MW00709883

EIGHTEENTH TEXAS SYMPOSIUM ON RELATIVISTIC ASTROPHYSICS AND COSMOLOGY

EIGHTEENTH TEXAS SYMPOSIUM ON RELATIVISTIC ASTROPHYSICS AND COSMOLOGY

"TEXAS IN CHICAGO"

CHICAGO, 15 – 20 DECEMBER 1996

EDITORS

ANGELA V OLINTO
The University of Chicago

JOSHUA A FRIEMAN
Fermi Laboratory

DAVID N SCHRAMM
The University of Chicago

World Scientific
Singapore • New Jersey • London • Hong Kong

Published by

World Scientific Publishing Co. Pte. Ltd.

P O Box 128, Farrer Road, Singapore 912805

USA office: Suite 1B, 1060 Main Street, River Edge, NJ 07661

UK office: 57 Shelton Street, Covent Garden, London WC2H 9HE

Library of Congress Cataloging-in-Publication Data
Texas Symposium on Relativistic Astrophysics (18th : 1996 : Chicago, Ill.)
　　Eighteenth Texas Symposium on Relativistic Astrophysics and
　Cosmology : "Texas in Chicago," December 15–20, 1996 / edited by
　Angela V. Olinto, Joshua A. Frieman, and David N. Schramm.
　　　　p.　　cm.
　　ISBN 9810234872
　　　1. Relativistic astrophysics -- Congresses.　　2. Cosmology -
- Congresses.　　3. Nucleosynthesis -- Congresses.　　4. Cosmic background
radiation -- Congress.　　5. Dark matter (Astronomy) -- Congresses.
6. Inflationary universe -- Congresses.　　I. Olinto, Angela V.
II. Frieman, Joshua A., 1959–　　╲　III. Schramm, David N.
IV. Title.
QB462.65.T489　　1996
523.01--dc21
　　　　　　　　　　　　　　　　　　　　　　　　　　　　98-16902
　　　　　　　　　　　　　　　　　　　　　　　　　　　　CIP

British Library Cataloguing-in-Publication Data
A catalogue record for this book is available from the British Library.

Printed in Singapore.

EIGHTEENTH TEXAS SYMPOSIUM ON RELATIVISTIC ASTROPHYSICS AND COSMOLOGY

Editors
Angela V. Olinto, Joshua A. Frieman, and David N. Schramm

Organizing Committee
J. Frieman (co-chair), D. Schramm (co-chair), F. Adams, G. Baym,
D. Duncan, J. Friedman, R. Kron, S. Meyer, A. Olinto, R. Rosner, M. Ulmer,
R. Bernstein, and C. Sazama

International Scientific Advisory Committee
V. Berezinsky, B. Cabrera, C. Cesarsky, L. da Costa, M. Davis, J. Ellis,
W. Freedman, M. Geller, R. Giacconi, J. Hartle, T. Kirsten, J. Mould, I.
Novikov, F. Pacini, T. Piran, T. Readhead, M. Rees, M. Ruderman, K. Sato,
R. Sunyaev, Y. Tanaka, K. Thorne, J. Trumper, A. Watson, D. Wilkinson

Texas Symposium International Organizing Committee
J. Audouze, J. D. Barrow, P. Bergmann, A. Cameron, J. Ehlers, L. Z. Fang,
E. J. Fenyves, R. Giacconi, J. C. Jones, M. Livio, L. Mestel, L. Motz, Y.
Neeman, I. Ozsvath, R. Ramaty, I. Robinson, R. Ruffini, B. Sadoulet, D.
Schramm, E. L. Schucking, G. Setti, M. M. Shapiro, G. Shaviv, L. C.
Shepley, J. J. Stachel, A. Trautman, V. Trimble, S. Weinberg, J. A. Wheeler,
and J. C. Wheeler

Financial assistance was received from:
Department of Energy
Fermi National Accelerator Laboratory
National Aeronautics and Space Administration
National Science Foundation
University of Chicago

Preface

Since 1963, the Texas Symposia have provided a biennial, peripatetic forum for forefront developments on a wide range of topics in relativistic astrophysics, from pulsars to string theory, from the birth of the Universe to the death of stars. In December, 1986, the Thirteenth Texas Symposium was held in Chicago. On the tenth anniversary of this meteorological experiment, "Texas in Chicago" returned. The Eighteenth Texas Symposium, with over 750 astrophysicists and physicists in attendance, was held at the historic Palmer House Hilton hotel in the Chicago Loop, December 15-19, 1996.

After welcoming remarks by Chicago Mayor Richard M. Daley, Jr., University of Chicago President Hugo Sonnenschein, and Fermilab Director John Peoples, the scientific program comprised five days of morning plenary talks and four afternoons of parallel sessions. The 26 plenary lectures, 230 parallel session talks, and 265 poster presentations attest to the scientific vitality of this interdisciplinary field. We thank the parallel session organizers for their excellent work in putting together their sessions: A. Konigl (AGNs,QSOs, and Jets), K. Sato (Big Bang Nucleosynthesis), J. R. Bond and G. Smoot (Cosmic Microwave Background), R. Kirshner (Cosmological Parameters), M. Turner (Dark Matter), E. Kolb (Early Universe), R. Giacconi and E. Wright (Future Projects and Instruments), J. Cronin and A. Watson (High Energy Cosmic and Gamma Rays), Y. Tanaka (Galaxy Clusters), R. Ellis (Galaxy Formation and Evolution), E. Fenimore and K. Hurley (Gamma Ray Bursts), R. Webster (Gravitational Lensing), A. Brillet and A. Giazotto (Gravitational Waves), F. Bouchet (Large-Scale Structure), K. Nomoto (Nuclear Astrophysics), E. Seidel (Numerical Relativity), D. Backer (Pulsars and Neutron Stars), D. York (QSO Absorption Line Systems), A. Ashtekar (Quantum Gravity and General Relativity), V. Berezinsky (Solar Neutrinos), W. Hillebrandt (Supernovae), E. van den Heuvel (X-Ray Binaries), and H. Bradt and N. Gehrels (X-Ray and Gamma-Ray Timing and Broad-band Spectroscopy).

During the Symposium, Stephen Hawking, Lucasian Professor at Cambridge University, delivered an enthusiastically received public lecture on Black Holes and Information Loss to an audience of 1,000 people at the Field Museum of Natural History. We are grateful to Prof. Hawking, and to the Adler Planetarium for organizing this event. The Symposium banquet was held in the Grand Ballroom of the Palmer House; the postprandial remarks of Robert Kirshner are regretfully omitted from these proceedings to avoid legal action.

The occasion of the Texas Symposium, with its large international audience, provided the opportunity for two 2-day conferences to be held just prior it. At the University of Chicago, the Symposium on Black Holes and Relativistic Stars was held in honor of S. Chandrasekhar, one of this century's most distinguished relativistic astrophysicists, who died in August, 1995. The 'Chandra' Symposium attracted 500 participants. At Fermilab, an informal workshop on the new field of Weak Gravitational Lensing was attended by more than 50 scientists.

Finally, we would like to thank all those who helped with the planning and logistical support which enabled the Symposium to run smoothly. The International Advisory Committee provided invaluable advice on the program, and the Organizing Committee helped put it in place. The Computing Division at Fermilab efficiently provided terminals and workstations. Graduate student volunteers from the University of Chicago and Northwestern and the staff at Fermilab and the University of Chicago helped with tasks from registration to poster mounting. We thank Roberta Bernstein, Joyce Rossi, and Patti Poole for many months of advance help and especially Cynthia Sazama of Fermilab for making the whole Symposium run smoothly. For help with these proceedings, we thank Colbey Harris, Jeff Bezaire, Clarice Assad, and The Print Lab. We are also grateful to the co-sponsoring institutions, Fermilab and the University of Chicago, and to the funding agencies for their support.

On the final day of the Symposium came the sad news of the passing of Carl Sagan, distinguished astrophysicist, author, and man of science. In his memory, we have included here a tribute from Peter Vandervoort of the University of Chicago.

Joshua A. Frieman, Editor and Co-chair
Angela V. Olinto, Editor
David N. Schramm, Editor and Co-chair

The death of Carl Sagan was publicly announced on the last day of the Symposium. Carl was well known to all of the Symposium participants for his writing, teaching, and other efforts to make science accessible to the public and especially to young people. His passing was a subject of reflection and discussion during the last hours of the Symposium. Younger participants spoke of the influence that he had on their own early interest in science. Carl's colleagues in his own areas of research had reason to respect his creativity and admire his vision. Those who had the good fortune to know him well during any period of his scientific career know how deeply he cared about people and about science. That combination made him a great teacher. Recalling the joy that he brought to the scientific enterprise, we dedicate this volume to his memory.

Peter Vandervoort

David N. Schramm

David N. Schramm
1945–1997

Exactly one year after the end of the 18th Texas Symposium, our co-editor and co-chair David N. Schramm died tragically when his twin-engine plane crashed near Denver, Colorado.

David was a prolific and much-honored scientist, whose visionary leadership helped guide the nascent field of particle astrophysics and early universe cosmology. A guiding light of the Texas Symposia for many years, Schramm's enthusiasm for science was infectious: he inspired those around him and fostered the careers of a large number of young astrophysicists. His enthusiasm for life was just as infectious, and he literally roped many of his colleagues into climbing mountains and lured them onto precipitous ski slopes.

David was one of the leading cosmologists of the present day, who made important contributions to big bang nucleosynthesis and other areas of nuclear and particle astrophysics. In books and lectures, he was an effective communicator of the excitement of science to the public and a respected advocate of the fundamental importance of scientific research.

Above all, David was a true friend whom we will greatly miss. We dedicate this volume to his memory.

Joshua A. Frieman
Angela V. Olinto

Contents

Part X. Galaxy Clusters   (*Y. Tanaka, Chair*)

ACTION IN BIG BANG NUCLEOSYNTHESIS

DAVID N. SCHRAMM

*University of Chicago, 5640 S. Ellis Avenue, Chicago,
IL 60637*

*NASA/Fermilab Astrophysics Center, Fermilab, Box 500, Batavia,
IL 60510, USA*

This review focuses on recent action involving one of the most important topics in physical cosmology: the light element abundances and their impact on the issue of dark matter. The agreement between the Big Bang Nucleosynthesis (BBN) predictions and the observed abundances is discussed. It is noted that the basic conclusions of BBN on baryon density are remarkably robust. However, detailed questions of ^{3}He evolution and potential ^{4}He systematics remain unresolved, but in no way do these issues lead to any doubt about the basic success of BBN itself. The recent extragalactic deuterium observations as well as the other light element abundances are discussed. The BBN constraints on the cosmological baryon density are reviewed and demonstrate that the bulk of the baryons are dark and also that the bulk of the matter in the universe is non-baryonic. Comparison of baryonic density arguments from Lyman-α clouds, x-ray gas in clusters, and the microwave anisotropy are made. It is shown that all are slightly more consistent with the higher end of the BBN baryon density regime and the lower extragalactic D/H values.

1 Introduction

Big Bang Nucleosynthesis and the light element abundances have undergone a recent burst of activity on many fronts. New results on each of the cosmologically significant abundances have sparked renewed interest and new studies. The bottom line remains: primordial nucleosynthesis has joined the Hubble expansion and the microwave background radiation as one of the three pillars of Big Bang cosmology. Of the three, Big Bang Nucleosynthesis probes the universe to far earlier times ($\sim$ 1 sec) than the other two and led to the interplay of cosmology with nuclear and particle physics. Furthermore, since the Hubble expansion is also part of alternative cosmologies such as the steady state, it is BBN and the microwave background that really drive us to the conclusion that the early universe was hot and dense.

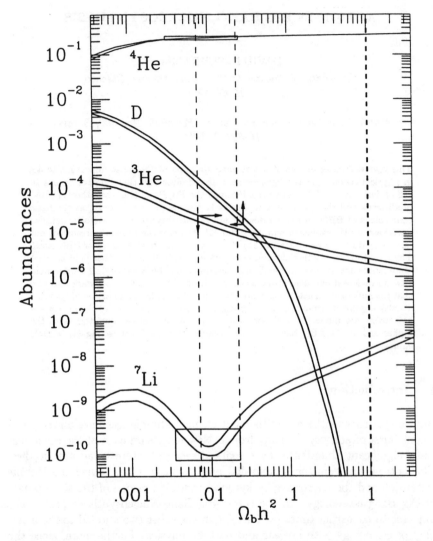

Figure 1: Figure 1. Big Bang Nucleosynthesis abundance yields versus baryon density (Ω_b) and $\eta \equiv \frac{n_b}{n_\gamma}$ for a homogeneous universe. ($h \equiv H_0/100$ km/sec/Mpc; thus, the concordant region of $\Omega_b h^2 \sim 0.015$ corresponds to $\Omega_b \sim 0.06$ for $H_0 = 50$ km/sec/Mpc.) Figure is from Copi, Schramm and Turner[8]. Note concordance region is slightly larger than Walker et al.[9] due primarily to inclusion of possible systematic errors on Li/H. The width of the curves represents the uncertainty due to input of nuclear physics in the calculation.

2 Overview

Although the extragalactic D/H observations have naturally attracted the most attention, it should not be forgotten that there are also recent heroic observations of ^{6}Li, Be and B, as well as ^{3}He and new ^{4}He determinations. Let us now briefly review the history, with special emphasis on the remarkable agreement of the observed light element abundances with the calculations. This agreement works only if the baryon density is well below the cosmological critical value.

It should be noted that there is a symbiotic connection between BBN and the 3K background dating back to Gamow and his associates, Alpher and Herman. The initial BBN calculations of Gamow's group[1] assumed pure neutrons as an initial condition and thus were not particularly accurate, but their inaccuracies had little effect on the group's predictions for a background radiation.

Once Hayashi[2] recognized the role of neutron-proton equilibration, the framework for BBN calculations themselves has not varied significantly. The work of Alpher, Follin and Herman[3] and Taylor and Hoyle[4], preceeding the discovery of the 3K background, and of Peebles[5] and Wagoner, Fowler and Hoyle[6], immediately following the discovery, and the more recent work of our group of collaborators[7,8,10,11,12,9,13,14] all do essentially the same basic calculation, the results of which are shown in Figure 1.

As far as the calculation itself goes, solving the reaction network is relatively simple by the standards of explosive nucleosynthesis calculations in supernovae, with the changes over the last 25 years being mainly in terms of more recent nuclear reaction rates as input, not as any great calculational insight, although the current Kawano code[14] is somewhat streamlined relative to the earlier Wagoner code[6]. In fact, the earlier Wagoner code is, in some sense, a special adaptation of the larger nuclear network calculation developed by Truran[15,16] for work on explosive nucleosyntheis in supernovae. With the exception of Li yields and non-yields of Be and B[17], the reaction rate changes over the past 25 years have not had any major affect (see Yang et al. [13] and Krauss and his collaborators[18,19], or Copi. Schramm, and Turner[8] for a discussion of uncertainties). The one key improved input is a better neutron lifetime determination[20,21]. There has been much improvement in the t(α,γ) ^{7}Li reaction rate, but as the width of the curves in Figure 1 shows, the ^{7}Li yields are still the poorest determined, both because of this reaction and even more because of the poorly measured ^{3}He (α,γ) ^{7}Be.

With the exception of the effects of elementary particle assumptions, to which we will return, the real excitement for BBN over the last 25 years has not really been in redoing the basic calculation. Instead, the true action is focused on understanding the evolution of the light element abundances and using that information to make powerful conclusions. In the 1960's, the main focus was on ^{4}He which is very insensitive to the baryon density. The agreement between BBN predictions and observations helped support the basic Big Bang model but gave no significant information, at that time, with regard to density. In fact, in the mid-1960's, the other light isotopes (which are, in principle, capable of giving density information) were generally assumed to have been made during the T-Tauri phase of stellar evolution[22], and so, were not then taken to have cosmological significance. It was during the 1970's that BBN fully developed as a tool for probing the universe. This possibility was in part stimulated by Ryter et al.[23] who showed that the T-Tauri mechanism for light element synthesis failed. Furthermore, ^{2}D abundance determinations improved significantly with solar wind measurements[24,25] and the interstellar work from the Copernicus satellite[26]. (Recent HST observations reported by Linsky et al.[27] have compressed the ^{2}D error bars considerably.) Reeves, Audouze, Fowler and Schramm[28] argued for cosmological ^{2}D and were able to place a constraint on the baryon density excluding a universe closed with baryons. Subsequently, the ^{2}D arguments were cemented when Epstein, Lattimer and Schramm[29] proved that no realistic astrophysical process other than the Big Bang could produce significant ^{2}D. This baryon density was compared with dynamical determinations of density by Gott, Gunn, Schramm and Tinsley[30]. See figure 2 for an updated H_0 - Ω diagram.

In the late 1970's, it appeared that a complimentary argument to ^{2}D could be developed using ^{3}He. In particular, it was argued[31] that, unlike ^{2}D, ^{3}He was made in stars; thus, its abundance would increase with time. Unfortunately, recent data on ^{3}He in the interstallar medium[32] has shown that ^{3}He has been constant for the last 5 Gyr. Thus, low mass stars are not making a significant addition, contrary to these previous theroetical ideas. Furthermore, Rood, Bania and Wilson[33] have shown that interstellar ^{3}He is quite variable in the Galaxy, contrary to expectations for a low mass star-dominated nucelus. However, the work on planetary nebulae shows that at least some low mass stars do produce ^{3}He. Nonetheless, the current observational situation clearly shows that arguments based on theoretical ideas about ^{3}He evolution should be avoided (c.f. Hata et al.[34]) where their "crisis" is really about ^{3}He problems, not BBN. Since ^{3}He now seems not to have a well behaved history, simple ^{3}He or ^{3}He + D inventory arguments are misleading at best. However, one is not free to go to arbitrary low baryon densities and high primordial D and ^{3}He, since processing of D and ^{3}He in massive stars also produces metals which are constrained[35,36] by the metals in the hot intra-cluster gas, if not the Galaxy.

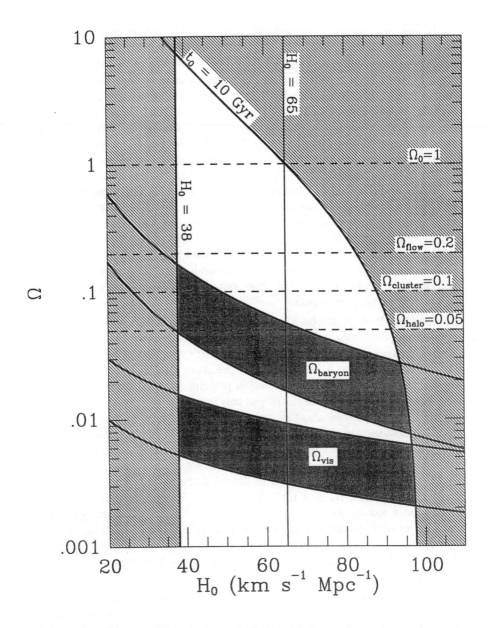

Figure 2: Figure 2. An updated version of $H_0 - \Omega$ diagram of Gott, Gunn, Schramm and Tinsley[30] showing that Ω_b does not intersect $\Omega_{VISIBLE}$ for any value of H_0 and that $\Omega_{TOTAL} > 0.1$, so non-baryonic dark matter is also needed.[65]

It was interesting that other light elements led to the requirement that ^{7}Li be near its minimum of ^{7}Li/H $\sim 10^{-10}$, which was verified by the Pop II Li measurements of Spite and Spite[37,38,39], hence yielding the situation emphasized by Yang et al.[13] that the light element abundances are consistent over nine orders of magnitude with BBN, but only if the cosmological baryon density, Ω_b, is constrained to be around 6% of the critical value (for $H_0 \simeq 50$ km/sec/Mpc). The Li plateau argument was further strengthened with the observation of ^{6}Li in a Pop II star by Smith, Lambert and Nissen[40]. Since ^{6}Li is much more fragile than ^{7}Li, and yet it survived, no significant nuclear depletion of ^{7}Li is possible[41,42,43]. This observation of ^{6}Li has now been verified by Hobbs and Thorburn[44], and a detection in a second Pop II star has been reported. Lithium depletion mechanisms are also severly constrained by the recent work of Spite et al.[45] showing that the lithium plateau is also found in Pop II tidally locked binaries. Thus, meridonal mixing is not causing lithium depletion. Recently Nollett et al.[46] have discussed how ^{6}Li itself might eventually become another direct probe of BBN depending on the eventual low energy measurement of the ^{2}D $(\alpha\gamma)$ ^{6}Li cross section on spectroscipy improvements for extreme metal-poor dwarfs.

Another development back in the 70's for BBN was the explicit calculation of Steigman, Schramm and Gunn[47] showing that the number of neutrino generations, N_ν, had to be small to avoid overproduction of ^{4}He. (Earlier work[4,48,49] had commented about a dependence on the energy density of exotic particles but had not done an explicit calculation probing N_ν.) This will subsequently be referred to as the SSG limit. To put this in perspective, one should remember that the mid-1970's also saw the discovery of charm, bottom and tau, so that it almost seemed as if each new detector produced new particle discoveries, and yet, cosmology was arguing against this "conventional" wisdom. Over the years, the SSG limit on N_ν improved with ^{4}He abundance measurements, neutron lifetime measurements, and with limits on the lower bound to the baryon density, hovering at $N_\nu \lesssim 4$ for most of the 1980's and dropping to slightly lower than 4 just before LEP and SLC turned on.[10,9,50,51] This was verified by the LEP results[52] where now the overall average is $N_\nu = 2.99 \pm 0.02$. A recent examination of the cosmological neutrino limit by Copi et al.[7] in the light of the recent ^{3}He and D/H work shows that the BBN limit remains between 3 and 4 for all reasonable assumption options.

The recent apparent convergence of the extra-galactic D/H measurements towards the lower values[53] D/H $\sim 3 \times 10^{-5}$ may eventually collapse the Ω_B band in figure 1 to a relatively narrow strip on the high Ω_B side (see arrows on Figure 2). However, such a collapse at present is probably a bit premature. It should be noted that this low D/H value also has important implications for galactic evolution since the present ISM value is not significantly lower than the high Z value. This would seem to favor infall models or models with variable initial mass functions to explain heavy element production in our Galaxy. In any case, it is clear that deuteronomy (the study of deuterium in the cosmos) is a success since: 1) deuterium is clearly cosmological as it is seen in low metalicity and high redshift Lyman-α clouds; 2) the primordial D/H is higher than the present ISM D/H, as predicted by theory; and 3) the range of values for primordial D/H, regardless of whether or not the high or low ones win out, is consistent with the range of expectations based on the other light nuclei.

One potential problem that the low D/H, high Ω_B solution raises is the fact that the central primordial ^{4}He mass fraction is ~ 0.23, rather than ~ 0.245, which the Tytler D/H[53] value would prefer for concordance. However, as Copi et al.[7] emphasize, systematic uncertainties in y cannot rule out such an excursion. But clearly we have to look carefully at ^{4}He. The recent work of Izo tov et al.[54] on $y \sim 0.24$ shows how uncertain the present situation is, but the resolution remains to be found.

The power of homogeneous BBN comes from the fact that essentially all of the physics input is well determined in the terrestrial laboratory. The appropriate temperature regimes, 0.1 to 1 MeV, are well explored in nuclear physics laboratories. Thus, what nuclei do under such conditions is not a matter of guesswork, but is precisely known. In fact, it is known for these temperatures far better than it is for the centers of stars like our sun. The center of the sun is only a little over 1 keV, thus, below the energy where nuclear reaction rates yield significant results in laboratory experiments, and only the long times and higher densities available in stars enable anything to take place.

3   Density of Baryons

The bottom line that emerges from the above discussion is that[7]

$$0.01 \lesssim \Omega_B h_0^2 \lesssim 0.025$$

where $h_0 \equiv H_0/100$ km/sec/Mpc. If the Tytler arguments on D/H do indeed hold up, then this will compress towards the high side, say $\Omega_B h^2 \sim 0.02 \pm 0.005$. Let us now compare with other ways of estimating Ω_b.

3.1   Lyman-α Clouds

Recent work by Bi and Davidsen[55], by Quashnock and Vanden Berk[56], and by Weinberg[57] also argues that the density of gas in the form of Lymon-α clouds at high redshift is consistent with the high end of the Big Bang Nucleosynthesis range on Ω_b. This would appear to resolve the long time problem of where are the "dark baryons." It is well known that $\Omega_{VISIBLE} \lesssim 0.01$, which, when compared to Ω_{BBN}, implies that the bulk of the baryons are not associated with stellar material. At least at high redshift this unseen material appears to have been found in these Lyman-α clouds. In conjunction with the Lyman-α clouds, it should also be noted that singly ionized helium is seen in the intergalactic gas, thus supporting the BBN fact that helium is primordial, and also supporting the point that significant numbers of baryons were between galaxies at high redshift[58,55].

3.2   Hot Gas in Clusters

Hot gas has been found in clusters of galaxies by ROSAT and ASCA. The temperature of the gas can be used to estimate the gravitational potential of the clusters if it is assumed that the gas is vivialized and purely supported by thermal pressure. Similarly, the intensity of the emission can be used to estimate the density of the gas. White et al.[59] have shown that the typical values for x-ray clusters yield a hot gas to total mass ratio M_{HOT}/M_{TOT} of about 0.2. If clusters are represenative of the universe as they would be in standard cold dark matter models, and if $\Omega_{TOTAL} = 1$, then $M_{HOT}/M_{TOT} = \Omega_b$. Figure 3 shows the range on M_{HOT}/M_{TOT} plotted as Ω_x on an $H_0 - \Omega$ diagram, showing that even with the spread in M_{HOT}/M_{TOT} in clusters as found by Mushotsky,[60] there is only a marginal overlap at very low H_0 and the high end of Ω_b from BBN. Clearly, higher values of Ω_{BBN} are favored by this argument. However, it should be noted that various systematic effects such as:

1) magnetic field pressure in clusters[61]
2) clumping of the hot gas regions[62]
3) admixture of hot dark matter

all go in the direction of increasing M_{TOTAL} and improving the overlap. Also, if clusters are not fair samples of the universe, then there is no need for concordance here. This latter point may be implied by the variations observed by Mushotsky, since, if clusters are fair samples, then they should all be giving the same answer. Another way to obtain concordance is if $\Omega_{TOTAL} < 1$! All of these caveats tell us that clusters do not represent any "baryon catastrophe" but they are important to continue to monitor, and it is clear that the overlap is better for higher Ω_{BBN} and requires less variation from the standard assumptions.

3.3   Microwave Anisotropies

The method with the most potential for checking Ω_b is the measurement of the acoustic peaks in the microwave background anisotropy at angular scales near

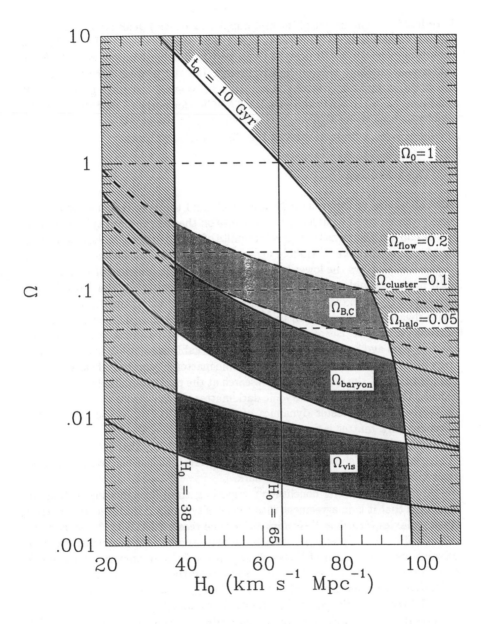

Figure 3: An Ω-H_0 diagram with the Ω_x added, showing the implied Ω in x-ray emitting gas in clusters if $\Omega_T = 1$ and if clusters are representative.

$1°$ or less[63,64]. The height of the first doppler (acoustic) peak for gaussian fluctuation models is directly related to $\Omega_B h^2$, thus a direct check on BBN. Current experiments at the south pole, at Sasketoon, and using balloons seem to favor values near the high side of the BBN range, but the present uncertainties are too large to make any strong statements. However, the next generation of satellites, NASA's MAP and ESA's PLANCK (formerly COBRAS/SAMBA) should be able to fix $\Omega_b h^2$ to better than 10% (if the sky is gaussian), which should provide a dramatic test of Big Bang Nucleosynthesis.

4 Dark Matter

The robustness of the basic BBN arguments and the new D/H measurements have given renewed confidence to the limits on the baryon density constraints. Let us convert this density regime into units of the critical cosmological density for the allowed range of Hubble expansion rates. This is shown in Figure 2. Figure 2 also shows the lower bound on the age of the universe of 10 Gyr from both nucleochronology and from globular cluster dating[65] and a lower bound on H_0 of 38 from extreme type IA supernova models with pure 1.4 $M_\odot$ carbon white dwarfs being converted to ^{56}Fe. The constraint on Ω_b means that the universe *cannot be closed with baryonic matter*. (This point was made over twenty years ago[66] and has proven to be remarkably strong.) If the universe is truly at its critical density, then nonbaryonic matter is required. This argument has led to one of the major areas of research at the particle-cosmology interface, namely, the search for non-baryonic dark matter. In fact, from the lower bound on Ω_{TOTAL} from cluster dynamics of $\Omega_{TOTAL} > 0.1$, it is clear that nonbaryonic dark matter is required unless $H_0 < 50$. The need for non-baryonic matter is strengthened on even larger scales[67]. Figure 2 also shows the range of $\Omega_{VISIBLE}$ and shows that there is no overlap between Ω_b and $\Omega_{VISIBLE}$. Hence, the bulk of the baryons are dark.

Another interesting conclusion[30] regarding the allowed range in baryon density is that it is in agreement with the density implied from the dynamics of single galaxies, *including their dark halos*. The recent MACHO [68] and EROS[69] reports of halo microlensing may well indicate that at least some of the dark baryons are in the form of brown dwarfs in the halo. However, Gates, Gyuk and Turner[70,71], and Alcock et al.[68] show that the observed distribution of MACHOs favors less than 50% of the halo being in the form of MACHOS, but a 100% MACHO halo cannot be completely excluded, yet.

For dynamical estimates of Ω, one estimates the mass from $M \sim \frac{v^2 r}{G}$ where v is the relative velocity of the objects being studied, r is their separation distance, and G is Newton's constant. The proportionality constant out front depends on orientation, relative mass, etc. For large systems such as clusters, one uses averaged quantities. For single galaxies v would represent the rotational velocity and r, the radius of the star or gas cloud. It is this technique which yields the cluster bound on Ω shown in Figure 2. It should be noted that the value of $\Omega_{CLUSTER} \sim 0.2$ is also obtained in those few cases where alignment produces giant gravitational-lens arcs. Recent work using weak gravitational lensing by Kayser[72] also supports large Ω. As Davis [67] showed, if the large scale velocity flows measured from the IRAS survey are due to gravity, then $\Omega_{IRAS} \gtrsim 0.2$. For $H_0 < 50$, $\Omega_{CLUSTER}$ already requires $\Omega_{TOTAL} > \Omega_{BARYON}$ and hence the need for non-baryonic dark matter.

An Ω of unity is, of course, preferred on theoretical grounds since that is the only long-lived natural value for Ω, and inflation[73,74] or something like it provided the early universe with the mechanism to achieve that value and thereby solve the flatness and smoothness problems. Note that our need for exotica is not dependent on the existence of dark galatic halos and that high values of H_0 increase the need for non-baryonic dark matter.

5   Acknowledgments

I would like to thank my collaborators, Craig Copi, Ken Nollett, Martin Lemoine, David Dearborn, Brian Fields, Dave Thomas, Gary Steigman, Brad Meyer, Keith Olive, Angela Olinto, Bob Rosner, Michael Turner, George Fuller, Karsten Jedamzik, Rocky Kolb, Grant Mathews, Bob Rood, Jim Truran and Terry Walker for many useful discussions. I would further like to thank H.G. Davidson, Poul Nissen, Jeff Linsky, Julie Thorburn, Doug Duncan, Lew Hobbs, Evan Skillman, Bernard Pagel and Don York for valuable discussion regarding the astronomical observations.

This work is supported by the NASA and the DoE(nuclear) at the University of Chicago, and by the DoE and by NASA grant NAG5-2788 at Fermilab.

References

1. R.A. Alpher, H. Bethe, and G. Gamow, *Phys. Rev.*, **73**, 803 (1948).
2. C. Hayashi, *Prog. Theor. Phys.* **5**, 22 (1950).
3. R.A. Alpher, J.W. Follin, and R.C. Herman, *Phys. Rev.*, **92**, 1347 (1953).
4. R. Taylor and F. Hoyle, *Nature* **203**, 1108 (1964)
5. P.J.E. Peebles, *Phys. Rev. Lett.* **16**, 410 (1966).
6. R. Wagoner, W.A. Fowler, and F. Hoyle, *Astrophys. J.* **148**, 3 (1967).

7. C.J. Copi, D.N. Schramm, and M.S. Turner, *Phys. Rev. Lett.* **55** (Feb. 15,1997).

8. C.J. Copi, D.N. Schramm, and M.S. Turner, *Science* **267**, 192 (1994).

9. T. Walker, G. Steigman, D.N. Schramm, K. Olive, H.S. Kang, *Astrophys. J.* **376**, 51 (1991).

10. K. Olive, D.N. Schramm, G. Steigman, and T. Walker, *Phys. Lett.* B **236**, 454 (1990).

11. D.N. Schramm and R.V. Wagoner, *Ann. Rev. of Nuc. Sci.* **27**, 37 (1977).

12. K. Olive, D.N. Schramm, G. Steigman, M. Turner, and J. Yang, *Astrophys. J.* **246**, 557 (1981) .

13. J. Yang, M. Turner, G. Steigman, D.N. Schramm, and K. Olive, *Astrophys. J.* **281**, 493 (1984).

14. L. Kawano, D.N. Schramm, and G. Steigman, *Astrophys. J.* **327**, 750 (1988).

15. J. Truran, Doctoral Thesis, Yale University (1965); J.W. Truran, A.G.W. Cameron, and A. Gilbert, *Can. Jour. of Phys.* **44**, 563 (1966).

16. J. W. Truran, A.G.W. Cameron, and A. Gilbert, *Can. Jour. of Phys.* **44**, 563 (1966).

17. D. Thomas, D.N. Schramm, K. Olive, and T. Walker, *Astrophys. J.* **415**, L 35 (1993).

18. L.M. Krauss and P. Romanelli, *Astrophys. J.* **358**, 47 (1990).

19. P. Kernan and L. Krauss, *Phys. Rev. Lett.* **72**, 3309 (1994).

20. W. Mampe, P. Ageron, C. Bates, J.M. Pendlebury, and A. Steyerl, *Phys. Rev. Lett.* **63A**, 593 (1989).

21. W. Mampe, P. Ageron, C. Bates, J.M. Pendlebury, and A. Steyerl, *Phys. Rev. Lett.* **63A**, 593 (1098); W. Mampe *et al.*, *JETP Lett.* **57**, 82 (1993).

22. W.A. Fowler, J. Greenstein, and F. Hoyle, *Geophys. J.R.A.S.* **6**, 6 (1962).

23. C. Ryter, H. Reeves, E. Gradstajn, and J. Audouze, *Astron. and Astrophys* **8**, 389 (1970).

24. J. Geiss and H. Reeves, *Astron. and Astrophys.* **18**, 126 (1971).

25. D. Black, *Nature* **234**, 148 (1971).

26. J. Rogerson and D. York, *Astrophys. J.* **186**, L95 (1973).

27. J. Linsky *et al.*, *Astrophys. J.* **402**, 694 (1993).

28. H. Reeves, J. Audouze, W.A. Fowler, and D.N. Schramm, *Astrophys. J.* **179**, 909 (1973).

29. R. Epstein, J. Lattimer, and D.N. Schramm, *Nature* **263**, 198 (1976).

30. J.R. Gott, III, J. Gunn, D.N. Schramm, and B.M. Tinsley, *Astrophys. J.* **194**, 543 (1974).

31. R.T. Rood, G. Steigman, and B.M. Tinsley, *Astrophys. J.* **207**, L57

(1976).

32. G. Gloeckler and J. Geiss, *Nature* **381**, 210 (1996).

33. R.T. Rood, T. Bania and J. Wilson, *Nature* **355**, 618 (1992).

34. N. Hata, R.J. Scherrer, G. Steigman, D. Thomas, T.P. Walker, S. Bludman, and P. Langacker, *Phys. Rev. Lett.* **75**, 3977 (1995).

35. C.J. Copi, D.N. Schramm, and M. Turner, *Astrophys. J.* **455**, L95 (1995).

36. S.T. Scully, M. Cassé, K.A. Olive, D.N. Schramm, J. Truran, and E. Vangioni-Flam, *Astrophys. J.* **462**, 960 (1996).

37. J. Spite and M. Spite, *Astron. and Astrophys.* **115**, 357 (1982).

38. R. Rebolo, P. Molaro, and J.Beckman, *Astron. and Astrophys.* **192**, 192 (1988).

39. L. Hobbs and C. Pilachowski, *Astrophys. J.* **326**, L23 (1988).

40. V.V. Smith, D.L. Lambert, and P.E. Nissen, *Astrophys. J.* **408**, 262 (1982).

41. K. Olive and D.N. Schramm, *Nature* **360**, 434 (1992).

42. G. Steigman, B. Fields, K. Olive, D.N. Schramm, and T. Walker, *Astrophys. J.* **415**, L35 (1993).

43. M. Lemoine, D.N. Schramm, J.W. Truran and C.J. Copi, *Astrophys. J.* in press (April 1, 1997).

44. L. Hobbs and J. Thorburn, *Astrophys. J. Lett.* **428**, L25 (1994).

45. M. Spite, P. Francois, P.E. Nissen, and F. Spite, *Astron. and Astrophys.* **307**, 172 (1996).

46. K. Nollett, M. Lemoine and D.N. Schramm, *Phys. Rev. C.* in press (1997).

47. G. Steigman, D.N. Schramm, and J. Gunn, *Phys. Lett.* B **66**, 202 (1977).

48. V.F. Schvartzman, *JETP Letters* **9**, 184 (1969).

49. P.J.E. Peebles, *Physical Cosmology* (Princeton University Press, 1971).

50. D.N. Schramm and L. Kawano, *Nuc. Inst. and Methods A* **284**, 84 (1989).

51. B. Pagel, in Proc. of 1989 Rencontres de Moriond (1990).

52. ALEPH, L3, OPAL, DELPHI results, 1993 Lepton-Photon meeting at Ithaca, NY (1993).

53. D. Tytler and C. Hogan, in *Proc. of the 18th Texas Symposium on Relativisitic Astrophysics*, in press (World Scientific, Singapore 1997).

54. Y.I. Izotov, T.X. Thuan, and V.A. Lipovetsky, *Astrophys. J.* **463**, 120 (1996).

55. H.G. Bi and A.F. Davidsen, *Astrophys. J.*, in press (1997).

56. J.M. Quashnock and D.E. Vanden Berk, *Astrophys. J.*, submitted (1997).

57. D.H. Weinberg, J. Miralda-Escué, L. Hernquist and N. Katz, preprint astro-ph/9801012 and *Astrophys. J.*, in press (1997).

58. P. Jakobson, A. Bokensberg, J.M. Deharveng, P. Greenfield, R. Jedrzewski, and F. Paresce, *Nature* **370**, 3 (1994).

59. S.D.M. White, J.F. Navarro, A.E. Evrard, and C.S. Frenck. *Nature* **366**, 261 (1993).

60. R. Mushotsky, in *Relativistic Astrophysics and Particle Cosmology: Texas PASCOS 92*, C.W. Akerlof and M.A. Srednicki, *Annals of the N.Y Academy of Sciences* **688**, 184 (1993).

61. P.P. Kronberg, *Rep. Prog. Phys.* **57**, 325 (1994).

62. R. Strickland and D.N. Schramm *Astrophys. J.* in press (1997).

63. G. Jungman, A. Kosowsky, M. Kamionkowski, D.N. Spergel, *Phys. Rev. Lett.* **76**, 1007 (1996); *Phys. Rev. D* **54**, 1332 (1996).

64. D. Scott and M. White, *Gen. Rel. and Grav.* **27**, 1023 (1995).

65. X. Shi, D.N. Schramm, D. Dearborn, and J.W. Truran, *Comments on Astrophys.* **17**, 343 (1995).

66. H. Reeves, W.A. Fowler, and F. Hoyle, *Nature* **226**, 727 (1970).

67. M. Davis in *Proc. 18th Texas Symposium on Relativistic Astrophysics*, in press (World Scientific, Singapore 1997).

68. C. Alcock *et al.*, *Nature* **365**, 621 (1993).

69. E. Aubourg *et al.*, *Nature* **365**, 623 (1993).

70. E. Gates, G. Gyuk, and M. Turner, *Phys. Rev. Lett.* **74**, 3724 (1995).

71. D. Bennett, *et al.* 1995, LLNL preprint.

72. N. Kayser in *Proc. of the Texas Symposium on Relativistic Astrophysics*, Munich, December 1994 (in press, 1995).

73. A. Guth, *Phys. Rev. D* **23**, 347 (1981).

74. A. Linde, *Particle Physics and Inflationary Cosmology* (Harwood Academic Publishers, N.Y., 1990).

Deuterium and Helium Absorption at High Redshift: Mapping the Abundance, Density and Ionization of Primordial Gas

Craig J. Hogan

Astronomy and Physics Departments, University of Washington
PO Box 351580, Seattle, WA 98195, USA

Spectra of quasars at high redshift with high resolution and high signal-to-noise allow in favorable circumstances detection of absorption by deuterium and ultimately measurement of its primordial abundance. Ultraviolet spectra of high redshift quasars allow measurement of absorption by the most abundant cosmic absorber, singly ionized helium, thereby mapping gas even in the most rarefied cosmic voids. These new techniques already provide significant constraints on cosmological models but will soon become much more precise.

1   Absorption by Primordial Gas

Soon after quasars were discovered in the 1960's it was realized (by Bahcall and Salpeter, Gunn and Peterson, Lynds and others) that their exceptional brightness and simple, smooth intrinsic spectra made their absorption spectra ideal probes of distant material along the line of sight— a unique opportunity to study material at great distances and early times in detail. By 1980 landmark papers by Sargent and collaborators had laid out a considerable detailed statistical knowledge of the absorption, with many important implications for cosmology. The statistical properties of the absorption firmly established that it is caused by cosmologically distributed foreground gas, mostly by clouds constituting an intergalactic or protogalactic population. Some fraction of the clouds with high column density and metal enrichment were identified with galaxies already in the process of chemical evolution. During the 1980's, catalogs of the highest column density "damped" absorbers (especially by Wolfe and collaborators) seemed to isolate a population that could be readily identified as progenitors of modern galaxies.

Theoretical ideas for the absorbing clouds initially explored a very large space of physically plausible populations embedded within the wide variety of extant galaxy formation scenarios. As the Cold Dark Matter paradigm for structure formation sharpened during the 1980's, Rees and others showed that within this picture the absorbing clouds are a natural accompaniment to galaxy formation; they are the condensations of the primordial baryonic material during its early stages of collapse into the dark matter potentials. It became clear that the study of the distribution and the enrichment of the gas through absorption would provide one of the richest and most precise ways of viewing directly the process of galaxy and structure formation and chemical evolution.

This promise has been realized in the last few years as the pace of progress has advanced quickly both in observations and interpretation. The Keck telescope now provides a qualitatively new type of data: spectra from $z \approx 3$ with high resolution (that is, better than the thermal widths of lines), with signal-to-noise of the order of 100. The Hubble and HUT telescopes provide another qualitatively new type of data: spectra from $z \approx 3$ down to below the He^+ Lyman-α line 304Å in the rest frame, at lower resolution but still comparable to the earlier work on HI. These two developments now allow (among many other things) the study of absorption not only by hydrogen and metal ions but also by the other important primordial elements in this high redshift gas, deuterium and helium.

At the same time, ideas about the cosmic gas distribution have sharpened quantitatively in recent years, due to hydrodynamic simulations which accurately predict the distribution and motion of matter in hierarchical models of galaxy formation (Cen et al. 1994, Hernquist et al. 1996, Miralda-Escudé et al. 1996, Rauch et al. 1997, Croft et al. 1997, Zhang et al. 1997; see also Bi and Davidsen 1997). Departing from earlier analytic models based on isolated clouds with symmetric geometries such as spheres and slabs, simulations of gravitational collapse from nearly uniform gas (with linear gaussian noise) into nonlinear structures produce dynamical systems with a complex geometry (poetically described as voids, pancakes, filaments, knots) and no sharp distinction between diffuse gas and clouds; similarly, the complex simulated absorption spectra reveal no sharp distinction between lines and continuum. It is now possible to use absorption spectra to apply statistical tests to CDM models which are in some ways cleaner than the traditional ones based on galaxies— since the gas distribution and ionization is computed accurately up to the point where optical depths become large, and do not depend much on uncertainties from star formation and extinction. The cleanest predictions concern the gas in the least dense regions, especially in the voids.

In this context the accessibility of deuterium and helium absorption opens up new types of cosmological tests. The primordial abundance of deuterium is a critical observational test of cosmological theory, both as a test of the basic Big Bang picture and as a measure of the cosmic baryon density, a central parameter of structure-formation models (Peebles 1966, Walker et al. 1991, Smith et al. 1993, Copi et al. 1995, Sarkar 1996, Hogan 1997). Even though deuterium is detected in the Galaxy, high redshift absorption probes its primordial abundance with better control over the effects of chemical evolution, and provides an opportunity to map the abundance in space and its evolution in time in different environments.

Helium absorption is also interesting for its primordial abundance, but more so for the insights it provides about the density, distribution and ionization of gas at high redshift. Simulated spectra reveal that gas in the most underdense regions, filling the bulk of the spatial volume, is so highly ionized that it produces absorption features with very low HI Lyman-α optical depth. The more abundant absorbing ion He^+ however produces optical depths of the order of unity even in these regions, so its absorption is easily detectable, mapping the distribution of cosmic baryons at the lowest densities.

2 Deuterium Abundance

There are now about eight plausible detections of extragalactic deuterium in the literature, reviewed recently for example in Hogan (1997). There is currently no case where the candidate deuterium feature can be identified positively as such: in every instance, the data could be interpreted as an HI cloud accompanied by another cloud blueshifted by 82 km/sec with a much smaller column density. In the cases where we had thought this was impossible based on the narrow widths of candidate D absorption (Rugers and Hogan 1996ab), new and better data now show that the features are not so narrow after all (Tytler, Burles and Kirkman 1996, and Cowie et al, private communication.) Indeed the new data shows that there must be at least some contamination by HI at the DI Lyman-α feature because the velocity centroid is slightly displaced from the bulk of the hydrogen as measured from the higher Lyman series lines. Thus, the evidence for a high deuterium abundance is not conclusive, but is based on the anecdotal accumulation of several high estimates, some of which are reported in the literature. I describe here briefly some recent work (Rugers and Hogan 1997) to seek a more reliable statistical estimate.

2.1 Interloper statistics from "Pretend" deuterium absorbers

In principle, because of the monotonic destruction of deuterium by chemical evolution, one should pay the most attention to the highest measured abundances. But contamination by hydrogen lines are bound to yield some spurious detections. Indeed, at some level there is always some contamination, since there is some hydrogen absorption at all redshifts— the problem is to quantify its effect on abundance estimates.

To estimate the contamination by hydrogen interlopers, we explore a statistical technique. We have assembled a control sample of "pretend" deuterium candidates associated with hydrogen lines. These candidates are selected the

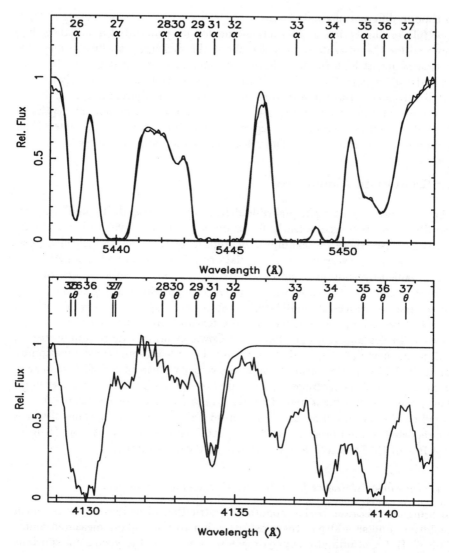

Figure 1: Lyman-α fit (top) for one of seven new deuterium candidates, at $z = 3.478401$ in Q1422+230. The main HI absorption, component 31, has redshift determined by the optically thin Lyman-θ fit (below); the DI feature, component 30, matches this redshift to within the fitting error (difference -3 ± 14 km/sec). Hydrogen contamination (fitted mostly with component 28) is superimposed on the deuterium feature, making the abundance fit very uncertain, $\log(D/H) = -3.53 \pm 0.45$. It can nevertheless be used in a statistical study.

same way the deuterium candidates are selected, in all respects but one: the control sample candidates are drawn from the red side of hydrogen absorbers, rather than the blue side where the real deuterium feature appears. To make a larger sample, the velocity bin to accept a pretend candidate is also larger than that for a deuterium candidate (i.e., $[-60, -100]$ km/sec rather than $[-82 \pm 1\sigma]$ km/sec). The pretend candidates are all interlopers, drawn from a population with the same statistical properties as the deuterium interlopers, including their joint correlations in velocity, column density and width. The properties of two samples can then be compared statistically, using the Doppler parameters and column densities from the line fits.

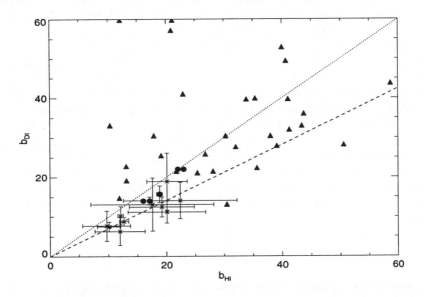

Figure 2: Fitted deuterium and hydrogen Doppler parameters of real deuterium candidates (a uniform sample represented by crosses and published values by filled circles), compared with those in the sample of pretend deuterium candidates, plotted as solid triangles; errors in the latter are omitted to enhance the clarity of the plot. The two lines represent the boundaries of the interval allowed for real deuterium, between the extremes of thermal and turbulent broadening. Unlike the pretend sample, all of the real deuterium candidates are consistent with the allowed ratio. For the new uniform sample, no cut was applied based on Doppler parameter.

For a fair comparison we have also assembled a uniform sample of deuterium absorbers from the same spectra as the pretend deuterium absorbers, based purely on redshift selection relative to HI features. There are seven new candidates with detectable deuterium features at the right redshift, in addition to the previously published candidates from Rugers and Hogan (1996ab). The reason that they have not been published is that they are not very convincing, or do not yield a precise abundance. One new example is shown in figure 1.

The fitted Doppler parameters (figure 2) show that the claimed deuterium detections are at least mostly real, even if individual cases are suspect. This is true both for the new candidates and for the previously published ones. The D features have statistically narrower profiles than the pretend candidates, consistent with the deuterium identification.

Since we believe the deuterium is statistically real, we can derive a statistical abundance. For the new sample of absorbers, (that is, the uniform sample excluding previously published values), we obtain $\langle \log(D/H) \rangle = -3.75 \pm 0.51$. The reduced χ^2 is 0.70, indicating consistency with a universal abundance. This is certainly not the case for the sample of pretend absorbers; even confining ourselves to those with high HI column (so as to be more directly comparable to the real ones), the abundance of "pretendium" is $\langle \log(P/H) \rangle = -3.2 \pm 2.6$, with a reduced χ^2 of 8.7. Other statistics also display differences. While the real deuterium candidates display consistency with a universal abundance (linear regression coefficient between $N(D)$ and $N(H)$ of 1.00), the pretend sample is a scatter plot (linear regression coefficient between $N(P)$ and $N(H)$ of 0.55.) These statistics are reflected in the scatter plot shown in figure 3. And the distributions of $N(D)$ and $N(P)$ also differ, in the sense that P is not as common as D— another way of saying that for the most part, interlopers may be there but statistically have a lower column density than the real deuterium.

2.2   Current situation and future prospects

Although deuterium is detected, its primordial abundance is still uncertain by an order of magnitude. At present, we have a very firm lower limit $D/H \geq 2 \times 10^{-5}$ on primordial deuterium, from Galactic measurements (Linsky et al. 1993, 1996), as well as from some quasar absorbers (e.g. Tytler et al. 1996). There is some evidence for a somewhat higher lower limit $(D/H \geq 4 \times 10^{-5})$ from quasar absorbers (Songaila et al. 1996). A higher abundance than this is not clearly required by the data, although there is some statistical evidence for it and there is no very strong evidence against it, since the low abundances are still found in just a few cases where deuterium may have been destroyed.

Better even than a much larger statistical study would be to find a system where a clear signature of deuterium can be proven, or where the interloper probability is small. A very promising possibility is a damped absorber in

Q2206-199, at $z = 2.559$, with a low metallicity and very narrow lines (Pettini et al. 1994). The interloper problem is much smaller here, both because of the very high column and the low redshift. The low redshift requires HST, but with STIS the entire Lyman series can be seen at once so this is a practical program, currently approved for cycle 7.

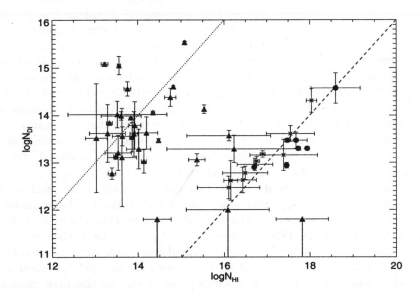

Figure 3: Column densities for fitted hydrogen components and their deuterium counterparts (crosses for the uniform sample, dots for other published values), together with the same quantities for the control sample of pretend candidates (triangles). The distributions are clearly different, as confirmed by statistical tests. Lines correspond to constant D/H.

3 Resolving the Helium Lyman-α Forest: Mapping Intergalactic Gas and Ionizing Radiation at $z \approx 3$

There is certainly HI absorption between the identified lines of the Lyman-α forest. As spectra of higher signal-to-noise ratio are obtained they reveal absorption of progressively lower optical depth. In HI however, even at the highest S/N so far available (about 100 at high resolution), the HI absorption does not yet fill redshift space. Limits on HI continuous optical depth (or "Gunn-Peterson effect") provide useful constraints on diffuse gas density, subject to uncertain quantities such as the ionizing radiation field (Giallongo et al. 1992,1994,1996).

Because of its higher ionization potential, the most abundant absorbing ion in the universe is not HI but He^+. Even in hard radiation fields near quasars it is more abundant by a factor η between 10 and 20. In intergalactic space the ratio is probably greater than 100, so in the emptiest void regions, He^+ produces much more easily detectable optical depths even at modest signal-to-noise. It is therefore the best tool for mapping the distribution of cosmic baryons at the lowest densities. Absorption by He^+ is also the most direct probe of the hard ultraviolet cosmic radiation field, which can be predicted from semiempirical models based on observed quasar and absorber populations (Haardt and Madau 1996). The spectral shape also influences other observables such as the ratio of CIV to SiIV (Songaila & Cowie 1996), so information from helium absorption allows information about relative C and Si abundances to be derived. In situations where the ionizing spectrum is known, such as the near proximity of a quasar, He^+ absorption can be compared to HI absorption to extract independent information about the primordial abundance of helium, an important test of Big Bang Nucleosynthesis.

The difficulty of course is that the Lyman-α transition of He^+ is at 304Å, requiring observations both from space and at high redshift. The first detection of cosmic He^+ absorption was made in Q0302 by Jakobsen et al. (1994) using the Hubble Space Telescope Faint Object Camera (FOC). They found an absorption edge and a large continuous optical depth at low resolution (10Å), $\tau > 1.7$, caused by a mixture of lines and truly diffuse He^+ Lyman-α absorption (Songaila et al. 1995). A similar observation has also been made of the $z = 3.185$ quasar PKS 1935-692, with a similar result (Tytler & Jakobsen 1996).

A significant improvement came from the Hopkins Ultraviolet Telescope (Davidsen et al. 1996), which can reach shorter wavelengths and hence lower redshift than HST, and also provides better resolution and wavelength calibration than FOC. Davidsen et al. observed the $z = 2.72$ quasar HS 1700+64 and found an He^+ edge close enough to the predicted redshift to rule out the possibility of foreground HI as an important contaminant. They also found that the flux below the edge is not consistent with zero, and measured accurately a mean optical depth, $\tau = 1.00 \pm 0.07$. The smaller absorption at lower redshift reflects the increasing ionization of He^+ with time, the thinning due to the expansion, and the conversion of diffuse gas into clouds.

New observations of Q0302 (Hogan et al. 1997) were made to improve both the wavelength calibration and resolution of the He^+ absorption, with enough sensitivity to correlate usefully with the HI absorption (see figure 4). We can now explore in detail the relative contributions of clouds and diffuse

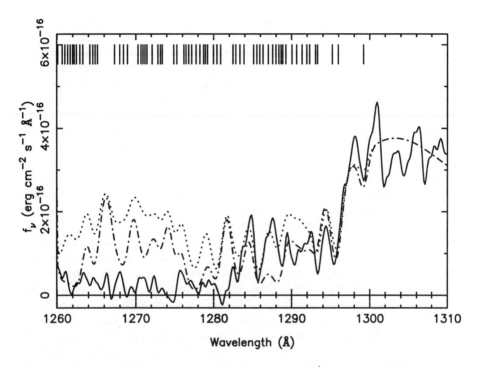

Figure 4: Detail of HST spectrum of 0302-003 near the He$^+$ edge. The HST spectrum is overlaid with a model spectrum predicted on the basis of the model distribution of HI derived from a Keck spectrum of the HI Lyman-α forest. Ticks indicate the fitted HI velocity components from the Keck spectrum. Doppler parameters and column densities from the fit were used to predict the He$^+$ absorption spectrum at the GHRS resolution. Two predictions are shown: dotted and dot-dash curves corresponding to He$^+$/HI ratios $\eta = 20$ and 100 respectively, both models assuming pure turbulent broadening, $b_{He^+} = b_{HI}$. The HST and Keck spectra appear to show corresponding absorption features, including the main He$^+$ edge. Near the quasar, a hard spectrum with $\eta = 20$ is probably sufficient to explain the absorption features entirely with clouds. A large z-filling optical depth is only allowed outside the proximity of the quasar, presumably because the diffuse gas nearer the quasar is doubly ionized. There is significant He$^+$ opacity ($\tau_{GP} > 1.3$) even at the redshift of the conspicuous HI Lyman-α forest void near 1266Å, indicating the presence of diffuse gas between identified HI absorbers. At the same time there is significant nonzero flux everywhere, setting a limit on the density of diffuse, z-filling gas.

gas, as well as measuring independently the ionizing radiation field and the helium abundance. We find significant He^+ absorption from HI clouds (with optical depth of the order of unity) but also comparable He^+ absorption even in redshift intervals where the best Keck spectrum reveals no detectable HI; thus we directly measure absorption attributable separately to both the clouds and the diffuse gas. Our spectrum suggests nonzero flux at all wavelengths, which constrains the ionizing background spectrum and leads to an upper limit on the density of diffuse gas. Absorption from gas near the quasar, where the incident spectrum is known approximately from the direct measurement of the quasar spectrum, allows independent constraints on the density and helium abundance of the gas.

The main new conclusions from the current data are: 1. The He^+ Lyman-α forest is detected, and indeed discrete clouds identified in HI are responsible for the main He^+ absorption edge in the spectrum; 2. The "diffuse" (redshift-space-filling) medium is also detected, and must have a low density ($\Omega \leq 0.01(h/0.7)^{-3/2}$) consistent with standard primordial nucleosynthesis and models of early gas collapse into protogalaxies; 3. The intergalactic ionizing spectrum is soft ($\eta \geq 100$), although the intergalactic helium is probably mostly doubly ionized by $z = 3.3$; 4. The helium abundance is within a factor of a few of standard Big Bang predictions, over a large volume of space at high redshift. We expect that these conclusions will be made more general and precise with the new 2-dimensional spectroscopic capability of HST/STIS.

Acknowledgments

Keck spectra used in this work were generously shared by L. Cowie and collaborators. I am grateful for many useful discussions with participants in the 1996 workshop on Nucleosynthesis at the Institute for Nuclear Theory in Seattle, funded by DOE. This work was supported by NASA and the NSF at the University of Washington.

References

1. Bi, H. and Davidsen, A. F. 1996, *ApJ*, in press
2. Cen, R., Miralda-Escudé, J., Ostriker, J. P., and Rauch, M. 1994, *ApJ* 437, L9.
3. Copi, C. J., Schramm, D. N., and Turner, M. S. 1995, Science 267, 192.
4. Croft, R. A. C., Weinberg, D. H., Katz, N. and Hernquist, L. 1997, *ApJ*, submitted (astro-ph 9611053)
5. Davidsen, A. F., Kriss, G. A. and Zheng, W., 1996 *Nature* **380**, 47.

6. Giallongo, E., Cristiani, S., Trevese, D., et al. 1992, *ApJ*, **398**, L9.
7. Giallongo, E., et al. 1994, *ApJ*, **425**, L1.
8. Giallongo, E., Cristiani, S., D'Odorico, S., et al. 1996, *ApJ*, to appear (astro-ph 9602026)
9. Haardt, F., and Madau, P. 1996, *ApJ* 461, 20
10. Hernquist, L, Katz, N., Weinberg, D. H., and Miralda-Escudé, J. 1996, *ApJ* 457, L51.
11. Hogan, C.J. 1997, in *Critical Dialogues in Cosmology*, ed. N. Turok (Princeton), in press (astro-ph 9609138)
12. Hogan, C. J., Anderson, S. F., and Rugers, M. H. 1997, *Astron.J.*, in press (astro-ph 9609136)
13. Jakobsen, P., et al., 1994, *Nature* **370**, 35.
14. Linsky, J. L. et al. 1993, *ApJ*, 402, 694
15. Linsky, J.L., Diplas, A., Wood, B. E., Brown, A., Ayres, T. R. and Savage, B. D. 1996, *ApJ*, 463,254
16. Miralda-Escudé, J., Cen, R., Ostriker, J. P., and Rauch, M. 1996, *ApJ* **471**, 582.
17. Peebles, P.J. E. 1966, *ApJ*, 146, 542.
18. Pettini, M., Smith, L. J., Hunstead, R. W., and King, D. L. 1994, *ApJ* 426, 79
19. Rauch, M., et al. 1997, *ApJ*, submitted (astro-ph 9612245)
20. Rugers, M. and Hogan, C. J., 1996a, *ApJ* **459**, L1.
21. Rugers, M. and Hogan, C. J., 1996b, *Astron. J.* **111**, 2135.
22. Rugers, M. and Hogan, C. J., 1997, in preparation.
23. Sarkar, S., 1996, Rep. Prog. Phys, 59, 1493
24. Smith, M. S., Kawano, L. H., and Malaney, R. A. 1993,*ApJ* 85, 219
25. Songaila, A. and Cowie, L. L., 1996, *Astron. J.*,112,335.
26. Songaila, A., Hu, E. M., and Cowie, L. L., 1995, *Nature* **375**, 124.
27. Songaila, A., Wampler, E.J., Cowie, L. L. 1996, *Nature*, submitted.
28. Tytler, D., Fan, X.-M. and Burles, S., 1996, *Nature* **381**, 207.
29. Tytler, D., Burles, S., and Kirkman, D. 1996, astro-ph 9612121
30. Tytler, D. and Jakobsen, P., 1996, manuscript in preparation.
31. Walker T. P., Steigman, G., Schramm, D. N., Olive, K. A. and Kang, H.-S. 1991, *ApJ* 376, 51
32. Zhang, Y., Anninos, P., Norman, M., and Meiksin, A. 1997, *ApJ*, submitted (astro-ph 9609194)

GRAVITATIONAL MICROLENSING SEARCHES AND RESULTS

C. ALCOCK

Lawrence Livermore National Laboratory, Livermore,
CA 94550, USA

Baryonic matter, in the form of Machos (MAssive Compact Halo Objects), might be a significant constituent of the dark matter that dominates the Milky Way. This article describes how surveys for Machos exploit the gravitational microlens magnification of extragalactic stars. The experimental searches for this effect monitor millions of stars, in some cases every night, looking for magnification events. The early results of these surveys indicate that Machos make up a significant fraction of the dark matter in the Milky Way, and that these objects have *stellar* masses. Truly *substellar* objects do *not* contribute much to the total. Additionally, the relatively high event rate towards the Galactic bulge seems to require that the bulge be elongated, and massive.

1 Introduction

The elucidation of how much dark matter exists in the universe, where this matter is located, and *what it is made of* is one of the most pressing issues facing astrophysics and cosmology at the end of the century. The broader parts of this question are addressed very ably elswhere in this volume[16]; additionally, there are very up to date research summaries. The purpose of this article is to point to these articles, and to introduce the gravitational microlens with sufficient completeness that the reader of this volume can understand the whole story.

Most of the mass of the Milky Way and similar galaxies is in some presently invisible form (see e.g., the review of Fich and Tremaine[5]). This "dark matter" cannot be in the form of normal stars or gas, which can readily be detected. Additionally, there is compelling evidence for much larger quantities of dark matter on larger scales in the universe.

Many candidates have been proposed to account for this dark matter[16]. These fall into two main classes: the particle-physics candidates such as massive neutrinos, axions or other weakly interacting massive particles (WIMPs)[14], and the astrophysical candidates, including substellar objects below the hydrogen burning threshold $\approx 0.08 M_\odot$ ('brown dwarfs'), or stellar remnants such as white dwarfs, neutron stars or black holes; these are generically known as massive compact halo objects (Machos). Such objects would be much too faint to have been detected in current sky surveys.

The searches for Machos described here (and all of the experimental searches for particle-physics candidates) are looking exclusively for objects located in the dark halo of the Milky Way. The total amount of dark matter in all galactic halos (located say, within 50 kpc of spiral galaxies) is approximately known from rotation curve data, and contributes to the mean density of the universe $\Omega \sim 0.05$: if $\Omega = 1$, either the halos must extend far beyond 50 kpc or there must be intergalactic dark matter.

Paczynski [11] suggested that Machos could be detected by their gravitational 'microlensing' of background stars. This indirect technique does not depend upon light emitted by the Machos. (It should be remembered, though, that the lensing objects in question are not required to be dark, merely significantly fainter than the source stars.)

A good technical review of this area of research is given by Paczynski [12], which is very complete on the theory and contains a thorough summary of the current experimental situation. There are also several very good articles in this volume which give excellent summaries of the current experimental situation with the two principal collaborations: Griest et al. [10] describes the current state of the MACHO Project, and Palanque-Delabrouille [13] describes the current state of the EROS Project.

2 The Gravitational Microlens

The simple gravitational lens comprises a point-like source of light (typically a star), a point-like massive deflector (the Macho), and an observer. It is characterized by an angular size θ_E, called the Einstein radius. If the target star lies directly behind the Macho, its image will be a ring of light of this angular radius. If the Macho is separated from the line of sight to the source by some finite angle, $\theta = b/D_d$ (where b is the physical distance of the Macho from the line of sight), the ring splits into two arcs. The combined light from the two images produced by the gravitational lens causes a net magnification

$$A(u) = (u^2 + 2)/u\sqrt{u^2 + 4} \qquad (1)$$

that depends only on the ratio of the separation to the Einstein radius, $u = \theta/\theta_E$. Note that $A > 1.34$ when $u < 1$, and $A \sim u^{-1}$ when $u \ll 1$. The Einstein radius is related to the underlying physical parameters by

$$\theta_E = \sqrt{\frac{4GMD_{ds}}{c^2 D_d D_s}} \qquad (2)$$

where M is the Macho mass, and D_d, D_s, and D_{ds} are the (observer-lens), (observer-source), and (lens-source) distances, respectively. The term 'microlensing' is used when θ_E is so small that the two images cannot be separated with current observing equipment, and the image doubling cannot be seen.

Frequently the related quantity $R_E = \theta_E D_d$ is referred to as the radius of the Einstein ring, where R_E is the physical size of the ring described above, measured at the location of the Macho. For a source distance of $50\,\text{kpc} \approx 10^{10}\,\text{AU}$ and a deflector distance of $10\,\text{kpc}$, the Einstein radius is $R_E \approx 8\sqrt{M/M_\odot}\,\text{AU}$. The coincidence between this scale and the orbit of the earth around the sun can be exploited in the study of microlenses, as will be described below.

The observable phenomenon, as the Macho moves at constant relative projected velocity $\mathbf{v}$, is the varying magnification of the star. This magnification is given as a function of time by $A[u(t)]$, where

$$u(t) = \sqrt{\omega^2(t - t_0)^2 + \beta^2} \tag{3}$$

and β is the impact parameter in units of the Einstein radius, t_0 is the epoch of maximum magnification, and ω^{-1} is the characteristic time (duration) of the event.

Fortunately, microlensing has distinctive signatures which can be used to discriminate it from intrinsic stellar variability:

- Since the optical depth is so low, only one event should be seen in any given star (optical depth is discussed below).

- The deflection of light is wavelength-independent; hence, the star should not change color during the magnification.

- The events should have lightcurves as predicted by the theory (above).

All these characteristics are distinct from known types of intrinsic variable stars; most variable stars are periodic or semi-regular, and do not remain constant for long durations. They usually change temperature and hence color as they vary, and they usually have asymmetrical lightcurves with a rapid rise and slower fall.

In addition to these individual criteria, if many candidate microlensing events are detected, there are further statistical tests that can be applied:

- The events should occur equally in stars of different colors and luminosities.

- The distribution of impact parameter u_{min} should be uniform from 0 to the experimental cutoff $u(A_{min})$.

- The event timescales and peak magnifications should be uncorrelated.

When a microlensing event is detected and its light curve is measured, one determines three parameters, ω, t_0, and β. Of these, only ω is related to the Macho's physical parameters:

$$\omega = \frac{v}{D_{\rm d}\theta_E}. \tag{4}$$

For microlensing of sources in the Large Magellanic Cloud (LMC), $\omega^{-1} \approx 70\sqrt{M/{\rm M}_\odot}$ days. One of the principal limitations of present experiments in this area, one which cannot be resolved with purely ground-based work, is that our uncertain knowledge of the quantities $D_{\rm d}$, $D_{\rm ds}$, and v for an observed microlensing event means that the uncertainty in the inferred mass of the Macho, M, spans more than an order of magnitude [9]. The mass is our only clue to the true nature of the Macho.

Parallax, obtained by the simultaneous observation of a microlens event from telescopes separated by a distance of order R_E, allows one to measure a second parameter [8],

$$\tilde{v} = (D_{\rm s}/D_{\rm ds})v. \tag{5}$$

(Generally one obtains one vector component of $\tilde{v}$, and the absolute magnitude of another.)

The complete solution of a microlensing event requires, in addition, the determination of θ_E. This is very difficult to achieve, since typically $\theta_E <$ milliarcsec. For a handful of cases in which the Macho passes directly in front of the source star (or nearly in front of the star), it will be possible to measure θ_E. This is possible because the magnification equation given above must be modified to take into account the finite angular size θ_S of the star. The modification depends principally upon the ratio θ_S/θ_E. (An important complication is introduced by the center to limb variation of brightness over the face of the star.) If θ_S can be estimated spectroscopically, one can in turn infer θ_E.

In the rare cases when one has ω, $\tilde{v}$, and θ_E, the Macho mass, distance, and transverse velocity can each be determined. For example,

$$M = \frac{c^2\tilde{v}\theta_E}{4G\omega}. \tag{6}$$

3 The "Macho Fraction" in the Galactic Halo

The gravitational microlens 'optical depth' is the quantity that probes directly the Macho fraction of the dark matter, since it is 'proportional' to the density of microlensing objects along the line of sight to the target stars. If the total mass density $\rho_{total}(D_d)$ is known, for instance along the line of sight to the Large Magellanic Cloud, then the experimental estimate of the optical depth τ along this line of sight yields the fraction of the total dark matter that is in the form of Machos.

The 'optical depth' τ for gravitational microlensing is defined as the probability that a given star is lensed with $u < 1$ ($A > 1.34$) at any given time, and is

$$\tau = \pi \int_0^{D_s} \frac{\rho(D_d)}{M} R_E^2(D_d)\, dD_d, \tag{7}$$

where ρ is the density in Machos. Since $R_E \propto \sqrt{M}$, while for a given ρ the number density of lenses $\propto M^{-1}$, the optical depth is independent of the individual Macho masses. Using the virial theorem, one finds that $\tau \sim (V/c)^2$, where V is the rotation speed of the Galaxy.

More detailed calculations [9] give an optical depth for lensing by halo dark matter of stars in the Large Magellanic Cloud of $\tau_{LMC} \approx 5 \times 10^{-7}$, under the assumptions that (1) all of the dark matter is in the form of Machos; and (2) the most naive model of the halo (spherically symmetric, small core radius) is correct. This very low value means that only one star in two million will be magnified by $A > 1.34$ at any given time. (Note that this estimate assumes that all of the dark matter is in Machos, and hence is a crude upper limit to the optical depth.)

Surveys for gravitational microlensing follow millions of stars photometrically in order to obtain event rates of a few per year against the Large Magellanic Cloud. The optical depth in principle can be estimated directly from the experimental data, once a statistically significant number of events has been recorded. This interpretation is complicated, in practice, by inefficiencies introduced by the irregular sampling, and by the very crowded star fields that must be observed in order to make the large number of photometric measurements.

The interpretation of microlens optical depths in terms of the fraction of the dark matter in the form of Machos is limited by our poor knowledge of ρ_{total}. The dark halo may be spherical or flattened [15], and if flattened it may or may not be aligned with the plane of the disk [6]. The core radius of the halo is not securely known. Also important is a complete understanding of the contribution of the disk and bulge to the total mass interior to the solar circle. This has turned out to be more significant than expected.

Improving our knowledge of ρ_{total} is clearly important. Improvement can be obtained by measuring the optical depth along many well separated lines of sight, combined with the measured rotation curve of the disk. The variation of

τ with location on the sky would provide us with a form of *tomography* of the dark matter distribution, which gives the missing shape information. The rotation curve data provide the normalization for ρ_{total}. To date, measurements have only been attempted along two well separated lines of sight, towards the Large Magellanic Cloud and towards the central bulge of the Milky Way. The latter line of sight does not probe the dark halo unambiguously, but is very useful in probing the structure of the Milky Way, and thus helping to determine ρ_{total}.

4   What has been learned so far

Four groups have reported detections of gravitational microlens events: the MACHO Project (Alcock *et al.*[3] and references therein), the EROS Project[4], the OGLE Project[17], and the DUO Project[1]. These projects follow tens of millions of stars over periods of years and have detected well over a hundred microlensing events; most of these have been seen by the MACHO Project. It is the results of the MACHO and EROS Projects that bear most directly on the dark matter story. The current situation with these two projects is summarized ably elsewhere in this volume by Griest *et al.*[10], and by Palanque-Delabrouille[13]). The most important conclusions drawn from this work are:

- Gravitational microlensing has been observed.

- Machos make up a significant fraction (f) of the dark matter ($20\% < f < 100\%$).

- The Machos appear to have *stellar* masses ($\sim 0.5 M_\odot$).

- True *substellar* objects do not comprise a significant fraction of the dark halo.

5   The problem with the underlying model of the Galaxy

The Macho *fraction* of the dark matter in the halo of the Milky Way is the most important single number that might be determined in the microlensing experiments. This number involves the ratio of the gravitational microlensing optical depth estimated experimentally to the value of that optical depth for a model of the mass distribution in the dark halo (which assumes that all of the dark matter is in Machos). This model introduces significant uncertainty into the result, as shown, for instance, in Alcock *et al.*[3]. This uncertainty arises in part because of inadequate knowledge of the rotation curve of the Galaxy, but also in large part because the *shape* of the dark halo is not well constrained by observation (in this regard, there is much that could be done, as shown by Sackett and Gould [15] and Frieman and Scoccimarro [6]).

Gates [7] and her collaborators have explored very large numbers of models in order to quantify this model uncertainty. In this way, they are able to incorporate the results of the microlensing surveys towards the Galactic bulge, and it becomes clear that significant progress in understanding the mass model of the Milky Way will be needed before a defintive result on the Macho fraction can be obtained.

6 The Macho Parallax Effect

Macho parallaxes merit a special section because they hold the promise greatly to improve on our understanding of gravitational microlensing. A measurement of Macho parallax provides an additional fit parameter that has bearing on the physics of an event, thus doubling the information value of the event. Specifically, one can obtain an estimate of $\tilde{v}$ by measuring an event from two different lines of sight [8]. From the ground one can measure Macho parallaxes only for events of exceptionally long duration, so that the acceleration of the earth around the sun significantly changes the projected velocity during the event. These long duration events are uncommon, but the effect has been measured in an event towards the Galactic bulge [2].

Ground-based work faces a severe limitation in measuring parallaxes: it is only for the very longest duration events that the earth will move far enough in its orbit during the event to make the parallax effect visible from the ground. This limits us to measurements made on events for which the Machos are in the galactic disk, and eliminates from useful study bulge and dark halo Machos (for which the events are of duration much shorter then six months, because of the high expected transverse velocities). For this vast majority of cases, the microlensing events will end before the earth has had a chance to move very far, so the parallax information will only be available if events can be observed simultaneously from the earth and a small satellite in a solar orbit of order ~ 1 AU away from the earth [8].

Satellite parallaxes would greatly advance the study of Machos. For events seen toward the Large Magellanic Cloud, measurement of $\tilde{v}$ would clearly distinguish between Galactic Machos ($\tilde{v} < 300 \, \mathrm{km \, s^{-1}}$) and those in the Large Magellanic Cloud ($\tilde{v} \sim 2000 \, \mathrm{km \, s^{-1}}$). For Galactic Machos, $\tilde{v} \sim v$, so that one could distinguish between Machos in the disk ($v \sim 50 \, \mathrm{km \, s^{-1}}$), the thick disk ($v \sim 100 \, \mathrm{km \, s^{-1}}$), and the halo ($v \sim 200 \, \mathrm{km \, s^{-1}}$). It would be possible to measure the transverse direction for at least some of these Machos, yielding additional information about their distribution.

7 Summary

This short review has only superficially covered the substantial progress made in this field since the seminal suggestion by Paczynski [11]. Much has been learned since gravitational microlensing changed from speculation to observation. In particular, we have discovered that substellar objects do not make a significant fraction of the dark matter in the Milky Way, but that objects of *stellar* mass ($0.5M_\odot$) do. These are the first quantitative statements to have been made about the composition of this elusive, but pervasive matter.

Acknowledgments

Everything I know about gravitational microlensing I have learned either from, or with, my colleagues in the Macho Project. Work performed at LLNL is supported by the DOE under contract W7405-ENG-48.

References

1. Alard., C., *et al.*, 1995 *The Messenger*, No. 80, p. 31.
2. Alcock, C., *et al.*, 1995, *ApJ*, **454**, L125.
3. Alcock, C., *et al.*, 1997, *ApJ*, in press (astro-ph/9606165).
4. Aubourg, E., *et al.*, 1993, *Nature*, **365**, 623.
5. Fich, M. and Tremaine, S. 1991 *Ann. Rev. Astron. Astrophys.*, **29**, 409.
6. Frieman, J., and Scoccimarro, R. 1994, *ApJ*, **431**, L23.
7. Gates, E., 1997, *elsewhere in this volume*.
8. Gould, A., 1994, *ApJ*, **421**, L75.
9. Griest, K., 1991, *ApJ*, **366**, 412.
10. Griest, K., *et al.*, 1997, *elsewhere in this volume*.
11. Paczynski, B., 1986, *ApJ*, **304**, 1.
12. Paczynski, B., 1996, *Ann. Rev. Astr. Ap.*, **34**, *in press*..
13. Palanque-Delabrouille, N., 1997, *elsewhere in this volume*.
14. Primack, J., Seckel, D., and Sadoulet, B., 1993, *Ann. Rev. Nuc. Part. Sci.*, **38**, 751.
15. Sackett, P., and Gould, A., 1993, *ApJ*, **419**, 648.
16. Turner, M., 1997, *elsewhere in this volume*.
17. Udalski, A., *et al.*, 1994, *ApJ*, **426**, L69.

GAMMA-RAY BURSTS: CHALLENGES TO RELATIVISTIC ASTROPHYSICS

MARTIN J. REES

Institute of Astronomy, Madingley Road, Cambridge, CB3 OHA, UK

Although they were discovered more than 25 years ago, gamma-ray bursts are still a mystery. Even their characteristic distance is highly uncertain. All that we can be confident about is that they involve compact objects and relativistic plasma. Current ideas and prospects are briefly reviewed. There are, fortunately, several feasible types of observation that could soon clarify the issues.

1 History

Astrophysics is a subject where the observers generally lead, and theorists follow behind. The topic of my talk is one where the lag is embarrassingly large. However, gamma-ray bursts raise issues which are certainly fascinating to everyone involved in relativistic astrophysics.

Even though the history of gamma-ray bursts dates back more than 25 years, we still know neither where nor what they are. The story started in the late 1960s, when American scientists at Los Alamos had developed a set of satellites aimed at detecting clandestine nuclear tests in space by the associated gamma-ray emission. Occasional flashes, lasting a few seconds, were indeed detected. It took several years before these were realised to be natural, rather than sinister phenomena, and in 1973 a paper was published by Klebesadal, Strong & Olson entitled *Observations of Gamma-ray Bursts of Cosmic Origin*. This classic paper reported 16 short bursts of photons in the energy range between 0.2 and 1.5 MeV, which had been observed during a three-year period using widely separated spacecraft. The burst durations ranged from less than 0.1 second up to about 30 seconds, but significant fine time-structure was observed within the longer bursts. The bursts evidently came neither from the Earth nor from the Sun, but little else was clear at that time.

It did not take long for the theorists to become enthusiastically engaged. At the Texas conference in December 1974, Ruderman (1975), gave a review of models and theories. He presented a long and exotic menu of alternatives that had already appeared in the literature, involving supernovae, neutron stars, flare stars, antimatter effects, relativistic dust, white holes, and some even more bizarre options. He noted also the tendency, still often apparent, for theorists to "strive strenuously to fit new phenomena into their chosen specialities".

In the 1970s and 1980s, data accumulated on gamma-ray bursts, due to a number of satellites. Particular mention should be made of the contributions by Mazets and his colleagues in Leningrad. Also important were the extended observations made by the Pioneer Venus Orbiter (PVO). The number of detected bursts rose faster than the number of models – a further index of progress is that some of the conjectures reviewed by Ruderman were actually ruled out.

During that period, three classes of models were pursued: those in which the bursts were respectively in the Galactic Disc (at distances of a few hundred parsecs), in the halo (at distances of tens of kiloparsecs), and at cosmological distances. The characteristic energies of each burst, according to these three hypotheses, are respectively 10^{37} ergs, 10^{41} ergs, and 10^{51} ergs. The most popular and widely-discussed option during the 1980s was that the bursts were relatively local, probably in our Galactic Disc, and due to magnetospheric phenomena or "glitches" on old neutron stars (defunct pulsars).

It was clear that there were two statistical clues which could in principle decide the location of gamma-ray bursts as soon as enough data had accumulated, and selection effects were understood. One was the number-versus-intensity of the events, which tells us whether they are uniformly distributed in Euclidean space, or whether we are in some sense seeing the edge of the distribution. The other is the degree of anisotropy.

There was already evidence that the counts of gamma-ray bursts were flatter than the classic Euclidean slope, since otherwise more faint bursts would have been detected by balloon experiments. This would not of course have been unexpected if the bursts were within the galaxy. However, the real surprise came with the launch, in April 1991, of the Compton Gamma Ray Observatory (GRO) satellite, whose Burst and Transient Source Experiment (BATSE) offered systematic all-sky coverage, with good sensitivity over the photon energy range 30 keV - 1.9 MeV. Data from BATSE have transformed the subject.

The most remarkable BATSE result is the unambiguous evidence that the bursts are highly isotropic over the sky. More than 1700 have now (December 1996) been recorded, and there is still no statistical evidence for any dipole or quadrupole anisotropy, nor for any two-point correlation (Briggs *et al.* 1997) The lack of any enhancement either towards the plane of the Galaxy, or towards the Galactic Centre, is a very severe constraint on the hypothesis that bursts come from the Galaxy. Note that they cannot be ultra-local objects within our galactic disc: this would naturally permit isotropy, but is ruled out by the flatness of the number counts. The "non-Euclidean" counts imply that the surveys are probing to distances where the sources are, for some reason, thinning out; the problem is to account for this by a hypothesis that is also consistent with the isotropy.

The experiments on GRO have produced evidence on the spectra and time structure of events. (For a recent review, see Fishman (1995) and references cited therein.) Despite the large variety, there is little doubt that gamma-ray bursts are a well-defined class of objects, distinguished spectrally from phenomena such as X-ray bursters, and also from the so-called "soft gamma repeaters" which have substantially softer spectra. Within this class, there are some apparent correlations. For instance, the shorter bursts tend to be stronger and to have somewhat harder spectra; the histogram plotting burst durations may have two peaks; and the counts deviate most from the Euclidean slope for the bursts with harder spectra (Kouveliotou et al 1996).

The manifest isotropy has tilted the balance of opinion strongly towards a cosmological interpretation of the classical gamma-ray bursts. I will concentrate on discussing the challenge posed to theorists by that model. But I will then mention, more briefly, types of halo model that are compatible with the isotropy since these cannot yet be definitively deemed irrelevant. In conclusion, I will list some observations which might in the near future settle the issue, or at least reduce the current level of perplexity. This talk (and the present written version) is intended as a general overview. Fuller details, and more extensive references, can be found in the papers from the special session on gamma-ray bursts, elsewhere in these proceedings, or in Hartmann (1996).

2 Models for "Cosmological" Bursts

If the bursts are cosmological, then the sub-Euclidean counts imply that the typical burst has a redshift z of order 1. The precise redshift distribution depends on how much evolution there is in the population. The mean redshift would be less, for instance, if the burst rate increased with cosmic time. However, we can confidently say that all but the very nearest of the observed bursts must have redshifts of at least 0.2. Otherwise evolution would need to be implausibly steep to explain the non-Euclidean counts, and nearby superclusters would show up in the distribution over the sky. (Since the bursts exhibit such a wide variety of time-structures, it would be astonishing if, by any measure, they were anywhere near being standard candles. Obviously, detailed interpretations of the counts depend on the luminosity function.)

The event rate per unit volume is very low if we are sampling a population out to cosmological distances. It is of order 10^{-5} per year per galaxy, in other words a thousand times less than the supernova rate in galaxies. The required energy release then amounts to 10^{51} ergs in a few seconds. (Both the estimates of the rates and of the energy per event would need to be adjusted in a straightforward way, of course, if the individual events were beamed in a small solid angle.)

3 "The trigger"

The total energy is not necessarily in itself a problem. After all, whenever a supernova goes off, the binding energy of a neutron star is released in a fraction of a second, and this amounts to 10^{53} ergs, a hundred times what is needed for the burst. But in a supernova most of this energy goes to waste as neutrinos; moreover, any impulsive electromagnetic release would not escape promptly, but would be degraded by adiabatic expansion of the envelope before, much later, it could leak out. So is it possible for some rare events to occur where the energy release can escape promptly, rather than being surrounded by an extensive opaque envelope? The most widely favoured possibility is coalescence of binary neutron stars (see, for example, Narayan, Paczynski and Piran 1992). Systems such as the famous binary pulsar will eventually coalesce, when gravitational radiation drives them together. The final merger, leading probably to the production of a black hole, happens in a fraction of a second (though the swallowing or dispersal of all the debris may take somewhat longer). The calculated event rates for such phenomena – and perhaps also for the coalescence of binaries consisting of a neutron star and a black hole, rather than two neutron stars – are uncertain but are probably high enough to supply the requisite rates of bursts.

4 Fireball and gamma-ray emission

How can the energy be transformed into some kind of fireball after such a coalescence event? There seem to be two options. The first is that some of the energy released as neutrinos is reconverted, when the neutrinos collide outside the dense core where they were produced, into electron-positron pairs or photons. The rate of this process depends on the square if the neutrino luminosity, and those simulations that have so far been carried out yield rather pessimistic estimates for the efficiency (Ruffert *et al.* 1996). The second option is that strong magnetic fields directly convert the rotational energy of the system into a directed outflow. This latter option requires that the magnetic fields be amplified to strengths of order 10^{15} Gauss. (Usov 1994; Thompson 1994)

The observed gamma rays seem to have a nonthermal spectrum. Moreover, they commonly extend to energies above 1 MeV, the pair production threshold in the rest frame. These facts together imply that the emitting region must be relativistically expanding. We draw this conclusion for two reasons. Firstly, if the region were indeed only a light second across or less, as would be implied by the observed rapid variability in the absence of relativistic effects, the total mass of baryons in the region would need to be below about 10^{21} grams in order that the electrons associated with the baryons should not provide a large opacity: the rest mass energy of the baryons would need to be 10 orders of magnitude less than that of the radiation energy in the same volume. Not only is this a remarkably low figure, implying that only 10^{-12} of the material from the compact objects is mixed up in the emitting region, but it would in any case imply a relativistic expansion. Quite apart from the baryon constraint, there is a second reason for invoking relativistic expansion. Larger source dimensions are required in order to avoid opacity due to photon-photon collisions (via $\gamma + \gamma \rightarrow e^+ + e^-$).

If the emitting region is expanding relativistically, then for a given observed variation timescale the dimension R can be increased by γ^2. The opacity to electrons and pairs is then reduced by γ^4, and the threshold for pair production, in our frame, goes up by $\sim \gamma$ from its "rest" value of ~ 1 MeV. A high γ will of course only be attained if the baryon loading is sufficiently low, such that the ratio of total energy to rest mass energy is larger than γ. A variety of models have been discussed. Best-guess numbers are, for an energy of 10^{51} ergs, a Lorentz factor γ in the range 10^2 to 10^3, allowing the rapidly-variable emission to occur at radii in the range 10^{14} to 10^{16} cms. The entrained baryonic mass would need to be below $10^{-6} M_\odot$ to allow these high relativistic expansion speeds.

Because the emitting region must be several powers of ten larger than the compact object that acts as "trigger", there is a further physical requirement: the original energy – whether envisaged as an instantaneous fireball or as a short-lived quasi-steady wind – would, during expansion, be transformed into bulk kinetic energy (with associated internal cooling). It must be re-randomised and efficiently radiated as gamma rays: this requires relativistic shocks. Impact on an external medium (or an intense external radiation field) would randomise half of the initial energy merely by reducing the expansion Lorentz factor by a factor of 2. Alternatively, there may be internal shocks within the outflow: for instance, if the Lorentz factor in an outflowing wind varied by a factor more than 2, then the shocks that developed when fast material overtakes slower material would be internally relativistic (Piran 1997 and references cited therein).

In the case of expansion into an external medium, the energy would be rethermalised after sweeping up external matter with rest mass $E/c^2\gamma^2$ (Rees & Mészáros 1992; Mészáros & Rees 1993). For $E = 10^{51}$ ergs and $\gamma = 10^3$, only

$10^{-9}M_\odot$ of external matter need be swept up. In an unsteady wind, if γ were to vary on a timescale δt, internal shocks would develop at a distance $\gamma^2 c \delta t$, and randomise most of the energy (eg Rees & Mészáros 1994). For instance, if γ ranged between 500 and 2000, on a timescale of δt second, internal shocks with Lorentz factors ~ 2 (measured in the frame of the mean $\gamma \simeq 1000$ outflow) would lead to efficient dissipation at $3 \times 10^{16}\delta t$ cms.

Another important consequence of relativistic outflow is that only material moving within an angle γ^{-1} of the line of sight contributes to what we observe. Observations cannot therefore tell us if bursts are highly beamed. Transverse pressure gradients are only effective on angles below γ^{-1}, so material ejected in widely differing directions behaves quite independently. There are already a variety of models in the literature discussing the radiation from shocks in expanding fireballs and relativistic winds (see Piran 1997 for a recent review). The parameters are uncertain, and the relevant physics, involving for instance the coupling between electrons and ions in relativistic shocks, is not sufficiently well developed to allow accurate modelling of the radiation (see, for instance, Gallant et al. 1992).

So how is the original energy channelled from the central object into the outflowing fireball or wind. Recent calculations by Ruffert et al., 1996, suggest problems with releasing neutrino energy efficiently enough, and on a short enough timescale, to allow production of a fireball. The options involving *magnetic* energy (cf Narayan, Paczynski & Piran 1992) are rather less quantitative, but I still believe they are more promising. As discussed by Usov, (1994), and Thompson (1994), a millisecond pulsar with a $\sim 10^{15}$ Gauss field would be slowed down in 1 second, its spin energy being dumped in a pair-dominated relativistic wind. As these authors and others have discussed, internal processes in such a wind could explain gamma rays with the observed spectrum and variability characteristics.

5   A "best buy" model

My personal favourite model (cf Meszaros and Rees 1997b) involves the toroidal debris from a disrupted neutron star orbiting around a black hole. If this debris contains a strong magnetic field, amplified perhaps by differential rotation, then an axial magnetically-dominated wind may be generated along the rotation axis, perpendicular to the plane of the torus. The advantage of this geometry is that it seems to offer the best chance of preventing baryon contamination, because the baryonic material would be precluded by angular momentum from getting near the axis without first falling into the black hole or being on a positive-energy trajectory.

Such a configuration could arise from capture of a neutron star by a black hole of less than $5M_\odot$, this mass limit being required because otherwise the neutron star would be swallowed before disruption. Alternatively, it could be the outcome of the merger of two neutron stars, where most of the mass collapses to a black hole, leaving some fraction of the original material in orbit around it. (cf Ruffert *et al.* 1996; Jaroszynski 1996)

The available energy in this model is the kinetic or gravitational energy of the neutron-star debris left behind in the torus, plus the spin energy of the hole itself (which, being the outcome of binary coalescence, is almost guaranteed to have a high angular momentum). Near the axis, we would expect maximal dissipation (from fields threading the hole or anchored in the torus) but minimum baryonic loading. The Lorentz factor would therefore be largest along the axis. Indeed, a narrow channel, essentially free of baryons, may carry a Poynting-dominated outflow, energised by the hole via the Blandford-Znajek process.

Along any given line of sight, the time-structure would be determined partly by the advance of jet material into the external medium, but probably even more by internal shocks within the jet, which themselves depend on the evolution and instabilities of the torus, from its formation to its eventual swallowing or dispersal. Even if the bursts were caused by a completely standardised set of objects, their appearance would be likely to depend drastically on orientation relative to the line of sight. Other phenomena as yet undiscovered – for instance some new class of X-ray or optical transient – may be attributable to gamma-burst sources viewed from oblique orientations.

6 Physics of the emission mechanism

We are a long way from a convincing model for what triggers gamma-ray bursts: coalescing compact binaries seem likely to be implicated, but we should remain open-minded to more exotic options. A precise description of the dynamics, along with the baryon content, magnetic field, and Lorentz factor of the outflow, might allow us to predict the gross time-structure. But even then we could not predict the intensity or spectrum of the gamma rays – still less answer key questions about the emission in other wavebands – without also having an adequate theory for particle acceleration in relativistic shocks. We need the answers to the following poorly-understood questions:

(i) Do relativistic shocks yield particle spectra that obey power laws? This is in itself uncertain: the answer probably depends on the ion/positron ratio, and on the relative orientation of the shock front and the magnetic field (e.g. Gallant *et al.* 1992).

(ii) In ion-electron plasmas, what fraction of the energy goes into the electrons?

(iii) Even if the shocked particles establish a power law, there must be a low-energy break in the spectrum at an energy that is in itself relativistic. But will this energy, for the electrons, be $\Gamma_s m_e c^2$, or $\Gamma_s m_p c^2$ (or even, if the positive charges are heavy ions like Fe, $\Gamma_s m_{Fe} c^2$?

(iv) Can ions be accelerated up to the theoretical maximum where the gyroradius becomes the scale of the system? If so, the burst events could be the origin of the highest energy cosmic rays (an interesting possibility addressed by other speakers at this conference)

(v) Do magnetic fields get amplified in shocks? This is relevant to the magnetic field in the swept-up external matter outside the contact discontinuity, and determines how sharp the external shock actually is (cf Mitra 1996)

(vi) Can radio emission be generated by a coherent process? If not, the usual surface brightness constraint implies that there would be little chance of detecting a radio "afterglow".

These questions, crucial for gamma ray bursts, are also relevant to other phenomena. For example, Lorentz factors of at least 10 (and probably electron-positron plasmas) exist in the compact components of strong extragalactic radio sources probed by VLBI.

If one is prepared to parametrise the uncertainties implicit in the above questions, predictions can be made of how the spectrum would evolve during a burst with simple time-structure. (eg Meszaros *et al.* 1994, Tavani 1996, Meszaros and Rees 1997). For a wide range of parameters, the associated X-rays would be above the threshold of small omnidirectional detectors such as those developed for the High Energy Transient Explorer (HETE) It was therefore a real setback to the subject – particularly to the prospect of using concurrent X-ray or UV emission to pinpoint the burst locations more accurately – when HETE failed to go properly into orbit.

After the main emission is over, the fireball material would continue to expand, with steadily-falling Lorentz factor, into the external medium. Associated optical emission may persist for hours or even days. This is long enough to allow an initial detection with BACODINE to be followed up by raster scans with a 1 m telescope, that could detect even emission down to 15th magnitude.

7 "Extended Halo" Models

In the interests of balance, I would like to make a few remarks about the alternative idea that the bursts are not from cosmological distances, but instead come from within our own galaxy. Classical gamma-ray bursts could be isotropic enough to be consistent with the BATSE data if they came from neutron stars ejected from our Galactic Disc at more than 700 km s^{-1}, which remained active, bursting sporadically, for long enough to allow them to reach distances of at least 100 kiloparsecs. They may either escape from the galaxy, or be on very extended bound orbits. These high velocity objects could be a special subset of pulsars. The typical velocities of the pulsars sampled in surveys may be as high as 400 km/sec (Lyne and Lorimer 1994, J. Taylor, these proceedings). Moreover, those that formed with higher kick velocities and/or with strong magnetic fields (and therefore short lifetimes) are under-represented in surveys; we cannot exclude the possibility that a high fraction of newly-formed pulsars are of such types. If we conservatively suppose that they are only a few percent of all pulsars, and form in our Galactic Disc at a rate of about 1 per thousand years, then each must produce 10^6 bursts, of typical energy 10^{41} ergs. (If the relevant objects formed at a rate of one per 100 years, the requirements placed on each would be ten times more modest.) Repetition would not necessarily be expected, since each neutron star could in principle continue bursting at a slow rate for more than a billion years. However, if the bursts came in groups, rather than being independent poissonian events, repetition would not be impossible.

8   Fitting the isotropy

Podsiadlowski has done detailed calculations of whether such a population can provide an isotropic distribution. He shows this is indeed possible for long-lived bursters whose orbits take them out beyond 100 kiloparsecs. An important feature of such orbits is that, because the galactic halo potential is not spherical (and may indeed be rather irregular at such large distances) objects do not conserve their angular momentum and therefore, even if they started off near the centre of our Galaxy, they need not return so close to the centre in later orbits. This effect helps to ensure greater isotropy. Another possibility, favoured by Lamb (1995), is that the typical objects have velocities above a thousand kilometres per second, and are escaping the Galaxy completely. In this case, the best fit is obtained if the bursts do not start until after a delay of around 10^7 years, by which time all neutron stars have reached distances of 30 kiloparsecs or more.

I think it is fair to say that such models need to be carefully tuned in order to fit the existing isotropy data, but that, though perhaps unappealing, they cannot be ruled out. The constraints on orbital parameters would be eased in alternative schemes where the neutron stars *formed* far out in the halo (being perhaps, as Woosley (1993) has discussed, relics of an early population of halo stars) rather than being ejected from the disc.

9 Mechanisms for halo bursts

If a "halo" model is to be taken seriously, there must be an acceptable mechanism for producing the succession of 10^{41} erg bursts, spread over a very long timescale. Two options have been proposed (Podsiadlowski, Rees & Ruderman 1995).

The first possibility is that the relevant subset of neutron stars start off with a super-strong ($\gtrsim 10^{15}$ Gauss) magnetic field. This field, penetrating the core of the neutron star, would gradually rise towards the surface through buoyancy effects, thereby causing stress in the crust. The timescale for the buoyancy is estimated to be at least 10^6 years. Acceptable models require that it be $\gtrsim 10^9$ years. The total stored energy is $\sim 10^{47}(B/10^{15}G)$ ergs. The energy depends linearly on B, rather than quadratically, because the field in the core is concentrated into tubes where its strength has a standard value of $\sim 3 \times 10^{15}G$.

The crust gets stretched as the field drifts outwards. The units in which energy is released depend on how much stress can build up in the crust, and what fraction is released when the crust cracks. This is a complicated problem in asteroseismology. However, a release of 10^{41} ergs per event is plausible, in which case the total stored magnetic energy would be sufficient to supply the requisite 10^6 events.

The second very different option for triggering halo bursts involves asteroidal impacts on to a neutron star. Each event requires, on energetic grounds, the impact of 10^{21} grams. The main problem with this idea is that such asteroidal or cometary bodies would be tidally disrupted too far out to give a sudden enough event. A possible solution is that the debris from the disrupted body squashes down the magnetic field, which then rebounds, generating high electric fields and thereby a pair cascade. Alternatively, the debris may form a disc which accumulates before triggering a sudden electromagnetic release when it couples its rotation to that of the neutron star.

The total impacting mass, to get enough bursts per star, must be 10^{27} grams. It is not impossible (especially now we know that planetary systems can exist around pulsars) that a neutron star could carry with it $\gtrsim 10^{27}$ gm of asteroidal debris. However, a larger reservoir, plus at least one large planet, is needed in order for enough of these planetesimals to be perturbed on to near radial orbits. We know that at least one pulsar has a planetary system. This fact, plus the evidence that even typical pulsars may have velocities of 400 km s^{-1}, suggests that models of this kind should not be dismissed. Whatever the bursts turn out to be, the primary trigger, and the efficient conversion of its energy into gamma rays, involve physical conditions that are extreme and unfamiliar.

10   How can we settle the debate?

There is no convincing and fully worked out model for the bursts on either the halo or the cosmological hypothesis. Neither option, however, seems to violate any cherished beliefs in physics or relativistic astrophysics. The issue is one of plausibility, and how one weighs different lines of evidence. The isotropy would be a natural consequence of the cosmological hypothesis. But the level of isotropy so far revealed by BATSE, which restricts any dipole or quadrupole anisotropy below the few per cent level and shows no evidence for clumping on smaller scales, could be accommodated in a halo hypothesis if high speed neutron stars were implicated.

In April 1995, the 75th anniversary of the Shapley/Curtis debate, there was an interesting debate in Washington on the location of gamma-ray bursts – a current issue offering some amusing parallels to the earlier controversy concerning the distances of the nebulae. The two main protagonists were Don Lamb and Bohdan Paczynski (a written version of the argument appears in Lamb (1995) and Paczynski (1995)). I had the privilege of acting as the moderator in this debate, perhaps because I was one of the few people who had not already taken a firm stance on one side of the issue or the other. There was an agreement among all participants that the issue would be settled only by more data. Indeed, there was a broad consensus on some particular tests that could be crucial, or at least highly suggestive. Among these might be the following.

Most valuable of all would be a firm identification of a burster with some other class of object. The stumbling block here is the poor positional accuracy of most gamma-ray detectors. BATSE itself has error circles of 1 or 2 degrees for the brightest bursts, and more than 5 degrees for the fainter ones. However, the locations of some bursts have been pinned down with a precision of minutes of arc or better by triangulation experiments involving deep space probes; this technique utilises the rapid time structure, which, when recorded and timed by detectors separated by 10 light minutes or more, allows accurate positioning. There is still no firm identification of any classical gamma-ray burst, though there are tantalising indications that some of the brighter bursts may be correlated with galaxies or clusters of galaxies, whose distances are not inconsistent with what is expected on the cosmological hypothesis. (It is disappointing, incidentally, that the failure of the recent Mars probe, which would have carried a small gamma-ray detector, means that we now lack the requisite deep-space network for obtaining accurate "triangulation".)

Even though the gamma-ray positional information is poor, one might be able to pin down the position of the sources more accurately if they displayed concurrent transient emission in some other waveband. Various projects have been undertaken in the optical and radio band. Ground-based observers can be notified of a BATSE event within a few seconds; a small telescope can then be rapidly slewed to seek an optical counterpart within less than a minute. No such counterparts have been detected, nor have radio searches yet yielded positional or timing coincidences. The likely strength of gamma-ray bursts in the optical or radio band is uncertain and highly model-dependent. Indeed, any detection in these wavebands would have the bonus that it would help to narrow down the range of possible models and emission mechanisms. However, most theories predict that there should be substantial spectral extension from gamma-rays down towards the X-rays, so it would seem less of a gamble to seek X-ray counterparts.

If the bursts have a local rather than cosmological origin, then, at some level, anisotropies over the sky would be bound to show up. A particularly crucial test would be feasible if bursts more than ten times fainter than those recorded by BATSE could be detected. It would then, according to the halo hypothesis, be feasible to detect bursts from the halo of Andromeda, and there should be a definite excess of weak events from that direction (Bulik & Lamb 1997, Ruszkowski & Wijers 1997). The lack of such a trend would severely embarrass halo models. A specific proposal has been made to look at a 10 degree field around Andromeda with 20 times the sensitivity of BATSE. But X-ray detectors are more readily available and more sensitive. For this reason, and also because the x-ray emission from bursts seems stronger than a straight extrapolation of the gamma-ray spectrum suggests (Preece et al. 1996), the best prospects for testing the halo model might be from long-duration observations of Andromeda and other nearby galaxies.

The cosmological interpretation of bursts would be confirmed, as Paczynski (1986) first pointed out, if there were evidence of gravitational lensing by an intervening galaxy. If a suitable galaxy lay along the line of sight to a cosmologically distant burst, radiation would reach us by two or more different paths, whose light travel times would differ typically by weeks or months. We would therefore detect two bursts from the same direction. Even though the positions could not be pinned down accurately, the elaborate time structure of each burst is highly distinctive, and if two bursts with identical "fingerprints" were detected from within the same error circle, this would be compelling evidence that they were actually separate gravitationally-lensed images of the same burst. (As a technical point, it should be noted that microlensing by stars or substellar objects would only introduce differences on millisecond timescales between the two burst profiles (Williams and Wijers 1997), and therefore would not vitiate this possibility.)

Unfortunately, the probability that a galaxy lies along a random line of sight to a high redshift object is below one per cent, the exact value depending of course on the presumed redshift of the burst. Moreover, because BATSE can only observe a given direction in the sky for about 40 per cent of the time it is more likely than not that, if a lensed event occurred, the recurrence would be missed because it would occur during dead time. Taking these effects into account, it is rather marginal whether we would expect BATSE to detect a single instance of this lensing before it dies, even if the bursts indeed come from cosmological distances. However, if we were lucky, such a double burst could clinch the cosmological interpretation.

A further issue which has figured strongly in the debate on the location of gamma-ray bursts concerns the existence or otherwise of spectral features attributable to cyclotron lines. This is a technical controversy which I will not enter here. However, its relevance lies in the fact that halo models involve neutron stars, where the magnetic fields are expected to be in the range such that cyclotron lines should be in the hard X-ray band. On the other hand, the fields in the emitting regions of cosmological fireballs or relativistic winds would not, even when relativistic effects are taken into account, give rise to such features.

The controversies in the Shapley-Curtis debate were settled within a few years. Our knowledge of extragalactic astronomy thereby made a forward leap, and astronomers moved on to address more detailed issues. I'm enough of an optimist to believe that it will be only a few years before we know where and perhaps even what, the gamma-bursters are. Even if this optimism is misplaced, I am completely sure that these mysterious phenomena will serve as a continuing challenge and stimulus to theorists, and will remain high the agenda of future Texas Conferences.

I thank my colleagues Josh Bloom, Peter Meszaros, Philipp Podsiadlowski, and Ralph Wijers for collaboration and discussion. I am also very grateful to the members of the BATSE team, especially Jerry Fishman and Jim Brainerd, for updating me on the observations and answering several queries.

1. Briggs, M. *et al.* ApJ, 1997 (in press).
2. Bulik, T, and Lamb, D.Q. Space Sci. Rev 1997 (in press).
3. Fishman, G.J. PASP, 107, 1145, 1995.
4. Gallant, Y.A., Hoshino, M., Langdon, A.B., Arons, J. and Max, C.E. ApJ, 391, 73, 1992.

47

5. Hartmann, D.H. A&AS, 120, 31, 1996.
6. Jaroszynski, M. A&A, 305, 839, 1996.
7. Klebesadal, R.W., Strong, I.B. and Olsen, R.A. ApJ Lett., 182, L85, 1973.
8. Kouveliotou, C. *et al.* Proc 3rd Huntsville Symposium on Gamma-Ray Bursts (AIP) (in press).
9. Lamb, D.Q. PASP, 107, 1152, 1995.
10. Lyne, A.G and Lorimer, D.R. Nature, 369, 127, 1994.
11. Mészáros, P. and Rees, M.J. ApJ, 405, 278, 1993.
12. Mészáros, P. and Rees, M.J. ApJ, 1997 (in press)
13. Mészáros, P., Papathanassiou, H. and Rees, M.J. ApJ, 432, 181, 1994.
14. Mitra, A. A&A, 313, L9, 1996.
15. Narayan, R., Paczynski, B. and Piran, T. ApJ Lett., 395, L83, 1992.
16. Piran, T. In *Unsolved Problems in Astrophysics* (ed. J. Bahcall and J.P. Ostriker) Princeton U.P. 1997
17. Paczynski, B. ApJ, 308, L43, 1986.
18. Paczynski, B. PASP, 107, 1167, 1995.
19. Podsiadlowski, P., Rees, M.J. and Ruderman, M. MNRAS, 273, 755, 1995.
20. Preece, R. *et al.* ApJ, 473, 310, 1996.
21. Rees, M.J. and Mészáros, P. MNRAS, 258, 41P, 1992.
22. Rees, M.J. and Mészáros, P. ApJ, 430, L93, 1994.
23. Ruderman, M. Ann. N.Y. Acad. Sci. 262, 164, 1975.
24. Ruffert, M., Janka, H.-T., Takahashi, K. and Schäfer, G. A&A 1996 (in press).
25. Ruszkowski, M. and Wijers, R.A.M.J. MNRAS 1997 (submitted).
26. Tavani, M. ApJ, 466, 768, 1996.
27. Thompson, C. MNRAS, 270, 480, 1994.
28. Usov, V.V. 1994, MNRAS, 267, 1035, 1994.
29. Williams, L.L.R. and Wijers, R.A.M.J. MNRAS 1997 (in press).
30. Woosley, S.E. ApJ, 45, 273, 1993.

EXTREMELY HIGH ENERGY COSMIC RAYS : AGASA RESULTS

M. NAGANO

Institute for Cosmic Ray Research, University of Tokyo
3-2-1 Midoricho, Tanashi, Tokyo, 188 JAPAN

Recent results on primary cosmic rays of energies above 1×10^{19} eV observed by the Akeno Giant Air Shower Array (AGASA) are summarized. It is most likely that extremely high energy cosmic rays are from diffuse sources distributed isotropically in the universe. However, some fraction of cosmic rays beyond 40EeV seem to come nearby sources composing double events within a limited space angle. The implications of these results for the origin and propagation through the intergalactic space are discussed.

1 Introduction

AGASA [1] is the Akeno Giant Air Shower Array covering over 100km^2 area in operation at Akeno village about 130km west of Tokyo, in order to study extremely high energy cosmic rays (EHECR) above 10^{19} eV. In this energy region, distinctive features in the energy spectrum and arrival direction distribution are expected. If the cosmic rays are of extragalactic origin, photopion production between cosmic rays and primordial microwave background photons becomes important at energies above 6×10^{19} eV with a mean free path of about 6Mpc. Therefore a cutoff in the spectrum may be observed around several times 10^{19} eV even if the primary cosmic ray energy spectrum extends beyond 10^{20} eV. This is called as the Greisen-Zatsepin-Kuzmin (GZK) cutoff [2]. In this senario, the expected arrival direction distribution of EHECR may be quite isotropic. On the other hand, if they are galactic origin, their expected arrival direction distribution is no more isotropic, since the gyroradius of protons of energy above 10^{19} eV exceeds the thickness of the galactic disc. Therefore a study of correlations of EHECR with the galactic structure and/or with the large scale structure of galaxies is very important.

So far the experiments in this energy region have been made at Volcano Ranch [3], Haverah Park [4], Narabrai [5], Yakutsk [6], Dugway [7] and Akeno [8], and the significances of evidence for the *GZK cutoff* and their isotropic arrival direction distributions have increased [9]. Therefore detection of a few$\times 10^{20}$ eV cosmic rays, by the Fly's Eye [10] and AGASA [11] well beyond the expected cutoff energy, has posed a puzzle concerning its origin. Recent observation of AGASA events with energies above 4×10^{19} eV, coming within a space angle of $2.5°$ from the direction of supergalactic plane [12] and a possible correlation of

Haverah Park data with supergalactic plane [13] should be also remarked.

As accerleration mechanism of these EHECR, the diffusive shock acceleration is most widely accepted and shocks at radio lobes of relativistic jets from Active Galactic Neuclei, termination shocks in local group of galaxies, etc. are discussed to be possible sites [14]. The acceleration above 10^{20}eV in dissipative wind models of cosmological gamma-ray burst is also proposed [15]. Since various difficulties are still anticipated in accelerating cosmic rays up to the highest observed energies, it is also discussed that these cosmic rays are possiblly the decay products of some massive particles produced at the collapse and/or annihilation of cosmic topological defects which could have been formed in a symmetry-breaking phase transiton in the early universe [16].

In this report recent results on AGASA experiment are described.

2 AGASA

The AGASA consists of 111 scintillation detectors of $2.2m^2$ area each, which are arranged with inter-detector spacing of about 1km over 100km^2 area. Muon detectors of various areas (2.4m$^2 \sim$ 10m^2 area) are installed in 27 sites out of 111. The two-way communication between the detectors and the central station for data transmission is carried out through two optical fiber cables. The triggering requirement is more than 5 fold coincidences of neighbouring detectors. The details of the AGASA are described in Chiba et al. [1] and Ohoka et al. [17]. The array is operation since 1991 and the exposure is about 800km^{2}year till the end of 1996. In the following analysis, data till the end of 1995 are included.

3 Energy Spectrum

In order to estimate the primary energy of giant air showers observed by the AGASA, the particle density at a distance of 600m from the shower axis (S(600)) is used as an energy estimator, which is known to be a good parameter [18]. The conversion factor from S(600) [per m^2] to primary energy E_0 [eV] is derived by simulation [19] as

$$E_0 = 2.0 \times 10^{17} \times S(600)^{1.0}. \tag{1}$$

The details of determination of arrival directions and S(600) are described in Yoshida et al.[20].

The differential and integral primary energy spectra above $10^{18.5}$eV are shown in figure 1 [21]. The bars represent statistical errors only and the error in the energy determination is about 30% above 10^{19}eV, which is estimated from

50

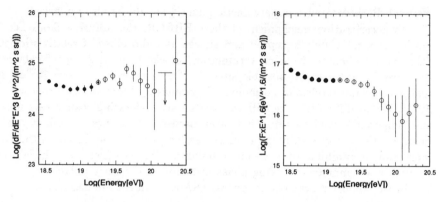

Figure 1: The differential and integral energy spectrum around the ankle. New data(○) are normalized to our pevious one (•) at $10^{19.25}$ eV. An upper limit at $10^{20.2}eV$ is 90%C.L.

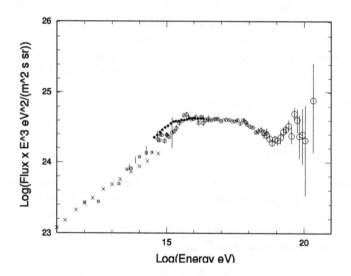

Figure 2: *Differential energy spectrum of primary cosmic rays. The energy spectrum determined by AGASA (large open circles) in figure 1 is normalized to that by Akeno A1(small open circles) by multiplying 0.9 in energy. Closed circles:Tibet, crosses:Aoyama-Hirosaki, squares:Proton satellite, Bars:JACEE.*

the analyzing the artificial showers simulated considering shower development fluctuation and experimental errors in each detector[20].

In order to estimate the systematic error of primary energy determined from (1), the whole energy spectrum determined at Akeno together with that in lower energy region is shown by open circles in figure 2, along with the energy spectrum determined by the direct observations[22,23,24] and the results determined at 4300m.a.s.l.[25].

The Akeno energy spectrum between $10^{14.5}$ and 10^{18}eV is based on the number spectrum of total charged particles (shower size, N_e) in EAS, determined by the '1km^2 Array'(A1)[26] which is in the southeast corner of the AGASA. A1 consists of 156 scintillation detectors, each 1m^2 in area, which are distributed over an area of about 1km^2 with detector spacing of 120m, and 30m in three regions of (90×90)m^2 area each where clusters of detectors were operated to observe the showers whose total number of particles around 10^6. The N_e is converted to energy[26] as

$$E_0 = 3.9 \times 10^{15} \times (\frac{N_e}{10^6})^{0.9}. \tag{2}$$

By using big showers hitting inside A1, the energies converted by equation (1) [E_{S600}] and (2)[E_{Ne}] are compared. The median value of $\frac{E_{S600}}{E_{Ne}} = 1.10$ and the dispersion is 45% for showers of median energies of $10^{18.1}$eV. This value agrees with the simulation result[19] which gives the error in energy estimation in individual arrays as about 20% and 40% around $10^{18.1}$eV.

The Akeno energy spectrum determined by using (2) coincides very well with that determined by Tibet group (closed circles in figure 2[25]) which may be the best one around knee region, since the shower sizes at Tibet altitude don't depend much on the shower development fluctuation and different primary composition. The small difference around 10^{15}eV between Akeno and Tibet spectrum may be due to the fact that some showers from heavy composition are attenuated considerably and can't be reached at Akeno level.

The energy spectrum in the highest energy region from four different experiments from Haverah Park[4], Yakutsk[6], Fly's Eye[7] and the present one agrees well within 20% in energy around 10^{19}eV, though the difference is large above 3×10^{19}eV due to low statistics of each experiments. It should be emphasized here that the agreement of the calorimetric method used by Fly's Eye and Yakutsk with the Monte Carlo based method used by Haverah Park and AGASA implies that the energy determination is rather good in these experiments.

From the above discussion, the systematic error in energy determination in Akeno experiment may not be very large and the energy spectrum shown in figure 1 may be used as a standard one for discussion of the origin of primary cosmic rays. Though a possibility of extending a power law spectrum without cutoff up to the observed highest energy still remains, the observed spectrum of figure 1 is well fitted to the expected ones which were simulated under an assumption of diffuse sources distributed isotropically in the universe or sources at very far distances [8,27]. If the sources are very far or uniform in universe, we may not observe any anisotropy in the arrival direction distribution of these events.

4   Arrival Direction

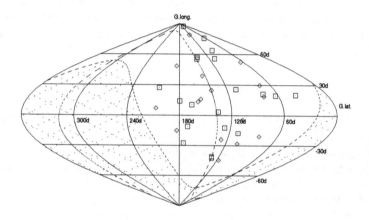

Figure 3: *The arrival direction distribution of 36 cosmic rays above* $4 \times 10^{19}\,eV$ *in galactic coordinates. Open squares represent events above* $5 \times 10^{19}\,eV$ *(20 events) and open diamonds between* 4×10^{19} *and* $5 \times 10^{19}\,eV$(16 events). *The dashed curve shows the supergalactic plane and sky not observed by AGASA due to a zenith angle cut at 45° is shown as cross-hatched area.*

The arrival direction distribution around 10^{19}eV is quite isotropic and no significant preference along the galactic or supergalactic plane was observed below $10^{19.6}$eV as discussed in Takeda et. al. [28]. The AGASA data are plotted in figure 3 for events with energy > 4×10^{19}eV [12]. It is seen that arrival direction of significant fraction of EHECR are uniformly distributed over the observable sky, supporting the interpretation of energy spectrum at the highest energy end. However, it should be also remarked that two pairs of showers, each clustered within a 2.5°, are observed among the 20 events

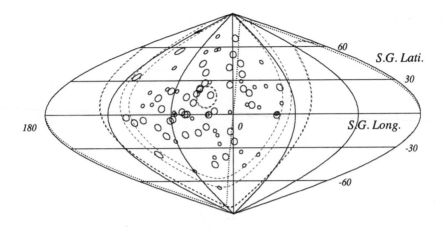

Figure 4: *Arrival directions with error circle of EHECR of four ground array experiments on supergalactic coordinates. Total number of events is 81.*

above 5×10^{19} eV, corresponding to a chance probability of 1.7%. If we refer to 36 showers with energies above 4×10^{19} eV, another pair is observed at high supergalactic latitude with 2.9% chance probability. It should be noted that two pairs of them are within 2.0° of the supergalactic plane.

There is no experimental evidence for the primaries of pair events to be gamma-rays. Since decay length of a neutron is about 1Mpc at 10^{20} eV, neutrons after being produced and escaping from the large magnetic field environment around a source must travel most of their way through intergalactic space as protons.

If the primary particles are protons, significant constraints are imposed on the scale, the strength and the direction of the magnetic field configuration and its turbulence in the interstellar and intergalactic space to explain pair events with angular separation of only a few degrees.

Recently the arrival direction distribution of EHECR above 4×10^{19} eV determined by four ground array experiments in the northern hemisphere is summarized [29] as shown in figure 4. In case of AGAGA, radii of error circle is 1.6° and in case of Haverah Park, Volcano Ranch and Yakutsk, they are 3.0°. There are three doublets and a triplet on very narrow region within ±2° from the supergalactic plane. The chance of observing doublets within ±10° of supergalactic plane is less than a few %.

5 Composition

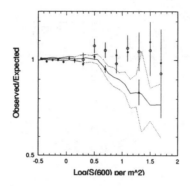

Figure 5: *$\rho_\mu(600)$'s divided by expected values are plotted as a function of S(600). The expected densities are extrapolations from the average relation determined below $10^{17.5}$ eV. An expectation from two composition model proposed by the Fly's Eye group is shown by dotted lines (68% C.L.) and a solid line (the average) respectively.*

From the comparison of the experimental results on the depth for the maximum shower development (X_{max}) from the Fly's Eye experiment [31] with the expected energy dependence of X_{max} for proton and iron primaries obtained from simulations, Gaisser et al. [32] interpreted the Fly's Eye result that the proportion of protons increases with increasing energy and is about 90% at 10^{19} eV.

At AGASA X_{max} can't be determined, however, the low energy muon density far from the core, which is related to the primary composition, can be estimated. If we use superposition model on nucleus-nucleus interaction, an elongation rate of total number of low energy muons (N_μ) is derived from

$$N_\mu = kA\left(\frac{E}{A}\right)^\alpha as \quad \frac{\delta ln N_\mu}{\delta ln E} = \alpha + (1-\alpha)\frac{\delta ln A}{\delta ln E} \qquad (3)$$

That is, the slope of N_μ vs E relation may change from 0.82 to 0.72 around $10^{17.5}$ eV, if the above Fly's Eye result is due to a change in the primary composition. In figure 5, the AGASA result [33] on muon density at 600m from the core ($\rho_\mu(600)$) and S(600) relation is plotted, along with the expected relation from Fly's Eye two composition model by a Monte Carlo simulation based on the MOCCA program [34,35,36]. The change of the composition above $10^{17.5}$ eV has not been detected beyond the experimental uncertainties and assumptions used in the simulation.

Walker and Watson [37] estimated elongation rate of X_{max} from the arrival time distribution of shower particles far from the core measured by water Čerenkov detectors and concluded that the main composition around 10^{19}eV becomes lighter than that in the lower energy region. In Auger Project [38], water Čerenkov detectors of 10m^2 area each (1.2m depth) are planned to deploy as a ground array. At Akeno, a prototype Auger water Čerenkov detector is installed and now in operation [39,40]. A detector with two scintillators sandwiching a lead plate of 1 cm thickness (leadburger) in total 12m^2 area has been also operational since September, 1994 [41]. These detectors will be helpful to determine arrival time distributions of muons, electrons and photons, separately and to estimate X_{max} of each shower in the next few years.

6 The highest energy events

There are two events whose energies exceed considerably the GZK cut-off energy. One is an AGASA event [11] and the other is a Fly's Eye event [10] observed by the optical method. As described before their sources can't be very far and must be within a few tens of Mpc. There are no candidate astronomical objects within a few tens of Mpc in the direction of the highest energy events which may be able to accelerate particles to more than 10^{20}eV and there is an apparent gap in the existing data between the highest energy events and other events.

These observations have led to suggestions that particles may be directly produced by decay from objects produced in phenomenon occurring on a higher energy scale [16,42]. For example, topological defects (TD's) left over from the phase transitions in the early universe, caused by the spontaneous breaking of symmetries, have been proposed as a candidate for production of extremely high energy cosmic rays. Such TD's are magnetic monopoles, cosmic strings, domain walls, superconducting strings etc. The remarkable result from this scenario is that bulk of cosmic rays above 10^{20}eV may be gamma-rays rather than protons. Also, there is expected to be a gap around 10^{20}eV and the energy spectrum is expected to extend up to the GUT scale $\sim 10^{25}$eV with a hard exponent such as 1.35 [43], which is based on the exponent of hadronization in QCD.

7 Conclusion

It is most likely that EHECR's are from diffuse sources distributed isotropically in the universe. However, some fraction of cosmic rays beyond 40EeV seem to come nearby sources composing double events within a limited space angle.

In order to confirm the above findings, it is quite important to increase data significantly. We are planning to continue the AGASA experiment for more than five years expecting that the HiRes Detector(phase II) [44], the Telescope Array Project [45] and the Auger Project [38], will be in their steady operation during that time.

Acknowledgments

I am grateful for the contributions of each AGASA collaborators. This work is supported in part by the Grant-in-Aid for Scientific Research No.06402006 from the Japanese Ministry of Education, Science and Culture.

References

1. N.Chiba et al., *Nucl. Instrum. Methods* A **311**, 388 (1992).
2. K.Greisen, *Phys. Rev. Lett.* **16**, 748 (1966); G.T.Zatsepin and V.A.Kuzmin, *Pisma Zh. Eksp. Teor. Fiz.* **4**, 144 (1966).
3. J.Linsley, Proc. 13th ICRC, Denver **5** (1973).
4. M.A.Lawrence, R.J.O.Reid and A.A.Watson, *J. Phys. G: Nucl. Part. Phys.* **17**, 733 (1991).
5. M.M.Winn, *J. Phys. G: Nucl. Part. Phys.* **12**, 653 (1986).
6. B.N.Afanasiev et al., *Proc. Tokyo Workshop on Techniques for the Study of the Extremely High Energy Cosmic Rays*, ed. Nagano (Inst. Cosmic Ray Research, Univ. of Tokyo, 1993) p.35.
7. D.J.Bird et al., *Ap. J.* **424**, 491 (1994).
8. S.Yoshida et al., *Astroparticle Phys.* **3**, 105 (1995).
9. M.Teshima, Invited, Rapporteur and Highlight Papers, *Proc. 23th ICRC, Calgary*, ed. D.A.Leahy, R.B.Hicks and D.Venkatesan (World Scientific, Singapore, 1993) p.257.
10. D.Bird et al., *Ap. J.* **424**, 491 (1995).
11. N.Hayashida et al., *Phys. Rev. Lett.* **73**, 3491 (1994).
12. N.Hayashida et al., *Phys. Rev. Lett.* **77**, 1000 (1996).
13. T.Stanev et al., *Phys. Rev. Lett.* **75**, 3056 (1995).
14. Various acceleration models are discussed in *Astrophysical Aspects of the Most Energetic Cosmic Rays*, ed. M.Nagano and F.Takahara (World Scientific, Singapore, 1990) pp.252-334.
15. E.Waxman, *Phys. Rev. Lett.* **75**, 386 (1995); M.Vietri, *Ap. J.* **453**, 883 (1995); M.Milgrom and V.Usov, *Ap. J.* **449**, L37 (1995);
16. Recent summary and references are described in P.Bhattacharjee, *Proc. of ICRR Symp. on Extremely High Energy Cosmic Rays : Astrophysics*

and Future Observatories ed. M.Nagano, (Inst. of Cosmic Ray Research, University of Tokyo., 1997) p.125.

17. H.Ohoka et al., *Nucl. Instrum. Methods* A , (1997) in print.
18. A.M.Hillas et al, *Proc. 12th ICRC, Hobart* **3**, 1001 (1971).
19. H.Y.Dai et al., *J. Phys. G: Nucl. Part. Phys.* **14**, 793 (1988).
20. S.Yoshida et al., *J. Phys. G: Nucl. Part. Phys.* **20**, 651 (1994).
21. T.Doi et al., *ICRR-Report-353-96-4* 1 (1996).
22. N.L.Grigorov et al, *Proc. 12th ICRC, Hobart* **5**, 1760 (1971).
23. K.Asakimori et al, *Proc. 23rd ICRC, Calgary* **2**, 25 (1993).
24. M.Ichimura et al, *Phys. Rev.* D **48**, 1949 (1993).
25. M.Amenomori et.al., *Ap. J.* **461**, 461 (1996).
26. M.Nagano et al., *J. Phys. Soc. Japan* **53**, 1667 (1984).
27. N.Hayashida et al., *Proc. of ICRR Symp. on Extremely High Energy Cosmic Rays : Astrophysics and Future Observatories* ed. M.Nagano, (Inst. of Cosmic Ray Research, University of Tokyo., 1997) p.17.
28. M.Takeda et al., *op. cit.* p.398.
29. Y.Uchihori et al., *op. cit.* p.50.
30. R.M.Baltrusaitis et al., *Nucl. Instrum. Methods* A **240**, 410 (1985).
31. D.Bird et al., *Phys. Rev. Lett.* **71**, 4301 (1993).
32. T.K.Gaisser et al., *Phys. Rev.* D **47**, 1919 (1993).
33. N.Hayashida et al., *J. Phys. G: Nucl. Part. Phys.* **21**, 1101 (1995).
34. A.M.Hillas, *Nucl. Phys.* B *(Proc. Suppl.)* **28**, 67 (1992).
35. J.W.Cronin, *University of Chicago preprint EFI 92-8* (1992).
36. B.R.Dawson, *Proc. Tokyo Workshop on Techniques for the Study of Extremely High Energy Cosmic Rays*, ed. M.Nagano (Inst. of Cosmic Ray Research, University of Tokyo, 1993) p.125.
37. R.Walker and A.A.Watson, *J. Phys. G: Nucl. Part. Phys.* **7**, 1297 (1981).
38. Pierre Auger Project Design Report, The Auger Collaboration (1995).
39. N.Sakaki and M.Nagano, *Proc. of ICRR Symp. on Extremely High Energy Cosmic Rays : Astrophysics and Future Observatories* ed. M.Nagano, (Inst. of Cosmic Ray Research, University of Tokyo 1997) p.402.
40. C.Pryke, *op. cit.* p.407.
41. K.Honda et al. submitted to *Phys. Rev.* D , (1996).
42. P.Bhattacharjee, *Phys. Rev.* D **40**, 3968 (1989).
43. P.Bhattacharjee, C.T.Hill, and D.N.Schramm, *Phys. Rev. Lett.* **69**, 567 (1992).
44. M.Al-Seady et al., *op. cit.* p.191.
45. N.Hayashida et al., *op. cit.* p.205.

HIGH-ENERGY GAMMA AND NEUTRINO ASTRONOMY

L. BERGSTRÖM

Department of Physics, Stockholm University
Box 6730, S-113 85 Stockholm, Sweden

An overview is given of high-energy gamma-ray and neutrino astronomy, emphasizing the links between the two fields. With several new large detectors just becoming operational, the TeV gamma-ray and neutrino sky will soon be surveyed with unprecedented sensitivity.

1 Introduction

These are exciting times for high-energy gamma ray and neutrino astronomy. During the last couple of years several sources of TeV gamma rays have finally been convincingly detected, after many years of marginal and sometimes erroneous claims of detection at higher energies in air shower arrays.

This healthy development of the field is due to the operation of several new large experimental facilities, in particular the CASA, HEGRA and Whipple experiments (for a summary of these experiments, see Ref. [1]).

In neutrino astronomy, the first sources beyond the Sun (and the transient SN 1987A) remain to be discovered. There are great expectations that this will happen soon, as new large neutrino telescopes are just about to become operational.

There are several areas of intersection between gamma ray and neutrino astronomy. By both probes one gets a view of violent astrophysical processes, and in contrast to charged cosmic rays the direction to the source is preserved. Most of the processes that give rise to high-energy neutrinos should also generate gamma rays, and vice versa. By studying both types of emission valuable information about the production mechanisms of these energetic particles can be obtained. Due to the difference in absorption (TeV gamma rays are absorbed on IR intergalactic photons, whereas neutrinos are unaffected), useful information on the intergalactic radiation field may be obtained if far-away sources are observed.

Besides the more "mundane" local processes creating gamma rays and neutrinos, such as cosmic ray collisions with interstellar gas and dust, or with the Earth's atmosphere, there are some very intriguing sources like the central parts of Active Galactic Nuclei (AGN) and some more exotic possibilities like radiation from nonbaryonic dark matter annihilations and from topological defects. While a discovery of the latter class of course would be quite remarkable, also non-discovery is useful to establish limits on the underlying particle physics theories.

2 High-Energy Gamma Rays

Traditionally, gamma ray astronomy has been divided into several subfields based on the energy range studied, from the MeV region all the way up to YeV (10^{21} eV). This is due to the fact that completely different experimental techniques are used, and also different physical processes are involved in the sources.

In fact, due to the overwhelming background of low-energy gamma rays produced in the atmosphere by the intense cosmic ray flux, it is necessary to use space detectors to detect gammas of energy below roughly 50 GeV. Above that energy, ground-based air Cherenkov telescopes of much larger area may be employed. With instruments on board the Compton-GRO satellite, notably the EGRET detector [2], data is now available up to 20 GeV. The EGRET catalog comprises a large number of supernova remnants and AGNs, but also many sources of unknown origin. An interesting new result is that the diffuse γ ray flux from the galactic center recently detected by EGRET seems to show shows some evidence of an excess at high energy which is not easily explained in conventional models [3].

At present, there is is an annoying gap in the energy range between around 20 and 250 GeV, above which energy the most advanced ground-based air Cherenkov telescopes become functional. The principle of these is to detect in optical mirrors the Cherenkov radiation caused by air showers initiated by the primary particles. Above around 10 TeV, some particles of the air showers penetrate all the way down to the surface (at least at mountain altitudes) and can be detected directly. In air shower arrays these cascades are sampled sparsely but over large areas.

The energy gap will most probably be filled from both sides the next few years as, e.g., both new space detectors (like GLAST [4]) and large solar power plant mirror arrays [5] are planned to be deployed.

The problem of establishing a signal from a gamma ray point source is highly nontrivial, since the cosmic ray flux, roughly 10^{-7} cm^{-2} s^{-1} sr^{-1} at 10 TeV, is much higher than any expected gamma ray flux. The low signal to noise was probably the reason for some seemingly erroneous claims of detection of galactic point sources in the 1980's, something that was rectified by the standard-setting CASA experiment [6].

The last two or three years, remarkable improvements in the imaging qualities and hadron rejection of air Cherenkov telescopes has finally resulted in solid detection of the first few TeV gamma ray point sources. The first one to be detected, with remarkably high statistics by the Whipple group [7], was the Crab nebula. (It was confirmed by several other groups like ASGAT, Themistocle, CANGAROO and TIBET.) The pulsar-driven Crab supernova remnant is such a solid TeV gamma ray source that it has become something of a standard candle for high-energy gamma ray astronomy today.

The jets of AGN had also been hypothesized as being possible TeV gamma ray sources, since there is a Lorentz boost for jets viewed head-on. A complication here is that for extragalactic sources, the optical depth of gamma rays may become non-negligible. A gamma ray traveling through the intergalactic medium will interact with a high cross section with photons of energy corresponding to an invariant mass just above the e^+e^- cross section. For a TeV photon, this means a sensitivity to IR photons at $\sim 2\,\mu$m. (Note that the cosmic microwave background cuts off PeV γ radiation at a fraction of a Mpc.)

Recently, a detailed analysis [8,9] using the most recent determinations of the optical and IR intergalactic background, has shown that TeV sources more distant than $z \sim 0.1$ should hardly be seen due to absorption. Recent observations seem to verify this general picture.

The first observation of TeV gammas from an AGN was made by the Whipple collaboration [10], who detected a signal from the blazar Mkn 421 at the 6σ confidence level. Recently, the HEGRA collaboration independently confirmed this source using two of their instruments [11]. Another blazar, Mkn 501, which is too weak a GeV source to be detected by EGRET, has recently been seen in TeV γs by both Whipple and HEGRA [12].

A most remarkable, rapid outburst of TeV γs from Mkn 421 was detected by the Whipple group on May 7th, 1996 [13]. With a doubling time of about one hour, the flux increased above the quiescent value by a factor of more than 50, making this source even brighter than the Crab in TeV γ radiation. In a second outburst about a week later, the flux increased by a factor of almost 25 in approximately 30 minutes. This type of violent variability on very short time scales is bound to severely strain current models, although interesting attempts have appeared [14].

At this Conference, new results were presented from the HEGRA collaboration [15], indicating that there may be a handful of additional TeV sources (in fact, even above 30 TeV) among the nearby ($z \lesssim 0.06$) EGRET sources. If this is confirmed, it should have interesting consequences for the intergalactic IR and optical background. This could give useful information on the mechanisms for early galaxy formation [9].

The origin of the high-energy radiation from AGNs is still unclear. It seems probable that shock acceleration is involved near the black hole or, for the blazar class, in the jet, but how particles are transferred to the outer regions as well as how they interact is still mysterious. In fact, it is not known whether leptons or hadrons are mainly responsible for energy transport near the accretion region. It is conceivable that electrons, interacting with ambient magnetic fields, create synchrotron radiation which in turn may be inverse-Compton scattered to high energies. These are the so-called SSC (synchrotron self-Compton) models[16], which work very successfully for a supernova remnant like the Crab. In another class of models [17], mainly hadrons (protons) are accelerated, which interact with the dense photon gas in the AGN central region or in a jet. In $p\gamma \rightarrow \pi + X$ reactions, high-energy neutrinos, electrons, positrons and gamma rays are created in the decay of pions. All particles except the neutrinos induce electromagnetic cascades which terminate at low energy. In particular, the X-ray flux may be used to put an upper bound on the neutrino rates in this class of models [18,19]. Although estimates are uncertain, it seems that the integrated rate from all AGNs may give a "diffuse" source of very high energy neutrinos which could be detectable in the new generation of neutrino telescopes like AMANDA.

It appears that if the recent detection of γs of more than 30 TeV from several blazars[15] is confirmed, it may lend credibility to the hadronic model[20]. A solid answer must, however, await a detailed analysis of time-correlated multi-waveband data and/or the findings from neutrino telescopes.

3 High-Energy Neutrinos

Neutrino astronomy was born with the first detection of solar neutrinos (too few to fit standard solar models) by R. Davis et al. in the 1960s, with the proof two decades later by the Kamiokande collaboration that the neutrino events really point back to the Sun. The solar neutrino problem is of course still one of the most intriguing indications we have for physics outside the Standard Model of particle physics[21]. The remarkable detection of neutrinos from SN1987A in the Kamiokande and IMB detectors (originally constructed to search for proton decays) has established neutrino astronomy as a useful branch of astrophysics. In addition, the observed neutrino rates from the SN1987A event has helped particle physicists to put limits on neutrino properties as well as on various hypothetical, weakly interacting particles. Indeed, neutrino astrophysics is one of the areas where the connections between astrophysics and particle physics are perhaps the strongest.

The first neutrino telescopes typically had effective areas of the order of one to a few hundred m^2. They have been followed by a new generation (MACRO, Super-Kamiokande) which approaches 10^3 m^2. Super-Kamiokande, for instance, is an extremely well-equipped and sensitive laboratory for all types of neutrino physics of energy from a few MeV upwards [22]. MACRO has recently published [23] its first measurement of the atmospheric neutrino flux above 1 GeV.

However, for TeV neutrino energies and above, all estimates indicate that the effective areas must be much larger to give a fair chance of detection [24]. Therefore, a new generation of very large telescopes has been developed, which sacrifice sensitivity of MeV neutrinos for large area (10^4 to 10^5 m^2 at present - the aim is for 1 km^2 within a few years) for multi-GeV neutrinos. (Typical thresholds are some tens of GeV.) A pioneer of this type was the deep ocean DUMAND experiment [25] outside Hawaii, which now seems to be discontinued at the prototype stage due to various technical problems related to the very demanding ocean environment. However, even with a small prototype, they were able to put some limits on cascades initiated by AGN neutrinos [26], showing the promise of this type of technique. In Europe, the ocean detector concept is being further investigated in the Mediterranean by the NESTOR [27] and ANTARES [28] collaborations, with a large-scale detector still being a couple of years ahead.

The Lake Baikal experiment [29] has become the first of the natural-water detectors to successfully detect atmospheric neutrinos, although only a few events so far in its 96-fold OM (optical module) array. The array is successively being expanded to 200 OMs, with 3/4 of that expected by the spring of 1997. It has the advantage over ocean detectors of being in fresh water, thus avoiding the high radioactive background from ^{40}K present in salt water. Also, the ice cover during winter months helps the logistics of the deployment substantially. However, bioluminescence is present and sedimentation necessitates regular cleaning of the optical modules. In addition, the relatively shallow depth (1300 m) means that a large background of downward atmospheric muons has to be fought. In is an impressive achievement of the Baikal group to have obtained the up/down rejection factor needed to detect upward-going muons.

In the deep under-ice US-German-Swedish detector AMANDA at the South Pole, none of these problems is present (although the maximum useable depth of around 2500 m still gives substantial downward-going muon flux). On the other hand, it was not clear before last year that the ice quality was good enough to deploy a large detector. In particular, a prototype deployed in 1994-95 at 800 to 1000 m depth showed severe degradation of timing resolution due to scattering on residual air bubbles at that depth. However, ice inbetween air bubbles was found to be remarkably clean, with absorption lengths in the near-UV being more than ten times longer than ever measured in laboratory ice [30].

In the 1995-96 season, 4 strings of 20 OMs each (20 m spacing between OMs) were deployed to 2000 m depth, and the scattering on bubbles was found to be absent (or at least two orders of magnitude smaller than at 800 m), permitting the first muons to be tracked [31]. In the soon finished, highly successful 1996-97 season, 6 additional strings have been deployed. Thanks to improvements in signal transmission, thinner twisted quad cables could be used permitting 36 OMs per string, with now 10 m separation between OMs. The average distance between nearest-neighbor strings in the 10-string detector is around 30 m. Of the 216 new OMs, only half a dozen have failed, giving the AMANDA collaboration the hope of soon having at its disposal a detector of around 10^4 m^2 for upward-going single muons, and much larger for cascades initiated, e.g., by electron neutrinos.

3.1 Sources of High-Energy Neutrinos

In a large detector, like the present AMANDA neutrino telescope, there will be a real chance to detect neutrinos from AGNs, if the models involving acceleration of hadrons are correct. Besides the "diffuse" integrated contribution from all AGNs, which could amount to several hundred events per km^2 per year [24], the blazars (i.e. AGNs with jets viewed nearly head-on) from the EGRET catalog will be promising objects to study. The fact that the TeV gamma ray sources seen by air Cherenkov telescope are all relatively nearby, whereas many stronger such EGRET sources are not seen in TeV gammas, has as its most natural explanation the intergalactic absorption of gamma rays. Thus there could be a large number of very intense neutrino sources awaiting discovery.

The fact that whenever hadrons are accelerated, both gamma rays and neutrinos will be produced through pion decay, means that models, e.g., for gamma ray bursts (GRBs), where hadronic fireballs are excited inevitably predict also neutrino radiation [32,33]. In the AMANDA detector, a trigger has been set up which can correlate an excess of neutrino events with satellite detection of a GRB. (A supernova trigger is also implemented.) As has been pointed out [33,34], if an extragalactic source of neutrinos is found, there are many interesting tests of neutrino properties (mass, mixings, magnetic moments etc) that can be made, which would supersede terrestrial tests and constraints from SN1987A by orders of magnitude.

If very-high energy (PeV) neutrinos from AGNs are present, a whole range of other exotic particle physics processes could be investigated as well (such as leptoquarks, multi-W processes etc [35]). An interesting process in addition is the resonant $\bar{\nu}_e + e^- \to W^-$ at around 6 PeV, which could give spectacular, background-free cascades in Cherenkov detectors. [36,35] In fact, for such high energies, the way to get a large effective detector volume may be to use the coherent radio wave radiation from the shower in the ice.[37] Prototype radio detectors have been deployed piggy-back on AMANDA strings this year.

3.2 Indirect Detection of Supersymmetric Dark Matter in Neutrino Telescopes

Supersymmetric neutralinos with masses in the GeV–TeV range are among the leading non-baryonic candidates for the dark matter in our galactic halo. One of the most promising methods for the discovery of neutralinos in the halo is via observation of energetic neutrinos from their annihilation in the Sun and/or the Earth[38,39,40]. (In some regions of parameter space, also detection in gamma rays in air Cherenkov telescopes through the unique signature of a line of narrow width, could be feasible[41].) Neutralinos do not annihilate into neutrinos directly, but energetic neutrinos may be produced via hadronization and/or decay of the direct annihilation products. These energetic neutrinos may be discovered by terrestrial neutrino detectors.

The prediction of muon rates is in principle straight-forward but technically quite involved: one has to compute neutralino capture rates in the Sun and the Earth, fragmentation functions in basic annihilation processes, propagation through the solar or terrestrial medium, charged current cross sections and muon propagation in the rock, ice or water surrounding the detector.

The neutralinos $\tilde{\chi}_i^0$ are linear combinations of the neutral gauginos $\tilde{B}$, $\tilde{W}_3$ and of the neutral higgsinos $\tilde{H}_1^0$, $\tilde{H}_2^0$, the lightest of which, called χ, is then the candidate for the particle making up (at least some of) the dark matter in the universe.

With Monte Carlo simulations one can consider the whole chain of processes from the annihilation products in the core of the Sun or the Earth to detectable muons at the surface of the Earth.

Unfortunately, no details about supersymmetry breaking are known at present, which means that a lot of parameters are undetermined. The usual strategy[39,40,38] is then to scan the parameter space of the minimal supersymmetric extension to the Standard Model.

The best present limits[42] for indirect searches come from the Baksan detector. The limits are $\Phi_\mu^{Earth} < 2.1 \times 10^{-14}$ cm^{-2} s^{-1} and $\Phi_\mu^{Sun} < 3.5 \times 10^{-14}$ cm^{-2} s^{-1} at 90% confidence level and integrated over a half-angle aperture of 30° with a muon energy threshold of 1 GeV. This has already allowed some models to be excluded[40]. A neutrino telescope of an area around 1 km^2, which is a size currently being discussed for a near-future neutrino telescope, would improve these limits by two or three orders of magnitude and would have a large discovery potential for supersymmetric dark matter.

Indirect dark matter searches and LEP2 probe complementary regions of the supersymmetric parameter space. Moreover, direct detection[38] is reaching a sensitivity that allows some models to be excluded[43], with somewhat different characteristics than those probed by the other methods. This illustrates a nice complementarity between direct detection, indirect detection and accelerator methods to bound or confirm the minimal supersymmetric standard model.

3.3   Establishing a Neutrino Signal from a Point Source

For neutralino detection, as well as for other physics objectives of neutrino telescopes, a problem will always be the irreducible background coming from atmospheric neutrinos. However, a typical signal will appear as a peak in the angular distribution; usually the energy distribution is different as well. The question of how the discovery potential depends on the angular and energy resolution has recently been investigated[44].

Due to the finite muon production angle, one would like to accept muons from a large enough solid angle around the point source to assure all the signal events are accepted. For example, the rms angle between the neutrino direction and the direction of the induced muon is $\sim 20°/\sqrt{E_\nu/10\,\text{GeV}}$. Furthermore, the muon typically carries half the neutrino energy, so the angular radius of the acceptance cone should be $\sim 14°/\sqrt{E_\mu/10\,\text{GeV}}$. The problem is of course that the a priori energy of signal neutrino events is unknown, so one has to optimize angular and energy acceptance according to varying hypotheses for the neutrino source.

A general covariance-matrix formalism has been set up[44] and applied to the specific example of neutralino annihilation in the Sun and Earth, for detectors with various values of angular and energy resolution. Comparing, e.g., the improvement by using a 3-parameter fit for the signal to the simple case of using just one bin up to a certain angle θ_{max} one finds that there could be an improvement of up to a factor of 2 at high masses. Although this application was for neutralino annihilation, the formalism[44] is general enough to be applicable for a generic point source. As large neutrino experiments now come on-line, we can expect successive improvements in their discovery potential.

4   Conclusions and Acknowledgments

With new windows to the universe, historically it has always been the case that unexpected discoveries have appeared. I have tried to summarize the status and expectations for high-energy gamma ray and neutrino astronomy.

66

Maybe the outcome will be different than predicted here, but it certainly will be interesting.

The author wishes to thank J.J. Aubert, V. Berezinsky, J. Edsjö, P.O. Hulth, J. Learned, H. Meyer, H. Rubinstein, C. Spiering and T. Weekes for useful discussions, and the organizers of "Texas in Chicago" for hospitality. This work was sponsored by the Swedish Natural Science Research Council.

References

1. J. Cronin, K.G. Gibbs and T.C. Weekes, *Ann. Rev. Nucl. Part. Sci.* **43**, 883 (1993).
2. http://cossc.gsfc.nasa.gov/cossc/EGRET.html
3. M. Mori, Ap. J. **478** (1997).
4. GLAST home page: http://www-glast.stanford.edu
5. R.A. Ong, these Proceedings.
6. http://hep.uchicago.edu/ covault/casa.html
7. P.T. Reynolds et al., Astrophys. J. **404**, 206 (1993).
8. F.W. Stecker, O.C. De Jager and M.H. Salamon, Ap. J. **390**, L49 (1992).
9. D. MacMinn and J.R. Primack, Space Sci. Rev. **75**, 413 (1996).
10. M. Punch et al., Nature **358**, 477 (1992).
11. D. Petry et al., Astron. Astrophys. **311**, L13 (1996).
12. J. Quinn et al., Ap. J. **456**, L83 (1996); S.M. Bradbury et al., astro-ph/9612058 (1996).
13. J.A. Gaidos et al., Nature **383**, 319 (1996); T.C. Weekes, these Proceedings.
14. W. Bednarek and R. Protheroe, astro-ph/9612073 (1996); A. Dar and A. Laor, astro-ph/9610252 (1996).
15. H. Meyer, these Proceedings.
16. C.D. Dermer, R. Schlickeiser and A. Mastichiadis, Astron. Astrophys. **256**, L27 (1992); A.A. Zdziarski and J.H. Krolik, Ap. J. **409**, L33 (1994).
17. P.L. Biermann and P.A. Strittmatter, Ap. J. **322**, 643 (1987); M.C. Begelman, B. Rudak and M. Sikora, Ap. J. **362**, 38 (1990); K. Mannheim, Phys. Rev. **D48**, 2408 (1993).
18. F.W. Stecker et al., Phys. Rev. Lett. **66**, 2697 (1991); (E) **69**, 2738 (1992).
19. V.S. Berezinsky and J.G. Learned, in *Proc. of the Workshop on High Energy Neutrino Astrophysics*, eds. V.J. Stenger, J.G. Learned, S. Pakvasa and X. Tata, World Scientific, 1992.
20. K. Mannheim, S. Westerhoff, H. Meyer and H.-H. Fink, Astron. Astrophys. **315**, 77 (1996).

21. J.N. Bahcall, these Proceedings.
22. Y. Totsuka, these Proceedings.
23. S. Ahlen et al., Phys. Lett. **B357** 481.
24. T. K. Gaisser, F. Halzen and T. Stanev, Phys. Rep. **258**, 173 (1995).
25. P.K.F. Grieder, in *Trends in Astroparticle Physics*, Stockholm, Sweden, 1994, eds. L. Bergström, P. Carlson, P.O. Hulth and H. Snellman, Nucl. Phys. (Proc. Suppl.) **B43** (1995) 265.
26. J. Bolesta, these Proceedings.
27. L. Resvanis, Europhys. News **23**, 172 (1992).
28. ANTARES home page: http://marcpl1.in2p3.fr/astro/astro.html
29. I.A. Belolaptikov et al., in *Trends in Astroparticle Physics*, Stockholm, Sweden, 1994, eds. L. Bergström, P. Carlson, P.O. Hulth and H. Snellman, Nucl. Phys. (Proc. Suppl.) **B43** (1995) 241.
30. L. Bergström et al., physics/9701025, Appl. Optics, in press (1997).
31. P.O. Hulth et al (AMANDA Collaboration), to appear in *Proc. of Neutrino 96*, eds. K. Enqvist, K. Huitu and J. Maalampi (World Scientific, Singapore, 1997).
32. F. Halzen and G. Jaczko, astro-ph/9602038 (1996).
33. E. Waxman and J.N. Bahcall, astro-ph/9701231 (1997).
34. J.G. Learned, in *Proc. of Neutrino 94*, eds. A. Dar, G. Eilam and M. Gronau, Nucl. Phys. (Proc. Suppl.) **38**, 484 (1995).
35. L. Bergström, R. Liotta and H. Rubinstein, Phys . Lett. **B276**, 231 (1992); N. Arteaga-Romero et al., hep-ph/9701339 (1997).
36. V.S. Berezinsky and A.Z. Gazizov, JETP Lett. **25**, 254 (1977); R. Gandhi, C. Quigg, M.H. Reno and I. Sarcevic, Astropart. Phys. **5**, 81 (1996).
37. G. Frichter, J. Ralston and D. Mackay, Phys. Rev. **D53**, 1684 (1996); P.B. Price, Astropart. Phys. 5, 43 (1996).
38. For a comprehensive review containing historical references, see G. Jungman, M. Kamionkowski, and K. Griest, Phys. Rep. **267**, 195 (1996).
39. V. Berezinsky, A. Bottino, J. Ellis , N. Fornengo, G. Mignola and S. Scopel, Astropart.Phys. 5 333 (1996).
40. L. Bergström, J. Edsjö, and P. Gondolo, hep-ph/9607237, Phys. Rev. D, in press (1997).
41. L. Bergström and J. Kaplan, Astrop. Phys. **2**, 261 (1994); G. Jungman and M. Kamionkowski, Phys. Rev. **D51**, 3121 (1995).
42. M.M. Boliev et al., in *TAUP 95*, Nucl. Phys. (Proc. Suppl.) **B48**, 83 (1996).
43. L. Bergström and P. Gondolo, Astropart. Phys. 5, 263 (1996).
44. L. Bergström, J. Edsjö and M. Kamionkowski, astro-ph/9702037.

ANISOTROPIES IN THE COSMIC MICROWAVE BACKGROUND: THEORY

SCOTT DODELSON

NASA/Fermilab Astrophysics Center, P.O. Box 500, Batavia, IL 60510, USA

Anisotropies in the Cosmic Microwave Background (CMB) contain a wealth of information about the past history of the universe and the present values of cosmological parameters. I ouline some of the theoretical advances of the last few years. In particular, I emphasize that for a wide class of cosmological models, theorists can accurately calculate the spectrum to better than a percent. The spectrum of anisotropies today is directly related to the pattern of inhomogeneities present at the time of recombination. This recognition leads to a powerful argument that will enable us to distinguish inflationary models from other models of structure formation. If the inflationary models turn out to be correct, the free parameters in these models will be determined to unprecedented accuracy by the upcoming satellite missions.

1 History

The Texas Symposium on Relativistic Astrophysics was held in Chicago ten years ago in 1986. David Wilkinson spoke about the cosmic microwave background. He undoubtedly made the point that the CMB provides us with some of the best evidence for the Big Bang. There was no evidence (and there still is no evidence) for any deviations from a black-body spectrum. And this is one of the primary predictions of the Big Bang.

Wilkinson devoted most of his talk to searches for anisotropies in the CMB. The fact that the CMB temperature is the same in all directions indicates that the universe was very smooth early in its history. However, cosmologists generally work within the framework of gravitational instability which says that small inhomogeneities early on grew via gravity into the large structures we see today. Thus, the CMB should *not* be perfectly isotropic; it should carry some imprint of those small, early inhomogeneities. Wilkinson compiled the upper limits on anisotropies from the experiments of the time. This compilation is reproduced in Figure 1, where I have taken the liberty of slightly changing his notation. In particular, it is convenient to expand the temperature on the sky in terms of spherical harmonics

$$\frac{T(\theta, \phi)}{T_0} = \sum_{l=0}^{\infty} \sum_{m=-l}^{l} a_{lm} Y_{lm}(\theta, \phi). \tag{1}$$

When we expand in this fashion, low l's correspond to anisotropies on large angular scales (the quadrupole is $l = 2$) while large l's correspond to anisotropies on small scales. The square of the coeffients of the Y_{lm}'s are known as the C_l's. These are extremely useful things becuase they can be calculated by theorists and measured by observers. A given experiment at angular scale l measures $\delta T_{rms} \sim [l(l+1)C_l/2\pi]^{1/2}$. The upper limits at the time correspond to $\delta T_{rms} \sim 50 - 200 \mu K$.

Wilkinson was obviously aware of the fact that these upper limits were tantalizingly close to the levels of anisotropies predicted by many theories. He ended his talk by saying, *"If the anisotropies are indeed just below current limits, as most of us feel they must be, the next few years should see this field turn from one of searching to one of studying."*

2 Experiments

How has the field progressed since Wilkinson's review? Figure 1 shows, along with Wilkinson's compilation, a recent compilation[1] of all experiments in the last two years. Starting with the COBE[2] detection in 1992 at the largest angular scales, there have been dozens of detections on a wide range of angular scales. These detections, as Wilkinson's quote makes clear, were anticipated based on typical models of structure formation. Although the details are not yet in, it is safe to say that gravitational instability theories predicted the level of anisotropies that are observed today.

Another feature of the detections is just now becoming evident. As one moves from low l to high l (from large scales to small scales), one sees evidence of a gradual rise in the the amplitude of the anisotropies. We will see shortly that this too is a prediction of some of the more popular models of structure formation.

How will the situation look ten years from now when results from the current crop of balloon-borne and ground-based experiments have come in, and the two satellite experiments (MAP[3] and PLANCK[4]) will have made all-sky maps? Figure 1 shows the expected error bars in the year 2006. There are several ways to represent the knowledge we will have at the time. First, it is important to note that, today, experiments are sensitive to a range of l's: thus, C_{500} for example is not measured by a given small scale experiment. Rather, each experiment measures a signal integrated over a wide range of l. This range is depicted by the horizontal error bars in Figure 1. In the future, we can continue to smooth over the l's in this fashion. Then the errors will be as shown on the far right in Figure 1. In order to see them on this graph, I have blown them up by a factor of 100! We will also have the ability by then, though, to determine each individual C_l. The expected errors on C_{500}

70

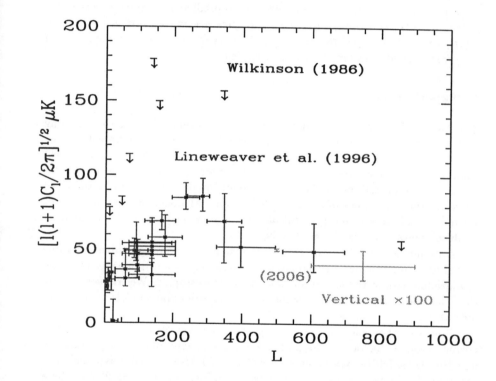

Figure 1: Observations of the CMB spectrum. Upper limits are those compiled by Wilkinson in 1986. Detections were compiled by Lineweaver based on results within the last two years. Anticipated error bars from satellites are also shown.

are shown in Figure 1. Either way you look at it, we will have an extraordinary amount of information in ten years. For the experimentalists, at least, it is clear that Wilkinson's prediction has come true. The field really has moved from searching to studying.

3   Theory

CMB theorists have also been very active over the last few years. First of all, for a wide range of models, we are confident that we can calculate[5] the anisotropy spectra – the C_l's – to an accuracy of better than a percent. Figure 2 shows the results of seven different groups who independently calculated the C_l's for a given model. This graph was made about two years ago, and the agreement has only gotten better since then. Not only can we calculate accurately, but we can also calculate quickly. Thanks to Seljak and Zaldariagga[6], in the time it has taken me to write this paragraph, we could have run off another set of C_l's.

We also have made great strides in the last few years understanding the bumps and wiggles in the theoretical curves. To understand the structure of these anisotropies, we need to review the thermal history of the universe. Recall that, early in the history of the universe, the temperature of the cosmic gas was very high. So, anytime a free electron and proton came together to form a hydrogen atom, a high energy photon immediately destroyed it. There was essentially no neutral hydrogen early on. This situation changed dramatically when the temperature dropped below 1/3 eV. After that time, there were not enough ionizing photons around. So almost all the free electrons and protons combined into neutral hydrogen. This had dramatic implications for the cosmic photons. As long as the electrons were free, they interacted with the photons via Compton scattering. After they combined into hydrogen, the photons travelled freely from the "surface of last scattering" to us today. So, for the purposes of the CMB, the universe is neatly divided into two epochs: Before Recombination when the photons and electrons behaved as a tightly coupled fluid and After Recombination when photons freestreamed. The mathematics of freestreaming is a little complicated, but the physics is completely trivial: it just requires us to trace the paths of free photons. So the physics behind the spectrum of anisotropies comes solely from the epoch Before Recombination.

It pays to reiterate that Before Recombination, the photons and electrons acted as a *fluid*. By this, I mean that it could be described by only its $l = 0$ component (as opposed to all the multipole moments that are needed to describe it today). This represents an immense simplification: instead of solving a infinite heirarchy of coupled differential equations for all the photon mo-

72

NOW !

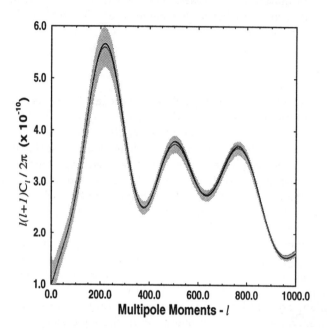

Figure 2: Seven different calculations of the CMB power spectrum for a model with adiabatic fluctuations. All lie within the shaded band which illustrates the minimum "cosmic variance" errors.

ments, we need solve for only one of the moments. The forces acting on this moment, let's call it δT, are pressure and gravity. These forces act in opposite directions. Pressure tends to smooth out any inhomogeneities (i.e. drives δT to zero) while gravity produces inhomogeneities. It is not surprising then that acoustic oscillations are set up in the medium. In fact, Hu & Sugiyama[7] have shown that this oscillation pattern is precisely the one imprinted in the C_l spectrum of Figure 2. A quantitative analysis shows that there are two possible modes[a] that can be excited in this fluid. In particular,

$$\delta T(\vec{x}, \eta) = \int d^3k e^{i\vec{k}\cdot\vec{x}} \left[A\cos[k\eta/\sqrt{3}] + B\sin[k\eta/\sqrt{3}] \right] \qquad (2)$$

---

[a]The modes look this simple only in the idealized case of zero baryons and pure matter domination. Accounting for baryons and other complications though does not alter the qualitative fact that there are two very distinct modes.

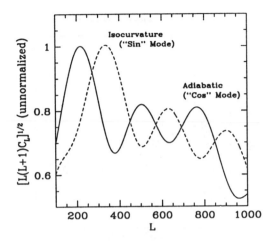

Figure 3: The power for two different theories, one of which (adiabatic) excites the cosine mode of acoustic oscillations the other (isocurvature) the sine mode. Inflationary predictions typically look like the adiabatic spectrum here. Defect models should have some of the features of the isocurvature spectrum, but the calculations at present are not yet believable.

where η is conformal time. Again, not surprisingly, the C_l spectrum today is radically different if the sine mode is excited than if the cosine mode is excited. Figure 3 shows that, as you would expect, the spectra are out of phase with each other.

It is clear from the present data shown in Figure 1 that we will shortly be able to tell which of the two theoretical curves in Figure 3 is more accurate. That is, we will soon know whether the sine or the cosine mode were excited in the early universe. This is extremely important because we expect the two most popular mechanisms of structure formation – inflation and topological defects – to excite different modes. Let me walk through this argument which has recently been clearly elucidated by Hu & White[8]. Any theory which respects causality necessarily requires that there be no correlations on very large scales (scales that have not been in causal contact with each other). This is equivalent to a boundary condition on δT; namely that the Fourier transform vanishes at $k = 0$. This means that only B in equation 2 can be non-zero. So topological

Adiabatic Models

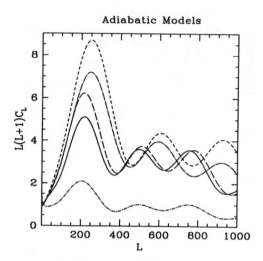

Figure 4: The spectra for adiabatic models with different sets of parameters. Varied are the cosmological constant, neutrino mass, spectral index, and Hubble constant.

defects, which of course obey causality, can be expected to excite the sine mode. Inflation is a theory which introduces correlations amongst scales that appear to be causally disconnected. Thus, inflationary models can, and most often do, excite the cosine mode.

There are several caveats to the above argument. First of all, the predictions for the adiabatic models depend on various cosmological parameters[9]: the slope of the primordial spectrum, the contribution from tensor modes, the Hubble constant, the baryon density, and several others. Thus the actual curves share some of the features of the curve labelled "Adiabatic" in figure 3, but the predictions are by no means unique (see figure 4). Fortunately there are some robust features of these curves which hold up even after allowing many parameters to vary. The second caveat is that we simply do not know for sure that defect theories follow the general isocurvature model. There have been a few calculations of the spectrum in defect models[10]. As one who is actively at work on one such calculation, I think it is fair to say that we have not yet reached agreement.

Assuming there are no major theoretical surprises, we can expect the experiments over the next several years to pick out whether toppological defects or inflation are correct. Once that issue is settled, it remains to pin down the cosmological parameters which impact upon the spectrum. One might think that since there are so many free parameters, they cannot all be determined simultaneously. Recent work has shown that this is *not* true. Figure 5 shows an example[11]: we let five parameters vary and show the error ellipses projected down onto a couple of two dimensional planes. The top figure shows that it is quite possible that by the year 2006, we will not be arguing about whether the Hubble constant is 50 or 100, but rather whether it is 50.5 or 50.0. The bottom figure shows that, in addition to the cosmological parameters, we should get a good handle on the inflationary parameters, thereby allowing us to distinguish amongst different inflationary models[12]. A number of groups[13] have varied even more parameters and all have reached the same general conclusion: the cosmological parameters will be pinned down to unprecedented accuracy by the satellite experiments.

4 Conclusions

About thirty years ago, Penzias and Wilson discovered the cosmic microwave radiation. This discovery convinced the vast majority of physicists that the Big Bang model was correct. In 1992, the COBE satellite discovered anisotropies in the CMB. The existence and amplitude of these anisotropies were *predicted* by theories which relied on gravitational instability to form structure. It is perhaps too early to know for sure, but I would guess that COBE's most enduring legacy will be its evidence that current models of structure formation are on the right track.

A number of cosmologists are beginning to speculate about what we will have learned in ten years after the next generation of balloon and ground based experiments and after the MAP and PLANCK satellites have flown. There is a very good chance these measurements will clearly distinguish between the two most popular models of structure formation: inflation and topological defects. Indeed, this could happen very soon. If a peak does indeed develop in the C_l spectrum at around $l \sim 200$, this will be strong evidence for inflation. If the general picture of inflation is verified in this manner, the fun will begin. It will then be possible to determine many of the cosmological parameters to unprecedented accuracy. Further, the experiments will contain so much information that it will be possible to distinguish amongst different inflationary models. This opens a window to study physics at energies that are twelve orders of magnitude higher than those probed by the largest accelerators.

Many people are fond of pointing out that something completely unex-

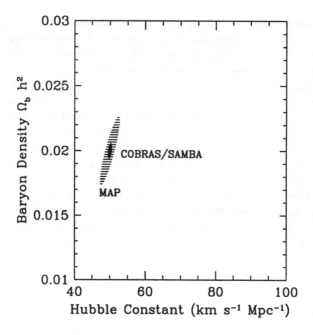

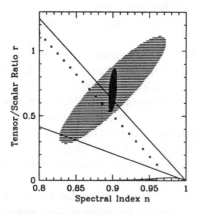

Figure 5: The estimated 95% contours for the MAP (larger ellipses in each case) and PLANCK (formerly COBRAS/SAMBA) satellites. In each case five variables – the amplitude of the scalar and tensor perturbations, the spectral index of the scalars, the baryon density, and the Hubble constant – are allowed to vary. The ellipses are the projections of the five dimensional ellipses onto the (Hubble constant,Baryon density) plane and the (spectral index,tensor/scalar ratio) plane. The points and lines in the $n - r$ plane correspond to the predictions of different inflationary models.

pected and confusing may turn up, thereby upsetting the possibility of any such determinations. Of course this is possible. Unexpected and confusing discoveries have rocked cosmology for decades. The "man-bites-dog" story in cosmology though is the one in which the confusion ends; this may well happen within the next ten years.

Acknowledgments

This work was supported in part by DOE and NASA grant NAG5-2788 at Fermilab.

References

1. C. H. Lineweaver *et al, astro-ph/9610133*.
2. G.P. Smoot *et al, Astrophys. J.* **396**, L1 (1992).
3. The MAP home page is http://map.gsfc.nasa.gov/.
4. The PLANCK (formerly COBRAS/SAMBA) home page is http://astro.estec.esa.nl/SA-general/Projects/Cobras/cobras.html.
5. P.J.E. Peebles and J.T. Yu, *Astrophys. J.* **162**, 815 (1970); M.L. Wilson and J. Silk, *Astrophys. J.* **243**, 14 (1981); J.R. Bond and G. Efstathiou, *Astrophys. J.* **285**, L45 (1984).
6. U. Seljak and M. Zaldariagga, *Astrophys. J.* **469**, 437 (1996).
7. W. Hu and N. Sugiyama, *Astrophys. J.* **444**, 489 (1995); F. Atrio-Barandela and A. G. Doroshkevich, *Astrophys. J.* **420**, 26 (1994); P. Nasel'skij and I. Novikov, *Astrophys. J.* **413**, 14 (1993); U. Seljak, *Astrophys. J.* **435**, L87 (1994); A. G. Doroshkevich. Ya. B. Zeldovich, and R. A. Sunyaev, *Sov. Astron.* **22**, 523 (1978); A. G. Doroshkevich, *Sov. Astron. Lett.* **14**, 125 (1988).
8. W. Hu and M. White, *Phys. Rev. Lett.* **77**, 1687 (1996); N. G. Turok, astro-ph/9607109 (1996).
9. J. R. Bond, R. Crittenden, R. L. Davis, G. Efstathiou and P. J. Steinhardt, *Phys. Rev. Lett.* **72**, 13 (1994).
10. R. G. Crittenden and N. G. Turok, *Phys. Rev. Lett.* **75**, 2642 (1995); A. Albrecht, D. Coulson, P. Ferreira, and J. Magueijo, *Phys. Rev. Lett.* **76**, 1413 (1996).
11. S. Dodelson, W. Kinney, and E. W. Kolb, astro-ph/9702166 (1997).
12. L. Knox and M. S. Turner, *Phys. Rev. Lett.* **73**, 3347 (1994).
13. L. Knox, *Phys. Rev.* **D52**, 4307 (1995); G. Jungman, M. Kamionkowski, A. Kosowsky , and D. N. Spergel, *Phys. Rev.* **D54**, 1332 (1996); S. Dodelson, E. I. Gates, and A. S. Stebbins, *Astrophys. J.* **467**, 10 (1996).

MEASUREMENTS OF THE COSMIC MICROWAVE BACKGROUND

L.A. PAGE

Dept. of Physics, Princeton University, PO Box 708, Princeton NJ 08544, USA
E-mail: page@pupgg.princeton.edu

The cosmic microwave background (CMB) is arguably our best cosmological observable. All precise and accurate measurements of its attributes serve to distinguish between cosmological models. Measurements of the absolute emission temperature as a function of frequency probe the history of cosmic energetics. Measurements of the anisotropy strongly constrain models of structure formation. In addition, for some cosmological models, key cosmological parameters such as Ω_0, Ω_B, and h may be extracted from the angular power spectrum of the anisotropy. In these notes, we review the status of measurements of both the absolute temperature of and temperature anisotropy in the CMB. Future directions are indicated and the upcoming satellite experiments are briefly discussed.

1 Introduction

The CMB is a powerful cosmological probe because essentially no steps separate what is measured from what is of cosmological import; what you see is what you get. The indistinguishability of the CMB from a Planck spectrum to an accuracy of 0.01% tells us that there were not any highly energetic cosmic processes that coupled to photons before $z \approx 10^3$. The near perfect shape of the spectrum is the strongest evidence to date that the universe went through a hot dense phase when everything that interacted with photons was in thermal equilibrium.

As Scott Dodelson explained in the preceeding talk, the CMB photons have free-streamed through the cosmos since last scattering off electrons some 100,000 years after the big bang.[a] In the epoch of last scattering, the primordial photon/baryon fluid flowed and oscillated in response to fluctuations in the gravitational potential that fostered galaxies and clusters of galaxies. The photons, which decoupled from the plasma in this epoch, now bring to us, in the form of an angular anisotropy, an imprint of the potential wells and a signature of the dynamics of the photon/baryon fluid's response to the wells. From the angular spectrum of the anisotropy, we determine the properties of the fluid and the potential fluctuations, thereby distinguishing among various possible mechanisms of structure formation.

---

[a]There are many models of structure formation. The following description stems from the popular cold dark matter model in a post-inflation universe with "standard" recombination. Some observations are incompatible with the simplest version of this model.

2 Measurements of the Spectrum

The spectrum of the CMB is as close to that of a blackbody as can currently be measured; no distortions have been detected. The FIRAS experiment aboard the *COBE* satellite measured the flux from the sky between $60 - 2880$ GHz. The team [1,2] finds $T_{CMB} = 2.728 \pm 0.004$ K (95% CL). One doubts that short of another satellite-based experiment this result will be matched or bettered at frequencies above 100 GHz. It is comforting that the UBC rocket experiment [3] gives a consistent result.

The error on the FIRAS result, 4 mK, is due entirely to systematic effects. It tells the reader how confident the authors are in the results; it is not a formal statistical error. In Figure 1, one sees that the lack of distortions is a robust conclusion; however, one must bear in mind that the quoted errors come from a *model* of the data. If the model is not correct, the results must be re-interpreted.

Before $z \approx 3 \times 10^6$, double Compton scattering and free-free emission maintain the thermal equilibrium of the CMB with the surroundings by creating photons. An energy input simply results in a hotter CMB temperature. Between $10^5 < z < 3 \times 10^6$, single Compton scattering, which conserves the number of photons, is the dominant scattering mechanism over most of the frequency spectrum. In this epoch, the CMB is in *statistical* equilibrium with its surroundings and the distribution is characterized by a chemical potential μ. (The quoted numbers are for the unitless chemical potential; the flux is $S_\nu(T, \mu) = 2h\nu^3/[\exp(h\nu/kT_{CMB} + \mu) - 1]$). At long wavelengths, free-free emission is still effective and "fills in the tail" of the distribution. For $z < 10^5$, hot electrons, which are neither in statistical nor in thermal equilibrium with the CMB, can inverse Compton scatter the CMB photons to produce a Compton y distortion. When there are relatively few scattering events, one may think of y as the average fractional energy change per scattering event times the average number of scatterings, or $y = 1/m_e c^2 \int [k(T_e - T_{CMB})] d\tau_e$ where T_e is the electron temperature and τ_e is the optical depth due to scattering [4]. Finally, if the universe is ionized at $z < 10^3$ then there may be enough free-free emission from the plasma to increase the photon occupation number at low frequencies. This distortion is parameterized by $Y_{ff} = (h\nu/kT)^2[T_{eff}(\nu) - T_{CMB}]/T_{CMB}$, with T_{eff} the plasma temperature. These and other distortions, along with their interpretation, are discussed in Sunyaev & Zel'dovich [4], Bartlett & Stebbins [5], Bond [6], and Danese *et al.* [7]

From FIRAS, Fixsen *et al.*[2] find $|y| < 1.5 \times 10^{-5}$ (95% CL) and $|\mu| < 9 \times 10^{-5}$ (95% CL). From these, Wright *et al.*[8] and Fixsen *et al.*[2] constrain energy injection in the early universe to $\Delta U/U < 7 \times 10^{-5}$ between $10^3 <$

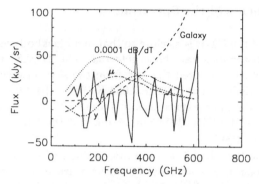

Figure 1. Residuals to a fit of the component of the FIRAS data that exhibits no spatial variation (monopole) outside a Galactic latitude of 5° from Fixsen *et al.*[2] The fit is to a Planck spectrum, a calibration correction term (labeled dB/dT), a model of the Galaxy, and either a μ or y distortion. After subtracting the first three of these, the solid line is obtained. The Galactic spectrum is shown scaled to 1/4 its value at the Galactic poles. The μ and y distortions are shown at the 95% CL values in the text. The intensity of the CMB at 165 GHz is 385 MJy/sr.

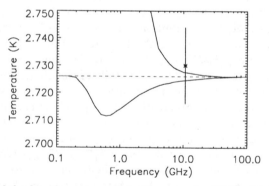

Figure 2. Plot of the low frequency distortions to the CMB. The flat line is for a perfect blackbody. The curve with a pronounced minimum near 700 MHz shows a μ distortion with $\mu = 9 \times 10^{-5}$, the *COBE* limit. The curve that begins to rise near 10 GHz is for a Y_{ff} distortion with $Y_{ff} = 1.5 \times 10^{-5}$. All data with error bars small enough to fit on this plot are shown. The measurement at 10.7 GHz comes from Staggs *et al.* [10] Note the three FIRAS data points near 100 GHz. This plot was adapted from a similar plot made by Al Kogut (http://ceylon.gsfc.nasa.gov/DIMES).

$z < 10^6$. From an analysis of the low frequency data, Bersanelli *et al.*[9] find $Y_{ff} < 1.9 \times 10^{-5}$ (95% CL).

The y distortion is manifest at high frequencies and it will be a long time before the FIRAS limit is improved. The limit on y strongly constrains alternative models of the origin of the CMB. For instance, if one tries to mimic a Planck spectrum with a superposition of multiple gray bodies, a y-distortion will result.

The signatures of any μ and Y_{ff} distortions are evident at low frequencies. While the current generation of experiments will just barely, if at all, improve on the FIRAS limits, they are paving the way for the next generation which may actually detect a distortion. Figure 2 shows a plot of the spectrum along with the μ and Y_{ff} distortion limits.

3 Anisotropy Measurements

The anisotropy[b] was first unambiguously measured by the DMR experiment aboard the *COBE* satellite[11] using 7° resolution full-sky maps at 30, 53, and 90 GHz. To date, these are still the cleanest and best checked data. When smoothed to a 10° resolution, the *rms* of the temperature fluctuations is about 30 μK. This is the canonically quoted value. If DMR had a 1/2° resolution, the *rms* would be closer to 90 μK. A linear combination of DMR's maps, optimized to accentuate the anisotropy, has a signal-to-noise of ≈ 2 per $10° \times 10°$ pixel. The two primary results derived from DMR's maps are:

1. From a fit of the data to a power spectrum parameterized by the spatial index and the quadrupole amplitude, the DMR team[12, 13, 14, 15, 16] finds $n_{DMR} = 1.21 \pm 0.3$ and $Q_{rms-PS} = 15.3^{+3.8}_{-2.8}$ μK. The quadrupole of the raw maps is slightly below Q_{rms-PS}, but not by a statistically significant amount. Note that for "standard CDM" one expects $n_{DMR} = 1.1$ instead of 1 because DMR probes the low-l tail of the acoustic peak. In this notation, the C_l are given by[17]:

$$C_l = \frac{4\pi}{5}Q_{rms-PS}^2 \frac{\Gamma[l + (n_{DMR} - 1)/2]\Gamma[(9 - n_{DMR})/2]}{\Gamma[l + (5 - n_{DMR})/2]\Gamma[(3 + n_{DMR})/2}$$ (1)

with $n_{DMR} = 1$ this reduces to $C_l \propto 1/l(l+1)$.

2. The data appear best described by Gaussian statistics[18]. At these large angular scales this is not surprising because even non-Gaussian processes

---

[b]The "dipole" term[2], with amplitude 3.372 mK and direction $(l, b) = (264.26°, 48.22°)$ comes from our motion with respect to the frame in which the CMB was emitted. This term is always subtracted before data are analyzed for primordial anisotropy.

at small scales, when averaged over a large enough volume, appear Gaussian. However, it is reassuring that the statistics we all assume have some basis in reality.

There are now three experiments that have observed the same fluctuations as DMR. They are FIRS[19], Tenerife[20], and most recently, FIRAS[21]. This last result is a tour-de-force in instrument design and analysis because FIRAS is an *absolute* experiment. From these results we begin to see the frequency spectrum of the anisotropy; all indications are that the fluctuations are thermal.

Many groups are working to measure the anisotropy. Though some are focussing on large angular scales and frequencies not observed with DMR, most concentrate on smaller angular scales. Table 1 contains a list of recent, current and planned experiments. It does not include the satellite experiments nor does it claim to be comprehensive; more details on each experiment may be found in[22].

To give a broad and almost un-biased sense of what the anisotropy data tell us, we take the compilation from Ratra[23] and compute the weighted mean of *rms* fluctuations per logarithmic interval in l in logarithmically spaced bins[22]. The results are shown in Figure 3.

The principle conclusion one should draw from Figure 3 is that there is a general rise in δT_l as one moves from the *COBE* scales to smaller angular scales. This is a stunning observation that was predicted long before the anisotropy was discovered. Has a peak to the spectrum been detected? Possibly, but it is still too early to say this with confidence. The errors are simply too large and there are too many systematic effects hidden in the data. In the analyses that give rise to Figure 3, an entire data set is reduced to give one measurement and one statistical error bar. When this is done, there is simply not enough signal-to-noise to quantify systematic effects that are lurking at the 1σ to 2σ level. On top of this, the analysis of these experiments is tricky; "new" effects are still being discovered by many groups. Finally, the inter-calibration of the experiments is uncertain to the 10%-15% level.

To improve on these results, a number of experimental and observational challenges must be met. For instance, greater knowledge of Galactic foreground emission is essential. (Because the CMB is the brightest broad-band diffuse emitter between about 1 and 500 GHz, this has not been a major problem so far.) New observational techniques are needed. The results that went into Figure 3 are primarily from difference measurements. Eventually we will want maps of the sky so that experiments are easily compared, foreground contamination is more easily identified, and powerful statistical tests can be performed. Interferometers offer one proven way to do this at low frequencies[24]

Table 1: Recently Completed, Current and Planned Anisotropy Experiments

Experiment	Resolution	Frequency	Detectors	Groups
ACE	0.2°	25-100 GHz	HEMT	UCSB
APACHE	0.33°	90-400 GHz	Bol	Bologna, Bartol Rome III
ARGO	0.9°	140-3000 GHz	Bol	Rome I
ATCA	0.03°	8.7 GHz	HEMT	CSIRO
BAM	0.75°	90-300 GHz	Bol	UBC, CfA
BEAST	0.2°	25-100 GHz	HEMT	UCSB
BOOMERanG	0.2°	90-400 GHz	Bol	Rome I, Caltech UCB, UCSB
CAT	0.17°	15 GHz	HEMT	Cambridge
CBI	0.0833°	26-36 GHz	HEMT	Caltech, Penn.
FIRS	3.8°	170-680 GHz	Bol	Chicago, MIT, Princeton, NASA/GSFC
HACME	0.6°	30 GHz	HEMT	UCSB
IAB	0.83°	150 GHz	Bol	Bartol
MAT	0.2°	30-150 GHz	HEMT/SIS	Penn, Princeton
MAX	0.5°	90-420 GHz	Bol	UCB, UCSB
MAXIMA	0.2°	90-420 GHz	Bol	UCB, Caltech
MSAM	0.4°	40-680 GHz	Bol	Chicago, Brown, Princeton, NASA/GSFC
OVRO 40/5	0.033°, 0.12°	15-35 GHz	HEMT	Caltech, Penn
PYTHON	0.75°	35-90 GHz	Bol/HEMT	Carnegie Mellon Chicago, UCSB
QMAP	0.2°	20-150 GHz	HEMT/SIS	Princeton, Penn
SASK	0.5°	20-45 GHz	HEMT	Princeton
SuZie	0.017°	150-300 GHz	Bol	Caltech
TopHat	0.33°	150-700 GHz	Bol	Bartol, Brown, DSRI,Chicago, NASA/GSFC
Tenerife	6.0°	10-33 GHz	HEMT	NRAL, Cambridge
Tenerife2	2.4°	90-270 GHz	Bol	Bartol
VCA	0.33°	30 GHz	HEMT	Chicago
VLA	0.0028°	8.4 GHz	HEMT	Haverford, NRAO
VSA	–	30 GHz	HEMT	Cambridge
White Dish	0.2°	90 GHz	Bol	Carnegie Mellon

84

(and eventually at higher frequencies) but other strategies are being developed as well.

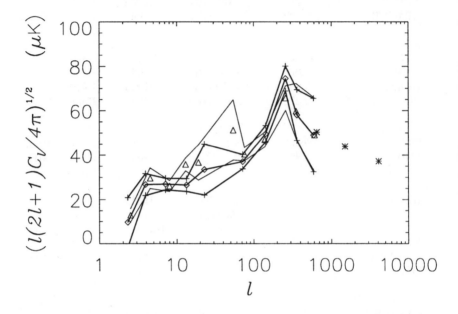

Figure 3. Angular spectrum of the anisotropy. The thick lines connect our estimates of the errors [22]. Bharat Ratra has used somewhat different criteria and obtained the results indicated by the thin lines. The asterisks near $l = 1000$ are upper limits. Clearly, the data indicate a rise in the spectrum from $l = 10$ to $l \approx 200$.

4 The New Satellite Experiments

Two satellite missions will endeavor to map the CMB anisotropy over the entire sky. The European space mission is called *PLANCK*.[c] The US mission, which is in the final design and definition phase, is called *MAP*. Because I am part of the *MAP* team, I will focus on it.

---

[c]Originally named *COBRAS/SAMBA*. It will use HEMT amplifiers and will carry cryogens to operate bolometers at 0.1 K. The planned frequency range is between 30 and 900 GHz with an angular resolution between 30′ and 4.4′. The first possible launch date is 2005. See web site for additional information.

MAP is based on HEMT amplifiers developed by Marian Pospieszalski at the National Radio Astronomy Observatory (NRAO). The radiometers are intrinsically polarization-sensitive and differential, similar in some regards to the successful *COBE*/DMR design. The instrument will span from 20 to 106 GHz in five frequency bands with an angular resolution ranging between 54' and 15'. The sensitivity per $0.3° \times 0.3°$ pixel (of which there are roughly 400,000 in the sky) will be about 35 μK. Because the instrument is passively cooled, it can in principle observe longer than the 27 month design life.

The primary goal of *MAP* is to make multi-frequency, high-fidelity, high-sensitivity maps of the sky. This requires extreme control of systematic effects. We believe the best vantage for these observations is L2, the Earth-Sun Lagrange point. At L2, the Sun, Earth, and moon are $\approx 90°$ out of the beams and the environment is essentially isothermal. ¿From work on DMR and balloons, the team has found that successful map production requires reference of one pixel to another over many directions and over many time scales. *MAP* plans to do this with the scan strategy shown in Figure 4.

¿From a high-quality map, one may not only obtain the power spectrum, but one may also compare the data to the results of other CMB experiments and to maps of the foreground emission at different frequencies. Also, a map gives the best data set for testing the underlying statistics of the fluctuations. For instance, we will be able to tell from the *MAP* data if the CMB is a Gaussian random field. Finally, with its high sensitivity and large scale coverage, the time-line data from *MAP* will be ideal for searching for transient radio emission.

The question of how well one can determine the parameters of cosmological models is still an active area of research. Early results suggest that Ω_0, Ω_B, h, etc. may be determined to a few percent accuracy [25]. One must bear in mind that these estimates are for only one class of models (standard CDM); we are braced for surprises. At any rate, if the anisotropy is normally distributed, the *MAP* CMB data will be cosmic variance limited up to $l \approx 600$ (assuming the foreground/radio source emission is successfully removed) and will probe multipoles up to $l \approx 1200$.

In the current schedule, the satellite design and definition will be complete by November 1997 and then the building will begin. *MAP* is scheduled for launch late in 2000. The *MAP* science team comprises Chuck Bennett (PI) at NASA/GSFC, Mark Halpern at UBC, Gary Hinshaw at NASA/GSFC, Norm Jarosik at Princeton, John Mather at NASA/GSFC, Steve Meyer at Chicago, Lyman Page at Princeton, Dave Spergel at Princeton, Dave Wilkinson at Princeton, and Ned Wright at UCLA. More information about *MAP*, the CMB, and other experiments may be obtained from http://map.gsfc.nasa.gov/.

86

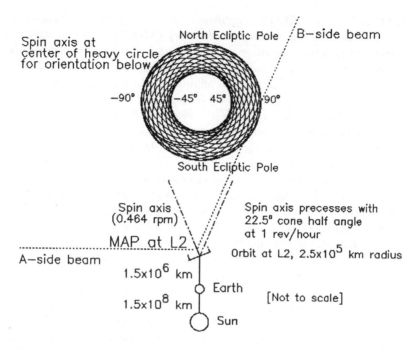

Figure 4. The *MAP* scan pattern for one hour of observation. The lines show the path for one side of a differential pair of beams. The other pair member follows a similar path, only delayed by 1.1 min. There are four principal time scales for the observations. The phase of the difference signal is switched by 180° at 2.5 KHz. The spacecraft spins around its symmetry axis with a 2.2 min period (bold circle) with cone opening angle of roughly 135°. This pattern precesses about the Earth-Sun line with a period of 60 minutes. Thus, in about 1 hour, over 30% of the sky is covered. Every six months, the whole sky is observed. Note that any pixel is differenced to many other pixels in many directions.

Acknowledgments

Conversations with Suzanne Staggs and Steve Meyer were especially useful in preparing this presentation. This work was supported by the NSF and a grant to L. Page from the David and Lucile Packard Foundation.

References

1. Mather, J. C., et al. 1994, Ap. J. ApJ, 420:439.
2. Fixsen, D. J., et al. 1996, ApJ, 473:576. (astro-ph/9605054).
3. Gush, H. et al., 1990, PRL, 65, 537.
4. Sunyaev, R. A. & Zel'dovich , Ya. B. Ann. Rev, Astron. Astrophys, 1980, 18:537.
5. Bartlett, J. G., Stebbins, A. 1991, ApJ, 371:8.
6. Bond, J. R., in *Cosmology and Large Scale Structure*, ed R. Schaeffer, 1995, Elsevier Science Publishers, Netherlands
7. Danese, L., Burigana, C. Toffolatti, L., De Zotti, G. & Franceschini, A. *The Cosmic Microwave Background: 25 Years Later*, 153, 1990, Kluwer Academic Publishers. Mandolesi & Vittorio (eds.)
8. Wright, E. L., et al. 1994, Ap. J. ApJ, 420:450.
9. Bersanelli, M. *et al.* 1994, Ap.J. 424:517.
10. Staggs, S. T., *et al.* 1996, ApJ 473:L1 (astro-ph/9609128).
11. Smoot, G. F. *et al.* 1992, Ap.J. 396:L1.
12. Bennett, C. *et al.* 1996, Ap.J. 464:L1-L4.
13. Kogut, A. *et al.* 1996, Ap.J. 464:L5-L9.
14. Górski K. *et al.* 1996, Ap.J. 464:L11-L15.
15. Hinshaw G. *et al.* 1996, Ap.J. 464:L17-L20.
16. Wright E. *et al.* 1996 Ap.J. 464:L21-L24.
17. Bond, J. R. & Efstathiou, G. 1987, MNRAS, 226, 655.
18. Kogut, A. *et al.* 1996, ApJ 464:L29-L33.
19. Ganga, K. M. *et al.* 1993, Ap.J. 432:L15-L18.
20. Hancock, *et al.* 1994, Nature, 367, 333.
21. Fixsen, D.J. *et al.* 1997 pre-print.
22. Page, L.A. 1997 Proceedings from the Ettorre Majorana school on Large Scale Structure.
23. Ratra, B. 1996. The original compilation was reported in Ratra & Sugiyama, 1995. This is available through astro-ph/9512157. Ratra has kept the list up-to-date and kindly supplied his more recent results.
24. Scott et al. 1996, Ap.J. 461:L1.
25. Jungman, G., Kamionkowski, M., Kosowsky, A. and Spergel, D. Phys. Rev. D, 1996, 54 1332.

COSMOLOGICAL MAGNETIC FIELDS

Angela V. OLINTO

University of Chicago,
5640 S. Ellis Ave, Chicago, IL 60637

During this talk, we follow the history of magnetic fields as the universe evolves from the early universe to the present. We review different scenarios for magnetogenesis in the early universe and follow the subsequent evolution of these fields as the universe recombines. We then focus on the role magnetic fields play after recombination in the formation of structure and the seeding of stellar and galactic fields. Cosmological magnetic fields in the intergalactic medium trace the turbulent history of the universe and may contain fossils of the early universe. We close by reviewing the challenges of observing cosmological magnetic fields and discuss the possibility of studying these fields with extremely–high-energy cosmic rays.

1 Introduction

While the existence of magnetic fields in galaxies and clusters of galaxies is well established, the origin of these fields and the existence of cosmological fields are still unknown. Today, galactic and cluster fields play important dynamical roles as they reach energy equipartition with the gas and the cosmic rays in these systems. These fields may have been seeded by cosmological fields that would permeate the space between galaxies today and affect extremely–high-energy cosmic rays (EHECRs). In the past, cosmological fields may have played a dynamical role in the early universe and as galaxies and stars formed.

Synchrotron maps and Faraday rotation measurements show significant magnetic fields on inter-cluster scales on unknown origin.[1] If there are large scale fields in the universe today, when and how did they form and what is their present structure and spectrum? Is there a component from pre-recombination times? Did these fields influence the formation of galaxies and stars? Are primordial fields the seeds of stellar, galactic, and cluster fields? If we measure fields today on scales larger than clusters, is that evidence for primordial fields or could it be due to pollution from galaxies?

Answering these questions will not only help understand the history of magnetic fields but it may also help understand the formation of the first stars and galaxies and the dynamics of the early universe. We may find that magnetic fields are as relevant to galaxies as they are to the Sun. At a minimum, the study of large scale magnetic fields will help follow the hydrodynamic evolution of the universe, as magnetic fields are fossils of shocks, outflows, winds, and/or early universe processes. In addition, large scale fields affect strongly the propagation of EHECRs. Studying this effect with future observations will help solve the mystery of the origin and nature of EHECRs.

In the following sections, we follow the history of magnetic fields from magnetogenesis in the early universe to the present and discuss possible ways of observing the unknown structure of cosmological magnetic fields. We start by discussing the motivation for early universe magnetogenesis based on the need for primordial fields to seed galactic dynamos. We then discuss the evolution and damping of magnetic fields up to recombination. Fields present at recombination can play a role on the formation of galaxies and stars by preserving fluctuations below the Silk mass and by generating density perturbations after recombination. As dynamical measurements of the mean density of the universe on large scales as well as studies of galaxy clustering indicate that the universe may have a low mean density ($\Omega \simeq 0.2$), the baryonic component of the universe increases in dynamical relevance and so do cosmological fields. Finally, magnetic fields on scales of clusters and larger play an important role on the origin and propagation of extremely–high-energy cosmic rays. As future experiments try to address the origin and nature of EHECRs, the structure of magnetic fields in the intergalactic medium around the Local Group will be simultaneously addressed.

2 Magnetic Fields in the Early Universe

Historically the study of magnetic field generation in the early universe was motivated by the search for the origin of galactic large scale fields. Galaxies have large scale coherent fields of about a few μG ($\equiv 10^{-6}$ G) including the Milky Way with $B_{MW} \simeq 3\mu$G. These galactic fields could have originated from a relatively large primordial seed field amplified by the collapse of the galaxy or by a much smaller seed field that is greatly amplified by a galactic dynamo. These two possibilities give very different estimates for the field needed to seed the galactic field. (Magnetohydrodynamic equations are linear in B, thus, seed fields are necessary for the growth of magnetic fields with time.)

If the galactic dynamo is efficient at amplification, the seed field can start as low as $B_{seed} \simeq 10^{-23}$ G. This value is estimated assuming that the galactic dynamo amplified the field present at the newly formed galaxy by a factor of $\sim e^{30}$ corresponding to about 30 dynamical timescales or complete revolutions since the galaxy formed. After the dynamo amplification factor is taken into account, the present galactic field requires that the newborn galaxy had a field of $\sim 10^{-19}$ G. Finally, we can get an estimate of the primordial field needed to explain present galactic fields by assuming that the newborn galaxy amplified a frozen-in cosmological field as the protogalactic primordial gas collapsed to present galactic densities. The collapse implies a $B \propto \rho^{2/3}$ scaling which gives primordial fields of about $B_{seed} \simeq 10^{-23}$ G for a density increase of 10^6 between the galactic density and the average density in the universe. If a galactic dynamo is not efficient at amplification the primordial field requirements increase to $B_{seed} \simeq 10^{-9} - 10^{-12}$ G.[2]

The success of galactic dynamo models remains elusive. Dynamo models work well for the Earth where fluctuations about the large scale field are small compared to the average field. This hierarchy of scales does not hold in stars and galaxies, making the dynamo problem more challenging. A compass would not be very useful on stars or galaxies since the fluctuation fields have similar or larger strengths to the large scale coherent field. Here we take the two extremes as far as primordial seeding is concern, with and without dynamo amplification.

With the above numbers, we can proceed to discuss early universe processes that can generate seed fields. A useful way of comparing the magnitude of fields at different times is to define the energy density in magnetic fields, $\rho_B \equiv B^2/8\pi$, in units of the background photon energy density, $\rho_\gamma \equiv \pi^2 g_* T^4/30$ where g_* is the number of degrees of freedom and T is the temperature of the cosmic background radiation (CBR). The present CBR temperature implies $\rho_\gamma \simeq (4\mu G)^2/8\pi$. Both energy densities redshift the same way with the expansion of the universe and can be compared in a comoving way ($B \propto a^{-2}$ and $\rho_B \propto a^{-4}$, where $a(t)$ is the cosmic scale factor). In these units the field strength required to seed galactic fields with an efficient galactic dynamo translates into $\rho_B \simeq 10^{-34}\rho_\gamma$, while the primordial seed without dynamo amplification requires $\rho_B \simeq 10^{-14}\rho_\gamma$.

Once a primordial field is generated, it redshifts with the expansion of the universe. Although viscous processes during decoupling damp magnetic fields (as we discuss in §2.3), the cosmological plasma is highly conductive and magnetic diffusion is usually insignificant. If we take the residual ionization of the universe today and make a conservative estimate of the scale of magnetic diffusion during the age of the universe, we find that the field would only diffuse on scales smaller than an $\sim$ A.U.[3] Here we discuss fields on scales of galaxies and larger which correspond to comoving scales of about 1 Mpc ($\simeq 10^{11}$ A.U.) or larger.

2.1 Primordial Magnetogenesis

A number of authors have proposed scenarios for generating seed fields in the early universe.[3-14] These models make use of the few out-of-equilibrium epochs of the pre-recombination era. Phase-transitions are usually necessary, with models ranging from inflation (when the universe had temperatures, $T \gtrsim 10^{16}$ GeV) to the QCD phase-transition era ($T \sim 100$ MeV).

Standard inflationary models give rise to insignificantly small vector perturbations in contrast to the observable scalar and tensor perturbations. The simplest inflationary models give $\rho_B \sim 10^{-104} \rho_\gamma$ which is too small to act as a seed field. Reasonable seed fields can be generated if one breaks electromagnetic gauge invariance,[6] or change the gravitational couplings.[7] String cosmology may also give rise to primordial fields[8] but the magnitude of the effect is still under debate[9].

Although somewhat contrived, inflation based models for seed field generation have the advantage that large scale fields can be easily generated due to the exponential expansion during vacuum domination. On the other hand, observations of the cosmic background radiation (CBR) seem to strongly constrain these models,[10] living little room for their viability.

Several scenarios have been proposed below the GUT scale such as during the Electroweak and the QCD transitions.[11,14] These are usually based on out-of-equilibrium processes during first or second order phase-transitions were shocks, interfaces, and turbulent motions may generate significant fields through battery and amplification mechanisms.

Models based on phase-transitions other than inflation are limited by causality to less power on large scales. In order to compare different models, we use the definitions in [15] where the fourier transform, $\tilde{\mathbf{B}}(\mathbf{k})$, of the magnetic field at a point in space, $\mathbf{B}(\mathbf{x})$, can be described by a scalar power spectrum $\tilde{B}^2(k)$ when $\mathbf{B}(\mathbf{x})$ is assumed to be a random, homogeneous, and isotropic field. The total magnetic energy density is then given by:

$$\rho_B \equiv \frac{1}{2} \int \tilde{B}^2(k) d^3k.$$

The spectrum generated by different models can then be described by power laws on lengthscales above the horizon, H_{pt}^{-1}, (or wavenumbers below $2\pi H_{pt}$) of the particular phase transition. During radiation domination, the horizon for any given phase-transition is much smaller than the galactic scales today, $H_{pt}^{-1} \ll$ Mpc, and the power law behavior is a good description of the spectrum on the scales of interest. (The last transition of interest is the QCD transition when the universe had a horizon that corresponds to a comoving scale of $H_{QCD}^{-1} \simeq 1$ pc today. Earlier transitions have much smaller horizon scales.) We can write

$$\tilde{B}^2(k) = Ak^n$$

where the constant A is related to the total magnetic field energy density, ρ_B, the power index, n, and the cutoff scale of the power law spectrum, $k_{\max}$ via

$A \simeq (n+3)8\pi\rho_B/k_{\max}^{n+3}$.

Causality limits the spectrum of fields generated at transitions during radiation dominated epochs to have spectral index $n \geq 0$, while inflationary models can, in principle, generate fields with negative spectral indexes, i. e., more power on large scales.

Independent of the specific mechanism for magnetic field generation, we can estimate the ability of phase transitions to seed galactic fields given the limit on the total energy available for magnetic fields, the power index limit due to causality, and the maximum cutoff scale given by the horizon. The maximum energy density in magnetic fields that any scenario can generate is $\rho_B = \rho_\gamma$. This is an overestimate since hydrodynamical processes usually lead to magnetic energies in equipartition with plasma motions and $\rho_B \simeq \rho_{\text{plasma}} v^2 \lesssim 0.1 \rho_\gamma$ for most phase transitions. This maximum energy together with the cutoff scales limit the amplitude of the spectrum for a given spectral index. For the electroweak transition the cutoff is $k_{EW} = 2\pi H_{EW} \simeq 2\pi/10\text{AU}$ while for the QCD transition the cutoff is $k_{QCD} = 2\pi H_{QCD} \simeq 2\pi/\text{pc}$)

The best case scenario for large scales fields generated in phase transitions is the white noise spectrum with $n = 0$.[4] In this case, we can estimate the maximum strength of the magnetic field on Mpc scales by choosing a window function that extracts the contribution of the field spectrum on these scales. Following [15], for $n = 0$ the average field on a scale l is:

$$\bar{B}(l) \leq \pi^{3/4} \sqrt{\frac{6\rho_B}{(lk_{\max})^3}}$$

This model-independent estimate gives for the electroweak transition a field on scales $l \simeq$ Mpc of $\bar{B}_{EW}\text{Mpc} \lesssim 10^{-22}$ G and for the QCD transition $\bar{B}_{QCD}\text{Mpc} \lesssim 10^{-16}$ G. Therefore, neither transitions can seed the galactic field without dynamo amplification since the generated fields are much smaller than the needed $B_{\text{seed}} \sim 10^{-9} - 10^{-12}$ G.

2.2 Evolution up to Recombination

After primordial fields are generated, they redshift with the expansion of the universe frozen into the plasma for most of the early universe's history. Although magnetic diffusion is insignificant, during certain epochs in the early universe magnetic field energy is converted into heat through the damping of magneto-hydrodynamic (MHD) modes.[16] This damping is caused by dissipation in the fluid due to the finite mean free path of photons or neutrinos. The result of these fluid viscosities is the efficient damping of MHD modes similar to the Silk damping of adiabatic density perturbations.

The evolution of MHD modes, such as fast and slow magnetosonic, and Alfvén waves, in the presence of viscous and heat conducting processes can be studied in both the radiation diffusion and the free-streaming regimes in the early universe. Fluid viscosities damp cosmic magnetic fields from prior to the epoch of neutrino decoupling up to recombination. Similar to the case of sound waves propagating in a demagnetized plasma, fast magnetosonic waves are damped by radiation diffusion on all scales smaller than the radiation diffusion length. The characteristic damping scales are the horizon scale at neutrino decoupling $(M_\nu \approx 10^{-4} M_\odot$ in baryons) and the Silk mass at recombination $(M_\gamma \approx 10^{13} M_\odot$ in baryons). In contrast, the oscillations of slow magnetosonic and Alfvén waves get overdamped in the radiation diffusion regime, resulting in frozen-in magnetic field perturbations. Further damping of these perturbations is possible only if before recombination the wave enters a regime in which radiation free-streams on the scale of the perturbation. The maximum damping scale of slow magnetosonic and Alfvén modes is always smaller than or equal to the damping scale of fast magnetosonic waves, and depends on the magnetic field strength and its direction relative to the wave vector.

The dissipation of magnetic energy into heat during neutrino decoupling weakens big bang nucleosynthesis constraints on the strength of magnetic fields present during nucleosynthesis. The observed element abundances require that the energy density in magnetic fields be less than one-third of the photon energy density during nucleosynthesis.[17,18] Even if processes prior to neutrino decoupling generate magnetic fields with initial energy density comparable to the photon energy density, neutrino damping causes the magnetic energy to decrease substantially relative to that of radiation by the time of nucleosynthesis. This ensures that most magnetic field configurations generated prior to neutrino decoupling satisfy big bang nucleosynthesis constraints.

Further dissipation before recombination constrains models in which primordial magnetic fields give rise to galactic magnetic fields or density perturbations. Since a sizable fraction of the energy in magnetic field fluctuations is erased up to the Silk scale, it becomes even more difficult to produce the observed galactic field without dynamo amplification. Models which generate primordial fields in sub-horizon scales during phase transitions are particularly constrained since these models have more power on small scales where damping is most efficient.

Although Alfvén and slow magnetosonic modes also undergo significant damping, their damping scales depend on the strength and the direction of the background magnetic field and are generally smaller than the damping scale for fast magnetosonic modes. Therefore, magnetic energy can be stored in Alfvén and slow modes on scales well below the Silk mass. The survival of these modes help the seeding of galactic fields by keeping magnetic energy on scales of interest to galaxy formation and may also be of significance to the formation of structure on relatively small scales. In particular, these modes may be responsible for fragmentation of early structures that may seed early star or galaxy formation.

2.3   Early Dynamos

An important issue in the evolution of cosmological magnetic fields is the nature of hydrodynamic flows as the universe recombines. If there are velocity flows present as the universe evolves, these flows may have enough helicity to set up early dynamos. Magnetic energy can be regained at the expense of the kinetic energy of the flow and primordial fields may be amplified to the necessary level to seed galactic fields.[12,13] Some simulations of pre-recombination MHD systems show dynamo behavior.[14,19] However, viscous damping due to photon and neutrino decoupling damp velocity flows as well as MHD modes through recombination. If velocity fields exist at recombination, the most likely scale for driving the flow will be close to the Silk mass. Since the couplings in this problem are non-linear, it is possible that the driving at large scales cascades down to small scales and back to large scales, therefore, fields may be amplified during and after recombination. Constrains on such pre-recombination flows should be studied in the light of recent CBR observations.

Alternatively, the collapse of density perturbations generated by inflation and imprinted in a non-baryonic cold dark matter component may generate the necessary seed field and amplification. Numerical simulations of the growth of density perturbations after recombination show the formation of shocks that can generate seed fields to be amplified by large scale dynamos.[20]

2.4   CBR and B

In the future, observations of anisotropies of the CBR by MAP and PLANCK will become precise enough to study in detail the acoustic Doppler peaks from sound waves at recombination and polarization. If magnetic fields of significant magnitude are present at recombination, they polarize the CBR photons[21] and generate signatures of the different MHD modes.[22] The precise signatures are likely to depend on the nature of the mode, since different modes evolve differently through recombination.[16]

3 Magnetic Fields in the late Universe

3.1 Galaxy Formation

While magnetic fields are recognized as central agents in regulating the dynamics of star formation and the general interstellar medium in galaxies, it is generally assumed that they do not play a significant role during the epoch of galaxy formation. However, if fields of the order of 10^{-9}G to 10^{-12}G are present in protogalactic clouds, this view should be questioned. Depending on the spectrum of such cosmological fields, structure formation can be initiated by magnetic fields.

Magnetic fields may act as seeds for density perturbations.[?] The evolution of density perturbations, peculiar velocities, and magnetic fields from recombination to the present, give rise to a steep density perturbations spectrum,[15] namely $P(k) \sim k^4$. This spectrum is too steep to account for the observed large-scale structure of the universe, thus, magnetic fields alone cannot reproduce the observed clustering on large scales. On the other hand, magnetic fields do generate small-scale structure shortly after recombination, even if the rms magnetic field on intergalactic scales is as small as 10^{-12} Gauss today. Thus, magnetic fields may provide a natural source of (scale-dependent) bias of the luminous baryonic matter with respect to the dark matter in the universe.

Another consequence of primordial magnetic fields is to add power to the density perturbation spectrum on small scales to that arising primordially, a welcome ingredient for models of structure formation which lack small-scale power, such as tilted cold dark matter, mixed dark matter, and hot dark matter.

3.2 Pollution of the IGM

When trying to measure the magnitude of cosmological magnetic fields today, the most significant observations are those of magnetic fields on the largest scales and away from virialized systems such as galaxies and clusters of galaxies. Thus, observations of magnetic fields in the intergalactic medium are the most helpful in understanding the origin of cosmological magnetic fields.

Intergalactic and intercluster magnetic fields of significant magnitudes have been observed,[1,24] but their interpretation is not unique. One of the problems that need to be addressed is the ability of galaxies to pollute the intergalactic medium and how can "pollution" fields be differentiated from primordial fields. Galactic outflows may pollute a significant fraction of the IGM[25] depending on different models of galaxy formation and cosmological evolution.

With the discovery of the highest energy cosmic rays and the possibility of studying the sources of these events in the future, the question of whether an extragalactic magnetic field exists outside of clusters has gained a new observational tool.

4 Observing Cosmological Magnetic Fields

4.1 Present Observations

Of great relevance to understanding the history of cosmological magnetic fields are observations of intergalactic magnetic fields and magnetic fields at high redshifts. Reports of Faraday rotation associated with high-redshift Lyman-α absorption systems (see, e.g., [1]) suggest that dynamically significant magnetic fields (of order μG) may be present in condensations at high redshift. Together with observations of strong magnetic fields in clusters,[24] these observations support the idea that magnetic fields play a dynamical role in the evolution of structure and maybe present throughout the universe.

Present Faraday rotation measurements of intergalactic fields using the emission from high-redshift quasars place limits of $\lesssim 10^{-9}$ G for fields with Mpc reversal scales and $\lesssim 10^{-11}$G for fields coherent on the present horizon scale. Future measurements may help determine the strength of galactic magnetic fields at high redshifts as well as the presence of significant fields in the IGM today.

4.2 Future Observations

The detection of extremely-high energy cosmic rays (EHECRs) has triggered considerable interest on the origin and nature of these particles. To date, more than 60 cosmic ray events with energies above $\sim 5 \ 10^{19}$ eV have been observed by experiments such as Haverah Park, Fly's Eye, and AGASA. The Fly's Eye experiment has recorded a $3 \ 10^{20}$ eV event, the highest energy event so far. These EHECRs most likely originate from extragalactic sources and their spectrum and spatial as well as temporal distributions are affected by the presence of extragalactic magnetic fields.

The study of EHECRs is closely related to the study of cosmological magnetic fields.[27] Charged particles of energies $\sim 10^{20}$ eV can be deflected significantly in cosmic magnetic fields. Thus, detailed information on the structure of extragalactic magnetic fields may be contained in the time, energy, and arrival direction distributions of charged extremely-high energy particles emitted from powerful discrete sources. In this context, the clustering among EHECRs suggested by recent AGASA data is very encouraging and its confirmation would have important consequences for understanding the nature and origin of cosmological magnetic fields as well as for EHECRs.[28]

Another signature of cosmological magnetic fields is in the shape of the spectrum of γ-rays secondaries to EHECRs. The γ-ray spectrum depends on synchrotron losses versus inverse-Compton regeneration of the electromagnetic cascade. This signature is most sensitive to extragalactic magnetic fields around the Faraday rotation limit of $\sim 10^{-9}$G.[26]

5 Conclusion

We followed some of the history of cosmological magnetic fields. At each step of this history, new questions arise. Magnetogenesis in phase transitions alone cannot generate galactic fields, but is there a pre-recombination dynamo or is the amplification done as galaxies form? Decoupling damps fast magnetosonic waves, but do Alfvén and slow magnetosonic modes play a role on scales below the Silk mass? Protogalactic shocks may generate small seed fields, but do these get amplified on large scales?

These are some of the questions that future observations of large scale magnetic fields will address. In addition to improving traditional observations, the study of extremely–high-energy cosmic rays will play an important in probing cosmological fields on a 50 Mpc volume around us.

Acknowledgments

We acknowledge the support of the US Department of Energy and the hospitality of the Aspen Center for Physics.

References

1. P. P. Kronberg, *Rep. Prog. Phys.* **57**, 325 (1994).
2. R.M. Kulsrud, in *Galactic and Intergalactic Magnetic Fields*, ed. R. Beck, P.P. Kronberg, & R. Wielebinski (Dordrecht: Kluwer), p. 527 (1990).
3. B. Cheng and A. V. Olinto, *Phys. Rev.* D **50**, 2421 (1994).
4. C. J. Hogan, *Phys. Rev. Lett.* **51**, 1488 (1983).
5. E. R. Harrison, *MNRAS* **147**, 279 (1970); E. R. Harrison, *MNRAS* **165**, 185 (1973); W. D. Garretson, G. B. Field, and S. M. Carroll, *Phys. Rev.* D **46**, 5346 (1992); A. D. Dolgov, *Phys. Rev.* D **48**, 2499 (1993); A. D. Dolgov and J. Silk, *Phys. Rev.* D **47**, 3144 (1993).
6. M. S. Turner and L. M. Widrow, *Phys. Rev.* D **30**, 2743 (1988).

7. B. Ratra, *Ap. J* **391**, L1 (1992); B. Ratra, *Phys. Rev.* D **45**, 1913 (1992);
8. M. Gasperini, M. Giovannini, and G. Veneziano, *Phys. Rev. Lett.* **75**, 3796 (1995).
9. D. Lemoine and M. Lemoine, *Phys. Rev.* D **52**, 1955 (1995);
10. D. Lemoine, PhD Thesis (1995)
11. T. Vaschaspati, *Phys. Lett.* B **265**, 258 (1991); R. H. Brandenberger, A.-C. Davis, A. M. Matheson, and M. Trodden, *Phys. Lett.* B **293**, 287 (1992); A. P. Martin and A.-C. Davis, *Phys. Lett.* B **360**, 71 (1995); T. W. B. Kibble and A. Vilenkin, *Phys. Rev.* D **52**, 679 (1995); J. Quashnock, A. Loeb, and D. N. Spergel, *Ap. J.* **344**, L49 (1989).
12. G. Baym, D. Bödecker, and L. McLerran, *Phys. Rev.* D **53**, 662 (1996);
13. G. Sigl, K. Jedamzik, and A. V. Olinto, *Phys. Rev.* D **55**, 4582 (1997).
14. K. Enqvist, P. Olensen, *Phys. Lett.* B **329**, 195 (1994).
15. E. Kim, A. V. Olinto, & R. Rosner, *Ap. J.*, 467 (1996).
16. K. Jedamzik, V. Katalinic, and A. V. Olinto, *Phys. Rev. D*, in press (1998).
17. P. Kernan, G. Starkman, and T. Vachaspati, Phys. Rev. D **54**, 7202 (1996).
18. B. Cheng, A. V. Olinto, D. Schramm, and J. Truran, Phys. Rev. D **54**, 4714 (1996).
19. A. Brandenburg, K. Enqvist, and P. Olesen, *Phys. Lett.* **B391**, 395 (1997).
20. R. Kulsrud, D. Ryu, R. Cen, and J. P. Ostriker, *Ap. J.* (1997).
21. A. Loeb and A. Kosowsky, *Ap..J.* **469** 1 (1996).
22. J. Adams, U. Danielsson, D. Grasso, and H. Rubinstein, *Phys.Lett.* **B388** 253 (1996).
23. I. Wasserman, *Ap. J.*, **224**, 337 (1978).
24. K.-T. Kim, P.C. Tribble, and P.P. Kronberg, *Ap. J.*, **379**, 80 (1991).
25. P. Kronberg, and H. Lesch, in "The Physics of Galactic Halos" eds. H. Lesch, R-J Dettmar, U. Mebold and R. Schlickeiser, Berlin: Akademie Verlag (1996).
26. M. Lemoine, A. V. Olinto, G. Sigl, and D. Schramm, *Ap. J. Lett.* (1997)
27. G. Sigl, A. V. Olinto, and M. Lemoine, *Phys. Rev. D* (1997).
28. S. Lee, A. V. Olinto, and G. Sigl, *Ap. J. Lett.* **455**, L1 (1995).

SOLAR NEUTRINOS: WHERE WE ARE

JOHN BAHCALL

Institute for Advanced Study, Princeton, NJ 08540

This talk compares standard model predictions for solar neutrino experiments with the results of actual observations. Here 'standard model' means the combined standard model of minimal electroweak theory plus a standard solar model. I emphasize the importance of recent analyses in which the neutrino fluxes are treated as free parameters, independent of any constraints from solar models, and the stunning agreement between the predictions of standard solar models and helioseismological measurements.

1 Introduction

Joe Taylor mentioned in his beautiful preceding review of pulsar phenomena that the discussion of pulsars has a long history at the Texas conferences. I want to add a footnote to his historical remarks: the subject of solar neutrinos has an even longer history in this context. Both Ray Davis and I gave invited talks [1,2] on solar neutrinos at the 2nd Texas Conference, which was held in Austin, Texas in December 1964.

Solar neutrino research has now achieved the primary goal that was discussed in the 1964 Texas Conference, namely, the detection of solar neutrinos. This detection establishes empirically that the sun shines by fusing light nuclei in its interior.

The subject of solar neutrinos is entering a new phase in which large electronic detectors will yield vast amounts of diagnostic data. These new experiments [3,4,5], which will be described by Professor Totsuka in the following talk, will test the prediction of the minimal standard electroweak theory [6,7,8] that essentially nothing happens to electron type neutrinos after they are created by nuclear fusion reactions in the interior of the sun.

The four pioneering experiments—chlorine [9,10] Kamiokande [11] GALLEX [12] and SAGE [13]—have all observed neutrino fluxes with intensities that are within a factors of a few of those predicted by standard solar models. Three of the experiments (chlorine, GALLEX, and SAGE) are radiochemical and each radiochemical experiment measures one number, the total rate at which neutrinos above a fixed energy threshold (which depends upon the detector) are captured. The sole electronic (non-radiochemical) detector among the initial experiments, Kamiokande, has shown that the neutrinos come from the sun, by measuring the recoil directions of the electrons scattered by solar neutrinos. Kamiokande has also demonstrated that the observed neutrino energies are consistent with the range of energies expected on the basis of the standard solar model.

Despite continual refinement of solar model calculations of neutrino fluxes over the past 35 years (see, e.g., the collection of articles reprinted in the book edited by Bahcall, Davis, Parker, Smirnov, and Ulrich [14]), the discrepancies between observations and calculations have gotten worse with time. All four of the pioneering solar neutrino experiments yield event rates that are significantly less than predicted by standard solar models.

This talk is organized as follows. I first discuss in section 2 the three solar neutrino problems. Then I review in section 3 the recent work by Heeger and Robinson [15] which treats the neutrino fluxes as free parameters and shows that the solar neutrino problems cannot be resolved within the context of minimal standard electroweak theory unless solar neutrino experiments are incorrect. Next I discuss in section 4 the stunning agreement between the values of the sound velocity calculated from standard solar models and the values obtained from helioseismological measurements. Finally, in section 5 I compare the success of the MSW neutrino mixing hypothesis with the success of the solar ^{3}He mixing hypothesis recently discussed by Cumming and Haxton[16].

See http://www.sns.ias.edu/~jnb for further information about solar neutrinos, including viewgraphs, preprints, and numerical data.

2   Three Solar Neutrino Problems

I will first compare the predictions of the combined standard model with the results of the operating solar neutrino experiments. By 'combined' standard model, I mean the predictions of the standard solar model and the predictions of the minimal electroweak theory. We need a solar model to tell us how many neutrinos of what energy are produced in the sun and we need electroweak theory to tell us how the number and flavor content of the neutrinos are changed as they make their way from the center of the sun to detectors on earth.

We will see that this comparison leads to three different discrepancies between the calculations and the observations, which I will refer to as the three solar neutrino problems.

Figure 1 shows the measured and the calculated event rates in the four ongoing solar neutrino experiments. This figure reveals three discrepancies between the experimental results and the expectations based upon the combined standard model. As we shall see, only the first of these discrepancies depends sensitively upon predictions of the standard solar model.

2.1 Calculated versus Observed Absolute Rate

The first solar neutrino experiment to be performed was the chlorine radio-chemical experiment, which detects electron-type neutrinos that are more energetic than 0.81 MeV. After more than 25 years of the operation of this experiment, the measured event rate is 2.55 ± 0.25 SNU, which is a factor ~ 3.6 less than is predicted by the most detailed theoretical calculations, $9.5^{+1.2}_{-1.4}$ SNU [17,18]. A SNU is a convenient unit to describe the measured rates of solar neutrino experiments: 10^{-36} interactions per target atom per second. Most of the predicted rate in the chlorine experiment is from the rare, high-energy ^{8}B neutrinos, although the ^{7}Be neutrinos are also expected to contribute significantly. According to standard model calculations, the *pep* neutrinos and the CNO neutrinos (for simplicity not discussed here) are expected to contribute less than 1 SNU to the total event rate.

This discrepancy between the calculations and the observations for the chlorine experiment was, for more than two decades, the only solar neutrino problem. I shall refer to the chlorine disagreement as the "first" solar neutrino problem.

2.2 Incompatibility of Chlorine and Water (Kamiokande) Experiments

The second solar neutrino problem results from a comparison of the measured event rates in the chlorine experiment and in the Japanese pure-water experiment, Kamiokande. The water experiment detects higher-energy neutrinos, those with energies above 7 MeV, by neutrino-electron scattering: $\nu + e \longrightarrow \nu' + e'$. According to the standard solar model, ^{8}B beta decay is the only important source of these higher-energy neutrinos.

The Kamiokande experiment shows that the observed neutrinos come from the sun. The electrons that are scattered by the incoming neutrinos recoil predominantly in the direction of the sun-earth vector; the relativistic electrons are observed by the Cherenkov radiation they produce in the water detector.

In addition, the Kamiokande experiment measures the energies of individual scattered electrons and therefore provides information about the energy spectrum of the incident solar neutrinos. The observed spectrum of electron recoil energies is consistent with that expected from ^{8}B neutrinos. However, small angle scattering of the recoil electrons in the water prevents the angular distribution from being determined well on an event-by-event basis, which limits the constraints the experiment places on the incoming neutrino energy spectrum.

The event rate in the Kamiokande experiment is determined by the same high-energy ^{8}B neutrinos that are expected, on the basis of the combined

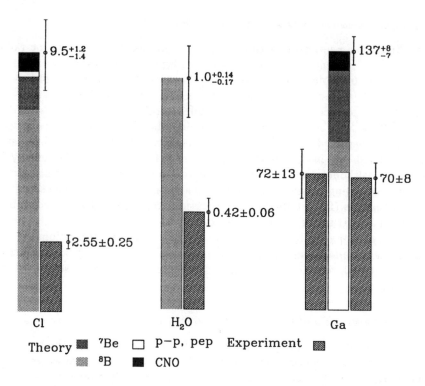

Figure 1: Comparison of measured rates and standard-model predictions for four solar neutrino experiments.

standard model, to dominate the event rate in the chlorine experiment. I have shown[19] that solar physics changes the shape of the ^{8}B neutrino spectrum by less than 1 part in 10^5 . Therefore, we can calculate the rate in the chlorine experiment that is produced by the ^{8}B neutrinos observed in the Kamiokande experiment (above 7 MeV). This partial (^{8}B) rate in the chlorine experiment is 3.2 ± 0.45 SNU, which exceeds the total observed chlorine rate of 2.55 ± 0.25 SNU.

Comparing the rates of the Kamiokande and the chlorine experiments, one finds that the net contribution to the chlorine experiment from the *pep*, ^{7}Be, and CNO neutrino sources is negative: -0.66 ± 0.52 SNU. The standard model calculated rate from *pep*, ^{7}Be, and CNO neutrinos is 1.9 SNU. The apparent incompatibility of the chlorine and the Kamiokande experiments is the "second" solar neutrino problem. The inference that is often made from this comparison is that the energy spectrum of ^{8}B neutrinos is changed from the standard shape by physics not included in the simplest version of the standard electroweak model.

2.3 Gallium Experiments: No Room for ^{7}Be Neutrinos

The results of the gallium experiments, GALLEX and SAGE, constitute the third solar neutrino problem. The average observed rate in these two experiments is 70.5 ± 7 SNU, which is fully accounted for in the standard model by the theoretical rate of 73 SNU that is calculated to come from the basic p-p and pep neutrinos (with only a 1% uncertainty in the standard solar model p-p flux). The ^{8}B neutrinos, which are observed above 7.5 MeV in the Kamiokande experiment, must also contribute to the gallium event rate. Using the standard shape for the spectrum of ^{8}B neutrinos and normalizing to the rate observed in Kamiokande, ^{8}B contributes another 7 SNU, unless something happens to the lower-energy neutrinos after they are created in the sun. (The predicted contribution is 16 SNU on the basis of the standard model.) Given the measured rates in the gallium experiments, there is no room for the additional 34 ± 4 SNU that is expected from ^{7}Be neutrinos on the basis of standard solar models.

The seeming exclusion of everything but p-p neutrinos in the gallium experiments is the "third" solar neutrino problem. This problem is essentially independent of the previously-discussed solar neutrino problems, since it depends strongly upon the p-p neutrinos that are not observed in the other experiments and whose calculated flux is approximately model-independent.

The missing ^{7}Be neutrinos cannot be explained away by any change in solar physics. The ^{8}B neutrinos that are observed in the Kamiokande experiment are produced in competition with the missing ^{7}Be neutrinos; the competition is between electron capture on ^{7}Be versus proton capture on ^{7}Be. Solar model explanations that reduce the predicted ^{7}Be flux generically reduce much more (too much) the predictions for the observed ^{8}B flux.

The flux of ^{7}Be neutrinos, $\phi(^7\text{Be})$, is independent of measurement uncertainties in the cross section for the nuclear reaction $^7\text{Be}(p, \gamma)^8\text{B}$; the cross section for this proton-capture reaction is the most uncertain quantity that enters in an important way in the solar model calculations. The flux of ^{7}Be neutrinos depends upon the proton-capture reaction only through the ratio

$$\phi(^7\text{Be}) \propto \frac{R(e)}{R(e) + R(p)}, \tag{1}$$

where $R(e)$ is the rate of electron capture by ^{7}Be nuclei and $R(p)$ is the rate of proton capture by ^{7}Be. With standard parameters, solar models yield $R(p) \approx 10^{-3}R(e)$. Therefore, one would have to increase the value of the ^{7}Be$(p, \gamma)^8$B cross section by more than 2 orders of magnitude over the current best-estimate (which has an estimated uncertainty of $\sim 10\%$) in order to affect significantly the calculated ^{7}Be solar neutrino flux. The required change in the nuclear physics cross section would also increase the predicted neutrino event rate by more than 100 in the Kamiokande experiment, making that prediction completely inconsistent with what is observed. (From time to time, papers have been published claiming to solve the solar neutrino problem by artificially changing the rate of the ^{7}Be electron capture reaction. Equation (1) shows that the flux of ^{7}Be neutrinos is actually independent of the rate of the electron capture reaction to an accuracy of better than 1%.)

I conclude that either: 1) at least three of the four operating solar neutrino experiments (the two gallium experiments plus either chlorine or Kamiokande) have yielded misleading results, or 2) physics beyond the standard electroweak model is required to change the neutrino energy spectrum (or flavor content) after the neutrinos are produced in the center of the sun.

3 "The Last Hope": No Solar Model

The clearest way to see that the results of the four solar neutrino experiments are inconsistent with the predictions of the minimal electroweak model is not to use standard solar models at all in the comparison with observations. This is what Berezinsky, Fiorentini, and Lissia[20] have termed "The Last Hope" for a solution of the solar neutrino problems without introducing new physics.

Let me now explain how model independent tests are made.

Let $\phi_i(E)$ be the normalized shape of the neutrino energy spectrum from one of the i neutrino sources in the sun (e.g., ^{8}B or $p - p$ neutrinos). I have shown [19] that the shape of the neutrino energy spectra that result from radioactive decays, ^{8}B, ^{13}N, ^{15}O, and ^{17}F, are the same to 1 part in 10^5 as the laboratory shapes. The $p - p$ neutrino energy spectrum, which is produced by fusion has a slight dependence on the solar temperature, which affects the shape by about 1%. The energies of the neutrino lines from ^{7}Be and pep electron capture reactions are also only slightly shifted, by about 1% or less, because of the thermal energies of particles in the solar core.

Thus one can test the hypothesis that an arbitrary linear combination of the normalized neutrino spectra,

$$\Phi(E) = \sum_i \alpha_i \phi_i(E), \qquad (2)$$

can fit the results of the neutrino experiments. One can add a constraint to Eq. (2) that embodies the fact that the sun shines by nuclear fusion reactions that also produce the neutrinos. The explicit form of this luminosity constraint is

$$\frac{L_\odot}{4\pi r^2} = \sum_j \beta_j \phi_j , \qquad (3)$$

where the eight coefficients, β_j, are given in Table VI of the paper by Bahcall and Krastev [21].

The first demonstration that the four pioneering experiments are by themselves inconsistent with the assumption that nothing happens to solar neutrinos after they are created in the core of the sun was by Hata, Bludman, and Langacker [22]. They showed that the solar neutrino data available by late 1993 were incompatible with any solution of equations (2) and (3) at the 97% C.L.

In the most recent and complete analysis in which the neutrino fluxes are treated as free parameters, Heeger and Robertson [15] showed that the data presented at the Neutrino '96 Conference in Helsinki are inconsistent with equations (2) and (3) at the 99.5% C.L. Even if they omitted the luminosity constraint, equation (3), they found inconsistency at the 94% C.L.

It seems to me that these demonstrations are so powerful and general that there is very little point in discussing potential "solutions" to the solar neutrino problem based upon hypothesized non-standard scenarios for solar models.

4 Comparison with Helioseismological Measurements

Helioseismology has recently sharpened the disagreement between observations and the predictions of solar models with standard (non-oscillating) neutrinos. This development has occurred in two ways.

Helioseismology has confirmed the correctness of including diffusion in the solar models and the effect of diffusion leads to somewhat higher predicted events in the chlorine and Kamiokande solar neutrino experiments [17]. Even more importantly, helioseismology has demonstrated that the sound velocities predicted by standard solar models agree with extraordinary precision with the sound velocities of the sun inferred from helioseismological measurements [18]. Because of the precision of this agreement, I am convinced that standard solar models cannot be in error by enough to make a major difference in the solar neutrino problems.

The physical basis for the helioseismological measurements was described beautifully by Joergen Christensen-Dalsgaard in his talk yesterday afternoon. I recommend the text of his discussion that you will also find in these proceedings. You will see in Joergen's article references to other papers about helioseismology that you can use to become better acquainted with the subject.

I will report here on some comparisons that Marc Pinsonneault, Sarbani Basu, Joergen, and I have done recently which demonstrate the precise agreement between the sound velocities in standard solar models and the sound velocities inferred from helioseismological measurements. These results are based upon an article that has since appeared in *Physical Review Letters*[18].

Since the deep solar interior behaves essentially as a fully ionized perfect gas, $c^2 \propto T/\mu$ where T is temperature and μ is mean molecular weight. The sound velocities in the sun are determined from helioseismology to a very high accuracy, better than 0.2% rms throughout nearly all the sun. Thus even tiny fractional errors in the model values of T or μ would produce measurable discrepancies in the precisely determined helioseismological sound speed

$$\frac{\delta c}{c} \simeq \frac{1}{2} \left(\frac{\delta T}{T} - \frac{\delta \mu}{\mu} \right) . \tag{4}$$

The remarkable numerical agreement between standard predictions and helioseismological observations, which I will discuss in the following remarks, rules out solar models with temperature or mean molecular weight profiles that differ significantly from standard profiles. The helioseismological data essentially rule out solar models in which deep mixing has occurred (cf. *PRL* paper[23]) and argue against unmixed models in which the subtle effect of particle diffusion–selective sinking of heavier species in the sun's gravitational field–is not included.

Figure 2 compares the sound speeds computed from three different solar models with the values inferred[24,25] from the helioseismological measurements. The 1995 standard model of Bahcall and Pinsonneault (BP)[17], which includes helium and heavy element diffusion, is represented by the dotted line; the corresponding BP model without diffusion is represented by the dashed line. The dark line represents the best solar model which includes recent improvements[26,27] in the OPAL equation of state and opacities, as well as helium and heavy element diffusion. For the OPAL EOS model, the rms discrepancy between predicted and measured sound speeds is 0.1% (which may be due partly to systematic uncertainties in the data analysis).

In the outer parts of the sun, in the convective region between $0.7R_\odot$ to $0.95R_\odot$ (where the measurements end), the No Diffusion and the 1995 Diffusion

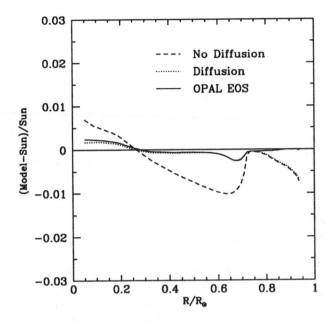

Figure 2: Comparison of sound speeds predicted by different standard solar models with the sound speeds measured by helioseismology. There are no free parameters in the models; the microphysics is successively improved by first including diffusion and then by using a more comprehensive equation of state. The figure shows the fractional difference, $\delta c/c$, between the predicted model sound speed and the measured[24,25] solar values as a function of radial position in the sun ($R_{\odot}$ is the solar radius). The dashed line refers to a model[17] in which diffusion is neglected and the dotted line was computed from a model[17] in which helium and heavy element diffusion are included. The dark line represents a model which includes recent improvements in the OPAL equation of state and opacities[26,27].

model have discrepancies as large as 0.5% (see Figure 2). The model with the Livermore equation of state [27], OPAL EOS, fits the observations remarkably well in this region. We conclude, in agreement with the work of other authors[28], that the OPAL (Livermore National Laboratory) equation of state provides a significant improvement in the description of the outer regions of the sun.

The agreement between standard models and solar observations is independent of the finer details of the solar model. The standard model of Christensen-Dalsgaard et al. [29], which is derived from an independent computer code with different descriptions of the microphysics, predicts solar sound speeds that agree everywhere with the measured speeds to better than 0.2%.

Figure 2 shows that the discrepancies with the No Diffusion model are as large as 1%. The mean squared discrepancy for the No Diffusion model is 22 times larger than for the best model with diffusion, OPAL EOS. If one supposed optimistically that the No Diffusion model were correct, one would have to explain why the diffusion model fits the data so much better. On the basis of Figure 2, we conclude that otherwise standard solar models that do not include diffusion, such as the model of Turck-Chièze and Lopez[30], are inconsistent with helioseismological observations. This conclusion is consistent with earlier inferences based upon comparisons with less complete helioseismological data[24,31,23], including the fact that the present-day surface helium abundance in a standard solar model agrees with observations only if diffusion is included[17].

Equation 4 and Figure 2 imply that any changes $\delta T/T$ from the standard model values of temperature must be almost exactly canceled by changes $\delta\mu/\mu$ in mean molecular weight. In the standard model, T and μ vary, respectively, by a factor of 53 and 43% over the entire range for which c has been measured and by 1.9 and 39% over the energy producing region. It would be a remarkable coincidence if nature chose T and μ profiles that individually differ markedly from the standard model but have the same ratio T/μ. Thus we expect that the fractional differences between the solar and the model temperature, $\delta T/T$, or mean molecular weights, $\delta\mu/\mu$, are of similar magnitude to $\delta c^2/c^2$, i.e. (using the larger rms error, 0.002, for the solar interior),

$$|\delta T/T|, \ |\delta\mu/\mu| \lesssim 0.004. \tag{5}$$

How significant for solar neutrino studies is the agreement between observation and prediction that is shown in Figure 2? The calculated neutrino fluxes depend upon the central temperature of the solar model approximately as a power of the temperature, Flux $\propto T^n$, where for standard models the exponent n varies from $n \sim -1.1$ for the $p-p$ neutrinos to $n \sim +24$ for the ^{8}B neutrinos[32]. Similar temperature scalings are found for non-standard solar models[33]. Thus, maximum temperature differences of $\sim 0.2\%$ would produce changes in the different neutrino fluxes of several percent or less, much less than required[34] to ameliorate the solar neutrino problems.

Figure 3 shows that the "mixed" model of Cummings and Haxton (CH)[16] (illustrated in their Figure 1) is grossly inconsistent with the observed helioseismological measurements. The vertical scale of Figure 3 had to be expanded by a factor of 2.5 relative to Figure 2 in order to display the large discrepancies with observations for the mixed model. The discrepancies for the CH mixed model (dashed line in Figure 3) range from +8% to −5%. Since μ in a standard solar model decreases monotonically outward from the solar interior, the mixed model–with a constant value of μ– predicts too large values for the sound speed in the inner mixed region and too small values in the outer mixed region. The asymmetric form of the discrepancies for the CH model is due to the competition between the assumed constant rescaling of the temperature in the BP No Diffusion model and the assumed mixing of the solar core (constant value of μ). We also show in Figure 3 the relatively tiny discrepancies found for the new standard model, OPAL EOS.

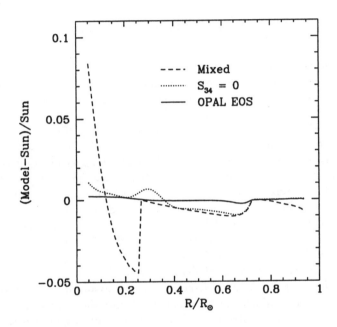

Figure 3: Non-standard solar models compared with helioseismology. This figure is similar to Figure 2 except that the vertical scale is expanded. The dashed curve represents the sound speeds computed for the mixed solar model of Cumming and Haxton [16] with ^{3}He mixing. The dotted line represents the sound speed for a solar model computed with the rate of the ^{3}He$(\alpha, \gamma)^7$Be reaction set equal to zero. For comparison, we also include the results for the new standard model labeled OPAL EOS in Figure 2.

More generally, helioseismology rules out all solar models with large amounts of interior mixing, unless finely-tuned compensating changes in the temperature are made. The mean molecular weight in the standard solar model with diffusion varies monotonically from 0.86 in the deep interior to 0.62 at the outer region of nuclear fusion ($R = 0.25R\odot$) to 0.60 near the solar surface. Any mixing model will cause μ to be constant and equal to the average value in the mixed region. At the very least, the region in which nuclear fusion occurs must be mixed in order to affect significantly the calculated neutrino fluxes [35,36,37,38,39]. Unless almost precisely canceling temperature changes are assumed, solar models in which the nuclear burning region is mixed ($R \lesssim 0.25 R_\odot$) will give maximum differences, δc, between the mixed and the standard model predictions, and hence between the mixed model predictions and the observations, of order

$$\frac{\delta c}{c} = \frac{1}{2} \left(\frac{\mu - <\mu>}{\mu} \right) \sim 7\% \text{ to } 10\%, \qquad (6)$$

which is inconsistent with Figure 2.

Are the helioseismological measurements sensitive to the rates of the nuclear fusion reactions? In order to answer this question in its most extreme form, we have computed a model in which the cross section factor, S_{34}, for the $^3\text{He}(\alpha, \gamma)^7\text{Be}$ reaction is artificially set equal to zero. The neutrino fluxes computed from this unrealistic model have been used [35] to set a lower limit on the allowed rate of solar neutrinos in the gallium experiments if the solar luminosity is currently powered by nuclear fusion reactions. Figure 3 shows that although the maximum discrepancies ($\sim 1\%$) for the $S_{34} = 0$ model are much smaller than for mixed models, they are still large compared to the differences between the standard model and helioseismological measurements. The mean squared discrepancy for the $S_{34} = 0$ model is 19 times larger than for the standard OPAL EOS model. We conclude that the $S_{34} = 0$ model is not compatible with helioseismological observations.

To me, these results suggest strongly that the assumption on which they are based—nothing happens to the neutrinos after they are created in the interior of the sun—is incorrect. A less plausible alternative (in my view) is that some of the experiments are wrong; this must be checked by further experiments.

5 ^{3}He Mixing versus MSW Mixing

It is instructive to compare the success of the hypothesis of ^{3}He solar mixing[16] with the success of the hypothesis of MSW neutrino mixing. I do so below.

•**Consistency With Solar Neutrino Experiments.** The ^{3}He mixing hypothesis is inconsistent at the 99.5% C.L. with solar neutrino experiments (a special case of the general result of Heeger and Robinson[15]); MSW mixing is consistent with the solar neutrino experiments (best value of χ^2 is less than one per degree of freedom).

•**Consistency With Helioseismology.** The ^{3}He mixing hypothesis is inconsistent with helioseismology; mixing of the solar core necessarily implies ~ 7% discrepancies with the measured solar sound velocities. The standard solar model with no free parameters predicts sound velocities that agree with the measured velocities to a rms accuracy of 0.2% in the solar core.

•**Free Parameters.** The ^{3}He mixing has 3 free parameters; the MSW mixing has 2 free parameters.

6   Discussion

The combined predictions of the standard solar model and the standard electroweak theory disagree with the results of the four pioneering solar neutrino experiments. The disagreement persists even if the neutrino fluxes are treated as free parameters, without reference to any solar model.

The solar model calculations are in excellent agreement with helioseismological measurements of the sound velocity, providing further support for the inference that something happens to the solar neutrinos after they are created in the center of the sun.

Looking back on what was envisioned in 1964, I am astonished and pleased with what has been accomplished. In 1964, it was not clear that solar neutrinos could be detected. Now, they have been observed in different experiments and the theory of stellar energy generation by nuclear fusion has been directly confirmed. Moreover, particle theorists have shown that solar neutrinos can be used to study neutrino properties, a possibility that we did not even consider in 1964. In fact, much of the interest in the subject stems from the fact that the four pioneering experiments suggest that new neutrino physics may be revealed by solar neutrino measurements. Finally, helioseismology has confirmed to high precision predictions of the standard solar model, a possibility that also was not imagined in 1964.

Acknowledgments

This research is supported in part by NSF grant number PHY95-13835.

References

1. John N. Bahcall, Proceedings of the 2nd Texas Symposium on Relativistic Astrophysics, December 1964, in *Quasars and High-Energy Astronomy*, eds. K.N. Douglas, E.L. Schucking, I. Robinson, J.A. Wheeler, A. Schild, and N.J. Woolf (Gordon and Breach, 1969) p. 321.

2. R. Davis, Jr., D.S. Harmer, and F.H. Nelly, Proceedings of the 2nd Texas Symposium on Relativistic Astrophysics, December 1964, in *Quasars and High-Energy Astronomy*, eds. K.N. Douglas, E.L. Schucking, I. Robinson, J.A. Wheeler, A. Schild, and N.J. Woolf (Gordon and Breach, 1969) p. 287.

3. C. Arpesella *et al.*, *BOREXINO proposal, Vols. 1 and 2*, eds. G. Bellini, R. Raghavan, et al. (Univ. of Milano, 1992).

4. M. Takita, in *Frontiers of Neutrino Astrophysics*, eds. Y. Suzuki and K. Nakamura (Universal Academy Press, 1993) p. 147.

5. A.B. McDonald, in *Proceedings of the 9th Lake Louise Winter Institute*, eds. A. Astbury *et al.* (World Scientific, 1994) p. 1.

6. S.L. Glashow, *Nucl. Phys.* **22**, 579 (1961).

7. S. Weinberg, *Phys. Rev. Lett.* **19**, 1264 (1967).

8. A. Salam, in *Elementary Particle Theory*, ed. N. Svartholm (Almqvist and Wiksells, 1968) p. 367.

9. R. Davis, Jr., *Phys. Rev. Lett.* **12**, 303 (1964).

10. R. Davis, Jr., *Prog. Part. Nucl. Phys.* **32**, 13 (1994).

11. Y. Suzuki, KAMIOKANDE collaboration, *Nucl. Phys.* B *(Proc. Suppl.)* **38**, 54 (1995).

12. P. Anselmann, et al., GALLEX collaboration, *Phys. Lett.* B **342**, 440 (1995).

13. J.N. Abdurashitov, *et al.*, SAGE Collaboration, *Phys. Lett.* B **328**, 234 (1994).

14. Eds. J.N. Bahcall, R. Davis, Jr., P. Parker, A. Smirnov, and R.K. Ulrich, *Solar Neutrinos: The First Thirty Years* (Addison Wesley, 1995).

15. K. H. Heeger and R.G.H. Robertson, *Phys. Rev. Lett.* **77**, 3720 (1996).

16. A. Cumming and W. C. Haxton. Phys. Rev. Lett, **77**, 4286 (1996).

17. J.N. Bahcall, M.H. Pinsonneault, *Rev. Mod. Phys.* **67**, 781 (1995).

18. J.N. Bahcall, M.H. Pinsonneault, S. Basu, and J. Christensen-Dalsgaard, *Phys. Rev. Lett.* **78**, 171 (1997).

19. J.N. Bahcall, *Phys. Rev.* D **44**, 1644 (1991).

20. V. Berezinsky, G. Fiorentini, and M. Lissia, *Phys. Lett.* B **365**, 185 (1996).

21. J.N. Bahcall and P.I. Krastev, *Phys. Rev.* D **53**, 4211 (1996).

22. N. Hata, S. Bludman, and P. Langacker, *Phys. Rev.* D **49**, 3622 (1994).

23. Y. Elsworth *et al.*, *Nature* **347**, 536 (1990).

24. S. Basu *et al.*, *Astrophys. J.* **460**, 1064 (1996).
25. S. Basu *et al.*, *Bull. Astron. Soc. India* **24**, 147 (1996).
26. C.A. Iglesias and F.J. Rogers, *Astrophys. J.* **464**, 943 (1996); D.R. Alexander and J.W. Ferguson, *Astrophys. J.* **437**, 879 (1994). The new OPAL opacities include more elements (19 rather than 12) and cover a wider range in temperature, density, and composition. The low temperature opacity tables include more opacity sources and a wider range of composition.
27. F.J. Rogers, F.J. Swenson, and C.A. Iglesias, *Astrophys. J.* **456**,902 (1996). Our previous equation of state [17] assumed that the plasma was fully ionized in the interior, and included the Debye-Huckel correction, relativistic effects, and degeneracy. The OPAL EOS is based on an activity expansion of the grand canonical partition function which does not require an *ad hoc* treatment of pressure ionization.
28. D.B. Guenther, Y.-C. Kim, and P. Demarque, *Astrophys. J.* **463**, 382 (1996); H. Shibahashi and M. Takata, *Publ. Astron. Soc. Japan* **48**, 377 (1996).
29. J. Christensen-Dalsgaard *et al.*, *Science* **272**, 1286 (1996).
30. S. Turck-Chièze and I. Lopez, *Astrophys. J.* **408**, 347 (1993).
31. J. Christensen-Dalsgaard, C.R. Proffitt, and M.J. Thompson, *Astrophys. J.* **403**, L75 (1993).
32. J.N. Bahcall and A. Ulmer, *Phys. Rev.* D **53**, 4202 (1996).
33. V. Castellani, S. Degl'Innocenti, G. Fiorentini, and M. Lissia, *Phys. Rev.* D **50**, 4749 (1994).
34. J.N. Bahcall and H.A. Bethe, *Phys. Rev. Lett.* **75**, 2233 (1990); M. Fukugita, *Mod. Phys. Lett.* A **6**, 645 (1991); M. White, L. Krauss, and E. Gates, *Phys. Rev. Lett.* **70**, 375 (1993); V. Castellani, S. Degl'Innocenti, and G. Fiorentini, *Phys. Lett.* B **303**, 68 (1993); N. Hata, S. Bludman, and P. Langacker, *Phys. Rev.* D **49**, 3622 (1994); V. Castellani *et al.*, *Phys. Lett.* B **324**, 425 (1994); J.N. Bahcall, *Phys. Lett.* B **338**, 276 (1994); S. Parke, *Phys. Rev. Lett.* **74**, 839 (1995); G.L. Fogli and E. Lisi, *Astropart. Phys.* **3**, 185 (1995); V. Berezinsky, G. Fiorentini, and M. Lissia, *Phys. Lett.* B **185**, 365 (1996). *Phys. Rev. Lett.* **77**, 4286 (1996).
35. J.N. Bahcall, *Neutrino Astrophysics* (Cambridge University Press, 1989).
36. D. Ezer and A.G.W. Cameron, *Astrophys. Lett.* **1**, 177 (1968).
37. J.N. Bahcall, N.A. Bahcall, and R.K. Ulrich, *Astrophys. Lett.* **2**, 91 (1968).
38. G. Shaviv and E.E. Salpeter, *Phys. Rev. Lett.* **21**, 1602 (1968).
39. E. Schatzman, *Astrophys. Lett.* **3**, 139 (1969); E. Schatzman and A. Maeder, *Astron. Astrophys.* **96**, 1 (1981).

FIRST RESULTS FROM SUPER-KAMIOKANDE

Y. TOTSUKA

Kamioka Observatory,
Institute for Cosmic Ray Research, University of Tokyo,
Higashimozumi, Kamioka,
Gifu, 506-12 Japan
(for the Super-Kamiokande Collaboration)

A 50,000 ton water Čerenkov detector, Super-Kamiokande, has been operational since April 1996. Observation is currently being made with the threshold total-energy of 5.6 MeV. Data taken for 102 days have been analyzed and the preliminary results on solar neutrinos were obtained. Based on the data with visible energies $E_{vis} > 7$ MeV, the ^{8}B neutrino flux was $2.51 \pm^{0.14}_{0.13} \pm 0.18 \times 10^6$ cm^{-2}s^{-1}. The data were divided in daytime and nighttime. The day and night fluxes agreed within statistical errors.

1 Introduction

Neutrinos are elusive and still mysterious particles. They interact only by weak forces and thus possess an enormous puch-through capability. Neutrinos are copiously produced in the core of a star by weak nuclear processes which are the source of energy and luminosity. Most of the neutrinos have energies less than 1 MeV but a small fraction of them are emitted with energies as large as 15 MeV by β-decay of rarely produced ^{8}B nuclei (half life 0.77 s).

The Sun is the star nearest to us and is indeed the only main-sequence star that can be seen with neutrino detectors. Since R. Davis had pioneered his work, the four experiments, Homestake, Kamiokande, SAGE and GALLEX [1], successfully observed the solar neutrinos, whose energies ranged from 0.23 MeV to 15 MeV. Their results thus confirmed that the Sun and in general the stars generate energies by nuclear processes. However when comparison was made quantitatively, their results strongly disagreed with what the standard solar model (SSM) predicts. In fact the Homestake, Kamiokande, SAGE and GALLEX experiments observed only 30 % ($E_\nu > 0.814$ MeV), 50 % ($E_\nu > 6.5$ MeV), 50 % ($E_\nu > 0.233$ MeV) and 50 % ($E_\nu > 0.233$ MeV) of the expected yields [2], respectively. It is of utmost importance to find what causes such large discrepancies.

The supernova is another astronomical object that is visible with neutrinos though its occurrence is very rare. The supernova SN1987A in the Large Magellanic Cloud (LMC) went off in February 1987 and the underground detectors, notably Kamiokande and IMB [3], recorded neutrino bursts and confirmed the

basic process underlying the explosion of type II supernovae, namely gravitational collapse of the central iron core. However the observed number of events
(11 in Kamiokande and 8 in IMB) were frustratingly few and did not allow us
to study in detail the explosion mechanism of the supernova.

Successful observations of solar and supernova neutrinos motivated us to
construct a new neutrino detector, Super-Kamiokande, much larger than the
current underground detectors which, compared with other nuclear and particle experiments, are already extremely massive. Super-Kamiokande began its
operation in April 1996.

The Super-Kamiokande experiment studies not only astrophysical neutrinos but also atmospheric neutrinos. It has been known for some time that the
ratio of observed μ to e in $E_{vis} \leq 1\,\text{GeV}$ which are produced by atmospheric
neutrinos is about 40 % lower than what one naively expects[4]. The ratio μ/e
is colsely related to $(\nu_\mu + \overline{\nu_\mu})/(\nu_e + \overline{\nu_e})$ and its theoretical uncertainties such
as the primary cosmic-ray flux largely cancel. Hence the small μ/e ratio, if
the results are indeed true, provides strong evidence for neutrino oscillations
of $\nu_\mu \rightarrow \nu_e$, ν_τ or ν_s (sterile neutrinos). It is of utmost importance to confirm
this atmospheric-neutrino anomaly with the Super-Kamiokande detector.

2 Super-Kamiokande Detector

The Super-Kamiokande detector is a 50,000 ton water Čerenkov detector located 1,000 m underground in the Kamioka zinc mine and about 150 m south
of the old Kamiokande facility. The excavation of a 65,000 m³ cavity began in
December 1991 and ended in June 1994. Lining of the cavity with stainless-
steel plates followed and a huge water tank of 39.3 mϕ × 42 mh was completed
in April 1995. It then took till December 1995 to mount 11,200 photomultiplier tubes (PMT) of 50 cmϕ in the inner detector and 1,800 PMTs of 20 cmϕ
in the outer detector. The tank was completely filled with pure water by the
end of March 1996 and the observation started in 1 April 1996 as originally
scheduled.

The detector consists of the inner and outer parts. The outer one serves as
a veto counter for incoming cosmic-ray muons as well as a shield layer against
γ-rays and neutrons coming from the rock. The inner detector is the central
32,000 ton of water mass viewed by 11,200 PMTs (50 cmϕ) and is completely
surrounded by the outer detector with a water layer of about 2.7 m thick. It
is this inner part that detects interactions of solar and atmospheric neutrinos
and hopefully supernova neutrinos in near future.

50,000 ton of water contained in the inner and outer detectors is continuously purified by means of de-ionization, fine filtering and degasification. Water

must be free of natural-radioactive elements, especially ^{222}Rn whose daughter nuclei ^{214}Bi β-decay and emit electrons of kinetic energy $T \leq 3.3$ MeV. Due to a finite energy resolution of the detector these electrons occasionally output signals equivalent to $E_{vis} \geq 6$ MeV (E_{vis} is the visible energy and equivalent to the electron *total*-energy) and hence are significant background to the solar-neutrino observation. Water should be transparent in order for Čerenkov photons to traverse on average 20 m of water and to reach PMTs on the wall without substantial losses. Rn concentration and water transparency have been monitored continuously and the current levels (as of October 1996) are less than 5 mBq/m^3 and 55 m for Rn concentration and average absorption length of Čerenkov light, respectively.

PMT signals are fed into ADC/TDC circuits of double-hit capability through 70 m coax-cables. The circuits are capable of recording double pulses with a minimum separation of 1 μs and thus are able to efficiently detect electrons from μ- decay. The data-acquisition system as well as the electronics were specially designed to cope with burst events of more than 40 kHz so that Super-Kamiokande should be kept alive in case of near-by supernovae.

Gain factors of all the PMTs were set to 10^7 and their variations measured within 7 % by means of a Xe-lamp system. The timing resolution is typically 2.9 ns r.m.s. at one photoelectron (p.e.) level.

3   Calibration

The Super-Kamiokande detector must be calibrated periodically to update the basic parameters, the most important of which are the energy estimator (ratio of the number of hit PMTs in the time interval of 50 ns to the deposited energy with corrections of water transparency and PMT photosensitive area) and the resolutions of position and angle determinations. To experimentally determine these parameters we have been using γ-rays of about 9 MeV emitted from Ni(n,γ)Ni reactions. The γ-source consists of a polyethylene vessel (19 cm $\phi \times$ 20 cm$h$) and 2.8 kg of Ni threads. A ^{252}Cf neutron source is placed at its center. The vessel is lowered at a predetermined position in the water. Water embeds the vessel when it descends and thermalizes the neutrons. The calibration with the Ni-γ system has been carried out on average every two weeks. The energy estimator was found to be stable with an accuracy better than 1 %.

Recently the calibration with a low-energy electron beam has been successfully carried out. A small electron linear acceleratior (LINAC) was installed in a tunnel near the Super-Kamiokande cavity. The electron beam of energies between 5 MeV and 16 MeV were transported in a magnetically-shielded vacuum pipe, bent by 90 deg downward and shot in the water through a 30 m vacuum pipe. At the exit window a thin plastic scintillation counter was placed and its signals were used to initiate the data-taking. The first calibration was carried out at the coordinate (-1237, -70.7, 1206) cm where (0,0,0) corresponds to the detector center. The data analysis then followed in exactly the same way as for the real data. Figure 1 shows the E_{vis} distributions for four electron energies of 5.866, 6.782, 8.637 and 15.966 MeV , where E_{vis} is expressed as the effective number of hit PMTs . They were found to agree with the Monte Carlo simulation within an accuracy of 1 %. The beam position was also successfully reconstructed. The systematic shift of the reconstructed position at 8.6 MeV, for example, was less than 10 cm and the resolution was about 60 cm and 35 cm for the longitudinal and transverse directions to the beam, respectively. The beam direction was also reconstructed and reproduced very well by the Monte Carlo simulation. Though the calibration results are still preliminary, we believe that the detector characteristics are well understood to perform the detailed study of solar neutrinos.

The trigger of the Super-Kamiokande detector is made by simply counting the number of hit PMTs within a gate time of 200 ns. Its efficiency was determined with the LINAC calibration system, which is shown in Fig.2 .

4   First Results on Solar Neutrinos

Low-energy data accumulated during 101.9 day of live-time have been analyzed for solar neutrinos. Preliminary results will be presented here.

The basic reaction for detecting solar neutrinos is neutrino-electron elastic scattering, $\nu_e + e \rightarrow \nu_e + e$. Its cross section is precisely known from electroweak theory. Elastic scattering by other neutrino species, $\nu_i + e \rightarrow \nu_i + e$, $i = \mu$ or τ also takes place but its cross section is only a sixth. It is this recoil electron that emits Čerenkov light and is detected. The electron is scattered in the forward direction if the neutrino energy is much larger than the electron mass, which is the case for the present experiment. This forward peak was used to pick up solar-neutrino signals from the isotoropic background.

The analysis approximately followed the following way which had been established in the old Kamiokande experiment:
(1) Eliminate obvious external noises such as electric noises (6.1×10^7 events left); (2) Eliminate low-energy events within 20μs after preceding muons to avoid contamination of electrons from μ-decay (5.5×10^7 events); (3) Space-reconstruct remaining low-energy events and determine the event positions and directions (4.7×10^7 events); (4) Estimate visible energies, E_{vis}, assuming events to be electrons; (5) Apply the fiducial-volume cut, namely pick up events within the central 22.5 kton of the inner detector to reduce incoming

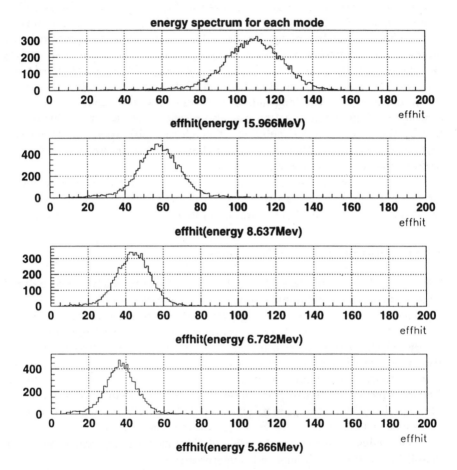

Figure 1: Distribution of the energy estimator (effective number of hit PMTs) for electrons of 5.866, 6.782, 8.637 and 15.966 MeV (preliminary).

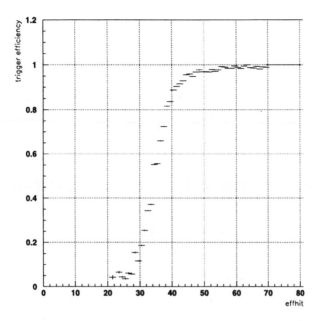

Figure 2: Trigger efficiency of the S-K detector at the position of the beam injection as a function of the effective number of hit PMTs, which is an energy estimator of electons. 35 hits correspond to electron tatal-energy of 5.51 MeV. This result was obtained with the LINAC calibration. Note the trigger efficiency slightly depends on event positions in the detector.

background γ-rays (4.2×10^6 events); (6) Pick up events with $7 \leq E_{vis} \leq 20\,\text{MeV}$ (1.1×10^5 events); (7) Eliminate spallation products by cosmic-ray muons, by taking advantage of temporal and spacial correlations of low-energy event and preceding muons (1.4×10^4 events). Now the basic data sample is ready: (8) Make the $\cos\theta_{sun}$ distribution, where θ_{sun} is the directional correlation of the event with respect to the Sun, $\theta_{sun} = 0$ being the forward direction. Solar-neutrino events ought to be accumulated near $\cos\theta_{sun} = 1$: (9) Count the number of excess events near $\cos\theta_{sun} = 1$.

Figure 3 shows the resultant angular distribution for $E_{vis} \geq 7\,\text{MeV}$. A forward peak is clearly visible above the flat background. The number of excess events in the peak is $1,005^{+55}_{-52}$. Based on this number, the total ^{8}B-neutrino flux is $2.51^{+0.14}_{-0.13} \pm 0.18 \times 10^6\,\text{cm}^{-2}\text{s}^{-1}$. The systematic error will soon be reduced significantly thanks to the precise LINAC calibration. This result is quite consistent with what Kamiokande obtained over 2,000 days of live-time.

However the absolute flux value is not a main issue any more. What is

120

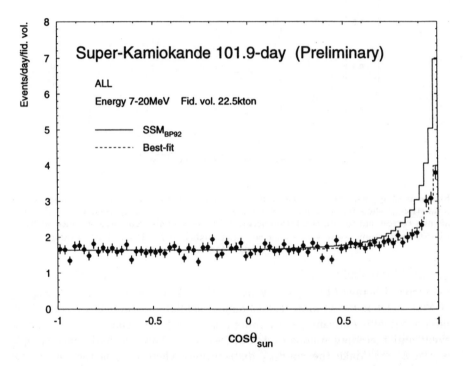

Figure 3: $\cos\theta_{sun}$ distribution for $E_{vis} > 7\,\text{MeV}$ after fiducial-volume and spallation cuts. Excess events near the forward direction are signals produced by solar neutrinos. Data correspond to 101.9 day live-time.

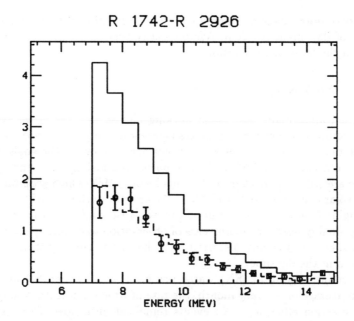

R 1742-R 2926

Figure 4: E_{vis} distribution of solar-neutrino events. The solid histogram is the prediction from Bahcall-Pinsonneault's theory [6], while the dashed one is the best fit with a scale factor of 0.441 to the theoretical prediction.

important is to find the true mechanism that causes the solar-neutrino problem. Super-Kamiokande intends to measure the precise shape of the ^{8}B neutrino spectrum independent of the absolute flux value. The data were binned by the visible energies, E_{vis} and the above procedures were repeated. Figure 4 shows the resultant spectrum together with the expected histogram. The shape itself agrees with the expectation within the statistical errors. Unfortunately the current statistics is not large enough to find a faint spectral distortion predicted by the small-angle solution of the MSW mechanism [5], which is thought to be one of the best candidate solutions to the solar-neutrino problem.

Another important measure is to study the difference in the ^{8}B fluxes between day and night. In fact the large-angle solution of the MSW mechanism [5] predicts a small excess in the nighttime which is caused by the regeneration of ν_e in the earth. We divided the data into day and night and analyzed them in the same way. The resultant ^{8}B fulxes are $2.30^{+0.18}_{-0.17} \pm 0.17 \times 10^6$ and $2.76^{+0.21}_{-0.20} \pm 0.20 \times 10^6 \, \mathrm{cm^{-2}s^{-1}}$ for day and night, respectively. We do not see the difference within the present statistical errors. Note that the systematic errors of both fluxes are highly correlated and some of them cancel when the ratio of day/night is taken. We obviously need more data to confirm or disconfirm the large-angle solution.

5   More Work Needed

The Super-Kamiokande detector is working almost as good as we expected. Various calibrations such as water transparency, energy estimator, etc. are being conducted well. However we have not achieved one of the goals, namely the threshold energy of 5 MeV. Water purity reached the low enough level but there are still large background events below 7 MeV which prohibit further study of solar neutrinos down to 5 MeV. We are still working hard to find what is the origin of these events and hope to resolve the problem in near future.

We probably need two more years of observation to definitely see if the ^{8}B neutrino spectrum is really distorted or if the flux in the nighttime is indeed larger than in the daytime.

The online supernova (SN) watch has been installed and SN alarms are issued on average once every month, the rate of which is adjustable depending on the detection efficiency. No events remained until now after inspecting the spacial uniformity of the alarm events: They were clusters of spallation products by hard interactions of muons. Currently Super-Kamiokande is 100 % efficient to find SNs at the distance as far as 100 kpc. Obviously the detector must be running all the time as the SN neutrino burst lasts for only tens of seconds. The current data-taking efficiency is about 95 %. The loss is due to frequent calibration work. It may eventually reach $97 \sim 98\%$ very soon.

The analysis of atmospheric neutrinos is going well and the first result will be presented in the summer this year with statistics several times larger than the total data obtained in Kamiokande.

In order to confirm the atmospheric-neutrino problem in a convincing way we will carry out the long-baseline neutrino-oscillation experiment between Super-Kamiokande and KEK from the beginning of 1999.

Proton decay is being and will be searched for as long as the Super-Kamiokande experiment continues, hopefully for more than 50 years. Our effort is especially focused on the decay mode $p \rightarrow \nu K^+$ which is predicted by the SUSY GUTs. In five years we will reach the sensitivity of 3×10^{33} years for this decay mode.

References

1. B.T. Cleveland *et al.*, *Nucl. Phys.* B (*Proc. Suppl.*) **38**, 47 (1995);

Y. Fukuda *et al.*, *Phys. Rev. Lett.* **77**, 1683 (1996);
J.N. Abdurashitov *et al.*, *Phys. Lett.* B **328**, 234 (1994);
P.Anselmann *et al.*, *Phys. Lett.* B **324**, 440 (1995).

2. J.N. Bahcall and M. Pinsonneault, *Rev. Mod. Phys.* **64**, 885 (1992), and **67**, 781 (1995);
   S. Turck-Chieze and I. Lopes, *Ap.J.* **408**, 347 (1993).

3. K. Hirata *et al.*, *Phys. Rev. Lett.* **58**, 1490 (1987), and *Phys. Rev.* D **38**, 448 (1988);
   R.M. Bionta *et al.*, *Phys. Rev. Lett.* **58**, 1494 (1987).

4. Y. Fukuda *et al.*, *Phys. Lett.* B **335**, 237 (1994).

5. S.P. Mikheyev and A.Y. Smirnov, *Sov. Jour. Nucl. Phys.* **42**, 913 (1985);
   L. Wolfenstein, *Phys. Rev.* D **17**, 2369 (1978).

6. First paper in reference 2.

GALAXY EVOLUTION: HAS THE "EPOCH OF GALAXY FORMATION" BEEN FOUND?

C. Steidel

Palomar Observatory, California Institute of Technology,
Caltech 105-24,
Pasadena, CA 91125

An enormous amount of progress has been made in the last year or two toward empirically constraining the evolutionary status of galaxies as a function of cosmic epoch. We now have a relatively consistent picture of the state of galaxy evolution for $z \lesssim 1$, and a first glimpse at $z > 2$. There are now several lines of argument, supported by independent data, that galaxy formation began in earnest between $z \sim 3$ and $z \sim 4$ and probably peaked near $z \sim 2$. Thus, the "epoch of galaxy formation" is now directly accessible to wholesale observations and a great deal of progress is to be expected in the very near future.

1 Introduction

The rate of change of the state of the whole field of galaxy formation and evolution is presently so large that there is little doubt that by the time these proceedings appear in print, much of what I will say will have been superseded by new results. It is a very exciting time, as new observational facilities and techniques are being brought to bear on fundamental questions concerning the nature of galaxies as a function of cosmic epoch, and for the first time it appears that significant empirical constraints on theories of galaxy and structure formation, at early epochs where different theories diverge most significantly, may be imminent.

I would characterize the current state of affairs in the following way: thanks to large, comprehensive surveys of field galaxies using multi-object spectrographs primarily on 4-meter class telescopes, and to the re-furbished Hubble Space Telescope, we now have a relatively consistent picture of the history of field galaxies since $z \lesssim 1$, or the last $\sim 60\%$ of the age of the universe. There are certainly many issues to be resolved, and there are differences of opinion as to how to interpret the empirical data, but overall there is remarkable qualitative agreement among various groups on the general history of galaxies since $z \sim 1$[1]. At the same time, the $z \gg 1$ universe is now very much open to empirical study for quite run–of–the–mill, representative objects, and while the dust has not yet settled and a great deal of new data is being obtained at the moment, there is a "thumb–nail" sketch of the history of global galaxy and star formation that is emerging from the qualitative consistency of a number of independent techniques for studying the high–redshift universe. In these written proceedings, due to limited space, I will very briefly summarize the overall results at $z < 1$ and highlight some of the very new results at very high redshifts.

2 Galaxies at $z \lesssim 1$

There is now a relatively large amount of information on field galaxies out to $z \sim 1$, ranging from comprehensive apparent magnitude–selected redshifts surveys[2,3,4], to detailed kinematic observations of individual galaxies[5,6] and morphological studies with the refurbished Hubble Space Telescope[7,8,10]. While there are differences in the details, the general picture of the evolution of the luminosity function of faint field galaxies is that objects of luminosity $\sim L^*$ and greater appear to have undergone only modest evolution since $z \sim 1$, whereas there appears to be a pronounced "steepening" of the faint–end slope of the luminosity function which can be interpreted as enhanced star formation in moderate–to–low luminosity galaxies. When galaxies are selected in the near–IR (or by gas cross-section), which is quite insensitive to the instantaneous star formation rate, there appears to be no significant change in the luminosity function out to $z \sim 1$ relative to local samples[4,11]. At the same time, there is a huge increase in the integrated luminosity density, particularly at blue and UV rest-wavelengths[12]. While kinematic studies of distant galaxies are extremely challenging to both observation and interpretation, it appears that at least some giant spiral galaxies at $z \sim 1$ have essentially the same circular velocities for their absolute blue luminosities as local spirals[5], while small, "faint blue galaxies" appear to be as much as 1.5 magnitudes brighter than local counterparts with the same circular velocities[6].

Morphologically, when galaxy number counts are segregated by HST morphology, there is approximately the expected number of objects that can be classified along the traditional Hubble sequence, but a huge excess of objects which would be classified as "peculiar"[7,8]. This large population of blue, peculiar objects grows more prominent with increasing apparent magnitude, and quickly dominates the number counts and overwhelms the traditional Hubble sequence objects. At very faint apparent magnitudes the Hubble classification scheme appears to be completely inadequate to describe most of the galaxies observed in extremely deep HST images[10], although this may be due to the fact that the galaxies are (on average) at high enough redshifts that observations in the observed–frame optical are sampling the unfamiliar far UV in the galaxies' rest frame.

While the jury is still out on *exactly* what is happening at $z < 1$, the points of general consensus appear to be:

• Massive galaxies (E/S0 and early spirals) are more or less in place and evolving relatively quiescently by $z \sim 1$, implying a much earlier "formation epoch" (although even this statement is not without controversy[13,14].)

• Star formation is a strongly increasing function of redshift, where the evolution in the luminosity density is carried mostly by late type spirals and irregular galaxies (less–massive systems?).

At around $z \sim 1$ it appears that objects that are intrinsically more massive (as reflected in their near-IR luminosities) are beginning to show signs of rapid star formation, while at lower redshifts most of the star formation is occurring in less–massive systems; apparently star formation "migrates" from large objects to small as one goes forward in time[4,15].

All of these points suggest that the galaxy formation timescale is very different depending on the mass of the galaxy, and that the most important epoch for the formation of today's massive and luminous galaxies is at high redshift. This apparent empirical situation may be at odds with many current theories of the galaxy formation process which involve a primarily bottom–up growth of structure (see also below).

3 Beyond $z \sim 1$

Given that the important epoch for the formation of at least a large fraction of the "massive" galaxies seems to be pushed to well beyond $z \sim 1$, it is clear that in order to directly observe the galaxy formation process observations at still earlier cosmic epochs are probably necessary. Searches for "proto-galaxies", or galaxies caught directly in the act of forming the bulk of their stars on a relatively short timescale, have a long history dating back to the 1960's. Over the intervening years, while searches continued, the prevailing theoretical expectations have changed considerably, to the point where the current picture is one in which the "formation" of a big galaxy is expected to be a protracted process in which a galaxy grows relatively slowly by successive mergers of smaller sub-units. Within this general "bottom–up" hierarchical scenario for structure formation, there would not necessarily be any singular time period in a galaxy's history which one could point to as the "formation" epoch, and perhaps no time when the galaxy would be especially bright and therefore easy to detect at large distances. That there appeared to be far too many galaxies in terms of sheer numbers as a function of apparent magnitude (the famous "faint blue galaxy" problem), and the lack of very high redshift galaxies in the spectroscopic samples, seemed to support a picture in which today's galaxies were in many pieces in the observable past. There has been recent support for this "sub–galactic fragment" picture [16] for high redshift galaxies from HST images in the field of a low–power radio galaxy, where a large surface density of small, Lyman–alpha emitting blobs is seen. At the same time, however, there are a handful of examples of apparent "field" objects (and many in high redshift clusters of galaxies[18]) which apparently formed the bulk of their stars even earlier than the early epoch at which they are observed, and have evolved relatively quiescently since that time.[17] It is certain that at some level structure in the universe developed hierarchically and that mergers of sub–galactic "fragments" must have occurred and were probably important to the formation of what we now call a galaxy; however, it is also evident that there are objects out there which formed early and, necessarily, fairly quickly, reaching conservative middle-age prior to a redshift of 1. It is not yet clear which is the dominant historical paradigm for which type (or what fraction) of present–day galaxy.

Until quite recently, examples of any object powered completely by star formation at redshifts beyond $z \sim 1$ could be counted on one hand, while tremendous progress was being made with the high redshift radio galaxies[19] and with high redshift QSOs[20]. It was clear that actually finding high redshift star–forming galaxies with traditional redshift surveys would be an inefficient and painstaking process given that "foreground" objects severely dominate even at fairly faint apparent magnitude, judging by the results of the field galaxy redshift surveys (also, there is the practical matter that beyond $z \sim 1.2$, essentially all of the prominent spectral features used to confirm galaxy redshifts move into the near-IR, where the background is orders of magnitude higher and in any case the features are beyond the wavelength range of multi-object spectrographs with CCD detectors). Clearly, a more subtle approach is necessary to isolate the high redshift objects from the dominant foreground, and one approach is to isolate some strong spectral feature that is essentially guaranteed to be present and which can be discerned via "ultra–low resolution spectroscopy" with relatively broad filters and imaging techniques. At high redshifts, the strongest such feature is the Lyman limit of hydrogen at 912Å (13.6 eV). An object with far–UV flux, due either to ongoing massive star formation or AGN activity, will exhibit a rather "hard" or "blue" spectrum up to the Lyman limit, whereupon a combination of intrinsic stellar energy distributions, H I gas in the galaxy itself, and intervening H I external to the galaxy effectively reduce the continuum to zero intensity. [21,22,23]. This "Lyman break" feature becomes accessible to ground-based telescopes for redshifts near $z \sim 3$, where it will appear in the center of the traditional U band. Figure 1 illustrates the basic idea. It has been demonstrated[24], based on limits on the numbers of objects with discontinuities in the observed U band, that more

128

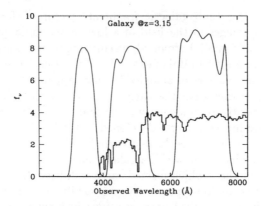

Figure 1: Schematic showing how a galaxy at $z \sim 3$ can be isolated by the presence of the Lyman break, using broad–band filters. The spectrum is that expected for a star forming galaxy after passing the model through a reasonable interstellar medium and the intergalactic medium at $z \sim 3$. The 3 filter bandpasses are the actual ones used in our ground based surveys for $z \sim 3$ galaxies. The resulting "colors" as measured using such filters are unlike any other astronomical object at any other redshift, allowing efficient isolation of such objects from the dominant foreground galaxies.

than 90% of the field galaxies to $R = 26$ magnitude have $z < 3$. However, it turns out that this leaves plenty of room for a substantial population of very high redshift galaxies hidden among the dominant foreground.

In view of this, about 6 years ago we began a program whose aim was to assess the evolutionary status of "normal" (i.e., run-of-the-mill) galaxies, in whatever unknown form they might take at such high redshifts. The approach was to look for, using the Lyman break expectations and deep imaging techniques, objects that we already knew were there because they had given rise to metallic line absorption in the spectra of background QSOs[21,22,25]. QSO absorbing galaxies are expected to be quite "representative" objects[11], at least if the $z \sim 3$ examples were similar in nature to those at somewhat smaller redshifts. The original motivation for this study was to credibly demonstrate that a simple photometric method could be used to identify high redshift galaxies, and the QSO absorbers were seen as a way to obtain "poor–man's spectroscopy" for objects which we guessed would always be too faint for actual confirming spectra. The method could then be extended to the general field with some confidence, and broad statistics on the surface density and luminosities of $z \sim 3$ galaxies would be accessible. However, we were being overly pessimistic, and in fact the Lyman break technique turns out to be extremely efficient for isolating (from ground based photometry) galaxies in the

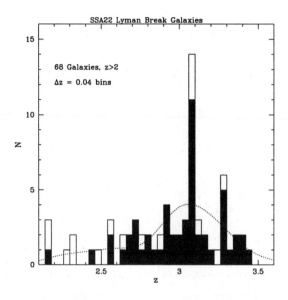

Figure 2: Plot showing the redshift distribution of Lyman break galaxies that have been spectroscopically confirmed in a single 9' by 18' area of sky. The differently shaded histograms reflect slightly different color selection criteria (see Steidel *et al.* 1997 for details); the dotted curve is the smoothed redshift selection function for the color selection corresponding to the un-shaded histogram, normalized to the same total number of galaxies. The "spike" at $z = 3.09$ is significant at the 99.9% confidence level.

redshift range $2.7 \leq z \leq 3.4$ (more or less as expected *a priori* based on models), and at the time of this writing more than 170 galaxies with $z > 2$ have been spectroscopically confirmed with the Keck telescopes [26,27,28,29]. There is a tremendous amount of astrophysics that can now be done with representative objects at very high redshifts, from the properties of the objects themselves (spectroscopic, kinematic, stellar populations, etc.) to global measurements of star formation activity and large scale structure as traced by galaxies. The Hubble Deep Field[30] has provided an excellent focal point for many studies because of the exceptionally deep photometry that can be used in a manner analogous to the ground based photometry for isolating the very high redshift objects[27,32,31,28] (in fact Lyman break objects can be found to lower redshifts because the HDF photometric system includes the F300W filter, which probes considerably farther to the UV than can be done from the ground), and the fact that extremely high quality, high resolution images are available for all of the objects falling within the $\sim 2\overset{.}{'}5$ field of view.

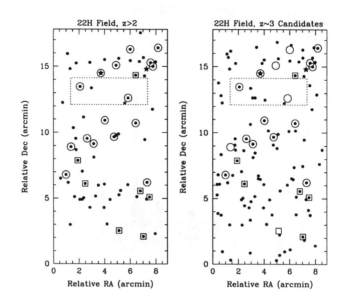

Figure 3: Plot showing the distribution of high redshift galaxies on the plane of the sky from the same field as in the previous figure. The rectangular dotted region corresponds to the area surveyed in another faint redshift survey conducted with the Keck telescope.[4] The left panel shows all objects with confirmed redshifts $z > 2$, and the second shows the overall distribution of photometrically selected $z \sim 3$ candidates. Objects comprising the "spike" at $z = 3.09$ are circled. The two "starred" objects are faint QSOs discovered using the same selection technique.

We have been engaged in trying exploit the efficiency with which high redshift galaxies can now be selected and confirmed to begin to "map out" the large scale distribution of galaxies at such early epochs. The idea is to try to provide a solid "data point" for the progress of the growth of structure at the earliest epoch at which galaxies can be observed directly. Toward this end, we have compiled photometric data in about 10 different high latitude fields, typically from 100–200 square arcminutes per field, in which we have identified many hundreds of candidate high redshift objects for follow–up spectroscopy. Experience has shown that with a single good night on the Keck 10m telescope with the Low Resolution Imaging Spectrograph[34] we can obtain confirming spectra of more than 40 $z \sim 3$ galaxies. While the work is ongoing, we have been able to obtain enough redshifts in one of our fields to enable meaningful statistical analyses of the redshift histogram, shown in Figure 2. Figure 3 shows the corresponding map of the field on the plane of the sky. At least one of the redshift "spikes" is highly significant, at $z \sim 3.09$, and preliminary results from some of our other fields indicates that such prominent structures are common at these high redshifts. Qualitatively, our "pencil beam" redshift survey at $z \sim 2.5 - 3.5$[35] appears quite similar to those at much smaller redshifts[37], and it is the case that a substantial fraction of star-forming galaxies are already organized into large structures as early as $z \sim 3$! Incidentally, the co-moving scale between the "peaks" in the high redshift histograms appears to be $\sim 100h^{-1}$ Mpc (for $q_0 = 0.1$), apparently as at all other redshifts[38,39,40]! We are currently working on quantitative analysis of the state of the large scale galaxy distribution at these high redshifts[35,36].

As another example of the kind of science that can now be addressed which would have seemed very remote a couple of years ago, it is possible to combine the ground–based results, which because of the wide area covered provide a much better sampling of the bright end of the luminosity function, with the Hubble Deep field data which are able to reliably detect much fainter Lyman break objects. Figure 4 shows the "star formation" or far–UV luminosity function at $z \sim 3$ from a such a pooling of data[41].

4   The Star Formation History of the Universe

In addition to isolating populations of high redshift objects, the Lyman break technique is also very powerful as a means of taking a census of the abundance of high redshift star forming objects, and thereby provides estimates of the total star formation activity as a function of time. By using different combinations of broad-band filters, one can count the co-moving numbers and luminosities of the Lyman break objects from $z \sim 2$ to $z > 4$, which has been done for the Hubble Deep Field[32]. Since the observed-frame optical samples the rest–frame far-UV, the observed fluxes from the galaxies can be converted directly into a massive star formation rate, which is directly proportional to the rate of production of metals, and given an assumed stellar initial mass function, a total star formation rate. The Lyman break galaxy data can then be combined with data from redshift surveys for $z \lesssim 1$ galaxies[12] to produce the total metal production rate versus time over essentially the whole history of the Universe. Of course, there are a number of uncertainties involved in such an analysis–for example the far–UV is particularly subject to extinction by dust, and so it is likely that some correction upward in the inferred metal production/star formation rate would be required. However, it appears that (at least for the

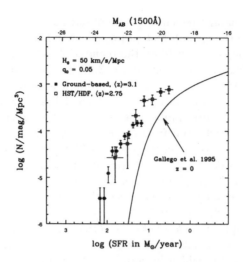

Figure 4: The "star–formation" luminosity function (from Dickinson *et al.* 1997) compiled by combining the Lyman break galaxies from our ground–based survey with the "F300W" break objects in the Hubble Deep Field. The conversion from UV luminosity to star formation rate (SFR) assumes a Salpeter initial mass function. Note the much higher number density of extremely high SFR objects at $z \sim 3$ relative to the curve for the local Universe.[9] Adopting a reasonable correction for internal extinction in the far-UV in the high redshift galaxies increases the inferred SFR by about a factor of 2.

galaxies that have actually been observed) reddening and extinction are likely to be relatively minor ($\sim$ a factor of 2-3 in far-UV flux)[27,41]. In addition, there are problems in that the metal production/star formation rates are estimated using slightly different indices at different redshifts. All of this being said, however, a very nice qualitative picture emerges in which the overall metal production rate, which is likely to trace the overall stellar formation rate, increases very rapidly between $z \sim 4$ and $z \sim 3$, peaks near $z \sim 2$, and decreases monotonically from $z \sim 1$ to the present time[32,41]. A schematic "cartoon" is shown in Figure 5 which roughly describes the star formation history of the Universe as deduced from our present knowledge of the emission from galaxies, as a function of cosmic time (rather than as a function of z).

It is still premature to take this kind of diagram overly seriously, but it is worth pointing out that the qualitative shape of this curve bears remarkable resemblance to the space density of bright QSOs versus time[20], and is very much consistent with the picture of the overall *chemical* history of the universe that comes from both statistical and detailed studies of QSO absorption line systems[43,44,45,46,47]. The fact that there is qualitative agreement among these

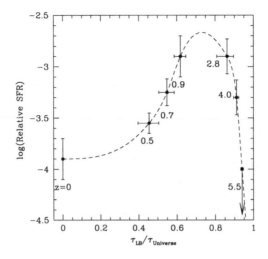

Figure 5: Schematic view of the relative star formation rate per unit comoving volume versus cosmic epoch. The points at $z \leq 1$ come from the CFRS redshift survey[12], while the $z > 2$ points are based on Lyman break objects in the Hubble Deep Field[32,33]. The relative placement of the $z < 1$ and $z > 2$ points will depend to some extent on the value of q_0, so the diagram should be taken as qualitative rather than absolutely quantitative.

very different techniques suggests that we may be converging on a working sketch of the global history of galaxy formation as a function of cosmic epoch.

Of course, what we'd really like to have is a series of "snapshots" at different cosmic epochs of the same galaxies in order to understand their detailed history. Even if we had a perfectly accurate version of the curve in Figure 5, it would not tell us about the history of individual galaxies that we see today. In the empirical universe, it is sobering that, in the end, only circumstantial connections can be made between "populations" of objects at different cosmic epochs, and it will always be difficult to be certain of comparing "apples to apples". However, we are now in a position where (at the very least) data *exist* or are accessible over about 90% of the age of the universe. It will be a monumental task to fill in the details of the rough sketch that now exists, but it is certainly encouraging that doing so does not seem so far out of reach.

134

Acknowledgments

I would like to thank M. Dickinson, M. Giavalisco, K. Adelberger, M. Pettini, and M. Kellogg for allowing me to present some results of collaborative work prior to publication. I would also like to thank the Alfred P. Sloan Foundation and the U.S. National Science Foundation for financial support.

References

1. R. S. Ellis, *ARAA* (1997), in press.
2. R.S. Ellis, M. Colless, T. Broadhurst, J. Heyl, and K. Glazebrook, *MN-RAS* **280**, 235 (1996).
3. S.J. Lilly, Le Févre, O., Crampton, D., Hammer, F., and Tresse, L., *ApJ* **455**, 50 (1995).
4. L. L. Cowie, A. Songaila, E. M. Hu, and J. G. Cohen,*AJ* **112**, 839 (1996).
5. N.P. Vogt *et al*, *ApJ* **465**, L15 (1996).
6. H.-W. Rix, P. Guhathakurta, M. Colless, and K. Ing, *MNRAS*, in press.
7. K. Glazebrook, R. Ellis, M. Colless, T. Broadhurst, and J. Allington-Smith, *MNRAS* **273**, 157 (1995)
8. S. Driver, R. Windhorst, R. Griffiths, *ApJ* **453**, 48 (1995).
9. J. Gallego, J. Zamorano, A. Aragon-Salamanca, and M. Rego, *ApJ* **455**, L1 (1995).
10. R. Abraham *et al*, *MNRAS* **273**, 157 (1996).
11. C. Steidel, M. Dickinson, and S. Persson, *ApJ* **437**, L75 (1994).
12. S. Lilly, O. LeFévre, F. Hammer, D. Crampton, *ApJ* **460**, L1 (1996).
13. G. Kauffmann, *MNRAS* **281**, 487 (1996).
14. D. Schade, S. Lilly, O. LeFèvre, F. Hammer, and D. Crampton, *ApJ* **464**, 79 (1996).
15. M. Giavalisco, C. Steidel, and F. Machetto, *ApJ* **470**, 189 (1996).
16. S. Pascarelle, R. Windhorst, W. Keel, and S. Odewhan, *Nature* **383**, 45 (1996).
17. J. Dunlop, J. Peacock, H. Spinrad, A. Dey, R. Jiminez, D. Stern, and R. Windhorst, *Nature* **381**, 581 (1986).
18. M. Dickinson, in *Galaxy Scaling Relations*, ed. L. da Costa, (Springer-Verlag), in press.
19. P. Mc Carthy, *ARAA* **31**, 639 (1993).
20. M. Schmidt, D. Schneider, J. Gunn, *AJ* **110**, 68 (1995).
21. C. Steidel and D. Hamilton, *AJ* **104**, 941 (1992).
22. C. Steidel and D. Hamilton, *AJ* **105**, 2017 (1993).
23. P. Madau, *ApJ* **441**, 18 (1995).

24. P. Guhathakurta, A. Tyson, and S. Majewski, *ApJ* **357**, L9 (1990).
25. C. Steidel, M. Pettini, and D. Hamilton, *AJ* **110**, 2519 (1995).
26. C. Steidel, M. Giavalisco, M. Pettini, M. Dickinson, and K. Adelberger, *ApJ* **462**, L17 (1996)
27. C. Steidel, M. Giavalisco, M. Dickinson, and K. Adelberger, *AJ* **112**, 352 (1996).
28. J. Lowenthal *et al*, *ApJ*, in press (1997).
29. C. Steidel *et al*, in preparation.
30. R. Williams *et al*, AJ **112**, 1335 (1996).
31. K. Lanzetta, A. Yahil, and A. Fernandez-Soto, *Nature* **381**, 759 (1996).
32. P. Madau, H. Ferguson, M. Dickinson, M. Giavalisco, C. Steidel, and A. Fruchter, *MNRAS* **283**, 1388 (1996).
33. P. Madau, in proc. of 7th Maryland Astrophysics Conference,*Star Formation Near and Far*, in press.
34. J.B. Oke *et al*, *PASP* **107**, 375 (1995).
35. C. Steidel, K. Adelberger, M. Dickinson, M. Giavalisco, and M. Kellogg, *ApJ*, submitted.
36. M. Giavalisco, C.Steidel, K. Adelberger, M. Pettini, and M. Dickinson, in preparation.
37. J. Cohen, D. Hogg, M. Pahre, and R. Blandford, *ApJ* **462**, L9 (1996).
38. T. Broadhurst, R. Ellis, D. Koo, and A. Szalay, *Nature* **343**, 726 (1990).
39. S. Landy, S. Shectman, H. Lin, R. Kirshner, A. Oemler, and D. Tucker, *ApJ* **456**, L1 (1996).
40. A. Szalay, this volume.
41. M. Dickinson *et al*, in preparation.
42. C. Steidel, W. Sargent, and A. Boksenberg, *ApJ* **333**, L5 (1988).
43. C. Steidel, *PASP* **104**, 843 (1992).
44. M. Pettini, R. Hunstead, L. Smith, and D. King, *ApJ*, submitted.
45. L. Lu, W. Sargent, T. Barlow, C. Churchill, and S. Vogt, *ApJS* **107**, 475 (1996).
46. L. Storrie-Lombardi, M. Irwin, and R. McMahon, *MNRAS* **282**, 1330 (1996).
47. M. Fall, S. Charlot, and Y. Pei, *ApJ* **464**, 43 (1996).

WALLS AND BUMPS IN THE UNIVERSE

Alexander S. SZALAY

Department of Physics and Astronomy

The Johns Hopkins University, Baltimore, MD 21218, USA

Observations of large scale structure indicate the presence of very large, wall-like superclusters. In their distributions there is a growing evidence, that there is a characteristic scale in excess of 100 h^{-1} Mpc. Other observations seem to imply only Gaussian fluctuations in the power spectrum. We describe how this apparent conflict can caused by differences in sampling strategy. If the baryon fraction of the background density is high, close to the upper limits from the D abundance, the characteristic scale can possibly be interpreted as a remnant of the acoustic oscillations at recombination. Such a model would have very specific predictions for the position and amplitude of the Doppler-peaks in the cosmic microwave background.

1 Introduction

The study of large scale structure is one of the most dynamically evolving areas of astrophysics today. Cosmology and large scale structure is growing into an accurate science and requires correspondingly more sophisticated methods of analysis. Twenty years ago the estimates of the fluctuation amplitude were about 10^{-3}, almost a factor of 100 off of today's measurements. Ten years ago we could only hope for high precision measurements of large scale structure, there were less than 5000 redshifts measured, and only a handful of normal galaxies with $z > 1$ were known. Computer models of structure formation had just begun to consider non-power-law spectra based on physical models like hot/cold dark matter. As a consequence there was considerable freedom in adjusting parameters in the various galaxy formation scenarios. In contrast, many of today's debates are about factors of 2 and soon we will be arguing about 10% differences. The shape of the primordial fluctuation spectrum, first derived from philosophical arguments[2,1], can now be quantified from detections of fluctuations in the CBR made by COBE[3]. The number of available redshifts is beyond 50,000, and soon we will have redshift surveys surpassing 1 million galaxies. N-body simulations are becoming more sophisticated, of higher resolution, and incorporating complex gas dynamics. The unprecedented number of new observations currently under way give us hope that over the next decade we will gain a clear understanding of the shape and evolution of the primordial fluctuation spectrum, understand from first principles how galaxies were formed, and make quantitative comparisons and tests to differentiate among the various galaxy formation scenarios.

2 Quantifying Large Scale Structure

Structure in the universe evolves from the initially small primordial fluctuations. These fluctuations can arise during an inflationary expansion or come from topological defects later. They grow in amplitude, due to gravitational instability, and the shape of the fluctuation spectrum is altered by different physical processes. The nature of the dark matter, whether hot or cold, believed to dominate the mass density of the universe, determines the shape of the power spectrum on small (< 100 Mpc) scales. On the other hand, the shape of the large scale part of the fluctuations (> 200 Mpc) remains remarkably unchanged, because no scale in the evolutionary process becomes this large.

The COBE measurements constrain both the amplitude and the initial spectrum of the fluctuations in this regime, and demonstrate extremely good agreement ($n = 1.1 \pm 0.4$)[4] with the Harrison- Zeldovich predictions of $P(k) = k^n$, with $n = 1$. These fluctuations are due to differences in the gravitational potential at the surface of last scattering[5], reflecting the state of the universe at a redshift of ≈ 1000. Galaxy surveys (at $z < 0.3$) are rapidly increasing in size, thus providing increasingly better measurements of the fluctuations on small scales (CfA slices[6], IRAS[7], APM[8], APM redshift surveys[9], LasCampanas[10]). One can use theoretical scenarios to evolve and extrapolate the large scale CBR measurements into the structure of the local universe, but the two regimes do not yet overlap directly.

The currently most popular scenario is the Cold Dark Matter dominated universe, where most of the mass is dark, interacting only via gravity, consisting of particles of such a large mass that their thermal motion is negligible. To match the observed clustering of galaxies without producing too large a velocity dispersion, the concept of 'biasing' has been invoked[11,12]: mass is converted into light only at the densest regions in the universe, creating a luminous component more clustered than the mass. This scenario, modulo a properly chosen initial normalization, has been remarkably successful over the last fifteen years.

The COBE measurements create a conflict with the minimal biased CDM model: if a Harrison-Zeldovich spectrum is assumed and the normalization is locked to COBE, then the biasing parameter must be unity to match the small scale part of the fluctuation spectrum, leading to very large small scale velocities. Several alternative models have been rapidly suggested. Gravity waves, which decay with time, may contribute to the largest scale modes observed by COBE and produce a 'tilt' of the spectrum[13]. Alternative scenarios invoke a large cosmological constant[14]. A mixture of cold and hot dark matter would also help [15].

What are the most important measurements we can make in order to differentiate between proposed models? Overlap between scales probed by CBR experiments and redshift surveys in the 'local' universe would place strong constraints on the power spectrum. Measurements of galaxy clustering on scales of 200-500 Mpc from well-sampled redshift surveys would tell us whether the gravity wave/tilted model is relevant, measure the bias factor, and determine the shape of the spectrum on scales where most of today's models differ but which are too small for COBE and beyond the scale of current galaxy measurements. For the same reason, many CBR experiments are probing 1-2 degree scales, corresponding to a comoving scale of about $\approx$ 120 Mpc. These experiments together with the redshift surveys will soon yield an unambigous answer.

3   Observing Walls

Several surveys have now found evidence for sharp, wall-like structures in the universe. The existence of such features is by no means unexpected, Zeldovich[18] predicted, that the generic features in a pressure-free gravitational collapse will be highly flattened 'pancakes'. Observational confirmation took a few years, Chincarini & Rood[16], and Gregory & Thompson[17] identified the excess of galaxies between Coma and A1367 with a supercluster, resembling a 'pancake'.

In 1980, Kirschner etal[19] identified the Bootes void, showing the first big 'void' in the galaxy distribution. A major breakthrough in our understanding of large scale structure came from the CfA 'slice' by deLapparent, Geller and Huchra[20], 6 degrees wide in declination but over 100 degrees in right ascension. In the region where the slice maps the universe, at a radial distance of $70h^{-1}$ Mpc, a distinct pattern appears: a 'Great Wall' containing hundreds of galaxies, connecting several of the known Abell clusters. Its tranverse spatial extent exceeds 100 by $50h^{-1}$ Mpc. The general trend has been profoundly summarized by Geller and Huchra[6]: 'all surveys have detected structures as large as they could ...'

If the universe were full of 'Great Walls', i.e. if they are typical of the very large scale structure, already from the surface density of galaxies one can get an estimate what would a 'fair sample' consist of. If we assume (in the extreme), that all bright galaxies are on these surfaces, with the surface density of $\mu = 0.4$ galaxies Mpc^{-2}, we can estimate the characteristic 'cell' size by requiring that the corresponding 'local' volume density of bright galaxies, $n = 0.01$ galaxies h^{-1} Mpc^{-3}, be approximately reproduced. Assuming spherical bubbles, and counting only half of the surface area, since the walls separate two volumes,

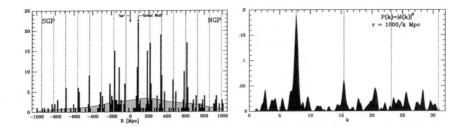

Figure 1: The redshift distribution of galaxies in the BEKS survey, comprised of two narrow pencilbeams towards the Galactic poles. The left hand figure shows the histogram of all galaxies, while the right hand side shows the 1D power spectrum. The big spike corresponds to the comoving scale of 128 h^{-1} Mpc.

the typical size of the voids is

$$\lambda = 2R = \frac{3\mu}{n} = 120h^{-1} \text{ Mpc}. \tag{1}$$

This gives us some idea what cell sizes can one expect in a universe dominated by 'Great Walls', derived solely from the observations.

In 1990 Broadhurst, Ellis, Koo & Szalay[21] (BEKS) published results from a redshift survey in two opposite pencilbeams. The angular diameter of the survey is 30', and the depth is about 0.5 in redshift, both at the North and South Galactic Poles. The combined surveys have a joint length in excess of $2000h^{-1}$ Mpc, considerably deeper than any other survey before. To compensate for the small physical size of the survey at low redshifts, data from two bright surveys in almost the same directions were used, resulting in a combined selection well approximated by a cylinder of constant comoving radius.

The Northern pencilbeam is in the CfA slice, and one can find the 'Great Wall' without much difficulty. Surprisingly, however, at very large radial distances one still cannot see a homogeneous distribution, rather most galaxies are in a few large 'spikes' along the line of sight, separated typically by more than $100h^{-1}$ Mpc. The simplest explanation was that further 'walls' were found, meaning that the 'Great Wall' is by no means unique, and that these structures contain a large percentage ($\approx 50\%$) of the galaxies.

4   Observing Bumps

Even more suprising was the fact, that in the one-dimensional Fourier transform of the redshift distribution, a highly significant peak was found at the wavelength of 128 h^{-1} Mpc, with a probability of $P < 3 \times 10^{-4}$. This observation prompted many debates, and even more exotic theories. The main question was, of course, whether the peak in the Fourier spectrum is just a random accident, or does this scale arise as a result of a physical process? Extending the BEKS survey to 9 pencilbeams, randomly distributed over a 6x6 degree region at both galactic poles, it was shown, that the cross-correlation signal stays strong up to about 60 h^{-1} Mpc transverse separation[22], indicating that the redshift spikes are indeed 'Great-Wall' like structures, also that the power spectrum peak was not due to a random alignment of small groups.

Several years later bigger redshift surveys became available. The LasCampanas survey[10], consisting of 6 slices of 450 h^{-1} Mpc depth, found evidence for statistically significant excess power on 100 h^{-1} Mpc scales[23]. A similar slice near the South Galactic Pole, the ESP project[24] confirms the BEKS spikes in the overlap region. Deeper surveys on the Keck telescope[25], and the CFRS[26] found evidence for the existence of sharp walls at $z = 1$. Excess power on $100^{+}h^{-1}$ Mpc scales is present at even higher redshifts, in QSO absorption systems[27] and in galaxies[28]. These high redshift observations are extremely important — if the bumps appear on the same comoving scale at much earlier Hubble times, then there is a built-in feature in the power spectrum!

5 Sampling strategies

If fluctuations in the universe are strictly Gaussian, their full statistical description is contained in the two-point correlation function or in its Fourier transform, the power spectrum. The phases of the individual Fourier components are random for such a process, and all high-order measures of clustering vanish. In this case, Kaiser[29] shows that measuring the redshifts of a small fraction of the galaxies is the most efficient way to measure the power spectrum on large scales. However, if the universe contains sharp large scale features like the "Great Wall," a sparsely sampled survey may fail to identify them because it is less sensitive to the higher-order correlations (equivalently, the phase correlations) that characterize such structures. A distribution with high-order clustering can be very different from a homogeneous, isotropic, Gaussian random field with the same second-order statistical properties. A sparse survey optimized to measure the two-point correlation function would miss the very real differences between these distributions.

Thus we can understand now how sparse sampling affects power spectrum estimators: in the ensemble average, the power spectrum as a second order statistic is invariant with respect to sampling. Averaging over an infinite number of finite size realizations, the correct power spectrum is recovered, even for

Voronoi foam, smoothed original

Voronoi foam, random phases

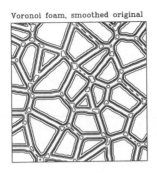 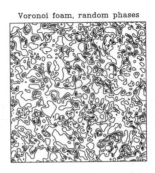

Figure 2: Two simple realizations of a two-dimensional universe, with identical second order statistical properties. The left hand figure is a two dimensional Voronoi foam, generated by the median surfaces between Poisson 'seeds' at the mean separation of 100 Mpc. In this simple toy model galaxies reside only on the walls of the foam, smoothed, so the walls have a finite thickness. The structure has a well defined second order statistic, but also has well correlated phases. This picture has been Fourier transformed, all the phases randomized, then transformed back again. The result is shown on the right hand side, with the same second order properties, but with a Gaussian distribution. It is easy to see, that placing well-sampled pencilbeams across both surveys will easily distinguish between the two, whereas a sparse sample drawn from the two realizations cannot differentiate.

a very low sampling rate. On the other hand, if there is a network of sharp 'walls' present, they are manifested as a set of sharp 'spikes' in Fourier space. These sharp spikes will vary from realization to realization, and in an ensemble average they will converge to the underlying power spectrum. Even though both scenarios converge to the true power spectrum in the infinite limit, it is much harder to detect the sharp Fourier spikes with a low sampling rate in a single local realization, like our nearby Universe. This is why the well-sampled pencilbeams may yield seemingly quite different results for the statistics of power spectrum amplitudes than wide angle sparsely sampled surveys.

5 Origin of the 100 Mpc Bumps

Here I would like to discuss, how can such 100 h^{-1} Mpc bumps arise in the power spectrum. It has been understood for a long time [31,32] that around recombination due to the high pressure in the photon-baryon plasma, fluctuations oscillate like sound waves. On smaller wavelengths, these oscillations damp, but on larger scales, near the horizon scale at recombinations, they may survive longer. The motion of baryons due to these sound waves gives rise to the Doppler-peaks in the CMB fluctuations. At the same time it was

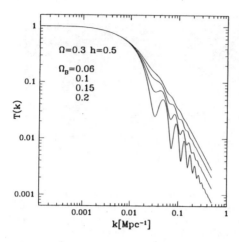

Figure 3: The transfer function for the fluctuations in a high baryon content universe. In order to get the dimensionless contribution to the variance, $k^3 P(k)$, this curve needs to be multiplied by k^4, for a Zeldovich spectrum. Note, that the scale of the first peak is at $k = 0.05$, corresponding to $125\ h^{-1}$ Mpc. This figure was kindly provided by Wayne Hu.

understood early on, that after recombination, as the sound speed approaches zero, an interference pattern may emerge, the so-called Sakharov-oscillations[33]. Sound waves which exactly fit inside the horizon will amplify, others with opposite phases will cancel. Since the horizon scale at recombination is very close to the regime of interest (between 100-200 h^{-1} Mpc, depending on $\Omega_0 h$), it is worth to consider what does it take for these sound-waves to have an appreciable effect on not only the CMB but on the galaxy distribution!

Since galaxy clustering is only affected by gravity, the fluctuations in the baryons due to the sound waves need to leave an imprint in the gravitational potential. This requires as high a baryon fraction as possible. How high can this number be? From observations of the primordial deuterium[34], the 1σ limit $\Omega_B h^2 \leq 0.025$, can be combined with reasonably low estimates of the Hubble constant, and values of $\Omega_B \approx 0.1$ are not unimaginable. At the same time, in order for a large imprint, Ω_t has to be low, in the range of $\Omega_t \approx 0.4$. Given the faint number counts of galaxies, this is again not an outrageous idea any longer. Calculations of W. Hu[35] indicate, that in these scenarios there will be several bumps in the power spectrum, the first at $kh = 0.05$. The amplitude for $\Omega_B = 0.1$, and $\Omega_t = 0.3$ is quite substantial, although nowhere as high as e.g. the BEKS or the LasCampanas detections.

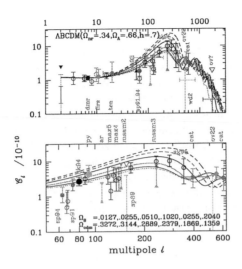

Figure 4: The angular multipole amplitudes of the CMB fluctuations for a family of models with high baryon density plotted with the results of recent small-scale anisotropy measurements. This figure is from Bond and Szalay (1997).

On the other hand, there are several other amplification mechanisms at work. The surveys measure the distribution of galaxies, while the above calculations refer to the linear fluctuations in all the mass. First of all, the formation of the pancakes is a highly non-linear process, which will amplify fluctuations, if there is a distinct scale associated with them. Second, the galaxy surveys are analysed in redshift space, thus infall on to the walls will enhance these structures, and will result in a further amplification. This effect will depend on the survey geometry: it is very important for pencilbeams, less so for slices, and spherical volumes. Even in the Las Campanas survey one can notice, that some of the walls curve to stay perpendicular to the line of sight – a consequence of redshift space enhancements. The high baryon content has another effect: the increased viscous damping will decrease the fluctuations on small scales. If this effect is too large, it cannot be compensated for by the bias factor. This provides a practical upper limit on how high Ω_B/Ω_t can become. The relevant parameter range ($h = 0.5 - 0.65, \Omega_B = 0.06 - 0.12, \Omega_t = 0.2 - 0.4$) has not been particularly well studied. We are currently undertaking such a linear analysis, combining the results with the possible nonlinear amplifications mechanisms, to sharpen these constraints further[36] (Bond and Szalay 1997).

144

7 Summary

In summary, there are several observations pointing to excess power on 100-130 h^{-1} Mpc scales, which manifests itself in a small number of sharp spikes in Fourier space. These reflect the presence of walls and voids on similar scales. The emergence of this comoving scale at high redshift implies that this is imprinted on the fluctuations. Such a scale occurs naturally at recombination. The Sakharov oscillations, remnants of the sound waves at that epoch may provide an intriguing explanation. In such scenarios the baryon content of the Universe must be high, the Hubble constrant low, and the Universe open. This family of models deserves further investigation, and it is just barely possible that the 100 Mpc bumps may be the first preview of the elusive Doppler peaks — a fascinating preview of further connections between the galaxy distribution and the Cosmic Microwave Background.

Acknowledgments

The author would like to acknowledge useful discussions with Dick Bond, Joe Silk, Lyman Page, Wayne Hu and Istvan Szapudi. The author is supported by NASA LTSA and a grant from the Seaver Institute.

References

1. Zel'dovich, Ya.B. 1972, *M.N.R.A.S.*, **160**, 1P.
2. Harrison, E.R. 1970, *Phys.Rev.D*, **1**, 2726.
3. Smoot, G.F. et al. 1992, *Ap.J. Lett.*, **396**, L1.
4. Gorski, K.M., Hinshaw, G., Banday, A.J., Bennett, C.L., Wright, E.L., Kogut, A., Smooth, G.F. & Lubin, P. 1994, *Ap.J. Lett.*, **430**, L89.
5. Sachs, R.K., & Wolfe, A.M. 1967, *Ap.J.*, **147**, 73.
6. Geller, M.J. & Huchra, J.P. 1989, *Science*, **246**, 897.
7. Saunders, W., Frenk, C., Rowan-Robinson, M., Lawrence, A. & Efstathiou, G. 1991, *Nature*, **349**, 32.
8. Maddox, S.J., Efstathiou, G., Sutherland, W.J. & Loveday, J. 1990, *M.N.R.A.S.*, **242**, 43P.
9. Loveday, J., Efstathiou, G, Peterson B.A. & Maddox, S.J. 1992, *Ap.J.*, **400**, L43.
10. Shectman, S.A., Landy, S.A., Oemler, A., Tucker, D., Lin, H., Kirschner, R.L. & Schechter, P.L. 1996, *Ap.J. Supp.*, **470**, 172.
11. Kaiser, N. 1984, *Ap.J. Lett.*, **284**, L9.

12. Bardeen, J.M., Bond, J.R., Kaiser, N., & Szalay, A.S. 1986, *Ap.J.*, **304**, 15.
13. Davis, R.L., Hodges, H.M., Smoot, G.F., Steinhardt, P.J., & Turner, M.S. 1992, *Phys. Rev. Lett.*, **69**, 1856.
14. Kofman, L., Gnedin, N., & Bahcall, N.A. 1993, *Ap.J.*, **413**, 1.
15. Klypin, A., Holtzman, J., Primack, J., & Regös, E. 1993, *Ap.J.*, **416**, 1.
16. Chincarini,G. & Rood, H.J. 30, *Ap.J.*, **206**, 1976.
17. Gregory,S.A. & Thompson, L.A. 784, *Ap.J.*, **222**, 1978.
18. Zeldovich,Ya.B. 1970, *Astron. Astrophys.*, **5**, 84.
19. Kirshner, R.P., Oemler, A.J., Schechter, P.L. & Shectman, S.A. 1983, *Astron.J.*, **88**, 1285.
20. de Lapparent, V., Geller, M.J., & Huchra, J.P. 1986, *Ap.J. Lett.*, **301**, L1.
21. Broadhurst, T.J., Ellis, R.S., Koo, D.C., & Szalay, A.S. 1990, *Nature*, **343**, 726.
22. Broadhurst,T.J., Ellis,R.S., Ellman,N.E., Koo,D.C. & Szalay,A.S. eds: H. McGillivray and C. Collins, *Proc. of ROE Meeting on Digital Sky Surveys*, **1992**, p.397.
23. Landy, S.D., Shectman, S.A., Lin H., Kirschner, R.P., Oemler, A.A. & Tucker, D. 1996, *Ap.J. Lett.*, **456**, 1L.
24. Vettolani, G., Zucca, E. etal 1997, *Astron. Astrophys.*, **in press**, .
25. Cohen, J.G., Cowie, L.L., Hogg D.W., Songalia, A., Blandford, R., Hu, E.M. & Shopbell, P. 1996, *Ap.J.*, **471**, 5.
26. Lilly, S.J., Tresse, L., Hammer, F., Crampton, D. & Le Fevre, O. 1996, *Astron.J.*, , in press.
27. Quashnock, J.M.;VanDen Berk, D.E., York, D.G 1996, *Ap.J.*, **472**, 69.
28. Steidel, C. 1997, this volume.
29. Kaiser, N. 1986, *M.N.R.A.S.*, **219**, 785.
30. Szapudi,I. and Szalay, A.S. 1996, *Ap.J.*, **459**, 504.
31. Peebles, P.J.E. 1968, *Ap.J.*, **153**, 1.
32. Sunyaev, R.A. & Zeldovich, Ya.B. 1970, *Astrophys. Space Sci.*, **7**, 3.
33. Sakharov, A. 1966, *JETP(english)*, **22**, 241.
34. Tytler, D. 1997, this volume.
35. Hu, W. *Ph.D. Thesis*, (The University of California, Berkeley, 1996).
36. Bond, J.R. & Szalay, A.S. 1997, *Nature*, in preparation.

Large Scale Flows

Luiz Nicolaci da Costa

European Southern Observatory, Karl-Schwarzschild Str.2 D-85748
Garching bei München, Germany

Observatório Nacional, Rua Gen. José Cristino 77,
Rio de Janeiro, R.J., Brazil

New measurements of the peculiar velocity field of spiral galaxies (SFI) are used to investigate the nature of the density and three-dimensional velocity fields. We find that the reconstruction based on the SFI sample yields a mass distribution and a velocity field that more closely resemble the redshift distribution of galaxies and the $IRAS$ predicted velocity field. Although a bulk flow of about 300 km s^{-1} is measured in a top-hat window 6000 km s^{-1} in radius, the flow is not uniform and has a smaller coherence length (3000 km s^{-1}) then previously claimed. A preliminary analysis of the data also suggests a low value of the parameter $\beta \lesssim 0.6$, favoring a low–Ω universe. In contrast to earlier results the constraints imposed on cosmological models from cosmic flows are consistent with those coming from galaxy clustering and the local abundance of clusters.

1 Introduction

Recent breakthroughs in the study of the high-redshift universe, allowing us to observe galaxies in the making for the first time, raise the question of whether or not it is worthwhile to continue the ongoing work in the $z \lesssim 0.05$ universe.

In my view, the answer is yes for several reasons. Available wide-angle, complete redshift surveys of the nearby universe like CfA2 [21] and SSRS2 [9] provide a dense and quasi three-dimensional view of the local distribution of galaxies, probing L_* galaxies out to scales of $\sim 100\ h^{-1}$ Mpc. Moreover, future surveys like SLOAN and 2dF will provide truly volume-limited samples on this scale. These samples are useful for evolutionary studies, analysis of high-order moments of the galaxy distribution and for comparison with N-body simulations. Perhaps more importantly, the volume mapped out by these local surveys is well-matched to that probed by redshift-distance surveys, from which the mass distribution can be reconstructed from the observed peculiar motion of galaxies, if gravitational instability drives the growth of structures. Therefore, studies of the nearby universe are unique because we can only expect to reconstruct the underlying mass distribution from peculiar motions locally, given the errors in the distance estimates.

Together, data from complete redshift surveys and from measurements of the peculiar motion of galaxies may allow the direct study of galaxy biasing models. Hopefully, we will soon be able not only to understand the relative distribution of galaxies and mass, but also the relation between internal properties of galaxies and the local density of galaxies and mass. With this information we may gain additional insight into the process of galaxy formation and evolution.

Peculiar motions also complement other ways of probing the mass content of the universe filling in the gap between gravitational lensing, probing galactic and cluster scales, and cosmic microwave background anisotropies probing scales $\gtrsim 100\, h^{-1}$ Mpc.

In section 2 we review how to measure distances and peculiar velocities, and discuss some of the results stemming from these studies. In section 3 we review the properties of the all-sky samples currently available pointing out some of their differences. In section 4 we present some results from the ongoing analysis of the SFI sample of field spirals. A brief summary is presented in section 5.

2   Overview

2.1   Measuring Peculiar Velocities

The radial component of the peculiar velocity of a galaxy is given by

$$u_{pec} = cz - R \tag{1}$$

where cz is the measured recessional velocity of the galaxy and R is the distance, expressed in km s^{-1}. The distance of a galaxy can be estimated using secondary distance indicators like the $D_n - \sigma$ relation for early-types[28] and for spirals the Tully-Fisher (TF) relation. Although these relations are a generic prediction of the virial theorem, they are empirically determined and provide important information about the process of galaxy formation.

On large scales, where we expect the linear theory to be valid, the three-dimensional peculiar velocity is related to the mass density fluctuations

$$\mathbf{v}(\mathbf{r}) = \frac{\beta}{4\pi} \int d^3 r \prime \delta_g \frac{(\mathbf{r}\prime - \mathbf{r})}{|\mathbf{r}\prime - \mathbf{r}|^3} \tag{2}$$

where $\beta = \Omega^{0.6}/b$. Here we have assumed that galaxies are related to the underlying mass distribution according to the simple linear bias model, where b is the biasing factor. The above relation together with the assumption of a curl-free flow allows one to derive the mass distribution from measurements of the radial component of the peculiar velocity for individual galaxies.

148

2.2 Previous Work

After the pioneering work of Rubin and collaborators [32], spirals [1] and ellipticals [35] were used to map out the local peculiar velocity field with the goal of estimating the value of the density parameter Ω based on the measured infall of galaxies towards Virgo.

The first estimates of Ω were made using the Virgocentric infall model

$$u_{pec} = \frac{1}{3}R\Omega^{0.6}\frac{\delta N}{N} \qquad (3)$$

based on linear theory. From measured values of u_{pec}, the distance R to Virgo and the overdensity of galaxies $\delta N/N$ measured from redshift surveys early estimates yielded $\Omega = 0.2$. Note that in this estimate there is the implicit assumption that galaxies trace matter, as the bias concept was not yet in vogue at the time.

Soon after these first estimates, it was recognized that the spherical infall model was too simplistic and that shear motion could have a considerable effect [5][36]. It was also obvious that there was a misalignment between the direction of Virgo and the CMB dipole suggesting that other sources of gravity pull on the Local Group. In fact, Tammann and Sandage [34] pointed out that in order to account for the Local Group motion measured from the CMB anisotropy and the estimates for the infall towards Virgo, the LG motion would require a component pointing towards the general direction of Hydra-Centaurus, an overdense region in the projected distribution of galaxies. This was perhaps the first indirect evidence of a large mass concentration in the region, subsequently confirmed by redshift surveys [8][16].

In order to address this question the local samples were extended by the 7 Samurai [28], who mapped the peculiar velocity field of early-type galaxies out to 8000 km s^{-1}, although the effective depth of the sample has been a point of controversy. Still the sample was the deepest all-sky sample available until recently. A startling result was that the preliminary analysis [15] indicated the existence of a bulk motion of ~ 600 km s^{-1} in the general direction of the Hydra-Centaurus complex, much larger than expected by the standard biased-CDM model, the most popular model at the time. Subsequent analysis [28] showed that the flow was more adequately described by postulating the existence of a large mass concentration, the Great Attractor (GA), at a distance of about 4500 km s^{-1} and at $l = 307°$ and $b = 9°$. Several authors were quick to point out that the Great Attractor model was also inconsistent with a high-bias model and suggested large values of Ω [3][28].

The 7 Samurai work was followed by less ambitious surveys which concentrated primarily in spiral galaxies in selected regions of the sky [7][38]. Nevertheless, the results of these surveys were equally disturbing. They suggested that the Perseus-Pisces (PP) complex, an impressive concentration of galaxies in the opposite direction of the GA, was itself moving towards the Local Group at about 450 km s^{-1}. A possible interpretation is that the local flow has a large coherence length ($\sim 100\ h^{-1}$ Mpc) and that density fluctuations exist on much larger scales[7]. Adding to the confusion was the observation[27] that Abell clusters out to 15,000 km s^{-1} were not at rest relative to the CMB restframe but exhibited a dipole motion with an amplitude of about 600 km s^{-1} in a direction which did not coincide with the LG dipole motion. Several surveys of distant clusters are now underway to verify this result[37].

3 New All-Sky Samples

This confusing state of affairs pointed out the need to generate more uniform and deeper redshift-distance surveys. Two such surveys have recently been completed: one of spirals of different types in the southern hemisphere (Mat92)[29] and the other consisting of over 1200 Sbc-Sc galaxies[22] in the region $\delta > -45°$ and $b > |10°|$ and about 800 spirals in the the direction of 24 clusters (SCI)[23]. Both surveys are based on I-band photometry and measurements of the rotational velocity from either optical rotation curves or the 21 cm line-width. These surveys are the basis of the most extensive all-sky catalogs currently available to map the peculiar velocity field.

One such catalog (Mark III) has been assembled[39] by combining all the available distance measurements in the literature. The relative contribution of the various samples to the final catalog can be found in the figures presented by Kolatt and collaborators[26], who show the projected distribution of galaxies within ±2000 km s^{-1} from the supergalactic plane. The second is the so-called SFI sample[11] which combines the Sbc-Sc galaxies measured north of $\delta = -45°$ with a subset of the Mat92 sample. This subset includes only Sbc-Sc galaxies, and the raw data were converted to a common system using about 300 galaxies measured by both groups. The resulting SFI sample is shown in figure 1. Comparison of this figure with those of Kolatt et al. immediately shows the remarkable uniformity in sky coverage and depth of the SFI sample.

We should also point out the ongoing effort to extend in depth the 7 Samurai sample of early-type galaxies. The ENEAR project[2], consisting of about 1500 galaxies covering the whole sky, will be of great value for an independent confirmation of the results obtained with spiral galaxies as it uses a different distance estimator and has different selection biases.

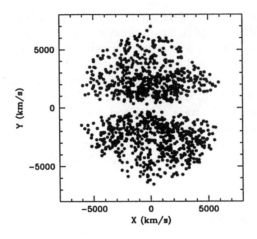

Figure 1: Projected distribution of SFI galaxies within 2000 km s^{-1} of the supergalactic plane, illustrating the uniformity of the sample.

4   Results

4.1   Reconstruction of the Mass and the Velocity Fields

We have used the SFI sample to reconstruct the mass and three-dimensional velocity fields [11] under the assumption that the peculiar motions are induced by the gravitational field associated with mass fluctuations, in which case the flow is fully described by a scalar potential [4], at least on large scales where linear theory is valid. Distances to the individual galaxies were estimated using the I-band direct Tully-Fisher relation derived by combining the data for 24 clusters [24] in the SCI sample. In order to correct for biases, which include the homogeneous and inhomogeneous Malmquist bias as well as sample selection bias, a Monte-Carlo approach has been used whereby the bias is directly estimated from the real space density field of galaxies and "observed" using the assumed TF relation and scatter as in the real data [19].

One of the main findings is that the derived mass distribution resembles the observed galaxy redshift distribution much more closely than any previous reconstruction [14]. In particular, voids in the galaxy distribution reflect real voids in the mass distribution [17]. Another remarkable feature is that velocity field along the supergalactic plane shows for the first time a bifurcation towards the GA and PP, much more consistent with the predicted *IRAS* velocity field,

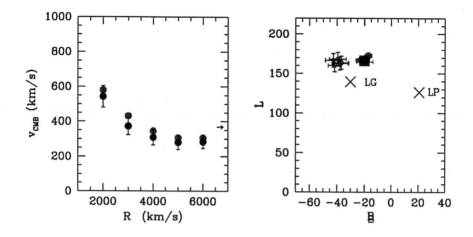

Figure 2: Amplitude (left panel) and direction (right panel) of the bulk motion as measured in top-hat windows of radius R using the SFI sample. The direction of the LG motion, the Lauer-Postman dipole and the amplitude and direction obtained from the Mark III catalog are also shown.

as expected if light traces matter.

4.2 Characteristics of the Velocity Field

From the reconstructed three-dimensional velocity field, the volume-weighted bulk motion in top-hat windows has been computed and the result is shown in figure 2. The values obtained using the Mark III catalog[14] are also indicated. For comparison, we also show the bulk velocity estimated from the maximum-likelihood fit to the radial component of the peculiar velocity measured for individual galaxies. The left panel shows the amplitude and the right panel the direction of the bulk motion in supergalactic coordinates, for different radius R. As it can be seen the bulk motion is a robust measurement both in amplitude (~ 300 km s^{-1}) and direction.

However, in contrast to earlier claims, we find no evidence for a large amplitude, uniform flow across the surveyed volume. This can be seen in figure 3 where we compare the bulk velocity for galaxies separated into the two hemispheres defined by the plane perpendicular to the bulk motion direction. From the figure we see that there is a considerable shear across the volume, with galaxies in the hemisphere that contains PP moving in the same direction as the general bulk motion but with a significantly smaller amplitude ~ 200 km s^{-1}.

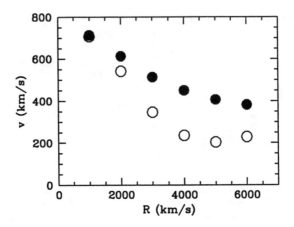

Figure 3: Comparison of the amplitude of the bulk motion dividing the volume into two hemispheres, defined by the plane perpendicular to the direction of the bulk motion.

The amplitude of the bulk is much higher close to the GA then at the location of PP, in the opposite side.

In order to determine the coherence length of the peculiar velocity field we have computed the velocity correlation tensor using a weighted least-square technique to estimate its parallel and perpendicular components assuming a homogeneous and isotropic flow [25]. From these components we have derived the velocity correlation function[31] $\xi_v = \Pi(r) + 2\Sigma(r)$ shown in figure 4. From the figure we find that the flow is characterized by a coherence length of about 3000 km s^{-1}, in marked contrast to earlier claims [7][38]. For comparison, also shown are the velocity correlations obtained from linear theory using a Γ-model power-spectrum. All curves have been computed with the same shape parameter ($\Gamma = 0.2$) but for different normalizations, with $\sigma_8\Omega^{0.6}$ varying from 0.4 to 0.8. We find that for values of the shape parameter consistent with those inferred from galaxy clustering data[10], the value of $\sigma_8\Omega^{0.6}$ must lie in the range 0.4 to 0.6, significantly smaller than the value obtained from the analysis of the Mark III data [40]. Similar low values of $\sigma_8\Omega^{0.6}$ have been obtained from an independent analysis based on the distribution of peculiar velocities of SCI clusters [6].

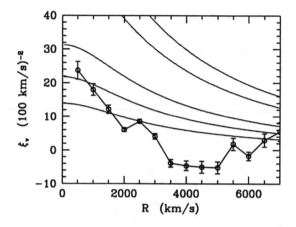

Figure 4: The velocity correlation tensor as measured from the SFI data compared to linear theory predictions based on the Γ-model power-spectrum. The shape parameter is taken to be $\Gamma = 0.2$ and the amplitude $\sigma_8 \Omega^{0.6}$ varies from 0.4 (bottom curve) to 0.8 (top curve) in 0.1 steps.

5   Constraints on $\beta = \Omega^{0.6}/b$

Analysis of cosmic flows have been extensively used to estimate the density parameter Ω, or more precisely the parameter β, on large scales. The parameter β can be derived either by comparing the reconstructed density field obtained from peculiar velocity measurements with the real space galaxy density field recovered from all-sky redshift surveys, or from a velocity-velocity comparison between the predicted *IRAS* velocity field and that directly observed. Regardless of the method previous results based on large scale flows have favored large values of β [33].

A preliminary determination of β from the SFI data has been carried out using the ITF method of Nusser and Davis [30]. The method is based on minimizing the scatter of the inverse Tully-Fisher relation for a model peculiar velocity field, described by a set of smooth orthogonal functions. This method is particularly useful for comparison between different data sets. Recently, it has been applied to the Mark III catalog [13] and the *IRAS* 1.2 Jy yielding values of β in the range 0.4-0.6. However, significant differences between the Mark III and *IRAS* fields were found, with the residual field exhibiting a coherent dipole. Using the ITF method to the SFI data, we find that the most likely value for β

154

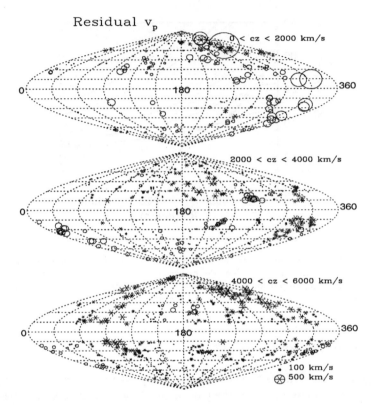

Figure 5: Sky projection of residuals of the ITF velocity field after subtraction of the *IRAS* predicted velocity field for $\beta = 0.6$ in galactic coordinates, as seen in the LG frame. The open symbols are points that are moving inward and the stars are points flowing outward.

lies in the same range but, as speculated earlier[11], leads to a much more robust determination because of the good agreement between the SFI and *IRAS* fields.

This can be seen in figure 5 where the residuals obtained from the comparison of *IRAS* ($\beta = 0.6$) and SFI peculiar velocity fields are shown. This figure should be compared to figures 12 and 13 of Davis *et al.* [13]. It is clear that there is a much better agreement between the *IRAS* and SFI peculiar velocity fields than in the case of the Mark III catalog. The value of β obtained from the ITF analysis is also consistent with that derived from the density-density comparison [20] which yields $\beta \sim 0.4$. We find no evidence in support to recent speculations that β may be scale dependent. Results from these analyzes will be reported in forthcoming papers.

6 Summary

The most important conclusion is that by using a homogeneous, all-sky sample of I-band TF distances of spiral galaxies, most of the inconsistencies of previous works on large-scale flows have been by and large resolved. In particular, the cosmological constraints imposed by cosmic flows seem to agree remarkably well with those stemming from studies of galaxy clustering, the local abundance of clusters and small-scale velocity field. Therefore one of the long standing arguments against an open universe may no longer be valid.

Acknowledgments

I would like to thank my collaborators W. Freudling, R. Giovanelli, M. Haynes, J. Salzer and G. Wegner in the SFI project for allowing me to present some results before their publication. My special thanks to GW for his careful reading of the manuscript. I would also like to thank Adi Nusser for his collaboration with the ITF method, and M. Davis and A. Dekel for useful discussions over the years.

References

1. Aaronson, M., Huchra, J., Mould, J., Schechter, P.L. and Tully, R.B. *ApJ* **258**, 64 (1982).
2. Alonso, M.V., Bernardi, M., da Costa, L.N., Freudling, W., Pellegrini, P., Wegner, G and Willmer, C. *in preparation*.
3. Bertschinger, E. and Juszkiewicz, R *ApJ* **334**, L59 (1988)
4. Bertschinger, E., Dekel, A., Faber, S. M., Dressler. A. and Burstein, D. *ApJ* **364**, 370 (1990).
5. Bushouse, H., Mellot, A., Centrella, J. and Gallagher, J. *MNRAS* **217**, 7p (1985).
6. Borgani, S., da Costa, L. N., Freudling, W., Giovanelli, R, Haynes, M.P., Salzer, J.J and Wegner, G. *ApJ submitted*.
7. Courteau, S., Faber, S., Dressler, A. and Willick, J.A. *ApJ* **412**, L51 (1993).
8. da Costa, L.N., Nunes, M.A., Pellegerini, P.S.S., Willmer, C.N.A. Chincarini, G. and Cowan, J.J. *AJ* **91**, 6 (1986)
9. da Costa *et al.* *ApJ* **424**, L1 (1996).
10. da Costa, L.N., Vogeley, M., Geller, M.J., Huchra, J.P. and Park, C. *ApJ* **437**, L1 (1994).

11. da Costa, L. N., Freudling, W., Wegner, G., Giovanelli, R, Haynes, M. P. and Salzer, J.J. *ApJ* **468**, L5 (1996)

12. da Costa *et al.* , *in preparation*.

13. Davis, M., Nusser, A. and Willick, J.A. *ApJ* **473**, 22 (1996).

14. Dekel, A. ARAA **32**, 371 (1994)

15. Dressler, A., Faber, S.M., Burstein, D., Davies, R.L., Lynden-Bell, D., Terlevich, R.J. and Wegner, G.*ApJ* **313**, L37 (1987).

16. Dressler, A. *ApJS* **75**, 241 (1991).

17. Elad, H. Piran, T. and da Costa, L. *MNRAS, in press* (1997).

18. Efstathiou, G., Bond, R. and White, S.D.M. *MNRAS* **258**, 1p (1992).

19. Freudling, W., da Costa, L. N., Wegner, G., Giovanelli, R, Haynes, M. P. and Salzer, J.J. *AJ* **110**, 920 (1995).

20. Freudling *et al.* , *in preparation*.

21. Geller, M.J. and Huchra, J. P. Science **246**, 897 (1989).

22. Giovanelli, R., Haynes, M.P., Salzer, J.J., Wegner, G., da Costa, L.N. and Freudling, W. *AJ* **107**, 2036 (1994).

23. Giovanelli,R., Haynes, M. P.,T. Herter,T., N. Vogt, N..Salzer, J. J., Wegner, G., da Costa, L. N. and Freudling, W. *AJ* **113**, 22 (1997).

24. Giovanelli,R., Haynes, M. P.,T. Herter,T., N. Vogt, N.. da Costa, L. N, Freudling, W., Salzer, J. J. and Wegner, G.*AJ* **113**, 53 (1997).

25. Groth, E.J., Juszkiewicz, R. and Ostriker, J.P. *ApJ* **346**, 558 (1989).

26. Kolatt, T., Dekel, A., Ganon, G. and Willick, J.A. *ApJ* **458**, 419 (1996).

27. Lauer, T.R. and Postman, M. *ApJ* **425**, 418 (1994)

28. Lynden-Bell, D. Faber, S.M., Burstein, D., Davies, R.L., Dressler, A. Terlevich, R.J. and Wegner, G. *ApJ* **326**, 19 (1988).

29. Mathewson, D.S., Ford, V.L. and Buchhorn, M. *ApJS* **81**, 413 (1992).

30. Nusser, A. and Davis, M. *MNRAS* **276**, 1391 (1995).

31. Peebles, P.J.E. Nature **327**, 210 (1987).

32. Rubin, V.C., Ford, W.K., Thonnard, N., Roberts, M.S. and Graham, J.A. *AJ* **81**, 687 (1976).

33. Strauss, M. A. and Willick, J.A. Physics Report **261**, 271 (1991).

34. Tammann, G.A. and Sandage, A. *ApJ* **281**, 31 (1985).

35. Tonry, J.J. and Davis, M. *ApJ* **246**, 680 (1981).

36. Villumsen, J. and Davis, M. *ApJ* **308**, 499 (1986).

37. Wegner, G., Colless, M., Baggley, G., Davies, R.L., Bertschinger, E., Burstein, D., McMahan, R.K. and Saglia, R.P. *ApJS* **106**, 1 (1996).

38. Willick, J.A., *ApJ* **351**, L5 (1990).

39. Willick, J.A., Courteau, S., Faber, S.M., Burstein, D. and Dekel, A., *ApJ* **446**, 12 (1995).

40. Zaroubi, S., Zehavi, I., Dekel, A., Hoffman, Y. and Kollat, T. preprint astro–ph/9610226 (1996).

NUMERICAL RELATIVITY

M.W. CHOPTUIK
Center for Relativity, Department of Physics, UT Austin
Austin TX 78712-1081, USA

I review recent progress in numerical relativity with emphasis on work relating to the coalescence of black hole binaries, the close limit in head-on black hole collisions, and critical phenomena in gravitational collapse.

1 Introduction

Research in numerical relativity continues at a brisk pace. Advances along many fronts—ranging from the development of new formalisms to the use of ever more powerful computational facilities—are rapidly increasing our ability to simulate those astrophysical scenarios where the dynamics of the relativistic gravitational field is important. In addition, numerical simulation continues to provide us with new physical insights into the phenomenology of strong-field gravitational dynamics.

2 Binary Black Hole Coalescence

2.1 Overview

The problem of binary black hole (BBH) coalescence is currently the focus of a large fraction of the active researchers in numerical relativity. It is widely believed that BBH mergers rank among the most copious generators of gravitational waves, and hopes for detection of such events by next-generation instruments such as LIGO run high. Conventional wisdom suggests that detailed waveform predictions (templates) from numerical simulations will be crucial for early detection of these events, and this has provided much of the impetus for the current flury of activity devoted to the BBH problem. Binary coalescence is also of fundamental physical importance *vis a vis* its central role in our understanding of the general-relativistic two-body problem. Finally, the BBH problem is very much a watershed calculation for numerical relativity; its successful solution requires machinery which should largely provide the means to generate truly general solutions of Einstein's equations. Much of the current BBH research is being carried out under the auspices of an NSF "Grand Challenge" grant, awarded to a multi-institution team of researchers led by R.A. Matzner [1].

2.2 Magnitude of the Computational Problem

Following Finn[2], the basic magnitude of the BBH computational problem may be estimated as follows. We assume that the binary consists of equal mass black holes, with a total mass M, initially in a circular orbit with diameter $\sim 6M$. The quadrupole radiation from this source has a wavelength of order $100M$ and to reliably read off the radiation, the outer boundary of the computational domain must be placed at least one wavelength from the source. If we assume that we need a finite-difference (FD) mesh spacing $\sim M/20$ to resolve the black holes themselves, and that we will use a uniform three-dimensional (3D) FD mesh, the total number of grid points used in the calculation will be $\sim 10^{10}$. Typical 3D codes require 50-100 floating-point numbers per grid-point, which, for standard 8-byte arithmetic, implies a total storage requirement of 10^{12}–10^{13} bytes!

CPU requirements are just as prodigious. For our fiducial binary, the estimated physical time for a couple of orbits, coalescence and ring-down is $\sim 500M$. This implies that a simulation will involve $\sim 10^4$ discrete time steps, each of which will require on the order of 5000 floating-point-operations (flops) per grid point. This gives a total of about 5×10^{17} flops/simulation or about a CPU-week on a Teraflop/s machine.

Although machines with capacities commensurate with these requirements should be available within the next three years, it is clear that we will not have dedicated access to such facilities and thus the reduction of the computational burden, particularly using adaptive mesh refinment techniques, is a high priority.

2.3 Formalisms for Numerical Relativity

The traditional approach to general-relativistic simulations has employed the ADM (3+1) formalism[3] in which the geometry of spacetime in viewed as the "time history" of the geometry of a spacelike hypersurface ("instant of time"). In the ADM approach, just as the geometry of spacetime is described by a 4-metric $^{(4)}g_{\mu\nu}$, so is the geometry of a spacelike hypersurface described by a 3-metric g_{ij} (Latin indices run over spatial values 1, 2, 3; Greek indices run over space-time values 0,1,2,3). In particular, the spacetime-displacement-squared can be written

$$
\begin{aligned}
^{(4)}ds^2 &= {}^{(4)}g_{\mu\nu}dx^\mu\,dx^\nu \\
&= -\alpha^2\,dt^2 + g_{ij}\left(dx^i + \beta^i\,dt\right)\left(dx^j + \beta^j\,dt\right)
\end{aligned}
\tag{1}
$$

where α (the lapse function) and β^i (the shift vector), which, in principle, can be chosen essentially arbitrarily, represent the 4-fold coordinate freedom of general relativity. Another tensor of prime importance in the ADM formalism is the extrinsic curvature (second fundamental form), K_{ij} which, loosely speaking, may be viewed as the "velocity" of g_{ij}:

$$K_{ij} = \frac{1}{2\alpha}\left(-\frac{\partial g_{ij}}{\partial t} + D_i\,\beta_j + D_j\,\beta_i\right) \qquad (2)$$

where D_i is the 3-covariant derivative (i.e. $D_i\,g_{ij} = 0$).

As is well known, the Einstein equations (restricted here to the case of vacuum):

$$G_{\mu\nu} = 0 \qquad (3)$$

naturally decompose into two sets. Four of the equations

$$G_{0\nu} = 0 \qquad (4)$$

do *not* involve time derivatives of the K_{ij}, and thus represent equations of constraint which must be satisfied at all times, including the initial time (see the next subsection). The remaining 6 equations are the evolution equations, *per se*, for the gravitational field. Written in first-order-in-time form, they are:

$$\frac{\partial g_{ij}}{\partial t} = -2\alpha K_{ij} + D_i\,\beta_j + D_j\,\beta_i \qquad (5)$$

$$\frac{\partial K_{ij}}{\partial t} = \mathcal{L}_\beta K_{ij} - D_i D_j \alpha + \alpha\left(R_{ij} - 2K_{ik}K^k{}_j + K_{ij}K\right) \qquad (6)$$

where R_{ij} is the 3-Ricci tensor, $K \equiv K^i{}_i$, and $\mathcal{L}_\beta$ is the Lie derivative along β^i. It is worth noting that the vast majority of numerical relativity codes which have been constructed over the past few decades have been based on the ADM formalism.

Successful as the ADM approach has been in numerical relativity, the general form of the ADM equations has been of considerable concern to many researchers. In particular, the ADM form of Einstein's equations is generically non-hyperbolic. Roughly speaking, a hyperbolic formulation of a system of evolution equations is one in which characteristic speeds are either 0 or the local light speed, so that information manifestly propagates along light cones. In addition, hyperbolicity generically allows the evolution equation to be cast into a so-called "flux conservative form":

$$\partial_t u + \partial_i F^i(u) = S(u) \qquad (7)$$

where the source term $S(u)$ does not involve any spatial derivatives of the dynamical variables, u. In the past few years there has been tremendous progress made in re-casting Einstein's equations in hyperbolic form [4]. Possible benefits of such an approach range from improved treatment of both inner and outer boundaries in black hole collisions to the *en masse* appropriation of the large body of advanced numerical techniques which have been developed (particularly in the context of computational fluid dynamics) for flux-conservative systems. It is still too early to tell whether the hyperbolic approaches will live up to their promise but it is abundantly clear that research in this area will be followed with intense interest by the numerical relativity community.

2.4 The Initial Value Problem for Two Black Holes

As mentioned above, four of Einstein's equations:

$$G_{0\nu} = 0 \tag{8}$$

are equations of constraint, which, in a given coordinate system, become a system of 4 coupled, non-linear elliptic equations in the dynamical variables $\{g_{ij}, K_{ij}\}$. Thus, even the problem of determining *initial data* for general-relativistic simulations is decidely non-trivial. Historically a key issue has been deciding which of the $\{g_{ij}, K_{ij}\}$ should be freely specified, and which should be determined via solution of the constraints. One general strategy, worked out in the 70's and early 80's by York and various collaborators [5], is based on ideas of conformal scaling and a "spin-decomposition" of K_{ij}. This approach allows the constraints to be cast as a system of quasi-linear elliptic equations for four potentials, $\{\psi, X^i\}$, which may viewed as the relativistic generalization of the familiar Newtonian gravitational potential. In many cases, the equations for the potentials X^i can be solved analytically, leaving only the equation for ψ (the Hamiltonian constraint) to be solved numerically. The state-of-the-art in Hamiltonian constraint solvers is quite advanced [5] and typically involves second-order finite-difference equations solved via a multi-grid algorithm. With a careful choice of coordinates and discretization (see Cook *et al*[5]), higher order accuracy can be achieved via Richardson extrapolation and, at least from the computational point of view, the initial value problem for two black holes can be considered solved. From a physical point of view, however, a crucial question remains unanswered: How do we generate *realistic* initial data for a black hole collision? In particular, we do not yet have a prescription for specifying data for a relatively tight binary (needed to minimize simulation cost) which accurately represents the late-time configuration of a binary which started in some well-separated,, quasi-Newtonian orbit.

2.5 Black Hole Excising Techniques

The fact that black hole spacetimes contain physical singularities has had a profound effect on numerical relativity. The traditional approach, pioneered by DeWitt and his students in the mid-70's, has been to use coordinate freedom (choice of lapse and shift) to "freeze" evolution near a physical singularity, while allowing the dynamics in regions far-removed from the singularity to be pushed ahead. Consider, for example, the case of a spherically-symmetric collapse of some initial matter distribution which leads to black hole formation. Then we can choose coordinates, r and t, such that the metric takes on a "time-dependent-Schwarzschild" form:

$$ds^2 = -\alpha(r,t)^2\,dt^2 + \left(1 - \frac{2m(r,t)}{r}\right)^{-1}\,dr^2 + r^2\,d\Omega^2. \qquad (9)$$

When black hole formation is imminent, one finds that near the Schwarzschild radius, $r = R_S$, the lapse function, $\alpha(r,t)$ rapidly drops to 0, effectively "freezing" the dynamics in the vicinity of the singularity. Unfortunately, one also finds that the radial metric function, $(1 - 2m/r)^{-1}$, and its spatial derivatives grow without bound for $r \approx R_S$. Thus, the evolution avoids the *physical* singularity, but at the price of generating a *coordinate* singularity. Moreover, from the point of view of numerical simulation, the coordinate singularity is just as pathological as the physical singularity, and a code using these coordinates will "crash" on a dynamical time-scale. Although a variety of other singularity-avoiding coordinate choices have been used in black hole studies, without exception they induce coordinate singularities which render them useless for the long-time evolution of black holes.

In the early 80's, Unruh struck on a possible solution to this serious problem by observing that since the interior of a black hole is, by definition, out of causal contact with the exterior spacetime, it should be possible to simply exclude the regions within event horizons from the computational domain. Further, since the event horizon cannot be located until the construction of the spacetime is complete, Unruh suggested that *apparent horizons* be used as inner boundaries for black hole calculations. (An apparent horizon is a closed 2-surface, defined at some instant at time, such that the outgoing light rays emanating from the surface "hover" at constant radius.) Thornburg[6] aggressively pursued this "excising" strategy in his thesis research on axisymmetric systems, but it was the spherically-symmetric work of Seidel and Suen[6] which first convincingly demonstrated the efficacy of the approach. Several other spherically-symmetric calculations[6] have furthered our confidence that the technique should be generically applicable, and all of the 3-dimensional codes currently being developed within the Grand Challenge project implement black-hole excising. Preliminary indications (mostly from the Postsdam/NCSA/Wash. U group) suggest that the approach is viable in the generic 3D case, and thus it seems very likely that excising (and related techniques) will have an enormous impact on our ability to carry out accurate, long-term black-hole integrations.

3 Head-on Black Hole Collisions: The Close Limit

The problem of two momentarily stationary, axisymmetric black holes which subsequently collide and coalesce has become a benchmark problem in numerical relativity. Initial data for the problem was constructed by Misner [7] in 1960; Smarr and Eppley [8] evolved the data in their famous studies in the mid 70's, and from the late 80's until the present time, researchers at NCSA have revisited and extended those pioneering computations. Advances in computer technology, coordinate conditions and numerical techniques have produced a steady improvement in the accuracy of the computations so that, currently, calculations can routinely produce expected waveforms with an uncertainty on the order of a few percent.

Recently, following an observation by Smarr, Price and Pullin [9] have formulated a perturbative alternative to full-blown numerical computations, which, for Misner and Misner-like data, has been spectacularly successful in reproducing the axisymmetric finite-difference calculations. Their "close-limit" approach is developed by first noting that there is a free parameter, μ_0 in the Misner data which controls the initial separation of the holes, and that for $\mu_0 \leq 1.4$, the data actually describe an "already-collided" system (i.e. a single black hole). Thus, for sufficiently small μ_0, the space-time geometry is nearly spherical and we can expect a perturbative expansion of the 4-metric to be sensible:

$$g_{\mu\nu} = g_{\mu\nu}^{(0)} + \epsilon h_{\mu\nu}^{(1)} + \epsilon^2 h_{\mu\nu}^{(2)} + \cdots \qquad (10)$$

Here, $g_{\mu\nu}^{(0)}$ is the metric of a Schwarzschild spacetime. As shown by Zerilli and Montcrief [10], the physical information in each ℓ-pole, and at each order in perturbation theory, can be encoded in a single "Zerilli function", $\psi_\ell^{(1)}, \psi_\ell^{(2)}, \ldots$. Each Zerilli function satisfies a relatively simple wave equation; for example, at first order we have

$$-\partial_t^2 \psi_\ell^{(1)} + \partial_{r^\star}^2 \psi_\ell^{(1)} + V_\ell(r^\star) \psi_\ell^{(1)} = 0 \qquad (11)$$

where $r^\star$ is the usual "tortoise" coordinate, and $V_\ell(r^\star)$ is an effective potential. Such 1+1 dimensional wave equations are much easier to integrate numerically than the full Einstein equations, and the results of such integrations readily yield estimates of pertinent physical quantities such as waveforms, power-law exponents and the total amount of energy radiated before the system settles down to a static, spherically symmetric final state. Price and Pullin's initial calculations have been applied to other configurations (see for example Abrahams and Price[9]), and very recently have been extended to second order[11] in ϵ, where the agreement with the NCSA results is even more striking.

4 Critical Phenomena in Gravitational Collapse

A final area which has recently seen considerable research activity concerns the threshold of black hole formation in models of gravitational collapse. Here I will only give a brief overview of this rapidly developing field; the interested reader should consult the references for futher information[12,13,14].

Black hole critical phenomena were first discovered in spherically-symmetric studies of the collapse of a massless scalar field (Choptuik, 1993[12]). Following a suggestion by Christodoulou, I considered families of collapse solutions each characterized by some control parameter, p, which governed the strength of the gravitational self-interaction of the scalar field. These families were constructed to "interpolate' between essentially flat spacetimes (for low values of the p) and black-hole spacetimes (for high values of p), such that for each family, there was a critical parameter value, $p^\star$, which demarked the threshold of black hole formation. Detailed empirical studies of many interpolating families revealed several intriguing features of the strong-field non-linear dynamics in the near-critical regime ($p \approx p^\star$). These included a type of exponential sensitivity to initial conditions, a novel form of self-similarity, and a scaling law for the masses of the black holes which formed in super-critical ($p > p^\star$) evolutions:

$$M_{\mathrm{BH}} \sim c_f |p - p^star|^\gamma \qquad (12)$$

Here, c_f is a family-dependent normalization factor, but the scaling exponent, $\gamma \sim 0.37$, is universal in the sense that it does not depend on the details of the initial data (i.e. on the specific interpolating family used).

These results strongly suggested that black-hole formation in the scalar-field model generically turned on at *infinitesimal* mass, and thus implied that the no-hole to hole transition was analogous to a second order phase transition (where the mass of the black hole plays the role of an order parameter). This provided a mechanism (admittedly largely of theoretical interest) by which arbitrarily small black holes, and regions of space-time with arbitrarily large curvature visible at infinity, could be created. Furthermore, on the basis of the numerical results one could reasonably conjecture that a precisely critical solution would contain a curvature singularity which was *not* surrounded by an event horizon, thus providing an example of naked singularity formation from a perfectly regular and "realistic" (i.e. not dust) initial matter distribution.

Since the original scalar field work, several other examples of critical solutions have been found—in most cases by employing the same general strategy of constructing an interpolating family and then searching for a critical point in parameter space. The models considered to date include:

- Axisymmetric, vacuum collapse of gravitational waves (Abrahams and Evans [12])

- Spherically-symmetric collapse of a radiation fluid (Evans and Coleman [12])

- Spherically-symmetric collapse of a Yang-Mills field (Choptuik *et al*[12])

- Extended models of spherically-symmetric scalar collapse (Hamade *et al*[12], Liebling and Choptuik [12], Hod and Piran [12])

These models all exhibit the "Type-II" (for second-order) features seen in the scalar calculations, and in the cases of the gravitational wave and radiation fluid, the mass-scaling exponents, γ, were found to close enough to the scalar value to prompt some speculation that the exponent in the mass-scaling law might be *truly* universal. However, Maison's [13] investigations of perfect fluid models with varying polytropic index, the Yang-Mills calculations and the extended-scalar simulations have all shown conclusively that this is not the case.

Considerable insight into the nature of these critical phenomena (at least in spherical symmetry) has been provided by the application of perturbation theory to precisely critical solutions [13]. The Type II critical solutions which have been found thus far all exhibit either continuous self-similarity or, more interestingly, discrete self-similarity (as first observed in the scalar collapse). In the former case, the precisely critical solution can be determined directly by assuming continuous self-similarity, reducing the equations of motion to a set of ODEs in the similarity variable, and then solving the ODEs to high accuracy via a shooting technique. One can then perform a linear-perturbation analysis about the critical background and examine the mode structure of the linear perturbations. For Type-II solutions one generically finds a *single* growing mode in perturbation theory, whose associated eigenvalue can be immediately related to the exponent, γ, in the mass-scaling law. The existence of the exactly self-similar solution, plus the existence of a single growing mode provides a unified theoretical understanding of the *universality* (family-independence) of the features seen in critical collapse. A similar analysis can be carried out for the case of discretely self-similar Type II solutions (see Gundlach[13]) where the basic picture of a locally unique critical solution with one unstable mode again emerges.

The Yang-Mills model is notable for possessing a second type of black-hole transition ("Type I" behaviour), where the critical solution is the $n = 1$ static Bartnik-Mckinnon[15] soliton, and black hole formation turns on at finite mass. Suitably constructed 2-parameter families exhibit a critical line in parameter space, and the two critical solutions co-exist at a point along this line. Finally, recent work by Hod and Piran[12] has verified Gundlach's[13] prediction that the scaling law (12) is not exact for scalar collapse, but rather has a low-amplitude periodic "wiggle" superimposed on it.

Acknowledgments

The author was supported in part by NSF PHY9318152.

References

1. See http://www.npac.syr.edu/projects/bh/ and links therein.
2. L.S. Finn, LANL preprint gr-qc/9603004 (1996)
3. C.W. Misner *et al*, *Gravitation*, (W. H. Freeman, San Francisco, 1973); J.W. York in *Sources of Gravitational Radiation*, (Cambridge University Press, Cambridge, 1979)
4. C. Bona *et al*, *Phys. Rev. Lett.* **75**, 600 (1995); A. Abrahams *et al*, *Phys. Rev. Lett.* **75**, 3377 (1995); M. Van Putten and D. Eardley, *Phys. Rev.* D **53**, 3056 (1996);
5. J.W. York and T. Piran, in *Spacetime and Geometry*, eds. R. Matzner and L. Shepley, (University of Texas Press, Austin, (1982); G.B Cook *et al*, *Phys. Rev.* D **47**, 1471 (1993)
6. J. Thornburg, University of British Columbia PhD Thesis (1993); E. Seidel and W-M. Suen, *Phys. Rev. Lett.* **69**, 1845 (1992); M. Scheel *et al*, *Phys. Rev.* D **51**, 4108 (1995); P. Anninos *et al*, *Phys. Rev.* D **51**, 5562 (1995); R. Marsa and M. Choptuik, *Phys. Rev.* D **54**, 4929 (1996)
7. C. Misner, *Phys. Rev.* **118**, 1110 (1960)
8. L. Smarr, in *Sources of Gravitational Radiation*, ed. L Smarr, (Cambridge University Press, Cambridge, 1979)
9. R. Price and J. Pullin, *Phys. Rev. Lett.* **73**, 3297 (1994); A. Abrahams and R. Price, *Phys. Rev.* D **53**, 1963 (1996); R. Gleiser *et al*, *Class. Quant. Grav* **13**, L117 (1996)

10. F. Zerilli, *Phys. Rev. Lett.* **73**, 3297 (1970); V. Montcrief, *Ann, Phys. (NY)* **88**, 323 (1974)

11. R. Gleiser *et al*, LANL preprint gr-qc/9609022 (1996)

12. M.W. Choptuik, *Phys. Rev. Lett.* **70**, 9 (1993); A.M. Abrahams and C.R. Evans, *Phys. Rev. Lett.* **70**, 2980 (1993); C.R. Evans and J.S. Coleman, *Phys. Rev. Lett.* **72**, 1782 (1994); C. Gundlach, *Phys. Rev. Lett.* **75**, 3214 (1995); M.W. Choptuik *et al*, *Phys. Rev. Lett.* **77**, 424 (1996); S.L. Liebling and M.W. Choptuik, *Phys. Rev. Lett.* **77**, 1424 (1996); R.S. Hamade *et al*, *Class. Quant. Grav* **13**, 2241 (1996); S. Hod and T. Piran, LANL preprint gr-qc/9606087 (1996); S. Hod and T. Piran, LANL preprint gr-qc/9606093 (1996);

13. T. Koike *et al*, *Phys. Rev. Lett.* **74**, 5170 (1995); D. Maison, *Phys. Lett.* B **366**, 82 (1995); E.W. Hirschmann and D. Eardley, *Phys. Rev.* D **52**, 5850 (1995); C. Gundlach and J.M. Martin-Garcia, *Phys. Rev.* D **54**, 7353 (1996); C. Gundlach, *Phys. Rev.* D **55**, 695 (1997); E.W. Hirschmann and D. Eardley, LANL preprint gr-qc/9511052

14. A. Strominger and L. Thorlacius, *Phys. Rev. Lett.* **72**, 1584 (1994); T. Chiba and M. Siino, Kyoto University preprint KUNS-1384, (1996); Y. Peleg it et al, LANL preprint gr-qc/9608040 (1996)

15. R. Bartnik and J. Mckinnon, *Phys. Rev. Lett.* **61**, 141 (1988)

BLACK HOLES, JETS, AND ACCRETION DISKS

MITCHELL C. BEGELMAN

JILA, University of Colorado, Boulder,
CO 80309-0440, USA

The prevailing paradigm for the energetic phenomena in active galactic nuclei and X-ray binaries — linking black holes, relativistic jets, and accretion disks — has been greatly strengthened in last the last few years by a host of spectacular observational discoveries, plus a few theoretical developments. I briefly describe some of the most dramatic new results.

1 Introduction

The advertised title of my talk — *AGNs, QSOs, and Jets* — is flawed in two respects, which is why I have decided to change it. First, it is redundant. If anyone at the start of 1996 still worried that radio-quiet QSOs differ fundamentally from Seyfert nuclei, i.e., garden-variety AGNs, in more than distance and luminosity, then the *HST/WFPC2* images of quasar host galaxies observed by Bahcall, Disney, and collaborators[1] should have laid those qualms to rest. Second, the original title was not general enough. It is becoming increasingly difficult for a theorist to discuss the central engines of AGNs without focusing on the generic features that make them so interesting: the conjunction of black holes, accretion disks, and jets. We now know that the latter trio also occurs, in scaled-down form, in X-ray binary systems (XRBs). A few of the developments I want to speak about concern the latter, so I am reserving the right to switch between AGNs and XRBs at will.

2 Searching for Black Holes

What is new in the search for black holes is an explosion of promising new techniques for finding them. The three "classical" techniques — binary mass functions for XRBs, velocity dispersions of stars and rotation curves of both stars and gas in galactic nuclei[2] — are becoming ever more refined, as they yield additional black-hole candidates. But the last couple of years has seen the appearance of at least four "new" techniques. Two of these — measurement of stellar proper motions in the Galactic Center, and Keplerian rotation traced by water masers — probe the region from about 10^3 to more than 10^5 Schwarzschild radii. While these new methods are greatly strengthening the case for massive dark objects, they are not testing the general relativistic description of black holes per se. But the other two methods — broad X-ray emission lines and "diskoseismology" — are potentially probing the regions at 10 Schwarzschild radii or less. These techniques offer the exciting promise of probing gravity in the "strong field" limit.

2.1 Stellar Proper Motions in the Galactic Center

High-resolution imaging spectroscopy of the Galactic Center star cluster at 2 μm had already established a fairly convincing case for a massive dark object of about 3 million $M_\odot$ on the basis of radial velocity dispersions.[3] But direct measurements of proper motions[4,5], made over 7 epochs using speckle methods, have moved the Galactic Center up to second place behind NGC 4258 (see below) as the most compelling black-hole detection in a galactic nucleus. Comparing 39 proper motions (19 with confidence levels exceeding 4σ) with 200 radial velocities spanning projected distances $0.03 - 0.3$ pc from SgrA*, Eckart and Genzel established 1) that the velocity dispersion in the central star cluster is isotropic; and 2) that the velocity dispersion steadily increases toward the putative Galactic Center. The former result removes much of the ambiguity afflicting previous applications of the Jeans equation to estimate the central mass, while the latter implies that the central mass density must exceed 6×10^9 $M_\odot$ pc^{-3}.

In an even more dramatic development, Eckart and Genzel have tentatively measured proper motions in the SgrA*(IR) star cluster, only 0.01 pc from SgrA*. The velocity dispersion has risen to 560 ± 90 km s^{-1} (which fits on the Keplerian extrapolation of the velocity dispersion measured further out), with one star clocked at over 1500 km s^{-1}! The corresponding density exceeds 10^{12} $M_\odot$ pc^{-3}.

2.2 H_2O Megamasers

It would seem that only a thin Keplerian disk could give a cleaner signal of a massive dark object than the stellar velocity dispersions mapped out in the Galactic Center. And that is precisely what has been found in the center of the nearby LINER galaxy NGC 4258, using a most unlikely diagnostic — maser emission from water molecules.[6,7] Precise measurements of positions, radial velocities, radial accelerations and proper motions of the maser spots lead to an amazingly consistent (indeed, overdetermined!) fit to a nearly edge-on, warped annulus extending from 0.13 pc to 0.26 pc from the nucleus. The rotation curve is fit by a Keplerian $r^{-1/2}$ law to within 0.3%, giving our most accurate black hole mass to date: 3.6×10^7 $M_\odot$.

Is NGC 4258 a fluke, or does it herald a new standard of evidence in the search for black holes? The fact that the disk must be nearly edge-on in order to observe the maser activity would make these systems somewhat rare. If the masers are pumped by X-rays emitted close to the accreting black hole, as seems likely, [8] then a warp is also necessary in order for the opaque disk to receive enough incident radiation. But as we shall see (§ 4.1), a warp may be the rule, not the exception, in accretion disks. Given these conditions, a standard thin accretion disk with enough mass flow to power the X-rays can account nicely for the scale and intensity of the maser emission. [9]

Indeed, water maser emission has been detected from nearly 20 AGNs. Although a simple kinematic signature may not be ubiquitous, at least one or two sources besides NGC 4258 show evidence for a thin disk. Most recently, the characteristic rotation curve of an annulus has been mapped out in the nucleus of the archetypal type 2 Seyfert galaxy NGC 1068. [10,11] The central mass is about 10^7 $M_\odot$, implying that this luminous AGN is radiating at close to its theoretical maximum, the Eddington limit.

2.3 Broad X-ray Lines

X-ray observations by the Japanese *ASCA* satellite may be providing the first direct evidence for disk-like flow close to the event horizon of a black hole. [12,13,14] *ASCA*'s superior spectroscopic capability has revealed that the iron Kα emission lines observed in some Seyfert 1 galaxies are extremely broad, with Doppler widths (at zero intensity) as high as $c/3$ in some cases. These lines are thought to arise from fluorescence of relatively cool ($\sim 10^6$ K or less), optically thick gas exposed to hard X-rays produced in an optically thin corona. If the fluorescing gas forms the inner part of an accretion disk orbiting a black hole, then the line profile should typically display a relatively narrow blue wing boosted in intensity by the radial Doppler shift, and a broad red wing shaped by the combination of gravitational redshift and transverse Doppler shift. The best studied case, MCG-6-30-15, shows exactly these features in its high- intensity state. Its rapid variability [15] indicates that the line is produced "close in."

Could a fortuitous combination of effects unrelated to the presence of a black hole conspire to produce the observed profiles? While it would be difficult to rule this out, simple models fall far short of explaining the data. [13] Yet the discovery that the Kα line profile in MCG-6-30-15 is strongly correlated with the X-ray continuum flux casts doubt on the simplest model fits to the profile. In particular, the intense blue wing disappears while the red wing is enhanced when MCG-6-30-15 drops into its low-intensity state. [15] These changes might be driven by changes in external illumination or ionization level of the accretion disk, and they may indicate that the black hole is spinning rapidly. More sophisticated models should take into account the emission from gas inside the innermost stable orbit of the accretion disk, which has been neglected to date.

2.4 Diskoseismology

While rapid variability has been a useful qualitative indicator of compactness, its role as a quantitative probe has seldom lived up to expectations. The rich phenomenology of quasi-periodic oscillations (QPOs) in X-ray binaries, with its interdependences among periods, spectra, and luminosities, evokes a complex set of gas dynamical processes in which the black hole's (or neutron star's) gravity is one among many uncertain factors. But the *Rossi X-ray Timing Explorer*'s discovery of a *stable* 67-Hz period in the black-hole candidate GRS 1915+105 [16] may prove to be a relatively clean probe of the black hole's mass and spin. [17]

"G−mode" oscillations of thin, Keplerian disks can exist only where their frequencies are smaller than the local epicyclic frequency for a nearly circular orbit. Due to the effects of general relativity, the epicyclic frequency has a maximum near the inner edge of an accretion disk around a black hole, and vanishes at the inner edge itself. This means that g−modes can be *trapped* in a narrow band near the inner edge of the disk. [18,19] If the disk is thin (but not too thin, $h/R \sim 0.1$), these modes could generate QPOs with coherence factors $Q \sim 20$, rms variability of about a percent (compared to the total X-ray luminosity), and a hard spectrum. [17] The 67-Hz QPO in GRS 1915+105 shows all of these characteristics. If this is the correct interpretation for the stable QPO, its frequency would constrain both the mass of the black hole and its Kerr spin parameter (a/m). Unfortunately, GRS 1915+105 does not have an independent mass function, so we can merely bracket the mass of the black hole between 10.6 $M_\odot$ (nonrotating) and 36.3 $M_\odot$ (maximally rotating). If the mass were measured independently using observations of the binary, one could then determine, for the first time, how fast a black hole is spinning.

3   The Speeds of Jets

Recent observations have strengthened the case for relativistic velocities in the jets emanating from the regions near accreting black holes. The principal indicators of flow at high Lorentz factor are superluminal expansion of compact features along a jet — essentially a light travel- time effect — and one-sidedness, which is thought to result from Doppler beaming along the direction of motion. This interpretation is corroborated by evidence that the "jetted" side is nearly always the one pointing towards us. Surveys of compact extragalactic radio sources continue to uncover new examples of both these effects, with apparent Lorentz factors approaching 30 or more in some cases. In the Galactic X-ray binaries GRO J1655-40 and GRS 1915+105, black-hole candidates which produce pairs of jets following outbursts, it has been possible to follow the evolution of *both* jets.[20,21,22] Measured asymmetries in brightness, length, and speed are fully consistent with the kinematic effects expected from a symmetric pair of relativistic outflows.

3.1 Gamma-ray Blazars

New evidence for relativistic flow comes from high-energy gamma-rays and rapid radio variability. The *EGRET* instrument on board *Compton Gamma-Ray Observatory* has shown that blazars — a class of AGNs with highly variable, polarized emission and compact, one-sided radio jets — often appear most luminous at GeV photon energies.[23] Two blazars have been found to vary rapidly at TeV energies,[24,25, 26] as well. Such energetic photons could not escape if they were produced too close to the black hole, or in any very compact region, since they would interact with the dense radiation field of lower-energy photons to produce electron–positron pairs. But the rapid variability of the gamma-rays suggests that they are produced in a confined space. The resolution of this paradox is almost certainly that the gamma-rays are produced in a jet flowing toward us with a Lorentz factor $\Gamma \sim 10$ or more. Light travel-time effects would compress the apparent variability timescale by a factor $\sim \Gamma^{-2}$, making the rapid variability compatible with the requirements for photon escape.

The gamma-rays presumably are produced by electrons (or pairs) in the jet Compton scattering background, or "seed," photons to higher energies. But it not clear where these seed photons come from. One source is synchrotron photons produced in the jet itself (and believed to be responsible for the lower-energy radiation from blazars).[27,28] Another is the ambient radiation which impinges on the jet either directly from the accretion disk[29] or after reprocessing by a diffuse scattering medium or emission-line clouds.[30,31] The relative importance of these two sources would vary from object to object, partly accounting for the observed wide range of spectral and temporal properties.

3.2 Intraday Radio Variability

The interpretation of Intraday Radio Variability (IDV), also a common feature of blazars, is less certain. Using the variability timescale to estimate a maximum source size, one obtains *apparent* radio brightness temperatures ranging up to 10^{17} K or more.[32] If the radio emission mechanism is incoherent synchrotron and the variability is intrinsic to the source, then avoiding (primarily) synchrotron self-absorption and (secondarily) catastrophic Compton cooling would require Lorentz factors larger than ~ 100 for the most highly variable sources ($\sim 10^3$ for the most extreme case reported[33]). But such high Lorentz factors come at a price: very low radiative efficiency.[34] Consequently, the jets producing the most extreme examples of IDV would have to carry implausibly large energy fluxes.

Is there a way out? Conceivably, the radiation mechanism could be coherent,[35] but then conditions would have to be just right in order for the intense radiation to escape without suffering induced scattering or absorption.[36] Also, there are lingering questions about whether the variability is entirely intrinsic, or could be due partially to interstellar scintillation. These doubts could be put to rest if the correlations among various wavebands are placed on a firmer footing. In any case, it seems likely that at least some of the rapid variability is intrinsic, and realistic limitations on the efficacy of coherent mechanisms would still require Lorentz factors of order 10.

3.3 Relativistic Proper Motions on Kiloparsec Scales

Do jets from AGNs retain their relativistic speeds as they travel through interstellar and intergalactic space? Or do they sweep up material and slow down? Many examples of large-scale one-sided jets in powerful (type FR II) radio sources are known. Now, relativistic proper motions of bright "knots" have been observed in the 2-kpc-long jet of M 87.[37] When combined with existing spectral and morphological data on the M 87 jet, these new observations allow one to constrain the jet's composition and interactions with its surroundings. According to one model,[38] the observed velocities (typically $\sim 0.5\ c$), are the pattern speeds of oblique shocks triggered by Kelvin–Helmholtz instabilities as the jet traverses the background medium. Estimates of the spacing and growth rate of the instability suggest that the jet contains little if any "cold" (i.e., nonrelativistic) plasma, may consist mainly of electron-positron pairs, and does not interact directly with undisturbed interstellar matter but rather traverses a low-density bubble, presumably cleared out by the expansion of the radio source itself. The Lorentz factor of the jet on kiloparsec scales is probably in the range $2 - 5$.

4 Accretion Disks

Whereas new, and often unanticipated, observational discoveries have been driving progress on black holes and jets, several theoretical developments have dominated progress on accretion disks. For the first time, three-dimensional magnetohydrodynamic models of Keplerian shear flows are beginning to provide a physical basis for the α−model of accretion disk viscosity. [39,40,41] Renewed interest in the role of energy advection in accretion disks has led to improved models and new insights into accretion flows with low radiative efficiency. [42,43,44]

4.1 Radiation-driven Warping and Precession

I will dwell briefly on a third development, Pringle's[45] discovery that accretion disks are unstable to warping driven by radiation pressure from the central luminosity source. Hints that direct radiation pressure or gas pressure from an X-ray heated wind could maintain an existing warp had previously appeared in the literature, [46,47,48] but Pringle was the first to demonstrate that disks with prograde warps (relative to the direction of rotation) are genuinely unstable at sufficiently large radii. The radial scale and growth time of the warp are compatible with thin accretion disk models for the maser disk in NGC 4258, which indeed has a mild warp. [49] Numerical models for the nonlinear growth of the instability, taking into account shadowing, suggest that the warping can become extreme, with the inner part of the disk effectively flipping over. [50] Radiation from the central source would then be confined to two cones with time-dependent opening angles, reminiscent of the "ionization cones" observed in Seyfert galaxies. [50,51]

Given enough time, warped disks could attain a steady state characterized by uniform precession. [49] Estimates of the precession timescale suggest that this may be the answer to the longstanding question of what causes the 164-day precession of the jets in SS 433 [52], the 35-day X-ray period in Hercules X-1 [53], and the precession thought to occur in many other X-ray binaries. [54] In these complicated systems, however, precession will not be driven solely by radiation torques — quadrupole gravitational torques must contribute as well, and will also break the symmetry (present in the linear theory of radiation-driven warping) between retrograde and prograde precession.

5 Concluding Remarks

¿From the many examples mentioned in this paper, it should be clear that the study of black-hole phenomenology is galloping ahead. A unified paradigm of AGNs and Galactic black-hole candidates, with its basic components of black hole, jets, and accretion disk, is more robust than ever before. Yet, such discoveries as maser disks in AGNs and Pringle's warping instability should

174

remind us that we are likely to encounter surprises for the foreseeable future.

Acknowledgments

Discussions with Martin Rees, Mike Nowak, Chris Reynolds, and Phil Maloney were particularly helpful in organizing this review. My work on AGNs and X-ray binaries is supported in part by NSF grants AST91-20599 and AST95-29175, and NASA grants NAG5- 2026 and NAGW-3838.

References

1. Bahcall, J. N., Kirkhakos, S., Saxe, D. H., & Schneider, D. P. 1996, ApJ (Letters), in press
2. Kormendy, J., & Richstone, D. 1995, ARA&A, 33, 581
3. Genzel, R., Thatte, N., Krabbe, A., Kroker, H., & Tacconi-Garman, L. E. 1996, ApJ, 472, 153
4. Eckart, A., & Genzel, R. 1996, Nature, 383, 415
5. Eckart, A., & Genzel, R. 1996, MNRAS, in press
6. Miyoshi, M., Moran, J., Herrnstein, J., Greenhill, L., Nakai, N., Diamond, P., & Inoue, M. 1995, Nature, 373, 127
7. Greenhill, L. J., Jiang, D. R., Moran, J. M., Reid, M. J., Lo, K. Y., & Claussen, M. J. 1995, ApJ, 440, 619
8. Neufeld, D. A., Maloney, P. R., & Conger, S. 1994, ApJ, 436, L127
9. Neufeld, D. A., & Maloney, P. R. 1995, ApJ, 447, L17
10. Greenhill, L. J., Gwinn, C. R., Antonucci, R., & Barvainis, R. 1996, ApJ, 472, L21
11. Greenhill, L. J., & Gwinn, C. 1997, in preparation
12. Mushotzky, R. F., Fabian, A. C., Iwasawa, K., Kunieda, H., Matsuoka, M., Nandra, K., & Tanaka, Y. 1995, MNRAS, 272, L9
13. Fabian, A. C., Nandra, K., Reynolds, C. S., Brandt, W. N., Otani, C., Tanaka, Y., Inoue, H., & Iwasawa, K. 1995, MNRAS, 277, L11
14. Tanaka, Y., et al. 1995, Nature, 375, 659
15. Iwasawa, K., et al. 1996, MNRAS, 282, 1038
16. Morgan, E. H., Remillard, R. A., & Greiner, J. 1996, IAU Circ. 6392
17. Nowak, M. A., Wagoner, R. V., Begelman, M. C., & Lehr, D. E. 1997, ApJ, (Letters), in press
18. Nowak, M. A., & Wagoner, R. V. 1992, ApJ, 393, 697
19. Nowak, M. A., & Wagoner, R. V. 1993, ApJ, 418, 187
20. Mirabel, I. F., & Rodriguez, L. F. 1994, Nature, 371, 46
21. Tingay, S. J., et al. 1995, Nature, 374, 141

22. Hjellming, R. M., & Rupen, M. P. 1995, Nature, 375, 464
23. Fichtel, C., et al. 1994, ApJS, 94, 551
24. Punch, M., et al. 1992, Nature, 358, 477
25. Quinn, J., et al. 1996, ApJ, 456, L83
26. Kerrick, A. D., et al. 1995, ApJ, 438, L59
27. Marscher, A., & Gear, W. 1985, ApJ, 198, 114
28. Ghisellini, G., & Maraschi, L. 1989, ApJ, 340, 181
29. Dermer, C. D., Schlickeiser, R., & Mastichiadis, A. 1992, A&A, 256, L27
30. Sikora, M., Begelman, M. C., & Rees, M. J. 1994, ApJ, 421, 153
31. Blandford, R. D., & Levinson, A. 1995, ApJ, 441, 79
32. Wagner, S. J., & Witzel, A. 1995, ARA&A, 33, 163
33. Kedziora-Chudczer, L., et al. 1996, submitted
34. Begelman, M. C., Rees, M. J., & Sikora, M. 1994, ApJ, 429, L57
35. Benford, G. 1992, ApJ, 391, L59
36. Coppi, P., Blandford, R. D., & Rees, M. J. 1993, MNRAS, 262, 603
37. Biretta, J. A., Zhou, F., & Owen, F. N. 1995, ApJ, 447, 582
38. Bicknell, G. V., & Begelman, M. C. 1996, ApJ, 467, 597
39. Brandenburg, A., Nordlund, A., Stein, R. F., & Torkelson, U. 1995, ApJ, 446, 741
40. Stone, J. M., Hawley, J. F., Gammie, C. F., & Balbus, S. A. 1996, ApJ, 463, 656
41. Balbus, S. A., Hawley, J. F., & Stone, J. M. 1996, ApJ, 467, 76
42. Narayan, R., & Yi, I. 1995, ApJ, 452, 710
43. Chen, X., Abramowicz, M. A., Lasota, J.-P., Narayan, R., & Yi, I. 1995, ApJ, 443, 61
44. Narayan, R., Yi, I., & Mahadevan, R. 1995, Nature, 374, 623
45. Prtingle, J. E. 1996, MNRAS, 281, 357
46. Petterson, J. A. 1977, ApJ, 216, 827
47. Iping, R. C., & Petterson, J. A. 1990, A&A, 239, 221
48. Schandl, S., & Meyer, F. 1994, A&A, 289,149
49. Maloney, P. R., Begelman, M. C., & Pringle, J. E. 1996, ApJ, 472, 582
50. Pringle, J. E. 1997, submitted
51. Begelman, M. C., & Bland-Hawthorn, J. 1997, Nature, 385, 22
52. Margon, B. 1984, ARA&A, 22, 507
53. Tananbaum, H., Gursky, H., Kellogg, E. M., Levinson, R., Schreier, E., & Giacconi, R. 1972, ApJ, 174, L143
54. Maloney, P. R., & Begelman, M. C. 1997, in Accretion Phenomena and Related Outflows, IAU Colloq. 163, ed. D. Wickramasinghe, L. Ferrario, & G. Bicknell, in press (San Francisco: Astron. Soc. Pacific)

SUPERNOVAE AND COSMOLOGY: LOW REDSHIFTS

ALEXEI V. FILIPPENKO

Department of Astronomy, University of California,
Berkeley, California 94720-3411, USA

I review the use of low-redshift ($z \lesssim 0.1$) supernovae for cosmological distance determinations. The data on SNe Ia suggest that $H_0 = 64 \pm 3$ km s^{-1} Mpc^{-1} if a range of peak luminosities is recognized, but arguments can be made for lower values (57 ± 4 km s^{-1} Mpc^{-1}) based on restricted subsets of objects. A different technique applied to SNe II yields $H_0 = 73 \pm 7$ km s^{-1} Mpc^{-1}. Supernovae demonstrate that the Hubble expansion is linear, that the bulk motion of the Local Group is consistent with the COBE result, and that the properties of dust in other galaxies are similar to those of dust in the Milky Way.

1 Introduction

Supernovae (SNe) come in two main varieties (see Filippenko 1997b for a review). Those whose optical spectra exhibit hydrogen are classified as Type II, while hydrogen-deficient SNe are designated Type I. The latter group is further subdivided according to the appearance of the early-time spectrum: SNe Ia are characterized by strong absorption near 6150 Å (now attributed to Si II), SNe Ib lack this feature but instead show prominent He I lines, and SNe Ic have neither the Si II nor the He I lines. SNe Ia are believed to result from the thermonuclear disruption of carbon-oxygen white dwarfs, while SNe II come from core collapse in massive supergiant stars. The latter mechanism probably produces most SNe Ib/Ic as well, but the progenitor stars previously lost their outer layers of hydrogen or even helium.

It has long been recognized that SNe may be very useful distance indicators for a number of reasons (Branch & Tammann 1992, and references therein). (1) They are exceedingly luminous objects, with absolute blue magnitudes averaging -19.7 for SNe Ia and -18.2 for SNe II (if $H_0 = 50$ km s^{-1} Mpc^{-1}). (2) "Normal" SNe Ia have relatively small dispersion among their peak magnitudes ($\sigma \lesssim 0.3$ mag). Also, the luminosity of some SNe II can be calibrated individually with high-quality data, despite showing wide variations among objects. (3) Our understanding of the progenitors and explosion mechanisms of SNe is on a reasonably firm physical basis. The results of recent models give good fits to the observed spectra and light curves, providing confidence that we are not far off the mark. (4) Little cosmic evolution is expected in the peak luminosities of SNe, and it can be modeled in detail. This makes SNe superior to galaxies as distance indicators. (5) One can perform *local* tests of various possible complications and evolutionary effects by comparing nearby SNe in different environments (elliptical galaxies, bulges and disks of spirals, galaxies having different metallicities, etc.).

Research on SNe over the past five years has amply confirmed these expectations, and has demonstrated their enormous potential as cosmological distance indicators. Although there are subtle effects that must indeed be taken into account, it appears that SNe may provide among the most accurate values of H_0, q_0, Ω_0, and Λ. As an example of the rapid pace of research, IAU Circulars 6270 and 6490 report the "batch" discovery of 11 SNe with $0.16 \lesssim z \lesssim 0.65$ (Perlmutter et al. 1995) and 17 SNe with $0.09 \lesssim z \lesssim 0.84$ (Suntzeff et al. 1996), respectively. The immediate goal of both teams is to measure q_0 with SNe Ia, and tantalizing preliminary results are discussed by Perlmutter et al. (1997). Distant SNe Ia have already provided evidence for cosmological time dilation through detailed analysis of light curves (Leibundgut et al. 1996; Goldhaber et al. 1997) and spectra (Riess et al. 1997).

Here I will concentrate on the use of relatively low-redshift SNe ($z \lesssim 0.1$) for measurements of H_0. In the next talk, my colleague Saul Perlmutter will describe the quest for q_0, Ω_0, and Λ with SNe Ia at higher redshifts.

2 The Expanding Photosphere Method

An atlas of light curves of SNe II is given by Patat et al. (1993). Although they exhibit bewildering variety, it is useful to subdivide the majority of early-time light curves ($t \lesssim 100$ days) into two relatively distinct subclasses (Barbon et al. 1979; Doggett & Branch 1985). The light curves of SNe II-L ("linear") generally resemble those of SNe I, with a steep decline after maximum brightness followed by a slower exponential tail, while SNe II-P ("plateau") remain within ~ 1 mag of maximum brightness for an extended period, when a recombination front is passing through the hydrogen-rich envelope and a reasonably well-defined photosphere is present. The peak absolute magnitudes of SNe II-P show a very wide dispersion (Young & Branch 1989), almost certainly due to differences in the radii of the progenitor stars. The light curve of SN 1987A, albeit unusual, was generically related to those of SNe II-P; the initial peak was very low because the progenitor was a blue supergiant, much smaller than a red supergiant (Arnett et al. 1989, and references therein).

Despite *not* being anything like "standard candles," SNe II-P (and some SNe II-L) are very good distance indicators. They are, in fact, "custom yardsticks" when calibrated with the "Expanding Photosphere Method" (EPM) described by Kirshner & Kwan (1974) and further developed by Wagoner (1981), Hershkowitz & Wagoner (1987), and others; see the complete history given by

Eastman *et al.* (1996). A variant of the famous Baade-Wesselink method for determining the distances of pulsating variable stars, this technique relies on an accurate measurement of the photosphere's effective temperature and velocity *during the plateau phase* of SNe II-P.

A brief synopsis of EPM is as follows. The radius (R) of the photosphere can be determined from its velocity (v) and time since explosion $(t - t_0)$ if the ejecta are freely expanding:

$$R = v(t - t_0) + R_0 \approx v(t - t_0),$$

where we have assumed that the initial radius of the star $(R_0$ at $t = t_0)$ is negligible relative to R after a few days. The velocity of the photosphere is determined from measurements of the wavelengths of the absorption minima in P-Cygni profiles of weak lines such as those of Fe II or, better yet, Sc II. (The absorption minima of strong lines like Hα form far above the photosphere.) The angular size (θ) of the photosphere, on the other hand, is found from the measured, dereddened flux density (f_ν) at a given frequency. We have

$$4\pi D^2 f_\nu = 4\pi R^2 \zeta^2 \pi B_\nu(T),$$

$$\text{so} \quad \theta = \frac{R}{D} = \left[\frac{f_\nu}{\zeta^2 \pi B_\nu(T)} \right]^{1/2},$$

where D is the distance to the supernova, $B_\nu(T)$ is the value of the Planck function at color temperature T (derived from broadband measurements of the supernova's brightness in at least two passbands), and ζ^2 is the flux dilution correction factor (basically a measure of how much the spectrum deviates from that of a blackbody, due primarily to the electron-scattering opacity).

The above two equations imply that

$$t = D\left(\frac{\theta}{v}\right) + t_0.$$

Thus, for a series of measurements of θ and v at various times t, a plot of θ/v versus t should yield a straight line of slope D and intercept t_0. This determination of the distance is independent of the various uncertain rungs in the cosmological distance ladder; it does not even depend on the calibration of the Cepheids. It is equally valid for nearby and distant SNe II-P.

An important check of EPM is that the derived distance be *constant* while the supernova is on the plateau (before it has started to enter the nebular phase). This has been verified with SN 1987A (Eastman & Kirshner 1989) and a number of other SNe II-P (Schmidt *et al.* 1992, 1994b). Moreover, the EPM distance to SN 1987A agrees with that determined geometrically through measurements of the brightening and fading of emission lines from the inner circumstellar ring (Panagia *et al.* 1991). It is also noteworthy that EPM is relatively insensitive to reddening (Schmidt *et al.* 1992; Eastman *et al.* 1996): an underestimate of the reddening leads to an underestimate of the color temperature T [and hence of $B_\nu(T)$ as well], but this is compensated by an underestimate of f_ν, yielding a nearly unchanged value of θ. Indeed, Schmidt *et al.* (1992) show that for errors in A_V [the visual extinction, or $\sim 3.1E(B-V)$] of 0–1 mag, one incurs an error in D of only $\sim$ 0–20%.

Of course, EPM has some caveats or potential limitations. A critical assumption is spherical symmetry for the expanding ejecta, yet polarimetry shows that SN 1987A was not spherical (e.g., Jeffery 1991), as do direct *Hubble Space Telescope (HST)* measurements of the shape of the ejecta (Pun 1996). This could be a severe problem for specific, highly asymmetric objects. On the other hand, the *average* distance derived with EPM for many SNe might be almost unaffected, given random orientations to the line of sight. (Sometimes the cross-sectional area will be too large, and other times too small, relative to spherical ejecta.) Indeed, comparison of EPM and Cepheid distances to the same galaxies shows agreement to within the expected uncertainties for a sample of 6 objects ($D_{\mathrm{Cepheids}}/D_{\mathrm{EPM}} = 0.98 \pm 0.08$; Eastman *et al.* 1996). The agreement will become somewhat worse, however, if the distances of Galactic Cepheids have previously been underestimated by $\sim$ 10% (as is rumored to be the case based on Hipparcos data).

Another limitation of EPM is that one needs a well-observed SN II-P in a given galaxy in order to measure its distance; thus, the technique is most useful for aggregate studies of galaxies, rather than for distances of specific galaxies in a random sample. Finally, knowledge of the flux dilution correction factor, ζ^2, is critical to the success of the method. Fortunately, the extensive grid of models recently published by Eastman *et al.* (1996) shows that the value of ζ^2 is mainly a function of T during the plateau phase of SNe II-P; it is relatively insensitive to other variables such as helium abundance, metallicity, density structure, and expansion rate. Also, it does not differ too greatly from unity during the plateau. There are, however, some differences of opinion regarding the treatment of radiative transfer and thermalization in expanding supernova atmospheres (e.g., Baron *et al.* 1996). The calculations are difficult and various assumptions are made, possibly leading to significant systematic errors.

The most distant SN II-P to which EPM has successfully been applied is SN 1992am at $z = 0.0487$ (Schmidt *et al.* 1994a). The derived distance is

$D = 180^{+30}_{-25}$ Mpc. This object, together with 15 other SNe II-P at smaller redshifts, yields a best fit value of $H_0 = 73 \pm 7$ km s^{-1} Mpc^{-1}, where the quoted uncertainty is purely statistical (Schmidt *et al.* 1994b; Eastman *et al.* 1996). A systematic uncertainty of $\sim \pm 6$ km s^{-1} Mpc^{-1} should also be associated with the above result. The main source of statistical uncertainty is the relatively small number of SNe II-P in the EPM sample, and the low redshift of most of the objects (whose radial velocities are substantially affected by peculiar motions). Doug Leonard and I are currently trying to remedy the situation with EPM measurements of additional nearby SNe II-P, as well as with Keck spectra of SNe II-P in the redshift range 0.1–0.3.

3   Type Ia Supernovae

3.1   Homogeneity and Heterogeneity

The traditional way in which SNe Ia have been used for cosmological distance determinations has been to assume that they are perfect "standard candles" and to compare their observed peak brightness with those of SNe Ia in galaxies whose distances have been independently determined (e.g., Cepheids), or with theoretical calculations of the peak luminosity of SNe Ia. This approach is well summarized in a review by Branch & Tammann (1992). The rationale is that SNe Ia exhibit relatively little scatter in their peak blue luminosity ($\sigma_B \approx 0.4$–0.5 mag; Branch & Miller 1993), and even less if "peculiar" or highly reddened objects are eliminated from consideration by using a color cut (e.g., $\sigma_B \approx 0.2$–0.3 mag, when one includes only those SNe Ia whose $B - V$ color at maximum brightness is within the range -0.25 to 0.25 mag; Vaughan *et al.* 1995). Moreover, the optical spectra of SNe Ia are usually quite homogeneous, if care is taken to compare objects at similar times relative to maximum brightness (Oke & Searle 1974, and references therein). Branch *et al.* (1993) estimate that over 80% of all SNe Ia discovered thus far are "normal."

From a Hubble diagram constructed with unreddened, moderately distant SNe Ia ($z \lesssim 0.1$) for which peculiar motions should be small and relative distances (as given by ratios of redshifts) are accurate, Vaughan *et al.* (1995) find that

$$< M_B(\text{max}) >= (-19.74 \pm 0.06) + 5 \log(H_0/50) \text{ mag.}$$

In a series of papers, Sandage *et al.* (1996) and Saha *et al.* (1996) combine similar relations with *HST* Cepheid distances to the host galaxies of six SNe Ia to derive $H_0 = 57 \pm 4$ km s^{-1} Mpc^{-1}, a value significantly higher than advocated by Sandage in his earlier studies (e.g., Sandage 1994). Alternatively, Nugent *et al.* (1995a) use the theory of SNe Ia to calculate their peak blue luminosity and find that $H_0 = 60^{+14}_{-11}$ km s^{-1} Mpc^{-1} (updated to 57 ± 11 km s^{-1} Mpc^{-1} in Nugent 1997). Despite its rather large uncertainty, this theoretical estimate suggests a fairly low value for H_0, in agreement with Sandage *et al.* (1996).

Recently, however, it has become clear that SNe Ia do *not* constitute a perfectly homogeneous subclass (see Filippenko 1997a). In retrospect this should have been obvious: the Hubble diagram for SNe Ia exhibits scatter larger than the photometric errors, the dispersion actually *rises* when reddening corrections are applied (under the assumption that all SNe Ia have uniform, very blue intrinsic colors at maximum brightness; van den Bergh & Pazder 1992; Sandage & Tammann 1993), and there are some significant outliers whose anomalous magnitudes cannot possibly be explained by extinction alone.

Spectroscopic and photometric peculiarities have been noted with increasing frequency in well-observed SNe Ia during the past decade. A striking case is SN 1991T; its pre-maximum spectrum did not exhibit Si II or Ca II absorption lines, yet two months past maximum the spectrum was nearly indistinguishable from that of a classical SN Ia (Filippenko *et al.* 1992b; Phillips *et al.* 1992; Ruiz-Lapuente *et al.* 1992). The light curves of SN 1991T were slightly broader than the SN Ia template curves, and the object was probably somewhat more luminous than average at maximum. The reigning champion of well-observed, peculiar SNe Ia is SN 1991bg (Filippenko *et al.* 1992a; Leibundgut *et al.* 1993; Turatto *et al.* 1996). At maximum brightness it was subluminous by 1.6 mag in V and 2.5 mag in B, its colors were intrinsically red, and its spectrum was peculiar (with a deep absorption trough due to Ti II). Moreover, the decline from maximum brightness was very steep, the I-band light curve did not exhibit a secondary maximum like normal SNe Ia, and the velocity of the ejecta was unusually low.

The photometric heterogeneity among extreme examples of SNe Ia is perhaps best demonstrated by Suntzeff (1996) with five objects having excellent $BVRI$ light curves, all scaled to the same maximum brightness. The differences in decline rates at *early* times are largest in B, followed by V; they are relatively minor at R and I, except for SN 1991bg. At *late* times, by contrast, the R and I light curves exhibit more scatter than the B and V curves. Two months past maximum, for example, SN 1991T and SN 1991bg differ by 1.3 mag in R (but very little in B) after normalizing to the same peak. Suntzeff (1996) goes on to show the difference in bolometric (near-UV through near-IR) luminosity of four SNe Ia as a function of time. At maximum brightness, SN 1991T was a factor of 5 more luminous than SN 1991bg, and this grew to a factor of 9 by one month past maximum.

3.2 Cosmological Uses

Although SNe Ia can no longer be considered perfect "standard candles," they are still exceptionally useful for cosmological distance determinations. Excluding those of low luminosity (which are hard to find, especially at large distances), most SNe Ia are *nearly* standard (Branch *et al.* 1993). Also, after many tenuous suggestions (e.g., Pskovskii 1977, 1984; Branch 1981), convincing evidence has finally been found for a *correlation* between light curve shape and luminosity. Phillips (1993) achieved this by quantifying the photometric differences among a set of nine well-observed SNe Ia using a parameter, $\Delta m_{15}(B)$, which measures the total drop (in B magnitudes) from maximum to $t = 15$ days after B maximum. In all cases the host galaxies of his SNe Ia have accurate relative distances from surface brightness fluctuations or from the Tully-Fisher relation. In B, the SNe Ia exhibit a total spread of ~ 2 mag in maximum luminosity, and the intrinsically bright SNe Ia clearly decline more slowly than dim ones. The absolute magnitude range is smaller in V and I, making the correlation with $\Delta m_{15}(B)$ less steep than in B, but it is present nonetheless.

Using SNe Ia discovered during the Calán/Tololo survey ($z \lesssim 0.1$), Hamuy *et al.* (1995, 1996b) confirm and refine the Phillips (1993) correlation between Δm_{15} and $M_{max}(B, V)$: it is not as steep as had been claimed. Apparently the slope is shallow at high luminosities, and much steeper at low luminosities; thus, objects such as SN 1991bg skew the slope of the best-fitting single straight line. The degree to which the relationship can be utilized to reduce the scatter in the Hubble diagram for SNe Ia is demonstrated by Hamuy *et al.* (1995, 1996b) for a set of SNe Ia that includes a color cut of $B_{max} - V_{max} \leq 0.20$ mag: after measuring the decline rate and making the appropriate correction for each object, the dispersion falls from 0.24 to 0.17 mag in B and from 0.22 to 0.14 mag in V. Their derived value for H_0 is 63.1 ± 3.4 (internal) ±2.9 (external) km s^{-1} Mpc^{-1}, 10–15% higher than their data would have implied had they used SNe Ia as perfect standard candles. [Note that Schaeffer (1996) claims $H_0 = 56 \pm 3$ km s^{-1} Mpc^{-1} from the Hamuy *et al.* relation, but my own average of the entries for the last four ("modern") SNe Ia in his Table 2 yields $H_0 = 60$ km s^{-1} Mpc^{-1}, excluding the seemingly discrepant U-band value for SN 1972E.]

In a similar effort, Riess *et al.* (1995a) show that the luminosity of SNe Ia correlates with the detailed shape of the light curve, not just its initial decline. They form a "training set" of light curve shapes from 9 well-observed SNe Ia having known relative distances, including very peculiar objects (e.g., SN 1991bg). When the light curves of an independent sample of 13 SNe Ia (the Calán/Tololo survey) are analyzed with this set of basis vectors, the dispersion in the V-band Hubble diagram drops from 0.50 to 0.21 mag, and the Hubble constant rises from 53 ± 11 to 67 ± 7 km s^{-1} Mpc^{-1}, comparable to the conclusions of Hamuy et al. (1995, 1996b). About half of the rise in H_0 results from a change in the position of the "ridge line" defining the linear Hubble relation, and half is from a correction to the luminosity of some of the local calibrators which appear to be unusually luminous (e.g., SN 1972E).

By using light curve shapes measured through several different filters, Riess et al. (1996) extend their analysis and objectively eliminate the effects of interstellar extinction: a SN Ia that has an unusually red $B - V$ color at maximum brightness is assumed to be *intrinsically* subluminous if its light curves rise and decline quickly, or of normal luminosity but significantly *reddened* if its light curves rise and decline slowly. With a set of 20 SNe Ia consisting of the Calán/Tololo sample and their own objects, Riess et al. (1996) show that the dispersion decreases from 0.52 mag to 0.12 mag after application of this technique. (Preliminary results with a very recent, expanded set of 44 SNe Ia indicate that the dispersion decreases from 0.42 mag to 0.17 mag; A. G. Riess 1996, private communication.) The resulting Hubble constant is 65 ± 3 km s^{-1} Mpc^{-1}, with an additional systematic uncertainty of ± 3 km s^{-1} Mpc^{-1}. Riess et al. (1996) also show that the Hubble flow is remarkably linear; indeed, SNe Ia now constitute the best evidence for linearity. Finally, they argue that the dust affecting SNe Ia is *not* of circumstellar origin, and show quantitatively that the extinction curve in external galaxies typically does not differ from that in the Milky Way (cf. Branch & Tammann 1992).

Riess et al. (1995b) capitalize on another use of SNe Ia: determination of the Milky Way Galaxy's peculiar motion relative to the Hubble flow. They select galaxies whose distances were accurately determined from SNe Ia, and compare their observed recession velocities with those expected from the Hubble law alone. The speed and direction of the Galaxy's motion are consistent with what is found from COBE (Cosmic Background Explorer) studies of the microwave background, and also with many plausible bulk flows expected to accompany observed density variations in the Universe. Their results are inconsistent with those of Lauer & Postman (1994), a conclusion that is reinforced with the recently expanded set of SNe Ia.

There are additional, potentially tight correlations that could enable one to calibrate even more accurately the peak luminosity of individual SNe Ia. Fisher et al. (1995) find that velocity of the red edge of the Ca II H&K absorption line at $t \gtrsim 60$ days in spectra of SNe Ia correlates reasonably well with absolute visual magnitude. Nugent et al. (1995b) discuss other spectral trends, and begin to explore their physical basis. For example, the ratio of the Si II absorption line at 5900 Å to that at 6150 Å increases with decreasing luminosity, as does the ratio of the two peaks on either side of the Ca II H&K absorption trough. The latter could be especially useful for calibrating the luminosity of SNe Ia at high redshifts ($\sim$ 0.5), since most of the optical spectrum is shifted to the infrared. Similarly, Branch *et al.* (1996a) find that the $U - B$ color at maximum brightness correlates with absolute magnitude; luminous SNe Ia generally exhibit the largest UV excess. Further work is required to see whether at least two independent methods yield the same corrections (within the uncertainties) to the derived peak luminosities of SNe Ia.

A more theoretical approach is adopted by Höflich & Khokhlov (1996), who calculate realistic light curves for various models of SN Ia explosions. They identify acceptable models by individually comparing the observed light curves (in at least two passbands) of 26 SNe Ia to their theoretical light curves. This procedure yields the distance and extinction of each SN Ia independent of all other calibrations (e.g., Cepheids); no objects are rejected on the grounds of peculiarity. The resulting value of H_0 is 67 ± 5 km s^{-1} Mpc^{-1}, consistent with the more empirical methods of Riess *et al.* (1996) and Hamuy *et al.* (1996b).

The advantage of systematically correcting the luminosities of SNe Ia at high redshifts rather than trying to isolate "normal" ones seems clear in view of recent evidence that the luminosity of SNe Ia may be a function of stellar population. If the most luminous SNe Ia occur in young stellar populations (Hamuy *et al.* 1995, 1996a; Branch *et al.* 1996a), then we might expect the mean peak luminosity of high-redshift SNe Ia to differ from that of a local sample. Alternatively, the use of Cepheids (Population I objects) to calibrate local SNe Ia can lead to a zero point that is too luminous. On the other hand, as long as the physics of SNe Ia is essentially the same in young stellar populations locally and at high redshift, we should be able to adopt the luminosity correction methods (photometric and spectroscopic) found from detailed studies of nearby samples of SNe Ia.

It is important to mention that not all workers agree with the procedures recommended by Hamuy *et al.* (1996b), Riess *et al.* (1996), and Höflich & Khokhlov (1996); see, for example, Sandage *et al.* (1996) and Branch *et al.* (1996a,b), who find that $H_0 \approx 57 \pm 4$ km s^{-1} Mpc^{-1}. A convenient summary of their concerns, many of which seem reasonable, is provided by Branch *et al.* (1996c). These must be further scrutinized, and tests should be conducted with independent samples of SNe Ia. At this time, I favor the empirical approach in which the luminosities of SNe Ia are adjusted according to the shapes of their light curves, preferably in multiple passbands (Riess *et al.* 1996), but it is best to keep an open mind on this issue.

4 Conclusions

The bottom line of the above discussions is that supernovae are potentially excellent cosmological probes; the current precision of individual distance measurements is roughly 5–10% — but a number of subtleties must be taken into account to obtain reliable results. Besides providing accurate estimates of the Hubble constant, SNe at $z \lesssim 0.1$ have been used to demonstrate that (1) the Hubble flow is definitely linear, (2) the bulk motion of the Local Group is consistent with that derived from COBE measurements, (3) the progenitors of SNe Ia do not generate significant dust in their vicinity, and (4) the dust in other galaxies is similar to Galactic dust.

Values for the Hubble constant obtained recently from SNe are 73 ± 7 km s^{-1} Mpc^{-1} with SNe II-P (EPM), and 64 ± 3 km s^{-1} Mpc^{-1} with SNe Ia empirically corrected for luminosity and reddening variations (or 67 ± 5 km s^{-1} Mpc^{-1} using the more theoretical approach of Höflich & Khokhlov 1996). With the quoted statistical uncertainties as well as possible systematic effects, the differences are not significant. Of course, since the empirical SN Ia method depends on the Cepheid calibration while EPM does not, the difference between the values derived from these two techniques will become greater when the new (rumored) Hipparcos results are adopted. Perhaps we live in a slightly underdense region of the Universe: EPM and other methods that favor $H_0 \approx$ 70–75 km s^{-1} Mpc^{-1} generally use galaxies with $z \lesssim 0.03$, while galaxies in the SN Ia analysis go out to $z \approx 0.1$. Given the uncertainties, this would not be inconsistent with the conclusion of Kim et al. (1997) that H_0(local; $z \lesssim 0.1)/H_0$(global; $z \gtrsim 0.35) \approx 1.0$, especially since a majority of the SNe Ia they used to determine the "local" value of H_0 have $z \gtrsim 0.03$ (i.e., larger distances than objects that yield $H_0 = $ 70–75 km s^{-1} Mpc^{-1}).

In view of the above discussion and my slight bias toward empirical (but still well substantiated) methods, here I choose the value of H_0 derived from SNe Ia by Riess et al. (1996) and Hamuy et al. (1996b): 64 ± 3 km s^{-1} Mpc^{-1}. The inverse of H_0 is consequently 15.3 ± 0.7 Gyr, an upper bound on the age of the Universe. If $\Omega_0 = 1$ and $\Lambda = 0$, the true age is 2/3 of this, or 10.2 ± 0.5 Gyr, probably in conflict with the ages of the oldest globular star clusters [15^{+5}_{-3} Gyr ($\sim 2\sigma$ limits); VandenBerg et al. 1996]. This problem has been discussed frequently (e.g., Bolte & Hogan 1995); perhaps the Universe really does have a low value of Ω_0. On the other hand, if the Hipparcos recalibration of the Galactic Cepheid distance (and hence luminosity) scale brings H_0 (as derived from SNe Ia) down to ~ 57 km s^{-1} Mpc^{-1} [i.e., $(2/3)H_0^{-1} \approx 11.5$ Gyr] and the globular cluster ages down to ~ 11 Gyr, there may not be any difficulty with a flat universe. This is especially true if H_0 is already 57 ± 4 km s^{-1} Mpc^{-1},

186

not including the Hipparcos results, as advocated by Sandage *et al.* (1996), Schaeffer (1996), and Branch *et al.* (1996a,b,c).

My research on supernovae is supported by NSF grant AST-9417213. I also thank the organizers of the Texas Symposium for partial travel funds. Useful comments on a draft of this paper were provided by Aaron Barth, David Branch, Ron Eastman, Doug Leonard, Peter Nugent, and Adam Riess.

References

1. W. D. Arnett, *et al.*, *ARAA*, 27, 629 (1989).
2. R. Barbon, F. Ciatti, & L. Rosino, *A&A*, 72, 287 (1979).
3. E. Baron, P. H. Hauschildt, & A. Mezzacappa, *MNRAS*, 278, 763 (1996).
4. M. Bolte & C. J. Hogan, *Nature*, 376, 399 (1995).
5. D. Branch, *ApJ*, 248, 1076 (1981).
6. D. Branch, A. Fisher, & P. Nugent, *AJ*, 106, 2383 (1993).
7. D. Branch & D. L. Miller, *ApJ*, 405, L5 (1993).
8. D. Branch, W. Romanishin, & E. Baron, *ApJ*, 465, 73 (1996a); erratum 467, 473.
9. D. Branch & G. A. Tammann, *ARAA*, 30, 359 (1992).
10. D. Branch, *et al.*, *ApJ*, 470, L7 (1996b).
11. D. Branch, *et al.*, in *Extragalactic Distance Scale: Poster Papers*, ed. M. Livio, *et al.* (Baltimore: STScI), p. 4 (1996c).
12. J. B. Doggett & D. Branch, *AJ*, 90, 2303 (1985).
13. R. G. Eastman & R. P. Kirshner, *ApJ*, 347, 771 (1989).
14. R. G. Eastman, B. P. Schmidt, & R. P. Kirshner, *ApJ*, 466, 911 (1996).
15. A. V. Filippenko, in *Thermonuclear Supernovae*, ed. P. Ruiz-Lapuente, *et al.* (Dordrecht: Kluwer), p. 1 (1997a).
16. A. V. Filippenko, *ARAA*, in press (1997b).
17. A. V. Filippenko, *et al.*, *AJ*, 104, 1543 (1992a).
18. A. V. Filippenko, *et al.*, *ApJ*, 384, L15 (1992b).
19. A. Fisher, *et al.*, *ApJ*, 447, L73 (1995).
20. G. Goldhaber, *et al.*, in *Thermonuclear Supernovae*, ed. P. Ruiz-Lapuente, *et al.* (Dordrecht: Kluwer), p. 777 (1997).
21. M. Hamuy, *et al.*, *AJ*, 109, 1 (1995).
22. M. Hamuy, *et al.*, *AJ*, 112, 2391 (1996a).
23. M. Hamuy, *et al.*, *AJ*, 112, 2398 (1996b).
24. S. Hershkowitz & R. V. Wagoner, *ApJ*, 322, 967 (1987).
25. P. Höflich & A. Khokhlov, *ApJ*, 457, 500 (1996).
26. D. J. Jeffery, *ApJ*, 375, 264 (1991).

27. A. G. Kim, *et al.*, *ApJ*, 476, L63 (1997).
28. R. P. Kirshner & J. Kwan, *ApJ*, 193, 27 (1974).
29. T. Lauer & M. Postman, *ApJ*, 425, 418 (1994).
30. B. Leibundgut, *et al.*, *ApJ*, 466, L21 (1996).
31. B. Leibundgut, *et al.*, *AJ*, 105, 301 (1993).
32. P. Nugent, *Ph.D. Thesis*, Univ. of Oklahoma (1997).
33. P. Nugent, *et al. Phys. Rev. Lett.*, 75, 394 (1995a); erratum 75, 1874.
34. P. Nugent, *et al. ApJ*, 455, L147 (1995b).
35. J. B. Oke & L. Searle, *ARAA*, 12, 315 (1974).
36. N. Panagia, *et al.*, *ApJ*, 380, L23 (1991).
37. F. Patat, *et al.*, *A&AS*, 98, 443 (1993).
38. S. Perlmutter, *et al.*, *IAUC* 6270 (1995).
39. S. Perlmutter, *et al.*, *ApJ*, in press (1997).
40. M. M. Phillips, *ApJ*, 413, L105 (1993).
41. M. M. Phillips, *et al.*, *AJ*, 103, 1632 (1992).
42. Yu. P. Pskovskii, *Soviet Astron.*, 21, 675 (1977); 28, 658 (1984).
43. C.-S. J. Pun, in *Science with the Hubble Space Telescope – II*, ed. P. Benvenuti, *et al.* (Baltimore: STScI), p. 379 (1996).
44. A. G. Riess, W. H. Press, & R. P. Kirshner, *ApJ*, 438, L17 (1995a).
45. A. G. Riess, W. H. Press, & R. P. Kirshner, *ApJ*, 445, L91 (1995b).
46. A. G. Riess, W. H. Press, & R. P. Kirshner, *ApJ*, 473, 588 (1996).
47. A. G. Riess, *et al.*, preprint (1997).
48. P. Ruiz-Lapuente, *et al.*, *ApJ*, 387, L33 (1992).
49. A. Saha, *et al.*, *ApJS*, 107, 693 (1996).
50. A. Sandage, *ApJ*, 430, 13 (1994).
51. A. Sandage & G. A. Tammann, *ApJ*, 415, 1 (1993).
52. A. Sandage, *et al.*, *ApJ*, 460, L15 (1996).
53. B. E. Schaefer, *ApJ*, 460, L19 (1996).
54. B. P. Schmidt, R. P. Kirshner, & R. G. Eastman, *ApJ*, 395, 366 (1992).
55. B. P. Schmidt, *et al.*, *AJ*, 107, 1444 (1994a).
56. B. P. Schmidt, *et al.*, *ApJ*, 432, 42 (1994b).
57. N. Suntzeff, in *Supernovae and Supernova Remnants*, ed. R. McCray & Z. Wang (Cambridge: Cambridge Univ. Press), p. 41 (1996).
58. N. Suntzeff, *et al.*, *IAUC* 6490 (1996).
59. M. Turatto, *et al.*, *MNRAS*, 283, 1 (1996).
60. D. A. VandenBerg, M. Bolte, & P. B. Stetson, *ARAA*, 34, 461 (1996).
61. S. van den Bergh & J. Pazder, *ApJ*, 390, 34 (1992).
62. T. E. Vaughan, *et al.*, *ApJ*, 439, 558 (1995).
63. R. V. Wagoner, *ApJ*, 250, L65 (1981).
64. T. R. Young & D. Branch, *ApJ*, 342, L79 (1989).

MEASURING COSMOLOGICAL PARAMETERS

Wendy L. FREEDMAN

Carnegie Observatories, 813 Santa Barbara St., Pasadena, CA 91101, USA

In this review, the status of measurements of the matter density (Ω_m), the vacuum energy density or cosmological constant (Ω_Λ), the Hubble constant (H_0), and ages of the oldest measured objects (t_0) are summarized. Measurements of the statistics of gravitational lenses and strong gravitational lensing are discussed in the context of limits on Ω_Λ. Three separate routes to the Hubble constant are considered: the measurement of time delays in multiply-imaged quasars, the Sunyaev-Zel'dovich effect in clusters, and Cepheid-based extragalactic distances. Globular-cluster ages plus a new age measurement based on radioactive dating of thorium in a metal-poor star are briefly summarized. Limits on the product of $H_0 t_0$ are also discussed. Finally, the areas where future improvements are likely to be made soon are highlighted; in particular, measurements of anisotropies in the cosmic microwave background. Particular attention is paid to sources of systematic error and the assumptions that underlie many of the measurement methods.

1 Introduction

Rapid progress is being made in measuring the cosmological parameters that describe the dynamical evolution and the geometry of the Universe. In essence, this is the first conclusion of this review. The second conclusion is that despite the considerable advances, the accuracy of cosmological parameters is not yet sufficiently high to discriminate amongst, or to rule out with confidence, many existing, competing, world models. We as observers still need to do better. Fortunately, there are a number of opportunities on the horizon that will allow us to do so.

In the context of the general theory of relativity, and assumptions of large-scale homogeneity and isotropy, the dynamical evolution of the Universe is specified by the Friedmann equation

$$H^2 = \frac{8\pi G \rho_m}{3} - \frac{k}{a^2} + \frac{\Lambda}{3}$$

where a(t) is the scale factor, $H = \frac{\dot{a}}{a}$ is the Hubble parameter (and H_0 is the Hubble "constant" at the present epoch), ρ_m is the average mass density, k is a curvature term, and Λ is the cosmological constant, a term which represents the energy density of the vacuum. It is common practice to define the matter density ($\Omega_m = 8\Pi G \rho_m / 3 H_0^2$), the vacuum energy density ($\Omega_\Lambda = \Lambda / 3 H_0^2$), and the curvature term ($\Omega_k = -k / a_0^2 H_0^2$) so that $\Omega_m + \Omega_\Lambda = 1$ for the case of a flat universe where k = 0. The simplest case is the Einstein-de Sitter model with $\Omega_m = 1$ and $\Omega_\Lambda = 0$. The dimensionless product $H_0 t_0$ (where t_0 is the age of the Universe) is a function of both Ω_m and Ω_Λ. In the case of the Einstein-de Sitter Universe

$$f(\Omega_m, \Omega_\Lambda) = H_0 t_0 = \frac{2}{3}$$

Bounds on several cosmological parameters are summarized in a plot of the matter density as a function of the Hubble constant from Carroll, Press and Turner (1992) (Figure 1) for both open ($\Lambda = 0$) and flat ($\Lambda \neq 0$) models. The dotted lines indicate the limits on the baryon density (Ω_b). Solid lines represent the expansion ages in Gyr. The grey boxes are defined by values of H_0 in the range of 40 to 100 km/sec/Mpc and $0.1 < \Omega_m < 0.2$. The left-hand plot illustrates the well-known "age" problem; namely that for an Einstein-de Sitter Universe ($\Omega = 1$, $\Lambda = 0$), H_0 must be less than ~45 km/sec/Mpc if the ages of globular clusters (t_0) are indeed ~15 billion years old. This discrepancy is less severe if the matter density of the Universe is less than the critical density, or if a non-zero value of the cosmological constant is allowed, as shown in the right-hand plot.

A number of issues that require knowledge of the cosmological parameters remain unresolved at present. First is the question of timescales ($H_0 t_0$) discussed above; possibly a related issue is the observation of red (if they are indeed old) galaxies at high redshift. Second is the amount of dark matter in the Universe. As discussed below, many dynamical estimates of the mass over a wide range of scale sizes are currently favoring values of Ω_m ~0.25±0.10, lower than the critical Einstein-de Sitter density. And third is the origin of large-scale structure in the Universe. Accounting for the observed power spectrum of galaxy clustering has turned out to be a challenge to the best current structure formation models.

Taking all of the data at face value, one way out of these current difficulties is to introduce a non-zero value of a cosmological constant, Λ. In fact, it is precisely the resolution of these problems that has led to a recent resurgence of interest in a non-zero value of Λ (*e.g.* Ostriker & Steinhardt 1995; Krauss & Turner 1995). Another means of addressing these issues (*e.g.* Bartlett *et al.* 1995) requires being in conflict with essentially all of the current observational measurements of H_0; from purely theoretical considerations, a very low value of H_0 (≤ 30) could also resolve these issues.

Ultimately we will have to defer to measurement as the arbiter amongst the wide range of cosmological models (and their very different implications) still being discussed in the literature. A wealth of new data is becoming available

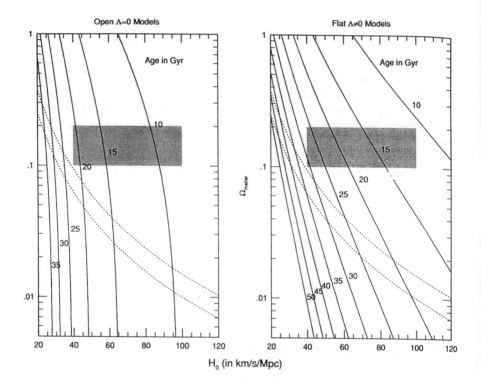

Figure 1: Ω_m versus H_0 from Carroll, Press and Turner (1992) (their Figure 10) for a) an open ($\Omega_\Lambda = 0$) Universe and b) a flat ($\Omega_\Lambda \neq 0$) Universe. See text for details.

and progress is being made in the measurement of all of the cosmological parameters discussed below: the matter density, Ω_m, the vacuum energy density, Ω_Λ, the expansion rate H_0, and age of the oldest stars t_0. The central, critical issues now are (and in fact have always been) testing for and eliminating sources of significant systematic error.

2 Ω_m – The Matter Density

Table 1 presents a summary of several different techniques for measuring the matter density of the Universe. These techniques have been developed over a wide range of scales, from galaxy ($\sim$100-200 kpc), through cluster (Mpc), on up to more global scales (redshifts of a few). The first part of the table lists Ω_m determinations that are independent of Ω_Λ; the second part lists Ω_m determinations that are not independent of Ω_Λ; and the third part of the table lists Ω_Λ determinations. In addition to listing the physical basis of the method, types of object under study, and values of Ω_m plus an estimated uncertainty, Table 1 makes explicit some of the assumptions that underlie each of these techniques. Although in many cases, 95% confidence limits are quoted, these estimates must ultimately be evaluated in the context of the validity of their underlying assumptions. It is non-trivial to assign a quantitative uncertainty in many cases, but in fact systematic effects may be the dominant source of uncertainty. Several of these assumptions and uncertainties are discussed further below. They include, for example, diverse assumptions about mass tracing light, mass-to-light ratios being constant, clusters being representative of the Universe, clumping of X-ray gas, non-evolution of type Ia supernovae, and the non-evolution of elliptical galaxies. For methods that operate over very large scales (gravitational lensing and type Ia supernovae), assumptions about Ω_Λ or Ω_{total} are currently required to place limits on Ω_m.

Since lower values of the matter density tend to be measured on smaller spatial scales, it has given rise to the suspicion that the true, global value of Ω_0 must be measured on scales beyond even those of large clusters, i.e., scales of greater than $\sim$100 Mpc (e.g., Dekel 1994). In that way, one might reconcile the low values of Ω_m inferred locally with a spatially flat Universe. However, recent studies (Bahcall, Lubin & Dorman 1995) suggest that the M/L ratios of galaxies do not continue to grow beyond a scale size of about $\sim$200 kpc (corresponding to the sizes of large halos of individual galaxies). In their Jeans analysis of the dynamics of 16 rich clusters, Carlberg et al. (1997) also see no further trend with scale. Hence, currently the observational evidence does not indicate that measurements of Ω_m on cluster size scales are biased to lower values than the true global value.

Table 1: SUMMARY OF Ω_m and Ω_Λ DETERMINATIONS

Sample	Method	Scale	Assumptions	Ω_m	Error
Ω_Λ Independent Methods					
Galaxies	dyn. M/L ratio	100 kpc	galaxies representative M/L constant	~0.1	
Clusters	dyn. M/L ratio	<few Mpc	clusters representative M/L constant	~0.2	
Clusters	X-ray M/L ratio	<few Mpc	hydrostatic eqm	~0.2	
Clusters	baryon fraction		clusters representative no clumping	0.3-0.5	
Clusters	morphology		model dept.	>0.3	
Local Group	Least action principle	1 Mpc	LG representative no external torques model uniqueness	~0.15	
Galaxies	Virial Theorem (pairwise velocities)	1-300 Mpc	mass-indept. biasing point masses	0.2-0.4	
Galaxies	Peculiar velocities	100 Mpc	biasing	>0.3	95%
Sample	Method	Scale	Assumptions	Ω_m	Error
Ω_Λ Dependent Methods					
Type Ia SNae	Hubble diagram	z<0.5	$\Omega_\Lambda = 0$ no evolution effects	0.88	90%
			$\Omega_{tot}=1$	>0.49	95%
Lensed QSO's	lensing statistics	global	$\Omega_\Lambda=0$ dark matter distrib. slow galaxy evolution dust small effect	>0.15	90%
6 lenses	strong lensing	global	$\Omega_\Lambda = 0$ model dependent	low Ω	
CMB	multipole analysis	global	CDM	0.3 - 1.5	
Sample	Method	Scale	Assumptions	Ω_Λ	Error
Type Ia SNae		z<0.5	$\Omega_{tot}=1$	<0.51	95%
Lensed QSO's	lensing statistics	global	$\Omega_{tot}=1$	< 0.66	95%
6 lenses	strong lensing	global	$\Omega_{tot}=1$	<0.9	95%
$H_0 t_0$	age discrepancy	100 Mpc	$H_0 >65$ $t_0 > 13$ Gyr	>0.5	66%

A brief description of several techniques for measuring the matter density is given below. These methods are discussed in the context of both their strengths and weaknesses, paying particular attention to the underlying assumptions. An excellent, recent, and more complete review on this topic is given by Dekel, Burstein & White (1997); also see Trimble (1987).

2.1   Galaxies and Clusters: Dynamical Measures & Mass-to-Light Ratios

The contribution of galaxies to the mass density can be determined by integrating the luminosity function per unit volume for galaxies and multiplying by an (assumed, constant) mean mass-to-light (M/L) ratio. The dynamical masses of galaxies can be determined from rotation curves for spiral galaxies, or the measurement of velocity dispersions and application of the virial theorem both for individual elliptical galaxies. The latter method can also be applied for groups and clusters of galaxies (as Zwicky did in the 1930's).

This method has several advantages. First it is conceptually simple and model-independent. Unlike some of the global techniques discussed below, this method is independent of both H_0 and Ω_Λ. However, there are a number of underlying assumptions. Most important is *the assumption that galaxies trace all mass*. In addition, there are implicit, underlying assumptions concerning the similarity of mass-to-light ratios in different systems (ignoring, for example, potential differences in initial mass functions, star formation histories, dark remnant populations, dust content, etc.) The estimates based on this method tend to yield low values of Ω_m of ≤ 0.25.

2.2   Dynamics of the Local Group

Peebles (1994) estimated Ω_m by calculating the orbits of galaxies in the Local Group based on observed radial velocities, positions, and distances. Shaya *et al.* (1995) extended this method to a catalog of galaxies within 3000 km/sec. Again this is a method that is conceptually straightforward and independent of H_0 and Ω_Λ. Moreover, since the galaxies are nearby, the errors in the distances are relatively small. However, only one (the radial) component of the motion is measured. This method too is based on *the assumption that galaxies trace mass*. It also assumes that external tidal influences and past mergers are not significant. Furthermore, the question of uniqueness is difficult to address. The estimates based on this method again give low values of Ω_m of ~ 0.15.

2.3   Cluster Baryon Fraction

This issue was discussed in detail by White *et al.* (1993) for the Coma cluster, and has been addressed now in many contexts by a number of authors (*e.g.*, White & Frenk 1991; White & Fabian 1995; Steigman & Felten 1995). The calculation goes as follows: First, the number density of baryons (Ω_b) can be determined based on the observed densities of light elements from big-bang nucleosynthesis. Hence, the fraction of baryons (f_b) measured in clusters of galaxies can be used to estimate of the overall matter density assuming

$$f_b = \frac{M_{gas}}{M_{TOT}} = \frac{M_b}{M_{TOT}} = \frac{\Omega_b}{\Omega_m}$$

There are four explicit assumptions made:
1) The gas is in hydrostatic equilibrium.
2) There is a smooth potential.
3) Most of the baryons in the clusters are in the X-ray gas.
4) The cluster baryon fraction is representative of the Universe.

If the gas is clumped or there is another source of pressure (magnetic fields or turbulence) in addition to the thermal pressure, the baryon fraction would be decreased and the matter density would be increased (Steigman & Felten 1995).

Recent measurements of X-ray clusters (*e.g.*, Loewenstein & Mushotsky 1996; White & Fabian 1995) indicate that the baryon fraction has a range of values from about 10->20%. The values for f_b tend to be smaller for small groups and in the inner regions of larger clusters. These results underscore the importance of ensuring that such measurements are made on large enough scales to be truly representative of the large-scale Universe as a whole.

Taken at face value, the cluster-baryon method estimates again favor low values of Ω_m. For $\Omega_b h^2 = 0.024 \pm 12\%$ (Tytler, this conference) relatively low values of $\Omega_m < 0.5$ are favored for the range of baryon fractions observed. The Tytler *et al.* 1997 baryon determination is at the high end of recent measures of this quantity (low end of the deuterium abundance measurements); lower baryon densities only serve to decrease the Ω_m estimates. (However, see the discussion by Bothun, Impey and McGaugh 1997; these authors suggest that perhaps low-surface-brightness galaxies could be source of most of the baryons in the Universe and that rich clusters are not representative of the overall baryon density.)

2.4 Peculiar Velocities: Density and Velocity Comparisons

On scales of $\sim$100 Mpc, the motions of field galaxies can be used to infer the mass density given independent distance information. These methods do not yield a measure of Ω_m directly, but rather yield the ratio $\beta = \Omega^{0.6}/b$ where b is the bias parameter (describing the relation between mass and light) over a scale of a few hundred km/sec. These methods are again insensitive to both H_0 and Ω_Λ. Several different approaches have been investigated. For more details, the reader is referred to Dekel (1994), Willick et al. (1997) and Dekel, Burstein and White (1997).

All methods make use of radial velocity catalogs and distances based on the Tully-Fisher relation. The analyses differ in detail and there are advantages and disadvantages to each type of approach. At the present time, the results from this type of technique have not yet yielded a consistent picture. Earlier analyses (e.g. Dekel et al. 1993) suggested large values of $\beta \sim 1.3$, and correspondingly rather high values of Ω (subject to assumptions about the value of b). More recently, the estimates of β have decreased somewhat (Dekel, Burstein & White 1997). At present, the results from different groups (e.g., Dekel, Willick, Davis and collaborators) appear to differ from the results of Giovanelli, Haynes, Da Costa and collaborators (see the contribution by Da Costa to this volume). Understanding the sources of the differences is clearly an important goal.

2.5 Galaxy Pairwise Velocities

Using the cosmic virial theorem, the relative velocity dispersion of galaxy pairs can be used to estimate the matter density (e.g. Davis & Peebles 1983). The Las Campanas Redshift Survey (Shectman et al. 1996) contains about 26,000 redshifts out to $\sim$30,000 km/sec and provides an excellent sample of galaxies not dominated by clusters. Davis (this conference) presented results based on this sample, concluding that relative galaxy pairs have a one-dimensional velocity dispersion of only 260 km/sec, implying $\Omega_m \sim 0.25$.

This method is very clean and conceptually simple; however, it again is limited by the assumption that bias is independent of scale. Moreover, Frenk (1997) argues that bulk velocity flows are not sensitive to Ω_m, and that the peculiar velocities are quite similar for a number of models with a range of values of Ω_m.

3 Ω_Λ and Ω_m Limits

The subject of the cosmological constant Λ has had a long and checkered history in cosmology. The reasons for skepticism regarding a non-zero value of the cosmological constant are many. First, there is a discrepancy of ≥ 120 orders of magnitude between current observational limits and estimates of the vacuum energy density based on current standard particle theory (*e.g.* Carroll, Press and Turner 1992). Second, it would require that we are now living at a special epoch when the cosmological constant has begun to affect the dynamics of the Universe (other than during a time of inflation). In addition, it is difficult to ignore the fact that historically a non-zero Λ has been dragged out prematurely many times to explain a number of other apparent crises, and moreover, adding additional free parameters to a problem always makes it easier to fit data. Certainly the oft-repeated quote from Einstein to Gamov about his "biggest blunder" continues to undermine the credibility of a non-zero value for Λ.

However, despite the very persuasive arguments that can be made for $\Lambda = 0$, there are solid reasons to keep an open mind on the issue. First, at present there is no known physical principle that demands $\Lambda = 0$. Second, unlike the case of Einstein's original arbitrary constant term, standard particle theory and inflation now provide a physical interpretation of Λ: it is the energy density of the vacuum (*e.g.*, Weinberg 1989). Third, *if* theory demands $\Omega_{total} = 1$, then a number of observational results can be explained with a low Ω_m and $\Omega_m + \Omega_\Lambda = 1$: a) for instance, the observed large scale distribution of galaxies, clusters, large voids, and walls is in conflict with that predicted by the (standard) cold dark matter model for the origin of structure (*e.g.* Davis *et al.* 1992; Peacock & Dodds 1994); and b) the low values of the matter density based on a number of methods as described in §2. In addition, the discrepancy between the ages of the oldest stars and the expansion age can be resolved. Perhaps the most important reason to keep an open mind is that this is an issue that ultimately must be resolved by experiment.

The importance of empirically establishing whether there is a non-zero value of Λ cannot be overemphasized. However, it underscores the need for high-accuracy experiments: aspects of the standard model of particle theory have been tested in the laboratory to precisions unheard of in most measurements in observational cosmology. Nevertheless, cosmology offers an opportunity to test the standard model over larger scales and higher energies than can ever be achieved by other means. It scarcely needs to be said that overthrowing the Standard Model (i.e., claiming a measurement of a non-zero value for Λ) will require considerably higher accuracy than is currently available.

What are the current observational limits on Ω_Λ? In the next sections, limits based on both the observed numbers of quasars multiply imaged by galaxy "lenses" and limits from a sample of strongly lensed galaxies are briefly discussed.

3.1 Gravitational Lens Statistics

Fukugita, Futamase & Kasai (1990) and Turner (1990) suggested that a statistical study of the number density of gravitational lenses could provide a powerful test of a non-zero Λ. Subsequently a number of studies have been undertaken (e.g. Fukugita & Turner 1991; Bahcall et al. 1992; Maoz et al. 1993; Kochanek 1993, 1996). The basic idea behind this method is simple: the number of gravitationally lensed objects is a very sensitive function of Ω_Λ. For larger values of Ω_Λ, there is a greater probability that a quasar will be lensed because the volume over a given redshift interval is increased. In a flat universe with a value of $\Omega_\Lambda = 1$, approximately an order of magnitude more gravitational lenses are predicted than in a universe with $\Omega_\Lambda = 0$ (Turner 1990). Thus, simply counting the numbers of gravitationally lensed quasars can provide a very powerful limit on the value of Ω_Λ. In practice, however, there are a number of complications: galaxies evolve (and perhaps merge) with time, even elliptical galaxies contain dust, the properties of the lensing galaxies are not well-known (in particular, the dark matter velocity dispersion is unknown), and the numbers of lensing systems known at present is very small ($\sim$ 20). Moreover, while the predicted effects are very large for $\Omega_\Lambda = 1$, because the numbers are such a sensitive function of Ω_Λ, it is very difficult to provide limits below a value of about 0.6, given these complicating effects.

Kochanek (1996) has recently discussed these various effects in some detail, and investigated the sensitivity of the results to different lens models and extinction. His best estimated limits to date are : $\Omega_\Lambda < 0.66$ (95% confidence) for $\Omega_m + \Omega_\Lambda = 1$, and $\Omega_m = 0.15$ (90% confidence) if $\Omega_\Lambda = 0$. Significant improvements to these limits could be made by increasing the size of the current lens samples.

3.2 Strong Gravitational Lenses

A number of strong (elliptical galaxy) gravitational lens systems are known that may offer the potential of constraining the value of Ω_m and Ω_Λ through modeling of the lens properties. This method is less sensitive to Ω_Λ than the statistics of lensing, and again it is sensitive to a number of possible systematic effects: possible perturbations by cluster potentials, uncertainties in the underlying properties of the lensing galaxies, and model-dependent corrections due to evolution. The objects are faint and the errors in the luminosities and velocity dispersions are potentially very significant. A recent analysis of 7 strong lenses has been undertaken by Im *et al.* (1996). Their current results yield $\Omega_\Lambda = 0.64^{+0.15}_{-0.26}$ (*i.e.*, this measurement sits almost at the end of the range *excluded* by Kochanek (1996) at 95% confidence. Im *et al.* exclude $\Omega_m = 1.0$ at 97% confidence.

3.3 Ω_m and Ω_Λ from Type Ia Supernovae

The use of type Ia supernovae for measuring cosmological parameters is covered elsewhere in this volume by Filippenko (nearby supernovae and determinations of H_0) and by Perlmutter (distant supernovae and Ω_m and Ω_Λ). Hence, these objects will not be discussed in much detail here, except to highlight their potential, and to summarize some of the main difficulties associated with them so that they can be compared relative to some of the other methods discussed in this review.

The obvious advantage of type Ia supernovae is the small dispersion in the Hubble diagram, particularly after accounting for differences in the overall shapes or slopes of the light curves (Phillips 1993; Hamuy *et al.* 1995: Reiss, Press & Kirshner 1997). In principle, separation of the effects of deceleration or a potential non-zero cosmological constant is straightforward, provided that (eventually) supernovae at redshifts of order unity can be measured with sufficient signal-to-noise and resolution against the background of the parent galaxies. The differences in the observed effects of Ω_m and Ω_Λ become increasingly easier to measure at redshifts exceeding ~ 0.5. In principle, the evolution of single stars should be simpler than that of entire galaxies (that have been used for such measurements in the past).

At the present time, however, it is difficult to place any quantitative limits on the expected evolutionary effects for type Ia supernovae since the progenitors for these objects have not yet been unequivocally identified. Moreover, there may be potential differences in the chemical compositions of supernovae observed now and those observed at earlier epochs. In principle, such differences could be tested for empirically (as is being done for Cepheid variables, for example). It is also necessary to correct for obscuration due to dust (although in general, at least in the halos of galaxies, these effects are likely to be small; a minor worry might be that the properties of the dust could evolve over time). In detail, establishing accurate K-corrections for high-redshift supernovae, measuring reddenings, and correcting for potential evolutionary effects will be challenging, although, with the exception of measurements of the cosmic microwave background anisotropies (discussed in §9 below), type

Ia supernovae may offer the best potential for measuring Ω_m and Ω_Λ.

The most recent results based on type Ia supernovae (Perlmutter *et al.* 1997 are encouraging, and they demonstrate that rapid progress is likely to be made in the near future. Currently, the published sample size is limited to 7 objects; however, many more objects have now been discovered. The feasibility of discovering these high-redshift supernovae with high efficiency has unquestionably been demonstrated (*e.g.* Perlmutter, this volume). However, systematic errors are likely to be a significant component of the error budget in the early stages of this program.

4   Summary of Current Ω_m and Ω_Λ Measurements

The results of the preceding sections on Ω_m and Ω_Λ are summarized graphically in Figure 2. The diagonal dashed line denotes a flat ($\Omega_m + \Omega_\Lambda = 1$) Universe. Plotted are the results from dynamical measurements (rotation curves, Local Group dynamics, galaxy velocity dispersions, X-ray clusters) that tend to give low values of $\Omega \sim 0.2\text{-}0.3$. In addition, the preliminary results from the Perlmutter *et al.* (1996) type Ia supernova search are plotted with quoted 1σ error bars, along with the 95% limits ($\Omega_\Lambda < 0.66$) on Ω_m and Ω_Λ from gravitational lens statistics from Kochanek (1996), shown as an arrow along the diagonal.

What can be concluded about the value of Ω? Given the available evidence and the remaining uncertainties, plus underlying assumptions at the present time, in my own view the data are still consistent with both an open and a flat Universe. This undesirable situation is very likely to be resolved in the near future with more accurate mapping of the anisotropies in the cosmic microwave background radiation (see §9). At this point in time, however, I believe that it is premature either to sound the death knell for ("standard") inflationary theories or to conclude contrarily that an open Universe is not a viable option.

5 H_0 – The Hubble Constant

Sandage (1995) likens the measurement of H_0 to a game of chess. In chess, only a grand master " experiences a compelling sense of the issue and the best move. This player "knows" by intuition which clues are relevant... In other words his or her intuition judges what is real in the game, what will or will not lead to contradiction, and what aspects of the data to ignore."

Although there are perhaps differences in philosophy and many different techniques for measuring H_0, its importance cannot be underestimated. Knowledge of H_0 is required to constrain the estimates of the baryon density from nucleosynthesis at early epochs in the Universe. The larger the value of

200

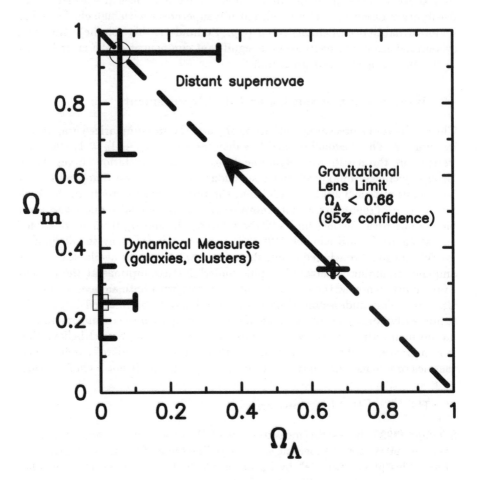

Figure 2: Summary of Omega Determinations. The dashed line corresponds to the case for a flat Universe: $(\Omega_m + \Omega_\Lambda = 1)$. See text for details.

H_0, the larger the component of non-baryonic dark matter is required, especially if the Universe has a critical density. The Hubble constant specifies both the time and length scales at the epoch of equality of the energy densities of matter and radiation. Both the scale at the horizon and the matter density determine the peak in the perturbation spectrum of the early universe. Hence, an accurate knowledge of the Hubble constant can provide powerful constraints on theories of the large-scale structure of galaxies. At present, large values of H_0 are problematic for the currently most successful models, those dominated by cold dark matter.

A value of H_0 to $\pm 1\%$ accuracy is still a goal far beyond currently available measurement techniques. However, if, for example a value of $H_0 = 70$ km/sec/Mpc were confirmed at $\pm 1\%$ (95% confidence), *and* the ages of the oldest objects in the Universe were confirmed to be >12 Gyr, then a number of issues would be brought into tight focus (and corresponding new problems raised!). A cosmological constant would be *required*, there would be no further debate over the need for non-baryonic dark matter, and at least the standard version of cold dark matter would be ruled out (conclusively).

The requirements for measuring an accurate value of H_0 are simple to list in principle, but extremely difficult to meet in practice. As discussed in more detail in Freedman (1997), in general, there are 4 criteria that need to be met for any method. First, the method should be based upon well-understood physics; second, it should operate well into the smooth Hubble flow (velocity-distances greater than 10,000, and preferably, 20,000 km/sec); third, the method should be based on a statistically significant sample of objects, empirically established to have high internal accuracy; and finally, the method needs to be demonstrated empirically to be free of systematic errors. This list of criteria applies both to classical distance indicators as well as to other physical methods (in the latter case, for example, the Sunyaev Zel'dovich effect or gravitational lenses). The last point requires that several distance indicators meeting the first three criteria be available, but the current reality is that, unfortunately, at the present time, an ideal distance indicator or other method meeting all of the above criteria does not exist. The measurement of H_0 to $\pm 1\%$ is not yet possible; however, recent progress (reviewed below) illustrates that a measurement to $\pm 10\%$ is now feasible.

5.1 *"Physical" versus "Astronomical" Methods*

There is a common (mis)perception that some methods for determining H_0 based on simple physical principles are free from the types of systematics that often affect distance indicators ("physical" versus "astronomical" methods).

However, the fact remains that aside from nearby geometric parallax measurements (d< 100 pc), *astrophysics enters all distance and H_0 determinations!* These methods include the gravitational lens time delay method, the Sunyaev Zel'dovich methods for clusters of galaxies, and theoretical modeling of type Ia and II supernovae.

For example, it is certainly true that the gravitational lensing method is premised on very solid physical principles (*e.g.* Refsdael 1964,1966; Blandford & Narayan 1992). Unfortunately, the astronomical lenses are not idealized systems with well-defined properties that can be measured in a laboratory; they are galaxies whose underlying (luminous or dark) mass distributions are not independently known, and furthermore they may be sitting in more complicated group or cluster potentials. A degeneracy exists between the mass distribution of the lens and the value of H_0 (*e.g.*, Kundić *et al.* 1997; Keeton and Kochanek 1997; Schechter *et al.* 1997. This is not a method based solely on well-known physics; it is a method that also requires knowledge of astrophysics. Ideally velocity dispersion measurements as a function of position are needed (to constrain the mass distribution of the lens). Such measurements are very difficult (and generally have not been available). Perhaps worse yet, the distribution of the dark matter in these systems is unknown. In a similar way, the Sunyaev-Zel'dovich method is sensitive to the clumping of X-ray gas, discrete radio sources, the projection of the clusters, and other astrophysical complications.

Hence the methods for measuring H_0 cannot be cleanly separated into purely "physical" and "astronomical" techniques. Rather, each method has its own set of advantages and disadvantages. In my view, it is vital to measure H_0 using a variety of different methods in order to identify potential systematic errors in any one technique. All methods require large, statistically significant samples. This is one of the current weakest aspects of the Sunyaev-Zel-dovich and gravitational-lens methods, for example, where samples of a few or only 2 objects, respectively, are currently available. In contrast, it is a clear disadvantage that many of the classical distance indicators (*e.g.*, the Tully-Fisher relation and at present, even the type Ia supernovae) do not have a well-understood physical basis. However, there are many cross-checks and tests for potential systematic effects that are now feasible and are being carried out for large samples of measured extragalactic distances (see §5.4 below). Assuming that systematic effects can eventually be understood and minimized, ultimately, the measurement of H_0 by a geometrical (or optical) technique at large distances will be crucial for establishing the reliability of the classical distance scale. For gravitational lenses, however, a considerable amount of work will be required to increase the numbers of systems with measured time delays, obtain velocity dispersion profiles for the faint lensing galaxies, constrain the lens models and test for other systematic effects, if this goal is to be reached.

Below, progress on H_0 measurements based on gravitational lenses, the Sunyaev Zel'dovich effect, and the extragalactic distance scale is briefly summarized.

5.2 Gravitational Lenses

Refsdael (1964, 1966) noted that the arrival times for the light from two gravitationally lensed images of a background point source are dependent on the path lengths and the gravitational potential traversed in each case. Hence, a measurement of the time delay and the angular separation for different images of a variable quasar can be used to provide a measurement of H_0. This method offers tremendous potential because it can be applied at great distances and it is based on very solid physical principles. Moveover, the method is not very sensitive to Ω_m and Ω_Λ. Some of the practical difficulties in applying this method have already been discussed in the previous section.

A number of new results based on this technique have recently appeared. Estimates of time delay measurements are now available for 2 systems: 0957 +561 (Kundić et al. 1997), and most recently, a new time delay has been measured for PG 1115 (Schechter et al. 1997; Keeton and Kochanek 1997).

In the case of 0957+561, progress has been made on several fronts. The time delay for this system has been a matter of some debate in the literature, with two different values of 410 and 536 days being advocated; extensive new optical data have now resolved this issue in favor of the smaller time delay (Δt=417±3 days (Kundić et al. 1997). Another large observational uncertainty has been due to the difficulty of measuring an accurate velocity dispersion for the lensing galaxy. Recent data from the Keck telescope have provided a new measurement of the velocity dispersion (Falco et al. 1997). In addition, there has been substantial progress in modeling this system (Grogin & Narayan 1996). Based on the new time delay and velocity dispersions measurements, and the model of Grogin and Narayan, Falco et al. have recently derived a value of $H_0 = $ in the range 62 - 67 ± 8 km/sec/Mpc for this system. The velocity dispersion in the lensing galaxy appears to decrease very steeply as a function of position from the center of the galaxy; further higher-resolution measurements will be required to determine the reliability of these faint measurements.

Schechter et al. 1997 have undertaken an extensive optical monitoring program to measure two independent time delays in the quadruply-imaged quasar PG 1115+080. They fit a variety of models to this system, preferring a solution that yields a value of $H_0 = 42$ km/sec/Mpc $\pm 14\%$ (for $\Omega = 1$). The model in this case consists of fitting isothermal spheres to both the lensing galaxy and a nearby group of galaxies. They also considered additional models that yield values of $H_0 = 64$ and 84 km/sec/Mpc. Keeton & Kochanek (1997) have considered a wider class of models. They stress the degeneracies that are inherent in these analyses; a number of models with differing radial profiles for the lensing galaxy and group, and with differing positions for the group, yield fits with chi-squared per degrees of freedom less than 1. They conclude that $H_0 = 60 \pm 17$ km/sec/Mpc (1-σ).

5.3   Sunyaev Zel'dovich Effect and X-Ray Measurements

The inverse-Compton scattering of photons from the cosmic microwave background off of hot electrons in the X-ray gas of rich clusters results in a measurable decrement in the microwave background spectrum known as the Sunyaev-Zel'dovich (SZ) effect (Sunyaev and Zel'dovich 1969). Given a spatial (preferably 2-dimensional) distribution of the SZ effect and a high-resolution X-ray map, the density and temperature distributions of the hot gas can be obtained; the mean electron temperature can be obtained from an X-ray spectrum. An estimate of H_0 can be made based on the definitions of the angular-diameter and luminosity distances. The method makes use of the fact that the X-ray flux is distance-dependent, whereas the Sunyaev-Zel'dovich decrement in the temperature is not.

Once again, the advantages of this method are that it can be applied at large distances and, in principle, it has a straightforward physical basis. As discussed in §5.1, some of the main uncertainties with this method are due to potential clumpiness of the gas (which would result in reducing H_0), projection effects (if the clusters observed are prolate, H_0 could be larger), the assumption of hydrostatic equilibrium, details of the models for the gas and electron densities, and potential contamination from point sources.

To date, a range of values of H_0 have been published based on this method ranging from ~25 - 80 km/sec/Mpc (*e.g.*, McHardy *et al.* 1990; Birkinshaw & Hughes 1994; Rephaeli 1995; Herbig, Lawrence & Readhead 1995). The uncertainties are still large, but as more and more clusters are observed, higher-resolution (2D) maps of the decrement, and X-ray maps and spectra become available, the prospects for this method will continue to improve. At this conference, Carlstrom reported on a new extensive survey of lenses being undertaken both at Hat Creek and the Owens Valley Radio Observatory. X-ray images are being obtained with ROSAT and X-ray spectra with ASCA.

5.4 The Cepheid-Calibrated Extragalactic Distance Scale

Establishing accurate extragalactic distances has provided an immense challenge to astronomers since the 1920's. The situation has improved dramatically as better (linear) detectors have become available, and as several new, promising techniques have been developed. For the first time in the history of this difficult field, relative distances to galaxies are being compared on a case-by-case basis, and their quantitative agreement is being established. Several, detailed reviews on this progress have been written (see, for example, the conference proceedings for the Space Telescope Science Institute meeting on the Extragalactic Distance Scale edited by Donahue and Livio 1997).

The Hubble Space Telescope (HST) Key Project on H_0 has been designed to undertake the calibration of a number of secondary distance methods using Cepheid variables (Freedman *et al.* 1994; Kennicutt, Freedman & Mould 1995; Mould *et al.* 1995). Briefly, there are three primary goals: (1) To discover Cepheids, and thereby measure accurate distances to spiral galaxies suitable for the calibration of several independent secondary methods. (2) To make direct Cepheid measurements of distances to three spiral galaxies in each of the Virgo and Fornax clusters. (3) To provide a check on potential systematic errors both in the Cepheid distance scale and the secondary methods. The final goal is to derive a value for the the Hubble constant, to an accuracy of 10%. Cepheids are also being employed in several other HST distance scale programs (*e.g.*, Sandage *et al.* 1996; Saha *et al.* 1994, 1995, 1996; and Tanvir *et al.* 1995).

In Freedman, Madore & Kennicutt (1997), a comparison of Cepheid distances is made with a number of other methods including surface-brightness fluctuations, the planetary nebula luminosity function, tip of the red giant branch, and type II supernovae. (Extensive recent reviews of all of these methods can be found in Livio and Donahue (1997); by Tonry; Jacoby; Madore, Freedman & Sakai; Kirshner). In general, there is excellent agreement amongst these methods; the relative distances agree to within ±10% (1-sigma). The use of both type Ia and type II supernovae for the purposes of determining H_0 are described in this volume by Filippenko.

The results of the H_0 Key Project have been summarized recently by Freedman, Madore & Kennicutt (1997); Mould *et al.* (1997); and Freedman (1997). For somewhat different views, see Sandage & Tammann (1997). The remarks in the rest of this section follow Freedman (1997). At this mid-term point in the HST Key Project, our results yield a value of $H_0 = 73 \pm 6$ (statistical) ± 8 (systematic) km/sec/Mpc. This result is based on a variety of methods, including a Cepheid calibration of the Tully-Fisher relation, type Ia

Table 2: SUMMARY OF KEY PROJECT RESULTS ON H_0

Method	H_0
Virgo	80 ± 17
Coma via Virgo	77 ± 16
Fornax	72 ± 18
Local	75 ± 8
JT clusters	72 ± 8
SNIa	67 ± 8
TF	73 ± 7
SNII	73 ± 7
$D_N - \sigma$	73 ± 6
Mean	73 ± 4

Systematic Errors	$\pm\,4$	$\pm\,4$	$\pm\,5$	$\pm\,2$
	(LMC)	([Fe/H])	(global)	(photometric)

Table 3: Current values of H_0 for various methods. For each method, the formal statistical uncertainties are given. The systematic errors (common to all of these Cepheid-based calibrations) are listed at the end of the table. The dominant uncertainties are in the distance to the LMC and the potential effect of metallicity on the Cepheid period-luminosity relations, plus an allowance is made for the possibility that the locally measured value of H_0 may differ from the global value. Also allowance is made for a systematic scale error in the photometry which might be affecting all software packages now commonly in use. Our best current weighted mean value is $H_0 = 73 \pm 6$ (statistical) ± 8 (systematic) km/sec/Mpc.

supernovae, a calibration of distant clusters tied to Fornax, and direct Cepheid distances out to ~ 20 Mpc. In Table 2 the values of H_0 based on these various methods are summarized.

These recent results on the extragalactic distance scale are very encouraging. A large number of independent secondary methods (including the most recent type Ia supernova calibration by Sandage *et al.* 1996) appear to be converging on a value of H_0 in the range of 60 to 80 km/sec/Mpc. The long-standing factor-of-two discrepancy in H_0 appears to be behind us. However, these results underscore the importance of reducing remaining errors in the Cepheid distances (*e.g.*, those due to reddening and metallicity corrections), since at present the majority of distance estimators are tied in zero point to the Cepheid distance scale. A 1-σ error of $\pm 10\%$ on H_0 (the aim of the Key Project) currently amounts to approximately $\pm$ 7 km/sec/Mpc, and translates into a 95% confidence interval on H_0 of roughly 55 to 85 km/sec/Mpc.

While this is an enormous improvement over the factor-of-two disagreement of the previous decades, it is not sufficiently precise, for example, to discriminate between current models of large scale structure formation, to resolve definitively the fundamental age problem, or to settle the question of a non-zero value of Λ. Before compelling constraints can be made on cosmological models, it is imperative to rule out remaining sources of systematic error in order to severely limit the alternative interpretations that can be made of the data. The spectacular success of HST, and the fact that a value of H_0 accurate to 10% (1-σ) now appears quite feasible, also brings into sharper focus smaller (10-15%) effects which were buried in the noise during the era of factor-of-two discrepancies. Fortunately, a significant improvement will be possible with the new infrared capability afforded by the recently augmented near-infrared capabilities of HST (the NICMOS instrument). Planned NICMOS observations will reduce the remaining uncertainties due to both reddening and metallicity by a factor of 3.

6 t_0 - Ages of the Oldest Stars

The ages of stars can be derived quite independently from the expansion age of the Universe (obtained by integrating the Friedmann equation), and have long been used as a point of comparison and constraint on cosmology; for example, globular cluster age-dating, nucleocosmochronology, and white-dwarf cooling estimates for the Galactic disk. The reader is referred to earlier reviews on these topics by Renzini (1991), Schramm (1989). For the purposes of this review, I briefly consider only two types of age determinations: those based on Galactic globular clusters, and a new estimate of the age based on a measurement of radioactive thorium in a metal poor Galactic halo star.

6.1   Globular Cluster Ages

There are also many excellent recent reviews covering in great detail the ages obtained for Galactic globular clusters (*i.e.,* from a comparison of observed color magnitude diagrams and theoretical evolution models). At the moment, there is a fairly broad consensus that Galactic globular clusters are most likely at least 14-15 Gyr old (*e.g.* Chaboyer *et al.* 1996; VandenBerg *et al.* 1996; Shi 1995).

It is not widely appreciated that *the largest uncertainty in the globular-cluster ages results from uncertainties in the distances to the globular clusters*, which currently are based on statistical parallax measurements of Galactic RR Lyrae stars or on parallaxes for nearby subdwarfs(*e.g.* Renzini, 1991; Chaboyer *et al.* 1996; VandenBerg *et al.* 1996). Although the ages of globular clusters are widely regarded as theoretically-determined quantities, in the process of determining ages, it is still necessary to interface theory with observation and transform the observed globular cluster magnitudes to bolometric luminosities (via an accurate distance scale). The subdwarf and RR Lyrae statistical parallax distance calibrations currently differ by about ~0.25-0.30 mag. Unfortunately, as emphasized by Renzini, small errors in distance modulus (0.25 mag or 13% in distance) correspond to 25% differences in age. Even with improved parallax measurements (for example, soon to be available from HIPPARCHOS), there are many subtle issues (*e.g.*, reddening, metallicity, photometric zeropoints) that combine to make it a very difficult problem to achieve distances to better than 5% accuracy.

As discussed previously in many contexts (*e.g.* Walker 1992; Freedman & Madore 1993; van den Bergh 1995, and most recently by Feast & Catchpole 1997), there is also currently a discrepancy in the Cepheid and RR Lyrae distances to nearby galaxies. If the Cepheid distances are correct, it would imply that the absolute magnitudes of RR Lyraes are brighter (by about 0.3 mag) than suggested by statistical parallax and Baade-Wesselink calibrations for Galactic RR Lyraes (*e.g.* see VandenBerg, Bolte & Stetson 1996 for a recent discussion). This brighter RR Lyrae calibration agrees well in zero point with that from Galactic subdwarfs. Based on the models of VandenBerg *et al.* 1997, applying this calibration (adopting $M_V(RR)=0.40$ mag) to the metal-poor globular cluster M92, results in an age of 15.8 ± 2 Gyr. If the fainter RR Lyrae distance scale is correct, the age derived for M92 based on these same recent models increases to ~19 Gyr. Alternatively, if the Feast & Catchpole calibration of Galactic Cepheids based on HIPPARCHOS parallaxes is correct, then the resulting RR Lyrae calibration is even brighter $(M_V(RR)=0.25$ at $[Fe/H] = -1.9)$, and the corresponding age for M92 would be reduced to about 13 Gyr (based on the same Vandenberg models). A new calibration of Galactic metal-poor subdwarfs, also based on new HIPPARCHOS parallaxes, appears to confirm these younger ages (Reid, private communication). It is interesting to note that while the distances to nearby galaxies have converged to a level where they no longer have a factor-of-two impact on the Hubble constant, subtle differences of only a few tenths of a magnitude in distance modulus can still have very significant impact on cosmology, through the ages determined from stellar evolution.

6.2 Thorium Ages

A new measurement of the age of a very metal poor star in the halo of our Galaxy has recently been made by Cowan *et al.* (1997), following a technique introduced by Butcher (1987). These authors make use of very high-resolution echelle spectra of CS22892-052, a star with a metallicity of only $[Fe/H] = -3.1$. They find that the observed abundances for stable elements in this star match the observed r-process elemental abundances observed in the Sun. However, for the radioactive element thorium, the abundance is down by a factor of 40 relative to solar. Allowing for the radioactive decay of thorium relative to (stable) europium yields a minimu age for this star of 15.2 ± 3.7 Gyr (1-sigma). If instead of europium alone, an average abundance for all r-process elements from Eu-Er is used, an age of 13.8 ± 3.7 Gyr results. This lower limit to the age is independent of any model of Galactic evolution (which only serve to increase the total age estimates for the Universe). It depends on both the decay rate and the initial abundance of thorium. Although the current sample is small (1 star!) and the uncertainties are correspondingly large, there is excellent promise for the future once the sample is enlarged. Methods like this one are particularly important because of the opportunity of having high-quality ages completely independent of the globular cluster age scale.

7 Remaining Issues for Measuring t_0

What are the ages of the oldest objects in the Universe? In this context, we need to keep in mind that it is currently only a useful working hypothesis that the Galactic globular clusters are representative of the oldest objects in the Universe (*e.g.* see Freedman 1995 for a more detailed discussion). Currently, the sample of objects for which direct (*i.e.* main-sequence-fitting) ages can be measured is limited to our own Galaxy and a small number of satellites around our own Galaxy. It is at least conceivable that in denser environments in the early Universe, star formation could have proceeded earlier than for Galactic globular clusters. At this time, there is no direct information with which to constrain the true dispersion in (or upper limit to) ages in environments outside the nearest galaxies in our own Local Group. There are, for example, no giant elliptical galaxies in the Local Group. Although considerable effort is now being invested in finding potential ways to lower the Galactic globular cluster ages, there is reason to keep in mind that the expansion-age discrepancy could potentially be even worse than is currently being discussed.

8 $H_0 t_0$

One of the most powerful tests for a non-zero cosmological constant is provided by a comparison of the expansion and oldest-star ages. To quote Carroll, Press and Turner (1990), "A high value of H_0 (>80 km/s/Mpc, say), combined with no loss of confidence in a value 12-14 Gyr as a *minimum* age for some globular clusters, would effectively prove the existence of a significant Ω_Λ term. Given such observational results, we know of no convincing alternative hypotheses."

In Figure 3, the dimensionless product of $H_0 t_0$ is plotted as a function of Ω. Two different cases are illustrated: an open $\Omega_\Lambda = 0$ Universe, and a flat Universe with $\Omega_\Lambda + \Omega_m = 1$. Suppose that both H_0 and t_0 are both known to $\pm 10\%$ (1-σ, *including systematic errors*). The dashed and dot-dashed lines indicate 1-σ and 2-σ limits, respectively for values of $H_0 = 70$ km/sec/Mpc and $t_0 = 15$ Gyr. Since the two quantities H_0 and t_0 are completely independent, the two errors have been added in quadrature, yielding a total uncertainty on the product of $H_0 t_0$ of $\pm 14\%$ *rms*. These values of H_0 and t_0 are consistent with a Universe where $\Omega_\Lambda = 0.8$, $\Omega_m = 0.2$. The Einstein-de Sitter model ($\Omega_m = 1$, $\Omega_\Lambda = 0$) is excluded (at 2.5σ).

Despite the enormous progress recently in the measurements of H_0 and t_0, Figure 3 demonstrates that significant further improvements are still needed. First, in the opinion of this author, *total* (including both statistical and systematic) uncertainties of $\pm 10\%$ have yet to be achieved for either H_0 or t_0. Second, assuming that such accuracies will be forthcoming in the near future for H_0 (as the Key Project, supernova programs and other surveys near completion), and for t_0 (as HIPPARCHOS provides an improved calibration both for RR Lyraes and subdwarfs), it is clear from this figure that if H_0 is as high as 70 km/sec/Mpc, then accuracies of significantly *better than* $\pm$ 10% will be required to rule in or out a non-zero value for Λ. (If H_0 were larger (or smaller), this discrimination would be simplified!)

9   Cosmological Parameters from Cosmic Microwave Background Anisotropies

One of the most exciting future developments with respect to the accurate measurement of cosmological parameters will be the opportunity to measure anisotropies in the cosmic microwave background to high precision. Planned balloon-born experiments (*e.g.*, MAX, MAXIMA, and Boomerang) will shortly measure the position of the first acoustic peak in the cosmic background anisotropy spectrum. Even more promising are future satellite experiments (*e.g.*, MAP to be launched by NASA in 2000, and the European COBRAS/SAMBA

H_0 and t_0 Measurements to $\pm 10\%$

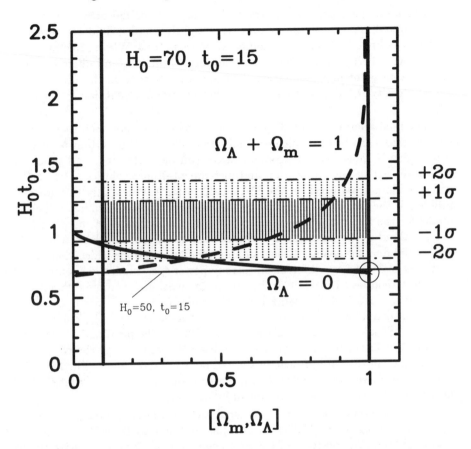

Figure 3: The product of $H_0 t_0$ as a function of Ω. The dashed curve indicates the case of a flat Universe with $\Omega_\Lambda + \Omega_m = 1$. The abscissa in this case corresponds to Ω_Λ. The solid curve represents a Universe with $\Omega_\Lambda = 0$. In this case, the abcissa should be read as Ω_m. The dashed and dot-dashed lines indicate 1-σ and 2-σ limits, respectively for values of H_0 = 70 km/sec/Mpc and t_0 = 15 Gyr in the case where both quantities are known to $\pm 10\%$ (1-σ). The large open circle denotes values of $H_0 t_0$ = 2/3 and $\Omega_m = 1$ (*i.e.*, those predicted by the standard Einstein-de Sitter model). Also shown for comparison is a solid line for the case H_0 = 50 km/sec/Mpc, t_0 = 15 Gyr.

mission, now renamed the PLANCK Surveyor mission, currently planned to be launched in 2005).

The underlying physics governing the shape of the anisotropy spectrum is that describing the interaction of a very tightly coupled fluid composed of electrons and photons before (re)combination (*e.g.*, Hu & White 1996; Sunyaev & Zel'dovich 1970). It is elegant, very simple in principle, and offers extraordinary promise for measuring cosmological parameters; (*e.g.*, H_0, Ω_0, and the baryon density Ω_b to precisions of 1% or better: Bond, Efstathiou & Tegmark 1997).

The final accuracies will of course (again) depend on how well various systematic errors can be controlled or eliminated. The major uncertainties will be determined by how well foreground sources can be subtracted, and probably to a lesser extent, by calibration and instrumental uncertainties. (PLANCK will provide a cross check of the MAP calibration.) Potentially the greatest problem is the fact that extracting cosmological parameters requires a specific model for the fluctuation spectrum. Currently the estimates of the precisions (*i.e.*, without systematic effects included) are based on models in which the primordial fluctuations are Gaussian and adiabatic, and for which there is no preferred scale. A very different anisotropy power spectrum shape is predicted for defect theories (Turok 1996), but these calculations are more difficult and have not yet reached the same level of predictive power. Important additional constraints will come from polarization measurements *e.g.*, Zaldarriaga, Spergel & Seljak 1997; Kamionkowski *et al.* 1997). The polarization data will provide a means of breaking some of the degeneracies amongst the cosmological parameters that are present in the temperature data alone. Furthermore, they are sensitive to the presence of a tensor (gravity wave) contribution, and hence will allow a very sensitive test of inflationary models.

Figure 4 shows a plot of the predicted angular power spectrum for cosmic microwave background (CMB) anisotropies reproduced from Hu, Sugiyama, & Silk (1997). The position of the first acoustic peak is very sensitive to the value of Ω_0, and, as noted by these authors, the spacing between the acoustic peaks in the power spectrum appears to provide a fairly robust measure of Ω_0. The accurate determination of other cosmological parameters will require the measurement of peaks at smaller (arcminute) angular scales. In general, the ratio of the first to the third peaks is sensitive to the value of value of H_0 (*e.g.*, Hu & White 1996). Excellent sky coverage is critical to these efforts in order to reduce the sampling variance.

Can the cosmological parameters be measured to precisions of $\leq 1\%$ with currently planned experiments as advertised above? I believe that both MAP and PLANCK are likely to revolutionize our understanding of cosmology. Ob-

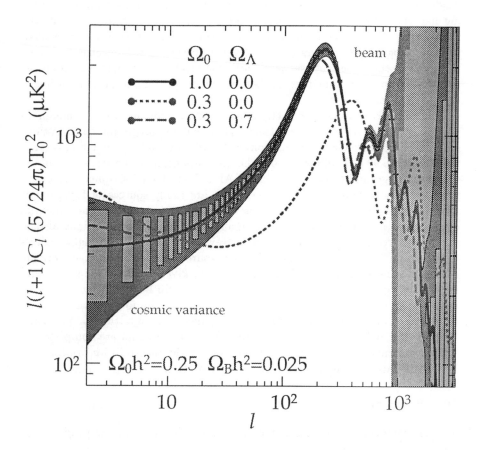

Figure 4: The angular power spectrum of cosmic microwave background anisotropies assuming adiabatic, nearly scale-invariant models for a range of values of Ω_0 and Ω_Λ (Hu, Sugiyama, and Silk 1997; their Figure 4). The C_l values correspond to the squares of the spherical harmonics coefficients. Low l values correspond to large angular scales ($l \sim \frac{200^\circ}{\theta}$). The position of the first acoustic peak is predicted to be at $l \sim 220\Omega_{TOT}^{-1/2}$, and hence, shifts to smaller angular scales for open universes.

servation of a Gaussian, adiabatic fluctuation spectrum would be a stunning confirmation of the "standard" cosmology. However, equally fundamental would be the case where the observed anisotropy spectrum resembles nothing like those for any of the various current theoretical predictions. In the former case, *if* foreground effects can be accounted for, then measurement of the cosmological parameters to these levels of precision will eventually follow. However, in the latter case, at least until the origin of the spectrum could be predicted from first principles, all bets would be off for the determination of cosmological parameters.

Can the foreground subtraction be accounted for accurately enough to yield final accuracies of 1% (or better)? There will be foreground contributions due to faint, diffuse Galactic emission. MAP will have 5 frequency bands ranging from 22 to 90 GHz allowing both the spectral and spatial distribution of the Galactic foreground to be measured. PLANCK will have 9 frequency channels from 30 GHz to 900 GHz. However, there are many sources of foregrounds whose subtraction is critical; perhaps the greatest unknown is the potential contribution from GHz radio sources, many of which could potentially also be variable sources. Deep 90 GHz radio surveys from the ground might address the question of how serious an issue such sources could be (Spergel, private communication). Although MAP will cover any given region of the sky several times, the signal-to-noise for an individual image will be insufficient to detect any but the brightest sources. In addition there will be foreground contributions due to diffuse emission from external galaxies, dust within galaxies, and bright infrared luminous galaxies. Until these experiments are completed, it will be difficult to assess whether these systematic uncertainties are likely to be small relative to the quoted formal uncertainties.

10 Summary

The current best measurements for the cosmological parameters yield:

$$\Omega_m \sim (0.2 - 0.4) \pm 0.1 \qquad (1\text{-}\sigma)$$
$$H_0 \sim (67 - 73) \pm 7 \text{ km/sec/Mpc} \qquad (1\text{-}\sigma)$$
$$t_0 \sim (14 - 15) \pm 2 \text{ Gyr} \qquad (1\text{-}\sigma)$$
$$\Omega_\Lambda < 0.7 \qquad (2\text{-}\sigma)$$

The low value for Ω_m and relatively high value for $H_0 t_0$ do not favor the standard Einstein-de Sitter ($\Omega_m = 1$, $\Omega_\Lambda = 0$) Universe; however, this model cannot be ruled out at high statistical significance. Moreover, systematic errors are still a source of serious concern. If the new HIPPARCHOS calibrations are confirmed, the ages of globular clusters may be as low as 10-12 Gyr. Rapid progress is expected in addressing these systematic effects; in particular new data from

HST, HIPPARCHOS, and MAP/PLANCK offer the enticing possibility that all of the cosmological parameters may soon be measured to unprecedented accuracies of ±1-5% within a decade. Let us hope that unexpected systematic errors will not continue to lurk (as they have done historically so many times before) in these future efforts to define the basic cosmological parameters.

Acknowledgments

It is a pleasure to thank the organizing committee for an extremely enjoyable and interesting conference, and for the opportunity to speak. The work presented on the Cepheid-based extragalactic distance scale (§5.4) has been done in collaboration with the Hubble Space Telescope Key Project team on the Extragalactic Distance Scale and I would like to acknowledge the contributions of R. Kennicutt, J.R. Mould (co-PI's), S. Faber, L. Ferrarese, H. Ford, B. Gibson, J. Graham, J. Gunn, M. Han, J. Hoessel, J. Huchra, S. Hughes, G. Illingworth, B.F. Madore, R. Phelps, A. Saha, S. Sakai, N. Silbermann, and P. Stetson, and graduate students F. Bresolin, P. Harding, D. Kelson, L. Macri, D. Rawson, and A. Turner. This work is based on observations with the NASA/ESA Hubble Space Telescope, obtained by the Space Telescope Science Institute, which is operated by AURA, Inc. under NASA contract No. 5-26555. Support for this work was provided by NASA through grant GO-2227-87A from STScI.

1. J. A. Bahcall, *et al.*, *Astrophys. J.* **387**, 56 (1992).
2. N. A. Bahcall, L. M. Lubin, and V. Dorman *Astrophys. J. Lett.* **447**, 81 (1995).
3. J. Bartlett, A. Blanchard, J. Silk, and M. S. Turner *Science,* **267**, 980 (1995).
4. M. Birkinshaw and J. P. Hughes *Astrophys. J.* **420**, 33 (1994).
5. R. Blandford and R. Narayan, *Astron. Rev. Astron. Astrophys.* **30**, 311 (1992).
6. J. R. Bond, G. Efstathiou, and M. Tegmark, *Mon. Not. Royal Astr. Soc.* **000**, 1997 (,) astro-ph/9702100, in press.
7. G. Bothun, C. Impey, and S. McGaugh, *Publ. Astr. Soc. Pac.* **000**, 000 (1997), astro-ph/9702156.
8. D. Tytler, S. Burles & D. Kirkman, *Science,* **000**, 000 (1997), astro-ph/9612121.
9. H. R. Butcher *Nature,* **328**, 127 (1987).
10. R. G. Carlberg, *et al.*, *Astrophys. J. Lett.* **000**, 000 (1997), astro-ph/9611204.
11. Carroll, Press, & Turner, *Astron. Rev. Astron. Astrophys.* **30**, 499 (1992).

216

12. B. Chaboyer, P. Demarque, P. J. Kernan and L. M. Krauss, *Science*, **271**, 957 (1996).
13. J. J. Cowan, A. McWilliam, C. Sneden, and D. L. Burris, *Astrophys. J.* **000**, 000 (1997), in press.
14. M. Davis and P. J. E. Peebles *Astrophys. J.* **267**, 465 (1983).
15. M. Davis, G. Efstathiou, C. S. Frenk, and S. D. M. White *Nature*, **356**, 489 (1992).
16. A. Dekel, *Astron. Rev. Astron. Astrophys.* **32**, 371 (1994)
17. A. Dekel *et al.*, *Astrophys. J.* **412**, 1 (1993)
18. A. Dekel, D. Burstein and S. D. M. White in *Critical Dialogs in Cosmology*, ed. N. Turok (World Scientific, 1997).
19. M. Donahue and M. Livio, eds. *The Extragalactic Distance Scale* (Cambridge University Press, 1997).
20. E. E. Falco, I. I. Shapiro, L. A. Moustakas, and M. Davis*Astrophys. J.*0000001997, astro-ph/9702152.
21. M. W. Feast and R. M. Catchpole, *Mon. Not. Royal Astr. Soc.* **000**, 000 (1997), press.
22. W. L. Freedman *et al.*, *Nature*, **371**, 757 (1994).
23. W. L. Freedman in *The Local Group: Comparative and Global Properties* eds. A. Layden, R. Smith, & J. Storm, (European Southern Observatory: Garching, 1995).
24. W. L. Freedman in *Critical Dialogs in Cosmology*, ed. N. Turok (World Scientific, 1997).
25. W. L. Freedman in *New Perspectives on Stellar Pulsation and Pulsating Variable Stars*, eds. J. M. Nemec and J. M. Matthews (1993).
26. W. L. Freedman, B. F. Madore & R. C. Kennicutt, in *The Extragalactic Distance Scale* eds. M. Donahue & M. Livio (Cambridge University Press, 1997).
27. C. Frenk, Aspen Winter Conference on Cosmology, 1997.
28. M. Fukugita, T. Tutamase and M. Kasai, *Mon. Not. Royal Astr. Soc.* **246**, 24p (1990).
29. M. Fukugita and E. Turner, *Mon. Not. Royal Astr. Soc.* **253**, 99 (1991).
30. N. A. Grogin and R. Narayan, *Astrophys. J.* **464**, 92 (1996).
31. M. Hamuy *et al.*, *Astrophys. J.* **109**, 1 (1995).
32. T. Herbig, C. R. Lawrence and A. C. S. Readhead *Astrophys. J. Lett.* **449**, 5 (1995)
33. W. Hu and M. White *Astrophys. J.* **441**, 30 (1996)
34. W. Hu, N. Sugiyama, and J. Silk *Nature*, **000**, 000 (1997), in press.
35. M. Im, R. E. Griffiths, and K. E. Ratnatunga *et al.*, *Astrophys. J.* **000**, 000 (1997), astro-ph 9611105.

217

36. M. Kamionkowski, A. Kosowskyi, and A. Stebbins, astro-ph/9609132.
37. C. R. Keeton and C. S. Kochanek, 1997, in preparation.
38. R. C. Kennicutt, W. L. Freedman & J. R. Mould, *Astron. J.* **110**, 1476 (1995).
39. C. S. Kochanek, *Mon. Not. Royal Astr. Soc.* **261**, 453 (1993).
40. C. S. Kochanek, *Astrophys. J.* **466**, 638 (1996).
41. L. Krauss and M. S. Turner, *Gen. Rel. Grav.*2711371995.
42. Kundić *et al.*, 1997, preprint.
43. M. Loewenstein and R. F. Mushotsky, *Astrophys. J. Lett.* **471**, 83 (1996).
44. I. M. McHardy *et al.*, *Mon. Not. Royal Astr. Soc.* **242**, 215 (1990).
45. D. Maoz *et al.*, *Astrophys. J.* **409**, 28 (1993).
46. J. R. Mould *et al.*, *Astrophys. J.* **449**, 413 (1995).
47. J. R. Mould *et al.*, in *The Extragalactic Distance Scale* eds. M. Donahue & M. Livio (Cambridge University Press, 1997).
48. J. P. Ostriker and P. Steinhardt, *Nature,* **377**, 600 (1995).
49. J. A. Peacock and S. J. Dodds, *Mon. Not. Royal Astr. Soc.* **267**, 1020 (1994).
50. P. J. E. Peebles, *Astrophys. J.* **429**, 43 (1994).
51. S. Perlmutter *et al.*, *Astrophys. J.* **000**, 000 (1997), in press.
52. M. Phillips, *Astrophys. J. Lett.* **413**, 105 (1993).
53. S. Refsdael, *Mon. Not. Royal Astr. Soc.* **128**, 295 (1964).
54. S. Refsdael, *Mon. Not. Royal Astr. Soc.* **132**, 101 (1966).
55. Y. Rephaeli, *Astron. Rev. Astron. Astrophys.* **33**, 541 (1995).
56. A. Reiss, W. Press, & R. Kirshner *Astrophys. J.* **000**, 000 (1997), in press.
57. A. Renzini, in *Observational Tests of Inflation,* ed. T. Shanks *et al.* (Dordrecht, Kluwer, 1991).
58. A. Saha *et al.*, *Astrophys. J.* **425**, 14 (1994).
59. A. Saha *et al.*, *Astrophys. J.* **438**, 8 (1995).
60. A. Saha *et al.*, *Astrophys. J.* **466**, 55 (1996).
61. A. R. Sandage, in *The Deep Universe*, eds. B. Binggeli and R. Buser (Springer-Verlag, New York, 1995).
62. A. R. Sandage and G. Tammann, in *Critical Dialogs in Cosmology*, ed. N. Turok (World Scientific, 1997).
63. A. Sandage *et al.*, *Astrophys. J. Lett.* **460**, 15 (1996).
64. P. Schechter *et al.*, *Astrophys. J. Lett.* **475**, 85 (1997).
65. D. N. Schramm, in *Astrophysical Ages and Dating Methods*, eds. E. Vangioni-Flam *et al.* (Edition Frontieres: Paris, 1989).
66. E. J. Shaya, P. J. E. Peebles and R. B. Tully, *Astrophys. J.* **454**, 15 (1995)

67. S. Shectman *et al.*, *Astrophys. J.* **470**, 172 (1996).

68. X. Shi *Astrophys. J.* **446**, 637 (1995)

69. G. Steigman and J. E. Felten, in *Proceedings of the St. Petersburg Gamow Seminar*, ed. A. M. Bykov and R. A. Chevalier (Dordrecht, Kluwer, 1995).

70. R. A. Sunyaev & Y. B. Zel'dovich, *Astrophys. & SS* **4**, 301 (1969)

71. R. A. Sunyaev & Y. B. Zel'dovich, *Astrophys. & SS* **7**, 3 (1970)

72. N. Tanvir *et al.*, *Nature*, **377**, 27 (1995).

73. V. Trimble, *Astron. Rev. Astron. Astrophys.* **25**, 425 (1987).

74. E. Turner, *Astrophys. J. Lett.* **365**, 43 (1990).

75. E. Turok, *Astrophys. J. Lett.* **473**, 5 (1996).

76. S. van den Bergh, *Astrophys. J.* **446**, 39 (1995)

77. D. A. VandenBerg, M. Bolte, and P. B. Stetson, *Astron. Rev. Astron. Astrophys.* **34**, 461 (1996)

78. D. A. VandenBerg, *et al.*, 1997, in preparation.

79. A. Walker, *Astrophys. J. Lett.* **390**, 81 (1992).

80. S. Weinberg, *Rev. Mod. Phys.*6111989.

81. S. D. M. White and C. S. Frenk, *Astrophys. J.* **379**, 52 (1991).

82. S. D. M. White and A. C. S. Fabian, *Mon. Not. Royal Astr. Soc.* **273**, 72 (1995).

83. S. D. M. White, J. F. Navarro, A. E. Evrard and C. S. Frenk, *Nature*, **366**, 429 (1993).

84. J. A. Willick *et al.*, *Astrophys. J. Suppl.* **000**, 000 (1997), in press.

85. M. Zaldarriaga, D. N. Spergel and U. Seljak, astro-ph/9702157.

HIGHLY IONIZED GAS IN ACTIVE GALACTIC NUCLEI

HAGAI NETZER

School of Physics and Astronomy, Tel Aviv University, Tel Aviv 69978, Israel

Absorption by highly ionized gas (HIG), in the 0.5–2 keV part of the spectrum, is a common property in AGN. This gas is ionized by the central radiation source and its temperature is close to the maximum allowed in a warm ionized gas, about 2×10^5 K. The HIG has recently been studied in a sample of Seyfert galaxies and the analysis shows that the X-ray ionization parameter cover an extremely small range, suggesting very similar conditions in all objects. At least 2/3 of all low luminosity AGN show the HIG component. There are indications that the HIG is moving out at a velocity of about 1000 km s^{-1} and that some ultraviolet absorption lines originate in the same location. A model is proposed in which bloated stars, or "seeds" in the broad line region are the origin of the HIG. The gas is blown out of the system by X-ray radiation pressure.

1 New X-ray observations and analysis of AGN

X-ray observations of many AGN show strong absorption features at energies of 0.5– 2 keV. The absorption is by highly ionized gas (hereafter HIG) which is dominated by OVII and OVIII absorption. Some observations and theoretical models of this component are given in Halpern (1984), Nandra & Pounds (1994), Reynold *et al.* (1995), George, Turner & Netzer (1995), Krolik & Kriss (1995), and Netzer (1996).

Recently, we (George *et al.* 1997) have undertaken a systematic study of 18 low luminosity AGN observed by ASCA and analyzed the HIG properties. The analysis includes fitting of the spectrum by various models, and comparison with photoionization calculations. The result is a uniformly analyzed data set that shows the distribution of column density, ionization parameter and covering fraction of the HIG. The analysis suggests that:

1. More than two thirds of low luminosity AGN show evidence of HIG.

2. The HIG show no preference in column density over the range $21.5 < \log N_H < 23.0$ cm^{-2}.

3. There is an extremely small range in ionization parameter and most HIG systems cluster around a value of $U_x=0.1$, where U_x is the X-ray ionization parameter (Netzer; 1996). At this ionization parameter, the electron temperature is about 2×10^5 K.

Except for point no. 1, that shows how common such systems are, point no. 3 is of great importance. The above mean value of U_x corresponds to a gas temperature just below the maximum temperature allowed on the cool branch of a thermally stable gas. Thus the HIG is as hot as possible to be on the the cool branch of the two-phase equilibrium curve and not hot enough to be on the hot branch. Part of this effect can be understood by noting that a larger U_x results in thermally unstable gas that can change to the stable solution on the hot branch. Such gas is likely to be completely transparent and escape the detection by ASCA. It is conceivable that the subgroup of AGN not showing evidence for HIG represent exactly this case. However, most objects are definitely on the cool branch and it is not clear what forces the HIG to attain the maximum allowed equilibrium temperature and not any temperature below it (i.e. much closer to the BLR gas temperature of about 10^4 K). Fitting and modeling the HIG is not sensitive to the gas density thus the analysis cannot help locate the HIG in the nucleus.

2   X-ray emission lines and ultraviolet absorption lines

The HIG must produce a large number of soft X-ray lines. The strongest transitions are due to H-like and He-like lines of oxygen, neon, magnesium, silicon and sulphur. L-shell lines of iron are also predicted to be strong (Netzer 1996). Calculations show that the strongest emission lines have typical equivalent widths, as measured against the unattenuated continuum, of about 50 eV for a covering factor, C_f of unity. This is at the limit of detection of present day instruments and proper line measurements must await the launch of AXAF and XMM.

As argued by Mathur and collaborators (e.g. Mathur, 1994), the HIG is also the likely origing of the weak (equivalent width of 1Å or less) ultraviolet absorption lines observed in numerous low luminosity AGN. Detailed calculations (Netzer 1996) show that HIG cloud with typical columns and ionization parameter, contain trace element of several lithium-like ions that can result in weak absorption lines of C IV λ1549 N V λ1240 and O VI λ1035. If this is confirm by simultaneous ultraviolet and X-ray observations, we can learn some more about the location of the HIG gas since the velocity of the UV absorbing gas is easily measurable.

3   HIG models

The HIG is an important new component which is very common in low luminosity AGN and has already been observed in several high luminosity objects. Yet, there is no clear idea about the location of the gas and hence its density and mass. One possible location is inside the BLR, where the density must be high and the HIG mass relatively small. Another suggestion (e.g. Krolik and Kriss; 1995) is that the gas is very far from the center, at a distance similar to the NLR. This must involve low density and large mass. Such very different ideas can be tested observationally, by looking for variable absorption and emission lines (e.g. George et $al.$; 1995, Reynolds et $al.$; 1995).

Simple scaling to the known BLR properties suggest the following relations: The HIG density is

$$N_{HIG} \simeq 3 \times 10^4 \, L_{44} R_{pc}^{-2} \ cm^{-3},$$

the physical thickness is

$$\Delta R_{HIG} \simeq 0.1 \, L_{44}^{-1} N_{22} R_{pc}^2 \ pc$$

and the mass is

$$M_{HIG} \simeq 10^3 \, N_{22} C_f R_{pc}^2 \ M\odot,$$

where N_{22} is the column density in $10^{22} \ cm^{-2}$ and L_{44} the ionizing luminosity in $10^{44} \ erg \ s^{-1}$. Given these properties, I propose yet another possible location for the HIG. The idea is that the HIG is the result of gas evaporation off "seeds" in the BLR. The seeds may be the high density cores of BLR clouds or bloated stars (Alexander and Netzer 1994) in the central star cluster. Such seeds loose mass and the outer boundary, or "wind", is made of lower density, transparent gas. This gas, which is exposed to the strong central radiation source, is being pushed by radiation pressure, reducing its density as it moves away from the center. At some stage it attains the HIG conditions, where the dominant absorbing species are OVII and OVIII and where the main driving force is radiation pressure by the central X-ray continuum. This model suggests that the HIG mass can be orders of magnitude large that the mass in the "traditional BLR", i.e. the BLR which is made of $10^{23} cm^{-2}$ column density clouds.

References

1. Alexander, T., & Netzer, H., 1994, MNRAS, 270, 781.
2. George, I., Turner, T.J., & Netzer, H., 1995, ApJLett, 438, L69.
3. George, I., Turner, T.J., Netzer, H., Mushotzky, R.F, & Nandra, P., 1997 (ApJ, submitted)
4. Halpern, J.P., 1984, ApJ, 281, 90
5. Krolik, J. & Kriss, G., 1995, ApJ, 447, 512.
6. Mathur, S., ApJLett, 431, L75.
7. Nandra, K., & Pounds, K.A., 1994, MNRAS, 268, 405.
8. Netzer, H., 1996, ApJ, 473, 781.
9. Reynolds, C.S., Fabian, A.C., Nandra, K., Inoue, H., Kunieda, H., & Iwasawa, K., 1995, MNRAS, 277, 901.

RECENT RESULTS ON BROAD ABSORPTION LINE QSOs

R. W. GOODRICH

The W. M. Keck Observatory,
65-1120 Mamalahoa Highway, Kamuela, HI 96743, USA

Optical surveys for QSOs generally find that roughly 10% of detected QSOs show broad, blue-shifted absorption lines characteristic of the so-called "BAL" QSOs, or BALQs. However, evidence from optical polarization, radio surveys, and X-ray measurements indicate that at least some BALQs are seen through heavy attenuation, hence are observed to be fainter than they would be from lines of sight which do not intersect the BAL region. This implies that the surveys are missing a significant number of BALQs, and hence they are not as rare as once thought, comprising more than $\sim$ 30% or more of the total QSO population.

1 Introduction

Broad-absorption line QSOs (BALQs) comprise a relatively rare class of optically quiet QSOs, at least in terms of observed numbers in optical surveys. [a] Typical optically-selected samples of QSOs show a few percent of the objects to be BALQs. Since these surveys are usually flux-limited, the absorption troughs in the BALQ spectra cause some of them to fall below the flux limit, even though they would otherwise qualify. Correcting for this effect, surveys typically find $\sim$ 10% of QSOs are BALQs, which is then the fraction of QSOs that are thought to be BALQs.

BALQs also show some other unusual characteristics compared to their supposed parent population of all optically-quiet QSOs. They show higher rest-frame UV polarizations, higher radio-to-optical flux ratios, and lower X-ray fluxes. Goodrich[1] pointed out that these characteristics can all be explained if at least some BALQs have highly attenuated views of the direct (unscattered, unpolarized) nuclear light. This contribution summarizes the most important points in the argument, and the most significant conclusion, that in reality 30% or more of QSOs are BALQs.

2 BALQs Have High Polarization

Stockman, Moore, and Angel[2] reported that 20% of BALQs have high polarization, $P > 3\%$. In contrast, non-BALQs almost never show $P > 1\%$; in fact 20% of non-BALQs have $P > 0.8\%$. Goodrich and Miller[3] argued that this implied that in high-P BALQs the direct, unpolarized view of the nuclear

---

[a]Throughout this paper I use the term "QSO" to refer only to radio quiet objects.

light is significantly attenuated compared to the lines of sight which would not show BAL characteristics. The scattered, polarized light is not so heavily attenuated, hence the net polarization in these objects is higher. If these same objects could be viewed along lines of sight missing the BAL region, they would appear as normal QSOs, and the direct view of the continuum source would swamp the scattered light, producing a low net polarization. A continuum attenuation by a factor of four at rest-frame UV wavelengths in BALQs is enough to roughly reconcile the polarization statistics.

3 BALQs Are "Radio-Moderate"

Francis, Hooper, and Impey[4] defined "radio-moderate" QSOs as those objects on the radio-loud end of the radio-quiet QSO distribution. They also noted that the few BALQs in their sample tended to lie within this radio-moderate class. They used the ratio of radio to optical fluxes to parametrize the class. Goodrich[1] pointed out that if some BALQs have attenuated optical-UV continua, their radio-to-optical flux ratio would appear to be higher than similar QSOs viewed along non-BAL lines of sight, thus explaining the observations of Francis *et al.* A continuum attenuation by a factor of three in BALQs is enough to explain the difference between BALQs and non-BALQs in their sample, similar to the factor derived from polarization statistics.

4 BALQs Are Weak in X-Rays

I have carefully used the word "attenuation" to describe a lack of direct photons along BAL lines of sight, relative to lines of sight that miss the BAL region. This attenuation does not necessarily arise in an absorption process, which implies that the photons are destroyed somehow. Instead the attenuation along BALQ lines of sight may be due to an asymmetric distribution of scatterers, which can preferentially remove photons from some lines of sight and redirect them to others. For example, a highly flattened oblate distribution of pure scatterers, such as electrons, will have higher optical depth along equatorial lines of sight (equivalent to the BALQ views). Photons which start out in the equatorial directions can escape more easily along lines parallel to the polar (symmetry) axis of the system, hence these photons will be removed from the equatorial lines of sight and added to the polar lines of sight.

Where there are free electrons there are also ions. These ions should absorb heavily on the soft X-rays, thus depressing the X-ray flux. X-ray spectra of a BALQ by Singh, Westergaard, and Schnopper[7] did indeed show a large absorption column, and subsequently Green and Mathur[5] (see also Green *et al.*[6])

showed that in general BALQs are underluminous compared to non-BALQs.

5 BALQs Are Common!

If some BALQs are observed to be fainter than they would be in the absence of an attenuating medium, then they must be underrepresented in flux-limited (i.e. nearly all) optical surveys for QSOs. Goodrich[1] showed that if *all* BALQs are attenuated by such a large factor, then the *majority* of QSOs must be BALQs. However, not *all* BALQs are required to be so heavily attenuated. At least 20% of them must have large attenuations in order to explain the polarization statistics, and this predicts that *at least* 1/3 of all QSOs are BALQs (as compared to the previously accepted value of 10%). Goodrich[1] further noted that at least 3/4 of all BALQs are attenuated, high polarization objects. Most of the attenuated BALQs are not found in surveys as the attenuation has forced their optical fluxes below the survey detection limits. Those high-P BALQs that we do see are among the most luminous of QSOs. Of course, the true fraction of BALQs translates directly into the covering factor of the broad-absorption region, and a covering factor of $> 30\%$ puts severe constraints on some theoretical models of the BAL phenomenon.

All of these pieces of evidence begin to give us clues as to the nature and location of the attenuating region, how it relates to the BAL region itself, and, ultimately, may shed light on the dichotomy between radio-loud and radio-quiet objects. Much more work needs to be done in order to shore up the statistics and follow up on the sometimes small differences seen between BALQs and non-BALQs (or even high-P BALQs and low-P BALQs).

Acknowledgements

I would like to acknowledge the support of HST grant GO-6766 in this work.

References

1. Goodrich, R. W., ApJ **474**, 606 (1997).
2. Stockman, H. S., Moore, R. L., & Angel, J. R. P., ApJ **279**, 485 (1984).
3. Goodrich, R. W., and Miller, J. S., ApJ **448**, L73 (1995).
4. Francis, P. J., Hooper, E. J., and Impey, C. D., AJ **106**, 417 (1993).
5. Green, P. J., and Mathur, S., ApJ **462**, 637 (1996).
6. Green, P. J., et al., ApJ **450**, 51 (1995).
7. Singh, K. P., Westergaard, N. J., and Schnopper, H. W., AandA **172**, L11 (1987).

Shock Excitation of Emission Lines in Active Galactic Nuclei

G.V. Bicknell[1], Z. Kuncic[1], Z. Tsvetanov[2], M.A. Dopita[3] and R. S. Sutherland[1]

[1] *ANU Astrophysical Theory Centre, Australian National University, Canberra, A.C.T. 0200, Australia*

[2] *Dept. of Physics and Astronomy, Johns Hopkins University, Baltimore, MD 21218, USA*

[3] *Mount Stromlo and Siding Spring Observatories, Weston PO, ACT, 2611, Australia*

1 Introduction

The dominant paradigm for the excitation of the narrow line region of many AGN, has for a long time, been photoionization by a central power-law. However, increasing attention has been paid to the possible role of shocks.[1,2,3,4] Here, we summarize further ideas relating to the relevance of shock models to Gigahertz Peak Spectrum (GPS) and Compact Steep Spectrum (CSS) radio sources and to the narrow line regions of Seyfert Galaxies.

2 Induced Compton Absorption in GPS and CSS Sources

It has been suggested[6] that the expansion of a jet-driven radio lobe drives a radiative bow shock into the dense ISM surrounding GPS and CSS sources. Free-free absorption by the ionized ISM produces a peak in the radio spectrum at $\nu \sim 0.1 - 1$ GHz. Here we note that induced Compton absorption[7] can also be important in defining the peak since the brightness temperatures of the components of GPS sources can be quite high, $T_\nu \sim 10^{9-10}$ K, (Stanghellini, pte. comm.). Induced Compton absorption sets in when $(kT_\nu/m_e c^2) \sim \tau_{\mathrm{T}}^{-1}$ where τ_{T} is the Thomson optical depth of the external screen. The value of τ_{T} is dominated by the ionized bow-shock precursor. Using estimates of precursor electron column density[4] for shocks in the velocity range $100 - 500$ km s^{-1}, we derive $\tau_T \approx 2.0 \times 10^{-2} V_3^{3.0}$ where $V_3 = v_{\mathrm{sh}}/1000$ km s^{-1}. This simply represents the electron opacity of the Stromgren column ionized by the post-shock radiation flux $\propto V_{\mathrm{sh}}^3$ and should extrapolate reasonably to velocities > 500 km s^{-1}. For a spectral index of α, induced Compton absorption implies that

$$\frac{\nu_{\mathrm{peak}}}{\nu_0} = \left[3.47 \times 10^{-3} \left(\frac{T_{\nu_0}}{10^9 \, \mathrm{K}} \right) V_3^{3.0} \right]^{1/(\alpha+2)} = 0.11 \left(\frac{T_{\nu_0}}{10^9 \, \mathrm{K}} \right)^{0.38} V_3^{1.15} \quad (1)$$

where T_{ν_0} is the brightness temperature at $\nu = \nu_0$ and for numerical values $\alpha = 0.6$. Equation (1) predicts $\nu_{\text{peak}} \sim 1\,\text{GHz}$ for $v_{\text{sh}} \sim 1000\,\text{km s}^{-1}$. This estimate depends only on the velocity and is independent of the density. However, for a given jet energy flux, F_E, the bow-shock velocity depends upon F_E/ρ_{a} where ρ_{a} is the ambient density. The velocities required are larger than those implied by the free-free absorption model and the implied densities and X-ray column depths ($\sim 10^{22}\,\text{cm}^2$) are smaller and may be relevant to the small number of GPS sources recently observed by Elvis and collaborators[8]. Thus, induced Compton absorption widens the range of densities over which a peak in the radio spectrum may be produced by gas external to the radio source.

3 The correlation between radio and optical line emission in Seyfert Galaxies

The correlation between radio and optical emission line luminosities for Seyfert galaxies[9] may be indicative of a correlation between jet luminosity and accretion disk luminosity[10] or it may suggest a more direct link through the excitation of the emission line plasma through mechanical energy transported by the jet. Indeed, if the jet energy flux is comparable to the energy dissipated in the accretion disk, then one expects comparable photoionization and mechanical excitation. On the other hand, it is sometimes argued[11] that Seyfert jet energy fluxes are insufficient to excite emission lines. This is usually based on a radio galaxy conversion factor between radio luminosity and jet energy flux. However, the ratio of monochromatic radio power from a jet-fed lobe to jet energy flux is[12]

$$\kappa_\nu \approx (a-2)\,C_{\text{syn}}(a)\,(\gamma_0 m_{\text{e}} c^2)^{(a-2)} \left[1 - (\gamma_1/\gamma_0)^{-(a-2)}\right]^{-1} f_{\text{e}}\, f_{\text{ad}}\, B^{(a+1)/2} t \quad (2)$$

where C_{syn} is a synchrotron parameter, the electron ditribution, $N(E) = KE^{-a}$ with lower and upper cutoffs cutoff at $E_{0,1} = \gamma_{0,1} m_{\text{e}} c^2$, the fraction of the energy in electrons and positrons is f_{e}, $f_{\text{ad}} \sim 0.5$ is an adiabatic factor, B is the magnetic field and t is the age of the source. The major factor here that can make κ_ν low is that the dynamical ages $t \sim 10^6$ yrs inferred for Seyfert galaxies[11] are much smaller than those assumed for radio galaxies, in particular FR1s. This is counterbalanced by the higher equipartition magnetic fields $B \sim 10^{-4}$ G. If the magnetic fields are sub-equipartition in Seyferts and/or if $f_{\text{e}} < 0.1$, then a value of $\kappa_\nu \sim 10^{-14} - 10^{-13}$ is feasible. The factor f_{e} may be low if the jet entrains substantial thermal material before it enters the lobe. A value $\kappa_\nu \sim 10^{-13}$ and a radio power $\sim 10^{23}$ W Hz^{-1} implies a jet energy flux $\sim 10^{43}$ ergs s^{-1}. Given that a substantial proportion of this energy can be converted into work done on the interstellar medium via shocks driven by the expansion of the lobe, the corresponding [OIII]λ5007 luminosity $\sim 5 \times 10^{41}$ ergs s^{-1}.

The sizes of Seyferts also relate to the jet energy flux. For a dentist-drill type model of the jet-driven lobe[13,6] the distance from core to hotspot, x_h is given by $x_h = x_0 \xi^{1/(5-\delta)}$ where the dimensionless parameter

$$\xi = \frac{(5-\delta)^3 \zeta^2}{18\pi(8-\delta)} \left(\frac{F_E t^3}{\rho_0 x_0^5} \right) \approx 0.26 \frac{(5-\delta)^3}{(8-\delta)} \frac{F_{E,43} t_6^3}{n_0 (x_0/\text{kpc})^5} t \tag{3}$$

Here, x_0 is an arbitrary scale length (taken to be a kpc here), the ambient density $\rho = \rho_0 (x/x_0)^{-\delta}$, n_0 is the Hydrogen density at a kpc, $10^{43} F_{E,43}$ ergs s^{-1} is the jet energy flux, t_6 Myr is the age and $\zeta \approx 2$ is the ratio of averaged hotspot pressure to cocoon pressure. In NGC 1068, the projected $x_h \sim 700$ pc and the predicted ambient density at a kpc is of order $1 - 10$ cm^{-3} for $F_{E,43} \sim 1$, $t_6 \sim 1$ and $\delta \sim 0 - 2$. The observed [SII] densities[14] are generally greater than $10^{2.5}$ cm^{-3} and are consistent with the above density estimate since the density in the [SII]-forming region of shocks is typically a factor of 100 higher than the pre-shock density.

4 Shock excitation of emission-line clouds in NGC 1068

The visual impression conveyed by the combined radio-optical images of the inner part of the narrow line region of NGC 1068[15] is that the jet is deflected by emission line clouds as it struggles to make progress to the kpc scale. Such deflections produce a shock in the jet which is visible as a knot and the increase in pressure, ΔP behind the oblique shock drives a shock into the cloud with a velocity $v_{\text{sh}} \approx (\Delta P/\rho)^{1/2}$. The luminosities of radiative shocks in [OIII] and Hα, including shocked and precursor emission are for $v_{\text{sh}} = 500 - 1000$ km s^{-1}

$$L(H\alpha) = 5.3 \times 10^{-3} n_H V_3^{2.41} A_{\text{sh}} \text{ ergs s}^{-1} \tag{4}$$

$$L([OIII]) = 2.3 \times 10^{-2} n_H V_3^3 A_{\text{sh}} \text{ ergs s}^{-1} \tag{5}$$

where n_H cm^{-3} is the pre-shock Hydrogen density, $V_3 = v_{\text{sh}}/1000$ km s^{-1}, as above and $A_{\text{sh}} \sim 10^{40}$ cm^2 is the shock area[6]. We have used HST FOC images to estimate the [OIII] and Hα luminosities from the various clouds and the implied Hydrogen density and shock velocities are given in the following table. The higher shock velocities are unreliable since the shock models are invalid beyond a thousand km s^{-1}. Nevertheless it is evident that shock velocities ~ 1000 km s^{-1} and Hydrogen densities $\sim 10 - 100$ cm^{-3} are involved. Such velocities are characteristic of the velocity dispersions of the NLR of Syefert

228

galaxies and NGC 1068 in particular. The driving pressures in the last column are consistent with the increase in nonthermal pressure at an oblique shock.

Cloud	$\log L(H\alpha)$ ergs s^{-1}	$\log L([OIII])$ ergs s^{-1}	$n_H A_{sh,40}$ cm^{-3}	V_{sh} km s^{-1}	P dyn cm^{-2}
NLR-C	40.14	40.71	500	800	7.0×10^{-6}
NLR-D	39.81	40.35	300	700	3.4×10^{-6}
NLR-F	39.85	40.57	60	1,400	2.8×10^{-6}
NLR-G	39.62	40.40	20	1,700	1.5×10^{-6}

References

1. R. S. Sutherland, G. V. Bicknell, and M. A. Dopita. **Ap. J.**, 414, 510 (1993)
2. S. M. Viegas and M. Contini. In B. M. Peterson, F.-Z. Chang, and A. S. Wilson, editors, *Emission Lines in Active Galaxies: New Methods and Techniques*, page 365, San Francisco, Astronomical Society of the Pacific, (1997)
3. A. M. Koekemoer and G. V. Bicknell. **ApJ**, submitted (1997)
4. M. A. Dopita and R. S. Sutherland. **ApJS**, 102, 161 (1996)
5. M. A. Dopita and R. S. Sutherland. **ApJ**, 455, 468 (1996)
6. G. V. Bicknell, M. A. Dopita, and C. P. O'Dea. **ApJ**, in press (1997)
7. R. A. Sunyaev. **Astrophys. Lett.**, 7, 19 (1970)
8. M. Elvis, S. Mathur, B. J. Wilkes, F. Fiore, P. Giommi, and P. Padovani. In B. M. Peterson, F.-Z Cheng, and A. S. Wilson, editors, *Emission Lines in Active Galaxies: New Methods and Techniques*, ASP Conference Series, page 236, San Francisco. Astronmical Society of the Pacific (1997)
9. M. Whittle. **MNRAS**, 213, 189 (1985)
10. H Falcke and P. L. Biermann. **A&A.**, 293, 665 (1995)
11. A. S. Wilson. In B. M. Peterson, F. Cheng, and A. S. Wilson, editors, *Emission Lines in Active Galaxies: New Methods and Techniques*, San Francisco (1997) Astronomical Society of the Pacific.
12. G. V. Bicknell, Z. Tsvetanov, M. A. Dopita, and R. S. Sutherland. **ApJ**, submitted (1997)
13. M. C. Begelman. In C. L. Carilli and D. A. Harris, editors, *Cygnus A: Study of a Radio Galaxy*, page 209, Cambridge. University Press (1996)
14. G. Cecil, J. Bland, and R. B. Tully. **ApJ**, 355, 70 (1990)
15. J. F. Gallimore, S. A. Baum, and C. P. O'Dea. **ApJ**, 464, 198 (1996)

RECENT PROGRESSES OF ACCRETION DISK MODELS AROUND BLACK HOLES

SANDIP K. CHAKRABARTI[a]

S.N. Bose National Center for Basic Sciences, Salt Lake, Calcutta 700091

Accretion disk models have evolved from Bondi flows in the 1950s to Keplerian disks in the 1970s and finally to advective transonic flows in the 1990s. We discuss recent progresses in this subject and show that sub-Keplerian flows play a major role in determining the spectral properties of black holes. Centrifugal pressure supported enhanced density region outside the black hole horizon produces hard X-rays and gamma rays by reprocessing intercepted soft photons emitted by the Keplerian disk terminated farther out from the black holes. Quasi-periodic oscillations can also be understood from the dynamic or thermal resonance effects of the enhanced density region.

1   Introduction

Matter accreting on galactic and extragalactic black holes need not be spherically symmetric Bondi flow or purely thin and Keplerian. In fact, since by definition, matter must enter into a black hole with radial velocity on the horizon similar to the velocity of light [2,3], it must be supersonic and hence sub-Keplerian [4]. If one assumes that matter forms a thin, subsonic, Keplerian disk very far away then it has to pass through at least one sonic point before the flow enters the black hole. Second, since just outside of the black hole, the infall time scale is much shorter compared to the viscous time scale (even when viscosity is high), angular momentum is roughly constant in the last few Schwarzschild radii and as a result, the centrifugal force increases rapidly enough as the flow approaches the black hole so as to form a centrifugal barrier behind which matter piles up. The resulting enhanced density region may be abrupt (if behind a shock), or smooth if the shock conditions are not satisfied.

2   Nature of Advective Solutions

Chakrabarti [1] classified all possible solutions of the inviscid adiabatic flows around black holes. Since angular momentum is likely to be almost constant close to a hole even for highly viscous flows, the inviscid solutions would be important even when viscosity is significant. It is observed that a large region of parameter space can produce standing shocks behind the centrifugal barrier. The classification is discussed in detail in Chakrabarti [3]. In presence of

---

[a] e-mail:chakraba@bose.ernet.in

viscosity, the advective disk may still produce shocks provided the viscosity parameter is less than some critical value [2,4]. The steady solutions with or without shocks in one and two dimensions have been tested to be stable by explicit numerical simulations [10,13]. In a large region of the parameter space, two saddle type sonic points are present, but the shock conditions are not satisfied. The numerical simulations [13] indicate that the shocks form nevertheless, but they are unstable, and oscillate back and forth. The period of oscillation depends on the specific angular momentum, but typically they can be around a fraction of a second for galactic black hole candidates and around a day for extragalactic massive black holes. Even when stable shock conditions are satisfied, shocks can be oscillatory when significant cooling effects are present [12] provided the cooling time scale in the post-shock region is comparable to the infall time scale in the pre-shock region.

As viscosity is added, the closed topologies in the Mach No. vs. radial distance space of the inviscid solutions open up and the solution joins with a Keplerian disk farther out, provided the accretion rate is large enough to have efficient emission from the disk. For low enough viscosity and the accretion rate the advective flow may join with a Keplerian disk very far away (forming a giant primarily rotating ion torus which surround the disk of size $10^{3-4}x_g$, see, g21-g41 topologies in Fig. 2a of Chakrabarti [4]), while for higher viscosity and accretion rate the Keplerian disk will come closer to the black hole (see, g13-g14 topologies in Fig. 2a of Chakrabarti [4]). In intermediate viscosities, these flows may have shocks. For low and intermediate viscosities the flow would definitely have a centrifugal barrier and consequent enhanced density region in between the black hole and the Keplerian disk. If viscosity varies with height, it is expected that all the three types of flows would be manifested in a single flow [7,5]. Chakrabarti [5] discussed in detail the nature of the multi-component advective disk. The boundaries of Keplerian and sub-Keplerian regions, as well as the accretion rates in different components will vary from case to case as well as from time to time. As the viscosity at the outer edge increases, more and more matter goes from sub-Keplerian component to the Keplerian component. The soft photons from the Keplerian flow cools the sub-Keplerian component (thermal Comptonization) and hence the inner edge of the Keplerian component also advances. Eventually, the sub-Keplerian component cools completely and the Keplerian disk advances till the last stable orbit. Quasi-spherical flows in between the horizon and the last stable orbit reprocesses the soft photons out of the Keplerian component by transferring the bulk momentum of the electrons to the photons [7] (bulk motion Comptonization) and as a result, long extended power law component is formed even in the very soft state.

3 Spectral Properties

When the accretion rate of the cooler Keplerian component ($\dot{m}_d$) is much smaller compared to that of the hotter sub-Keplerian component ($\dot{m}_h$), the soft photons emitted from the Keplerian component are unable to cool the hot electrons of the later component by thermal Comptonization processes. Thus, predominantly hard X-rays are produced with little or no soft bump. Generally, this soft bump may not be observed since the soft X-rays may be further reprocessed by the extended atmosphere of the disk. The X-ray spectral index α ($F(\nu) \sim \nu^{-\alpha}$) is around $0.5 - 0.8$ and $\dot{m}_d$ is typically less than $0.1 - 0.3$ when $\dot{m}_h = 1$ depending on Models[5]. As $\dot{m}_d$ is farther increased in comparison to $\dot{m}_h$ (which may be due to sudden increase of viscosity in the flow [sudden capture of magnetic clouds from the companion, for instance] which converts some of the sub-Keplerian flow into Keplerian [6]), the soft photons of the Keplerian disk cools the electrons of the sub-Keplerian component catastrophically and the spectra consists of only the soft bump without any extended power law tail. This is called soft state by some observers and this may happen for Keplerian rate of around $0.3 - 0.7$ or so when $\dot{m}_h = 1$ depending on models. One may also see broken power law in this state because of the contribution from both thermal and bulk motion effects. As $\dot{m}_d$ is farther increased, the optical depth in the centrifugal pressure supported enhanced density region becomes high and the flow drags most of the photons (which were supposed to be emitted in between $3x_g$ and $1x_g$. The sub-Keplerian flow farther out effectively looses its identity as it cools down completely to a temperature a little above the corresponding Keplerian component [7]). However, a fraction of photons energized by infalling matter due to bulk motion Comptonization can still come out. Chakrabarti & Titarchuk [7] (see also Ebisawa et al. [9] for details) showed that the power law hard component (with energy spectral slope $\alpha \sim 1.5 - 2.0$) can easily explain the behavior of the very soft state of the black hole candidates. Recent computation [14] shows that the power law extends till almost 1MeV. This power law component is absent for both neutron stars as well as naked singularities[8]. Thus, the presence of power law component in very soft states has enabled observers to distinguish a black hole from a neutron star very easily. Although the black hole horizons permit energetic matter to be swallowed directly and therefore, for a given accretion rate, black hole accretion could be less luminous than the neutron star accretion [8], argument based on total luminosity cannot be full proof, since there could be any number of other physical effects (such as bipolar outflows which carry away energy and matter) confusing the situation.

232

4 Quasi-Periodic Oscillations

When Quasi-Periodic Oscillations (QPOs) were discovered in black hole candidates a few years ago, it was surprising, since black holes have neither hard surfaces nor any anchored magnetic fields. It now appears, that the QPOs could be manifestation of the time-dependent solutions of the same set of equations which produced Keplerian and sub-Keplerian flows. As discussed in §2 above, simulations from a large region of parameter space showed that either because of resonance between inflow and outflow time-scales, or between inflow and cooling time scale, the enhanced density region oscillates with frequency very similar to QPO frequency. Furthermore, these oscillating regions intercept different amount of soft photons from the Keplerian component and produces hard X-rays of significant amplitude as is observed. Mechanisms based on acoustic oscillations or some such possibilities are incapable of producing significant modulations.

5 Acknowledgments

The author thanks Goddard Space Flight Center and USRA for partially supporting the cost of participation at the conference.

1. S.K. Chakrabarti, *Astrophys. J.*, **347**, 365 (1989).
2. S.K. Chakrabarti, Theory of Transonic Astrophysical Flows, World Scientific: Singapore (1990).
3. S.K. Chakrabarti, *MNRAS*, (Nov. 1st issue) (1996).
4. S.K. Chakrabarti, *Astrophys. J.*, **464**, 664 (1996).
5. S.K. Chakrabarti, *Astrophys. J.*, (in press) (1997).
6. S.K. Chakrabarti and D. Molteni, *MNRAS*, **272**, 80 (1995).
7. S.K. Chakrabarti, and L.G. Titarchuk, *Astrophys. J.*, **455**, 623 (1995).
8. S.K. Chakrabarti, and S. Sahu, *Astron. Ap*, (in press), (1997).
9. K. Ebisawa, L. Titarchuk, and S.K. Chakrabarti, *PASJ*, **48**, 1 (1996).
10. D. Molteni, G. Lanzafame, and S.K. Chakrabarti, *ApJ*, **425**, 161 (1994).
11. D. Molteni, D. Ryu, and S.K. Chakrabarti, *Astrophys. J.* (Oct 10th issue), (1996).
12. D. Molteni, H. Sponholz, and S.K. Chakrabarti, *Astrophys. J.*, **457**, 805 (1996).
13. D. Ryu, S.K. Chakrabarti, and D. Molteni, *Astrophys. J.*, (Jan. 1st) (1997).
14. L.G. Titarchuk, Proc. 2nd INTEGRAL Workshop "The Transparent Universe" Eds. C. Winkler et al. (in press).
15. S.N. Zhang, et al. 1997, *Astrophys. J.* (submitted).

TeV Gamma Rays from AGN

M. CATANESE

Dept. of Physics and Astronomy, Iowa State Univ., Ames, IA 50011, USA

Representing the Whipple Collaboration[a]

We report on recent observations of AGN with the Whipple Observatory γ-ray telescope. Particular emphasis is placed on results for Mrk 421 and Mrk 501. These include two episodes of $\lesssim$ hour-scale variability from Mrk 421 and multi-wavelength observations of Mrk 421 during a flare in 1995.

1 Introduction

The dominant radiation from the *blazar* class of active galactic nucleus (AGN) is widely believed to arise from relativistic jets viewed at small angles to their axes[1]. The broadband emission appears to consist of a low energy part, most likely synchrotron radiation from relativistic electrons within the jet, which extends from radio to optical-ultraviolet wavelengths (and even to X-rays in some BL Lac objects), and a high energy part, whose origin is unknown, which can extend to γ-ray energies. With EGRET's detection of more than 50 blazars[2], γ-ray observations of AGN have become important for the study of these objects . However, an important constraint on emission models is the endpoint of the energy spectra, and since no EGRET-detected AGN has shown evidence of a spectral cutoff up to 30 GeV, observations at higher energies are needed.

Imaging atmospheric Čerenkov telescopes (IACTs) detect γ-rays at energies above 300 GeV indirectly by focussing the Čerenkov light from air showers onto an array of photomultiplier tubes (PMTs). IACTs have huge effective areas ($\gtrsim 5 \times 10^8$ cm^2), limited by the size of the Čerenkov light pool rather than the telescope aperture. They are thus sensitive to very low fluxes of γ-rays and very short-term variations in AGN emission.

The Whipple Observatory γ-ray telescope[3] consists of a 10m diameter optical reflector focussed onto an array of 109 PMTs. The energy threshold of

[a]D.A. CARTER-LEWIS, F. KRENNRICH, F.W. SAMUELSON, J. ZWEERINK (Dept. of Physics and Astronomy, Iowa State Univ., Ames, IA 50011, USA); J.H. BUCKLEY, J. QUINN, T.C. WEEKES (F.L. Whipple Observatory, Harvard-Smithsonian CfA, P.O. Box 97, Amado, AZ 85645, USA); J.P. FINLEY, J.A. GAIDOS, R.W. LESSARD, G.H. SEMBROSKI, R. SRINIVASAN, C. WILSON (Dept. of Physics, Purdue Univ., Lafayette, IN 47907, USA); C.W. AKERLOF, M.S. SCHUBNELL (Randall Lab. of Physics, Univ. of Michigan, Ann Arbor, MI 48109); P. BOYLE, J. BUSSÓNS GORDO, D.J. FEGAN, J.E. MCENERY (Physics Dept., University College, Dublin 4, Ireland); A. BURDETT, A.M. HILLAS, A.J. RODGERS, H.J. ROSE (Dept. of Physics, Univ. of Leeds, Leeds, LS2 9JT, Yorskshire, England, UK); M.F. CAWLEY (Physics Dept., St. Patrick's College, Maynooth, County Kildare, Ireland); R.C. LAMB (Space Radiation Lab, Caltech, Pasadena, CA 91125)

234

the results presented here is ~350 GeV and the telescope obtains a 5σ excess from the Crab Nebula in one-half hour of on-source observations.

2 Whipple γ-ray observations

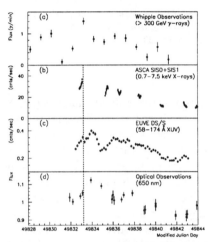

Figure 1. Lightcurves of Mrk 421 at (a) γ-ray, (b) X-ray, (c) extreme UV, and (d) optical wavelengths (1995 April 26 corresponds to MJD 49833).

Despite a systematic, on-going survey of nearby BL Lacs and observations of several EGRET-detected AGN[4], only Mrk 421[5] and Mrk 501[6] have so far been definitively detected at very high energies (VHE, E$\gtrsim$300 GeV). These are the two closest known BL Lacs and are among the brightest BL Lacs at X-ray wavelengths. Mrk 421 is an EGRET source, while Mrk 501 is not[2]. Both have recently been confirmed as TeV sources by observations with the HEGRA Čerenkov telescopes[7,11]. The mean VHE fluxes of Mrk 421 and Mrk 501 are approximately 30%[5] and 7%[6] that of the Crab Nebula, respectively. Day-scale flux variations have been detected at TeV energies for both objects[9,6] and the magnitude of the variations is quite large. Mrk 421[10] and Mrk 501[6] have peak measured fluxes 30 and 7 times their means, respectively.

Multi-wavelength observations of Mrk 421 taken in 1995 April-May revealed apparent correlations at VHE γ-ray, soft X-ray, extreme UV (XUV), and optical wavelengths[9] (Fig. 1). EGRET observations conducted during this time resulted only in an upper limit. The X-ray and VHE γ-ray flares show no apparent time lag while the XUV and optical flares appear to lag by one day. Optical polarization measurements taken at the same time also show a correlation with a one day time lag[9].

In May of 1996, two γ-ray flares were observed which are unprecedented for their magnitude and short duration[10] (Fig. 2). The first produced the largest VHE γ-ray flux (by a factor of >5) ever recorded and had a doubling time time of $\approx$1 hour. The VHE flux had returned to its quiescent level by the following night, implying a decay time of <24 hours. The second γ-ray flare had a duration of <1 hour, implying doubling and decay times of only $\approx$15 minutes.

3 Conclusions

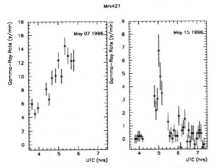

Figure 2. VHE γ-ray lightcurves of Mrk 421 for 1996 May 7 (left) and 1996 May 15 (right). The Crab Nebula flux in these units is 1.5γ/min.

The variability of the γ-ray flux from Mrk 421 and Mrk 501 requires that the emission region is compact. Doppler boosting may therefore be required to reduce the expected high opacities from γ-γ pair production. The correlated variability seen in Mrk 421 in 1995 allows restrictive limits to be set on the Doppler beaming factor and the magnetic field in the emission region, and provides stringent tests of blazar emission models [11]. The lack of detected VHE emission from other blazars could result, for the more distant sources, from intergalactic IR fields attenuating the VHE γ-rays [12]. Also, losses from interaction with fields near the source [13] and, in electron progenitor models, limitations of the emission mechanism [14] may be important in some objects.

Acknowledgments

We acknowledge the technical assistance of Kevin Harris and Emmet Roache. This work is supported by grants from the U.S. Department of Energy and from NASA, by PPARC in the UK, and by Forbairt in Ireland.

References

1. R.D. Blandford and A. Königl, *Astrophys. J.* **232**, 34 (1979).
2. D.J. Thompson *et al.*, *Astrophys. J. Suppl.* **101**, 259 (1995).
3. M.F. Cawley *et al.*, *Exp. Astron.* **1**, 173 (1990).
4. A.D. Kerrick *et al.*, *Astrophys. J.* **452**, 588 (1995).
5. M. Punch *et al.*, *Nature* **358**, 477 (1992).
6. J. Quinn *et al.*, *Astrophys. J.* **456**, L83 (1996).
7. D. Petry *et al.*, *Astron. & Astrophys.* **311**, L13 (1996).
8. S.M. Bradbury *et al.*, *Astrophys. J.*, submitted (1997).
9. J.H. Buckley *et al.*, *Astrophys. J.* **472**, L9 (1996).
10. J.A. Gaidos *et al.*, *Nature* **383**, 319 (1996).
11. J.H. Buckley *et al.*, *Adv. in Space Sci.*, in press (1997).
12. F.W. Stecker *et al.*, *Astrophys. J.* **415**, L71 (1993).
13. C.D. Dermer and R. Schlickeiser, *Astrophys. J. Suppl.* **90**, 945 (1994).
14. M. Sikora *et al.*, *Astrophys. J.* **421**, 153 (1994).

DEUTERIUM ABUNDANCE IN THE LOCAL ISM AND POSSIBLE SPATIAL VARIATIONS

JEFFREY L. LINSKY, BRIAN E. WOOD

JILA, University of Colorado and NIST,
Boulder, CO 80309-0440 USA

Excellent HST/GHRS spectra of interstellar hydrogen and deuterium Lyman-α absorption toward nearby stars allow us to identify systematic errors that have plagued earlier work and to measure accurate values of the D/H ratio in local interstellar gas. Analysis of 12 sightlines through the Local Interstellar Cloud leads to a mean value of D/H = $(1.50 \pm 0.10) \times 10^{-5}$. Whether or not the D/H ratio has different values elsewhere in the Galaxy is a very important open question.

An accurate measurement of the D/H abundance ratio in the local interstellar medium (LISM), $(D/H)_{LISM}$, and an assessment of possible spatial variations in the Galaxy are required for two important reasons. First, this information provides a lower limit to the primordial D/H ratio, $(D/H)_{prim}$, which constrains the critical density of baryons Ω_B. Second, $(D/H)_{LISM}$ is the end result of an incompletely understood complex set of Galactic chemical evolution processes. Comparison of the value of $(D/H)_{LISM}$ with D/H ratios characteristic of the protosolar nebula and elsewhere in the Galactic disk and halo will test our understanding of stellar evolution, stellar mass loss, interstellar physics, and the rate of infall and chemical composition of halo gas. At this symposium the precise value of $(D/H)_{LISM}$ has acquired greater importance as the previously announced high value of $(D/H) = 2 \times 10^{-4}$ in the absorption spectrum toward Q0014+813 is spurious (Tytler *et al.* 1996). Thus $(D/H)_{LISM}$ is much closer to $(D/H)_{prim}$ than some authors have thought, and our understanding of Galactic chemical evolution is tested by a measurement of the smaller difference between $(D/H)_{prim}$ and $(D/H)_{LISM}$.

Lyman line absorption is the most reliable technique for inferring the present local value of D/H, because the Sun is surrounded by a cloud of warm, partially-ionized gas (Lallement *et al.* 1995) with very few molecules and the ionization and adsorption on to grains is nearly the same for H and D. Lyman-α line absorption by H and D can be observed with high S/N in HST/GHRS spectra toward nearby stars as the line separation is 81 km s^{-1}, provided that the hydrogen column density is not so large as to obliterate the D line $(N_{HI} < 10^{18.7}$ cm$^{-2})$. The launch of the Far Ultraviolet Spectroscopic Explorer (FUSE) in 1998 will allow us to observe the higher Lyman lines to extend this method to more distant lines of sight (LOS).

Ferlet *et al.* (1996) have reviewed D/H measurements obtained with the Copernicus, IUE, and GHRS instruments. Since the analysis of new LOS observed with the GHRS is proceeding very rapidly, the conclusions that one can now draw from the data have changed. The previous studies of Lyman line absorption toward both hot and cool stars with the Copernicus and IUE satellites left a confused picture in which the uncertainties in D/H for individual LOS were large and the possiblilty of spatial variations in D/H by a factor of 2 or larger was consistent with the data. The flood of beautiful new GHRS spectra has changed this picture dramatically. The first clear indication of this paradigm shift was the measurement of D/H = $(1.60^{+0.14}_{-0.19}) \times 10^5$ for the Capella LOS (Linsky *et al.* 1995). Since this GHRS result lies outside of the published error bars for all previous results for this LOS to a bright star, the older results are likely unreliable because of systematic errors.

GHRS echelle spectra have far higher S/N and spectral resolution than Copernicus and IUE. Since the core of Lyman-α is highly saturated, high S/N and spectral resolution are critical for inferring the H column density which is more uncertain than the D column density. This alone may explain much of the previous scatter in the D/H values. Another critical issue is the presence of many velocity components in the LOS. Two or more velocity components are often observed even for short LOS. Ultra-high resolution spectra of the NaI and CaII lines for many LOS show many closely spaced narrow velocity components (e.g., Welty *et al.* 1996). Components with column densities orders of magnitude smaller than the main absorber would not be detected in metal lines but could be optically thick in the H Lyman-α line and change the inferred H column densities if not taken into account. For example, the analysis of GHRS spectra of the nearest (1.3 pc) stars, α Cen A and B (Linsky & Wood 1996), required a second absorption component in the LOS with a column density 1/1000 that of the main component, a red shift of about 4 km s^{-1}, and a high temperature (30,000 K). The most likely explanation is a "hydrogen wall" around the Sun created by charge exchange reactions between the inflowing interstellar gas and the solar wind leading to a pileup of neutral hydrogen atoms (cf. Baranov and Malama 1995). Given the highly saturated nature of the main component, the inclusion of this extra component raises the inferred D/H ratio from $\approx 6 \times 10^{-6}$ to $(1.2 \pm 0.7) \times 10^{-5}$. We have now identified hydrogen walls around other stars. Another potential source of systematic error is the unknown stellar Lyman-α emission line, but we have learned to minimize this problem by observing high radial velocity stars and spectroscopic binary systems at opposite quadratures.

Figure 1 shows the derived D/H ratios for 12 stars with interstellar radial velocities indicating that their LOS pass through the Local Interstellar Cloud

238

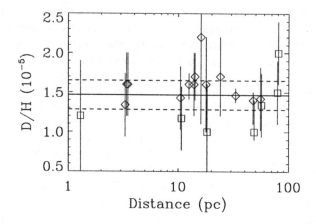

Figure 1: D/H ratios for interstellar gas toward all nearby stars observed with the GHRS. Diamond symbols are for gas in the LIC and square symbols are for other warm clouds.

(LIC) and 7 stars with LOS that pass through other warm clouds. The mean value for the LIC is $(D/H) = (1.50 \pm 0.10) \times 10^{-5}$ and the 1σ error bars for all 12 data points are consistent with the errors in the mean value. The data for other clouds are more scattered with $(D/H) = (1.28 \pm 0.36) \times 10^{-5}$ and the scatter may indicate real D/H variations. The figure shows the mean relation for all data points, $(D/H) = (1.47 \pm 0.18) \times 10^{-5}$. These results will be described in detail elsewhere. We conclude that the value of D/H in the tiny region of the Galaxy occupied by the LIC is now known, but we are just beginning to sample more distant lines of sight. STIS and FUSE data will allow us to study D/H further out in the Galactic disk and in the halo.

References

1. D. Tytler, S. Burles, & D. Kirkman, *Ap.J.* in press.
2. R. Lallement *et al.*, *A&A* **304**, 461 (1995).
3. R. Ferlet *et al.* in *Science with the HST-II*, ed. B. Benvenuti *et al.* (Space Telescope Science Institute, Baltimore, 1996).
4. J.L. Linsky *et al.*, *Ap.J.* **451**, 335 (1995).
5. D. Welty *et al.*, *Ap.J. Suppl.* **106**, 533 (1996).
6. J.L. Linsky & B.E. Wood, *Ap.J.* **463**, 254 (1996).
7. V.B. Baranov & Y.G. Malama, *JGR* **100**, 14755 (1995).

THE PRIMORDIAL HELIUM ABUNDANCE: AN UPDATE

Evan D. Skillman

Astronomy Department, University of Minnesota
116 Church St. SE, Minneapolis, MN 55455, USA

I review the current status on attempts to determine the primordial helium abundance. In particular, I review some of the most recent determinations and the current emphasis on evaluation of systematic effects. I hold a very optimistic outlook, as I think that the remaining areas of uncertainty (both theoretical and observational) will be addressed by programs which are already in progress.

1 Background

Since the original suggestion by Peimbert & Torres-Peimbert [1] the preferred method for determining the primordial helium abundance consists of measuring the He/H ratio in HII regions of ever decreasing metallicity and extrapolating this relationship back to zero metallicity. Over the last two decades, improvements in this technique have come from increasing the sample of low metallicity HII regions, decreasing the statistical uncertainties associated with the emission line ratios measured from the HII regions spectra, improvements in the understanding of the atomic physics and radiative transfer necessary to convert the emission line ratios into a helium abundance, and a better appreciation of possible systematic effects and their limits.

The work of Pagel et al. [2] represents a landmark study in this field. They presented new data, reanalyzed and homogenized data from the literature, and provided a detailed description of the methodology required. They found that the relationships between He/H and metallicity (using with both O/H and N/H) were linear to within the uncertainties. They then derived a value of the primordial helium abundance of 0.228 ± 0.005, or an upper limit of 0.242 (95% confidence interval) accounting for "reasonably likely" systematic effects.

Pagel et al. also noted that HII regions with Wolf-Rayet features in their spectra may have their abundances altered by the mass loss processes associated with the stars in the exciting cluster. However, Olive & Steigman [3] found no statistical evidence (from a study of the data set of Pagel et al.) in support of this hypothesis. Recently, Kobulnicky & Skillman [4] pointed out that the absence of low metallicity HII regions with Wolf-Rayet features in their spectra skews the comparison with normal HII regions. Constraining the comparison to the metallicity range in which Wolf-Rayet features are found, there is no evidence of a difference.

While the desired precision for the primordial helium abundance is better than 2%, the He/H determinations for the individual HII regions observed by Pagel et al. typically carry uncertainties in the range of 3 to 8%. Skillman et al. [5] obtained multiple observations of a very low metallicity HII region in the nearby dwarf galaxy UGC 4483 and showed that it is possible to determine the relevant emission line ratios with accuracies of order 2%, and spatially resolved spectra showed no evidence of the presence of neutral helium.

2 Recent Contributions

Recently Izotov et al. [6,7] have added new observations of HII regions over a large range in metallicity and have derived a new value of the primordial helium abundance, using exclusively their own data [7]. The resultant value of 0.243 ± 0.003 is significantly higher than that of Pagel et al. They attribute the bulk of the difference to the use of the improved He emissivities of Smits [8] and the collisional excitation rates calculated by Kingdon & Ferland [9].

Olive, Skillman, & Steigman [10] have shown that the new data of Izotov et al. are in excellent agreement with the previous abundance determinations of Pagel et al. They further show that the changes in the atomic data are negligible (as had been stated previously in the literature [8,9]). The main reason for the increased value derived by Izotov et al. is the rejection of observations of I Zw 18 (the HII region with the lowest measure metallicity [11]) from their analysis and a resulting lack of low metallicity data points.

3 Systematic Errors and Philosophical Differences

Given the quality of the data obtained so far and the attention to statistical and systematic uncertainties, one might consider that the primordial helium abundance determination is in good shape. In fact, there appears to be considerable concerns in the community about the currently published values. Since there is a lack of concordance with the recently obtained *low* values of D/H [12], it has been inferred that helium abundance determinations suffer from rather large systematic uncertainties. It has further been suggested that the best upper limit is derived by taking the extrapolated value, adding 2σ of statistical uncertainty, and adding (linearly!) every imaginable systematic uncertainty at its maximum possible value [13].

Olive, Skillman, & Steigman [10] have attempted to determine the best estimate given all the available data (0.234 ± 0.002). A thorough discussion of the systematic effects and their known limits is given therein, and a semi-empirical analysis supports the estimate of 0.005 given by Pagel et al. (resulting in a"firm" 2σ upper bound of 0.244). Both of the above approaches to systematic errors leave something to be desired. To best estimate the total systematic error, we will need to estimate not only the amplitude, but also the *distribution*, of each of the various uncorrelated sources of systematic errors.

4 Future Work

In order to significantly improve on the current status, we must:

(1) Increase the number of measurements of new regions with extremely low metallicities. This is being done by working with other groups that have been obtaining spectra of galaxies for the purposes of measuring large scale structure (Terlevich et al. in prep.).

(2) Use only measurements with uncertainties of 3% or less. With these measurements we will be able to establish the intrinsic dispersion in the He/H vs. O/H,N/H relationships and test for several classes of systematic errors.

(3) Continue to improve our knowledge of the basic atomic data on He and the relevant collisional and radiative transfer effects.

Acknowledgments

I am grateful for partial support from a NASA LTSARP grant No. NAGW-3189, and I wish to thank my collaborators for their encouragement.

References

1. M. Peimbert and S. Torres-Peimbert, ApJ **193**, 327 (1974).
2. B. E. J. Pagel, E. A. Simonson, R. J. Terlevich, and M. G. Edmunds, MNRAS **255**, 325 (1992).
3. K. Olive and G. Steigman, ApJS **97**, 49 (1995).
4. H. A. Kobulnicky and E. D. Skillman, ApJ **471**, 211 (1996).
5. E. D. Skillman, R. J. Terlevich, R. C. Kennicutt Jr., D. R. Garnett, & E. Terlevich, ApJ **431**, 172 (1994).
6. Y. I. Izotov, T. X. Thuan, and V. A. Lipovetsky, ApJ **435**, 647 (1994).
7. Y. I. Izotov, T. X. Thuan, and V. A. Lipovetsky, ApJ, in press, (1997).
8. D. P. Smits, MNRAS **278**, 683 (1996).
9. J. Kingdon and G. Ferland, ApJ **442**, 714 (1995).
10. K. Olive, E. D. Skillman, and G. Steigman, ApJ, in press, (1997).
11. E. D. Skillman and R. C. Kennicutt Jr., ApJ **411**, 655 (1993).
12. D. Tytler, X.-M. Fan, S. Burles, Nature **381**, 207 (1996).
13. D. Sasselov and D. S. Goldwirth, ApJ **444**, L5 (1995).

THE LITHIUM PLATEAU IN METAL-POOR STARS: CURRENT STATUS AND IMPLICATIONS ON BBN

SUCHITRA C. BALACHANDRAN

*Department of Astronomy, University of Maryland,
College Park, MD 20742, USA*

1 Introduction

The value of the primordial lithium abundance provides yet another constraint on the standard Big Bang model and is thus a quantity of much interest. This fragile element is easily destroyed in Population I stars; the surface layers of the Sun, for example, are depleted in lithium by a factor of 100 compared to their original (meteoritic) composition, presumably by circulation to the hotter interior. Given this evidence, the possibility of extracting the primordial value from the oldest, metal-poor stars in the Galaxy was not seriously considered until the remarkable discovery that these stars had lithium abundances about a factor of 10 larger than solar (Spite and Spite[1]). By some means not as yet understood, metal-poor stars are able to preserve lithium far more easily than their Population I counterparts. More significantly, metal-poor stars have nearly the same lithium abundance over a large effective temperature range (5500 - 6300 K) and over 1 dex spread in metallicity. Have metal-poor stars preserved *all* their initial lithium? And, if so, was all of it produced in the Big Bang?

It is sometimes claimed that the Population II plateau must be the primordial value because it is easily explained by the absence of lithium depletion in the standard stellar models. The concern over this interpretation may be understood in the context of the Population I data. In contrast to the Population II plateau, Population I stars exhibit a large variety of abundance patterns: the lithium 'dip' in the F stars and the decline in abundance with age in cooler stars coupled with the large spread seen even in very young clusters (see conference proceedings edited by D'Antona [2], Crane [3], and Spite & Pallavicini [4]). The solar depletion is not predicted by the standard models; the base of the surface convective zone is not hot enough to destroy lithium and "extra mixing" must be invoked. The large spread in lithium seen at each mass and age has suggested the role of other parameters, the most investigated of which is rotation. So far, none of the models, which include diffusion and a variety of rotationally-driven mixing processes, explains all of the observations adequately. It is therefore premature to assume that metal-poor stars have escaped the clutches of all of the non-standard mixing mechanisms that are so active in a metal-rich environment. Some non-standard models are able to produce a uniform depletion of lithium in metal-poor stars, albeit by an appropriate (but not necessarily unreasonable) adjustment of some free parameters. Given the overall inadequate understanding of mixing in Population I stars, much of the recent observational work has justifiably concentrated on determining whether the plateau stars have depleted any lithium.

2 Some Recent Results

Two approaches have been followed. First, using largely increased data samples, recent studies have attempted to determine whether there is an intrinsic dispersion in the lithium plateau larger than can be accounted for by observational and analysis errors. Such variations would require either depletion or enhancement from the star's initial abundance, the magnitude of which must then be ascertained before the primordial value can be determined. Studies by Deliyannis et al. [5], Thorburn [6] and Ryan et al. [7] suggested that the dispersion in lithium as a function of temperature was larger than observational error and pointed to a slope in the abundance as a function of temperature and metallicity. Deliyannis et al. examined the data in the observational plane (Li I equivalent width versus b-y color) in order to avoid errors in transformation to the theoretical abundance versus temperature plane. However they failed to account for the fact that b-y is not only temperature but also metallicity dependent. Thus the 20 % dispersion which they discovered in the plateau does *not* translate to as large a dispersion in the lithium abundance as they interpreted. Both Thorburn and Ryan et al. derived temperatures from colors. Citing underestimates in their temperature and equivalent width errors, the magnitude of the scatter in their studies was challenged by Spite et al. [8]. By a careful analysis using 3 different temperature determination techniques, they showed that extreme care was needed to fully constrain the sources of the error in the abundance analysis. However, since their sample was small, a statistical study of the plateau could not be performed. The recent work by Bonifacio and Molaro [9] provides a clear and definitive result. Rather than using color-temperature calibrations, they adopted temperatures from the extensive infrared flux method determinations of Alonso et al. [10]. Bivariate fits to the lithium plateau in temperature and metallicity showed the dispersion to be no larger than observational error. A small slope with temperature remains.

Detection of the more fragile ^{6}Li isotope in metal-poor stars has been considered to be a clear indicator that the extent of ^{7}Li depletion must be negligible. ^{6}Li has only been detected in *one* star by two independent groups

244

[11,12]. Measurements in two other stars are controversial. Detection in the first is unexpected because the star has depleted ^{7}Li. In the second, ^{6}Li is expected but not seen. In addition to the difficulty of the measurement, these ambigious results may reflect our incomplete understanding of the stellar atmosphere to the required accuracy. Additional ^{6}Li results are awaited with interest.

3 Conclusions

It is now clear that the metal-poor plateau is essentially flat and uniform. Once again it is urged that this is consistent with the standard stellar models and must thus represent the primordial value [9]. One crucial point must be noted. In the absence of mixing, gravitational settling is inevitable. Models predict that settling will result in a sharp drop in lithium at the hot end of the plateau. The flat plateau disputes this prediction. Settling may be curtailed by mixing. Does such mixing affect the rest of the plateau?

To the cosmologist I would urge patience. It is difficult to solve the Population II problem in isolation but persistence in the Population I problem will reveal many details about the inner workings of stars which, in addition to solving many lithium puzzles, may shed light on ^{3}He production at later evolutionary stages, yet another key parameter for BBN.

References

1. Spite, F., & Spite, M. 1982, *A&A*, 115, 357
2. D'Antona, F. 1990, *MSAIT*, 61, No.1
3. Crane, P. 1995 *The Light Element Abundances, ESO Astrophysics Symposia*
4. Spite, F. & Pallavicini, R. 1995, *MSAIT*, 66, No. 2
5. Deliyannis, C. P., Pinsonneault, M. H., & Duncan, D. K. 1993, *ApJ*, 414, 740
6. Thorburn, J. A. 1994, *ApJ*, 421, 318
7. Ryan, S. G., Beers, T. C., Deliyannis, C. P., & Thorburn, J. A. 1996, *ApJ*, 458, 543
8. Spite, M., Francois, P., Nissen, P. E., & Spite, F. 1996, *A&A*, 307, 172
9. Bonifacio, P. & Molaro, P. 1997, *A&A*, in press.
10. Alonso, A., Arribas, S., & Martinez-Roger, C. 1996, *A&AS*, 117, 227
11. Smith, V. V., Lambert, D. L., & Nissen, *ApJ*, 408, 262
12. Hobbs, L. M. & Thorburn, J. A. 1994, *ApJ*, 428, L25

NEUTRINO PHYSICS AND THE PRIMORDIAL ELEMENTAL ABUNDANCES

CHRISTIAN Y. CARDALL AND GEORGE M. FULLER

Department of Physics, University of California, San Diego, La Jolla,
CA 92093-0319, USA

Limits can be placed on nonstandard neutrino physics when big bang nucleosynthesis (BBN) calculations employing standard neutrino physics agree with the observationally inferred primordial abundances of deuterium (D), ^{3}He, ^{4}He, and ^{7}Li. These constraints depend most sensitively on the abundances of D and ^{4}He. New observational determinations of the primordial D and/or ^{4}He abundances could force revisions in BBN constraints on nonstandard neutrino physics.

1 Big Bang Nucleosynthesis and Neutrino Physics

The primordial elemental abundances depend on $(n/p)_{\rm WFO}$, the ratio of neutrons to protons at "weak freeze-out." Weak freeze-out occurs when the expansion rate of the universe exceeds the rates of the reactions that interconvert neutrons and protons. The ^{4}He abundance depends strongly on $(n/p)_{\rm WFO}$, while the abundances of D, ^{3}He, and ^{7}Li have a weaker dependence on this quantity.

The expansion rate can be parametrized by an "effective number of neutrinos," N_ν, which represents the relativistic degrees of freedom in addition to those contributed by photons and electron-positron pairs. N_ν also indirectly parametrizes any phenomenon that affects the ^{4}He abundance by altering $(n/p)_{\rm WFO}$ (even though the effect may be on the $n \leftrightarrow p$ interconversion rates rather than through the expansion rate).

The primordial elemental abundances also depend on the baryon-to-photon ratio η. The ^{4}He yield depends somewhat weakly on η, while the yields of the other light elements have a strong dependence on η. The strength of the dependence of the primordial elemental abundance yields on N_ν and η can be visualized from the figures in Refs. [1,2].

Many investigators over the years have studied the effects of nonstandard neutrino physics on BBN. These have included, for example, the number of neutrinos; neutrino mixing with sterile neutrinos; adding a mass, lifetime, and various decay products to the tau neutrino; and net cosmic lepton number, i.e. endowing the neutrino seas with chemical potentials.

Standard big bang nucleosynthesis calculations assume three massless neutrinos with zero net cosmic lepton number and no other relativistic degrees of freedom. When a range of baryon-to-photon ratio η is obtained for which the calculated primordial elemental abundances agree with those inferred from observations[3], nonstandard neutrino physics can be constrained. In the absence of a concordant range of η, one can either reasses the observational errors or claim nonstandard neutrino physics (or some other physics) as a solution to bring the calculated abundaces into line with the primordial abundances inferred from observations. It should be noted that nonstandard neutrino physics primarily affects $(n/p)_{\text{WFO}}$, and could therefore change ^{4}He rather easily. But more extreme neutrino physics would have to be introduced to significantly adjust the D, ^{3}He, and ^{7}Li abundances, and then fine tuning would be required to avoid unacceptable changes in the calculated ^{4}He abundance yield[1].

2 The Observational Situation

The observationally inferred abundances of the light elements produced in BBN have continued to be the subject of discussion and observational effort. Of particular interest are deuterium (D), which provides the most sensitive measure of the baryon density; and ^{4}He, which constrains N_ν. Also of interest is the primordial abundance of ^{7}Li, which provides a cross-check to the range of η determined by D and ^{4}He. The primordial abundance of ^{3}He is less useful for the purpose of determining η due to its complicated history of chemical and galactic evolution.

While inferences of the primordial D/H ratio formerly were based on "local" measurements in the solar system and interstellar medium, observations of quasar absorbtion systems (QAS) have recently provided determinations of D/H in a much more pristine environment. Initially there was a strong dichotomy in inferences of D/H, with claimed values differing by an order of magnitude[1]. However, at this conference the situation appears to be coming to a resolution, with the low value of D/H now favored[4]. It is also worth pointing out that the higher baryon density implied by low D/H is advantageous from other cosmological/astrophysical considerations[1,2].

The inferred primordial abundance of ^{4}He has also been the subject of recent action. A commonly accepted range of the ^{4}He abundance[5] does not provide a value of η concordant with that inferred from the low values of D/H inferred from QAS[1], for N_ν=3. (As it turns out, the primordial abundance of ^{7}Li inferred from the "Spite plateau" does not agree with low D/H for N_ν=3 either.) One group of observers has claimed[6] a higher range of ^{4}He abundance, but this analysis may have some difficulties[5]. It is not unreasonable, however, that systematic errors in the determinations of ^{4}He and ^{7}Li could allow for concordance at low D/H for N_ν=3.

3 An Example: Mixing with Sterile Neutrinos

Neutrino mixing with "sterile" (SU(2) singlet) neutrinos during the epoch of BBN, which has been studied by many authors [7], provides an example of how constraints on nonstandard neutrino physics depend on the primordial abundances of D and ^{4}He. Two kinds of mixing with sterile neutrinos have been suggested to solve neutrino puzzles: $\nu_e \leftrightarrow \nu_s$ for the solar neutrino problem, and $\nu_\mu \leftrightarrow \nu_s$ for the atmospheric neutrino problem.

Neutrino mixing with steriles does two things to increase primordial ^{4}He: first, it effectively brings another degree of freedom into thermal contact; and second, it depletes the ν_e population, reducing the $n \leftrightarrow p$ interconversion rates. The demand that ^{4}He not be overproduced places limits on mixing with steriles. The constraint comes from upper bounds on D/H and the ^{4}He mass fraction. This is because the largest D/H (and hence the minimum allowed η) allows the largest N_ν for a given ^{4}He abundance.

In constraining mixing with steriles, previous studies assumed, for example, $\eta > 2.8 \times 10^{-10}$ based on estimates of primordial (D+^{3}He)/H derived from "local" measurements. Until this conference, it appeared that discordant deuterium determinations could be indicating a minimum value of η that was either lower or higher than this. As an example of the consequences of a possible change in the minimum value of η, Cardall and Fuller published a calculation [7] in which the $\nu_\mu \leftrightarrow \nu_s$ solution to the atmospheric neutrino problem was *assumed*, and the resultant ^{4}He mass fraction was plotted as a function of D/H. For the purported "high" values of D/H, it appeared that the $\nu_\mu \leftrightarrow \nu_s$ solution to the atmospheric neutrino problem might be allowed. For low D/H, this mixing is even more strongly ruled out.

References

1. C. Y. Cardall and G. M. Fuller, Astrophys. J. **472**, 435 (1996)
2. G. M. Fuller and C. Y. Cardall, Nucl. Phys. B (Proc. Suppl.) **51B**, 71 (1996)
3. C. J. Copi, D. N. Schramm, and M. S. Turner, Science **267**, 192 (1996)
4. C. Hogan, these proceedings; D. Tytler, these proceedings.
5. K. A. Olive, E. Skillman, and G. Steigman, astro-ph/9611166; E. Skillman, these proceedings.
6. Y. I. Izotov, T. X. Thuan, and V. A. Lipovetsky, Astrophys. J., submitted (1996).
7. C. Y. Cardall and G. M. Fuller, Phys. Rev. D **54**, 1260 (1996); and references therein.

ADDITIONAL ENERGY AT THE EPOCH OF PRIMORDIAL NUCLEOSYNTHESIS

MERAV OPHER[a], REUVEN OPHER[b]

Instituto Astronômico e Geofísico - IAG/USP, Av. Miguel Stéfano, 4200
CEP 04301-904 São Paulo, S.P., Brazil

We derive the electromagnetic spectrum in the epoch of primordial nucleosynthesis, based on the *Fluctuation-Dissipation Theorem*. Our description includes thermal and collisional effects in a plasma. We show that it differs from the blackbody spectrum in vacuum that is usually assumed. We estimate the additional energy that exists in the plasma, besides the energy of the blackbody in vacuum. For example, at the beginning of the primordial nucleosynthesis, at $T = 0.8\ MeV$, the additional energy is $1\%\rho_\gamma$, where ρ_γ is the photon energy density.

1 Introduction

It is usually assumed in standard Big Bang Nucleosynthesis calculations that the primordial plasma was a homogeneous plasma and that the electromagnetic field was a blackbody spectrum in vacuum. In the energy density calculation, for example, the energy density of the photons is given by the energy density of the blackbody spectrum in vacuum. In this manner, the Primordial Universe is treated as an ideal gas: *collective effects* are assumed to be negligible. However, a plasma differs from an ideal gas due to correlations.

A plasma, even in thermal equilibrium, has fluctuations, that is, physical variables such as temperature, density and electromagnetic fields, fluctuate. The electromagnetic fluctuations are described by the *Fluctuation-Dissipation Theorem*[1,2]. The intensity of such fluctuations is highly dependent on how the plasma is described, for example, on the dissipation mechanisms present in it. It is necessary to describe the plasma in the most complete way. Cabell and Tajima[3,4,5] studied the magnetic field fluctuations for a cold plasma description with a constant collision frequency and for a warm plasma in a collisionless description.

We present a model that includes in the same description collisional and thermal effects. By using the *Fluctuation-Dissipation Theorem* relations, we derive the electromagnetic spectrum in a plasma. In Section II we present the expression of the electromagnetic fluctuations and our model, and in Section III the results and conclusions.

---

[a]email: merav@orion.iagusp.usp.br
[b]email: opher@orion.iagusp.usp.br

2 Electromagnetic Fluctuations and BGK Collision Term

The spectrae of fluctuations of the magnetic and electric fields, for an isotropic plasma, given by the *Fluctuation-Dissipation Theorem* [1,2] are

$$\frac{\langle E^2 \rangle_{\mathbf{k}\omega}}{8\pi} = \frac{\hbar}{e^{\hbar\omega/T} - 1} \frac{Im\ \varepsilon_L}{|\ \varepsilon_L\ |^2} + 2 \frac{\hbar}{e^{\hbar\omega/T} - 1} \frac{Im\ \varepsilon_T}{|\ \varepsilon_T - \left(\frac{kc}{\omega}\right)^2\ |^2} \tag{1}$$

$$\frac{\langle B^2 \rangle_{\mathbf{k}\omega}}{8\pi} = 2 \frac{\hbar}{e^{\hbar\omega/T} - 1} \left(\frac{kc}{\omega}\right)^2 \frac{Im\ \varepsilon_T}{|\ \varepsilon_T - \left(\frac{kc}{\omega}\right)^2\ |^2} . \tag{2}$$

The first term of Eq. (1) is the longitudinal electric field fluctuations and the second term is the transverse electric field fluctuations.

In order to have a description that includes thermal effects as well as collisional effects, we use the Vlasov equation in first order [6] with the BGK collision term [7]. The transverse dielectric permittivity obtained is:

$$\varepsilon_T(\omega, \mathbf{k}) = 1 + \sum_\alpha \frac{\omega_{p\alpha}^2}{\omega^2} \left(\frac{\omega}{\sqrt{2}kv_\alpha}\right) Z\left(\frac{\omega + i\eta_\alpha}{\sqrt{2}kv_\alpha}\right) , \tag{3}$$

where α is the label for each species of the plasma, v_α is the thermal velocity for each species, η_α is the collision frequency and $Z(z)$ is the Fried & Conte function [8]. With this transverse dielectric permittivity we obtain the electromagnetic spectrum, that is, the sum of the magnetic and the electric field transverse spectrae. In Figure 1, we plot the electromagnetic spectrum (divided by the normalization $S_0 = \omega_{pe}^2 k_B T/c^3$) vs ω/ω_{pe}, where ω_{pe} is the plasma electron frequency. We use an electron-positron plasma at $T = 0.8\ MeV$ and $n_e = 1.1 \times 10^{31}\ cm^{-3}$, that is, the plasma at the beginning of the Primordial Nucleosynthesis era. The dotted curve is our model compared to the blackbody spectrum in vacuum (the solid curve).

3 Results and Conclusions

The electromagnetic spectrum obtained behaves like a blackbody spectrum in vacuum for high frequencies. However, for low frequencies, it is distorted. It has more energy than the blackbody spectrum in vacuum. To estimate the additional energy, the longitudinal energy also has to be added. The longitudinal energy is obtained from the longitudinal electric field spectrum with the longitudinal dielectric permittivity obtained from our description. Calculating the additional energy, for example, for $T = 0.8\ MeV$, we obtain that the additional energy is $1\%\rho_\gamma$, where ρ_γ is the photon energy density.

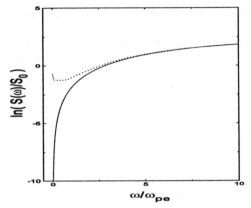

Figure 1: The electromagnetic spectrum for the electron-positron plasma at $T = 0.8$ MeV and $n_e = 1.1 \times 10^{31}$ cm^{-3}. The dotted curve is our model and the solid curve is the blackbody spectrum in vacuum.

Acknowledgments

The authors would like to thank Swadesh Mahajan for useful suggestions, especially concerning the BGK collision term. M.O. would like to thank the Brazilian agency FAPESP for support and R.O. would like to thank the Brazilian agency CNPq for partial support.

References

1. A.G. Sitenko, *Electromagnetic Fluctuations in Plasma* (Academic Press, NY, 1967).
2. A.I. Akhiezer, I.A. Akhiezer, R.V. Plovin, A.G. Sitenko, and K.N. Stepanov, *Plasma Electrodynamics* (Pergamon Press, Oxford, 1975).
3. S. Cabell, and T. Tajima, *Phys. Rev.* A, **46**, 3413 (1992).
4. T. Tajima, and S. Cable, *Phys. Fluids* B, **4**, 2338 (1992).
5. T. Tajima, S. Cable, K. Shibata, and R.M. Kulsrud, *ApJ*, **390**, 309 (1992)
6. S. Ichimaru, *Basic Principles of Plasma Physics* (Addison-Wesley, Redwood City, CA, 1992).
7. P.C. Clemmow, and J.P. Dougherty, *Electrodynamics of Particles and Plasmas*, (Addison-Wesley, Redwood City, CA, 1990).
8. B.D. Fried, and S.D. Conte, *The Plasma Dispersion Function* (Academic Press, NY, 1961).

INTRODUCTION & OVERVIEW: CMB SESSIONS

G.F. SMOOT

Lawrence Berkeley National Laboratory,
University of California, Berkeley CA 94720, USA

This is a very exciting time for the CMB field. It is widely recognized that precision measurements of the CMB can provide a definitive test of cosmological models and determine their parameters accurately. At present observations give us the first rough results but ongoing experiments promise new and improved results soon and eventually satellite missions (MAP and COBRAS/SAMBA now named Planck) are expected to provide the requisite precision measurements. Other areas such as observations of the spectrum and Sunyaev-Zeldovich effect are also making significant progress.

1 Introduction

There has long been anticipation that cosmic microwave background (CMB) radiation would provide significant information about the early Universe due to its early central role and its general lack of interaction in the later epochs.

1.1 COBE

Though there have been many observations of the CMB since its discovery by Penzias and Wilson [1] in 1964, the Cosmic Background Explorer satellite, COBE, provided two watershed observations. The first key observation is that the CMB is extremely well described by a black-body spectrum [2,3]. This observation of the CMB thermal origin strongly affirms the hot Big Bang predictions and tightly constrains possible energy releases, ruling out explosive and other exotic structure formation scenarios. The spectral measurement of the temperature as a blackbody and the dipole anisotropy as its derivative provides the basis of knowledge of the spectral shape of the higher order CMB anisotropies and thus a means to separate them from the foregrounds.

The detection of primordial CMB anisotropies [4] is the second key from COBE. The COBE large angular scale map of the microwave sky and detection of intrinsic anisotropies provides support for the gravitational instability picture and thus a link to large scale structure, an anchor point in the magnitude of fluctuations, an impetus and guidance to the field as a whole.

1.2 Theory

In addition to the increased attention to observations and the development of experimental techniques, a major thrust in the field has been the improvement in theory and ideas for extracting information from the data. Theoretical work has provided ideas and means to calculate the observable effects of various cosmological scenarios from standard Cold Dark Matter (sCDM) and its variants, various inflationary models through to nearly a good description of topological defects (an area still in active development). Other work has lead to a better physical intuition of the mechanisms involved [5,6,?].

The inverse problem of extracting the scenario and the appropriate cosmological parameters including error estimates is an active area following the pioneering paper by Llyod Knox[7] and a seminal paper by Jungman et al.[8]. The understanding that one can extract cosmological parameters with accuracy is now driving the excitement in the field in equal measure with the knowledge that the CMB anisotropies are there and are observable.

2 Current Observations

Since the COBE DMR detection of anisotropy, over a dozen groups have reported anisotropy detections and some interesting upper limits. The current power spectrum observations are summarized in the left panel of Figure 1.

For comparison the right panel shows the predicted anisotropy power spectrum for a number of different cosmological models.

3 Future and CMB Sessions

The CMB field is extremely active and exciting precisely because of the combination of rapid observational and theoretical development with definitive space missions in the coming decade and of the expectation that those observations will provide us with accurate determination of cosmological parameters. The great progress and interest are shown by the coverage in the invited talks by Scott Dodelson[9], Lyman Page[10], and Neil Turok[11] and in the talks in these two CMB sessions as well as posters and discussions by the meeting attendees.

A large total effort is necessary to achieve these lofty aims. The many people in the field must work together in a combination of competition and collaboration. Collaboration is necessary because the tasks are large and difficult as well as subtle. Competition is necessary as many people, especially the large number of young and very excellent scientists, must continue to establish their careers and accomplishments and because some competition is useful to

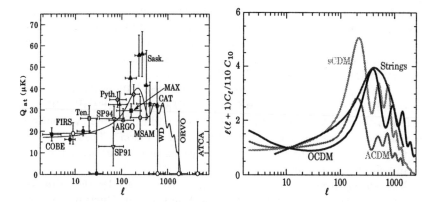

Figure 1: **Left Panel:** Current CMB anisotropy power spectrum observations. The horizontal axis ℓ is Legrendre polynomial number and is inversely proportional to the angular scale. The vertical axis is the RMS (rather than power) temperature variation for that angular scale. The solid curve is the calculated anisotropy for sCDM. **Right Panel:** Theoretical CMB anisotropy power spectrum for various models. The vertical axis is the temperature anisotropy power for that angular scale.

keep eveyone on their toes. However, it is important to keep in mind that our ultimate goal is the accurate testing and determination of cosmological models and parameters and that transcends any one group. The total resources that are and will be allocated to this effort are temendous - both in financial and human capital terms. It is everone's responsibility to work towards this ultimate goal and especially for the senior scientists to promote this and set a tone of friendly and helpful competitive cooperation. The rapid open sharing of data is a good standard but more is needed.

In addition to this cooperation a number of things must fall into place. The first is outstanding instrumentation and techniques for making the observations. Quality instrumentation is particularly important for the forthcoming satellite missions as they are costly in terms of money and mission opportunity.

The next is an understanding of how to process the data and then to turn the calibrated data into maps, power spectra, and other useful forms. This is intimately linked to understanding both the instruments and the foregrounds: galactic and extragalactic including the SZ effect. Work is needed in this area and has begun.

Finally the calibrated separated data must be used in the inverse problem to test cosmological models and recover their best-fitted parameters.

All of these efforts require a significant advancement in their technology. The talk by John Carlstrom [12] on making Sunyaev-Zeldovich maps using new

technology HEMT (high electron mobility transistor) amplifiers is an example
of the advances that can be made in a field with new techniques/technologies.
We have only recently come to appreciate the computorial complexity of utilizing a million-plus pixel map. Already a number of approaches are being tried
and programs being developed to address these issues.

The existence of already ongoing programs, e.g. the balloon-borne instruments: BOOMERANG/MAXIMA, MSAM/TOPHAT, HACME/BEAST, and
the interferometers: VCA, CBI, VSA, will provide additional and appropriate
data to test these techniques in a very short time horizon. These plus the
pressure from the future missions and observations provide us with both an
exciting and challenging field.

Acknowledgments. This work is supported in part by the Director,
Office of Energy Research, Office of High Energy and Nuclear Physics,
Division of High Energy Physics of the U.S. Department of Energy under
Contract No. DE-AC03-76SF00098.

References

1. A.A. Penzias and R.Wilson *Ap. J.* **142**, 419 (1965).
2. J. Mather *et al Ap. J.* **420**, 420 (1994).
3. D. Fixsen *et al Ap. J.* **424**, 1 (1996). astro-ph/9605054
4. G.F. Smoot. *et al, Ap. J.* **396**, 1 (1992).
5. W. Hu & M. White *Ap. J.* **xx**, 1 (1996). astro-ph/9609105, 9609079, 9604166
6. A. Albrecht *et al, Ap. J.* **xx**, 1 (1996). astro-ph/9612017
7. L. Knox *Phys. Rev.* D **D52**, 4307 (1995). astro-ph/9504054
8. Jungman *et al, Phys. Rev.* D **54**, 1332 (1996). astro-ph/9512139
9. S. Dodelson, These Proceedings "CMB Theory" 1997.
10. L. Page, These Proceedings "CMB Experiments" 1997.
11. N. Turok, These Proceedings "Early Universe" 1997.
12. J. Carlstrom, These Proceedings "SZ Observations" 1997.

THE MAXIMA AND BOOMERANG EXPERIMENTS

S. Hanany

The Center for Particle Astrophysics, University of California, Berkeley

The MAXIMA and BOOMERanG Collaborations
California Institute of Technology
Center for Particle Astrophysics, University of California, Berkeley
IROE-Firenze
Queen Marry and Westfield College
University of California, Berkeley
University of California, Santa Barbara
University of Rome

MAXIMA and BOOMERanG are balloon borne experiments designed to map the cosmic microwave background anisotropy power spectrum from $l = 10$ to $l = 900$ with high resolution. Here we describe the program of observations, the experiments, their capabilities, and the expected performance.

1   The MAXIMA/BOOMERanG Program

Observations of the cosmic microwave background anisotropy (CMBA) sky will be carried out by two balloon borne observatories, MAXIMA and BOOMERanG. We plan three consecutive flights in the next few years: a 24 hour north American flight of the BOOMERanG payload, a single night flight of MAXIMA, and a $\sim$ 7 day circumpolar flight of BOOMERanG in Antarctica. Throughout the program the CMBA power spectrum will be measured in the range $10 < l < 900$ with $\Delta l \leq 60$. Both experiments employ arrays of high sensitivity bolometric detectors in multiple frequencies. Figure 1 shows the focal plane array configurations of the three flights, and Table 1 lists the l space coverage, frequency bands, expected sky coverage and signal to noise per pixel given our expected instrument sensitivity and scan strategy. The payloads have been described recently by Debernardis *et al*[1], Hanany *et al*[2], and Lee *et al*[3].

2   Focal planes and Scan Strategies

The BOOMERanG focal plane for the north American flight stresses high optical efficiency and large sky coverage with modest spatial resolution. During CMBA observations the entire focal plane is swept in azimuth in a full 360 degree rotation at a rate of 1 RPM. The elevation is kept constant at 45 degrees. Data will be recorded in a total power mode, however the large

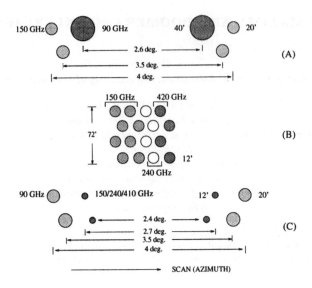

Figure 1: The focal plan configuration for (A) BOOMERanG north America, (B) MAXIMA and (C) BOOMERanG's Antarctic flight. The circles indicate the relative position and FWHM of the beams projected onto the sky. Each circle also represents a single frequency photometer in the instruments' focal plane, except for the 12' beams in case (C) in which the circles represent a multi frequency photometer. All the focal planes are scanned in azimuth as described in the text.

separation of pixels on the focal plane allows for the synthesis of multiple window functions if differencing of pixels proves to be required. By comparing our scan pattern during a typical night in April 1997 to The DIRBE 240 μm map of dust emission, and assuming a dust spectrum of $\nu I_\nu = \nu^{1.5} B_\nu (18K)$, we find that most of the scanned region has an RMS dust emission $\leq 4\,\mu K$ at 90 GHz.

The BOOMERanG focal plane for the Antarctic flight will have large spectral coverage and small beams. The long integration time together with the scan strategy will yield high signal to noise per pixel, in a low dust emission region. The high signal to noise will allow excellent identification and discrimination of systematic errors. During observations the focal plane is scanned 50 degrees peak to peak in azimuth at a rate of 0.7 deg/sec. The elevation is slowly changed between 33 and 65 degrees. The gondola will be pointed away from the sun, and during observations all direct paths between the sun and the payload are intercepted by sun shields. The entire region of observation is scanned during a single day, and the observation is then repeated throughout the $\sim$ 150 hours of flight. Comparison with IRAS and DIRBE map indicates that almost the entire region scanned in this flight has an exceptionally low

Table 1: The table summarizes the expected performance of the different flights in the program, given expected instrument noise, and scan strategy. 10, 5, and 150 hours of integration are assumed for the BOOMERanG north America, MAXIMA, and BOOMERanG long duration, respectively.

Flight	l space coverage	ν (GHz)	pixels	ΔT/pixel (μK)
BOOMERanG North America	$10 < l < 400$ $\Delta l = 25$	90	30,000	25
		150	120,000	55
MAXIMA	$60 < l < 900$ $\Delta l = 60$	150	26,000	24
		240	26,000	66
		410	26,000	
BOOMERanG Long Duration	$40 < l < 900$ $\Delta l = 60$	90	24,000	9
		150	66,000	16
		240	66,000	25
		410	66,000	

dust emission.

The MAXIMA focal plane employs high sensitivity 100 mK bolometers with high efficiency single frequency photometers. During observations the focal plane will be swept in azimuth in a superposition of two motions: a 4 deg/sec. triangular wave modulation with an amplitude of 6 degrees, achieved by motion of the primary mirror, and a slower slew rate of the entire gondola over 50 degrees peak to peak. For half of the observing period ($\sim$ 2.5 hours) the elevation will be fixed on the North Celestial Pole, and for the other half the elevation will be fixed at 50 degrees. Part of the region of observation will be scanned in both halves of the flight, assisting in systematic error rejection and map reconstruction.

1. P. Debernardis *et al*, Proceedings of the XXXI Moriond Conference, "Cosmic Microwave Background Anisotropies", in publication.
2. S. Hanany *et al*, Proceedings of the XXXI Moriond Conference, "Cosmic Microwave Background Anisotropies", in publication; astro-ph/ 9609098.
3. A. Lee *et al*, Proceedings of the XXXI Moriond Conference, "Cosmic Microwave Background Anisotropies", in publication.

The Very Compact Array

Mark Dragovan

*The Enrico Fermi Institute, The University of Chicago,
5640 S. Ellis Ave, Chicago, IL 60637, USA*

(For the VCA Collaboration [1])

The Very Compact Array is an interferometric array under construction designed
to image the anisotropy of the Cosmic Microwave Background on intermediate
angular scales.

1 Introduction

The Cosmic Microwave Background (CMB) has a wealth of information about
the origin and evolution of the Universe encrypted in its signal. Its frequency
spectrum is that of a blackbody at 2.7K, confirmation of the prediction made
by the standard big bang model. Its angular power spectrum contains information on the structure that existed at decoupling. Anisotropy in the matter
distribution at this time left imprinted upon the CMB a small anisotropy
($\Delta T/T \sim 10^{-5}$) in the angular distribution of microwave power. The amplitude and spatial distribution of the anisotropy are directly related to the
conditions in the early Universe which gave rise to the onset of structure
formation. All current models of structure formation predict specific power
spectra and morphologies for the CMB anisotropy. Determining the angular
power spectrum of the fluctuations, their distribution function, and imaging
the CMB anisotropy at intermediate angular scales (15' to 1°) is the goal of
this experiment, the Very Compact Array (VCA).

The advent of low–noise, broadband millimeter–wave amplifiers has made
interferometry a particularly attractive technique for detecting and imaging low
contrast emission, such as anisotropy in the CMB. An interferometer directly
measures the Fourier transform of the intensity distribution on the sky. By
inverting the interferometers output, images of the sky are obtained which
include angular scales determined by the size and spacing of the individual
array elements. Interferometry has successfully imaged the Sunyaev-Zeldovich
effect in distant clusters of galaxies at small angular scales using the existing
OVRO and BIMA arrays [2].

An interferometric system offers several desirable features: 1) An interferometer *directly* measures the power spectrum of the sky, in contrast to
differential or total power measurements. 2) An interferometer is intrinsically

stable since only correlated signals are detected; difficult systematic problems that are inherent in total power and differential measurements are absent in a well designed interferometer. 3) An interferometer can be designed for continuous coverage of the CMB power spectrum with the resolution determined by the number of fields imaged (the VCA is sensitive to scales from 0.25 to 1.15°). 4) With narrow band channels (1 GHz) and sufficiently wide total bandwidth (26 GHz to 36 GHz), the spectral index of the imaged emission can be determined and used to distinguish between true CMB fluctuations and galactic foregrounds [3].

Two interferometers are under construction in the US: 1) The Very Compact Array (VCA), a 13 element array of densely packed feedhorns arranged for continuous coverage over angular scales from 0.25° to 1.15°, and 2) The Cosmic Background Imager (CBI), currently under construction at Caltech, an array similar to the VCA, but with larger, Cassegrain elements arranged for continuous coverage over angular scales from 4' to 20'. The two instruments are designed to complement each other; together they span a factor of 17 in angular scale, and will sample the same region of the southern sky. In a three month observing season at the Pole the VCA will image a 2500 deg^2 region. This will give an excellent determination of the CMB power spectrum and, in the context of the popular CDM inflationary theory, will constrain $\Omega \equiv \rho/\rho_c$ to better than 20%. The ultimate goal is to image $\sim 25\%$ of the entire sky with the VCA and determine $\Omega, h, \Omega_b h^2$, and n to better than 10% [4].

2 Overview of Technical Approach

The interferometer consists of 13 scalar feed horns arranged in a three-fold symmetric extremely closed packed configuration which fills $\sim$50% of the aperture area to provide maximum brightness sensitivity. The feed horns are followed by low-noise HEMT amplifiers operating at 26 – 36 GHz with noise temperatures of $\sim$10 K. The amplified signals are then filtered and down-converted to 10 GHz wide IFs. The receivers are contained in a single cryostat cooled by two closed-cycle refrigerators; the horn, RF amplifier, filter, mixer, and first IF amplifier are all cooled to $\sim$ 15 K. An analog correlator performs the complex multiplications for the 78 baselines in 10 bands 1 GHz wide. The low noise, large bandwidth, and reliability of the HEMT amplifiers make them ideally suited to this experiment. Scalar feed horns offer the lowest sidelobes of any antenna system and also allow the elements to be densely packed with minimum crosstalk.

3   Observing strategy

The sensitivity of the VCA allows a 3° FWHM field to be imaged with 0.25° to 1.15° pixels and a sensitivity ranging from 4 to 10 μK per pixel in only 24 hours. This fast imaging speed makes it possible to mosaic large regions of the sky [5]. The CMB power spectrum can be recovered with channel widths set by the number of fields mosaiced. Mosaicing will allow the VCA to resolve any features found in the power spectrum of the CMB within $l = 150$ to 700.

4   The Antarctic Polar Site

The high altitude plateau at the South Pole is an exceptional site [6]. The low atmospheric opacity and smooth continuous airflow are ideal conditions for large field imaging. Long periods of excellent transparency and low sky noise have been used to great advantage by the Python experiment operating at 90 GHz [7]. The atmospheric emission is unlikely to lead to structure in an image made from integrations lasting several hours or more [8]. The low opacity and steady atmospheric conditions at the pole are ideal for imaging the CMB.

5   Summary

A Very Compact Array is under construction to image the anisotropy of the CMB at intermediate angular scales. Together with the Cosmic Background Imager, the two instruments will image the microwave sky with high signal to noise on angular scales ranging from 4' to 1.5°. Determining the power spectrum of the CMB anisotropy, as well as the statistical properties of its distribution function, are primary goals of modern experimental cosmology and are the scientific focus of the VCA and CBI interferometers.

1. J. Carlstrom, M. Dragovan, N. Halverson, W. Holzapfel, (Univ. of Chicago), J. Cartwright, S. Padin, T. Pearson, A.C. Readhead, M. Shepard, (Caltech), M. Joy (MSFC), S. Meyers (U Penn).
2. J. Carlstrom, these proceedings.
3. Brandt et al. Ap.J. **424**, 1 (1994).
4. White et al., in preparation (1997).
5. Cornwell, T. J., et al. Astron. Astrophys. **271**, 697 (1993).
6. Appl.Opt. **29(4)**, 463 (1990); Appl.Opt. **33(6)**, 1095 (1994); Ap.J. **476**, 428 (1997).
7. Ap.J. **427**, L67 (1994); Ap.J. **453**, L1 (1995); Ap.J. **475**, L1 (1997).
8. Church, S.E. MNRAS **272**, 551 (1995).

SUNYAEV ZELDOVICH OBSERVATIONS WITH THE OVRO AND BIMA ARRAYS

J. E. CARLSTROM, L. GREGO, W. L. HOLZAPFEL

Dept. of Astronomy & Astrophysics, University of Chicago, Chicago, IL 60637

M. JOY

Space Science Laboratory, NASA/MSFC, Huntsville, AL 35812

We present high signal to noise images of the Sunyaev Zeldovich Effect toward distant clusters. The data were obtained with low-noise HEMT amplifier cm-wave receivers mounted on the OVRO and BIMA mm-wave arrays. A total of eighteen clusters have been imaged successfully. We also discuss planned improvements to the instrument and the type of observations enabled.

1   Introduction

The Sunyaev Zeldovich Effect (SZE) is a small spectral distortion of the 3 K cosmic microwave background radiation resulting from inverse-Compton scattering of the microwave photons as they pass through the hot X–ray emitting gas in the extended atmospheres of clusters of galaxies. The spectral distortion appears as a decrement in the CMB brightness at frequencies below 218 GHz (> 1.4 mm) and as an increment at higher frequencies (see review by Sunyaev & Zel'dovich 1980). Although the magnitude of the effect is small ($\leq$ 1 mK at radio wavelengths) its promise of providing an independent estimate of the Hubble constant and of the peculiar motions of distant clusters of galaxies has continued to motivate increasingly sensitive observations since the effect was predicted 25 years ago.

The observable quantity $\Delta T/T_{CMB}$ is directly proportional to the electron density integrated along the line of sight weighted by the electron temperature. This leads to the remarkable fact that the magnitude of the observed quantity is independent of the distance to the cluster; it is only dependent on physical conditions within the cluster gas. Thus, sensitive SZE observations are potentially powerful probes of the high-z universe.

Thermal emission from the hot cluster gas is observed directly at X-ray wavelengths with a strength proportional to the square of the electron density integrated along the line of sight. It is the different dependences on electron density of the X-ray and SZE observables which allow a determination of the distance to the cluster, and therefore the Hubble constant. Such a distance determination is only dependent on the physics of the cluster gas. Systematics in this method for deriving the Hubble constant are therefore completely independent from those in other methods.

2 Observations and Results

The observations were obtained by outfitting the OVRO and BIMA mm-wave arrays with low-noise cm-wave receivers. The system as installed at OVRO and the first images obtained are discussed in Carlstrom, Joy and Grego (1996). The key advantage of the system is the ability to use the power of interferometric techniques and low-noise receiver technology to image large angular scales (up to $2'$) at cm-wavelengths.

In Figure 1 we present SZE images toward 8 clusters. The rms noise is typically 15 to $30\mu K$. Each cluster was observed for an average of ~ 6 transits, roughly 45 hours.

Preliminary estimates of the Hubble constant for the clusters CL0016+16 and MS0451 give values in the range 50 - 65 km s^{-1}/Mpc.

3 Future Plans

We are expanding the system from 6 to 9 telescopes and increasing the sensitivity of the receivers. The imaging speed should be increased by a factor of 4 or more. With this system we hope to determine the number density of high-z clusters by surveying a degree size region. The added sensitivity will also allow us to expand our targeted cluster observations. Lastly, it will allow higher quality images of clusters which can be used to better constrain the physical conditions within the cluster gas.

Acknowledgments

We gratefully acknowledge M. Pospieszalski for the Ka–band HEMT amplifiers. We thank the entire OVRO and BIMA staffs, in particular J. Lugten, S. Padin, R. Plambeck, S. Scott and D. Woody. JEC acknowledges support from an NSF-YI award and the David and Lucile Packard Foundation.

References

1. Carlstrom, J., Joy, M., and Grego, L. 1996, Ap.J. Letters, 456, L75.
2. Sunyaev, R. A. and Zel'dovich, Ya. B. 1980, ARAA, 18, 537.

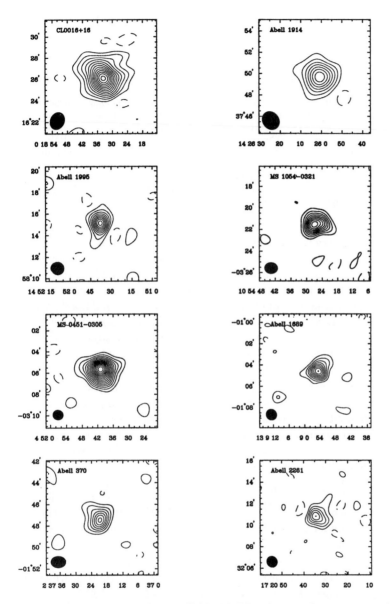

Figure 1: A sample of our Sunyaev Zeldovich Effect images. Contours are multiples of -2σ. The beam FWHM is shown by the filled black symbol in the lower left corner of each panel. Data for the top two images were obtained with BIMA; data for the lower six images were obtained with OVRO.

FOREGROUNDS AND CMB MEASUREMENT ACCURACY

F.R. BOUCHET

Institut d'Astrophysique, CNRS, 98 bis Boulevard Arago, Paris, F-75014, FRANCE

In order to predict the accuracy of the determination of the cosmological parameters of a given experiment, one assumes model power spectra for the CMB and their expected error bars. On the other hand, many foregrounds also contribute to the microwave fluctuations. The reachable accuracy will thus depend critically on the ability of different experiments to separate the true CMB fluctuations from those of other sources. Here I outline a foreground model introduced earlier and estimate the resulting error bars when wiener filtering is used to analyse the data. I compare with other estimates obtained by only combining detector noise and angular resolution figures for the experiment.

To model the galaxy [1], I use the spectral indexes $\alpha = -0.9, -0.16$ for the synchrotron & free-free emission, and the dust is assumed at 18K with an emissivity $\propto \nu^2$. But the emissions from dust and free-free are partially correlated [2], and comparisons with other tracers of the free-free emission show that the HI correlated free-free emission accounts for most (at least 50%), and maybe most, of that component. It is assumed that there may be a second, HI uncorrelated, component accounting for 5% of the total dust and 50% of the free-free emission. In conjunction with analyses of spatial templates (e.g. IRAS, DIRBE, Haslam 408 MHz), this leads to the following angular power spectra $\ell C_\ell^{1/2} = c\,\ell^{-1/2}\ \mu K$, with $c_{sync} = 2.1$, $c_{HI-U} = 8.5$ and $c_{HI-C} = 20.6$ at 100 GHz, where $HI - U$ and $HI - C$ stand for Dust+Free-free components, uncorrelated and correlated (respectively) with HI (which implies $c_{free} = 13.7$, and $c_{dust} = 13.5$). These normalisations should be appropriate for the "best" half of the sky, at scales $\ell \gtrsim 10$. For the unresolved background from radio-sources and infrared galaxies (i.e. once 3σ fluctuations are removed), I use the estimate by Hivon et al. [3] $\ell C_\ell^{1/2} = \frac{2.07}{\nu^5} \sinh^2\left(\frac{\nu}{113.6}\right) \left[\nu + \frac{10^{-15}\,\nu^8}{[1+(\nu/2500)^8][1+(\nu/1500)^2]}\right]^{1/2}$ K, with ν in GHz, which yields $\ell C_\ell^{1/2} = 2.2 \times 10^{-3}\ \ell\ \mu K$ at 100 GHz, although it might seriously underestimates the contribution from radio-sources. To evaluate the contribution from the Sunyev-Zeldovich effect from clusters of galaxies, I analysed maps [4] of compton parameter y and found that $\ell C_\ell^{1/2} = c_{ysz}\ell(1 + \frac{\ell}{100})^{-1}\ \mu K$ is a fair representation at $\ell < 1000$ (the white noise cutoff arising from the assumed cluster profiles) with $c_{ysz} = 0.02$ at 100 GHz.

Figure 1 shows the expected microwave landscape in this foreground model. Note that taking into account the Sunyaev-Zeldovich effect of clusters moves to higher frequencies the best location for CMB measurement.

"Wiener" filtering can be used [56] to separate these different components in multi-frequency, multi-resolution experiments. As I showed earlier [1], this allow predicting a "quality factor" $Q_p = <\hat{x}_p^2(\ell)>/<x_p^2(\ell)>$, where $\hat{x}_p$ is the wiener estimate of the spatial template of the "p" process, x_p, and $0 < Q_P < 1$. A low value of Q_p indicates that the information collected by the experiment is not sufficient, and Wiener filtering sets this mode ℓ of that process p to nearly zero in order to minimize the error in the template recovery.

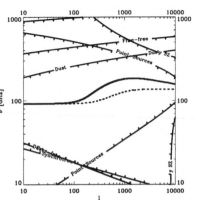

Figure 1: Contour levels of the different components in the angular scale–frequency plane. The levels indicate when the power spectra of the various components reach $\sim (3\mu K)^2$ - one tenth of the level of the COBE detection at low ℓ; tickmarks show the downhill direction. Thus the central area is the region where a CMB signal at COBE level will be 10 times bigger than any other component. The central solid & dashed line shows the locus of weakest overall contamination by all foregrounds when the Sunyaev-Zeldovich effect of clusters is respectiv included ot not.

Let us now try to define an experiment with angular resolution and noise characteristics such that it gives *without foregrounds* the same quality factor than the real experiment in presence of foregrounds. To that effect, consider the idealized experiment which maps directly the CMB, with a Fourier beam profile $w_B(\ell)$, and a detector noise characterized by its power spectrum C_N $(= \sigma_i^2 \Omega_i$, if σ_i stands for the 1σ ΔT sensitivity per field of view Ω_i). In the Gaussian case, the relative error on the CMB power spectrum is then [7] $\Delta \equiv \frac{\Delta C_{CMB}(\ell)}{C_{CMB}(\ell)} = \sqrt{\frac{2}{2\ell+1}}\left[1 + C_N(\ell)w_B(\ell)^{-2}\right]$. In this ideal case, one precisely finds $Q_{ideal} = w_B^2 C_{CMB}/(w_B^2 C_{CMB} + C_N)$, or equivalently $C_N w_B^{-2}(\ell) = 1/Q_{ideal}(\ell) - 1$. Thus the CMB "quality factor" of the real experiment $Q_{(p=CMB)}$ tells us which $\Delta(\ell)$ to consider to estimate the final noise contribution to the CMB power spectrum *once both the foregrounds and all channels properties are taken into account* (and the noise error is greater than cosmic variance when $Q \leq 1/2$).

In order to compare this estimate, Δ_Q, with more direct estimates of Δ by $1/\Delta = 1/\sum_{\nu_1 \leq \nu \leq \nu_2} C_N(\nu)w_B^{-2}(\nu)$ (at each ℓ), the next figure shows their relative difference with Δ_Q in two illustrative case, the baseline MAP and COBRAS/SAMBA projects[a]. In all cases, the error in the estimate goes to

---

[a]MAP: FWHM = 54, 39, 31.8, 23.4, 17.4 arcmin & C_N = 20.7, 18.9, 12.8, 12.3, 9.1 μK.deg at respectively 22, 30, 40, 60 & 90 GHz. COSA: FWHM = 30, 18, 12, 12, 10.3, 7.1,

zero at small ℓ, when cosmic variance dominates over the noise. If only the best channel (dashes) of the experiment is retained (i.e. $\nu_1 = \nu_2 = 90$ & 143 GHz for MAP & COBRAS/SAMBA), the error is always over estimated, i.e. proper use of all channels by Wiener filtering does much better than this naïve estimator would suggest. The error in estimating the noise level is much decreased if all the channels between 55 and 300 GHz are retained in the sum (dash-dots). There is essentially no error made in the MAP case if all 3 channels at 40, 60 & 90 GHz are retained, while retaining $35 \leq \nu \leq 400$ GHz in the COBRAS/SAMBA case is somewhat optimistic at very high ℓ.

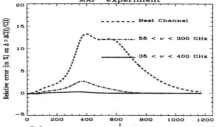 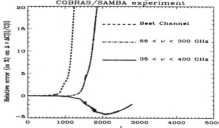

Of course, error estimates using Wiener's "quality factor" are only as good as the modeling of the foregrounds is. Clearly the situation regarding foregrounds emission might be quite more complicated than assumed (e.g. some free-free may locally be emitted at higher temperature, more point sources, etc...). Still, this modelling does give confidence that the future generations of satellite will unveil the CMB power spectrum with unprecedented accuracy.

Acknowledgments: I am grateful to my RUMBA collaborators for allowing me to use some of our unpublished material.

References

1. F. R. Bouchet, R. Gispert, F. Boulanger, and J.-L. Puget. Proceedings of the 30th Moriond meeting, Editions Frontières, 1996.
2. A. Kogut et al. ApJL **464**, L5 (1996).
3. E. Hivon, B. Guiderdoni, and F. R. Bouchet . same Proceedings as 1. Frontières, 1996.
4. N. Aghanim, A. De Luca, F. R. Bouchet, R. Gispert, and J.-L. Puget . *A&A, submitted*, 1996.
5. F. R. Bouchet, R. Gispert, and J.-L. Puget. In AIP Conference Proceedings 348, pages 255–268, 1995.
6. M. Tegmark and G. Efstathiou. MNRAS **281**, 1297 (1996).
7. Lloyd Knox. Phys Rev D **52**, 4307 (1995).

4.4, 4.4, 4.4 arcmin, $C_N = 12$, 6.9, 8.9, 21.8, 0.50, 0.57, 2.14, 13.6, 739 μK.deg at respectively 31.5, 53, 90, 125, 143, 217, 353.0, 545, 857 GHz.

CMB Error Forecasts for Cosmic Parameters

J. Richard Bond

CITA, University of Toronto, Toronto, ON M5S 3H8, CANADA

For theorists, cosmological parameter estimation was, is and always will be the raison d'etre for studying the CMB. The general parameter space for inflation-inspired models and plausible priors restricting freedom in this space, on theory or data grounds, is discussed. With current CMB data, e.g., DMR and SK95, the parameter sequences probed must be strongly constrained to get meaningful limits. Bayesian forecasts for future experiments, from long duration balloons (LDBs) through the satellites MAP and then Planck, are presented. These indicate a growing number of uncorrelated combinations of cosmological parameters may be determined to ± 0.1 and sometimes ± 0.01 accuracy. Errors on such combinations are more instructive than errors on H_0, Ω_0, Ω_b, Ω_{vac}, determined by marginalizing over all other parameters, though forecasts of these are given as well.

1 Cosmic Parameters

Parameters describing the theoretical angular temperature power spectra C_ℓ include early universe ones defining the initial conditions and those characterizing the transport of radiation through photon decoupling to the present. The early universe parameters could be very complex indeed, with non-Gaussian theories requiring an infinite number of N-point correlation functions to define them; for Gaussian theories, only the power spectra for the modes present are required, but these could have complicated hills and valleys arbitrarily located, depending e.g., upon the potential for the inflaton field in inflation models. Even with a gently varying inflaton potential, we still have: power spectrum amplitudes, $\{\mathcal{P}_\Phi(k_n), \mathcal{P}_{is}(k_n), \mathcal{P}_{GW}(k_n)\}$, at some normalization wavenumber k_n for the modes present (adiabatic scalar, possibly isocurvature scalar, gravity-wave or tensor); 'tilts' $\{\nu_s(k), \nu_{is}(k), \nu_t(k)\}$, nearly constant, though perhaps including a logarithmic correction, e.g., $d\nu_s(k_n)/d\ln k$. (The usual $n_s \equiv 1 + \nu_s$.)

The radiative transport, dependent upon physical processes, hence on physical parameters, can be conceptually split into two phases: the physics appropriate to the determination of the photon distribution through decoupling and the transport from then to now. At the time of decoupling, the important parameters are the densities of various types of matter present then, the expansion rate, the sound speed, and the damping rate; all of these are only dependent upon the density combinations $\Omega_j h^2$, where $j = b, cdm, hdm, \gamma, er\nu, \ldots$ refers to baryons, cold dark matter, hot dark matter, and the various relativistic particles present then (e.g., photons and relativistic neutrinos). $\Omega_{er} h^2 = (\Omega_\gamma + \Omega_{er\nu}) h^2$ depends upon the CMB temperature (now well determined), the number of relativistic neutrinos, and upon the density in decaying particle products, if any. The post-decoupling Hubble parameter only depends upon $\Omega_{nr}h^2 = (\Omega_B + \Omega_{cdm} + \Omega_{hdm})h^2$ (if the massive neutrinos were nonrelativistic then) and $\Omega_{er}h^2$. The Hubble parameter now is a derived quantity from the transport view, $h \equiv (\sum_j \Omega_j h^2)^{1/2}$ (in units of $100\,\mathrm{kms}^{-1}\mathrm{Mpc}^{-1}$). The abundance of primordial hydrogen (or helium), could also be considered to be a parameter to be determined, but constraints on it, on T_{cmb}, etc. are incorporated using prior probabilities.

The transport to an angular structure now from the post-decoupling spatial pattern of $\Delta T/T$ depends on the cosmological angle-distance relation, hence on $\Omega_{nr}h^2$, $\Omega_{vac}h^2$ and $\Omega_k h^2$, where $\Omega_{vac}h^2$ parameterizes the vacuum (or cosmological constant) energy density and $\Omega_k \equiv (1 - \Omega_{tot})$ the energy associated with the mean curvature of the universe. The angular pattern we observe, though, also depends upon the change of the gravitational metric in time between post-decoupling and the present, which breaks the $\Omega_{vac}h^2$, $\Omega_k h^2$ angle-distance degeneracy, but only at low redshift and for low ℓ, limiting the accuracy to which they can be simultaneously determined, irrespective of experimental precision. Many parameters could also be needed to describe the ionization history of the universe; here we just use the Compton optical depth τ_C from a reheating redshift z_{reh} to the present, assuming full ionization.

The parameter count is thus at least 17; even if an isocurvature contaminant is dropped as too baroque, 15 remain. In inflation, the ratio of gravitational wave power to scalar adiabatic power is $\mathcal{P}_{GW}/\mathcal{P}_\Phi \approx (-11\nu_t)/(1 - \nu_t/2)$, apart from small corrections, so one could argue for a parameter count reduction by one, as we do here. Estimates of errors on a smaller 9 parameter inflation set for the MAP and Planck (formerly COBRAS/SAMBA) satellites are given in the Table, based on quadratic expansions of log likelihoods about the maximum likelihood, with details given in Knox 1995, Phys. Rev. D48, 3502; Jungman et al. 1996, Phys. Rev. D54, 1332; Bond 1996, Les Houches Lectures, Elsevier Science Press; Bond, Efstathiou & Tegmark 1997, astro-ph/9702100; Zaldarriaga, Seljak & Spergel 1997, astro-ph/9702157.

For a given model, the $\mathcal{P}_\Phi(k_n)$, etc. are uniquely related to late-time power spectrum measures of relevance for the CMB, such as the quadrupole, $\mathcal{C}_2^{1/2}$, or the total bandpower for an ex periment, $\langle \mathcal{C}_\ell \rangle_B^{1/2}$, or to large scale structure observations, such as the rms density fluctuation level on the $8\,h^{-1}$ Mpc (cluster) scale, σ_8. The parameter error estimations depend upon which amplitude variable we choose. It is highly correlated with e^{τ_C} and a combination of the two is better determined. Orthogonal uncorrelated parameter eigenmodes for the log likelihood are even better. To characterize the tensor amplitude, we use $r_{ts} = \mathcal{C}_2^{(T)}/\mathcal{C}_2^{(S)}$.

Param	DMR	NCP +DMR	LDB +DMR	MAP (3ch)	Planck LFI(3)	Planck HFI(4)
$\overline{C}_{D\ell}/10^{-15}$	950	7	1.5	1.5	.10	.0033
f_{sky}	.65	.006	.028	.65	.65	.65
beams: ℓ_s	17	270	400	540	810	1970
ℓ_s		17	17	385	580	1225
ℓ_s				255	385	800
ℓ_s						560
cut: ℓ_{cut}	3	23	12	2	2	2
Orthogonal Parameter Combinations within ε						
$\varepsilon < 0.01$	0/9	0/9	1/9	2/9	3/9	5/9
$\varepsilon < 0.1$	1/9	2/9	4/9	6/9	6/9	7/9
Single Parameter Errors from Marginalizing Others						
$\delta\langle C_\ell\rangle_B^{1/2}/\langle C_\ell\rangle_B^{1/2}$	.07	.03	.022	.018	.019	.015
δn_s	.20	.05	.19	.06	.01	.006
δr_{ts}			.83	.34	.13	.09
$\delta\Omega_b h^2/\Omega_b h^2$			.20	.08	.016	.006
$\delta\Omega_{nr}h^2/h_0{}^2$			.32	.16	.04	.015
$\delta\Omega_{vac}h^2/h_0{}^2$			.82	.44	.14	.05
$\delta\Omega_{m\nu}(h)^2/h_0{}^2$			.24P	.07	.04	.02
τ_C			.30	.21	.17	.16
$\delta h/h$			.29	.14	.05	.02
$\delta\sigma_8/\sigma_8$			.28	.26	.21	.17
$\delta\mathcal{P}_\Phi^{1/2}/\mathcal{P}_\Phi^{1/2}$			.23	.22	.17	.15
$\delta\Omega_k h^2/h_0{}^2$			.14	.03	.008	.002

Table 1: Sample DMR, NCP, LDB, MAP, Planck parameter errors, based on the standard CDM model as true theory. The errors are of course sensitive to which model we choose. Details are in Bond, Efstathiou & Tegmark, $astro-ph/9702100$. $\overline{C}_{D\ell}$ is the channel-summed noise, assumed to be uniform for all maps. P means a prior probability constraint controls that parameter's value. Relative errors are quoted for h, $\Omega_b h^2$, and the amplitudes $\langle C_\ell\rangle_B^{1/2}$, σ_8 and $\mathcal{P}_\Phi^{1/2}(k_n)$ (the early universe gravitational potential amplitude for scalar adiabatic perturbations). The $\Omega_{vac}h^2$ numbers are determined with $\Omega_k h^2$ fixed, and the $\Omega_k h^2$ numbers are determined with $\Omega_{vac}h^2$ fixed, because of the angle-distance near-degeneracy; the other parameters are insensitive to fixing either, or neither. NCP denotes earth-based observations covering a $9°$ radius patch, similar to the 1994-95 Saskatoon observations. The LDB parameters are based upon ten days of observing with the bolometer-based TopHat experiment with 65% of a $24°$ diameter patch assumed usable. The usable sky fraction for DMR, 65%, was adopted for the satellites. For MAP, (home page $http://map.gsfc.nasa.gov$), one year observing and current (proposal-modified) beams have been assumed and the highest 3 frequency channels (90, 60, 40 GHz) have been used. (The 30 and 22 GHz channels will be more contaminated with bremsstrahlung and synchrotron emission.) For Planck, ($http://astro.estec.esa.nl/SA-general/Projects/Cobras/cobras.html$), 14 months of observing and current (proposal-modified) values are used. The Hemt-based LFI is significantly improved; the 100, 65, 44 GHz were used, but not the 30 GHz channel. For the bolometer-based HFI, 100, 150, 220 and 350 GHz were used, assuming the 550 and 850 GHz channels will be highly dust-contaminated. The DMR and DMR+NCP numbers should be compared with current constraints: $\langle C_\ell\rangle_{dmr}^{1/2} = 1.03 \pm 0.07 \times 10^{-5}$ for the best-fit primordial index, which is $n_s = 1.02 \pm 0.24$, errors in agreement with the table. Adding SK95 data gives 1.1 ± 0.1, errors twice as large as predicted because of calibration uncertainties (Bond & Jaffe, $astro-ph/9610091$).

UNCORRELATED MEASUREMENTS OF THE CMB POWER SPECTRUM

MAX TEGMARK [a]

Institute for Advanced Study, Princeton, NJ 08540, USA; max@ias.edu

A. J. S. HAMILTON

JILA and Dept. of Astrophysical, Planetary and Atmospheric Sciences, Box 440, Univ. of Colorado, Boulder, CO 80309, USA; ajsh@dark.colorado.edu

We describe how to compute estimates of the power spectrum C_ℓ from Cosmic Microwave Background (CMB) maps that not only retain all the cosmological information, but also have uncorrelated error bars and well-behaved window functions. We apply this technique to the 4-year COBE/DMR data.

Accurate future measurements of the angular power spectrum C_ℓ of the CMB would allow us to measure many key cosmological parameters with unprecedented accuracy [1]. It has recently been shown [2] how to compute a vector of power spectrum estimates $\mathbf{y}$ retaining all the cosmological information from a CMB map, and whose mean and covariance is given by

$$\langle \mathbf{y} \rangle = \mathbf{Fc},$$
$$\langle \mathbf{yy}^t \rangle - \langle \mathbf{y} \rangle \langle \mathbf{y}^t \rangle = \mathbf{F}. \tag{1}$$

Here $\mathbf{c}$ is the vector of true power coefficients, i.e., $c_\ell = C_\ell$, so the window function matrix and the covariance matrix are one and the same, equaling $\mathbf{F}$, the *Fisher information matrix* [2]. There are infinitely many ways of producing uncorrelated power estimates [3]. Making a factorization $\mathbf{F} = \mathbf{MM}^t$ for some matrix $\mathbf{M}$, the new power estimates in the vector defined by

$$\hat{\mathbf{c}} \equiv \mathbf{M}^{-1}\mathbf{y} \tag{2}$$

will be uncorrelated, since $\langle \hat{\mathbf{c}}\hat{\mathbf{c}}^t \rangle - \langle \hat{\mathbf{c}} \rangle \langle \hat{\mathbf{c}}^t \rangle = \mathbf{I}$. $\langle \hat{\mathbf{c}} \rangle = \mathbf{M}^t\mathbf{c}$, so the new window function matrix will be $\mathbf{M}^t$. However, whereas the window functions of the original power estimates (the rows of $\mathbf{F}$) are always well-behaved (they are always non-negative [2], and are generally quite narrow, as shown in the top panel of Figure 1), there is no guarantee that the same will hold for the new window functions. As was recently shown [3], however, one generally obtains beautiful window functions if one requires $\mathbf{M}$ to be lower-triangular, in which case $\mathbf{F} = \mathbf{MM}^t$ corresponds to a Cholesky decomposition. For the COBE/DMR case

---

[a]Hubble Fellow

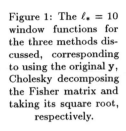

Figure 1: The $\ell_* = 10$ window functions for the three methods discussed, corresponding to using the original **y**, Cholesky decomposing the Fisher matrix and taking its square root, respectively.

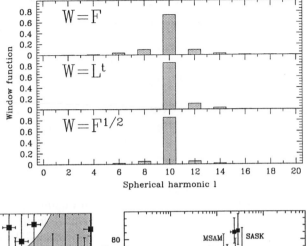

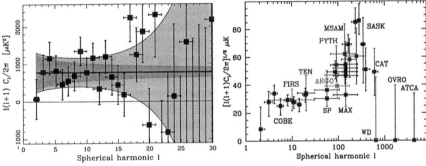

Figure 2: The power spectrum observed by COBE/DMR alone (left) and binned into 8 bands and compared with other experiments (right).

with a "custom" [4] Galaxy cut, this gives the narrow and non-negative window functions in the middle panel of Figure 1, with side lobes only to the right. Similarly, one could obtain window functions with side lobes only to the left by chosing **M** upper-triangular. A third choice, which is the one we recommend, is choosing **M** *symmetric*, which we write as $\mathbf{M} = \mathbf{F}^{1/2}$. The square root of the Fisher matrix is seen to give beautifully symmetric window functions (Figure 1, bottom) that are not only non-negative, but also even narrower than the original (top), which has roughly the bottom profile convolved with itself.

Figure 2 (left) shows the power spectrum extracted from the 4 year COBE data [4] with the minimum-variance method [2] and de-correlated with $\mathbf{M} = \mathbf{F}^{1/2}$. The error bars are for a flat 18 μK spectrum. These 29 data points thus contain all the cosmological information from COBE, distilled into 29 mutually exclusive (uncorrelated) and collectively exhaustive (jointly retaining all infor-

272

Table 1: The COBE power spectrum $\delta T \equiv [\ell(\ell+1)C_\ell/2\pi]^{1/2}$ in μK.

Band	ℓ_*	$\langle \ell \rangle$	$\Delta\ell$	δT	-1σ	$+1\sigma$
1	2	2.1	0.5	8.5	0	24.5
2	3	3.1	0.6	28.0	17.7	35.5
3	4	4.1	0.7	34.0	26.8	40.0
4	5-6	5.6	0.9	25.1	18.5	30.4
5	7-9	8.0	1.3	29.4	25.3	33.0
6	10-12	10.9	1.3	27.7	23.2	31.6
7	13-16	14.3	2.5	26.1	20.9	30.5
8	17-30	19.4	2.8	33.0	27.6	37.6

mation about cosmological parameters) chunks. To reduce scatter, these have been binned into 8 bands in Table 1 and Figure 2 (right). The data and references for the other experiments plotted can be found in recent compilations[5,6].

This method can readily be applied to other CMB experiments[7] as well as galaxy surveys[3,8]. In comparison, previous CMB power spectrum estimation methods[9,10,11,12] all had the drawback of giving correlated errors.

Support for this work was provided by NASA through a Hubble Fellowship, #HF-01084.01-96A, awarded by the Space Telescope Science Institute, which is operated by AURA, Inc. under NASA contract NAS5-26555. The COBE data sets were developed by the NASA Goddard Space Flight Center under the guidance of the COBE Science Working Group and were provided by the NSSDC.

References

1. G Jungman et al., *Phys. Rev. D* **54**, 1332 (1997).
2. M Tegmark, preprint astro-ph/961117 (1996).
3. A J S Hamilton, preprint astro-ph/9701009, *MNRAS*, in press (1997).
4. C L Bennett et al., *ApJ* 464, L1 (1996).
5. C Lineweaver et al., preprint astro-ph/9610133 (1997).
6. G Rocha & S Hancock, preprint astro-ph/9611228 (1997).
7. L Knox et al., in these proceedings.
8. A J S Hamilton, preprint astro-ph/9701008, *MNRAS*, in press (1997).
9. G Hinshaw et al., *ApJL* **464**, L17 (1996).
10. E L Wright et al., *ApJ* 464, L21 (1996).
11. M Tegmark, *ApJL* **464**, L35 (1996).
12. K M Górski, preprint astro-ph/9701191 (1997).

DATA COMPRESSION FOR CMB EXPERIMENTS

A.H. JAFFE

CfPA, 301 LeConte Hall, University of California, Berkeley, CA 94720

L. KNOX & J.R. BOND

CITA, 60 St. George St., Toronto, ON, M5S 3H8 CANADA

We discuss data compression for CMB experiments. Although "radical compression" to C_ℓ bands, via quadratic estimators or local bandpowers, potentially offers a great savings in computation time, they do a considerably worse job at recovering the full likelihood than the the signal-to-noise eigenmode method of compression.

We model a CMB observation at a pixel $p = 1 \ldots N_p$, as $\Delta_p = s_p + n_p$, where s and n represent the contribution of the CMB signal and the noise, respectively, to the observation. The signal is given by $s_p = \mathcal{F}_{p\ell m} a_{\ell m}$; $a_{\ell m}$ is the spherical harmonic decomposition of the temperature and $\mathcal{F}$ encodes the beam and any chopping strategy of the experiment.

We also assume that both the signal and noise contributions are described by independent, zero-mean, gaussian probability distributions, with correlation matrices given by $\langle s_p s_{p'} \rangle = C_{Tpp'}$ and $\langle n_p n_{p'} \rangle = C_{npp'}$, so $\langle \Delta_p \Delta_{p'} \rangle = C_{Tpp'} + C_{npp'}$; here, $C_T = C_T(\theta)$ is calculated as a function of θ, the parameters of the theory being tested in the likelihood function, and the noise matrix can include the effect of constraints due to, e.g., average or gradient removal. For Gaussian theories of adiabatic fluctuations, θ is typically the cosmological parameters; alternately they could be some set of phenomological parameters such as the value of the temperature power spectrum C_ℓ in some bands.

With this notation, the likelihood function is

$$\mathcal{L}_\Delta(\theta) = P(\Delta|\theta) = \frac{\exp\left[-\frac{1}{2}\Delta_p \left(C_T(\theta) + C_n\right)_{pp'}^{-1} \Delta_{p'}\right]}{(2\pi)^{N_p/2}|C_T(\theta) + C_n|^{1/2}} \qquad (1)$$

Calculating this requires extensive manipulations on the total correlation matrix $C_T(\theta) + C_n$ over extensive portions of the parameter space. Herein lies the problem: in order to calculate the determinant factor in the denominator requires time of $O(N_p^3)$.

We can state the problem as follows: find some functions of the data, $f_i(\{\Delta_p\})$ such that the new likelihood, $\mathcal{L}_f(\theta) = P(f|\theta) \simeq \mathcal{L}_\Delta(\theta) = P(\Delta|\theta)$ and $\mathcal{L}_f$ is "easier to calculate" in some appropriate sense. The definition of "$\simeq$" in this expression is crucial. If we take it to mean "having the same variance," the requirement reduces to the definition of "lossless" involving the Fisher matrix.[1] Note that this requirement will only be adequate very near the maximum of the distribution, i.e., where the Gaussian approximation is appropriate. For large data-sets, with high signal-to-noise, this will presumably be an adequate description; elsewhere (including those parts in the "tails" of an otherwise high-S/N experiment's window function that may be most interesting), it will not necessarily obtain that the derived confidence limits and parameter estimates will be the same.

If we can find a basis in which the matrices C_T and C_n are diagonal, the likelihood computation simplifies from matrix manipulations to $O(N)$ sums and products. First, we whiten the noise matrix using the transformation provided by its "Hermitian square root," and apply the same transformation to C_T, diagonalizing this in turn with the appropriate matrix of eigenvectors, $C_T \to \mathrm{diag}(\mathcal{E}_k)$, in units of $(S/N)^2$. We then transform the data into the same basis, $\Delta \to \xi$, in units of (S/N). In this basis, the noise and signal have diagonal correlations and $\langle \xi_k^2 \rangle = 1 + \mathcal{E}_k$. Now, the likelihood function is a simple product of one-dimensional uncorrelated gaussians in ξ_k.[2]

Modes with low $\mathcal{E}_k$ are linear combinations of the data which probe the theory poorly. Thus, they are ideal candidates for removal in a data compression scheme. Removal of these low-S/N modes is the optimal way to compress the data: for a given number of modes, it removes the most noise and least signal. Of course, the modes will change as the shape of the theory used for the covariance matrix C_T changes. In that case, we choose some fiducial theory (here, standard, untilted CDM) that represents the data moderately well, calculate its S/N-eigenmodes, and choose some number of modes to retain after compression. For other theories, these modes will not be optimal—we will have removed more signal and less noise. Nonetheless, this method does quite well even far from the fiducial theory, as we see in the Figure, which shows the likelihood for a parameterization of a standard CDM power spectrum with amplitude, σ_8, and a scalar tilt, n_s (so the primordial $P(k) \propto k^{n_s}$). Unfortunately, finding this basis is $O(N^3)$; even performing this operation once for a megapixel dataset will be prohibitively expensive.

Most experiments report their results in the form $[\hat{C}_\ell \pm \delta C_\ell]$; this encourages the use of a simple curve-fitting approach to parameter estimation[3]: form the obvious quantity $\chi^2(\theta) = \sum_\ell \left[\hat{C}_\ell - C_\ell(\theta) \right]^2 / (\delta C_\ell)^2$ where $C_\ell(\theta)$ is the predicted power spectrum for the theoretical parameters θ. In practice, the power spectrum is usually reported as a flat bandpower over some ℓ band with an appropriate window function, but the procedure remains the same.

Now, just do the usual fast χ^2-minimization for the parameters. This is

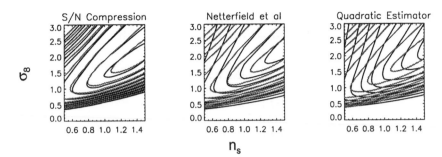

Figure 1: Contours of constant likelihood ratio (at $\delta 2 \ln \mathcal{L} = 1, 4, 9, \ldots$) for the SK94-5 data, parameterized by σ_8 and n_s, the primordial spectral slope, for different methods of data compression. The light-colored contours in each panel show the uncompressed likelihood.

a very "radical" approximation to the full likelihood: it assigns an uncorrelated gaussian distribution to the power spectrum: $\mathcal{L}(\theta) \propto \exp(-\chi^2/2)$. In the figure, we show confidence intervals for the SK95 experiment with the full likelihood and compare them to those obtained with 1) the S/N compression discussed above; 2) χ^2 using flat bandpowers as reported by Netterfield et al[4]; 3) χ^2 using a quadratic estimator of C_ℓ in bands of ℓ, modified somewhat to minimize covariance[5]. The peaks are nearby, and the contours are similar in the "amplitude" direction (σ_8), but less good in the "shape" (n_s) direction. Far from the peak, all of the χ^2 methods do quite poorly. The location of the peak is determined by the gross shape of the power spectrum, so the details of the calculation do not matter; this is especially true when combining the results of experiments probing very different scales, such as COBE/DMR and SK95. Away from the peak, the shape of the spectrum within the experimental windows, the covariances of the errors, and the non-Gaussian shape all contribute to the determination of which theories are more highly disfavored.

References

1. M. Tegmark, `astro-ph/9611174`, *Ap. J.*, submitted, 1996.
2. J.R. Bond & A. Jaffe, *Phys. Rev. Lett.*, submitted, 1997; A. Jaffe & J.R. Bond, in preparation, 1997.
3. C. Lineweaver, these proceedings.
4. C.B. Netterfield, M.J. Devlin, N. Jarosik, L. Page, & E.J. Wollack, *Ap. J.* **474**, 47 (1997).
5. L. Knox, J.R. Bond & A. Jaffe, these proceedings; M. Tegmark & A. Hamilton, these proceedings.

CONSTRAINING COSMOLOGICAL PARAMETERS
WITH CMB MEASUREMENTS

C.H. LINEWEAVER

Observatoire Astronomique de Strasbourg
11 rue de l'Université
67000 Strasbourg, France
charley@astro.u-strasbg.fr

The current enthusiasm to measure fluctuations in the CMB power spectrum at angular scales between 0.1 and 1° is largely motivated by the expectation that CMB determinations of cosmological parameters will be of unprecedented precision. In such circumstances it is important to estimate what we can already say about the cosmological parameters. In two recent papers (Lineweaver *et al.* 1997a & 1997b) we have compiled the most recent CMB measurements, used a fast Boltzmann code to calculate model power spectra (Seljak & Zaldarriaga 1996) and, with a χ^2 analysis, we have compared the data to the power spectra from several large regions of parameter space. In the context of the flat models tested we obtain the following constraints on cosmological parameters: $H_o = 30^{+13}_{-9}$, $n = 0.93^{+0.17}_{-0.16}$ and $Q = 17.5^{+3.5}_{-2.5}\,\mu$K. The n and Q values are consistent with previous estimates while the H_o result is surprisingly low.

1   Method

With two new CMB satellites to be launched in the near future (MAP $\sim$ 2001, Planck Surveyor $\sim$ 2005) and half a dozen new CMB experiments coming on-line (see contribution of Lyman Page to this volume), it is important to keep track of what we can already say about the cosmological parameters. In Lineweaver *et al.* (1997a) we considered COBE-normalized flat universes with $n = 1$ power spectra. We used predominantly goodness-of-fit statistics to locate the regions of the $H_o - \Omega_b$ and $H_o - \Lambda$ planes preferred by the data. In Lineweaver *et al.* (1997b) we obtained χ^2 values over the 4-dimensional parameter space $\chi^2(H_o, \Omega_b, n, Q)$ for $\Omega = 1$, $\Lambda = 0$ models. Projecting and slicing this 4-D matrix gives us the error bars around the minimum χ^2 values. Here we summarize several of our most important results.

2   Results and Discussion

One of the difficulties in this analysis is the 14% absolute calibration uncertainty of the 5 important Saskatoon points which span the dominant adiabatic peak in the spectrum (Figure 1). We treat this uncertainty by doing the analysis three times: all 5 points at their nominal values ('Sk0'), with a 14%

Figure 1. CMB Data
A compilation of 24 of the most recent measurements of the CMB angular power spectrum. Models with $h = 0.30$ and $h = 0.75$ are superimposed (both are $\Omega = 1$, $\Omega_b = 0.05$, $n = 1$ $Q = 18$ μK models). The dotted line is a 5th order polynomial fit to the data. The low-h value is favored. MAP and Planck Surveyor are expected to yield precise spectra for $\theta_{FWHM} \gtrsim 0°.3$ and $\theta_{FWHM} \gtrsim 0°.2$ respectively (see angular scale marked at the top). Figure from Lineweaver *et al.* (1997a).

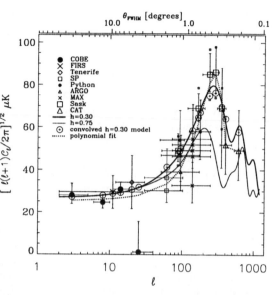

increase ('Sk+14') and a 14% decrease ('Sk-14'). Sk+14 and Sk-14 are indicated by the small squares in Figure 1 above and below the nominal Saskatoon points. Leitch *et al.* (1997) report a preliminary relative calibration of Jupiter and CAS A implying that the Saskatoon calibration should be $-1\% \pm 4\%$. Reasonable χ^2 fits are obtained for Sk0 and Sk-14.

In the context of the flat models tested, our χ^2 analysis yields: $H_o = 30^{+13}_{-9}$, $n = 0.93^{+0.17}_{-0.16}$ and $Q = 17.5^{+3.5}_{-2.5}$ μK. The H_o result is shown in Figure 2. The n and Q values are consistent with previous estimates while the H_o result is surprisingly low. For each result, the other 3 parameters have been marginalized over. This H_o result has a negligible dependence on the Saskatoon calibration, i.e., lowering the Saskatoon calibration from 0 to -14% does not raise the best-fitting H_o in flat models. The inconsistency between this low H_o result and $H_o \sim 65$ results will not easily disappear with a lower Saskatoon calibration. Our results are valid for the specific models we considered: $\Omega = 1$, CDM dominated, $\Lambda = 0$, Gaussian adiabatic initial conditions, no tensor modes, no early reionization, $T_o = 2.73$ K, $Y_{He} = 0.24$, no defects, no HDM.

There are many other cosmological measurements which are consistent with such a low value for H_o (Bartlett *et al.* 1995, Liddle *et al.* 1996). For example, we calculated a joint likelihood based on the observations of galaxy cluster baryonic fraction, big bang nucleosynthesis and the large scale density fluctuation shape parameter, Γ. We obtained $H_o \approx 35^{+6}_{-5}$.

I am grateful to my collaborators D. Barbosa, A. Blanchard and J. G. Bartlett and I acknowledge support from NSF/NATO post-doctoral fellowship

278

9552722.

References

1. C.H. Lineweaver, D. Barbosa, A. Blanchard & J.G. Bartlett *A & A*, in press astro-ph/9610133 (1997a).
2. C.H. Lineweaver, D. Barbosa, A. Blanchard & J.G. Bartlett *A & A*, submitted astro-ph/9612146 (1997b).
3. U. Seljak & M. Zaldariaga *Ap. J.*, 469, 437 (1996).
4. E. Leitch *et al.* in preparation (1997).
5. A. Liddle *et al. M.N.R.A.S.*, 281, 531, (1996).
6. J.G. Bartlett, A. Blanchard, J. Silk, & M.S. Turner Science, 267, 980 (1995).

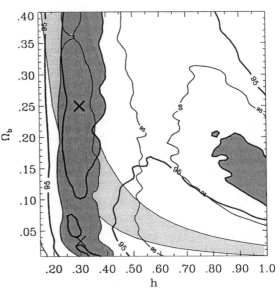

Figure 2. Constraints on Hubble's Constant The dark grey areas denote the regions of parameter space favored by the CMB data. They are defined by $\chi^2_{min} + 1$ for Sk0 and Sk-14 (minima marked with thick and thin 'x' respectively). '95' denotes the $\chi^2_{min} + 4$ contours for Sk0 (thick) and Sk-14 (thin). The light grey band is from big bang nucleosynthesis ($0.010 < \Omega_b\, h^2 < 0.026$). The parameters n and Q have been marginalized. In the H_o result quoted, we neglect the region at $H_o \sim 100$ with $\Omega_b \sim 0.15$. This figure shows clearly that lowering the calibration by 14% *does not* favor higher values of H_o (Figure from Lineweaver *et al.* 1997b).

THE INVERSE PROBLEM IN CMB ANISOTROPIES

M. WHITE

Enrico Fermi Institute, 5640 South Ellis Ave, Chicago
IL 60637, USA

In this talk I review some of the major themes in work I have done with Wayne Hu over the last year. These studies focus on the tale told by the CMB spectrum taken as a whole. In particular, the acoustic pattern, which arises from forced oscillations in the photon-baryon fluid before recombination, leaves a distinct signature from which we may measure cosmological parameters and begin to reconstruct the cosmological model.

1 Introduction

The CMB community has made a great deal of progress on the "forward problem" of CMB anisotropies, *i.e.* predicting the properties of the CMB given a specific cosmological model and values of the cosmological parameters, for classes of models loosely based on inflationary CDM. Numerical solution of the coupled Boltzmann, fluid and Einstein equations is now routinely carried out with a precision of 1% or better, as determined by comparing different codes, with equations solved in different gauges using different algorithms. Some of the questions which need to be addressed to obtain this level of precision are listed in [1] along with many references to the original literature.

Despite the large amount of work on the "foward problem", little work has been done on the "inverse problem" of CMB anisotropies. That is, how does one determine the cosmological parameters and the model for structure formation (independently?) from an observed CMB spectrum? Some first steps along that road are described in [2,3]. In those papers we have tried to establish a framework for thinking about the inverse problem. We have identified which features of the CMB spectra are model *in*dependent and which can be used to learn about the model for structure formation.

The production of CMB anisotropies is a linear process ($\Delta T/T \sim 10^{-5}$). Given a source of anisotropies, e.g. the gravitational potential Φ, one multiplies by Green's functions to find the power spectrum of the radiation today. Since this is a problem in linear response theory, the Greens functions, or transfer functions, form the desired set of "physical elements" for understanding the spectra. They are supplemented by an understanding of the source, which can have contributions from initial conditions (e.g. inflation), a time varying source (e.g. defects) and the feedback due to the baryon-photon fluid itself ($\Phi_{b\gamma}$).

The CMB anisotropy spectrum bears the imprint of physics which affects

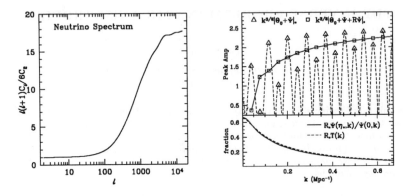

Figure 1: *Left:* The Neutrino Anisotropy Spectrum, from Ref. 1. The position of the rise is determined by matter-radiation equality, the height of the small-angle plateau is $\sim 4.2^2 \times$ the Sachs-Wolfe value. *Right:* The offset betwen the peaks in the CMB spectrum, induced by baryon drag, compared to the matter transfer function. Triangles indicate the RMS temperature, squares the potential envelope once the baryon drag is removed.

the anisotropy before horizon crossing (the acoustic phase), at horizon crossing (the potential envelope, discussed below), at last scattering (baryon drag and diffusion damping) and between last scattering and the present (the integrated Sachs-Wolfe effect [4,5]). The dependence on the super-horizon scale behavior of the anisotropy can be used to test inflation [6] and to distinguish it from models of structure formation based on defects [7,8,9] (see also [10]). The damping tail may be used as a model independent measure of spatial curvature [5,2,3].

Both the phase of the acoustic oscillations (important for distinguishing defects from inflation) and the potential envelope rely on feedback regulated by $\Phi_{b\gamma}$. The phase is treated in detail in [2,6], here I will concentrate on the potential envelope. Consider an adiabatic fluctuation which crosses the horizon at early times when the universe is radiation (photon) dominated. As the perturbation crosses the horizon it feels a gravitational potential Φ and falls into the potential well, becoming overdense. Eventually pressure resists the infall, halting the collapse and causing $\Phi \simeq \Phi_{b\gamma}$ to decay. This correlates the perturbation maximum with Φ decay. Thus photons no longer lose energy climing out of potential wells. The decay also causes space to contract, blueshifting the photons. In combination this enhances the small scale CMB anisotropy by 2Φ. The observed temperature is thus $-\Phi/3$ from the Sachs-Wolfe effect [4,11] plus 2Φ, or an enhancement of a factor [a] 5 over the large-angle $-\Phi/3$. This

---

[a]The presence of anisotropic stress changes the relation between space curvature and gravitational potentials, reducing this to ~ 4.

can be most clearly seen in the spectrum of *neutrino* anisotropy (Fig. 1). For the photons the power spectrum oscillates around this potential envelope due to baryon drag (see Fig. 1 and below) and is damped due to photon diffusion. The product of these envelopes forms the observed spectrum.

I end with one amusing connection between large-scale structure and the cosmic microwave background [3]. Due to the presence of baryon drag, the difference between compressions and rarefaction peaks in the RMS temperature can be used to measure the matter transfer function. Baryons add inertia to the baryon-photon fluid enhancing compressions (infall into potential wells) over rarefactions. The difference between odd and even peaks is thus proportional to the baryon-photon momentum density ratio at last scattering (R_*) and the gravitational potential Ψ at last scattering. If the former is known (e.g. from BBN) the latter can in principle be measured and compared with the gravitational potential measured today from large-scale structure (Fig. 1). While this connection is interesting, it is doubtful this would be a competitive way to measure $T(k)$ in practice!

Acknowledgments

The work reported here was done in collaboration with Wayne Hu.

References

1. W. Hu, D. Scott, N. Sugiyama, M. White, PRD **52**, 5498 (1995), astro-ph/9505043.
2. W. Hu, M. White, ApJ **471**, 30 (1996), astro-ph/9602019.
3. W. Hu, M. White, to appear in **ApJ**, astro-ph/9609079; W. Hu, M. White, submitted to **PRD**, astro-ph/9702170
4. R.K. Sachs, A.M. Wolfe, ApJ **147**, 73 (1967).
5. W. Hu, M. White, A&A **315**, 33 (1996), astro-ph/9507060.
6. W. Hu, M. White, PRL **77**, 1687 (1996), astro-ph/9602020.
7. R.G. Crittenden, N.G. Turok, PRL **75**, 2642 (1995), astro-ph/9505120.
8. N.G. Turok, PRD **54**, 3686 (1996), astro-ph/9604172; N.G. Turok, ApJ **473**, L5 (1996), astro-ph/9606087; N.G. Turok, PRL **77**, 4138 (1996), astro-ph/9607109.
9. W. Hu, D. Spergel, M. White, PRD **55**, 3288 (1997), astro-ph/9605193.
10. M. White, D. Scott, Comments Astrophys. **18**, 289 (1996), astro-ph/9601170.
11. M. White, W. Hu, to appear in **A&A**, astro-ph/9609105.

CMB POWER SPECTRUM ESTIMATION

L. Knox, J. R. Bond

Canadian Institute for Theoretical Astrophysics, 60 St. George St., Toronto, ON M5S 3H8, CANADA

A. H. Jaffe

Center for Particle Astrophysics, 301 LeConte Hall, University of California, Berkeley, CA, 94720

We explore power spectrum estimation in the context of a Gaussian approximation to the likelihood function. Using the Saskatoon data, we estimate the power averaged through a set of ten filters designed to make the power estimates uncorrelated. We also present an improvement to using the window function, W_l, for calculating bandpower estimates.

Estimates of parameters, θ_p, from data will in general have correlated errors, ϵ_p; $C_{pp'}^{\mathrm{P}} \equiv \langle \epsilon_p \epsilon_{p'} \rangle$ is not necessarily diagonal. A Taylor expansion of the log of the likelihood function, $\ln \mathcal{L}$, about the values of the parameter that maximize it, θ_p^*, identifies C^{P} with the inverse of the second derivative of $\ln \mathcal{L}$. The expectation value of this second derivative is an important quantity known as the curvature matrix or Fisher matrix, F:

$$F_{pp'} \equiv -\langle \frac{\partial^2 \ln \mathcal{L}(\theta_p)}{\partial \theta_p \partial \theta_{p'}} \rangle = \frac{1}{2} \mathrm{Tr} \left[(C_T + C_n)^{-1} \frac{\partial C_T}{\partial \theta_p} (C_T + C_n)^{-1} \frac{\partial C_T}{\partial \theta_{p'}} \right] \quad (1)$$

where C_T and C_n are the theory and noise covariance matrices, $\langle \Delta \Delta^T \rangle = C_T + C_n$, where Δ is the data (notation is more thoroughly explained in [1]). Since $\frac{\partial^2 \ln \mathcal{L}(\theta_p)}{\partial \theta_p \partial \theta_{p'}}$ is approximately equal to its expectation value, the parameter covariance matrix, C^P, is approximately the inverse of the Fisher matrix.

Knowing C^P is useful for two different purposes. Firstly, it is necessary if power spectrum estimation is to be used as a means of "radical data compression". We have tried this with the Saskatoon data[2] and the parametrization $C_l = q_B C_{l,B}^{\mathrm{cdm}}$, where $C_l \equiv l(l+1)C_l/(2\pi)$ and $C_{l,B}^{\mathrm{cdm}}$ refers to standard, untilted cdm normalized to $\sigma_8 = 1$ and restricted to ℓ within band B. Having estimated q_B for ten contiguous evenly spaced bands from $\ell = 19$ to $\ell = 499$ we can approximate the likelihood function of θ, where θ is some other parametrization of the spectrum:

$$-2 \ln \mathcal{L}(\theta) \simeq \chi^2(\theta) = (q_A(\theta) - q_A)\, F_{AB}\, (q_B(\theta) - q_B) \quad (2)$$

where $q_B(\theta) = \langle C_l(\theta) \rangle_B / \langle C_l^{\mathrm{cdm}} \rangle_B$ and the brackets mean a logarithmic average across band B. See[1] for how well this Gaussian approximation works when complicated slightly by a marginalization over calibration uncertainty.

The second use of C^P (or equivalently the Fisher matrix) is for the visual presentation of the power spectrum. We can plot linear combinations of power averaged through a set of filters where the filters are designed to produce uncorrelated estimates. One particularly useful set of filters comes from Cholesky decomposition[3], which is simply LU decomposition for a symmetric matrix; find L such that $F = LL^T$. Notice now that L^{-1} does diagonalize F since $L^{-1}F(L^{-1})^T$ is equal to the identity matrix. See L in the left panel of the figure. The transformation affects the parameters by taking $\mathbf{q}$ to $\mathbf{Q} = L^T\mathbf{q}$. To convert the estimate of each Q_β into a bandpower estimate in band β, $\langle C_l \rangle_\beta$ (plotted in the right panel of the figure), divide it by the sum over the filter function, $f_B^\beta = L_{B\beta}/\langle C_l^{\mathrm{cdm}} \rangle_B$. To find the bandpower prediction of another theory, $\langle C_l^t \rangle_\beta$, average it over the filter function: $\langle C_l^t \rangle_\beta = \sum_B f_B^\beta \langle C_l^t \rangle_B / \sum_B f_B^\beta$. Note that f_B^β is playing the role of W_l/l in the usual bandpower procedure[4]. One can use $F^{1/2}$ instead of L which has been done with the COBE data[5].

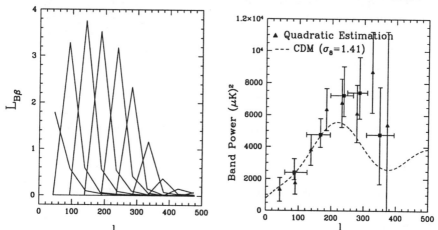

Left panel: Cholesky decomposition of the Fisher matrix. For each value of β the values of $L_{B\beta}$ at the ℓ value corresponding to the center of band B, are connected by straight lines. Right panel: Estimates of the power spectrum from the SK data set as given by the observing team (pentagons), and the quadratic estimator (triangles). The quadratic estimates are uncorrelated because they are estimates of power averaged over the filters f_B^β. The power estimates in the highest three bands have been averaged together.

We estimated $\mathbf{q}$ with a quadratic estimator, which can be derived from a Gaussian approximation to the likelihood function, which the central limit theorem tells us is good in the limit of large data sets. The Gaussian approximation is equivalent to truncating the Taylor series expansion of $\ln \mathcal{L}(\theta + \delta\theta)$ after the 2nd order term. Doing so allows us to solve for $\delta\theta$ that maximizes the likelihood:

$$\delta\theta_p = \frac{1}{2}(F^{-1})_{pp'}\mathrm{Tr}\left[\Delta\Delta^T(C_T + C_n)^{-1}\frac{\partial C_T}{\partial\theta_{p'}}(C_T + C_n)^{-1} - (C_T + C_n)^{-1}\frac{\partial C_T}{\partial\theta_{p'}}\right]$$

Note that due to the matrix inversions this is an order of n^3 operation, where n is the number of pixelized data points. Approximations to the weights $(C_T + C_n)^{-1}$ are necessary to make this estimator practical for very large data sets.

If we restrict ourselves to map-making experiments with no constraint removals, and make the parameters the C_l's then this reduces to the quadratic estimator independently advocated by M. Tegmark[6].

The dependence of the right hand side on θ_p suggests an iterative approach. The estimation of σ_8 for standard cdm took only three iterations starting from $\sigma_8 = 1$ and converging on $\sigma_8 = 1.41\pm0.08$ (c.f. 1.43 ± 0.08 via the full likelihood analysis). Therefore, for the power estimates shown in the figure, we used the quadratic estimator with C_T for $\sigma_8 = 1.41$ standard cdm.

The bandpower expected from a given theory, C_l, for a given experiment with window function W_l is usually calculated by $(\sum_l C_l W_l/l)/\sum_l W_l/l^4$. The optimal (minimum variance) filter, however, is not W_l/l but instead $\sum_{l'} F_{ll'}$ where the parameters of this Fisher matrix are C_l and it is evaluated with values of C_l consistent with the data and/or prior information. Heuristically, this is an inverse variance weighting. We recommend that observers reporting single bandpowers also report these filters which will improve the bandpower method as a means of "radical compression". In the limit that C_T is proportional to the identity matrix the two filters are equivalent: $\sum_{l'} F_{ll'} \propto W_l/l$.

Acknowledgments

We would like to thank Andrew Hamilton for a useful conversation.

References

1. A. H. Jaffe, L. Knox and J. R. Bond, these proceedings.
2. B. Netterfield, M. J. Devlin, N. Jarosik, L. Page, & E. J. Wollack, Ap. J. **474**, 47 (1997).
3. A. J. S. Hamilton, astro-ph/9701008; astro-ph/9701009.
4. J. R. Bond, Astrophys. Lett. and Comm. **32**, 63 (1995).
5. M. Tegmark and A. J. S. Hamilton, these proceedings
6. M. Tegmark, astro-ph/9611174

RECONSTRUCTING THE PRIMORDIAL POWER SPECTRUM

E. GAWISER

Physics Department, University of California, Berkeley, Berkeley, CA, 94720

Cosmological models predict transfer functions by which primordial density perturbations develop into CMB anisotropy and Large-Scale Structure. We use the current set of observations to reconstruct the primordial power spectrum for standard CDM, ΛCDM, open CDM, and standard CDM with a high baryon content.

1 Introduction

The combination of Cosmic Microwave Background (CMB) anisotropy measurements and Large-Scale Structure observations has caused dissatisfaction with the standard Cold Dark Matter (sCDM) cosmogony, leading some to advocate a "tilt" of the primordial power spectrum away from scale-invariant ($n = 1$) to $n = 0.8 - 0.9$.[1] Other CDM cosmogonies have not commonly been tilted but their agreement with the data might also improve. Because the primordial power spectrum is an inherent set of degrees of freedom in all CDM cosmogonies, we adopt a set of models and find the best-fit primordial power spectrum for each. This allows us to determine if the reconstructed primordial power spectra show any common features across the set of currently preferred cosmogonies.

2 The Primordial Power Spectrum

The initial density perturbations in the universe are believed to have originated from quantum fluctuations during inflation or from active sources such as topological defects. Defect models have not yet been calculated to high precision. Inflationary models predict rough scale-invariance; the shape of the inflaton potential leads to tilting as well as variation of the degree of tilt with spatial scale. The assumption of scale-invariance[2] is no longer acceptable because the data are now accurate enough to reveal the predicted deviations from $n = 1$. We adopt a parameterization of the primordial power spectrum as a polynomial in log-log space versus wave number k:

$$\log P_p(k) = \log A + n \log k + \alpha(\log k)^2 + \dots \qquad (1)$$

3 Data Analysis

Each cosmogony has transfer functions, T(k) and C_{lk}. CMB anisotropies are given by

$$C_l = \frac{1}{8\pi} \sum_k d \log k \, C_{lk} \, P_p(k),$$ (2)

where C_{lk} is the radiation transfer function after Bessel transformation into ℓ-space. The matter power spectrum is

$$P(k) = \frac{2\pi^2 c^3}{H_0^3} T^2(k) \, P_p(k).$$ (3)

We can predict the value of σ_8 using

$$\sigma_R^2 = \frac{2}{\pi^2} \int d \log k \, W^2(kR) \, k^3 \, P(k)$$ (4)

where W(kR) is a top-hat window function on the $R = 8h^{-1}$Mpc scale.

We use the CMB anisotropy observations catalogued in Scott, Silk and White[3] plus recent additions from Saskatoon[4], CAT[5], and the COBE 4-year data[6]. Peacock and Dodds[7] provide a careful compilation of the matter power spectrum, to which we add measurements of the power spectrum from peculiar velocities[8] and σ_8 from clusters[9]. For a given $P_p(k)$, we compare predictions with observations using the χ^2 statistic. We vary the coefficients of $P_p(k)$ given in Equation 1 to find the best fit for each cosmogony.

Table 1: Values of cosmological parameters for our models.

Model	h	Ω_{baryon}	Ω_{matter}	Ω_Λ
standard CDM	0.50	0.05	1.0	0.0
high-B sCDM	0.50	0.10	1.0	0.0
Λ CDM	0.65	0.04	0.4	0.6
Open CDM	0.65	0.04	0.4	0.0

4 Results

Our results are preliminary, but we have a qualitative understanding of the reconstructed primordial power spectra for each cosmogony. If we restrict $P_p(k)$ to scale-invariant ($n = 1$, $\alpha = 0$), the ΛCDM and OCDM models are a good fit to the data (meaning χ^2 per degree of freedom is about 1). The sCDM variants are poor fits, although high-B sCDM is better. Assuming only scale-free $P_p(k)$ ($\alpha = 0$), the sCDM models prefer $n = 0.8$. The ΛCDM and OCDM models choose a slight tilt ($n = 1.1$) but their agreement with the data only improves marginally, with ΛCDM fitting the CMB anisotropy data better than OCDM. Allowing the scalar index n to "run" yields interesting results; the ΛCDM and OCDM models prefer $\alpha=0$, so they remain good fits. The two sCDM models become good fits, with $n = 1.3$ on COBE scales and running by $\alpha = -0.1$. There is nothing to be gained by fitting additional terms of $P_p(k)$, as all four cosmogonies already fit the data well.

5 Conclusions

Assumptions about the primordial power spectrum make a tremendous difference in testing theories of structure formation. Allowing the power-law index to run makes standard CDM a good fit to the data, despite the apparent superiority of low Ω_{matter} models when restricted to a scale-invariant $P_p(k)$. Combining Large-Scale Structure observations with CMB anisotropy data gives us a long lever arm in k-space with which to reconstruct the primordial power spectrum. With the next generation of observations, we hope that our technique will prove powerful enough to either discredit inflation or reconstruct the inflaton potential.

Acknowledgments

This research is being performed in collaboration with Joe Silk and Martin White. E. Gawiser gratefully acknowledges the support of a National Science Foundation fellowship.

References

1. M. White *et al*, *Mon. Not. R. astr. Soc.* **276**, L69 (1995).
2. P.J.E. Peebles, *Ap. J.* **162**, 815 (1970), E.R. Harrison, *Phys. Rev.* D 1, 2726 (1970), Y.B. Zeldovich, *Mon. Not. R. astr. Soc.* **160**, 1P (1972).
3. D. Scott, J. Silk, and M. White, *Science* **268**, 829 (1995).
4. C.B. Netterfield *et al*, *Ap. J.* **474**, 47 (1997).
5. P.F. Scott *et al*, *Ap. J.* **461**, L1 (1996).
6. G. Hinshaw *et al*, *Ap. J.* **464**, L17 (1996).
7. J.A. Peacock and S.J. Dodds, *Mon. Not. R. astr. Soc.* **267**, 1020 (1994).
8. T. Kolatt and A. Dekel, *Ap. J.* **479**, 592 (1997).
9. P.T.P. Viana and A.R. Liddle, *Mon. Not. R. astr. Soc.* **281**, 531 (1996).

CMB Anisotropy due to Cosmic Strings: Large Angular Scale

B. Allen[a], R. R. Caldwell[b], E. P. S. Shellard[c], A. Stebbins[d], S. Veeraraghavan[e]

[a] *Department of Physics, University of Wisconsin - Milwaukee,*
P. O. Box 413, Milwaukee, Wisconsin 53201
[b] *Department of Physics and Astronomy, University of Pennsylvania,*
Philadelphia, PA 19104
[c] *University of Cambridge, DAMTP, Silver Street,*
Cambridge CB3 9EW, United Kingdom
[d] *NASA/Fermilab Theoretical Astrophysics Center,*
P.O. Box 500, Batavia, Illinois 60510
[e] *University of Manchester, NRAL, Jodrell Bank,*
Macclesfield, SK11 9DL, United Kingdom

We present results based on our simulation of the large angle CMB anisotropy induced by cosmic strings [1]. We have evolved a network of cosmic strings in order to generate full-sky CMB temperature anisotropy maps. Using 192 maps, we have computed the anisotropy power spectrum for $\ell \leq 20$. By comparing with the observed temperature anisotropy, we set the normalization for the string mass per unit length μ to be $G\mu/c^2 = 1.05^{+0.35}_{-0.20} \times 10^{-6}$, which is consistent with all other observational constraints on cosmic strings. We demonstrate that the large angle anisotropy pattern induced by strings is consistent with a Gaussian random field on large angular scales.

1 Cosmic String Simulation

Our goal in this project is to determine the large angular scale CMB anisotropies predicted by cosmic strings. These predictions are compared to the measurements of large scale anisotropies recorded by the COBE satellite. Because the predicted temperature perturbations are proportional to the dimensionless quantity $G\mu/c^2$, where G is Newton's constant and c is the speed of light, one may constrain μ, the cosmic string mass per unit length.

In order to simulate the network of cosmic strings, we have adapted the Allen-Shellard simulation [2] to evolve from a redshift $z = 100$ in a spatially-flat, $\Omega = 1$, dust-dominated Universe. In order to ensure that the anisotropy pattern is unaffected by simulation boundary effects, the simulation volume at $z = 0$ has side length $2H^{-1}$. Furthermore, a large dynamic range for the simulation was obtained by decreasing the number of segments used to represent the string network, while maintaining a good representation of the network on the angular scales of interest. Tests of this method are described elsewhere [1,3].

2 Methods of Computation

The temperature pattern for a single observer located at x_{obs} is computed using a discretized version of the integral equation

$$\frac{\Delta T}{T}(\hat{n}, x_{\text{obs}}) = \int d^4x \, G^{\mu\nu}(\hat{n}, x_{\text{obs}}, x)\Theta_{\mu\nu}(x). \tag{1}$$

Here, $\hat{n}$ is a direction on the discretized celestial sphere, $\Theta_{\mu\nu}$ is the string stress-energy tensor supplied by the numerical simulation, and the Green's function $G^{\mu\nu}$ is described in Stebbins & Veeraraghavan [4]. We have used the ansatz of local compensation [4] for the perturbations of the matter distribution at the beginning of the simulation. We have generated three independent realizations of the cosmic string network. For each realization we have computed the fractional CMB temperature perturbation, $(\Delta T/T)(\hat{n}, x_{\text{obs}})$, in 6144 pixel directions $\hat{n}$ on the celestial sphere of 64 observers distributed uniformly throughout each simulation volume. This scheme gives $\Delta T/T$ smoothed on about $3.5°$. The maps have subsequently been smoothed with an approximately $10°$ FWHM beam in order to facilitate comparison with COBE.

3 Results

We have normalized the cosmic string mass per unit length μ by matching estimates of $C(0°, 10°)$ from COBE-DMR [5] with our predictions. Using semi-analytic methods we have extrapolated our results for $\widehat{C}(0°, 10°)$, the standard estimator for the ensemble-averaged correlation function, from $z = 100$ to $z = 1100$ at recombination. We obtain $\widehat{C}(0°, 10°) = 103 \pm 24 \pm 20(G\mu/c^2)^2$ where the ± 24 gives the cosmic variance between the different observers in all three of our simulations, and the ± 20 is a conservative systematic error on the simulation and techniques. Hence, normalization to COBE yields

$$G\mu/c^2 = 1.05^{+0.35}_{-0.20} \times 10^{-6} \tag{2}$$

for the cosmic string mass per unit length.

We summarize the results of our simulations with the following figures. In Figure 1 we show the average rms anisotropy in each of the three simulations, smoothed with an effective $10°$ beam versus the maximal redshift z. We see that the rms anisotropy increases as the effects of more strings are included, with increasing z. In Figure 2 we show the mean multipole moments $\widehat{C}_\ell$ for $\ell \leq 20$, obtained by averaging over the 192 observers. The error bars give the rms variation. For $\ell > 20$ the errors due to finite pixel size and gridding effects are larger than 10%.

290

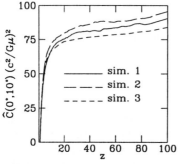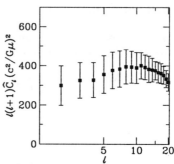

In Figure 3 we show the distribution $f[q_\ell]$ of values of the quantity $q_\ell \equiv (2\ell + 1)C_\ell/\widehat{C}_\ell$ for the 64 observers in a single simulation. A Kolmogorov-Smirnov test supports the hypothesis that q_ℓ is χ^2–distributed with $2\ell + 1$ degrees of freedom with a high level of significance. Hence, the CMB anisotropy pattern predicted by cosmic strings is consistent with a Gaussian random field on large angular scales. Other tests of the anisotropy patterns[1] have supported this conclusion.

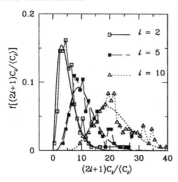

We believe that our estimate of μ is the most accurate and reliable to date. We refer the reader to Allen et al. [1,3] for more details.

Acknowledgements

The work of RRC was supported by the DOE at Penn (DOE-EY-76-C-02-3071).

References

1. B. Allen, R. R. Caldwell, E. P. S. Shellard, A. Stebbins, S. Veeraraghavan, Phys. Rev. Lett. **77**, 3061 (1996).
2. B. Allen and E. P. S. Shellard, Phys. Rev. Lett. **64**, 119 (1990).
3. B. Allen et al. (to be published).
4. A. Stebbins and S. Veeraraghavan, Ap. J. **365**, 37 (1990).
5. A. Banday et al., COBE Report No. 96-04, astro-ph/9601065, 1996; C. Bennett et al., COBE Report No. 96-01, astro-ph/9601067, 1996.

Non-Gaussian Spectra in the Cosmic Microwave Background

Pedro G. Ferreira [a]

Center for Particle Astrophysics, 301 Leconte Hall,
University of California, Berkeley CA 94720, USA

We propose a set of new statistics which can be extracted out of the angular distribution of the Fourier transform of the temperature anisotropies in the small field limit. They quantify generic non-Gaussian structure and complement the power spectrum in characterizing the sampled distribution function of a data set.

Recent years have seen scattered attempts at quantifying non-Gaussianity in cosmological data sets. It has become clear that there are two main lines of attack. One can focus on speculated sources of non-Gaussianity and design statistics which can best discriminate between them and Gaussian counterparts. This has been the approach in, for example, the study of cosmic defects such as strings or texture. The other, less prejudiced, approach is to define a general framework with which one can quantify deviations from Gaussianity. The main example of this strategy has been the n−point formalism which has been applied in a variety of settings [1]. The redundancy and inefficiency of such a method makes it pressing to look for viable alternatives.

This second approach is in general too ambitious. One must define restrictions to make any form of quantitative analysis tractable. One can do this by establishing requirements. We shall describe a formalism [2] which

- preserves information, i.e. given N independent pixels this will supply N quantities, one of which is the power spectrum

- is defined in Fourier space, and therefore tailored for high resolution, small field, interferometric measurements

In this setting it makes sense to consider the Fourier transform:

$$\frac{\Delta T(\mathbf{x})}{T} = \int \frac{d\mathbf{k}}{2\pi} a(\mathbf{k}) e^{i\mathbf{k}\cdot\mathbf{x}} \tag{1}$$

A Gaussian probability distribution function of the complex $a(\mathbf{k}_i)$ in a ring of fixed $|k|$ is given by:

$$F[a(\mathbf{k}_i)] = \frac{1}{(2\pi\sigma^2)^{m_k}} \exp\left(-\frac{1}{2\sigma_k^2} \sum_{i=1}^{m_k} |a(\mathbf{k}_i)|^2\right) \tag{2}$$

---

[a]In collaboration with João Magueijo (Imperial College)

where we have $2m_k = f_{sky}(2k+1)$ independent modes (f_{sky} is the fraction of the sky covered). Defining $a(\mathbf{k}_i) = \rho_i e^{i\phi_i}$ we can work in terms of m_k moduli ρ_i and m_k phases ϕ_i. The $\{\rho_i\}$ may be seen as Cartesian coordinates which we transform into polar coordinates. These consist of a radius r plus $m_k - 1$ angles $\tilde{\theta}_i$ given by

$$\rho_i = r \cos \tilde{\theta}_i \prod_{j=0}^{i-1} \sin \tilde{\theta}_j \qquad (3)$$

with $\sin \tilde{\theta}_0 = \cos \tilde{\theta}_{m_k} = 1$. In terms of these variables the radius is related to the angular power spectrum by $C(k) = r^2/(2m_k)$. In general the first $m_k - 2$ angles $\tilde{\theta}_i$ vary between 0 and π and the last angle varies between 0 and 2π. However because all ρ_i are positive all angles are in $(0, \pi/2)$. In order to define $\tilde{\theta}_i$ variables which are uniformly distributed in Gaussian theories one may finally perform the transformation on each $\tilde{\theta}_i$:

$$\theta_i = \sin^{N_k - 2i}(\tilde{\theta}_i) \qquad (4)$$

so that for Gaussian theories one has:

$$F(r, \theta_i, \phi_i) = \frac{r^{N_k - 1} e^{-r^2/(2\sigma_k^2)}}{2^{m_k - 1}(m_k - 1)!} \times 1 \times \prod_{i=1}^{m_k} \frac{1}{2\pi} \qquad (5)$$

The factorization chosen shows that all new variables are independent random variables for Gaussian theories. r has a $\chi_{N_k}^2$ distribution, the "shape" variables θ_i are uniformly distributed in $(0, 1)$, and the phases ϕ_i are uniformly distributed in $(0, 2\pi)$.

The variables θ_i define a non-Gaussian shape spectrum, the *ring spectrum*. They may be computed from ring moduli ρ_i simply by

$$\theta_i = \left(\frac{\rho_{i+1}^2 + \cdots + \rho_{m_k}^2}{\rho_i^2 \cdots + \rho_{m_k}^2} \right)^{m_k - i} \qquad (6)$$

They describe how shapeful the perturbations are. If the perturbations are stringy then the maximal moduli will be much larger than the minimal moduli. If the perturbations are circular, then all moduli will be roughly the same. This favours some combinations of angles, which are otherwise uniformly distributed. In general any shapeful picture defines a line on the ring spectrum θ_i. A non-Gaussian theory ought to define a set of probable smooth ring spectra peaking along a ridge of typical shapes.

We can now construct an invariant for each adjacent pair of rings, solely out of the moduli. If we order the ρ_i for each ring, we can identify the maximum moduli. Each of these moduli will have a specific direction in Fourier space; let $\mathbf{k}_{max}$ and $\mathbf{k}'_{max}$ be the directions where the maximal moduli are achieved. The angle

$$\psi(k, k') = \frac{1}{\pi}\text{ang}(\mathbf{k}_{max}, \mathbf{k}'_{max}) \tag{7}$$

will then produce an inter-ring correlator for the moduli, the *inter-ring spectra*. This is uniformly distributed in Gaussian theories in $(-1, 1)$. It gives us information on how connected the distribution of power is between the different scales.

We have therefore defined a transformation from the original modes into a set of variables $\{r, \theta, \phi, \psi\}$. The non-Gaussian spectra thus defined have a particularly simple distribution for Gaussian theories. We shall call perturbations for which the phases are not uniformly distributed localized perturbations. This is because if perturbations are made up of lumps statistically distributed but with well defined positions then the phases will appear highly correlated. We shall call perturbations for which the ring spectra are not uniformly distributed shapeful perturbations. This distinction is interesting as it is in principle possible for fluctuations to be localized but shapeless, or more surprisingly, to be shapeful but not localized. Finally we shall call perturbations for which the inter-ring spectra are not uniformly distributed, connected perturbations. This turns out to be one of the key features of perturbations induced by cosmic strings. These three definitions allow us to consider structure in various layers. White noise is the most structureless type of perturbation. Gaussian fluctuations allow for modulation, that is a non trivial power spectrum $C(k)$, but their structure stops there. Shape, localization, and connectedness constitute the three next levels of structure one might add on. Standard visual structure is contained within these definitions, but they allow for more abstract levels of structure.

References

1. P.J.E. Peebles, *The Large Scale Structure of the Universe*, Princeton University Press, (1980)
2. P. G. Ferreira, J. Magueijo, *Phys. Rev.* D, accepted, (1997)

FOREGROUND CONTAMINATION AROUND THE NORTH CELESTIAL POLE

ANGÉLICA DE OLIVEIRA-COSTA

Princeton University, Department of Physics, Jadwin Hall, Princeton, NJ 08544;
angelica@pupggp.princeton.edu
Institute for Advanced Study, Olden Lane, Princeton, NJ 08540

AL KOGUT

Hughes STX Corporation, Laboratory for Astronomy and Solar Physics, Code 685,
NASA/GSFC, Greenbelt MD 20771; kogut@stars.gsfc.nasa.gov

MARK J. DEVLIN, C. BARTH NETTERFIELD, LYMAN PAGE

EDWARD J. WOLLACK

Princeton University, Department of Physics, Jadwin Hall, Princeton, NJ 08544;
page@pupggp.princeton.edu

We cross-correlate the Saskatoon Q-Band data with different spatial template maps
to quantify possible foreground contamination. We detect a correlation with the
Diffuse Infrared Background Experiment (DIRBE) 100 μm map, which we interpret
as being due to Galactic free-free emission. Subtracting this foreground power
reduces the Saskatoon normalization of the Cosmic Microwave Background (CMB)
power spectrum by roughly 2%.

1 INTRODUCTION

One of the major concerns in any Cosmic Microwave Background (CMB) analysis is to determine if the observed signal is due to real CMB fluctuations
or due to some foreground contaminant. At the frequency range and angular
scale of the Saskatoon experiment [1,2], there are two major potential sources
of foreground contamination: diffuse Galactic emission and unresolved point
sources. The diffuse Galactic contamination includes three components: synchrotron and free-free radiation, and thermal emission from dust particles [3].
Although from a theoretical point of view, it is possible to distinguish these
three components, there is no emission component for which both the frequency
dependence and spatial template are currently well known [4]. The purpose of
this paper is to use the Saskatoon data to estimate the Galactic emission at
degree angular scales.

Figure 1: Saskatoon and template maps. For all maps, the temperatures are shown in coordinates where the North Celestial Pole is at the center of a circle of 15° diameter, with RA=0 at the top and increasing clockwise. The nine panels show the Saskatoon Ka Band map (Ka), the Saskatoon Q Band map (Q) and the full Saskatoon (Ka + Q Band) map (All), the 408 MHz Haslam survey (Ha), the 1420 MHz Reich & Reich survey (RR), the point source template (PS), and the DIRBE 240, 140 and 100 μm maps.

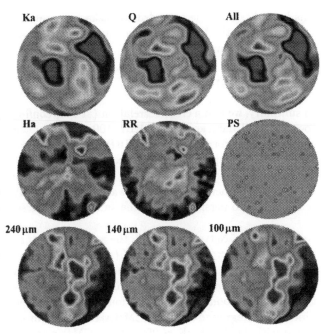

2 DATA ANALYSIS

We based our analysis on the 1994-1995 data from Saskatoon experiment [1,2,5]. We cross-correlate the Saskatoon Q-Band data with two different synchrotron templates: the 408 MHz survey [6] and the 1420 MHz survey [7]. To study dust and free-free emission, we cross-correlate the Saskatoon data with the Diffuse Infrared Background Experiment (DIRBE) sky map at wavelength 100 μm [8]. In order to study the extent of point source contamination in the Saskatoon data, we cross-correlate it with the 1Jy catalog of point sources at 5 GHz [9]. The templates used in this analysis, as well as the Saskatoon data, are shown in Figure 1.

The synchrotron templates, as well as the point source template, are found to be uncorrelated with the Saskatoon data. The DIRBE far-infrared template show a correlation, indicating a detection of signal with common spatial structure in the two data sets. Kogut et al.[4,10] detect a positive correlation between the DIRBE far-infrared maps and the DMR maps at 31.5, 53, and 90 GHz, which they identify as being the result of a free-free emission. Assuming that this hypothesis can be extended to Saskatoon scales, we argue that the correlation between the DIRBE template and the Saskatoon data is most likely due to free-free contamination[11].

3 CONCLUSIONS

In summary, we find a cross-correlation (at 97% confidence) between the Saskatoon Q-Band data and the DIRBE 100 μm map. The *rms* amplitude of the contamination correlated with DIRBE 100 μm is $\approx$ 17 μK at 40 GHz. We argue that the hypothesis of free-free contamination at degree angular scales is the most likely explanation for this correlated emission. Accordingly, the spatial correlation between dust and warm ionized gas observed on large angular scales seems to persist down to the smaller angular scales.

As reported by Netterfield et al.[2], the angular power spectrum from the Saskatoon data is $\delta T_\ell = 49^{+8}_{-5}$ μK at $l=87$ (corresponding to *rms* fluctuations around 90 μK on degree scales). This value of δT_ℓ is a much higher signal than any of the contributions from the foreground contaminants cited above, and shows that the Saskatoon data is not seriously contaminated by foreground sources. Since the foreground and the CMB signals add in quadrature, a foreground signal with 17μK /90μK $\approx$ 20% of the CMB *rms* only causes the CMB fluctuations to be over-estimated by $\sqrt{1 + 0.20^2} - 1 \approx$ 2%.

References

1. E.J. Wollack et al., *ApJ*, **476**, 440 (1997).
2. C.B. Netterfield et al., *ApJ*, **474**, 47 (1997).
3. R.B. Partridge in *3K:The CMBR* (Cambridge University, GB, 1995).
4. A. Kogut et al., *ApJ*, **460**, 1 (1996).
5. M. Tegmark et al., *ApJL*, **474**, 77 (1997).
6. C.G.T. Haslam et al., *A&A*, **100**, 209 (1981).
7. P. Reich & W. Reich, *A&A*, **74**, 7 (1988).
8. N.W. Boggess et al., *ApJ*, **397**, 420 (1992).
9. H. Kühr et al., *A&A*, **45**, 367 (1981)
10. A. Kogut et al., *ApJL*, **464**, 5 (1996).
11. A. de Oliveira-Costa et al., *ApJL*, submitted.

Constraints on Compact Hyperbolic Spaces from COBE

J. Richard Bond, Dmitry Pogosyan & Tarun Souradeep
CITA, University of Toronto, Toronto, ON M5S 3H8, CANADA

The (large angle) COBE DMR data can be used to probe the global topology of our universe on scales comparable to and just beyond the present "horizon". For compact topologies, the two main effects on the CMB are: [1] the breaking of statistical isotropy in characteristic patterns determined by the photon geodesic structure of the manifold and [2] an infrared cutoff in the power spectrum of perturbations imposed by the finite spatial extent. To make a detailed confrontation of these effects with the COBE maps requires the computation of the pixel-pixel temperature correlation function for each topology and for each orientation of it relative to the sky. We present a general technique using the method of images for doing this in compact hyperbolic (CH) topologies which does not require spatial eigenmode decomposition. We demonstrate that strong constraints on compactness follow from [2] and that these limits can be improved by exploiting the details of the geodesic structure for each individual topology ([1]), as we show for the flat 3-torus and selected CH models.

Flat or open FRW models adequately describe the observed average properties of our Universe. Much recent astrophysical data suggest the cosmological density parameter Ω_0 is < 1.[1] In the absence of a cosmological constant, this would imply a hyperbolic 3-geometry for the universe. There are numerous theoretical reasons, however, to favour compact topologies (reviewed in ref. 2). To reconcile this with a flat or hyperbolic geometry, compact models can be constructed by identifying points on the standard infinite FRW space by the action of certain allowed discrete subgroups of isometries, Γ. The FRW spatial hypersurface is then the "universal cover", tiled by copies of the compact space.[a] Dynamical chaos arising from the resulting complex geodesic structure in CH spaces has been proposed as an explanation of the observed homogeneity of the universe.[3]

Any unperturbed FRW spacetime will have an isotropic cosmic microwave background (CMB) regardless of global topological structure. However, the topology does affect the observed CMB temperature fluctuations $\Delta T/T$ and thus can be tested. At large angular scales, $\Delta T/T$ is dominated by the Sachs-Wolfe effect: $\Delta T/T \propto \Phi$, where Φ is the gravitational potential, appropriately smoothed to take into account the COBE beam. It is usually computed

---

[a] Any point x of the compact space has an image $x_i = g_i x$ in each "Dirichlet" domain on the universal cover, where $g_i \in \Gamma$. Compact hyperbolic manifolds (CHM) are described by discrete subgroups of the proper Lorentz group. A census of CHMs and software (SnapPea) for computing the generators of Γ for any CHM is freely available.[4] CHMs can be classified in terms of V/d_c^3, where V is the volume and $d_c = H_0^{-1}/\sqrt{1 - \Omega_0}$ is the curvature radius.[5]

by decomposing into eigenmodes of the 3-Laplacian for the space, $e.g.$ just plane waves for flat universes. We avoid the nontrivial task of finding these eigenmodes for a CH space by using the fact that the correlation function $C_c \equiv \langle \Phi(\mathbf{x}) \Phi(\mathbf{x}') \rangle$ in a compact space can be expressed as a sum over the correlation function C_u in its universal covering space between the images of $\mathbf{x}$ and $\mathbf{x}'$:

$$C_c(\mathbf{x}, \mathbf{x}') = \lim_{N \to \infty} \frac{1}{N} \sum_{i=0}^{N} \sum_{j=0}^{N} C_u(g_i \mathbf{x}, g_j \mathbf{x}'). \qquad (1)$$

The g_i are ordered in increasing displacement and g_0 is the identity. This procedure can be applied to any compact space provided the set of elements $\{g_i\}$ of the symmetry group is known. For a flat gravitational potential perturbation spectrum, $C_u(\mathbf{x}, \mathbf{x}') \propto \int d \ln(\beta^2 + 1) \left[\sin(\beta r) / (\beta \sinh r) \right]$, where r is the proper separation between $\mathbf{x}$ and $\mathbf{x}'$.

The prescription (1) applies separately to each "spectral mode", $C_c(\beta, \mathbf{x}, \mathbf{x}')$, defined as the integrand in $C_c(\mathbf{x}, \mathbf{x}') \propto \int d \ln(\beta^2 + 1) C_c(\beta, \mathbf{x}, \mathbf{x}')$. This decomposition is useful since the contribution to the ℓth multipole in the CMB angular correlation function comes predominantly from scales $\beta_\ell \approx \ell d_c / \mathcal{R}_H$ where $\mathcal{R}_H$ is the "angle-diameter distance" to the last scattering surface. In all the CH models that we have studied, compactness leads to a cutoff in $C_c(\beta, \mathbf{x}, \mathbf{x}')$ at scales at least four times the circum-radius of the space, $R_>$, $i.e.$, for $\beta < \beta_{cut} \approx (\pi/2) R_>^{-1}$ (see Fig. 1). Thus the ℓth multipole will be strongly suppressed if $\beta_{cut} > \beta_\ell$. We consider a model to contradict the COBE data if the $\ell \leq 4$ multipoles are strongly suppressed, translating to a criterion that the two parameters of the problem, $R_>/d_c$ and $\mathcal{R}_H/d_c$, a function only of Ω_0, have to satisfy: $R_> > (\pi/8) \mathcal{R}_H$. Fig. 1 shows our constraints in the $\Omega_0 - R_>$ parameter plane. We have studied some representative CH models (dots in Fig.1) in detail by confronting the full predicted statistics of the CMB anisotropy pattern with the all-channel COBE map.[6] We find that the exact likelihood falls steeply once the cutoff wavelength is reduced to a size comparable to the horizon, and the disallowed region in Fig. 1 determined by the simple β_{cut} criterion is strongly ruled out.

We conclude from Fig. 1 that the bounding scale $R_>$ of the universe cannot be much smaller than $\mathcal{R}_H$. This makes problematical topological explanations of galaxy and quasar distributions,[2,8] but there may be cases for which $R_>$ is large yet some closed geodesics are much shorter. Full statistical analysis using the COBE maps tightens the limits in all spaces considered so far.[6] When we apply our full statistical method to flat (equal-sided) 3-tori, we improve upon previous limits [7] on the size of the torus, $d_T \gtrsim 4 H_0^{-1}$ (95% CL), or $R_> \gtrsim \sqrt{3} \mathcal{R}_H$, much stronger than the conservative β_{cut} criterion given above because

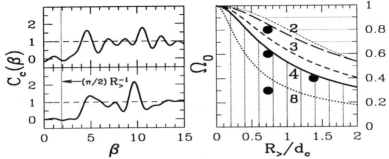

Figure 1: **left panel:** Examples of $C_c(\beta, \mathbf{x}, \mathbf{x}')$ for zero pixel separation in a CH model with $R_> = 0.83d_c$. Normalization is such that $C_u(\beta) = 1$. The infrared cutoff is present for $\mathbf{x} \neq \mathbf{x}'$ as well. **right panel:** Ω_0–$R_>$ constraints on CH models. Models in the region below the labeled lines have no power at multipoles $\ell \leq 2, 3, 4, 8$, respectively. In the upper shaded region, the compact space's volume exceeds the volume of a sphere of radius $\mathcal{R}_H$. Models in the lower shaded region fail our cut criterion with $\ell=4$. The large dots denote some examples for which a full statistical comparison with the DMR data has shown the topologies to be incompatible. This indicates the breaking of isotropy can lead to models outside of the lower shaded region being ruled out.

of the powerful breaking of statistical isotropy in the torus case. For compact hyperbolic universes, the low values of Ω_0 suggested by various astrophysical observations, $\lesssim 0.4$, are excluded for all the known [4] spaces. However, for $\Omega_0 \approx 1$, the multifaceted topology makes CH models more compatible with the COBE data than flat $\Omega_0 \equiv 1$ torus models.

References

1. Dekel, A., Burnstein, D., & White, S.D.M., 1996, in *Critical Dialogues in Cosmology*, ed. N.Turok, World Scientific.
2. Lachieze-Rey, M. & Luminet, J.-P. 1995, Phys. Rep. **25**, 136
3. Cornish, N.J., Spergel, D.N. & Starkman, G.D. 1996, Phys.Rev.Lett. **77**, 215
4. Weeks, J.R., *SnapPea: A computer program for creating and studying hyperbolic 3-manifolds*, University of Minnesota Geometry Center.
5. Thurston, W.P. 1979, *The Geometry of 3-Manifolds*, lecture notes, Princeton University; Thurston, W.P. & Weeks, J.R. 1984, Sci. Am. (July), 108
6. Bond, J.R., Pogosyan, D. & Souradeep, T. 1997, preprint
7. Sokolov, I.Y. 1993, JETP Lett. **57**, 617; Starobinsky, A.A. 1993, JETP Lett. **57**, 622; Stevens, D., Scott, D. & Silk, J. 1993, Phys. Rev. Lett. **71**, 20; de Oliveira Costa, A. & Smoot, G.F. 1995, Ap.J. **448**, 477; de Oliveira Costa, A., Smoot, G.F. & Starobinsky, A.A. 1996, Ap.J. **468**, 457
8. *e.g.*, Fagundes, H.V. 1989, Ap.J. **338**, 618; Roukema, B.F. 1996, preprint

A Search for Correlations between the Microwave and the X-ray Backgrounds

S. Boughn

Institute for Advanced Study, Princeton, NJ 08540
Department of Astronomy, Haverford College, Haverford, PA 19041

R. Crittenden

Canadian Institute for Theoretical Astrophysics, Toronto, ON M5S 3H8

N. Turok

DAMTP, University of Cambridge, Cambridge UK

In universes with significant curvature or a cosmological constant, some of the anisotropy of the cosmic microwave background (CMB) is created very recently and is correlated with the local matter density (RS or ISW effects). We examine the prospects of using the cosmic X-ray background (CXB) as a probe of the local matter density and obtain a constraint on the cosmological constant from the CXB/CMB cross-correlation function. Assuming that the X-ray map is dominated by large scale structure (which implies that the X-rays trace matter with a bias factor of $b_X \simeq 2$), we obtain a limit of $\Omega_\Lambda \lesssim 0.6$. Tighter constraints may be set with higher resolution maps of the CXB and CMB.

The recent observations of anisotropies in the microwave background [1,2] have been widely interpreted as the result of temperature fluctuations at the surface at last scattering, originating at high redshift ($z > 1000$). However, in a variety of cosmological models, a significant portion of the anisotropies are produced much more recently. The Rees-Sciama (RS) effect describes fluctuations which arise from non-linear effects on small scales[3]. In open universes or in universes with a significant cosmological constant, fluctuations are produced at late times by linear perturbations via the integrated Sachs-Wolfe (or ISW) effect[4]. These recent fluctuations result from time variations in the gravitational potential and are correlated with the nearby matter density ($z < 4$). Observing such correlations would result in important constraints on the mean matter density, Ω_m, as well as the cosmological constant, Ω_Λ, and offers a rare oppurtunity to observe CMB anisotropies as they are being produced[5,6].

Implementing this, however, requires a probe of the matter density at moderate redshifts ($z < 4$). Possible probes include radio galaxies and quasars; a number of large scale surveys of these objects are currently underway. Another probe is the hard (2-10 keV) X-ray background which appears to be produced primarily by active galactic nuclei[7] and so should reflect the mass distribution on large scales. In these proceedings, we investigate the cosmological limits which result from cross correlating the HEAO1 A2 2-10 keV X-ray map[8] with the four year COBE DMR map of the cosmic microwave background[1].

The combined HEAO data for a six month scan in 1977-8 were used to produce a full-sky CXB map with an effective angular resolution diameter of 3.3° FWHM. The Galactic plane was cut from the CMB and CXB maps and dipole moments were fit and removed from both maps. In addition, bright point sources were removed from the the X-ray map. The dimensionless cross correlation function, $\langle XT(\theta)\rangle/\sigma_T\sigma_X$, is shown in Figure 1a where σ_T and σ_X are the measured rms fluctuations in the two maps. The error bars plotted are from instrument noise and photon shot noise only. This correlation is consistant with a previous measurement at $\theta = 0°$ using the same data[9]. Correlations of the ROSAT All Sky Survey with COBE show somewhat more structure, presumably due to contamination from the Galaxy at lower (0.5-2.0 keV) X-ray energies[10].

Also shown for comparison are theoretical expectations for flat, CDM models with a range of values for the cosmological constant. As is shown in Figure 1b, there is a large variance in the theoretical predictions, which results primarily from accidental correlations between the X-ray map and the portion of the CMB fluctuations which was produced at the surface of last scattering. This 'cosmic' variance is highly correlated between neighboring angular bins and is the dominant source of uncertainty.

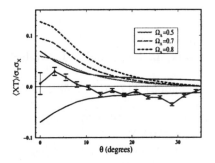 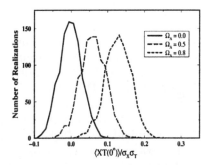

Figure 1. The left figure shows the calculated cross correlation between the CMB and CXB maps, compared to the theoretical expectations. The solid black lines show the rms fluctuation (cosmic variance) in the absence of correlations. The right figure shows a histogram of the zero lag cross correlation resulting from 1000 realizations of the different theories.

The theoretical predictions plotted in the figure assume that fluctuations in the CXB, σ_X, result entirely from large scale structure. Under this assumption, one can calculate the theoretical X-ray bias, b_x, which relates fluctuations in the X-ray emissivity to fluctuations in the matter density, $\delta\epsilon_x/\epsilon_x = b_x \delta\rho/\rho$. For the models we have considered, this implies an X-ray bias of $b_x \simeq 2$, which is consistent with estimates based on the acceleration of the Local Group.[11] However, it is not yet known how much of σ_X is due to the Poisson noise of unresolved sources or to other possible foregrounds. Therefore, our measurement of the X-ray bias should be considered an upper limit. To interprete Figure 1a for the case in which σ_X is contaminated (i.e. $b_x < 2$), one should reduce the theoretical curves in proportion to the inferred bias.

The observations are consistant with there being no correlation and, therefore, no cosmological constant. If there is no substantial contamination by unresolved sources, then for this class of models, $\Omega_\Lambda \leq 0.6$ at the 95 % confidence level. This constraint is weakened if some fraction of the X-ray fluctuations does not arise from large scale structure. In the future, higher resolution maps of the CMB and CXB should provide even better constraints.

Acknowledgments

The HEAO1 A2 data and map generating software were kindly provided to us by Keith Jahoda. Much of the data analysis was done at Princeton University and we thank Ed Groth for software and computing support. This work was supported in part by NASA, the NSF, and the Monell Foundation.

References

1. C.L. Bennett et al., Astrophysical J. **464**, L1 (1996).
2. C.B. Netterfield, M.J. Devlin, N. Jarosik, L. Page, and E.J. Wollack, Astrophysical J. **474**, 47 (1997).
3. M.J. Rees and D.W. Sciama, Nature (London) **217**, 511 (1968).
4. R. K. Sachs and A.M. Wolfe, Astrophysical J. **147**, 1 (1967).
5. R. G. Crittenden and N. Turok, Phys. Rev. Lett. **76**, 575 (1996).
6. M. Kamionkowski, Phys. Rev. **D54**, 4169 (1996).
7. For example, see A. Comastri, G. Setti, G. Zamorani and G. Hasinger, Astron. Astrophys. **296**, 1 (1995) and references therein.
8. E. Boldt, Phys. Rept. **146**, 215 (1987)
9. A.J. Banday, K.M. Gorski, C.L. Bennett, G. Hinshaw, A. Kogut, and G.F. Smoot, Astrophysical J. **468**, L85 (1996).
10. R. Kneissl, R. Egger, G. Hasinger, A. Soltan and J. Trumper, Astron. Astrophys., in press (1997).
11. T. Miyaji and E. Boldt, Astrophysical J. **353**, L3 (1990).

PHYSICS OF PHOTON DIFFUSION FOR ACTIVE SOURCES

R.A. Battye

Theoretical Physics Group, Blackett Laboratory, Imperial College
Prince Consort Road, London SW7 2BZ. U.K.

The physics of photon diffusion for the case of active sources is very different to that of passive sources. Fluctuations created just before the time of last scattering allow anisotropy to be created on scales much smaller than allowed by standard Silk damping. We develop a formalism to treat this effect and investigate its consequences using simple examples.

Understanding the implications of active sources models, such as topological defects, on the cosmic microwave background is very exciting area of research, since accurate measurements of the CMB over a wide range of scales will soon be available, enabling us to constrain or rule out whole classes of theories, with profound implications for our understanding of physics at high energies. Here, we present a summary of a study of a particular aspect of this subject : the effects of photon diffusion or 'Silk damping'[1]. We will extend previous work on passive[2] and active[3,4] theories. A more thorough discussion of this subject is currently in press[5].

The crucial period for understanding these effects is that just before last scattering. During this epoch the acoustic oscillations in the photon-baryon fluid are damped by the increasing mean free path of the photons. Active sources create fluctuations after the onset of this regime, which will receive less damping than those created before it. In particular, those created just before the time of last scattering will receive virtually no damping at all. If perturbations are created on all scales above the core size, as is thought to be the case for defect models, then it will be possible for anisotropy on small angular scales to remain to the present day. We will see that it is not sufficient to model these effects with a simple multiplication by an exponential suppression factor across all scales. Rather it requires careful consideration of the time at which fluctuations are created. Simple estimates will show that there are potentially important modifications to peak heights and also power-law suppression at small angular scales, rather than exponential.

In the region of angular scales of interest, the anisotropy can be estimated by investigating the behaviour of the intrinsic component of the anisotropy Θ_0, since it is this which is imprinted on the microwave background at last scattering, leading to Doppler or Sakharov peaks. An equation for $\widehat{\Theta}_0 = \Theta_0 + \Phi$

which includes the effects of photon diffusion at first order is

$$\ddot{\Theta}_0 + \left(\frac{\dot{R}}{1+R} + \frac{8}{27} k^2 \frac{1}{\dot{\kappa}} \frac{1}{1+R} \right) \dot{\Theta}_0 + \frac{1}{3} k^2 \frac{1}{1+R} \hat{\Theta}_0 = H(\eta) = \frac{1}{3} k^2 \left(\frac{\Phi}{1+R} - \Psi \right) \tag{1}$$

where $R = 3\rho_b/4\rho_\gamma$, $\dot{\kappa}$ is the differential optical depth due to Thomson scattering and Φ, Ψ are the gauge invariant gravitational potentials. Intuitively, one think of this as a forced-damped harmonic oscillator. Obviously, even in this very simple analogy no forcing effect can be damped until the it actually takes place.

One can solve (1) for large k using the WKB approximation. If one also ignores the transient solution, which can be thought of as being due to passive fluctuations, then $\hat{\Theta}_0$ is given by

$$[1 + R(\eta)]^{1/4} \hat{\Theta}_0(\eta) = \frac{\sqrt{3}}{k} \int_0^\eta d\eta' F(k, \eta, \eta') \tag{2}$$

where

$$F(k, \eta, \eta') = [1 + R(\eta')]^{3/4} e^{-k^2/k_s^2(\eta, \eta')} \sin [k r_s(\eta) - k r_s(\eta')] H(\eta'), \tag{3}$$

$r_s(\eta)$ is the sound horizon distance and $k_s^{-1}(\eta_2, \eta_1)$ is the Silk damping length, below which photon diffusion removes anisotropy,

$$r_s(\eta) = \frac{1}{\sqrt{3}} \int_0^\eta \frac{d\eta'}{\sqrt{1+R(\eta')}}, \quad k_s^{-2}(\eta_2, \eta_1) = \frac{4}{27} \int_{\eta_1}^{\eta_2} \frac{d\eta'}{\dot{\kappa}(\eta')} \frac{1}{1+R(\eta')}. \tag{4}$$

One can see immediately that the effect of photon diffusion is not an exponential suppression across all scales. Instead, the exponential suppression occurs inside the integral, with the damping scale being time dependent.

The effects of damping can be investigated by evaluating the power spectrum of of $\hat{\Theta}_0$ using a numerical integration routine and the simple structure functions designed to represent the qualitative behaviour of, for example, strings or textures. Here, we just present the results for a string structure function in figs. 1(a) and 1(b). Using a linear scale and a log-log scale illustrates the two effects of including this modified damping. One can see that the second, fourth and sixth peak heights are increased by 13%, 25% and 44% respectively and the tail at large $x_*(= k\eta_*/\sqrt{3})$ is no longer exponential, rather it is power law. The observable consequence of this effect seems to be a modulation of peak heights and a power-law tail, very similar to the effects of baryon drag[6]. In that case, the imperfect coupling between baryons and photons is responsible for an effective baseline shift. Here, what we are seeing is the sum of two

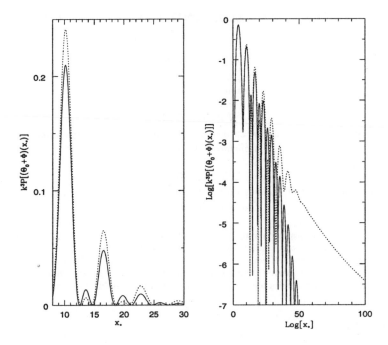

Figure 1: A comparison between the standard damping formalism (solid line) and the modified damping formalism (dotted line) for a string-like structure function : (a) on a linear scale and (b) on a log-log scale. Note that the linear scale starts at $x_* = 8$ and hence ignores the first peak.

components, one which is exponential, due to perturbations created before the onset of last scattering, and the other which is power law, due to perturbations created during the surface of last scattering.

References

1. J. Silk [1968], *Ap. J.* **151**, 459.
2. W. Hu & N. Sugiyama [1995], *Ap. J.* **444**, 489.
3. J. Magueijo *et al* [1996], *Phys. Rev. Lett.* **76**, 2617.
4. J. Magueijo *et al* [1996], *Phys. Rev.* **D54**, 3727.
5. R.A. Battye [1996], IMPERIAL/TP/96-96/46, astro-ph/9610197.
6. W. Hu & N. Sugiyama [1996], *Ap. J.* **471**, 542.

CALIBRATING CEPHEIDS AND THE DISTANCE SCALE

M.A. HENDRY

Department of Physics and Astronomy, University of Glasgow,
Glasgow G12 8QQ, UK

N.R. TANVIR

Institute of Astronomy, University of Cambridge, Madingley Road,
Cambridge CB3 OHA, UK

S.M. KANBUR

Department of Physics and Astronomy, University of Glasgow,
Glasgow G12 8QQ, UK

We discuss the calibration of multicolour Cepheid period luminosity (PL) relations. We describe a composite calibration procedure in which the PL slopes and zero points are determined self-consistently from Cepheids in several different galaxies. We also discuss the merits of an iterative 'inverse' PL fitting algorithm, and the implications of our results for the determination of the Hubble constant.

1   Introduction

In a recent article Tanvir [1] reviewed the use of Cepheids as extragalactic distance indicators and investigated the *robustness* of the V and I band Cepheid period luminosity (PL) relations in the LMC, which have become the 'benchmark' relations adopted in determining extragalactic distances (Madore & Freedman [2]; hereafter MF91). By extending this sample to a total of 53 Cepheids from the literature, Tanvir concluded that the PL relations constructed from the larger sample displayed no significant systematic differences from the fiducial relations presented in MF91. This suggested that the 'slide fitting' technique usually adopted in deriving distances to external galaxies – whereby the slopes of the V and I band PL relations in the external galaxy are assumed identical to those determined for the LMC sample – is unlikely to introduce a significant systematic error in the derived distance estimates.

Notwithstanding this, the number of Cepheids in external galaxies now detected with HST exceeds by at least a factor of ten in size the LMC calibrating sample. In this contribution we introduce a new, maximum likelihood, fitting procedure which determines simultaneously slopes and zero-points for the V and I band PL relations, and relative distance moduli and extinction coefficients, to a composite sample of Cepheids in several different galaxies. Thus, the assumption of PL slopes exactly equal to the LMC values is relaxed, and the impact of sampling error on the determination of the best-fit PL relations is reduced since they are estimated using the entire sample of Cepheids.

2 Method

This composite fitting procedure was first presented in an earlier paper [3], although we have now extended it to the multicolour case which allows simultaneous estimation of the extinction in each galaxy. We do no more than summarise the main points of the method here and will describe it in detail in a forthcoming paper.

We form a composite linear system of equations describing the V and I band PL relations. A likelihood function is then constructed for the observed apparent V and I band magnitudes, with the PL slopes and zero points, relative distance moduli and extinction coefficients as parameters. This system of equations is then solved by standard techniques and error estimates obtained for the fitted parameters from Monte Carlo simulations. The number of free parameters is reduced by making the usual assumption of a constant ratio for V and I band extinction. We also include a correction for the bias introduced by correlated observational errors on magnitudes and periods, extending an earlier univariate treatment [4]. The statistical formalism which we adopt is very similar to that used recently to analyse SNIa light curves [5].

3 Results

We have applied our composite fitting procedure to a number of different Cepheid samples comprising both LMC and HST data. Using a composite sample of 22 LMC stars [6] and a further 76 Cepheids in IC4182, NGC5253 and NGC4536, we obtain best-fit relations

$$M_V = -(2.73 \pm 0.07) \log P - (1.44 \pm 0.09) \tag{1}$$

$$M_I = -(2.99 \pm 0.07) \log P - (1.88 \pm 0.10) \tag{2}$$

which are completely consistent with those of MF91. When the LMC sample size is increased to include additional Cepheids from the literature [1] no significant discrepancy with MF91 is evident. This confirms the statistical robustness of the LMC calibration.

In order to address the possible impact of observational selection effects we have extended our composite fitting procedure to derive, iteratively, maximum likelihood fits to the *inverse* PL relations – i.e. minimising the residuals on log period at a given magnitude. We have shown that, in the case where the selection function is separable in period and magnitude, the fitted inverse PL relations will yield unbiased relative cluster distance moduli provided the same period range is selected in the LMC and external galaxies. We applied this procedure to composite samples of LMC and HST Cepheids, and again find that the relative distance moduli are not significantly changed from those obtained using the fiducial relations of MF91. We are testing the validity of assuming a separable selection function via Monte Carlo simulations of HST observations.

Our results indicate that recent determinations of the Hubble constant based on the Cepheid distance scale to e.g. SNIa [5] are not significantly affected by sampling error or uncorrected systematic bias in the LMC calibration.

We have extended our composite fitting procedure to more than two colours. We find from simulations that fitting simultaneously to the V, I and H band relations yields estimates of the extinction coefficients which are typically a factor of two more accurate than in the two colour case. The use of H band observations obtained with the NICMOS camera on HST will be an important step in further securing the reliability of the Cepheid distance scale.

References

1. N.R. Tanvir, in *The Extragalactic Distance Scale*, eds. M.Livio, M. Donohue and N. Panagia (CUP, Cambridge 1997)
2. B. Madore, W.L. Freedman, PASP **103**, 933 (1991)
3. M.A. Hendry and S.M. Kanbur, in *Mapping, Modelling and Measuring the Universe*, eds. P. Coles, V.J. Martinez and M.J. Pons-Borderia (ASP Conf. Series 94, 1996)
4. M. Akritas and M. Bershady, Ap.J. **470**, 706 (1996)
5. A.G. Riess, W. Press and R.P. Kirshner, Ap.J. **473**, 88 (1996)
6. N.R. Simon and T.S. Young, preprint, (1996)

MICROLENSING AND THE COMPOSITION OF THE GALACTIC HALO

Evalyn Gates[b], Geza Gyuk[d], and M.S. Turner[a,b,c]

[a] NASA/Fermilab Astrophysics Center, Fermi National Accelerator Laboratory, Batavia, IL 60510-0500

[b] Department of Astronomy & Astrophysics, Enrico Fermi Institute, University of Chicago, Chicago IL 60637-1433

[c] Department of Physics, Enrico Fermi Institute, University of Chicago, Chicago IL 60637-1433

[d] Scuola Internazionale Superiore di Studi Avanzati, Trieste, Italy

By means of extensive galactic modeling we study the implications of the more than 100 microlensing events that have now been observed for the composition of the dark halo of the Galaxy. Based on the currently published data, including the 2nd year MACHO results, the halo MACHO fraction is less than 60% in most models and the likelihood function for the halo MACHO fraction peaks around 20% - 40%, consistent with expectations for cold dark matter models.

Gravitational microlensing provides a valuable tool for probing the baryonic contribution to the dark matter in the halo of our Galaxy. However, even with precise knowledge of the optical depths toward the LMC and bulge, it would still be difficult to interpret the results because of the large uncertainties in the structure of the Galaxy. As it is, small number statistics for the LMC lead to a range of optical depths further complicating the analysis. Detailed modeling of the Galaxy is essential to drawing reliable conclusions.

The values of the parameters that describe the components of the Galaxy are not well determined; in order to understand these uncertainties we explore a very wide range of models that are consistent with all the data that constrain the Galaxy. We consider two basic models for the bulge, a triaxial model with the long axis oriented at an angle of about 10° with respect to the line of sight toward the galactic center, and an axisymmetric model. The bulge mass is not well determined, and we take $M_{\text{Bulge}} = (1 - 4) \times 10^{10} M_\odot$. For the disk component we consider a double exponential distribution and take the sum of a "fixed," thin luminous disk and a dark disk with varying scale lengths $r_d = 3.5 \pm 1\,\text{kpc}$, and thicknesses $h = 0.3\,\text{kpc}$, and $1.5\,\text{kpc}$. We also consider a model where the projected mass density varies as the inverse of galactocentric distance. We constrain the local projected mass density of the dark disk to be $10 M_\odot \leq \Sigma_{\text{VAR}} \leq 75 M_\odot\,\text{pc}^{-2}$. The dark halo is assumed to be comprised of two components, baryonic and non-baryonic, whose distributions are independent. We first assume independent isothermal distributions for the

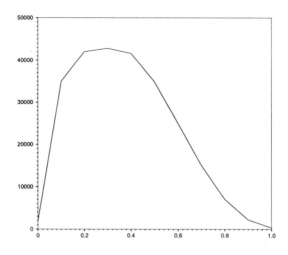

Figure 1: The number of viable models as a function of halo MACHO fraction.

MACHOs and cold dark matter, with core radii varying between 2 and 20 kpc. Since there are indications from both observations[1] and CDM simulations[2] that halos are significantly flattened, we also consider models with an axis ratio $q = 0.4$ (E6 halo) for both the baryonic and non-baryonic halos. While flattening does affect the local halo density significantly, increasing it by roughly a factor of $1/q$, it does not affect the halo MACHO fraction significantly[3]. Finally, we consider the possibility that the MACHOs are not actually in the halo, but instead, due to dissipation, are more centrally concentrated in a spheroidal component.

We then require that the following observational constraints be satisfied: circular rotation speed at the solar circle ($r_0 = 8.0\,\mathrm{kpc} \pm 1\,\mathrm{kpc}$) $v_c = 220\,\mathrm{km\,s^{-1}} \pm 20\,\mathrm{km\,s^{-1}}$; peak-to-trough variation in $v(r)$ between $4\,\mathrm{kpc}$ and $18\,\mathrm{kpc}$ of less than 14%; local escape velocity $v_{\mathrm{ESC}} > 450\,\mathrm{km\,s^{-1}}$ and circular rotation velocity at $50\,\mathrm{kpc}$, $180\,\mathrm{km\,s^{-1}} \leq v_c(50\,\mathrm{kpc}) \leq 280\,\mathrm{km\,s^{-1}}$. We also impose constraints from microlensing, both toward the bulge and toward the LMC. In calculating the optical depth toward the bulge, we consider lensing of bulge stars by disk, bulge and halo objects; for the LMC we consider lensing of LMC stars by halo and disk objects. We adopt the following constraints based upon microlensing data: (a)[4,5] $\tau_{\mathrm{BULGE}} \geq 2.0 \times 10^{-6}$ and (b)[6,7] $0.4 \times 10^{-7} \leq \tau_{\mathrm{LMC}} \leq 4 \times 10^{-7}$.

We summarize here our main results; details of the analysis can be found in ref. 8.

- The implications of the second year MACHO results for the halo MACHO fraction are shown in Figure 1. Incorporating the full set of constraints we find that the halo MACHO fraction is less than 60% in most models and peaks at a value of 20% − 40%. However, there are a small number of allowed models with a halo MACHO fraction greater than 80%. In addition to having a smaller total halo mass, these models all require an optical depth toward the LMC of greater than 2.5×10^{-7}, and most have $\tau_{LMC} \geq 3.5 \times 10^{-7}$.

- Bulge microlensing provides a crucial constraint to galactic modeling and eliminates many models. It all but necessitates a bar of mass at least $2 \times 10^{10} M_\odot$ and provides additional evidence that the bulge is bar-like. Because of the interplay between the different components of the Galaxy, the bulge microlensing optical depth indirectly constrains the MACHO fraction of the halo. On the other hand, LMC microlensing only constrains the MACHO fraction of the halo.

- Viable models with no MACHOs in the halo (where the LMC optical depth is due to a thick disk or spheroidal component population) are difficult unless $\tau_{LMC} \lesssim 2.0 \times 10^{-7}$.

This work was supported in part by the DOE (at Chicago and Fermilab) and the NASA (at Fermilab through grant NAG 5-2788).

1. P. Sackett, H. Rix, B.J. Jarvis, K.C. Freeman, *Astrophys. J.* **436**, 629 (1994).
2. C.S. Frenk, S.D.M. White, M. Davis, and G. Efstathiou, *Astrophys. J.* **327**, 507 (1988); J. Dubinski and R.G. Carlberg, *Astrophys. J.* **378**, 496 (1991).
3. E. Gates, G. Gyuk, and M.S. Turner, *Astrophys. J. Lett.* **449**, L123 (1995).
4. A. Udalski et al., *Astrophys. J.* **426**, L69 (1994); *ibid*, in press (1994); *Acta Astron.* **43**, 289 (1993); *ibid* **44**, 165 (1994); *ibid* **44**, 227 (1994).
5. C. Alcock et al., *Astrophys. J.* **445**, 133 (1995).
6. C. Alcock et al., *Phys. Rev. Lett.* **74**, 2867 (1995); C. Alcock et al., astro-ph/9606165.
7. E. Aubourg et al., *Nature* **365**, 623 (1993).
8. E. Gates, G. Gyuk, and M.S. Turner, *Phys. Rev. Lett.* **74**, 3724 (1995); E. Gates, G. Gyuk, and M.S. Turner, *Phys. Rev.* **D53**, 4138 (1996).

SUMMARY OF WIMP SEARCHES

N.J.C. SPOONER

Department of Physics, University of Sheffield, Hicks Building,
Hounsfield Road, Sheffield S3 7RH

A brief overview is presented of the state of experiments searching for dark matter in the form of Weakly Interacting Massive Particles by direct techniques. An outline is included of the various detector strategies adopted, recent improvements in the detector techniques, recent results and of future prospects for more sensitive detectors.

Progress towards WIMP detectors capable of sensitivity close to that needed to detect neutralinos at the predicted interaction rates has recently taken several important steps. A fuller overview of these can be found in several recent international workshops devoted to the subject[1,2,3]. Various different detector technologies have been adopted by the active groups but there has developed general commonality in the drive to obtain: i) active discrimination of the nuclear recoil events expected from WIMP interactions from residual ambient electron background and ii) increased active mass (scale-up). Table 1 attempts to summarise the principle active experiments recently run or presently running. Table 2 summarises various new planned experiments (as known to the author).

The strategies adopted to achieve points i) and ii) above can essentially be divided into two camps: low temperature bolometric detectors and scintillation detectors. Ge ionization detectors (see Table 1) are also being operated by several groups and have set useful limits (in particular by the Heidelberg/Moscow collaboration[4]) but although these can in principle be scaled up to many Kgs mass there is no proven technique of recoil discrimination and hence no direct means of identifying a WIMP signal, except via the small predicted annual modulation. Meanwhile recoil discrimination by bolometric and scintillation techniques have now both been proven. Various techniques are possible in the former technology, including the use of Superheated Superconducting Granules[1] (SSG) in which nuclear recoils cause fewer grain "flips" than electrons of the same energy, but the simultaneous detection of ionization and phonon energy in a semiconductor (Ge and Si) at mk temperatures has so far proved to be the most effective. In particular the CDMS experiment at Stanford[1,3] is now running several Ge and Si crystals of 60 and 100 g with recoil discrimination sufficient for dark matter experiments. No physics data has so far been obtained (at time of writing) but is expected soon. Several thermal experiments using alternative targets but without recoil discrimination are also now running. In particular the Milan group TeO_2 experiment[1] (primarily for $\beta\beta$ decay).

Table 1: Wimp experiments running or recently run

Experiment	Site	Target	~q.f.	Disc	Technology
Heidelberg/Moscow	Gran Sasso	Ge	0.25	No	Ionization
USC-PNL-Zaragoza-TANDAR	Canfranc Sierra Grande	Ge	"	No	Ionization
USC-PNL-Zaragoza	Soudan ββ	Ge	"	No	Ionization
Neuchatel-Caltech-PSI	St.Gottard	Ge	"	No	Ionization
UKDMC	Boulby	NaI(Tl) NaI	0.3(Na) 0.08(I)	Yes	Scintillation
DAMA-NaI (Rome)	Gran Sasso	NaI(Tl)	"	Yes	Scintillation
ELEGANTSV (Osaka)	Kamioka	NaI(Tl)	"	Yes	Scintillation
Saclay	Frejus	NaI(Tl)	"	Yes	Scintillation
USC-PNL-Zaragoza	Canfranc	NaI(Tl)	"	Yes	Scintillation
DAMA-Xe (Rome)	Gran Sasso	lq Xe	0.2-0.65	Yes	Scintillation
CDMS (CfPA)	Stanford	Ge, Si	>0.95	Yes	Thermal
EDELWEISS (France)	Frejus	Ge, Al_2O_3	>0.95	Yes	Thermal
Milan	Gran Sasso	TeO_2	0.93	No	Thermal
Tokyo	Osaka	LiF	>0.9	No	Thermal
Amherst-UCB	various	mica		yes	Track

Some of the most rapid progress has recently been made with scintillator technology, in particular with low background NaI(Tl). Recoil discrimination is possible here via use of differences in the scintillation pulse shape. Careful control of temperature, purity and doping has allowed useful discrimination to be obtained even at electron recoil energies well below 10 keV[5]. Furthermore, several groups have now proved operation of high mass arrays (>100 kg[1,2,3]). A disadvantage of these detectors is that the proportion of recoil energy detectable relative to electrons of the same energy (the q.f.) is relatively low compared to thermal detectors (see Table 1) making low energy thresholds difficult to achieve. Nevertheless, these detectors are presently setting the most stringent limits, particularly for spin-coupled candidates. The UKDMC group recently achieved a factor >x50 improvement of previous Ge limits[6] (see fig. 1).

Table 2: WIMP experiments proposed

Experiment	Site	Target	Disc	Technology
Osaka-Tokushima	Oto Cosmo	Ca_2F	Yes	Scintillation
ZEPLIN (UKDMC)	Boulby	Lq Xe	Yes	Ion-Scint
SIMPLE (CERN-Paris-Lisbon)	100 mwe	F	Yes	Superheated Droplet (SDD)
Montreal-Chalk River	0 mwe	F, Cl	Yes	SDD
SALOPARD	Canfranc	Sn	Yes	SSG
ORPHEUS	70 mwe	Sn	Yes	SSG
CREST (Munich-Oxford)	Gran Sasso	Al_2O_3	No	Thermal

314

The scintillator groups are planning upgraded detectors with inorganic halides (particularly NaI(Tl) and CaF$_2$(Eu)). However, it has now been demonstrated that Liquid Xe scintillator has potentially much greater discrimination power[3]. In this case a scintillation signal is collected in combination with ionization. Xe has the advantage of a higher (though still uncertain) q.f. value and higher A, which makes the target sensitive to coherent interactions and complementary to NaI. The UKDMC and Rome groups are actively developing a new generation of WIMP detectors based on Lq Xe. Other new detector techniques are also planned (see Table 2) including Superheated Droplet Detectors in which discrimination is predicted to be good enough not to need operation in a deep site[1]. However, there is an increasing interest in the possibility of detectors with sensitivity to the direction of nuclear recoils. This potentially would give a definitive signal for WIMP interactions, for which there is a well predicted forward:back asymmetry in the flux as the Earth moves through the dark matter halo (see various authors in ref. 1).

In conclusion there has recently been rapid progress towards improving the sensitivity of WIMP searches and this has enabled new limits to be set. But more importantly two significant technologies as applied to the field, scintillators and ionization/thermal detectors, are now well understood and there are clear routes to improving both towards sensitivity capable for a first detection of WIMPs.

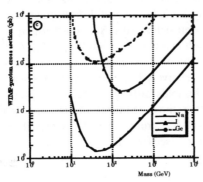

Figure 1: Spin dependent WIMP-proton cross section from the UKDMC 6 kg NaI(TL) detector[6]

References

1. see authors in Proc. *IDM96*, Sheffield, UK, 8-12th Sept. 1996
2. see authors in Proc. *Int. Work. Aspects of Dark Matter*, Heidelberg, Germany, 16-20th Sept. 1996
3. see authors in Proc. *2nd RESCEU Symposium*, Tokyo, 26-28th Nov. 1996
4. M. Beck et al., *Nucl. Phys. B (Proc. Suppl.)* 35, 150 (1994)
5. N.J.C. Spooner *et al.*, *Astrop. Phys.*, Vol. 5, No. 3 58 (1996)
6. P.F. Smith *et al.*, *Phys. Lett.* B379, 299 (1996)

A LARGE-SCALE SEARCH FOR DARK-MATTER AXIONS

C.A. HAGMANN, D. KINION, W. STOEFFL, K. VAN BIBBER
Physics and Space Technology Directorate,
Lawrence Livermore National Laboratory, 7000 East Avenue,
Livermore, CA 94550, USA

E.J. DAW, J. MCBRIDE, H. PENG, L.J. ROSENBERG, H. XIN
Department of Physics, Massachusetts Institute of Technology,
77 Massachusetts Avenue, Cambridge, MA 02139 USA

J. LAVEIGNE, P. SIKIVIE, N. S. SULLIVAN, D. B. TANNER
Department of Physics, University of Florida, Gainesville, FL 32611 USA

D.M. MOLTZ, J. POWELL, J. CLARKE
Lawrence Berkeley National Laboratory, 1 Cyclotron Road,
Berkeley, CA 94720 USA

F.A. NEZRICK, M.S. TURNER
Fermi National Accelerator Laboratory, P.O. Box 500,
Batavia, IL 60510-0500 USA

N.A. GOLUBEV, L.V. KRAVCHUK
Institute for Nuclear Research of the Russian Academy of Sciences,
60th October Anniversary Prospekt 7a, 117 312 Moscow, Russia

Early results from a large-scale search for dark matter axions are presented. In this experiment, axions constituting our dark-matter halo may be resonantly converted to monochromatic microwave photons in a high-Q microwave cavity permeated by a strong magnetic field. Sensitivity at the level of one important axion model (KSVZ) has been demonstrated.

1. Introduction

Axions result from a natural and minimal solution to the strong-CP problem as proposed by Peccei and Quinn. [1,2] The mass of the axion and all its couplings are inversely proportional to a large, unknown symmetry-breaking scale, i.e. f_a^{-1}; however its relic abundance from the time of the Big Bang goes as $f_a^{7/6}$, making extremely light axions good Cold Dark Matter candidates. Nominally the open mass range for axions is $10^{-(6-3)}$ eV, with the lower bound arising from the requirement that axions not overclose the Universe, and the upper bound from the effect axion cooling would have had on the neutrino signal from SN1987a. Such axions were long thought to be 'invisible' due to their extremely weak couplings, but as shown by Sikivie[3], halo axions could be detected by their resonant conversion to a monochromatic microwave signal in a high-Q cavity resonator placed in a strong magnetic field. The condition for resonant conversion is that the mass of the axion equal the frequency of the cavity, i.e. $m_a c^2 = h\nu$. The cavity is tuned in small steps, and at each value of central frequency the power spectrum is

315

evaluated. The integration time at each frequency is determined by the desired signal-to-noise according to Dicke's radiometer equation, $(S/N) = (P_s/kT) \cdot (t/\Delta v)^{1/2}$, where P_s is the expected signal power, T the total noise temperature, t the integration time, and Δv the signal bandwidth. Two early implementations of this concept established technical feasibility, but fell two to three orders of magnitude short in the required power sensitivity. [4,5]

2. Description of the Experiment

The design strategy for the experiment to reach cosmological sensitivity was to push on two fronts. First, for a given axion model (which specifies $g_{a\gamma\gamma}^2$), the axion-to-photon conversion power is mostly determined by the scale of the magnet, i.e. $P(a\text{-}>\gamma) \sim B^2 \cdot V \cdot Q$, where B is the magnetic field strength, V the volume, and Q the quality factor of the cavity. On the other hand, the search rate to achieve a certain model sensitivity with a given S/N additionally depends on the noise temperature of the experiment, $(1/v) \cdot (dv/dt) \sim B^4 \cdot V^2 \cdot T^{-2} \cdot Q$, where the total noise temperature of the experiment $T = T_{phys} + T_{elec}$. Therefore the magnet represents a large scale-up from the first generation experiments (the 6 ton Nb-Ti coil has a clear-bore diameter of 60 cm, a length of 100 cm, and a central field of 7.9 T), while the overall system noise profits from steady developments in HEMT performance and a suitably low physical temperature (the NRAO double-balanced amplifier has a minimum noise temperature of $T_{elec} \sim 3.9K$, and the cavity's physical temperature from pumped ^{4}He is $T_{phys} \sim 1.3K$). The quality factor Q for the copper-plated stainless steel cavity is already near the theoretical limit determined by the anomalous skin depth, which at optimal coupling in the 750 MHz region is $Q_L = Q/3 \sim 100,000$. In addition to the medium resolution data channel appropriate for the expected virialized component of the halo axions ($\Delta v/v \sim 10^{-6}$), we have instrumented a high resolution channel ($\Delta v/v \sim 10^{-11}$), to search for the possible ultra-fine structure predicted by Sikivie et al.[6] This latter possibility is based on both CDM being dissipationless, and the late-infall CDM not having experienced sufficient oscillation periods through the center of the galactic center to scatter gravitationally and thoroughly erase its initial phase-space distribution.

3. The Data

Typically each frequency bin is covered by 5-7 overlapping spectra, with an integration time of 70 seconds each. This is then repeated three times in sweeps separated by several weeks to reduce systematics; the data are then combined. Small gaps due to unwanted TE- or TEM-mode crossings with the TM_{010} mode of interest are filled by shifting the avoided crossings by running with the cavity flooded with superfluid helium. The distribution about the mean of the power spectrum is confirmed to be Gaussian out to +/- 4σ. Preliminary data for the frequency range 670 - 800 MHz is shown in Figure 1 below. In this figure, the signal power for KSVZ axions is assumed to fall into one channel (125 Hz), leading to a very large signal-to-noise, $S/N \sim (10\text{-}15)$. More realistically, a fully virialized axion signal would be about 6 channels wide, and S/N would be correspondingly reduced by about $6^{1/2}$ to $\sim (4\text{-}6)$. Peak search techniques include both the most robust method of excess power, as well as optimal filtering.

317

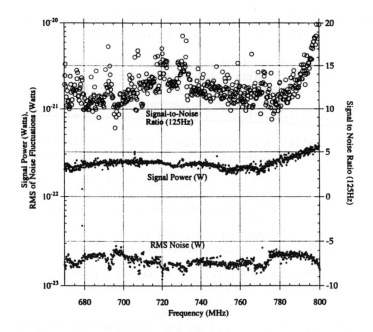

Figure 1. Noise, expected signal power (KSVZ) and *S/N* for the frequency range 670-800 MHz (2.77-3.3 μeV). As the entire spectrum contains more than 10^6 channels, only one out of every 2000 points has been plotted.

Research towards a DC SQUID-based amplifier is underway, aimed at an upgrade to ~300 mK noise temperature, which would permit sensitivity to DFSZ axions.

Acknowledgements

This work was performed under the auspices of the U.S. Department of Energy under contracts no. W-7405-ENG-48, DE-AC03-76SF00098, DE-AC02-76-CH03000, DE-FC02-94ER40818, DE-FG05-86ER40272 and FG02-90ER-40560.

References

1. R.D. Peccei and H. Quinn, *Phys. Rev. Lett.* **38** (1977) 1440 and *Phys. Rev.* **D16** (1977) 1791.
2. S. Weinberg, *Phys. Rev. Lett.* **40** (1978) 223; F. Wilczek, *Phys. Rev. Lett.* **40** (1978) 279.
3. P. Sikivie, *Phys. Rev. Lett.* **51** (1983) 1415
4. S. DePanfilis *et al.*, *Phys. Rev. Lett.* **59** (1989) 839; W.U. Wuensch *et al.*, *Phys. Rev.* **D40** (1989) 3153
5. C. Hagmann *et al.*, *Phys. Rev.* **D42** (1990) 1297
6. P. Sikivie and J. Ipser, *Phys. Lett.* **B291** (1992) 288

FIRST DATA FROM THE CRYOGENIC DARK MATTER SEARCH (CDMS) EXPERIMENT.

T. SHUTT, P.D. BARNES, JR., A. DASILVA, R.J. GAITSKELL, S.R. GOLWALA, B. PRITYCHENKO, R.R. ROSS, B. SADOULET, D. SEITZ, W. STOCKWELL, R. THERRIEN, T.L. TRUMBULL, S. WHITE
Center for Particle Astrophysics; and
Deptartment of Physics, UC Berkeley, Berkeley CA 94720;

J. EMES, E.E. HALLER, W.B. KNOWLTON, A. SMITH, G. SMITH, J.D. TAYLOR
Lawrence Berkeley National Laboratory, 1 Cyclotron Rd., Berkeley CA 94720;

P. BRINK, B. CABRERA, B. CHUGG, R.M. CLARKE, A. DAVIES, B. L. DOUGHERTY, K.D. IRWIN, S.W. NAM, M.J. PENN
Department of Physics, Stanford University, Stanford CA 94305;

D.S. AKERIB, R. SCHNEE, T. PERERA
Department of Physics, Case Western Reserve University, Cleveland, Ohio 44106;

D. BAUER, D.O.CALDWELL, D. HALE, A. SONNENSCHEIN, S. YELLIN
Deptarment of Physics, Univerisity of California, Santa Barbara, 93106;

B.A. YOUNG
Deptartment of Physics, Santa Clara University, Santa Clara CA 95053

V. KUZMINOV, V. NOVIKOV
Baksan Neutrino Observatory, Institute for Nuclear Research, Russian Academy of Science, Moscow, Russian Federation

The CDMS experiment is searching for WIMP dark matter in our galaxy with novel cryogenic detectors which reject background photons. The experiment began first data collection in the summer and fall of 1996; here we describe first data from a 60 g germanium and 100 g silicon detectors.

Direct detection of WIMP dark matter with particle detectors is a formidable experimental challenge [1]. Previous experiments [2-4] did not have the sensitivity to see particles predicted by minimal supersymmetric models [5] because of radioactive backgrounds. Here we describe the Cryogenic Dark Matter Search (CDMS), an experiment based upon a new detector technology with excellent background rejection capability.

The detectors consists of a target of germanium or silicon in which particles interact, creating both ionization and phonons, which we simultaneously measure. This technique allows us to distinguish WIMPs (which scatter elastically off nuclei) from background photons (which interact with electrons) since nuclear recoils are less ionizing than electron recoils by a factor of roughly 3.

Two complementary phonon measurement techniques are employed in the experiment. One [6] is a calorimetric measurement of the few μK/keV temperature rise in a germanium crystal using neutron transmutation doped Ge thermistors. The

318

first results reported here were obtained with a 62 g target; subsequent data runs will use detectors with equal performance but 165 g targets. The top and bottom faces of the disk-shaped target crystal have electrodes to collect the ionization. One electrode is segmented into two equal-volume regions. The baseline FWHM resolutions are 500 eV in phonons and 1.5 keV in ionization. These detectors are called Berkeley Large Ionization and Phonon based-detectors, or BLIPs.

The other technique utilizes a large array of sensors covering most of one face of the crystal. Each sensor has aluminum phonon collection pads with quasiparticle traps connected to tungsten transition edge sensors, and are called W/AL QET's [7, 8]. The signals are fast ($\approx$10 μsec rise time vs. a few ms for BLIPs), and give mm-scale position information. The current device has a 100 g silicon target, and resolutions of 7 keV and 3 keV in phonons and charge. These detectors are called Fast Large Ionization and Phonon based detectors, or FLIPs.

The detectors were calibrated with ^{60}Co photon and ^{252}Cf neutron sources (neutrons, like WIMPs, scatter on nuclei). In the 15-30 keV range BLIP has better than a 99% rejection of photons with a 95% acceptance of nuclear recoils. This photon rejection calibration is limited by electrons which land in a "dead layer" of some 10-30 μm beneath the electrodes where part of their charge is lost. FLIP has less of a dead layer, so that even with its lower energy resolution, it also achieved 99% photon rejection from 30-60 keV, with a 75% nuclear recoil acceptance.

The detectors are housed in modular "tower" packages which sit in an essentially all-copper, low-radioactivity extension to a dilution refrigerator surrounded by a nearly complete shield [9]. The CDMS experiment is located in a 10.5 m deep tunnel on the Stanford campus. At this depth cosmic-ray muons create a background of photon and neutrons. The outermost layer of the shield is thus a plastic scintillator active muon veto, with a measured efficiency of 99.6%. Inside this is a 15 cm-thick Pb photon shield, and a 25 cm thick polyethylene neutron moderator which reduces the energy of neutrons sufficiently that they can deposit <1 keV in germanium or silicon.

We have recently completed a data-taking run with the 60 g germanium BLIP and the 100 g silicon FLIP detectors discussed above. In figure 1 we show the measured background of all particles before application of the muon veto, and potential nuclear recoil anti-coincident with muons. The background photon rates (not shown in figure 1) are roughly 3 and 5 events/keV/kg/day in the BLIP and FLIP, with counting exposures of 1.4 and 0.52 kg-days, respectively. No lines from U or Th decay contamination have been seen. Preliminary Monte Carlo estimates of the neutron rate in coincidence with muons are, if anything, somewhat above the observed rate of nuclear recoils.

In BLIP there is a population of low energy electron events which limit our sensitivity to nuclear recoils. We have identified a small amount of ^{210}Pb, a beta emitter, in solder on a component near the detector which is consistent with the observed rate. Contamination by ^{40}K (from, e.g., fingerprints) or ^{210}Pb from airborne radon could also contribute to the beta background. In the next run we expect to reduce this rate greatly by removing the identified ^{210}Pb, and encasing a

320

new 165 g BLIP detector in a 1 mm thick Cu shield. We are also working to minimize this dead layer.

In the FLIP data there are three potential nuclear recoil events between 30 and 60 keV, corresponding to rate of 0.2 events/kg/keV/day. The beta rate in BLIP is lower in the inner charge collection electrode, so we use only this data (giving a fiducial mass of 30 g), and employ a 50% nuclear recoil acceptance cut which minimizes the beta acceptance. We obtain a rate of 0.4±0.2 events/kg/keV/day between 17.5 and 50 keV with 0.33 kg-days exposure, as shown in figure 1. Also shown is the spectrum from the Oroville experiment [2] re-scaled to recoil energy: the energy scale is increased and the rate decreased by a factor of three. These first data sets, obtained with 60 and 100 g detectors in a little over a month, are already competitive with the final results of previous experiments which had exposures on the order of 1 kg·year.

Once we have sufficiently reduced our beta background, a total detector mass of 1 kg (e.g., six 165 g detectors) counting for 100 days will improve existing WIMP limits by roughly 70, and, more importantly, will probe minimal supersymmetric WIMP models for the first time. Finally, preparations are underway for a second phase of the experiment, CDMS II, which will be located deep underground in the Soudan mine in Minnesota. At this deep site our goal is to improve our WIMP sensitivity by another factor of 30.

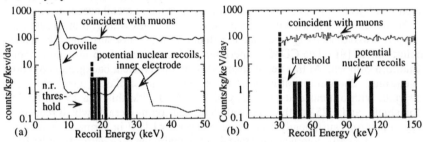

Figure 1.   Energy spectra in 62 g germanium (a) and 100 g silicon (b) detectors.

This work was supported by the Center for Particle Astrophysics, an NSF Science and Technology Center under Cooperative Agreement No. AST-912005, and by the DOE under Contract No. DE-ACO3-76SF00098 and Award Nos. DE-FG03-91ER40618 and DE-FG03-90ER40569.

1 J.R.Primack,D.Seckel, B.Sadoulet, *Ann.Rev. Nucl. Part. Sci.* **38**, 751-807 (1988).
2 D. O. Caldwell, *et al.*, *Phys. Rev. Lett.* , 510 (1988).
3 M. Beck, *et al.*, *Phys. Lett.* **B336**, 141-146 (1994).
4 E. Garcia, *et al.*, *Phys. Rev.* **D51**, 1458-64 (1995).
5 G. Jungman, M. Kamionkowski and K. Griest, *Phys.Repts.* **267**,195–373 (1996).
6 T. Shutt, *et al.*, *Phys. Rev. Lett.* **69**, 3531 (1992); *ibid* p. 3452.
7 K. D. Irwin, *Applied Physics Letters* **66**, 1988 (1995).
8 B. Cabrera,*et al.*, *Nuclear Physics B*, **51B**, pp. 294-303 (1996).
9 A. DaSilva, et al., *Nucl. Inst. Meth.* **A 354**, 553-559 (1995).

FINAL RESULTS FROM EROS 1

N. PALANQUE-DELABROUILLE

CE-SACLAY DAPNIA/SPP,
91191 Gif sur Yvette, FRANCE

EROS COLLABORATION

The final results from the first phase of the EROS search for gravitational microlensing of stars in the Magellanic Clouds by unseen deflectors are presented. The search is sensitive to events with time scales between 15 minutes and 200 days, corresponding to deflector masses in the range $10^{-7} \, M_\odot$ to a few $M_\odot$. Two events were observed that are compatible with microlensing by objects of mass $\sim 0.1 \, M_\odot$. By comparing the results with the expected number of events for various models of the Galaxy, we conclude that machos in the mass range $[10^{-7}, 0.02]$ $M_\odot$ make up less than 20 % (95 % C.L.) of the Halo dark matter.

1   Introduction

To search for compact objects in the Halo of our Galaxy using microlensing techniques, EROS monitored stars in two targets, the Large and the Small Magellanic Clouds. Possible deflectors can have masses ranging from $10^{-7} \, M_\odot$ to about a solar mass. Since the typical duration of an event is related to the mass of the deflector by $\Delta t \propto 70\sqrt{M/M_\odot}$ days, the above mass range implies durations from an hour to a few months. Assuming a given halo, many events of short duration are expected if it is filled with low-mass deflectors, thus requiring a very good time sampling. In contrast, very few events will be expected if the deflectors are heavy and many more stars will need to be monitored. The time sampling, however, will not be critical since the events will be of fairly long duration. So as to be sensitive to both ends of the mass spectrum, EROS conducted two observing programs:

- one using a 16 CCD camera mounted on a 40 cm telescope to search for short time scale microlensing events ($\Delta t <$ a few days i.e. $10^{-7} \lesssim m/M_\odot \lesssim 10^{-3}$); it covered a field of $\sim 0.5 \deg^2$ centered either on the bar of the LMC or the center of the SMC, with one point per color every 20 minutes;

- one with 25 $\deg^2$ Schmidt photographic plates (taken once per night, in both colors) for longer time scales ($10^{-4} \lesssim m/M_\odot \lesssim 1$). The plate data covers a large portion of the LMC.

We then analyse the 250 000 light curves from the CCD data and the $6 \, 10^6$ light curves from Schmidt plates in both colours.

2 Data analysis

To avoid excluding non-standard events such as those involving multiple lenses or sources, or those affected by the finite size of the source or additional light from unresolved stars (blending), the analysis contains *no* constraint on the shape of the light curve. We first search for a simultaneous fluctuation in the red and blue light curves of a given star. The main cuts, described herafter, are based on uniqueness: because of the low predicted optical depth ($\lesssim 5.10^{-7}$), the probability of a measurable microlensing effect to happen twice on the same star in a few years time is negligible. We thus require the main fluctuation to be much more statistically significant than the second one. This excludes most flat light curves on which there are only statistical fluctuations; first and second most important ones would be of similar amplitude. Periodic variables are also excluded by this cut. Most variables, however, are removed by asking that there be no correlation between the red and the blue light-curves, apart from the main bump. Some known long time scale variable stars[a] have amplification periods longer than what we are sensitive to (a couple years or so) and are therefore not removed by the previous cuts. Since they populate specific regions of the color-magnitude diagram, we only consider, in the analysis, stars that do not belong to those regions. This cut on the color-magnitude diagram only removes a small fraction of the stars monitored.

The detection efficiencies are obtained by processing through the same analysis software a sample of observed light curves with a random theoretical microlensing shape superimposed (finite source size effect and blending are taken into account in the simulated shape). With a chosen model of the Galaxy (disk and halo), one can then compute the expected number of events.

No microlensing event was identified in the CCD data.[1] Two events compatible with microlensing were identified from the Schmidt plate data,[2,3] with time scales of 23 and 29 days, corresponding to deflectors in the mass range $[0.01 - 1]\, M_\odot$.

3 Constraints on the Halo

From the comparison of the expected and observed number of events, we obtain an upper limit on the fraction of the Halo that can be composed of compact objects. Figure 1 presents 95% CL limits for a standard halo model, including all the EROS data and two microlensing candidates.[4] As indicated in the figure, the EROS analysis excludes the possibility that objects in the mass range $[10^{-7} - 0.02]\, M_\odot$ contribute to more than 20% to the mass of the halo. This

---

[a]very bright (and quite red) red giants, and very bright main sequence stars

323

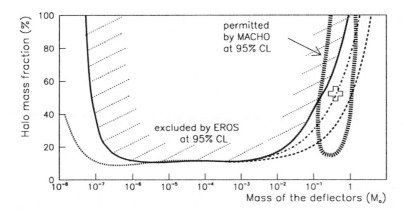

Figure 1: Exclusion diagram at 95% CL for a standard halo model with all EROS data, assuming a delta function for the mass distribution of the deflectors. For the CCD program, the influence of blending and finite size effects are shown (the dotted line on the left is the limit without those effects). Limits are shown for 0, 1 or 2 candidates (dashed, mixed or full line) assumed to be actual microlensing. The cross is centered on the area allowed at 95% CL by the MACHO experiment, assuming 6 microlensing events.

limit can be brought down to 10% for objects in the range $[5\ 10^{-7}-2\ 10^{-3}]\ M_\odot$. Since it is rather mass independent, this limit is valid for any mass function that peaks within this interval. Our results do not contradict the positive signal of the MACHO collaboration,[5] also plotted on the same figure.

The first phase of the EROS program has thus set very stringent limits on the contribution of low-mass objects to the Halo dark matter. Statistics for long time scale events, however, are still very low. The recently upgraded EROS program is now dedicated to high masses by monitoring a total of $80\ \mathrm{deg}^2$ on the LMC and SMC. EROS 2 also probes disk dark matter by monitoring about $100\ \mathrm{deg}^2$ toward the Galactic Center and spiral arms.

References

1. E. Aubourg et al, A&A **301**, 1 (1995).
2. E. Aubourg et al, Nat **365**, 623 (1993).
3. R. Ansari et al, A&A **314**, 94 (1996).
4. C. Renault et al, (submitted to A&A) astro-ph/9612102.
5. C. Alcock et al, (submitted to ApJ) astro-ph/9606165.

DARK OBJECTS' MASSES AND THE HALO STRUCTURE UNCERTAINTY

D. MARKOVIC

Theoretical Astrophysics Center
Juliane Maries Vej 30, Copenhagen DK-2100, Denmark

We estimate the accuracy of the massive halo objects' (MHO or 'Macho') mass function inference from microlensing events observed toward the Large Magellanic Cloud (LMC). It is assumed that one has access only to ground-based observations (which is currently the case) without the advantage of, e.g., parallax measurements. We also investigate quantitatively the effects of uncertainty of the MHOs' spatial distribution and kinematics on the determination of their mass function and the total mass of the MHO halo.

Microlensing of sources in the Large Magellanic Cloud has already yielded important information on the composition of the galactic halo. A statistical analysis[1] of the presently available data suggests that dark massive halo objects (MHOs) account for a significant portion of the total halo mass. In addition, their most likely masses seem to be in the range 0.1-$0.6 M_\odot$. This indicates that MHOs form a distinct, large population of old objects (e.g. white dwarves?) that cannot be easily extrapolated from any familiar stellar population. In order to place tighter constraints on the nature and origin of these objects, one should be able to extract a more accurate quantitative measure of their population, in particular their mass function $p_\mu(\mu)$ ($\mu \equiv M/M_\odot$).

The currently large uncertainty of MHOs' mass function stems in part from the still poor, small-number statistics, but also from our ignorance regarding the objects' spatial distribution and kinematics. In this chapter we will study these two main sources of error and evaluate prospects for an accurate determination of p_μ.

For simplicity, we will restrict our discussion to spherically symmetric halo models. An often used one is the isothermal sphere with an isotropic velocity dispersion, constant throughout the halo and the density profile well approximated by $\rho(r) = \rho_o \left(a^2 + R_\odot^2\right) / \left(a^2 + r^2\right)$, where $a \approx 5\,\mathrm{kpc}$ and $R_\odot = 8.5\,\mathrm{kpc}$ is the Sun - galactic centre distance. [This profile for the total (baryonic + non-baryonic matter) halo density yields the observed roughly flat galactic rotation curve.] We will use shorthand 'CIS' for this sphere with a core.

The MHO mass distribution, however, need not follow that of the total halo mass. One possibility is that the massive objects may be more concentrated toward the galactic centre like, e.g., the blue horizontal branch field stars of the outer halo. For this population we observe[2] $\rho \sim r^{-3.4}$ and a velocity dispersion

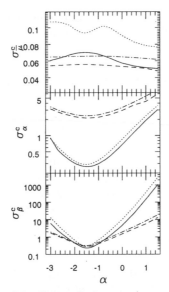

Figure 1: The dependence of the Cramer-limit errors (normalised to 100 events; $\bar{\mu} = 0.4$) on α for a 'broad' mass range, $\beta = 2$, are shown as the solid [$\varepsilon(T) = $ const] and dotted [MACHO-type detection efficiency, $\varepsilon(T) \neq$const.] lines. For a 'narrow' mass range, $\beta = 1$, the errors are shown as the dashed and dot-dashed lines for the respective detection efficiences.

that changes from $\beta_v \equiv 1 - \sigma_\theta^2/\sigma_r^2 > 0$ at smaller radii to $\beta_v < 0$ at larger distances. We will call this halo the 'concentrated sphere' (CS).

At first we make the important assumption that the halo structure, given by MHOs' density profile and kinematics, is known (we choose the CIS halo) and simulate the statistical inference of the mass function parameters $\mathbf{c} \equiv \{\bar{\mu}, \alpha, \beta\}$ [$p_\mu \propto \mu^\alpha$, $\bar{\mu} \equiv \int \mu\, p_\mu d\mu$, $\beta \equiv \log_{10}(\mu_{\max}/\mu_{\min})$] from the durations T of events. We find[3] that in the vicinity of $\alpha = -1.5$ the slope α and range β can be determined from $N \gtrsim 100$ events.[4] However, as we shift away from $\alpha = -1.5$, the error of determination grows very rapidly (see Fig. 1). [As detailed Monte-Carlo simulations show, the so-called Cramer-limit values,[3] shown in Fig. 1, are relatively inaccurate estimates of actual errors when the errors are large. Still, large values of Cramer errors imply large actual errors.] In particular, for any positive slope α one needs $N > 1000$ events for a reliable inference. A quantity that can be accurately ($\lesssim 30\%$ error) inferred at $N \gtrsim 100$ with *any* slope is $\bar{\mu}$. We find that our results do not depend strongly (Fig. 1) on whether the detection efficiency is T-independent or of the type presently available [1] for the MACHO project's microlensing searches.

In order to quantify the effects of the halo structure uncertainty, we will perform the following 'experiment:' we will assume that a specific halo model (the 'real' one) describes the actual halo accurately enough. We will then estimate how much the inferred parameters **c** of the mass function could be shifted *systematically* from the real values $\mathbf{c}_o$ if instead of fitting the event duration distribution function $P_o(T)$ based on the 'real' halo, we use the one, $P(T)$, based on a 'false' (yet plausible) halo model.

As an example, we choose CS as the 'real' halo and use the singular isothermal isothermal sphere (SIS), $\rho \sim r^{-2}$, for the inference. We find[3] that, although α and β are relatively weakly affected by the halo model ambiguity for $-2 \lesssim \alpha \lesssim 0$, the inferred $\bar{\mu}$ is on the average about 60% larger than the real value . It is important that unless $N \gtrsim 10^3$, the statistics based only on event durations does not allow us to distinguish between the two halos; the differences between them are swamped by the statistical noise.

The results of this 'experiment' indicate that the uncertainty of the inferred average mass will be difficult to reduce below about the factor of 2. A similar conclusion, with similar magnitudes of relative errors, holds for the MHOs' local density ρ_o and the total MHOs' mass M_{tot} between the solar orbit and LMC. The unresolvable (at $N < 10^3$) ambiguity is characteristic of a broad range of halo models that one might choose as the 'real' ones instead of CS, although the ensuing systematic errors may differ from those obtained in the specific case of the CS/SIS ambiguity. These results are supported by our maximum likelihood simulations,[3] where the halo model is treated as unknown and its parameters are varied together with those of the mass function: only at about $N \gtrsim 10^3$ do the errors in $\bar{\mu}$, ρ_o and M_{tot} fall below 50%.

Although the large requisite number of events does not support optimism regarding an accurate inference of MHOs' masses and their halo fraction on the basis of LMC events only, probing the halo in various directions, e.g. through microlensing of M31 sources, might help discriminate between different halo models. A considerable improvement should come from gaining more information from individual events by, e.g., parallax measurements.[3]

This research was generously supported by Danmarks Grundforsknings-fond through its establishment of the Theoretical Astrophysics Center.

References

1. C. Alcock *et al.*, 1996, astro-ph/9606165
2. J. Sommer-Larsen, C. Flynn & P.R. Christensen, *MNRAS* **271**, 94 (1994)
3. D. Markovic & J. Sommer-Larsen, to appear in *MNRAS*, astro-ph/9609187; and references therein.
4. Compare S. Mao and B. Paczynski, 1996, astro-ph/9604002

HALO DARK CLUSTERS OF BROWN DWARFS AND MOLECULAR CLOUDS

FRANCESCO DE PAOLIS and PHILIPPE JETZER

Paul Scherrer Institute, Laboratory for Astrophysics, CH-5232 Villigen PSI, and Institute of Theoretical Physics, University of Zurich, Winterthurerstrasse 190, CH-8057 Zurich, Switzerland

GABRIELE INGROSSO

Dipartimento di Fisica and INFN, Università di Lecce, CP 193, I-73100 Lecce

MARCO RONCADELLI

INFN, Sezione di Pavia, Via Bassi 6, I-27100 Pavia

The observations of microlensing events in the Large Magellanic Cloud suggest that a sizable fraction ($\sim$ 50%) of the galactic halo is in the form of MACHOs (Massive Astrophysical Compact Halo Objects) with an average mass $\sim 0.27 M_\odot$, assuming a standard spherical halo model. We describe a scenario in which dark clusters of MACHOs and cold molecular clouds (mainly of H_2) naturally form in the halo at galactocentric distances larger than 10-20 kpc.

1 Introduction

A central problem in astrophysics concerns the nature of the dark matter in galactic halos, whose presence is implied by the flat rotation curves in spiral galaxies. As first proposed by Paczyński[1], gravitational microlensing can provide a decisive answer to that question[2], and since 1993 this dream has started to become a reality with the detection of several microlensing events towards the Large Magellanic Cloud[3,4]. Today, although the evidence for MACHOs is firm, the implications of this discovery crucially depend on the assumed galactic model. It has become customary to take the standard spherical halo model as a baseline for comparison. Within this model, the mass moment method yields an average MACHO mass[5] of 0.27 $M_\odot$. Unfortunately, because of the presently available limited statistics different data-analysis procedures lead to results which are only marginally consistent. For instance, the average mass reported by the MACHO team is $0.5^{+0.3}_{-0.2}$ $M_\odot$. Apart from the low-statistics problem – which will automatically disappear from future larger data samples – we feel that the real question is whether the standard spherical halo model correctly describes our galaxy[6]. Besides the observational evidence that spiral galaxies generally have flattened halos, recent determinations of the disk scale length, the magnitude and slope of the rotation at the solar position indicate that our galaxy is best described by the maximal disk model. This conclusion is further strengthened by the microlensing results towards the galactic centre, which imply that the bulge is more massive than previously thought. So, the expected average MACHO mass should be smaller than within the standard halo model. Indeed, the value $\sim 0.1~M_\odot$ looks as the most realistic estimate to date and suggests that MACHOs are brown dwarfs.

2   Mass moment method

The most appropriate way to compute the average mass and other important properties of MACHOs is to use the method of mass moments developed by De Rújula et al. [7]. The mass moments $< \mu^m >$ are related to $< \tau^n > = \sum_{events} \tau^n$, with $\tau \equiv (v_H/r_E)T$, as constructed from the observations ($v_H = 210$ km s^{-1}, $r_E = 3.17 \times 10^9$ km and T is the duration of an event in days). We consider only 6 out of the 8 events observed by the MACHO group during their first two years. In fact, the two disregarded events are a binary lensing and one which is rated as marginal. The ensuing mean mass is [a] $< \mu^1 > / < \mu^0 > = 0.27~M_\odot$, assuming a standard spherical halo model.

Although this value is marginally consistent with the result of the MACHO team, it definitely favours a lower average MACHO mass. For the fraction of the local dark mass density detected in the form of MACHOs, we find $f \sim 0.54$, which compares quite well with the corresponding value ($f \sim 0.45$) calculated by the MACHO group in a different way.

3   Formation of dark clusters

A major problem concerns the formation of MACHOs, as well as the nature of the remaining amount of dark matter in the galactic halo. We feel it hard to conceive a formation mechanism which transforms with 100% efficiency hydrogen and helium gas into MACHOs. Therefore, we expect that also cold clouds (mainly of H_2) should be present in the galactic halo. Recently, we have proposed a scenario [8] in which dark clusters of MACHOs and cold molecular coulds naturally form in the halo at galactocentric distances larger than 10-20 kpc, with the relative abundance possibly depending on the distance.

The evolution of the primordial proto globular cluster clouds (which make up the proto-galaxy) is expected to be very different in the inner and outer parts of the Galaxy, depending on the decreasing ultraviolet flux (UV) from the centre as the galactocentric distance R increases. In fact, in the outer halo no substantial H_2 depletion should take place, owing to the distance suppression of

---

[a]When taking for the duration T the values corrected for "blending", we get as average mass 0.34 $M_\odot$. We thank D. Bennett for driving our attention on this point.

the UV flux. Therefore, the clouds cool and fragment - the process stops when the fragment mass becomes $\sim 10^{-2} - 10^{-1}$ $M_{\odot}$. In this way dark clusters should form, which contain brown dwarfs and also cold H_2 self-gravitating cloud, along with some residual diffuse gas (the amount of diffuse gas inside a dark cluster has to be low, for otherwise it would have been observed in the radio band).

We have also considered several observational tests for our model [8,9]. In particular, a signature for the presence of molecular clouds in the galactic halo should be a γ-ray flux produced in the scattering of high-energy cosmic-ray protons on H_2. As a matter of fact, an essential information is the knowledge of the cosmic ray flux in the halo. Unfortunately, this quantity is unknown and the only available information comes from theoretical considerations. Nevertheless, we can make an estimate of the expected γ-ray flux and the best chance to detect it is provided by observations at high galactic latitude. Accordingly, we find a γ-ray flux (for $E_\gamma > 100$ MeV) $\Phi_\gamma(90^0) \simeq 1.1 \times 10^{-6}$ ϵf photons cm^{-2} s^{-1} sr^{-1} (ϵ is a unknown parameter which takes into account the degree of confinement of the cosmic rays in the halo, whereas f stands for the fraction of halo dark matter in the form of cold molecular gas). This flux should be compared with the measured value for the diffuse background of $0.7-2.3 \times 10^{-5}$ photons cm^{-2} s^{-1} sr^{-1}. Thus, there is at present no contradiction with observations. Furthermore, an improvement of sensitivity for the next generation of γ-ray detectors will either discover the effect in question or yield more stringent limits on ϵf.

References

1. B. Paczyński, Astrophys. J. **304**, 1 (1986).
2. A. De Rújula, Ph. Jetzer and E. Massó, Astron. Astrophys. **254**, 99 (1992).
3. C. Alcock et al., Nature **365**, 621 (1993); astro-ph 9606165.
4. E. Aubourg et al., Nature **365**, 623 (1993).
5. Ph. Jetzer, Helv. Phys. Acta **69**, 179 (1996).
6. F. De Paolis, G. Ingrosso and Ph. Jetzer, Astrophys. J. **470**, 493 (1996).
7. A. De Rújula, Ph. Jetzer and E. Massó, Mont. Not. R. Astr. Soc. **250**, 348 (1991).
8. F. De Paolis, G. Ingrosso, Ph. Jetzer and M. Roncadelli, Phys. Rev Lett. **74**, 14 (1995); Astron. Astrophys. **295**, 567 (1995); Comments on Astrophys. **18**, 87 (1995); Astrophys. and Space Science **235**, 329 (1996); Int. J. Mod. Phys. **D5**, 151 (1996).
9. F. De Paolis, G. Ingrosso, Ph. Jetzer, A. Qadir and M. Roncadelli, Astron. Astrophys. **299**, 647 (1995).

The Minimum Total Mass of MACHOs and Halo Models of the Galaxy

Takashi Nakamura

Yukawa Institute for Theoretical Physics, Kyoto University, Kyoto 606, Japan

Yukitoshi Kan-ya and Ryouichi Nishi

Department of Physics , Kyoto University , Kyoto 606, Japan

If the density distribution $\rho(r)$ of MACHOs is spherically symmetric with respect to the Galactic center, it is shown that the minimal total mass M_{min}^{MACHO} of the MACHOs is $1.7 \times 10^{10} M_\odot \tau_{-6.7}^{\text{LMC}}$ where $\tau_{-6.7}^{\text{LMC}}$ is the optical depth (τ^{LMC}) toward the Large Magellanic Cloud (LMC) in the unit of 2×10^{-7}. If $\rho(r)$ is a decreasing function of r, it is proved that M_{min}^{MACHO} is $5.6 \times 10^{10} M_\odot \tau_{-6.7}^{\text{LMC}}$. Several spherical and axially symmetric halo models of the Galaxy with a few free parameters are also considered. It is found that M_{min}^{MACHO} ranges from $5.6 \times 10^{10} M_\odot \tau_{-6.7}^{\text{LMC}}$ to $\sim 3 \times 10^{11} M_\odot \tau_{-6.7}^{\text{LMC}}$. For general case, the minimal column density $\Sigma_{min}^{\text{MACHO}}$ of MACHOs is obtained as $\Sigma_{min}^{\text{MACHO}} = 25 M_\odot \text{pc}^{-2} \tau_{-6.7}^{\text{LMC}}$. If the clump of MACHOs exist only halfway between LMC and the sun, M_{min}^{MACHO} is $1.5 \times 10^9 M_\odot$. This shows that the total mass of MACHOs is smaller than $5 \times 10^{10} M_\odot$, i.e. $\sim 10\%$ of the mass of the halo inside LMC, either if the density distribution of MACHOs is unusual or $\tau^{\text{LMC}} \ll 2 \times 10^{-7}$. The details of the results can be found in ApJ. 473: L99-L102 (1996).

1 Spherically Symmetric Halo Models and the Minimal Total Mass of MACHOs

We assume that the density distribution function $\rho(r)$ of MACHOs is a function of the galactocentric radius r. The optical depth τ^{LMC} toward LMC is given by

$$\tau^{\text{LMC}} = \frac{4\pi G}{c^2} \int_0^{D_s} x(1 - \frac{x}{D_s})\rho(r)dx, \tag{1}$$

$$r^2 = R_0^2 - 2R_0\eta x + x^2, \tag{2}$$

and

$$\eta = \cos b \cos l, \tag{3}$$

where D_s, l, b and R_0 are the distance to LMC (50kpc), the galactic longitude and latitude of LMC and the galactocentric radius of the sun (8.5kpc), respectively. In Eq. (1) we assumed the threshold $u_T = 1$ for simplicity.

Equation (1) is rewritten as

$$\tau^{\text{LMC}} = \int f(x)dm, \tag{4}$$

$$f(x) = \frac{G}{c^2} \frac{x(1 - \frac{x}{D_s})}{r^2 \frac{dr}{dx}}, \tag{5}$$

and

$$dm = 4\pi r^2 \rho(r)\frac{dr}{dx}dx. \tag{6}$$

For LMC, $f(x)$ is infinite at $x = x_c \equiv R_0\eta = 0.153R_0$ so that the minimal total mass M_{min}^{MACHO} of MACHOs is zero for any given τ^{LMC} if MACHOs are distributed in an infinitesimally thin shell at $r = r_c \equiv \sqrt{1 - \eta^2}R_0$. However this is wrong. Since the angular size of LMC is $\sim 10° \times 10°$, M^{MACHO} is minimized if MACHOs are distributed in a shell at $r = r_c$ with width d given by

$$d = \frac{10\pi}{180}R_0\eta = 227\text{pc}. \tag{7}$$

It is easy to show that M_{min}^{MACHO} is given by

$$M_{min}^{MACHO} = \frac{c^2\tau^{LMC}}{3G} \frac{(r_c + d)^3 - r_c^3}{2\sqrt{2r_cd + d^2}(R_0\eta - \frac{R_0^2\eta^2 + 6r_cd + 3d^2}{3D_s})}, \tag{8}$$

$$= 1.7 \times 10^{10}M_\odot\tau_{-6.7}^{LMC}, \tag{9}$$

where $\tau_{-6.7}^{LMC}$ is τ^{LMC} in the unit of 2×10^{-7}. This shows that in principle M^{MACHO} can be only $\sim 3\%$ of the total mass of the halo inside LMC. However the density distribution function of MACHOs in this case is very peculiar so that we calculate M^{MACHO} for more realistic $\rho(r)$ to know more realistic M_{min}^{MACHO}. We consider two models;

1)Polytropic Model

$\rho(r)$ is given by polytrope of index N and the radius R_p.

2) α Model

$\rho(r)$ is given by

$$\rho(r) = \frac{\rho_0}{(1 + \frac{r^2}{R_a^2})^\alpha}. \tag{10}$$

This model is similar to the beta model of the cluster of galaxies with core radius R_a.

For polytropic models it is found that M_{min}^{MACHO} ranges from $5.6 \times 10^{10}M_\odot\tau_{-6.7}^{LMC}$ for N=0 to $7.8 \times 10^{10}M_\odot\tau_{-6.7}^{LMC}$ for N=3. For N=4 and 4.5, M^{MACHO} is greater than $1.0 \times 10^{11}M_\odot\tau_{-6.7}^{LMC}$ and the minimum does not exist for $R_p < D_s$. Under the assumption that $\rho(r)$ is a decreasing function of r, it can be proved that M^{MACHO} is minimized when $\rho(r)$ is constant. Therefore M_{min}^{MACHO} is $5.6 \times 10^{10}M_\odot\tau_{-6.7}^{LMC}$ if $\rho(r)$ is a decreasing function. For α models it is found that M_{min}^{MACHO} ranges from $1.3 \times 10^{11}M_\odot\tau_{-6.7}^{LMC}$ for $\alpha = 1.5$ to $8.7 \times 10^{10}M_\odot\tau_{-6.7}^{LMC}$ for $\alpha = 6$.

2 Axially Symmetric Halo Models and the Minimal Total Mass of the MACHOs

There are several suggestions that the Galactic halo is not spherically symmetric so that we study here axially symmetric halo models and calculate $M^{\rm MACHO}$. We consider two models; 1) Exponential Disk Model defined by

$$\rho(R, Z) = \rho_0 \exp(-\frac{R}{R_d} - \frac{|Z|}{Z_d}), \tag{11}$$

where R_d and Z_h are scale heights.
2) Elliptical Model defined by

$$\rho(R, Z) = \frac{\rho_0}{(1 + \frac{R^2}{a^2} + \frac{Z^2}{c^2})^\alpha}, \tag{12}$$

where a and c describe the ellipticity of the equidensity surface. For exponential disk models we found that $M_{min}^{\rm MACHO}$ is $\sim 1.0 \times 10^{11} {\rm M}_\odot \tau_{-6.7}^{\rm LMC}$ for $0.5 < Z_h/R_d < 1.0$ and it increases with the decrease of Z_h/R_d for $Z_h/R_d < 0.5$. For elliptical models we found that $M_{min}^{\rm MACHO}$ is $8.9 \times 10^{10} {\rm M}_\odot \tau_{-6.7}^{\rm LMC}$ at a=10kpc and c=6kpc for $\alpha = 2.5$ and is $7.08 \times 10^{10} {\rm M}_\odot \tau_{-6.7}^{\rm LMC}$ at a=22kpc and c=13.2kpc for $\alpha = 6.0$. For large α, $M_{min}^{\rm MACHO}$ converges, similarly to the α models.

3 Discussions

A deep north Galactic pole proper motion survey suggests that the halo is not dynamically mixed but contains a significant fraction of stars with membership in correlated stellar streams. If MACHOs are also dynamically unmixed, it is possible that the density distribution function is neither spherically nor axially symmetric but completely inhomogeneous. In such a case what we can say from the microlensing events toward LMC is the minimal column density $\Sigma_{min}^{\rm MACHO}$ of MACHOs. Since in Equation (1), $x(1 - x/D_s) < D_s/4$, $\Sigma_{min}^{\rm MACHO}$ is given by

$$\Sigma_{min}^{\rm MACHO} = 25 {\rm M}_\odot {\rm pc}^{-2} \tau_{-6.7}^{\rm LMC}. \tag{13}$$

Similar to Equation (7), the linear size of the clump of MACHOs should be larger than $174{\rm pc}(x/{\rm kpc})$ where x is the distance to the clump of MACHOs. For $x = D_s/2$, $M_{min}^{\rm MACHO}$ is $1.5 \times 10^9 {\rm M}_\odot$. If this is the case, the optical depth toward the Small Magellanic Cloud will be quite different and the inhomogeneity of the density distribution of MACHOs can be checked.

In conclusion, it is shown that the total mass of MACHOs becomes smaller than $5 \times 10^{10} {\rm M}_\odot$, i.e. $\sim 10\%$ of the mass of the halo inside the LMC, either if the density distribution of MACHOs is unusual or $\tau^{\rm LMC} \ll 2 \times 10^{-7}$.

HALO WHITE DWARFS AND THE HOT INTERGALACTIC MEDIUM

B. D. FIELDS, G. J. MATHEWS

University of Notre Dame, Notre Dame, IN 46635, USA

D. N. SCHRAMM

University of Chicago, Chicago, IL 60637, USA

We describe the formation of baryonic remnants in the halo along with hot intergalactic gas. In this scenario, the mass and metallicity of hot gas in the Local Group relates directly to the production of baryonic remnants during the collapse of galactic halos. We construct a schematic but self-consistent model in which early bursts of star formation lead to a large remnant population in the halo. These bursts also produce an outflow of stellar ejecta into the halo, Local Group, and ultimately, the intergalactic medium. This study suggests that an optimum value of 40% (and a 2σ upper limit of 77%) of the halo mass could be in the form of $0.5 M_\odot$ white dwarfs without violating any observational constraint. Thus the microlensing objects in the halo may predominantly be white dwarfs.

1 Introduction

We describe here a model[1] which relates recent observations of dark matter in the Galactic halo, and hot intergalactic gas in groups and clusters of galaxies. Recently, the presence of dark matter in our Galactic halo has been directly confirmed by microlensing observations towards the LMC[2,3]. Present estimates of the lensing object's mass, $m \sim 0.5 M_\odot$, is suggestive of white dwarfs[2]. On the other hand, hot intergalactic gas is found to be ubiquitous in clusters[4], and has recently been observed in groups[5]. This X-ray gas is metal enriched, which implies that some of the gas has undergone significant processing through massive stars, and subsequent ejection.

Our model attempts to connect these observations. The existence of a significant halo remnant population would imply that large stellar processing occurred in the past. The presence of metals in the hot gas also requires stellar processing, as well as a mechanism for the ejecta to escape to the intergalactic medium. Therefore, we posit that there were strong bursts of star formation in the early Galaxy. Most of the stars are now dead: the remnants are MACHOs; the ejecta were lost in galactic wind, and became the hot intergalactic medium. The model is presently viable, but is close to the observational limits on several fronts and should be confirmed or ruled out soon.

2 The Model

We model the Local Group evolution schematically, in the spirit of a hierarchical clustering scenario. Namely, we establish a hierarchy of three mass scales: (1) protogalactic clouds, (2) galactic halos, and (3) the group itself. Each mass scale corresponds to a primordial density fluctuation which must overcome the cosmological expansion. Thus the dynamics of each component comes from a collapse model, namely the behavior of a spherical overdensity[7]. Star formation is a key ingredient of the model; stars provide significant heating, as well as nucleosynthesis products. We also allow for different sources of gas heating and cooling.

The hierarchical scales are self-similar. At the smallest scale, the clouds contain stars, as well as both hot and cold gas. The halos include the clouds as sub-components, as well as their own component of hot gas and stars ejected from clouds. The group includes the two galaxies as a sub-component, as well as hot gas ejected from the halos. While the clouds contain only baryons, we introduce a component of non-baryonic dark matter at the halo and group levels.

The different mass scales interact via several mechanisms. One of these is merging, which reduces the number of clouds while increasing their average mass. Also, at each mass scale, gas heating leads to evaporation of the hottest gas particles whose thermal velocity exceeds the local escape velocity. This process is very efficient: there is a significant outflow from all levels, with mass loss even from the group itself (this leads to hot intergroup gas as a significant component of the dark baryons). The wind also serves to remove material after a single stellar processing and so prevents recycling which would otherwise lead to overproduction of metals and helium.

A remnant-rich halo scenario such as ours requires that the halo initial mass function (IMF) was biased away from low mass ($\lesssim 1M_\odot$) stars[8]. We parameterize the IMF as a log-normal form[9], with a centroid[9] at $2.3M_\odot$ and the width at the present-day value[10].

3 Results

To sketch the basic results, we summarize the mass and metal budgets. The clouds merge to form the proto-disk and bulge; starting with a mass $10^6 M_\odot$ and ending with $8.1 \times 10^{10} M_\odot$. The halos begin with a mass $1.35 \times 10^{12} M_\odot$, of which 81% is baryonic. The final galaxy mass is $5.0 \times 10^{11} M_\odot$, of which 50% is baryonic (40% of the dark halo is baryonic). The remnants are mostly white dwarfs, with 12% neutron stars. Thus there is a net loss of $8.5 \times 10^{11} M_\odot$

of gas from the galaxies into the IGM. The group itself begins with a mass of $5.7 \times 10^{12} M_\odot$, and ends with a mass $4.3 \times 10^{12} M_\odot$, 18% of which is baryonic. Of the baryonic group mass, $2.9 \times 10^{11} M_\odot$ resides in hot ($T \simeq 0.3$ keV) gas, a level below but close to the ROSAT limits. The group as a whole loses $1.4 \times 10^{12} M_\odot$ of gas to the intergalactic medium medium; thus about 65% of the initial baryonic mass in the group is ejected later into intergalactic space. This amount of hot (ionized) material is consistent with Gunn-Peterson limits on the intergalactic medium. A similar analysis shows that the gas and star metallicities[6] are reasonable; the luminosities are also acceptably low but near detection.

4   Observational Tests

We have shown that one may a plausible model of galaxy evolution that relates the hot gas seen in galaxy aggregates to a halo population of remnants. Since we link these phenomena, confirmation of one in the Local Groups would imply, in our model, the existence of the other. Further, signatures of the model are within reach. Unless the Local Group is anomalous, then the expected hot gas mass is too cool to provide a source of the diffuse X-ray background, as suggested by [11]. However, such gas may still be observable, and further work on such systems is crucial. The luminosity of halo white dwarfs should also be directly detectable with further pencil beam and wide angle observations. It is intriguing as well that several edge-on galaxies have an observed infrared halo [12], as one would expect from cooled white dwarfs. This model, and the white dwarf halo scenario in general, is eminently testable.

References

1. B.D. Fields, G.J. Mathews, and D.N. Schramm (1997) Ap.J., in press.
2. C. Alcock, these proceedings.
3. N. Palanque-Delabrouille, these proceedings.
4. R.F. Mushotzky, these proceedings.
5. J.S. Mulchaey, et al., Ap.J. 456, 80 (1996).
6. D. Ryu, K.A. Olive, and J. Silk, Ap.J. 353, 81 (1990).
7. G.J. Mathews, and D.N. Schramm, Ap.J. 404, 468 (1993).
8. J. Silk, Phys. Reports 227, 143 (1993).
9. F.C. Adams and G. Laughlin, Ap.J. 468, 586 (1996)
10. G.E. Miller, and J.M. Scalo, Ap.J.S., 41, 513 (1979).
11. Y. Suto, et al., Ap.J. 461, L33 (1996) .
12. M.D. Lehnert, and T.M. Heckman, Ap.J., 462, 651 (1996).

MICROLENSING BY NONCOMPACT OBJECTS

A.F. ZAKHAROV

Institute of Theoretical and Experimental Physics,
B. Cheremushkinskaya, 25, 117259, Moscow, Russia,
e-mail: zakharov@vitep5.itep.ru

The microlensing of the distant stars by neutralino stars has been considered. The neutralino stars have been considered in the recent paper of Gurevich and Zybin, moreover, the stars have been suggested to be regard as a component, carrying a major component of dark matter. The optics of these gravitational microlenses, namely, the gravitational lens equation, its solutions, magnifications, critical and caustic curves, light curves are analyzed by using a clear approximation.

The first results of observations of microlensing which were presented in the papers of several groups [1-3] have discovered a phenomenon, predicted in the papers [1,3]. A matter of the gravitational microlens is unknown till now, although the most widespread hypothesis assumes that they are compact dark objects as brown dwarfs. Nevertheless, they could be presented by another objects, in particular, an existence of the dark objects consisting of the supersymmetrical weakly interacting particles (neutralino) has been recently discussed in the papers [6,7]. The authors shown that the stars could be formed on the early stages of the Universe evolution and to be stable during cosmological timescale. We consider microlensing of a distant star by a neutralino star in framework of a rough model which is rather clear and we obtain analytical expressions. Of course, a more exact model of influence of the gravitational field of neutralino star may be considered [6,7], nevertheless, we think that the qualitative estimation of the effect was considered correctly. The geometric optics is used in the model which will be considered below and effects connected with diffraction and mutual interference of the images and analyzed in the papers [8-10] are neglected.

We approximate the density of distribution mass of a neutralino star in the form [11,12],

$$\rho_{NeS}(r) = \rho_0 \frac{a_0{}^2}{r^2}, \qquad (1)$$

where r - the current value of a distant from stellar center, ρ_0 - mass density of a neutralino star for distance a_0 from a center, a_0 - "radius" neutralino star. The dependence is approximation of the dependence which has been considered in the papers [6,7], namely

$$\rho_{NeS}(r) = K r^{-1.8},$$

but our model is more simple. If we normalize distances in the lens plane and in the source plane using R_0, namely if we introduce the variables $y = \eta D_s/(a_0 R_0 D_d)$, $x = \xi/(a_0 R_0)$, then the lens equation has quite clear form [11,12],

$$y = x - \frac{x}{|x|},\qquad(2)$$

where

$$R_0 = \frac{2\pi\rho_0 a_0}{\Sigma_{cr}} = \frac{M}{a_0{}^2}\frac{8\pi GD}{c^2},\quad D = \frac{D_d D_{ds}}{D_s},\qquad(3)$$

D_s is a distance from the source to the observer, D_d is a distance from the gravitational lens to the observer, D_{ds} ia a distance from the source to the gravitational lens, vectors (η, ξ) define a deflection on the plane of the source and the lens, respectively.

It is easy to see, that the equation of a lens in the dimensionless form identifies with the lens equation for the model of galactic mass distribution corresponding to a isothermal sphere [13] (because the expression (1) define the mass density distribution for a singular isothermal sphere). Therefore we have following expressions for magnification

$$\mu_{NeS}(y) = \left\{ \begin{array}{ll} 1 + \frac{1}{y}, & \text{for} \quad y \geq 1 \\ \frac{2}{y}, & \text{for} \quad 0 < y < 1. \end{array} \right.\qquad(4)$$

We recall that in a case when the gravitational microlens is a point gravitating body (Schwarzschild lens) then the magnification defined by the following expression [13,14]

$$\mu_S(y) = \frac{y^2 + 2}{y\sqrt{y^2 + 4}}.\qquad(5)$$

Thus, the difference between the magnification of Schwarzschild lens and neutralino star is an essential factor which distinguishes these objects.

Finally, we recall some papers where there was a detailed description of some properties of our model. Some criteria were investigated to distinguish compact and noncompact microlenses [15]. A geometrical optics, magnification, critical and caustic curves for some parameters of our model. were discussed in the paper [11], light curves for different parameters of the model of a neutralino star were analyzed in detail in the paper [12], an influence of Galactic mass on microlensing by noncompact bodies have been investigated [16]. Similarly to Chang - Refsdal model [13], caustic curves are astroids and the magnification near cusp singularities is calculated using expressions of the paper[17] (the magnification near fold singularities is well-known[13]). We remark that the average caustic size is very small therefore a probability to intersect the caustic curve is small also.

Acknowledgments

I appreciate A.V. Gurevich for useful discussions. It is a pleasure to thank M.V. Sazhin for all the work done in collaboration in this subject. The work was supported in part by Russian Foundation for Fundamental Research (Grant No. 96-02-17434). I acknowledge prof. M. Turner for his attention to the paper. I would like to thank organizers of 18th Texas Symposium for very enjoyable and successful meeting in Chicago. I also thank them for providing financial assistance that facilitated my attendance.

References

1. C. Alcock *et al.*, *Nature* **365**, 621 (1993).
2. E. Aubourg *et al.*, *Nature* **365**, 623 (1993).
3. A. Udalski *et al.*, *Astrophys. J.* **426**, L69 (1994).
4. A.V. Byalko, *Astron. Zhurn.* **46**, 998 (1969).
5. B. Paczinsky, *Astrophys. J.* **304**, 1 (1986).
6. A.V. Gurevich and K.P. Zybin, *Phys. Lett.* A **208**, 276 (1995).
7. A.V. Gurevich , K.P. Zybin and V.A. Sirota, *Phys. Lett.* A **214**, 322 (1996).
8. A.F. Zakharov, *Astron. Astrophys. Trans.* **5**, 85 (1994).
9. A.F. Zakharov, *Astron. Lett.* **20**, 359 (1994).
10. A.F. Zakharov and A.V. Mandzhos, *Journ. Exper. & Theor. Phys.* **104**, 3249 (1993).
11. A.F. Zakharov and M.V. Sazhin, *Journ. Exper. & Theor. Phys.* **110**, 1921 (1996).
12. A.F. Zakharov and M.V. Sazhin, *Journ. Exper. Theor. Phys. Lett.* **63**, 894 (1996).
13. P. Schneider, J. Ehlers and E.E. Falco, *Gravitational Lenses*, Springer, Berlin - Heidelberg - New York, 1992.
14. A.F. Zakharov, *Siberian Phys. Zhurn.* **4**, 38 (1995).
15. A.F. Zakharov, *Astron. & Astrophys.* (in press).
16. A.F. Zakharov and M.V. Sazhin, *Astron. Lett.* (in press).
17. A.F. Zakharov, *Astron. & Astrophys.* **293**, 1 (1995).

POSSIBLE EVIDENCE FOR PRIMORDIAL BLACK HOLES

DAVID B. CLINE

Department of Physics and Astronomy, Box 951547
University of California, Los Angeles
Los Angeles, CA 90095-1547, USA

We present evidence here from the study of a distinct sub class of very short, very hard gamma-ray bursts (GRBs) that $m \sim 10^{14}$ gm primordial black holes (PBHs) may exist in the local Galaxy environment. Some possible "smoking gun" observations that could prove this conjecture are listed here. The discovery of PBHs, while providing a new window on the early Universe, would also be of enormous importance in general.

1   Concept of the Search

We list the main ingredients to the method of the search here, which serves as a brief summary of this paper:

1. Primordial black holes with $m \sim 10^{14}$ gm can be produced in the early Universe; while $\Omega_{PBH} < 10^{-8}$, there can still be many in the vicinity of the Solar System.

2. A burst of photons could be produced as $T_{PBH} \sim T_{QGP}$ (QGP, quark–gluon phase), from the Hawking radiation.

3. We questioned: Are there any GRB events that could come from PBH evaporation?

4. Do these events have the characteristics expected from a homogeneous isotropic local source?

5. Are there other features of the events that are similar or that suggest a unique GRB population?

6. Is there a plausible mechanism that could produce these PBHs in the early Universe?

2   Evaporation of PBHs at $T = T_{QG}$

Ever since the theoretical discovery of the quantum-gravitational particle emissions from black holes by Hawking,[1] there have been many experimental searches for high-energy γ-ray radiation from PBHs (which would be formed in the early Universe) entering their final stages of extinction. The violent final stage evaporation or explosion is the striking result of the expectation that the PBH temperature is inversely proportional to the PBH mass, *i.e.*, $T = T_{PBH} \approx 100$ MeV (10^{15} g/m_{PBH}), since the black hole becomes hotter as it radiates more particles and eventually can attain extremely high temperatures.

Based on previous calculations and numerous direct observational searches for high-energy radiation from an evaporating PBH, we might conclude that it is not likely that we can single out such a monumental event. However in a previous work,[2] we pointed out a possible connection between very short GRBs observed by the BATSE team and a PBH evaporation caused by the emission of very low-energy γ-rays. If we want to accept this possibility, we may have to modify a way of calculating the particle emission spectra from an evaporating PBH, in particular at near the quark–gluon plasma (QGP) phase transition temperature at which the T_{PBH} arrives eventually. We assume that a first-order phase transition occurs at the QGP temperature (Fig. 1A). Tables 1 and 2 give more information about this model.

In the previous work, we briefly discussed that including the QGP effect around the evaporating PBH at the critical temperature might drastically change the resulting γ-ray spectrum. The QGP interactions around the evaporating PBH form an expanding hadronic- (mostly pions) matter fireball. Shortly after the decays of pions, the initial hadronic fireball converts to a fireball with mixtures of photons, leptons, and baryons. The photons could be captured inside this fireball until the photon optical depth becomes thin enough for the photons to escape as a very-short (orders of milliseconds duration) GRB.

Lacking a full description of the manner in which PBHs "explode," we must resort to phenomenology. We believe it is unlikely that the standard QCD framework can be used for PBHs with a temperature of about 100–200 MeV.[2] This is precisely the region where there does not seem to be an adequate description available.[4] However, the simple Hagedorn model is also likely excluded. We have studied a mixed model and used it to help get some insight into the general properties of the final stages of PBH

341

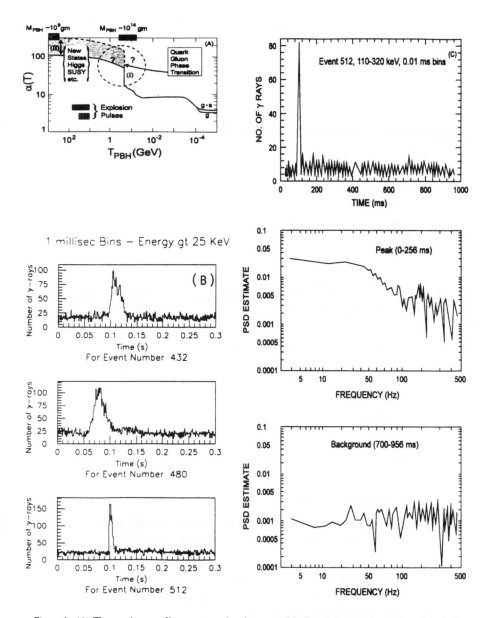

Figure 1: (A) The running coupling constant showing a possible first-order phase transition of the QGP temperature. (B) Time profile refit of the BATSE 3B events using the TTE data. (C) Fast Fourier analysis of one of the events analyzed here, showing the PSDs.

Table 1: Mass loss near the QGP transition temperature., T_{QGP}

Mass Loss Rate:

$$\frac{dm}{dt} = -\frac{\alpha(m)}{m^2} = \dot{m}$$

near $T = T_{QGP}$. $\left[\text{NOTE:} \quad \tau_0 = \frac{m_0^{-3}}{\alpha(m_0)} \right]$

$$\frac{d}{dt}[\dot{m}] = \frac{d}{dT}\left(-\frac{\alpha}{m^2}\right)$$

$$\frac{d\dot{m}}{dT} = \left(\frac{m_0}{\tau_0}\right)\left[+\frac{1}{m}\left(\frac{dM}{\alpha T}\right) - \frac{1}{\alpha}\left(\frac{d\alpha}{dT}\right)\right]_{T=T_{QGP}}$$

$$d\left(\dot{m}\right) = -\left(\frac{m_0}{\tau_0}\right)\left[-\frac{1}{\alpha}\left(\frac{d\alpha}{dT}\right)\right]_{T=T_{QGI}} dT \quad \text{- "explosion" for } \frac{1}{\alpha}\frac{d\alpha}{dT} \to \alpha \text{ at QGPT}$$

assume first-order transition.

Table 2: Fireball from the QGP transition (simple model).

$L_{QGP} \sim 5 \times 10^{34}$ ergs/s $T_{QGP} \geq 160$ MeV
$\sim L_{PBH}$ $\sim T_{PBH}$

$$r_s \sim \left\{ \frac{L_{PBH}}{4\pi\,T_{PBH}}\left[\frac{T}{S(T)}\right] \right\}^{1/2} \quad \text{(from Ref. 3)}$$

where S is the total entropy.

A simple radiation-dominated model (*i.e.*, $\pi^0 \to \gamma\gamma$) would give
$r_s \sim 10^9$ cm $\to \tau \sim \theta$ (100 ms) , rise time $\ll$ 100 ms.

Thus, expect GRB from fireball to have
- Very fast rise time ($\leq$ ms) and to be model-dependent (a guess),
- Durations of $\sim$ 100 ms for an order of magnitude,
- Low energy photons ($\sim$ 1 – 10 MeV),
- Hard spectrum.

evaporation. In addition, we believe it is essential to study unusual cosmic events, such as GRBs, to possibly identify unusual behavior that could be characteristic of PBH evaporation. We have described a class of GRBs that are of short durations, which could yield further evidence concerning this hypothesis. Another important test is $V/V_{max} \sim \frac{1}{2}$ combined with a Galactic Coordinate plot of the events, which is isotropic. We note that one observed short GRB has a fine time structure of $\sim 100 \ \mu s$.[5]

While quantitative calculations should be dependent on a particle physics model (or energy injection mechanism to the fireball), we may set up general criteria for the PBH evaporation as a fireball to be seen at an order of pc from the earth. Since the BATSE detector's observed fluences of $\approx 10^{-7}$ ergs/cm^2, we require the distance to the PBH fireball (R_{PBH}) and the release of total γ-ray energy (L_{PBH}^{γ}) during a short period of time to be $L_{PBH}^{\gamma}/4\pi R_{PBH}^2 \geq 10^{-7}$ ergs/cm^2. In fact, the total γ-ray energy from a PBH evaporation is closely related to the PBH mass at QGP transition temperature (roughly $T_{QGP} \geq 160$ MeV) as $L_{PBH}^{\gamma} = \kappa m_{PBH}$, where κ is a QGP model-dependent constant and calculable given a detailed parameter of a particle physics model. When the PBH surface temperature approaches to T_{QGP}, rapid interactions between the emitted quarks and gluons by the Hawking process at the PBH horizon ($\sim 2 \ m_{PBH}$) may result in a local thermal equilibrium and, thereafter, form an expanding ultra-relativistic QGP fireball. More detailed conditions for a QGP fireball formation and its opacity at T_{QGP} are to be presented elsewhere (work in progress). Subsequently the QGP fireball converts to a dense-matter fireball with the mixtures of pions, baryons, leptons, and radiation at a distance above the PBH horizon. It is expected that, because of a high degree of interaction between particles in the high-temperature-matter fireball (mostly pions), a local thermal equilibrium is obtained at a temperature, T.

3   The Evidence from the GRB Events

We present in Fig. 1B, the time profile of three of the BATSE 3B events analyzed here (Table 3); note the similarity of these events. We have also carried out a fast Fourier transform of 7 of the 11 events presented in Table 3 (with an analysis of one of the events shown in Fig. 1C). The remarkable similarity of these events suggests a common origin. Figure 2A–C provides information on the spatial and hardness distribution of the events. Note the very short time structures of these events in Table 3. Table 4 lists the various tests for the PBH hypothesis that we have made and the consistency. For additional information, see Ref. 6.

While there could be other origins for these events other than PBH evaporation, we believe that this analysis suggests that the theoretical study of the evaporation at the QGP transition could be useful. We also note that this transition is assumed to occur

Table 3: Hardness ratio versus duration (BATSE 3B)

Trigger No.	Duration (s)	T_{90} (s)	Hardness Ratio
1453	0.006 ± 0.0002	0.192	6.68 ± 0.33
512	0.014 ± 0.0006	0.183	6.07 ± 1.34
207	0.030 ± 0.0019	0.085	6.88 ± 1.93
2615	0.034 ± 0.0032	0.028	5.43 ± 1.16
3173	0.041 ± 0.0020	0.208	5.35 ± 0.27
2463	0.049 ± 0.0045	0.064	1.60 ± 1.55
432	0.050 ± 0.0018	0.034	7.46 ± 1.17
480	0.062 ± 0.0020	0.128	7.14 ± 0.96
3037	0.066 ± 0.0072	0.048	4.81 ± 0.98
2132	0.090 ± 0.0081	0.090	3.64 ± 0.66
799	0.097 ± 0.0101	0.173	2.47 ± 0.39

at $t_{universe} \sim 10^{-6}$ s, which is a very short time in the early Universe, and this is similar to what we assume here.

3 Possible Smoking Gun for PBH Observation

In Table 4, we summarize the basic evidence for PBHs. In this work, all tests are positive so far. We list the possible concentrations of PBHs in this Galaxy, based on different assumptions concerning the clustering, in Table 5. Figure 4(D) gives some limits on other searches for PBH evaporation, as well as the results presented here.

While we have presented circumstantial evidence for PHGs, this can hardly be considered a proof for the observation of such a fundamental class of objects in Nature. So what would constitute a smoking gun proof? The following are some ideas:

1. If V/V_{max} could be measured reliably for the GRB with $\Delta\tau < 100$ ms and were exactly 1/2, this would prove that the GRB population has no edge – or is homogeneous. By elimination, all other explanations of these events, except a local powerful GRB site (*i.e.*, PBH), would be eliminated, provided no clustering in the Galactic Plane was observed.

2. Unique single-event tests could also prove that the parent of the GRB must be nearby. For example, suppose a GRB is shown to be at the same exact angular location of a nearby quiet star. Thus, it must have originated in front of the star,

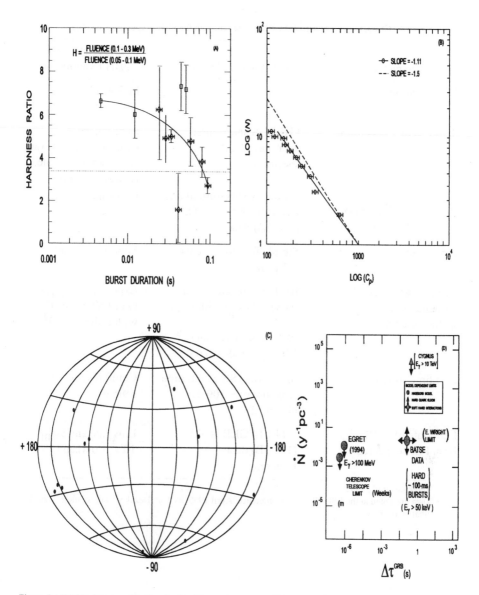

Figure 2: (A) Hardness vs duration for the 11 events presented here; note the correlation. (B) The $\ell n(N)$ vs $\ell n(C_p)$ plot for the 11 events. They are fully consistent with the expectation of an Euclidian space density, as would be expected for a local source. (C) Galactic coordinate of the events. Note that there is no concentration in the Galactic Dish or center. (D) Various limits on PBH density as described in Ref. 6.

Table 4: Characteristics of selected GRBs consistent with the PBH evaporation hypothesis.

Event Characteristics	GRB Events (Selected)	Expectation for PBH Evaporation
Time duration	~100 ms	In fireball model, $\Delta\tau \approx 200$ ms
Hardness in γ spectrum	PHEBUS short-burst data[a] have hard spectrum	Expect hard γ spectrum, but exact value not calculable. However, in pure Hawking Process, $\langle H \rangle \sim 250$
$H = F\,\dfrac{(0.1-0.3)}{(0.05-0.1)}\,\text{MeV}$	Hardest γ spectrum of any GRB $\langle H \rangle \cong 6$ for BATSE (1B–3B) data	
Time history	Simple for most events – 1 peak	Simple – 1 peak
$\ln(N) - \ln(C_p)$ test for population spatial structure	BATSE 3B: corrected $\ln(N) - \ln(C_p)$ with slope $= -3/2$ PHEBUS short events:[b] $V/V_{max} = 0.48 \pm 0.05$	Expected: $V/V_{max} = $ ½ or $\ln(N) - \ln(C_p)$ with a slope $= -3/2$
Fine time structure	In one BATSE event, time structure of ~ 100 μs observed[c]	Could reveal size of source
Limit rate of GRB from PBH expected	~ 11/614 GRB is ~2% (1B–3B) data; Low rate	Low rate and $\Omega_{PBH} \sim 10^{-7}$ – perhaps 10/y

[a]Terekhov *et al.*, 1995.
[b]Data presented at the ESLAB–ESA Conference, April 1995, by J. P. Dezalay from footnote a above.
[c]Bhat *et al.*, 1992.

Table 5: Estimated detection efficiencies for GRBs from a PBH.

N_{PBH} (pc^3)	Model of PBH Distribution	N (No Detection Assumed) (pc$^3\cdot$y)	N (Detection Limit)
$\sim 10^4 - 10^5$	γ background of the Universe ($\Omega_{PBH} \sim 10^{-8}$)	$\sim 10^5 - 10^4$	Diffuse γ spectrum (input)
$\sim 10^{10}$	"Mild" galactic concentration	~ 2	Could only be detected if Hagedorn model is correct (EGRET)
$\sim 10^{12}$	"Reasonable" galactic concentration	$\sim 10^{10}$	Can be detected in the mixed model (BATSE)
$\sim 10^{15}$	"Extreme" galactic concentration	$\sim 10^5$	Most likely can be consistent if hard QCD model (possible detection with Air Shower detector, $E_\gamma \gg$ TeV)

otherwise the photons would have been absorbed. If the star were known to be at a distance of a few parsec (or if a neutron star were in between, which is very unlikely), this could prove the conjecture!

I wish to thank W. Hong, D. Sanders, and M. Sanders for collaborative work on this subject.

References

1. S. W. Hawking, *Commun. Math. Phys.* **43**, 199 (1975).
2. D. B. Cline and W. P. Hong, *Astrophys. J.* **401**, L-57 (1992).
3. B. Carter *et al.*, *Astron. & Astrophys.*, **52**, 427 (1976).
4. P. N. Bhat et al., *Nature* **359**, 217 (1992).
5. F. Halzen et al, *Nature* **353**, 807 and references therein (1991).
6. D. Cline, D. Sanders, and W. Hong, "Further Evidence for Some Gamma Ray Bursts Consistent with Primordial Black Hole Evaporation," UCLA report #UCLA-APH-0091-1/97 (submitted to *Astrophys. J.*).

PRIMORDIAL BLACK HOLES AS DARK MATTER

M.R.S. HAWKINS

Royal Observatory, Blackford Hill, Edinburgh EH9 3HJ, Scotland

The idea that Jupiter mass primordial black holes are the main constituents of dark matter, and betray their presence by microlensing the light from quasars has been discussed in a number of recent papers. Here we present new data from a gravitationally lensed double quasar which shows unambiguous evidence for microlensing. It is argued that this must either be caused by a population of compact bodies along the line of sight sufficient to make up the critical density, or the halo of the lensing galaxy must be almost entirely composed of microlensing bodies. Either way this supports the idea that dark matter is in the form of planetary mass primordial black holes.

1 Introduction

The evidence for dark matter in the form of Jupiter mass compact bodies has been discussed in some detail recently [1,2]. The main line of argument is that the dark matter betrays its presence by microlensing quasars, and the consequent variations dominate quasar light curves and can be distinguished fom any intrinsic variability. The characteristic mass of the microlensing objects can be calculated from the timescale of variation [2], and also from the distribution of amplitudes [3], and is found to be about $10^{-3} M_\odot$. The cosmological mass density of the lenses Ω_λ can also be estimated from the observation that all lines of sight appear to be microlensed [4]. The implication is that the lens density must be around the critical density, $\Omega_\lambda = 1$. Baryon synthesis constraints imply that the lenses must be non-baryonic, and the most plausible candidates would appear to be primordial black holes created in the quark/hadron phase transition [5].

In this paper we concentrate on the microlensing of pairs of gravitationally lensed quasar images. Here, the question of whether the variation is caused by microlensing is not generally in dispute, but rather the extent to which the system can be seen as representative of a random line of sight. The analysis will focus on a newly discovered quasar pair, both components of which appear to be strongly microlensed.

2 The Double Quasar Q2138-431AB

When quasars were selected for the large variability study referred to in the last section [2], one of the constraints for inclusion in the sample was that the image should be round. This was a useful way of eliminating quasar candidates with images contaminated by foreground stars or galaxies, but had the additional consequence of removing any gravitationally lensed double quasars which would appear elongated.

It was only about three years ago that a search was carried out specifically to look for gravitationally lensed systems, according to well-defined criteria. The candidates were required to have a major to minor axis ratio of 1.5 or more and to show an excess of ultra-violet light, $U - B < -0.4$. Preliminary spectroscopy showed that a large fraction of the 23 candidates were quasars.

The first system to be studied in detail was Q2138-431AB, in position 21h 38m 06.7s -43° 10' 50" (1950) which turned out to be two quasars at the same redshift separated by 4.5 arcsecs. A detailed analysis of the system has been carried out which will be published elsewhere. Some specific details are given here:

Redshift $z = 1.64$, velocity difference $\delta v = 0 \pm 115$ km.sec^{-1}
Separation $= 4.5$"
B magnitude $m_A = 19.8$, $m_B = 21.0$
Amplitude $\delta m_A = 1.1$, $\delta m_B = 0.6$

In summary, the redshifts are in close agreement and the spectra are almost identical. The probability that two separate quasars lie so close together is remote and the system appears to be a strong candidate for gravitational lensing. However, the photographic plates showed no obvious sign of a lensing galaxy which would have to have a mass around $10^{12} M_\odot$ to produce the observed image separation. An extensive deep multicolour search with a CCD camera still failed to reveal the presence of a lensing galaxy, and we can now put a conservative limit of $R > 23$ for any object in the vicinity of the quasar images. This means that any lensing object is likely to have a mass to light ratio of more than $800 M_\odot/L_\odot$ with an absolute lower limit of $200 M_\odot/L_\odot$.

The extensive plate material available in the field containing Q2138-431 means that light curves covering 15 years can be measured for both images. These are plotted in Fig. 1 and it will be seen that although both quasars vary strongly on a timescale of about 5 years, their light curves are quite different. Since the expected light delay time for this system is about 6 months there is no way that the variations can be intrinsic to the parent quasar. If the quasar images are part of a lensed system the only plausible explanation for the observed variations is that they are the result of microlensing by planetary mass bodies along the line of sight, possibly in the undetected lensing galaxy.

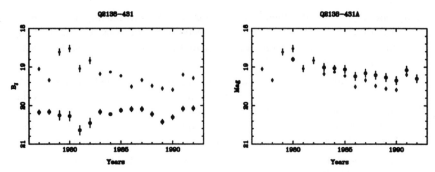

Figure 1: Light curves for the double quasar Q2138-431. The left panel is for the two components in the B_J passband, and the right panel for the A component in B_J and R.

3 Cosmological Implications

Q2138AB is only the second gravitationally lensed system to be monitored long enough to see the characteristic variations of most other quasars [2]. The other system, Q0957+561 is also known to be microlensed [6], but the amplitude is small and the structure of the light curve hard to interpret. Q2138-431 on the other hand shows large amplitude variations in both components. If as argued in the previous section the system is gravitationally lensed, the uncorrelated nature of the light curves imply that they are caused by microlensing. There appear to be two possibilities. Either the microlensing bodies are the Jupiter mass primordial black holes discussed by Hawkins [2], and are distributed along the line of sight, or they reside in the invisible lensing galaxy. In the latter case for the quasar image to be split there must be a critical mass density between the two images. For a high probability of microlensing an optical depth of around unity (critical density) is required in microlensing bodies. Thus for the observed continuous microlensing, the entire mass of the galaxy (including dark matter) must be in the form of microlensing bodies. Even if the stars form 10% of the galaxy mass, they will only cause occasional microlensing events. Thus if the lens galaxy is also responsible for the microlensing it must be composed almost entirely of microlensing bodies. Either way these observations provide new support for dark matter in the form of Jupiter mass primordial black holes.

References

1. M.R.S. Hawkins, *Nature* **366**, 242 (1993).
2. M.R.S. Hawkins, *MNRAS* **278**, 787 (1996).
3. P. Schneider, *A&A* **279**, 1 (1993).
4. W. Press, J. Gunn, *ApJ* **185**, 397 (1973).
5. M. Crawford, D. Schramm, *Nature* **298**, 538 (1982).
6. R.E. Schild, *ApJ* **464**, 125 (1996).

SDSS SEARCH FOR STELLAR MASS BLACK HOLES

Andrew F. Heckler

Astrophysics Group, Ohio State University
Columbus, OH 43210, USA

Edward W. Kolb

Fermi National Accelerator Laboratory,
Batavia, IL 60510, USA

We propose a strategy for searching for isolated stellar mass black holes in the solar neighborhood with the Sloan Digital Sky Survey. Due to spherical accretion of the ISM, an isolated black hole is expected to emit a roughly flat spectrum from the optical down to the far infra-red. We find that the Sloan Survey will be able to detect isolated black holes, in the considered mass range of 1–$100M_\odot$, out to a few hundred parsecs, depending on the local conditions of the ISM. We also find that the black holes are photmetrically distinguishable from field stars and they have a photometry similar to QSOs. They can be singled out from QSO searches because they have a spectrum with no emission lines. If no black hole candidates are found, important limits can be placed on the local density of black holes and the halo fraction in black holes, especially for masses greater than about $20M_\odot$.

As pointed out by several talks on MACHO results, a significant fraction of the halo *may be* made up of MACHOS with masses greater than a solar mass. Presumably, these would be black holes, and since microlensing searches are not very efficient at searching for MACHOS greater than a few solar masses, the question is: is there a way to determine if a significant fraction of the halo consists of stellar mass black holes? The answer, based on the work of Heckler and Kolb[1] is "quite possibly, yes", and as a bonus, one can also possibly find black holes right in our back yard, within about 100pc.

Isolated black holes may be detected because they emit radiation as they accrete the inter-stellar medium (ISM). The bremsstrahlung luminosity is very weak, but if inter-stellar magnetic fields are included in the accretion process, the resulting synchrotron luminosity can be quite high, with emission efficiencies as large as $0.1\dot{M}c^2$, and coincidentally, peaks in the optical[2,3].

The upcoming Sloan Digital Sky Survey (SDSS), which is an optical survey, therefore, offers an excellent opportunity to search for isolated black holes in our solar neighborhood in a systematic manner. Through planned QSO searches the SDSS will serendipitously collect both photometric and spectroscopic data on viable black hole candidates. Assuming they are bright enough to be observed (see table 1), the color-color diagram of Figure 1 reveals that accreting black holes occupy only a small region of color space, and the black

M_{BH}	g-magnitude		
	$(n = 1\,\mathrm{cm}^{-3})$	$(n = 0.1\,\mathrm{cm}^{-3})$	(loc. bubble)
$1\ M_\odot$	22.05	26.89	37.02
$10\ M_\odot$	14.55	19.39	29.52
$100\ M_\odot$	7.38	11.89	22.02

Table 1: The g-magnitudes of black holes at a distance of 10 pc, for various black hole masses M_{BH} and ISM densities n. For the local bubble, we set $n = 0.05\,\mathrm{cm}^{-3}$ and $T_{\mathrm{ISM}} = 10^5\mathrm{K}$. Otherwise we set $T_{\mathrm{ISM}} = 10^4\mathrm{K}$, and for all cases $V = 0$. To find magnitudes at other distances, add $5\log(D/10\mathrm{pc})$ to the values in the table. The SDSS limit is about $m \simeq 22$.

hole locus is easily distinguishable from the locus of main sequence stars, and shares the same region of color space as QSOs. This is a fortuitous result because the SDSS will take spectra of all QSO candidates. Since the black hole spectra will have no emission or absorbtion lines, they will be "easily" distinguishable from quasars. Naturally, featureless spectra could be other ob-

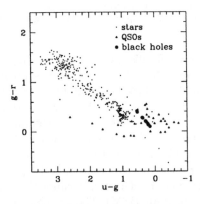

Figure 1: Color-color diagram for stars and QSOs, taken from Trevese et al.[4], and the expected colors of isolated black holes accreting the ISM. Notice that the black hole locus is distinct from the main-sequence stellar locus, and lies within the QSO locus. The SDSS QSO search will consider objects "bluer" (below and to the right) than the main-sequence stars to be QSO candidates.

jects such as BL-lacs, DC dwarfs, or simply quasars with low signal to noise, so one must use other criteria to further single out black hole candidates. For example, BL-lac object are bright in the radio band, whereas accreting black holes are not. One can also examine candidates in the infra-red band, look for time variability, which has two time scales: 10^{-4}s and on a scale of orders day to weeks, and one may detect proper motion.

Because of the relatively high luminosity of black holes with mass $M > 10M_\odot$, there are important consequences if no black holes are found in the SDSS. For example, Figure 2 shows the maximum amount black holes can contribute to the dark matter halo, if they are not detected in SDSS, assuming the SDSS is 100% efficient at finding black holes down to a limiting r magnitude of 20. One should note that the black hole luminosity is sensitive to its velocity,

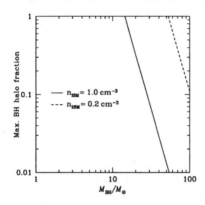

Figure 2: Maximum halo mass fraction in black holes if no black hole candidates are found with the SDSS, for two values of the ISM density. For $M \gtrsim 10M_\odot$, the SDSS is sensitive to black holes beyond the local bubble, where one expects $\bar{n}_{\rm ISM} \sim 1\,{\rm cm}^{-3}$.

and only a small fraction of halo black holes will have a relative velocity small enough to be sufficiently luminous. These conclusions therefore depend upon the halo model; here we have assumed a local maxwell distribution with average velocity of 270 km/s. If the halo is co-ratating with the disk, the limits can become much more constraining. Finally, note that the SDSS is sensitive to $M \gtrsim 10M_\odot$, which very complimentary to the microlensing searches.

We especially thank Brian Yanny, Heidi Newberg, and Rich Kron for providing the data on the star and QSO colors, and for providing useful discussions on both the data and on the SDSS QSO search program. This work was supported in part by the DOE and by NASA (NAG5-2788) at Fermilab.

1. A.F. Heckler and E.W. Kolb, *Ap.J. Lett.*, **472**, L85 (1996).
2. Shvartsman, V. F. *Soviet Astr.-AJ*, **15**, 377 (1971).
3. Ipser, J. R. and Price, R. H. *Ap.J.*, **255**, 654, (1982).
4. Trevese, D., Kron, R.G., Majewski, S.R., Bershady, M.A. and Koo, D.C. *Ap.J.*, **433**, 494, (1994).

IS GALAXY DARK MATTER A PROPERTY OF SPACETIME?

M. CARMELI

Department of Physics, Ben Gurion University, Beer Sheva 84105, Israel

Abstract

We describe the motion of a particle in a central field in an expanding universe. Use is made of a double expansion in $1/c$ and $1/\tau$, where c and τ are the speed of light and the Hubble time. In the lowest approximation the rotational velocity is shown to satisfy $v^4 = \frac{2}{3}GMcH_0$, where G is Newton's gravitational constant, M is the mass of the central body (galaxy) and H_0 is the Hubble constant. This formula satisfies observations of stars moving in spiral and elliptical galaxies, and in accordance with the familiar Tully-Fisher law.

1 Introduction

Equations of motion have a direct relevance to the problem of the existance of the galaxy dark matter. As is well known, observations show that the fourth power of the rotational velocity of stars in some galaxies is proportional to the luminousity of the galaxy (Tully-Fisher law), $v^4 \propto L$. Since the luminousity, by turn, is proportional to the mass M of the galaxy, $L \propto M$, it follows that $v^4 \propto M$, independent of the radial distance of the star from the center of the galaxy, and in violation to Newtonian gravity. Here came the idea of galaxy dark matter or, alternatively, modification of Newton's gravity.

In this paper we show how a careful application of general relativity theory gives an answer to the problem of motion of stars in galaxies in an expanding universe. If Einstein's general relativity theory is valid, then it appears that the galaxy halo dark matter is a property of spacetime and not some physical material. The situation resembles that existed at the beginning of the century with respect to the problem of the advance of the perihelion of the planet Mercury which general relativity theory showed that it was a property of curvature.

2 Geodesic Equation

The equation that describes the motion of a simple particle is the geodesic equation. It is a direct result of the Einstein field equations $G_{\mu\nu} = \kappa T_{\mu\nu}$ $\left(\kappa = 8\pi G/c^4\right)$. The restricted Bianchi identities $\nabla_\nu G^{\mu\nu} \equiv 0$ imply the covariant conservation law $\nabla_\nu T^{\mu\nu} = 0$. When volume-integrated, the latter yields the geodesic equation. To obtain the Newtonian gravity it is sufficient to assume the approximate forms for the metric: $g_{00} = 1 + \frac{2\phi}{c^2}$, $g_{0k} = 0$ and $g_{kl} = -\delta_{kl}$, where $k, l = 1, 2, 3,$

and ϕ a function that is determined by the Einstein field equations. In the lowest approximation in $1/c$ one then has [1]

$$\frac{d^2 x^k}{dt^2} = -\frac{\partial \phi}{\partial x^k}, \tag{1}$$

$$\nabla^2 \phi = 4\pi G \rho, \tag{2}$$

where ρ is the mass density. For a central body M one then has $\phi = -GM/R$ and Eq. (2) yields, for circular motion, the first integral

$$v^2 = GM/R, \tag{3}$$

where v is the rotational velocity of the particle.

3 Hubble's Law

The Hubble law asserts that faraway galaxies recede from each other at velocities proportional to their relative distances, $\mathbf{v} = H_0 \mathbf{R}$, with $\mathbf{R} = (x, y, z)$. H_0 is the universal proportionality constant (at each cosmic time). Obviously the Hubble law can be written as $\left(\tau = H_0^{-1} \right)$

$$\tau^2 v^2 - \left(x^2 + y^2 + z^2 \right) = 0, \tag{4}$$

and thus, when gravity is negligible, cosmology can be formulated as a new special relativity with a new Lorentz-like transformation [2-4]. Gravitation, being a nonlinear theory, however, does not permit global linear relations like Eq. (4) and the latter has to be adopted to curved space. To this end one has to modify Eq. (4) to the differential form and to adjust it to curved space. The generalization of Eq. (4) is, accordingly,

$$g'_{\mu\nu} dx^\mu dx^\nu = 0, \tag{5}$$

with $x^0 = \tau v$ and $g'_{\mu\nu}$ is a new metric. Since the universe expands radially (it is assumed to be homogeneous and isotropic), it is convenient to use spherical coordinates $x^k = (R, \theta, \phi)$ and thus $d\theta = d\phi = 0$. We are still entitled to adopt coordinate conditions, which we choose as $g'_{0k} = 0$ and $g'_{11} = g'^{-1}_{00}$. Equation (5) reduces to

$$\frac{dR}{dv} = \tau g'_{00}. \tag{6}$$

This is Hubble's law taking into account gravitation, and hence dilation and curvature. When gravity is negligible, $g'_{00} \approx 1$ thus $\frac{dR}{dv} = \tau$ and by integration, $R = \tau v$ or $v = H_0 R$ when the initial conditions are chosen appropriately.

356

4 Phase Space

As is seen, the Hubble expansion causes constraints on the structure of the
universe which is expressed in the phase space of distances and velocities, exactly the observables. The question arises: What field equations the metric
tensor $g'_{\mu\nu}$ satisfies? We *postulate* that $g'_{\mu\nu}$ satisfies the Einstein field equations
in the phase space, $G'_{\mu\nu} = KT'_{\mu\nu}$, with $K = \frac{8\pi k}{\tau^4}$, and $k = \frac{G\tau^2}{c^2}$. Accordingly,
in cosmology one has to work in both the real space and in the phase space.
Particles follow geodesics of both spaces (in both cases they are consequences
of the Bianchi identities). For a spherical solution in the phase space, similarly
to the situation in the real space, we have in the lowest approximation in $1/\tau$
the following: $g'_{00} = 1 + \frac{2\psi}{\tau^2}$, $g'_{0k} = 0$ and $g'_{kl} = -\delta_{kl}$, with $\nabla^2\psi = 4\pi k\rho$. For a
spherical solution we have $\psi = -kM/R$ and the geodesic equation yields

$$\frac{d^2 x^k}{dv^2} = -\frac{\partial\psi}{\partial x^k}, \tag{7}$$

with the first integral

$$\left(\frac{dR}{dv}\right)^2 = \frac{kM}{R} \tag{8}$$

for a rotational motion. Integration of Eq. (8) then gives

$$R = \left(\frac{3}{2}\right)^{2/3} (kM)^{1/3} v^{2/3}. \tag{9}$$

Inserting this value of R in Eq (3) we obtain

$$v^4 = \frac{2}{3} GMcH_0. \tag{10}$$

Acknowledgments

It is a pleasure to thank Y. Ne'eman for many discussions, illuminating remarks
and much encouragements. Thanks are also due to G. Erez, B. Carr, O. Lahav
and N. van den Bergh for useful remarks.

References

1. M. Carmeli, *Classical Fields: General Relativity and Gauge Theory* (John
Wiley, New York, 1982), chap. 6.
2. M. Carmeli, *Found. Phys.* **25**, 1029 (1995).
3. M. Carmeli, *Found. Phys.* **26**, 413 (1996).
4. M. Carmeli, *Cosmological Special Relativity: The Structure of Space, Time
and Velocity* (World Scientific, Singapore, 1997).

DARK MATTER: A CHALLENGE TO STANDARD GRAVITY OR A WARNING?

PHILIP D. MANNHEIM

Department of Physics, University of Connecticut, Storrs, CT 06269-3046
mannheim@uconnvm.uconn.edu

We suggest that the conventional need for overwhelming amounts of astrophysical dark matter should be regarded as a warning to standard gravity rather than as merely a challenge to it, and show that the systematics of galactic rotation curve data can just as readily point in the direction of the equally covariant conformal gravity alternative. In particular we identify an apparent imprint of the Hubble flow on those data, something which while quite natural to conformal gravity is not at all anticipated in the standard gravitational paradigm.

While the main thrust of current research is to meet the challenge of determining the precise nature and form of the dark matter which is widely thought to dominate the universe on large distance scales, nonetheless, given so startling a requirement as this, it is not inappropriate to ask whether this very need for dark matter, and in such copious proportions, might not instead be a warning that the standard Newton-Einstein theory might not in fact be the right one for gravity. Now while the standard theory was first established on solar system distance scales, for the moment its extrapolation to much larger distances is precisely just that, since there is not yet a single independent verification of Newton's Law of Gravity on galactic or larger distance scales which does not involve an appeal to dark matter, to thus show the complete circularity of the very reasoning which leads to dark matter in the first place. Thus at the present time observation can only mandate that gravity be a covariant theory whose metric reproduces the familiar Schwarzschild systematics on solar system distance scales, with (as noted long ago by Eddington) this actually being readily achievable in theories of gravity of order higher than the standard second order one. Since such higher order theories would however then also depart from Schwarzschild at larger distances [1] their phenomenological candidacy is immediate. In fact, motivated by the underlying conformal invariance of the three other fundamental interactions, Mannheim and Kazanas considered the candidacy of one explicit higher order theory, viz. fourth order conformal gravity, and found [2] that outside of a static, spherically symmetric system such as a star the exact metric takes the form $-g_{00} = 1/g_{rr} = 1 - 2\beta^*/r + \gamma^* r$, to thus precisely recover the Schwarzschild metric on small enough distances while both generalizing it and departing from it on much larger ones.

While higher order theories of gravity such as conformal gravity thus yield potentials which then dominate over the Newtonian one at large distances just as desired, the very fact that they do so entails that one is now no longer able to ignore the potentials due to distant matter sources outside of individual gravitational systems such as galaxies. Thus once we depart from the second order theory, we immediately transit into a world where we have to consider effects due to matter not only inside but also outside of individual systems, and we are thus led [3,4,5] to look for both local and global imprints on galactic rotation curve data, this being a quite radical (and quite Machian) conceptual departure from the standard purely local Newtonian world view.

To isolate such possible global imprints, it is instructive [3,4,5] to look at the centripetal accelerations of the data points farthest from galactic centers. In particular, for a large set of galaxies whose rotation curve data are regarded as being particularly characteristic of the pattern of deviation from the luminous Newtonian expectation that has so far been obtained, it was found that these farthest centripetal accelerations could all be parameterized by the universal three component relation $(v^2/R)_{last} = \gamma_0 c^2/2 + \gamma^* N^* c^2/2 + \beta^* N^* c^2/R^2$ where $\gamma_0 = 3.06 \times 10^{-30}$ cm^{-1}, $\gamma^* = 5.42 \times 10^{-41}$ cm^{-1}, $\beta^* = 1.48 \times 10^5$ cm, and where N^* is the total amount of visible matter (in solar mass units) in each galaxy. Since the luminous Newtonian contribution is decidedly non-leading at the outskirts of galaxies, we thus uncover the existence of two linear potential terms which together account for the entire measured departure from the luminous Newtonian expectation (not only for these farthest points but even [4,5] for all the other (closer in) data points as well in fact). Now while the $\gamma^* N^* c^2/2$ term can immediately be identified as the net galactic contribution due to the linear $\gamma^* c^2 R/2$ potentials of all the N^* stars in each galaxy, the inferred $\gamma_0 c^2/2$ term is on a quite different footing since it is independent of the mass content N^* of each of the individual galaxies. Moreover, since numerically γ_0 is found to have a magnitude of order the inverse Hubble radius, we can thus anticipate that it must represent a universal global effect generated by the matter outside of each galaxy (viz. the rest of the matter in the universe), and thus not be associated with any local dynamics within individual galaxies at all.

The emergence of the $\gamma_0 c^2/2$ term, a term which may be thought of as being a universal acceleration, immediately raises some questions. First, if it is an acceleration, then with respect to which frame is the acceleration - and no matter which particular one (the Hubble flow itself being the only apparent covariant possibility), how could it possibly be universal for each and every galaxy. And moreover, how could γ_0 be related to the Hubble radius at all since the Hubble parameter is not a static, time independent quantity. As we shall see, conformal gravity provides [3,4,5] answers to all these questions. Specifically, it was noted quite early on [2] that the general coordinate transformation $r = \rho/(1 - \gamma_0 \rho/4)^2$,

$t = \int d\tau/R(\tau)$ transforms the metric $ds^2 = (1+\gamma_0 r)c^2 dt^2 - dr^2/(1+\gamma_0 r) - r^2 d\Omega$ into $ds^2 = \Lambda(\rho, \tau)\{c^2 d\tau^2 - R^2(\tau)(d\rho^2 + \rho^2 d\Omega)/(1 - \rho^2 \gamma_0^2/16)^2\}$ (where the conformal factor $\Lambda(\rho, \tau)$ is given by $(1 + \rho\gamma_0/4)^2/R^2(\tau)(1 - \rho\gamma_0/4)^2)$, to yield a metric which is conformal to a Robertson-Walker metric with scale factor $R(\tau)$ and explicitly negative 3-space scalar curvature $k = -\gamma_0^2/4$, with this metric in fact being none other than that explicitly found [6,7] in conformal cosmology where only an open universe with explicitly negative k is realizable. Now, in a geometry which is both homogeneous and isotropic about all points, any observer located at the center of any comoving galaxy can serve as the origin for the coordinate ρ. Thus in his own local rest frame each such comoving observer is able to make the above coordinate transformation with the use of his own particular ρ, to then find that in his own frame the entire Hubble flow then acts as a universal linear potential coming directly from the spatial curvature of the universe; with galactic test particles then experiencing a universal acceleration $\gamma_0 c^2/2$ which is explicitly generated by the rest of the matter in the universe. Thus we see that rather then being an acceleration with respect to the Hubble flow (a non-relativistic notion), explicitly because of relativity, the universal acceleration in fact emerges as an intrinsic property of the Hubble flow itself as seen in each comoving observer's rest frame. Moreover, with it also emerging as the spatial curvature, it is then also a nicely time independent quantity, with the numerical determination of γ_0 given above actually yielding an explicit value for both the sign and magnitude of the spatial curvature of the universe, something which years of intensive work has yet to accomplish in the standard theory. Thus to conclude, it would appear from our analysis that there is something explicitly global at play in galactic dynamics, something not only quite suggestive of conformal gravity but also seemingly somewhat foreign to the standard gravitational paradigm. This work has been supported in part by the Department of Energy under grant No. DE-FG02-92ER40716.00.

References

1. P. D. Mannheim and D. Kazanas, Gen. Relativ. Gravit. **26**, 337 (1994).
2. P. D. Mannheim and D. Kazanas, Astrophys. J. **342**, 635 (1989).
3. P. D. Mannheim, "Cosmology and Galactic Rotation Curves", preprint UCONN 95-07, November 1995.
4. P. D. Mannheim, "Are Galactic Rotation Curves Really Flat?", preprint UCONN 96-04, May 1996, Astrophys. J. (in press).
5. P. D. Mannheim, "Local and Global Gravity", preprint UCONN 96-09, November 1996, Found. Phys. (in press).
6. P. D. Mannheim, Astrophys. J. **391**, 429 (1992).
7. P. D. Mannheim, "Conformal Cosmology and the Age of the Universe", preprint UCONN 95-08, December 1995.

DILATONS AS DARK MATTER CANDIDATES

RAINER DICK

Department of Physics, University of Munich, 80333 Munich, Germany

Instantons and axionic domain boundaries induce a dilaton potential at the QCD scale and yield a relation between the dilaton mass and the axion mass. It is found that the dilaton can complement the axion as a CDM component.

1. Dilatons are excitations of a scalar field ϕ which couples to gauge fields and axions through terms $\exp(\frac{\phi}{f_\phi})F^2$ and $\exp(-2\frac{\phi}{f_\phi})\partial a \cdot \partial a$, respectively. Such scalars are a genuine prediction of string theory and any theory involving a compactification scale, and the particular ratio of decay constants assumed in the two coupling functions is a prediction from strong/weak–coupling duality in field theory and string theory[1,2].

If the four–dimensional metric is directly induced from higher dimensions, then the dilaton couples also exponentially to the curvature scalar, implying a Brans–Dicke type theory of gravity with a Brans–Dicke parameter $\omega_{BD} \simeq 1$. Gasperini and Veneziano had pointed out that cosmology might dispense with an initial singularity in the Brans–Dicke framework and studied production of a relic dilaton background in this scenario[3]. On the other hand, solar system tests of gravity restrict Brans–Dicke couplings of massless scalars to values $\omega_{BD} > 500$, and it is apparent that a Brans–Dicke scalar with $\omega_{BD} \simeq 1$ must be very heavy to comply with gravitational light bending and time delay in the solar system.

An alternative scenario assumes that the four–dimensional metric is only conformally related to the metric induced from higher dimensions, whence low energy string theory could include standard Einstein gravity. This theory would not imply any Brans–Dicke type coupling and the initial singularity would not be removed (at least not on the level of four–dimensional field theory), whence the dilaton could be light. Since a light axion is a leading contender for cold dark matter in the universe[4] strong/weak coupling duality provides a strong indication for an additional light dilaton component accompanying the axion, and it was pointed out in[5] that instantons and axions together induce an effective dilaton potential

$$V(\phi) = \frac{m_\phi^2 f_\phi^2}{6}\left(2\exp(\frac{\phi}{f_\phi}) + \exp(-2\frac{\phi}{f_\phi})\right)$$

where the dilaton and instanton parameters are related via $m_\phi f_\phi \simeq m_a f_a$.
2. Unbroken strong/weak coupling duality at low energies would imply $f_\phi \simeq$

 f_a, and the dilaton and the axion would behave very similar concerning their contribution to the energy density of the universe. On the other hand, if f_ϕ is of the order of the reduced Planck mass $m_{Pl} = (8\pi G)^{-1/2} = 2.4 \times 10^{18}\text{GeV}$ as predicted by string theory, then the dilaton appears to be much lighter than the axion and oscillation dominance would start at a temperature $T_\phi \simeq 10^{-1}T_a$, i.e. coherent oscillations of the dilaton would start to contribute to the energy density of the universe after onset of the axion oscillations. In this setting the dilaton would still be non–relativistic with a velocity $p_\phi/m_\phi \simeq 10^{-2}$. However, such a large decay constant would imply overclosure of the universe[6] unless the dilaton misalignment at T_ϕ satisfies $\sqrt{\langle\phi^2\rangle} \leq 10^{-4}m_{Pl}$, implying either inhomogeneous inflation, or a strong impact of the tiny mass on $\sqrt{\langle\phi^2\rangle}$.

3. Superficially the tiny mass of the dilaton required by strong/weak coupling duality and implied by the previous findings seems to violate fifth force constraints derived by the assumption that couplings of the dilaton to mass terms would induce a Yukawa type potential in the gravitational sector. However, the dilaton would not couple directly to mass terms of basic leptons and quarks, which are generated at the weak scale far below any string or compactification scale. Effective couplings to meson and nucleon masses can't be estimated yet, since a recent investigation revealed that the dilaton has a profound impact on interaction potentials in gauge theory[7], which is not adequately described by the usual approximation of an effective local coupling. It turns out that the dilaton either regularizes the Coulomb potential to $(r+r_\phi)^{-1}$, with $r_\phi \sim f_\phi^{-1}$, or it induces a potential proportional to r.

References

1. A. Shapere, S. Trivedi and F. Wilczek, *Mod. Phys. Lett.* **A6**, 2677 (1991).
2. J.H. Schwarz, *Lectures on superstring and M–theory dualities*, hep–th/9607201.
3. M. Gasperini and G. Veneziano, *Phys. Rev.* **D50**, 2519 (1994).
4. E.W. Kolb and M.S. Turner, *The Early Universe*, (Addison–Wesley, Redwood City, 1990).
5. R. Dick, *Stabilizing the dilaton through the axion*, LMU–TPW–96/22, to appear in *Mod. Phys. Lett.* **A**.
6. M.S. Turner, *Phys. Rev.* **D33**, 889 (1986).
7. R. Dick, *The Coulomb potential in gauge theory with a dilaton*, LMU–TPW–97/02, hep–th/9701047.

FREE FLOATING BROWN DWARFS IN THE PLEIADES

J.E. GALLEGOS, R. REBOLO, M.R. ZAPATERO OSORIO,
E.L. MARTIN
Instituto de Astrofísica de Canarias, E-38200 LA LAGUNA, TENERIFE, SPAIN

In a new search for brown dwarfs in the Pleiades cluster, we have found an interesting object with optical and NIR characteristics that place it in the brown dwarf domain in the I vs. R-I and K vs. K-I diagrams. Spectroscopic observations determine that this object, Teide Pleiades 2, has radial velocity, H_α and spectral type consistent with membership in the cluster. We estimate its mass at 70±15 Jupiter masses. This new object adds to the previously discovered bona fide brown dwarfs in the cluster, confirming a continued rise in the mass function beyond the substellar limit.

1 The Optical R,I Survey and Near-infrared Photometry

The Pleiades cluster has specific characteristics which make it a very good place to find free-floating brown dwarfs (BDs). At its distance (125 pc) and age (70–120 Myr), BDs are still bright and detectable with current instrumentation. In fact the first free-floating BDs have been discovered in the Pleiades (Rebolo, R., *et al.*, 1995, Basri, G., *et al.*, 1996, Rebolo, R., it et al., 1996). The present survey[a] covers an area of 616 $arcmin^2$ with completeness magnitudes of R=20 and I=19; which increases by a factor of 2 the area surveyed by our group (Zapatero-Osorio et al. 1997a), although with less sensitivity. Follow-up infrared photometry can discriminate cluster members from field very cool stars contaminating optical surveys (Zapatero-Osorio et al. 1997b). Our JHK observationsObservations made with the 2.2 m telescope at Calar Alto Observatory. confirm that Teide 2 lies in the expected sequence of single BDs in the Pleiades (see Table 1).

2 Spectroscopy

The likelihood of membership to the Pleiades cluster for the objects which meet the optical and infrared conditions, is assessed via the study of their photospheric features, H_α emission, radial velocity and consistency of their spectral types and I-band magnitudes with known cluster members. A kinematic constraint on membership comes from the fact that isolated brown dwarfs are thought to be formed as independent condensations in a manner similar to

---

[a]Observations made with the IAC80 telescope operated by the Instituto de Astrofsica de Canarias at Teide Observatory.

Table 1: Photometric and spectroscopic data for the least massive Pleiades objects.

Object	I	R - I	I - K	H_{alpha}	$V_r(kms^{-1})$	Sp T
Teide 1	18.80	2.74	3.69	4.5±1.0	5.2±2.0	M8
Calar 3	18.73	2.54	3.79	6.5±1.0	1.0±8.0	M8
PPl 15	17.80	2.25	3.66	11.5	4.6±4.0	M6.5
Teide 2	17.82	2.23	3.26	10.0±3.0	3.7±6.0	M6

low mass stars, and hence they should also share the bulk motion of the cluster. The radial velocity for our object is $3.7\pm6.0kms^{-1b}$, consistent with the known radial velocities of members of the Pleiades which are in the range of $0 - 14kms^{-1}$. Another criterion of membership to the cluster is the presence of H_α in emission. All known Pleiades objects with spectral type in the range M4-M8 show H_α in emission with an equivalent width larger than 3 Å. Teide 2 is well located in the I-magnitude - spectral type sequence of the cluster and presents an H_α equivalent width of $10.0\pm3.0\overset{\circ}{A}{}^{c}$ (Gallegos, J.E., et al., 1997).

3 Discussion

From the RIJHK photometry and spectral type we have determined the bolometric luminosity and effective temperature of Teide 2, and comparing with theoretical evolutionary tracks we obtained a mass of 70±15 Jupiters, slightly below the substellar mass limit. This new object adds to the two BDs found in our previous, similarly deep, CCD survey (Zapatero Osorio et al. 1997a) and supports our claims that free-floating substellar objects could be quite numerous. The mass function appears to keep rising from very low-mass stars to BDs, at least until objects of about 70-50 Jupiter masses. The area so far covered by our CCD surveys represents only some 2% of the total cluster area and it will be desirable to extend it significantly. In any case, it is clear that star formation does not stop at the substellar limit. The BD contribution to the total mass of the Pleiades is probably a few percent or less, and if it is representative of the galactic disk, it would be consistent with current ideas that there is no need for missing mass in it. Since some amount of free-floating BDs

---

[b]Obtained with observations made with the Keck II telescope at the W.M. Keck Observatory, which is operated jointly by the University of California and the California Institute of Technology.

[c]Spectra obtained with the William Herschel Telescope, operated on the island of La Palma by the Royal Greenwich Observatory in the Spanish Observatorio del Roque de los Muchachos of the IAC, Spain.

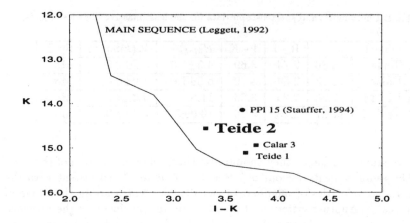

Figure 1: K-band vs. K-I diagram for Pleiades substellar objects.

are actually present in the Pleiades, which is where they are easier to detect, they remain an interesting kind of objects for the dark matter problem.

Acknowledgments

We would like to thank Gibor Basri for the colaboration with the Keck observations.

References

1. Basri, G., Marcy, G.W., Graham, J.R., *ApJ.* **458**, 600 (1996).
2. Gallegos, J.E., *et al.*, 1997, in preparation.
3. Leggett, S.K., ApJS **82**, 351 (1992).
4. Rebolo, R., Zapatero-Osorio, M.R., Martin, E.L., *Nature* **377**, 129 (1995).
5. Rebolo, R., *et al.*, *ApJ.* **469**, L53-L56 (1996).
6. Stauffer, J.R., Hamilton, D., Probst, R.G., *ApJ.* **108**, 155 (1994).
7. Zapatero-Osorio, M.R., Rebolo, R., Martin, E.L., *A&A.* **317**, 164 (1997a).
8. Zapatero-Osorio, M.R., Rebolo, R., Martin, E.L., *A&A.* , (1997b), in press.

DISCRIMINATING LIQUID-XENON DETECTOR FOR WIMPs SEARCH

DAVID B. CLINE AND HANGUO WANG

Department of Physics and Astronomy, Box 951547
University of California Los Angeles
Los Angeles, CA 90095-1547, USA

The search for SUSY cold dark-matter particles is of great importance. We describe a powerful discriminating liquid-xenon detector. A 2-kg detector (ICARUS-WIMPs) is now in the CNR Torino Mt. Blanc Underground Laboratory and a 20-kg detector is being planned by the UK-UCLA group. This set of detectors can, in principle, cover most of the SUSY discovery space.

1   Dark Matter--SUSY or Not?

The latest evidence for dark matter in the Universe has been reviewed recently at two University of California Los Angeles (UCLA) symposiums.[1] Remarkably, even in the 1920s some evidence had been found and of course in the 1930s, F. Zwicky provided perhaps the first definitive evidence for dark or non-luminous matter in Galaxies.[2]

While no one knows the exact cause of dark matter, there is a reasonable likelihood that new elementary particles play some role in this phenomenon. Of all of the current ideas in this regard, many feel supersymmetry is the most "natural." Our viewpoint is to take the SUSY model seriously and to see what level of detection and discrimination is required to observe such particles. While even the SUSY model is not fully predictive, it would appear to be better than other even more ad hoc models. The project described here grew out of the ICARUS project to construct a massive "electronic bubble chamber" using liquid argon.[3] The first stage of this project, the construction of a 600-ton detector for Hall C at the Gran Sasso, is now approved.

2   Rates for a SUSY Dark-Matter Detector

There are many estimates for the cross section of SUSY-WIMPs with various targets. We believe this illustrates the difficulty, as well as the promise, for the search for SUSY-WIMPs. In this report, we follow the recent work of Nath and Arnowitt[4] (and the references cited therein). Figure 1 shows the limits on the rate of interactions (per

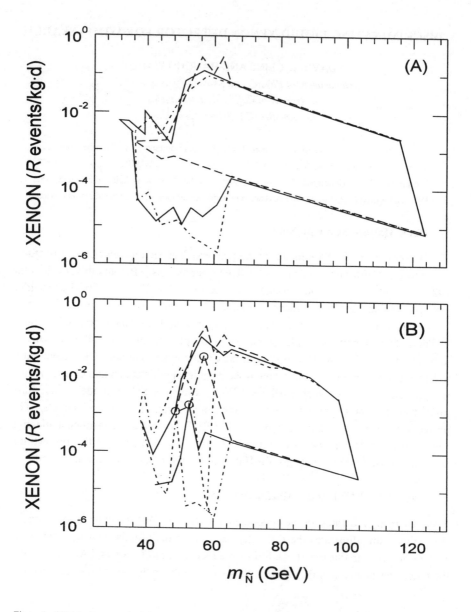

Figure 1: (A) Maximum and minimum curves of event rates for Xe as functions of neutralino mass when μ < 0 and all other parameters (m_0, A_t, $\tan \beta \leq 20$) run over the allowed ranges ($m_t = 175$ GeV, where m_t is the physical mass for $\Omega > 0.1$). The $b \to s\gamma$ constraint is not imposed. (B) Same as (A) but for $\Omega > 0.22$.

kg/d) as a function of the approximate neutralino mass (the gluino mass is expected to be approximately the same) for values of $\mu \lesssim 0$.[4] Without getting into the details of the assumptions in this calculation, we note that the range of rates goes from a few events / kg·d to 10^{-5} events/kg·d. Although the results are for Ge and Pb, we expect similar results for liquid Xe. These results, if taken at face value, suggest that the detection of SUSY-WIMPs could be very difficult, requiring large detectors of certainly 100 kg and possibly tons of detector. In this case, the rejection of background is even more important.

3 The ICARUS Liquid-Xenon Studies

In 1992, a subgroup of the ICARUS team started the study of liquid Xe for the purpose of WIMP detection. The first report of this work was given in 1992 at Waseda University and published in the proceedings of the conference[5] Figure 2A shows the initial experimental setup. Table 1 presents a schematic view of the reason that liquid Xe is potentially an excellent WIMP detector. The scope traces in Fig. 2B provide the essential discrimination method. The ratio of primary to secondary scintillation light is very sensitive to the initial ionization of the source; the γ and α particles are clearly separated. In addition, the pulse shapes provide discriminations against background. Results of detailed tests of the discrimination method can be seen in Fig. 2C,D[6]. More recently, this group has constructed a larger detector (Fig. 3A) and carried out very detailed tests of the discrimination methods (Fig. 3B).

Table 1: Signature and background in liquid xenon.

Recoil Nuclei
- Heavily ionizing particle
- High recombination, hence
- Mainly scintillation light is produced

Radioactivity
- Minimum ionizing particle
- Low recombination, hence
- Both charge and light are produced

In liquid Xe
- Both charge and light are visible
- This provides an efficient way for signal-to-background rejection

Moreover, in Xe
- No long-lived natural isotopes are present
- Xe^{127} has longest decay time (≈ 36 d)

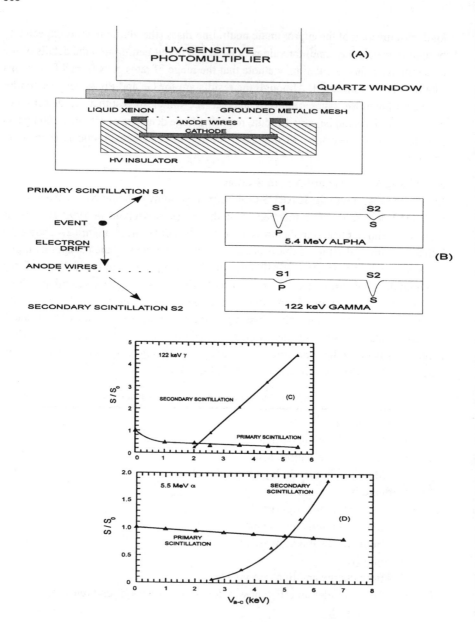

Figure 2: (A) Geometry of liquid-xenon test chamber, (B) observed primary and secondary scintillation signals showing S1/S2 >> 1 for α events and << 1 for γ events, (C) variations of the secondary scintillation intensity as a function of V_{a-c} for photons and (D) for α particles. (From Ref. 6.)

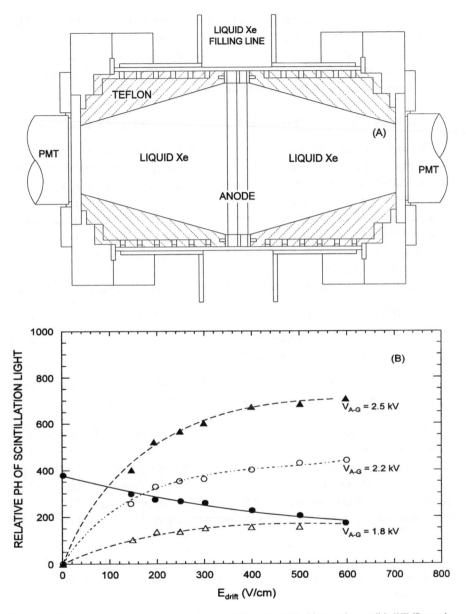

Figure 3: (A) A 2-kg detector that has been constructed for tests at Mt. Blanc and a possible WIMP search, and (B) variations of the secondary scintillation intensity as a function of E_{drift} and V_{A-G} for photons.

A successful test of the detection of a recoil Xe nucleus using neutron scattering has been recently carried out, and it shows clear evidence that SUSY-WIMPs will give a strong, unique signal on a discriminating liquid-Xe detector.[6] The 2-kg detector shown in Fig. 3A will be installed at the Mt. Blanc Underground Laboratory (UL) to perform a first search for SUSY-WIMPs using this tchnique.

4   The Proposed ZEPLIN Project to Definitively Search for SUSY Dark Matter[a]

To allow lower limits to be reached, it is essential to develop methods of differentiating the desired nuclear recoil events from γ- and β-decay backgrounds. At the same time, it is desirable to develop techniques capable of being substantially scaled up in target mass.   The need for targets in the 100-1000-kg region would arise in particular in searches for the 5% "annual modulation" of any true dark-matter signal (due to the Earth's motion combined with the solar motion through the Galaxy).   Large-mass targets would also be needed for heavier WIMP masses (> 100 GeV), because of the correspondingly smaller flux of such particles.

Liquid Xe satisfies all of the above requirements for a dark matter detector because:

1.   It is available in sufficiently large quantities with high purity.

2.   It scintillates via two mechanisms, which are stimulated to different extents by nuclear-recoil and background electron-recoil events.

3.   Its natural form consists of isotopes with and without nuclear spin, so it is suitable as a detector for both spin-independent and -dependent interactions.

The larger nuclear mass of Xe also makes it a better match to heavier WIMPs but, at the same time, the larger nuclear radius introduces a significant form-factor correction unless the energy threshold is low (1-10 keV).   Efficient light collection is, therefore, of prime importance in a liquid-Xe detector.   Figure 4 shows a schematic of the proposed ZEPLIN detector, for which the proposed location be England.

There are two distinct approaches to discriminating nuclear-recoil events in liquid xenon:

1.   Analyzing the total scintillation pulse shape or, at low energy, the individual photon arrival times, which will differ significantly for nuclear- and electron-recoil events;

---

[a]This section is adapted from a recent report[7] by the ZEPLIN group.

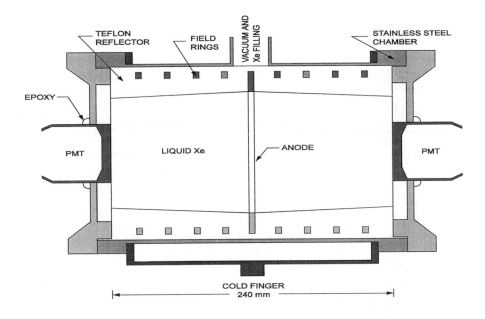

Figure 4: Conceptual design for ZEPLIN system, showing inner proportional zone and outer shielding zone (total length, 40 cm), with PMT for collecting proportional scintillation light.

2. Applying an electric field to prevent recombination and measuring (A) the primary scintillation and (B) the ionization component by drifting and producing "secondary scintillation."

Figure 5 shows the limits we hope to reach with the initial 20-kg detector. The ICARUS xenon detector operating at the Mt. Blanc UL could also reach favorable limits if the background can be kept under control and if a long operating period is utilized.

5   Recent Progress in the Test Liquid-Xenon Detector

During the first part of 1996, several advances have been made in the development of the ICARUS–WIMP liquid-Xe detector:

1. A detailed study of the background from low-energy gammas has been conducted.

2. Neutron-induced events in the presence of large backgrounds have been observed directly.

3. A detailed study of the rise-time distribution for Xe events and backgrounds has been made.

372

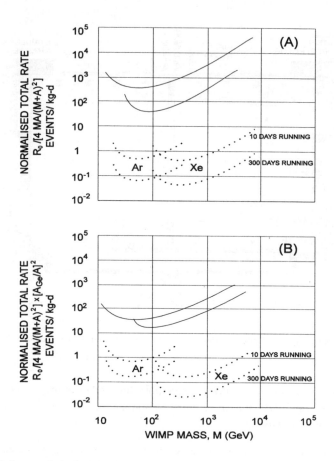

Figure 5: Dark matter limits for (A) spin-dependent and (B) spin-independent interactions. Typical existing limits are shown for Ge ionization detectors plus recent (1994) improvements using NaI detectors with pulse shape discrimination [from UK data (Gran Sasso data is similar)]. In each diagram, the lower pair of curves show estimated limits vs running time for this proposal, using primary/secondary scintillation in liquid Xe at 10-keV energy threshold. The (Xe-doped) Ar case shows the advantage of data from both Xe and Ar targets.

The study of the discriminating liquid-Xe detector continues to indicate that this may be one of the best methods to use to detect SUSY–WIMPs.

6 Status of the 2-kg Detector at the CNR Torino Mt. Blanc UL

The 2-kg ICARUS–WIMPs detector has now been installed at the CNR Torino Mt. Blanc UL and first tests are underway. Our goal is to search for WIMPs to the level of 10^{-1} events/kg·d during the next year or so.

Acknowledgments

We wish to thank the members of the ICARUS and ICARUS–WIMP groups (P. Picchi), as well as members of the proposed ZEPLIN group[7] (P. Smith in particular); my thanks also to P. Nath and R. Arnowitt for discussions on the theory of SUSY–WIMP detection and for Fig. 1.

References

1. See the proceedings of "Sources of Dark Matter in the Universe," Santa Monica, 1996, ed. D. Cline, *Nucl. Phys.* B (PS), **51B** (in press); and L. H. Aller and V. Trimble, *Sources of Dark Matter in the Universe*, ed. D. Cline (World Scientific, Singapore, 1994) p. 3 and p. 9.
2. F. Zwicky, *Ap. J.* **86**, 217 (1937).
3. CERN–UCLA–INFN Group, ICARUS Proposal (1993) unpublished.
4. P. Nath and R. Arnowitt, *Phys. Rev. Lett.* **74**, 4592 (1995).
5. D. Cline, *Nucl. Instrum. Methods* A **327**, 178–186 (1993).
6. P. Benetti et al., *Nucl. Instrum. Methods* A **327**, 203–206 (1993).
7. J. Park, M. Atac, D. B. Cline, H. Wang, and P. F. Smith, *Sources of Dark Matter in the Universe*, ed. D. Cline (World Scientific, Singapore, 1994) p. 288.

ACCURATELY DETERMINING INFLATIONARY PERTURBATIONS

ANDREW R. LIDDLE and IAN J. GRIVELL

Astronomy Centre, University of Sussex,
Brighton BN1 9QH, Great Britain

Cosmic microwave anisotropy satellites promise extremely accurate measures of the amplitude of perturbations in the universe. We use a numerical code to test the accuracy of existing approximate expressions for the amplitude of perturbations produced by single-field inflation models. We find that the second-order Stewart–Lyth calculation gives extremely accurate results, typically better than one percent. We use our code to carry out an expansion about the general power-law inflation solution, providing a fitting function giving results of even higher accuracy.

1 Motivation

Two newly approved satellites which will observe microwave background anisotropies, MAP and COBRAS/SAMBA, promise to measure the amplitude of these anisotropies at the percent level. Recent theoretical work has demonstrated that these anisotropies can be predicted from a given spectrum of adiabatic perturbations, to an accuracy of one percent or better[1]. The most promising theory for the generation of adiabatic perturbations is cosmological inflation, and here we address the question of whether or not it is possible to predict the spectrum from an inflationary model at a similar level of accuracy, in order to allow one to take full advantage of forthcoming observations. Full details can be found in Ref. 2, which also discusses gravitational waves.

We shall only consider models with a single scalar field ϕ, moving in an arbitrary potential $V(\phi)$. This is not to say that accurate calculations cannot be done in more general models, but there typically they have to be done on a case-by-case basis, whereas the single-field models can be analyzed simultaneously. Our aim is to compute the amplitude $\mathcal{P}_{\mathcal{R}}(k)$ of the curvature perturbation $\mathcal{R}$ on a given fixed comoving scale k, where the precise terminology is defined in Ref. 2. From this one could also compute the spectral index, its rate of change, etc. Typically though the corrections we discuss to the standard results are observationally negligible except for the amplitude itself.

2 Framework

The best calculational framework is that introduced by Mukhanov[3] and exploited by Lyth and Stewart.[4,5] The only knowledge we require of the background (homogeneous) evolution is of the combination $z = a\dot{\phi}/H$, where a is the scale factor and H the Hubble parameter. The perturbation can be expressed by a gauge-invariant potential $u = -z\mathcal{R}$. Finally, we use conformal time τ. Then a fourier mode of the perturbation u obeys the equation

$$\frac{d^2 u_k}{d\tau^2} + \left(k^2 - \frac{1}{z}\frac{d^2 z}{d\tau^2}\right) u_k = 0 \tag{1}$$

Typically we don't know the background evolution $z(\tau)$, so we don't even know the equation we are trying to solve. Fortunately though it possesses two asymptotic regimes where the solution can be found:

$$u_k = \frac{1}{\sqrt{2k}} e^{-ik\tau} \qquad k \gg aH \tag{2}$$

$$u_k \propto z \qquad k \ll aH \tag{3}$$

The former is the flat-space limit and includes the appropriate quantum normalization. The latter is the growing mode, and corresponds to $\mathcal{R}_k$ constant. Both regimes are independent of $z(\tau)$, so the amplitude of perturbations only depends on the transition regime. We can therefore expand $z(\tau)$ about the time τ where $k = aH$.

This expansion of the background evolution can be carried out using the slow-roll expansion.[6] Inflation is described using as a fundamental quantity the Hubble parameter as a function of the scalar field value, $H(\phi)$. From its derivatives one defines a series of slow-roll parameters, the first two being

$$\epsilon = \frac{m_{\mathrm{Pl}}^2}{4\pi}\left(\frac{dH/d\phi}{H}\right)^2 \quad ; \quad \eta = \frac{m_{\mathrm{Pl}}^2}{4\pi}\frac{d^2 H/d\phi^2}{H} \tag{4}$$

The slow-roll approximation demands that all these are small. They can readily be computed for a given $V(\phi)$.

3 Results

We solve the mode equation numerically, in order to compare with existing calculations. Figure 1 shows four panels, testing the standard slow-roll approximation result $\mathcal{P}_{\mathcal{R}}^{1/2} = H^2/2\pi|\dot{\phi}|$, the improved second-order result of Stewart and Lyth,[5] an approximation based on using the exact power-law inflation result,[4,2] and finally a new expression we derived[2] by expanding about a general power-law solution.

We conclude that it is possible to obtain very accurate results numerically, and also that anyway the Stewart–Lyth calculation[5] is already extremely accurate, being better than two percent accurate in all the parameter region currently favoured by observations.

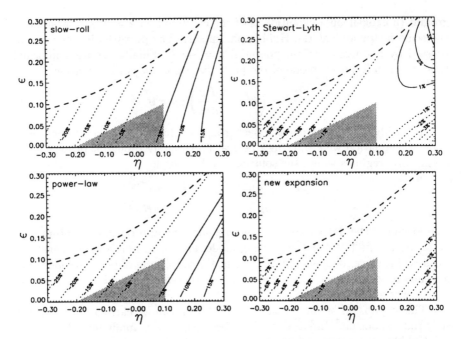

Figure 1: A comparison of four analytic expressions with the exact numerical result, with the contours showing the relative error. Each point in the ϵ–η plane corresponds to a different inflation model. Above the dashed line is a forbidden region (inflation ends too quickly), while the shaded area indicates those models currently preferred by observational data constraining the slope of the spectrum and amount of gravitational waves.

Acknowledgments

The authors were supported by the Royal Society.

References

1. W Hu, D Scott, N Sugiyama and M White, *Phys. Rev.* D **52**, 5498 (1995); U Seljak and M Zaldarriaga, *Astrophys. J.* **469**, 437 (1996).
2. I J Grivell and A R Liddle, *Phys. Rev.* D **54**, 7191 (1996).
3. V F Mukhanov, *Pis'ma Zh. Eksp. Teor. Fiz.* **41**, 402 (1985), *Zh. Eksp. Teor. Fiz.* **94**, 1 (1988).
4. D H Lyth and E D Stewart, *Phys. Lett.* B**274**, 168 (1992).
5. E D Stewart and D H Lyth, *Phys. Lett.* B**302**, 171 (1993).
6. A R Liddle, P Parsons and J D Barrow, *Phys. Rev.* D **50**, 7222 (1994).

NON-GAUSSIAN COSMOLOGICAL PERTURBATIONS AND INFLATIONARY GOLDSTONE MODES [a]

MARTIN BUCHER[1,2][b] and YONG ZHU[1][c]

[1] *Department of Physics, Princeton University*
Princeton, NJ 08544

[2] *Institute for Theoretical Physics, State University of New York*
Stony Brook, NY 11794

The simplest inflationary models predict a spectrum of almost scale-free, Gaussian, adiabatic primordial density perturbations. However it has been known that other types of density perturbations as well are possible from inflation. We discuss the non-Gaussian perturbations that result from exactly massless Goldstone modes disordered in a precise and predictable way during inflation. After inflation ends, these Goldstone modes reorder themselves in a self-similar manner described by a scaling solution, somewhat analogous to the evolution of topological defects after a symmetry-breaking phase transition. But unlike topological defects, the subsequent density perturbations generated by Goldstone modes disordered during inflation can be calculated by evaluating loop Feynman diagrams rather than resorting to numerical simulation. We compute the phase dispersion of the density perturbations generated in this model during the radiation-dominated epoch finding it to be small, which implies the existence of well-defined Doppler peaks in the multipole moments of the CMB anisotropy.

Many models of inflation predict a Gaussian, adiabatic spectrum of primordial density perturbations. However, it has been known for quite some time that non-Gaussian perturbations from inflation are possible as well. In this contribution, an abridged version of a longer paper,[1] which contains more complete references to the prior literature, we investigate one such possibility: density perturbations generated by Goldstone modes disordered during inflation. We assume an *exact* global symmetry G broken to a smaller group H *before* inflation. In this case there exist exactly massless Goldstone modes, described by an order parameter field taking values in the coset space G/H. During inflation these Goldstone modes are disordered in a precise and predictable way by the dynamics of the inflationary expansion. All vestiges of initial conditions are erased. After inflation these Goldstone modes order themselves in a self-similar way, the only relevant scale at a particular time being the Hubble length at that time. The ordering process after the end of inflation is

---

[a]Talk presented by Martin Bucher
[b]E-mail: bucher@insti.physics.sunysb.edu Current Address: Institute for Theoretical Physics, State University of New York, Stony Brook, NY 11794-3840
[c]E-mail: zhu@puhep1.princeton.edu

remarkably similar to what happens in the proposed field ordering scenarios for structure formation (e.g., cosmic strings, global monopoles, textures, etc.) in which field ordering takes place after a cosmological phase transition. In both cases the field ordering is described by a *scaling* solution. The only qualitative difference lies in the character of initial correlations on superhorizon scales. In the inflationary case correlations exist on arbitrary large scales; in field ordering scenarios there is a finite correlation length. However, one expects this difference to affect only the details of the scaling solution and not its essential character.

For the particular case in which a $U(1)$ global symmetry is broken completely before inflation, the evolution of the Goldstone mode during and after inflation takes a particularly simple form. In this case, the coset space is the circle S^1. In the field ordering case in which the same pattern of symmetry breaking occurs after a cosmological phase transition, the subsequent evolution is highly nonlinear and nontrivial to compute because global $U(1)$ cosmic strings are inevitably produced in large numbers. But in the inflationary case, in which the Goldstone field is disordered during inflation, cosmic strings can nucleate only in loops (of size of order the Hubble length during inflation in order not to recollapse), and the production of these loops can be exponentially suppressed by choosing a Mexican hat potential sufficiently stiff in the radial direction. With the production of loops suppressed, the nontrivial topology of S^1 can be ignored, so that the circle S^1 becomes the entire real line. In this case the Goldstone field is simply a free massless scalar field, and whatever its intial state may have been, at the end of inflation its state is well approximated by the Bunch-Davies vacuum.

The evolution of the Goldstone mode may be continued beyond the end of inflation straightforwardly by expanding in plane wave modes and modifying the time dependence to reflect the change in evolution of the scale factor $a(t)$. The contribution to the stress-energy from the Goldstone mode is given by

$$\Theta_{\mu\nu} = (\partial_\mu\phi)(\partial_\nu\phi) - \frac{1}{2}g_{\mu\nu}[g^{\alpha\beta}(\partial_\alpha\phi)(\partial_\beta\phi)]. \tag{1}$$

Müller and Schmid[2] have previously studied various correlations of $\Theta_{\mu\nu}$ for a free massless scalar field initially disordered by inflation. The Goldstone mode generates cosmological density perturbations by exciting the dominant forms of matter in the universe, taken in our calculation as a perfect fluid with $p = \frac{1}{3}\rho$ (i.e., radiation). One writes down the 'scalar' Einstein equations with $T_{\mu\nu}^{(total)} = T_{\mu\nu}^{(rad)} + \Theta_{\mu\nu}$ treating $\Theta_{\mu\nu}$ as a source and first solving for the homogeneous solutions in the absence of a source. From these homogeneous solutions Green's functions are constructed and cosmological perturbations at later times are expressed as a linear integral transform of the source $\Theta_{\mu\nu}$ at earlier times. For cosmological perturbations of a given co-moving wavenumber $\mathbf{k}$, Goldstone modes of all wavenumbers contribute because the source stress-energy is *quadratic* in the massless scalar field, rather than *linear* as is the case for the more usual adiabatic perturbations from simpler inflationary models. As in field ordering scenarios, for cosmological perturbations of wavenumber $\mathbf{k}$, the dominant contribution occurs during horizon crossing. Since unlike for the usual adiabatic perturbations, the decaying mode does not have a chance to decay away significantly—because the sourcing occurs late, at horizon crossing—one expects both the decaying and growing modes to be present. Quantitatively, the statistical ensemble may be characterized by the covariance matrix

$$C = \begin{pmatrix} \langle\psi(\mathbf{k},g)\psi(-\mathbf{k},g)\rangle & \langle\psi(\mathbf{k},g)\psi(-\mathbf{k},d)\rangle \\ \langle\psi(\mathbf{k},d)\psi(-\mathbf{k},g)\rangle & \langle\psi(\mathbf{k},d)\psi(-\mathbf{k},d)\rangle \end{pmatrix} \tag{2}$$

where g and d denote the growing and decaying modes, respectively. For ordinary perturbations from inflation, $C \sim \begin{pmatrix} 1 & 0 \\ 0 & 0 \end{pmatrix}$, because the decaying mode is completely absent. In this case there is no phase dispersion. The definite phase is responsible for the oscillatory character of the Doppler peaks. On the other hand, for a completely random phase one instead has $C \sim \begin{pmatrix} 1 & 0 \\ 0 & 1 \end{pmatrix}$, which would completely wash out the Doppler peaks. For the inflationary Goldstone modes we found that there is little phase dispersion: the large dominates the smaller one by approximately a factor of 50, suggesting the presence of well-defined Doppler peaks. Computing the matrix C involves evaluating one-loop Feynman diagrams.

The covariance matrix C is only one of many correlation functions that can be computed exactly in this simple non-Gaussian model by evaluating a finite number of Feynman diagrams. By straightforwardly extending these techniques, one can compute three-point and higher order correlations.

Acknowledgments We would like to thank A. Linde, G. Sterman, N. Turok, A. Vilenkin, and F. Wilczek for useful discussions. This work was supported in part by the David and Lucile Packard Foundation and by National Science Foundation grant PHY 9309888.

1. M. Bucher and Y. Zhu, (astro-ph 9610223) (1996).
2. H. Müller and C. Schmid, (gr-qc 9401020) (1994); H. Müller and C. Schmid, (gr-qc 9412022) (1994); H. Müller and C. Schmid, (gr-qc 9412021) (1994).

CLASSICAL AND QUANTUM COSMOLOGY WITH THE COMPLEX SCALAR FIELD

I.M. KHALATNIKOV

L. D. Landau Institute for Theoretical Physics, Russian Academy of Sciences, Kosygin Street, Moscow, 117940, Russia and Tel Aviv University, Mortimer and Raymond Suckler Institute of Advanced Studies, Ramat Aviv, 69978, Israel

A.YU. KAMENSHCHIK

Nuclear Safety Institute, Russian Academy of Sciences, 52 Bolshaya Tulskaya, Moscow, 113191, Russia

A.V. TOPORENSKY

Sternberg Astronomical Institute, Moscow University, Moscow, 119899, Russia

The cosmological model with complex scalar self-interacting inflaton field non-minimally coupled to gravity is studied. The different geometries of the Euclidean classically forbidden regions are represented. The instanton solutions of the corresponding Euclidean equations of motion are found by numerical calculations. Interpretation of obtained results and their connection with inflationary cosmology is discussed.

It is widely recognized that inflationary cosmological models give a good basis for the description of the observed structure of the Universe. Most of these models include the so called inflaton scalar field possessing non-zero classical average value which provides the existence of an effective cosmological constant on early stage of the cosmological evolution. In the series of recent papers [1,2,3,4,5] it was suggested to consider the cosmological models with the complex scalar field $\phi = x \exp(i\theta)$, that is equivalent to the inclusion into the theory the new quasi-fundamental constant – charge of the Universe Q. Appearance of this new constant essentially modifies the structure of the Wheeler-DeWitt equation. Namely, the form of the superpotential $U(a, x)$ displays now a new and interesting feature: the Euclidean region, i.e. the classically forbidden region where $U > 0$, is bounded by a closed curve in the minisuperspace (x, a) for a large range of parameters. Thus, in contrast with the picture of "tunneling from nothing" and with the "no-boundary proposal" for the wave function of the Universe we have Lorentzian region at the very small values of cosmological radius a and hence a wave can go into Euclidean region from one side and outgo from the other. These new features of the model require reconsideration of traditional scheme and give some additional possibilities.

Here we would like to dwell in more detail on the investigation of the cosmological models with the complex scalar field non-minimally coupled to gravity. This model described in our recent work [4] and in more detail in the forthcoming papers [5]. We shall consider the model with the following action:

$$S = \int d^4x \sqrt{-g} \left(\frac{m_{Pl}^2}{16\pi}(R - 2\Lambda) + \frac{1}{2}g^{\mu\nu}\phi_\mu^*\phi_\nu + \frac{1}{2}\xi R\phi\phi^* - \frac{1}{2}m^2\phi\phi^* - \frac{1}{4!}\lambda(\phi\phi^*)^2 \right).$$
(1)

Here ξ is the parameter of non-minimal coupling λ is the parameter of the self-interaction of the scalar field, Λ is cosmological constant, m is the mass of the scalar field. The complex scalar field ϕ can be represented in the form $\phi = x \exp(i\theta)$. We shall consider the minisuperspace model with the spatially homogeneous variables a (cosmological radius), x and θ. Taking into account that the variable θ is cyclical one can assume that the corresponding conjugate momentum is frozen and does not subject to quantization [1,2]. In this case the only influence of the variable θ on the dynamics of the system consists in the appearance of the new constant of the model: classical charge Q which is defined as $Q = a^3x^2\dot{\theta}$.

Writing down the super-Hamiltonian constraint in minisuperspace one can get the superpotential

$$U(a, x) = a \left(\frac{m_{Pl}^2}{16\pi}(6 - 2\Lambda a^2)) + 3\xi x^2 - \frac{Q^2}{a^4x^2} - \frac{1}{2}m^2x^2a^2 - \frac{1}{24}\lambda x^4a^2 \right).$$
(2)

Resolving equation $U = 0$ we can get the form of the Euclidean region in the plane of minisuperspace variables (a, x). It is interesting to compare the form of these regions for different values of parameters in the action (1). In the simplest case then $Q = \Lambda = \lambda = \xi = 0$ we have non-compact Euclidean region bounded by hyperbolic curve. Inclusion of the cosmological term $\Lambda \neq 0$ implies the closing of the Euclidean region "on the right". Inclusion of the non-zero classical charge of the scalar field $Q \neq 0$ implies the closing of the Euclidean region "on the left" and we have obtained "banana-like" structure of this region [1,2]. After the inclusion of the small term describing the non-minimal coupling between scalar field and gravity ($\xi \neq 0$) we obtain the second Euclidean region in the upper left corner of our picture. This new region is non-compact and unrestricted from above. While increasing the value of the parameter ξ this second Euclidean region drops down and at some value of ξ joins with the first banana-like Euclidean region. The boundary of this unified region is partially convex, partially concave and after further increasing of ξ it becomes convex. After inclusion of self-interaction of the scalar field $\lambda \neq 0$ we can have, depending on the values of the parameters Q, λ, ξ and m, various geometrical configurations of the Euclidean regions. we have three options. First, we can have only one closed banana-like Euclidean region; second, one can have two non-connected Euclidean regions : banana-like one and "bag-like" Euclidean region with an infinitely long narrow throat; third, one can have one open above bag-like Euclidean region which again has an infinitely long narrow throat. Thus, we have seen that inclusion of the charge Q, non-minimal coupling $\xi \neq 0$ and self-interaction of the scalar field implies a large variety of possible geometries of Euclidean regions in minisuperspace.

Equations of motion for our minisuperspace system are rather cumbersome [4,5] and can be studied numerically. Numerically integrating the Euclidean equations of motion we can investigate the question about the presence of instantons between the solutions of these equations in Euclidean region. Under instantons we shall understand solutions of Euclidean equations of motion which have vanishing velocities on the boundaries of Euclidean region and correspond to extrema of action (and, hence, to peaks in probability distributions for the birth of the inflationary Universe with definite initial values of a and x).

The distinguishing feature of the model with non-minimal coupling consists in the fact that in the case of the one open Euclidean region whose boundary is partially convex partially concave we have two instantons - corresponding to local maximum and local minimum of absolute value of Euclidean action. These instantons correspond to the opportunities to get probability peaks for Hartle-Hawking and tunneling wave functions of the Universe respectively.

In turn, in the model with minimal coupling where we have only one instanton in banana-like Euclidean region one can get probability peak only for Hartle-Hawking wave function [3].

This research was supported by Russian Foundation for Basic Research via grants No 96-02-16220, No 96-02-17591.

References

1. I.M. Khalatnikov and A. Mezhlumian, *Phys. Lett.* A **169**, 308 (1992).
2. I.M. Khalatnikov and P. Schiller, *Phys. Lett.* B **302**, 176 (1993).
3. L. Amendola, I.M. Khalatnikov, M. Litterio and F. Occhionero, *Phys. Rev.* D **49**, 1881 (1994).
4. A.Yu. Kamenshchik, I.M. Khalatnikov and A.V. Toporensky, *Phys. Lett.* B **357**, 36 (1995).
5. A.Yu. Kamenshchik, I.M. Khalatnikov and A.V. Toporensky, *Int. J. Mod. Phys.* D, to appear; I.M. Khalatnikov and A.Yu. Kamenshchik, *Phys. Rep.*, to appear.

DENSITY PERTURBATIONS FROM HYBRID INFLATION

DAVID WANDS

School of Mathematical Studies, University of Portsmouth,
Mercantile House, Hampshire Terrace, Portsmouth, PO1 2EG, United Kingdom

In models of hybrid inflation where the density during inflation is dominated by a constant false-vacuum energy, the spectrum of density perturbations has a positive tilt. In the limit that the energy density becomes constant, the amplitude and tilt of density perturbations can be calculated exactly, without recourse to the usual slow-roll approximation. If the phase transition required to end inflation is slow, it may result in the production of primordial black holes.

Inflation is an alluringly simple mechanism by which to produce a smooth, almost homogeneous universe. Accelerated expansion drives the universe towards an isotropic, spatially flat universe, while quantum fluctuations about the classical backgrounds field can produce the small density perturbations required for structure formation[1]. Despite the simplicity of the basic idea, the range of observable predictions from different models, particularly the perturbation spectra, is wider than is sometimes implied.

The archetypal picture of inflation has become the chaotic inflation scenario proposed by Linde[2] driven by a scalar field slowly rolling down a simple polynomial self-interaction potential, $V(\phi)$. In this model inflation naturally comes to an end when the dimensionless slope, parameterised by $\epsilon \equiv (M_{\rm Pl}^2/16\pi)(V'/V)^2$, becomes large as $V(\phi) \to 0$. While ϵ is small and the field is quasi-massless ($|V''| \ll H^2$) the field perturbations generated during inflation can be calculated using the slow-roll approximation. The curvature perturbation on comoving hypersurfaces, $\zeta = H\delta t = H\delta\phi/\dot{\phi}$, is conserved on super-horizon scales for adiabatic perturbations, and hence we can calculate the density perturbation when modes re-enter the horizon during the matter-dominated era as[1]

$$\delta_H^2 \equiv \left(\frac{\delta\rho}{\rho}\right)_{k=aH}^2 = \left(\frac{2}{5}\zeta\right)^2 \simeq \frac{32}{75\pi\epsilon}\frac{V}{M_{\rm Pl}^4}, \qquad (1)$$

and the spectral index is defined by $n \equiv 1 + (d\ln\delta_H^2/d\ln k)$, where k is the comoving wavenumber. Microwave background anisotropies and also large-scale structure observations imply that $\delta_H^2 \sim 10^{-10}$. For conventional inflation models (e.g., $V(\phi) = m^2\phi^2/2$) the slow-roll calculations imply that when these scales left the horizon during inflation we have $\epsilon \sim 0.1$ (smallish, but not very small) and hence $V \sim (10^{16}\text{GeV})^4$. Because V decreases with time, and ϵ

increases, δ_H becomes smaller as smaller scales leave the horizon and we have a red spectrum of density perturbations with n less than (but quite close to) unity.

This simple picture is dramatically altered if the energy density during inflation is dominated not by the inflaton's own energy density, but by some false-vacuum energy density which does not go to zero as the inflaton rolls to the minimum of its potential. If the potential energy $V = M_F^4 + f(\phi)$ remains almost constant, then the dimensionless slope ϵ can become almost arbitrarily small as the inflaton rolls to its minimum. The slow-roll density perturbations in Eq. (1) then imply $V \simeq M_F^4 \ll (10^{16}\text{GeV})^4$ for $\epsilon \ll 1$. The only constraint on the energy density during inflation comes from the requirement that one can reheat to a sufficiently high temperature to recover the standard hot big bang model after inflation[a]. The low energy scale during inflation has important consequences for building inflation models in realistic particle theories. In the simplest false-vacuum dominated model, where $V = M_F^4 + m^2\phi^2/2$, suitable density perturbations are obtained for scales, $m \sim m_{\text{SUSY}} \sim 1\text{TeV}$ and $M_F \sim \sqrt{M_{\text{Pl}} m_{\text{SUSY}}} \sim 10^{11}\text{GeV}$, which appear in supergravity models. Equally important is the fact that the trajectory, $\Delta\phi$, during the final stages of inflation is much less than the Planck scale in these models[4,7].

The amplitude of density perturbations, given in Eq. (1), increases on small scales as V remains approximately constant, while ϵ decreases during inflation. The tilt of the spectrum, $n - 1 \simeq 2V''/3H^2$, is positive, although for a wide range of parameters it will be indistinguishable from zero[4]. In the limit that $\epsilon \to 0$ we can take $V = M_F^4 + m^2\phi^2/2$ and we we have a massive scalar field in de Sitter spacetime. This is one of a handful of inflation models for which we can calculate the exact spectrum of field fluctuations and the resulting density perturbations without resorting to the usual slow-roll approximation. The exact result[8] for the tilt, valid even for $m^2 \sim H^2$, is $n = 1 + 2r$, where $r = 3/2 - \sqrt{9/4 - m^2/H^2} \to m^2/3H^2$ as $m^2/H^2 \to 0$.

So far I have presented a model of inflation which never ends. The key ingredient in hybrid inflation which allows inflation to end is a phase transition in a second field. The simplest example is provided by the potential

$$V(\phi, \psi) = \frac{\lambda}{4}\left(\psi^2 - M^2\right)^2 + \frac{1}{2}\gamma\phi^2\psi^2 + \frac{1}{2}m^2\phi^2. \qquad (2)$$

The false-vacuum state is $\psi = 0$ where $V = \lambda M^4/4 + m^2\phi^2/2$. The coupling between the inflaton, ϕ, and the second field, ψ, stabilises the false-vacuum

---

[a]Note that if $M_F \ll 10^{16}\text{GeV}$ then the amplitude of gravitational waves produced during inflation is much less than in conventional inflaton-dominated inflation and their contribution to the microwave background anisotropies will be negligible.

for $\phi^2 > \phi_c^2 \equiv \lambda M^2/\gamma$. If λ is of order unity then the transition to the true-vacuum state, where $\psi = M$ and $V = 0$, takes less than a Hubble time and inflation effectively ends once ϕ reaches ϕ_c [3,4].

The first consequence of ending inflation by a phase transition is that one may produce topological defects depending upon the symmetry of the vacuum manifold. Cosmic strings or textures would be important for structure formation or the microwave background limits only if the energy scale were close to the top of its allowed range, $V \sim (10^{16}\text{GeV})^4$ [4]. However monopoles or domain walls might rule out models at lower energy scales too.

Recently Randall $et\ al$ [5] have pointed out that if ψ itself has a very flat potential, typical of moduli fields in supergravity models, then there is the possibility that there may be a significant amount of inflation during the transition itself. This results in a new class of inflating topological defects [6]. Although the defects do not necessarily undergo a large amount of inflation, the density perturbations associated with the cores of the defects could be large, producing a population of primordial black holes [5,6]. It is important to treat this as a full two-field problem where both fields are evolving and the standard single-field results (e.g., adiabatic density perturbations) may not apply [9,6]. While there are severe constraints on the number of black holes with initial mass greater than about 10^6g, an interesting possibility arises if black holes with $m < 10^6$g are formed. Even a small initial number density can soon dominate the energy density of the universe, with or without conventional reheating. The black holes themselves will then evaporate and provide a novel non-equilibrium reheating mechanism [6].

References

1. A.R. Liddle and D.H. Lyth, *Phys. Rep.* **231**, 1 (1993).
2. A.D. Linde, *Phys. Lett.* B **129**, 177 (1983).
3. A.D. Linde, *Phys. Lett.* B **259**, 38 (1991); *Phys. Rev.* D **49**, 748 (1994).
4. E.J. Copeland, A.R. Liddle, D.H. Lyth, E.D. Stewart and D. Wands, *Phys. Rev.* D **49**, 6410 (1994).
5. L. Randall, M. Soljačić and A.H. Guth, *Nucl. Phys.* B **472**, 377 (1996).
6. J. García-Bellido, A. Linde and D. Wands, *Phys. Rev.* D **54**, 6040 (1996).
7. D.H. Lyth, preprint hep-ph/9609431 (1996).
8. J. García-Bellido and D. Wands, *Phys. Rev.* D **53**, 5437 (1996).
9. M. Sasaki and E.D. Stewart, *Prog. Theor. Phys.* **95**, 71 (1996); J. García-Bellido and D. Wands, *Phys. Rev.* D **54**, 7181 (1996).

GENERATION OF GRAVITATIONAL WAVES AND SCALAR PERTURBATIONS IN INFLATION WITH EFFECTIVE Λ-TERM AND T/S STORY

V.N.LUKASH and E.V.MIKHEEVA

Astro Space Center of P.N.Lebedev Physical Institute,
Profsoyuznaya Street 84/32, Moscow, 117810, Russia

We argue that gravitational wave contribution to the cosmic microwave background anisotropy at angular scale $\sim 10^0$ may exceed 50 % for some models of hybrid inflation producing standard cosmology with the density perturbation slope $n \simeq 1$.

1 Introduction

Two types of metric perturbations $h_{\alpha\beta}$ generating at inflation driven by scalar field, can contribute to large-scale cosmic microwave background anisotropy through the SW effect[1]. They are the Scalar (density) and Tensor (gravitational waves) perturbations which are presented as follows: $h_{\alpha\beta}/2 = A\delta_{\alpha\beta} + B_{,\alpha\beta} + G_{\alpha\beta}$. The first and second terms correspond to the scalar mode whereas the third one is for the tensor one ($G_\alpha^\alpha = G_{\alpha,\beta}^\beta = 0$). Their contribution into $\Delta T/T$ may be standardly separated over S and T parts:

$$\left\langle \left(\frac{\Delta T}{T}\right)^2 \right\rangle = \sum_\ell \langle a_\ell^2 \rangle = \sum_\ell \left(\langle a_\ell^2 \rangle_S + \langle a_\ell^2 \rangle_T \right) = S + T. \tag{1}$$

The available experimental data are not sufficient yet to recognize and find gravitational wave fraction in $\Delta T/T$ (that could be possible, in principle, by means of a joint analysis of the dependence of the temperature fluctuation spectrum on scale, the amplitude of Dopper peak, etc., which is the matter of future observations). Nevertheless, today we can evaluate T/S theoretically for various inflationary models to test the possible effect and see how large it may deviate from model to model and which particular properties of the model correlate with large T/S. Actually, we construct a counterexample to the common prejudice that *"T/S is negligible for the Harrison-Zel'dovich spectrum"*, introducing here a general class of models which face the opposite conclusion: *"T/S may be about or even larger than 1 for $n \simeq 1$".*

2 Generation of cosmological perturbations in inflationary model with effective Λ-term

A lot of inflation models has been discussed in the literature[2] for T/S. The result was that large T/S could be achieved at the expense of the rejection from the Harrison-Zeldovich spectrum: $T/S \geq 1$ for $n_S \leq 0.8$, the "red" spectra. Thus, it would be interesting to study possible T/S in models producing "blue" ($n_S > 1$) spectra of density perturbations. Here we present a general class of models based on the only scalar field with the potential:

$$V = V_0 + \frac{m^2 \varphi^2}{2}, \tag{2}$$

where V_0 and m are constants, V_0 describe the effective (metastable) Λ-term. The mechanism of its decay is not fixed here and may be arbitrary, for example, with help of another scalar field like in the hybrid inflation[3].

It is clear that the inflaton dynamics in the potential (2) has two regims and, consequently, the spectra of cosmological perturbations has two asymptotics. The regims are separated by the critical value of field φ_{cr} at which the first term is equal to the second $\varphi_{cr}^2 = 2V_0/m^2$.

At the first stage (we call it "red" asymptotic — by the form of the density perturbation spectrum) φ is large and the massive term dominantes the evolution. This is similar to the case of well-known chaotic inflation, so it is easy to write down the spectra of the perturbations[4,5] as follows ($c = \hbar = 8\pi G = 1$):

$$q_k^r = \frac{m\varphi^2}{4\sqrt{6}\pi}, \quad h_k^r = \frac{m\varphi}{2\sqrt{3}\pi}, \tag{3}$$

For $\varphi < \varphi_{cr}$ the constant term predominates. Here we obtain the "blue" asymptotic spectrum of density perturbations:

$$q_k^b = \frac{H_0 c_n^2 \varphi_{cr}^2}{4\pi\varphi}, \quad h_k^b = \frac{H_0}{\sqrt{2}\pi}, \tag{4}$$

where $H_0 = (V_0/3)^{1/2}$, $c_n^2 = 2^{1/2-n_S/2}(7 - n_S)\Gamma(2 - n_S/2)(3\sqrt{\pi})^{-1}$, $\Gamma(x)$ is gamma function, q coincides with A in comoving frame and h_k^2 is the spectrum of $G_{\alpha\beta}G^{\alpha\beta}$. φ in (3), (4) is taken at horizon crossing $k = a(\varphi)H(\varphi)$, a and $H = (V/3)^{1/2}$ are scale and Hubble factors, respectively. Note that we have the following relation between the spectrum slope at the "blue" asymptotic n_S and φ_{cr}: $(n_S - 1)(7 - n_S) = 24\varphi_{cr}^{-2}$, $\varphi_{cr}^2 \geq 8/3$, $1 < n_S < 4$.

As for gravitational waves their spectrum remains universal for any k, but the density perturbation spectrum requires approximation (fitting both

388

asymptotics and becoming exact in the slow-roll limit):

$$q_k = \frac{H_0\varphi_{cr}}{4\pi y}(1+y^2)^{1/2}(c_n^2+y^2), \quad h_k = \frac{H_0}{\sqrt{2}\pi}(1+y^2)^{1/2}, \tag{5}$$

where $y = \varphi/\varphi_{cr}$. It is easy to see that the ratio of tensor to scalar spectra achieves its maximum at $\varphi \simeq \varphi_{cr}$, where the minimum of density perturbation power occurs:

$$\frac{h_k}{q_k} = \frac{2\sqrt{2}}{\varphi_{cr}}\frac{y}{c_n^2+y^2} \leq \frac{\sqrt{2}}{c_n\varphi_{cr}} = D_n, \tag{6}$$

where $D_2 = 0.8$, $D_3 = 1$. We may suppose that here T/S gets its maximum as well.

T/S was calculated for potential (2) with DMR COBE window function. We obtained that T/S function is a two-parametric one and depends on n_S and k_{cr}, the latter corresponds to φ_{cr}. However, the earlier introduced approximation formula[2], $T/S \simeq -6n_T$, holds in our case as well if both T/S and local $|n_T|$ are taken at their maxima. We found a phenomenological relation between n_T and n_S: $|n_T| \simeq (n_S - 1)/4$, thus, another formula for T/S can be proposed for this model: $T/S|_{max} \simeq 1.5(n_S - 1)$.

3 Conclusions

Finally, we conclude: The inflation model predicting $T/S > 1$ has been constructed. A property of this model is "blue" spectrum of density perturbations for scales $k \gg k_{cr}$. However in the region, where $T/S \geq 1$, $k \sim k_{cr}$ and therefore the locally observed spectrum of cosmological perturbations is close to scale-invariant ($n \simeq 1$).

The work was supported partly by the Russian Foundation for Fundamental Research (project code 96-02-16689-a) and COSMION (cosmomicrophysics).

References

1. R.K. Sachs and A.M. Wolfe, *ApJ* **147**, 73 (1967).
2. J.E.Lidsey *et al.*, astro-ph/9508078.
3. A. Linde, *Phys. Rev.* D **49**, 748 (1994).
4. L.F. Abbott and M.B. Wise, *Nucl. Phys.* B **244**, 279 (1984).
5. V.N. Lukash, in *Cosmology and Gravitation II*, ed. M.Novello (Editions Frontieres, 1996).

PREHEATING IN FRW SPACETIMES

D. BOYANOVSKY[a], D. CORMIER[b], H.J. DE VEGA[c] R. HOLMAN[b], A. SINGH[d] and M. SREDNICKI[d]

(a) Department of Physics and Astronomy, University of Pittsburgh, Pittsburgh, PA. 15260, U.S.A.

(b) Department of Physics, Carnegie Mellon University, Pittsburgh, PA. 15213, U. S. A.

(c) Laboratoire de Physique Théorique et Hautes Energies Université Pierre et Marie Curie (Paris VI)

We study the time evolution of quantum fields in the $O(N)$ model with an eye towards understanding the phenomenon of *preheating*, that is, particle production due to spinodal instabilities and/or parametric amplification in the early universe. We find that the particle production as well as the late time behavior of the zero mode are governed by the existence of Goldstone modes. Similar effects occur in the $N = 1$ model in the Hartree approximation.

1 Introduction

Treating a field theory as an initial value problem rather than a boundary value one, such as is done when computing S-matrix elements, can give rise to startling new phenomena. In particular copious and explosive particle production can occur during the evolution of the zero mode of a quantum field.

This is both a non-perturbative as well as a non-equilibrium phenomenon; it is nonperturbative due to the fact that an initial energy density of order m^4/λ, where m is a mass scale in the theory and λ is a quartic coupling, is redistributed into $\mathcal{O}(1/\lambda)$ quanta. It is clearly non-equilibrium since it is the growth of quantum fluctuations as a function of time that drives the particle production.

In this talk we describe some work we have done on this phenomenon of preheating including the effects of backreaction of the quantum fluctuations produced as the zero mode of the field evolves [2].

2 The Non-Equilibrium $O(N)$ Model in an FRW Spacetime

We use the closed time path (CTP) formalism, adapted to the $O(N)$ vector model to do our calculations [1,3]. The $O(N)$ model is well suited to the task at hand since the leading term in $1/N$ already contains an infinite resummation of Feynman diagrams. This then allows for a *controllable* nonperturbative approximation to the theory.

The relevant equations found this way describe the evolution of the zero mode η of the σ component of the $O(N)$ multiplet as well as that of the $N-1$ "pion" fields. In the large N limit it is the fluctuations of these fields that is important.

Given these equations we can solve them for some cosmologically relevant situations. We will consider the case of a radiation dominated universe and treat both slow-roll initial conditions for the zero mode η as well as chaotic initial conditions.

2.1 Slow-Roll Initial Conditions

In this case, we take $\eta(\tau_0) \simeq 0$. Because the second derivative of the potential is negative at the origin, some of the modes display spinodal instabilities which drive the production of particles as the zero mode evolves.

For a small initial Hubble parameter, compared to the renormalized mass, the zero mode tends to follow the results found in Minkowski space [1]. It makes one excursion out to nearly the classical turning point and then tries to return to its initial value. However, by this time, the expansion of the universe has redshifted enough energy from the zero mode that it cannot come back to the origin and begins to execute damped oscillations about the tree-level minimum at $\eta = 1$. As it makes that excursion, the quantum fluctuatiuons grow as does the number of Goldstone particles produced, though at late times the fluctuations tend to zero. The distribution of Goldstone modes as a function of comoving wave number k is typically far removed from a Bose-Einstein distribution.

2.2 Chaotic Initial Conditions

In this case, the zero mode samples both minima for a period of time until enough energy is lost both due to redshift as well as to particle production to restrict it to small oscillations about one of the tree level minima, in this case the one at $\eta = 1$.

We find that the effective mass of the zero mode, quickly settles down to a value near zero as expected from Goldstone's theorem.

Changing the initial value of $\eta(\tau_0)$ will change the number of times both ground states are sampled. If the initial Hubble parameter is increased, for fixed $\eta(\tau_0)$, the energy of the zero mode is redshifted away faster, so that fewer large amplitude oscillations are executed and fewer particles per *physical* volume are produced.

3 Conclusions

We have generalized our previous results on the quantum dynamics of the $O(N)$ model in Minkowski space to FRW universes. In particular, we find that preheating can still occur in an expanding universe, though the actual amount of particle production depends sensitively on the size of the initial Hubble parameter relative to the mass of the field.

We also showed that the late time dynamics is determined by a sum rule enforcing Goldstone's theorem. This drives the fluctuations to zero, allowing the zero mode to find one of the tree level minima. This indicates that the symmetry is broken at late times[4]. We see no evidence of symmetry restoration.

Acknowledgments

D.B. thanks the N.S.F. for support under grant awards: PHY-9302534 and INT-9216755. R. H. and D. C. were supported by DOE grant DE-FG02-91-ER40682. A.S and M.S were supported by the N.S.F under grant PHY-91-16964.

References

1. D. Boyanovsky, H. J. de Vega, R. Holman, D.-S. Lee and A. Singh, Phys. Rev. **D51**, 4419 (1995);D. Boyanovsky, M. D'Attanasio, H. J. de Vega, R. Holman and D. S. Lee, Phys. Rev. **D52**, 6805 (1995); D. Boyanovsky, M. D'Attanasio, H. J. de Vega and R. Holman, Phys. Rev. **D52**, 6809 (1996), and references therein.
2. D. Boyanovsky, D. Cormier, H. J. de Vega, R. Holman, A. Singh, M. Srednicki, "Preheating in Friedmann-Robertson-Walker Universes", submitted to Physical Review Letters, (1996) hep-ph/9609527.
3. D. Boyanovsky, H.J. de Vega, R. Holman, J.F.J. Salgado, hep-ph/9608205, Phys. Rev. D. **54** 7570 (1996).
4. D. Boyanovsky, H.J. de Vega, R. Holman and J. F. J. Salgado, astro-ph/9609007, to appear in the Proceedings of the Paris Euronetwork Meeting 'String Gravity'.

A STUDY OF THE EFFECTS OF PREHEATING

I. TKACHEV

Department of Physics, The Ohio State University, Columbus, OH 43210
and
Institute for Nuclear Research of the Academy of Sciences of Russia
Moscow 117312, Russia

We review results of a numerical study of resonant inflaton decay in a wide range of realistic models and parameters. Wide enough parametric resonance can withstand the expansion of the Universe, though account for the expansion is very important for determining precisely how wide it should be. For example, the effective production of particles with mass ten times that of the inflaton requires very large values of the resonance parameter q, $q > 10^8$. For these large q, the maximal size of produced fluctuations is significantly suppressed by back reaction, $\langle X^2 \rangle \sim 10^{-10} M_{\text{Pl}}^2$. We discuss some physical implications of our results.

Recently the scenario of inflaton decay and reheating of the Universe after inflation has been revised considerably, starting from the observation [1] that this decay can naturally be in a regime of a wide parametric resonance. Wide resonance can be effective despite the expansion of the Universe. During this explosive process (called preheating [1]) particles heavier than inflaton can be created. This opens possibilities for many interesting and important effects to occur, including non-equilibrium phase transitions [2] and generation [3] of baryon asymmetry in old GUT frameworks (provided GUT is B-L non-conserving).

Particle content of the theory assumed to be as follows. First, there is an inflaton field which we denote as ϕ. Second, there are products of inflaton decay. We denote them as X and the mass of corresponding quanta is m_X. The possibility of the inflaton decay into X-quanta assumes that there is interaction of, say, the form $g^2 X^2 \phi^2/2$. In reality, there can be many channels for the inflaton decay and the final answer is the sum over all species. Third, the model allows for the possibility of symmetry breaking with an order parameter Φ. The X-particles couple to the order parameter, so their mass depends upon it, $m_X = m_X(\Phi)$. The part of the total potential (we break it into parts as $V = V_1 + V_2 + \ldots$) relevant for the symmetry breaking is $V_1 = -\frac{1}{2}\mu^2\Phi^2 + \frac{1}{2}\alpha X^2\Phi^2$.

If X-particles are abundantly created, the effective mass of Φ-filed can be written as $m_{\text{eff}}^2 = -\mu^2 + \alpha\langle X^2 \rangle$. The symmetry is restored if $m_{\text{eff}}^2 > 0$, or $\langle X^2 \rangle > \mu^2/\alpha$. In thermal equilibrium with temperature T one would have $\langle X^2 \rangle = T^2/12$. The main observation [2] is that right after the inflaton decay but long before equilibrium is established, $\langle X^2 \rangle$ is anomalously large. The strength of symmetry restoration in this non-equilibrium state can exceed that one in thermal equilibrium by many orders of magnitude.

Let us illustrate this point neglecting the expansion of the Universe. Energy density conserves in this case and energy density stored in X-filed after decay is equal to energy density in initial inflaton oscillations, $\rho \sim m_i^2 M_{\rm Pl}^2$. Typical energy of X-quanta is of order of the inflaton mass, $E \sim m_i$. We find $\langle X^2 \rangle \sim \rho/E^2 \sim M_{\rm Pl}^2$, i.e typical scale for the strength of symmetry restoration is Plankian. This estimate is encouraging but oversimplified. First, the expansion is very important: while parametric resonance develops, fields are red-shifted despite the explosive character of the process. Second, the inflaton not always decays completely during the resonance stage, and due to various back-reaction effects the fast decay can stop at much smaller values of $\langle X^2 \rangle$. The resulting picture is very model dependent, and due to complexity requires numerical study, model by model.

Number density of created particles can be easily determined if $\langle X^2 \rangle$ is known. For massive X, which are created mostly non-relativistic, we have $n_X = \langle X^2 \rangle/m_X$. Suppose the baryon number and CP are violated in decays of X particles (i.e. X are GUT leptoquarks) and they are created in sufficient number during parametric resonance. This opens a possibility of BAU generation at preheating [3,2].

Clearly, the quantity of interest is the maximum value of $\langle X^2 \rangle$ which can be achieved in inflaton decay. Here I review some results for different models.

1. The simplest one is pure $\lambda \phi^4$ model where the inflaton decays due to self-interaction. Fully non-linear calculation of the decay, which includes all rescattering and back-reaction processes [4], gives $\langle (\phi - \phi_0)^2 \rangle_{\rm max} = 10^{-7} M_{\rm Pl}^2$.

2. Massless inflaton decaying into X-particles, $V_2 = \frac{1}{4}\lambda\phi^4 + \frac{1}{2}g^2 X^2\phi^2$. Decay is very inefficient if X are massive [5]. In a conformally invariant case, typically [6] $\langle X^2 \rangle_{\rm max} \approx 10^{-7}(100/q)^{3/2} M_{\rm Pl}^2$ for $q \gtrsim 100$, where $q \equiv g^2/4\lambda$. Rescattering effects suppress $\langle X^2 \rangle$ as an inverse power of q [5,6,7].

3. The most complicated and efficient is the case with massive inflaton, $V_2 = \frac{1}{2}m_i^2\phi^2 + \frac{1}{2}g^2 X^2\phi^2$. Inflaton can decay even if X is considerably heavier than inflaton itself. A good idea of how $\langle X^2 \rangle_{\rm max}$ depends upon mass $m_\chi \equiv m_X/m_i$ and q (where now $q \equiv g^2\phi^2(0)/4m_i^2$) in an expanding universe can be obtained in the Hartree approximation. It is plotted in Fig. 1 by solid lines [5]. We see that to have resonance at all in an expanding universe, even for massless X-particles, one needs $q \gtrsim 10^3$. Required q rapidly grow with m_χ. E.g., to create X-particles sufficiently heavy to be promising for baryogenesis, $m_\chi \approx 10$, one already needs $q \sim 10^8$. For creation of massless particles expansion becomes unimportant and $\langle X^2 \rangle_{\rm max}$ becomes saturated by back-reaction if $q \gtrsim 10^4$. The Hartree approximation grossly overestimates $\langle X^2 \rangle_{\rm max}$ in this regime (however, this approximation can be good for X being a multicomponent field,

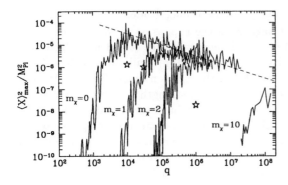

Figure 1: Variance $\langle X^2 \rangle$ as a function of q.

such as a large multiplet of $O(N)$ model). Results of fully non-linear lattice calculations [6], which include all effects but are more expensive, are presented in Fig. 1 by stars. Data points are presented for massless X at the moment when zero mode of inflaton oscillations completely decayed. Extrapolating these results to $q \approx 10^8$ we anticipate [6] $\langle X^2 \rangle_{\text{max}} \approx 10^{-10} M_{\text{Pl}}^2$.

Fig. 1 is a starting point for calculations of BAU and symmetry restoration effects at and after preheating.

Acknowledgments

This work was supported by DOE Grant DE-AC02-76ER01545 at Ohio State.

References

1. L. Kofman, A. Linde, A. Starobinsky, Phys. Rev. Lett. **73**, 3195 (1994).
2. L. Kofman, A. Linde, A. Starobinsky, Phys. Rev. Lett. **76**, 1011 (1996); I. I. Tkachev, Phys. Lett. **B376**, 35 (1996).
3. E. W. Kolb, A. D. Linde, A. Riotto, Phys. Rev. Lett. **77**, 4290 (1996); M. Yoshimura, hep-ph/9605246.
4. S. Khlebnikov and I. Tkachev, Phys. Rev. Lett. **77**, 219 (1996).
5. S. Khlebnikov and I. Tkachev, hep-ph/9608458.
6. S. Khlebnikov and I. Tkachev, hep-ph/9610477.
7. T. Prokopec and T. G. Roos, hep-ph/9610400.

WHAT IS FOUND UPON DEFROSTING THE UNIVERSE AFTER INFLATION [a]

A. RIOTTO

NASA/Fermilab Astrophysics Center,
Fermilab National Accelerator Laboratory, Batavia, Illinois 60510-0500, USA

At the end of inflation the universe is frozen in a near zero-entropy state with energy density in a coherent scalar field and must be "defrosted" to produce the observed entropy and baryon number. Baryon asymmetry may be generated by the decay of supermassive Grand Unified Theory (GUT) bosons produced non-thermally in a preheating phase after inflation, thus solving many drawbacks facing GUT baryogenesis in the old reheating scenario.

1 Prologo

Before going into details, let me explain the origin of the term "defrost" and its derivatives [1]. After inflation the universe appears slightly boring: no particles around, zero-entropy density, no thermal bath and all the energy density stored in the inflaton scalar field. It is clear that the universe must undergo a sort of phase transition from this state to produce the observed entropy and a thermal bath of particles. This process was denominated "defrosting" by R. Kolb. However, our collaborator A. Linde felt rather uncomfortable with this denomination. He wrote (quote):"... It is a cool idea and I am sure that Rocky will use this word in one of his brilliant talks. However, I really have severe problems with it. First of all, I associate it with the frozen chicken breasts, which I cannot stand...". In spite of Andrei's opinion, I decided to adopt the term "defrosting" anyway when Rocky dropped in my office and, with a very strong Chicago-Mafia accent, asked me how would I have translated the expression *"I break your legs"* in italian [b].

2 GUT Baryogenesis at Preheating

In models of slow-roll inflation, the universe is dominated by the potential energy density of a scalar field known as the *inflaton*. Inflation ends when the kinetic energy density of the inflaton becomes larger than its potential energy density. At this point the universe might be said to be frozen: any initial entropy in the universe was inflated away, and the only energy was in

---

[a]Talk given at the 18th *Texas Symposium on Relativisitc Astrophysics*, December 15-20, Chicago, Illinois.

[b]That sounds something like *" Ti spezzo le gambe"* in italian.

cold, coherent motions of the inflaton field. Somehow this frozen state must be transformed to a high-entropy hot universe by transferring energy from the inflaton field to radiation.

In the simple chaotic inflation model the potential is assumed to be $V(\phi) = M_\phi^2 \phi^2 / 2$, with $M_\phi \sim 10^{13}$GeV in order to reproduce the observed temperature anisotropies in the microwave background. In the old reheating (defrosting) scenario, the inflaton field ϕ is assumed to oscillate coherently about the minimum of the inflaton potential until the age of the universe is equal to the lifetime of the inflaton. Then the inflaton decays, and the decay products thermalize to a temperature $T_F \simeq 10^{-1}\sqrt{\Gamma_\phi M_\mathrm{P}}$, where Γ_ϕ is the inflaton decay width, and $M_\mathrm{P} \sim 10^{19}$ GeV is the Planck mass.

In supergravity-inspired scenarios, gravitinos have a mass of order a TeV and a decay lifetime on the order of 10^5s. If gravitinos are overproduced after inflation and decay after the epoch of nucleosynthesis, they would modify the successful predictions of big-bang nucleosynthesis. This can be avoided if the temperature T_F is smaller than about 10^{11} GeV (or even less, depending on the gravitino mass).

In addition to entropy, the baryon asymmetry must be created after inflation. There are serious obstacles facing any attempt to generate a baryon asymmetry in an inflationary universe through the decay of baryon number (B) violating bosons of Grand Unified Theories. The most tedious problem is the low value of T_F in the old scenario. Since the unification scale is expected to be of order 10^{16}GeV, B violating gauge and Higgs bosons (referred to generically as "X" bosons) probably have masses greater than M_ϕ, and it would be kinematically impossible to produce them directly in ϕ decay or by scatterings in a thermal environment at temperature T_F.

However, reheating may differ significantly from the above simple picture[2]. In the first stage of reheating, which was called "preheating" effective dissipational dynamics and explosive particle production even when single particle decay is kinematically forbidden. A crucial observation for baryogenesis is that even particles with mass *larger* than the inflaton mass, $M_X \sim 10\, M_\phi$, may be produced during preheating[2] by coherent effects provided that a coupling to the inflaton field of the type $|X|^2 \phi^2$ is present. A fully non-linear calculation of the amplitude of perturbations $\langle X^2 \rangle$ at the end of the broad resonance regime has been recently done[3] revealing that $\langle X^2 \rangle$ may be as large as $10^{-10}\, M_\mathrm{P}^2$. Since the value of the inflaton field ϕ_i at the beginning of preheating is of order of $10^{-2}\, M_\mathrm{P}$, one may assume that the first step in reheating is to convert a fraction $\delta \sim 10^{-4}$ of the inflaton energy density into a background of baryon-number violating X bosons. They can be produced even if the reheating temperature to be established at the subsequent stages of reheating is much smaller than M_X. Here we see a significant departure from the old scenario. In the old picture production of X bosons was kinematically forbidden if $M_\phi < M_X$, while in the new scenario it is possible as the result of coherent effects. The particles are produced out-of-equilibrium, thus satisfying one of the basic requirements to produce the baryon asymmetry. The next step in reheating is the decay of the X bosons. We assumed that the X decay products rapidly thermalize [1]. It is only after this point that it is possible to speak of the temperature of the universe. Moreover, we assumed that decay of an X-$\overline{X}$ pair produces a net baryon number B/ϵ (where ϵ is the CP-violating factor), as well as entropy [1]. We have numerically integrated the Boltzmann equations describing the temporal evolution of the baryon number, of the energy density of X-particles and of the inflaton field. Since the number of X bosons produced is proportional to δ, the final asymmetry is proportional to δ and we have noted [1] that $B/\epsilon \sim 10^{-9}$ can be obtained for δ as small as 10^{-6}.

Our scenario is based on several assumptions about the structure of the theory, but the feeling is that baryon number generation may be relatively efficient. Within uncertainties of model parameters, the value of ϵ, etc., the present $B \sim 10^{-10}$ may arise from GUT baryogenesis after preheating. Of course, additional work is needed to implement the ideas discussed above in the context of a more realistic model and a complete numerical analysis able to decribe the dynamics from the end of inflation down to the final baryon asymmetry production through the preheating era is urgently called for [4].

Acknowledgments

I would like to thank my collaborators A. Linde and R. Kolb from whom I am still learning so much. I would also like to thank Eleonora Riotto for entertaining discussions. This work is supported by the DOE and NASA under Grant NAG5-2788.

1. *GUT baryogenesis after preheating*, E.W. Kolb, A. Linde and A. Riotto, *Phys. Rev. Lett.* **77**, 4290 (1996).
2. L. A. Kofman, A. D. Linde and A. A. Starobinsky, *Phys. Rev. Lett.* **73**, 3195 (1994).
3. S. Yu. Khlebnikov and I.I. Tkachev, hep-ph/9610477.
4. E.W. Kolb, A. Riotto and I.I. Tkachev, in preparation.

THREE SCALE MODEL OF COSMIC STRING EVOLUTION

E.J. COPELAND

School of Chemistry, Physics and Environmental Sciences
University of Sussex,
Brighton, BN1 9QH, England

An analytical approach to describe the evolution of a network of cosmic strings is described. Including gravitational back reaction terms leads to new scaling solutions which differ from those previously obtained in numerical simulations. They have interesting cosmological implications.

Cosmic strings are particle physics candidates which could explain the observed large scale structure in the Universe, due to the fact that they have a non-trivial coupling to gravity – $G\mu \sim 10^{-6}$, where $\mu \sim \eta^2 \sim 10^{22}$ gcm^{-1} is the mass/length. Formed in a GUT phase transition the network evolves, long strings intercommute forming loops which then lose their energy through gravitational radiation, and the network reaches a scaling solution in which the characteristic length scales increase in proportion to the horizon distance (for a review see [1]). In this talk I will describe an analytical approach to understanding this evolution based on work published in [2,3].

Consider a network of long strings and loops. On a random segment of string choose a particular length scale l and consider the probability distribution for the end-to-end distance (or *extension*) $\mathbf{r}$: $p[\mathbf{r}(l)]d^3\mathbf{r}$ is the probability that a randomly chosen segment of length l will have extension $\mathbf{r}$ within the small volume $d^3\mathbf{r}$. A rate equation for $p[\mathbf{r}(l)]$, can be written where the terms represent the effects of stretching (due to the universal expansion), of gravitational radiation (back-reaction), of long-string intercommuting and of loop production. The goal is straightforward, to establish the nature of the solutions. For all except the smallest values of l, it is reasonable to assume that p is a Gaussian,

$$p[\mathbf{r}(l)] = \left(\frac{3}{2\pi K(l)}\right)^{3/2} \exp\left(-\frac{3}{2}\frac{\mathbf{r}^2}{K(l)}\right), \tag{1}$$

where $K(l)$ is the mean square extension, $K(l) = \overline{\mathbf{r}^2} = \int d^3\mathbf{r}\, \mathbf{r}^2 p[\mathbf{r}(l)]$.

The leading terms for very large and very small l may be written in the form

$$K(l,t) \approx 2\bar{\xi}(t)l, \qquad l \gg t, \tag{2}$$

$$\approx l^2 - \frac{l^3}{3\zeta(t)}, \qquad l \ll t. \tag{3}$$

It is also convenient to introduce the characteristic inter-string distance ξ defined by $\xi^2 = \frac{V}{L}$, where a length L of string exists within a comoving volume V. It is easiest to analyse the rate equations in terms of the dimensionless parameters $\gamma, \bar{\gamma}$ and ϵ defined by $\gamma = 1/H\xi$, $\bar{\gamma} = 1/H\bar{\xi}$, $\epsilon = 1/H\zeta$, where $H = \dot{R}/R$ is the Hubble parameter; $R \propto t^{1/p}$ is the scale factor in a flat FRW background, with $p = 2$ in the radiation era and $p = \frac{3}{2}$ in the matter era. The rate equations are given in [2,3], and I will briefly discuss the results below. Neglecting gravitational backreaction terms there exists a *transient* scaling regime – the one accessible to the numerical simulations. In [3] we find, that in the radiation (RD) and matter (MD) dominated regimes,

$$5.8 < \gamma_{\text{transient,RD}} < 6.5, \qquad 6.5 < \bar{\gamma}_{\text{transient,RD}} < 11,$$
$$2.1 < \gamma_{\text{transient,MD}} < 2.8, \qquad 1.7 < \bar{\gamma}_{\text{transient,MD}} < 3.0, \qquad (4)$$

which compares well with the numerical simulations of [4], $\gamma_{\text{transient,RD}} = 7.2 \pm 1.4$, $\gamma_{\text{transient,MD}} = 2.8 \pm 0.7$.

When the backreaction terms are included a *full* scaling regime is reached where $\epsilon \gg \gamma_{\text{full}} \sim \bar{\gamma}_{\text{full}}$. Briefly we find

$$\gamma_{\text{full,RD}} = 3 \text{ to } 6, \qquad \bar{\gamma}_{\text{full,RD}} = 2 \text{ to } 10,$$
$$\gamma_{\text{full,MD}} = 1.4 \text{ to } 2.6, \qquad \bar{\gamma}_{\text{full,MD}} = 0.5 \text{ to } 2.7, \qquad (5)$$

with $\epsilon_{\text{full}} \sim O(10^4)$ at full scaling. A second intriguing regime leads to effectively just one scale $\gamma \sim \bar{\gamma} \sim \epsilon \sim O(1)$. It is seen to occur when the chopping efficiency is high and will be discussed in detail elsewhere.

Scaling solutions are all well and good, but what are there implications for cosmology? For example, how far away is a cosmic string? What is most important here is the scaling value of γ in the recent, matter-dominated epoch. With $H_0 = 2/3t = 100h$ km s^{-1}Mpc^{-1} $(0.5 < h < 1)$, the mean inter string distance today would be $(\gamma_{\text{full,MD}} H_0)^{-1} = (1200 \text{ to } 2000)h^{-1}$ Mpc, with the nearest string to us being a distance say half that, *i.e.* in the range $z = 0.2$ to 0.5. The mean distance between loops can then be obtained, the present distance being dependent on $\gamma, \bar{\gamma}$ is $\hat{n}^{-1/3} \simeq 200h^{-1}$ Mpc. At the present time, the mean length of a loop is $\bar{l} \sim 200h^{-1}$ kpc with a corresponding mass $M_{\text{loop}} = \mu\bar{l} \sim 4 \times 10^{12}M_{\odot}$, i.e. surviving loops are *significant* objects gravitationally.

The spectrum of gravitational waves emitted by cosmic strings is constrained by the nucleosynthesis limit on relativistic particle species, as well as the pulsar timing on a stochastic gravitational wave background. At the time of nucleosynthesis Ω_{gr} can be constrained by using the bound on light neutrino species, $\Omega_{\mathrm{gr}}(t_{\mathrm{nuc}}) \leq 0.02$. This in turn constrains $G\mu$:

$$\mu_6[A_{\mathrm{RD}}(\kappa)\nu + \nu_{\mathrm{ls}}] < 15, \tag{6}$$

where $A_{\mathrm{RD}}(\kappa) = \frac{2}{3}\frac{\kappa^{3/2}-1}{\kappa-1}$, $\nu \sim 0.7c\bar{\gamma}\gamma^2/p^3$ and $2 < \kappa < 10$. The bound is not at all restrictive for any values of ν and κ in the expected range.

The pulsar timing provides an even tighter bound on the amount of allowed gravitational radiation from strings, with [3]

$$\mu_6 < 0.2\sqrt{\frac{\kappa-1}{\kappa}}\left(\frac{10}{\kappa}\right)\left(\frac{0.1}{\nu}\right)^{3/2}\frac{\Gamma_2^{1/2}}{h^{7/2}}, \tag{7}$$

$\Gamma_2 \sim 100$. Note the strong dependence on ν, κ and h. For $h = 0.5$ the condition is not particularly restrictive, though it puts bounds on ν and κ.

There are two other obvious consequences that should be mentioned, although I have not got time to go into them. The primordial density fluctuations which lead to the generation of anisotropies in the microwave background radiation and the large scale structure in the Universe all depend on the value of the scaling solutions. Since I have presented new solutions these will effect the normalisation of the model when these processes are calculated. Details are presented in [3].

In this talk, I have tried to motivate the need for a complementary analytical approach to the vast amount of work on cosmic strings which are numerical in origin. The biggest problem is to find a way of estimating the parameters which determine the effect on the small scale length of the gravitational back-reaction and loop formation, respectively.

Acknowledgements

I am very grateful to my collaborators Tom Kibble and Daren Austin for their support.

References

1. A. Vilenkin and E.P.S. Shellard,*Cosmic Strings and other Topological Defects*, (Cambridge: Cambridge University Press, 1994).
2. D. Austin, E.J. Copeland and T.W.B. Kibble, *Phys. Rev.* D **48**, 5594 (1993).
3. D. Austin, E.J. Copeland and T.W.B. Kibble, *Phys. Rev.* D **51**, R2499 (1995).
4. D.P. Bennett and F.R. Bouchet, *Phys. Rev.* D **41**, 720 (1990).

DEFECTS IN STRING COSMOLOGY

Ruth GREGORY
Centre for Particle Theory, University of Durham,
South Road, Durham, DH1 3LE, UK

I present results concerning the metric of an isolated self-gravitating abelian-Higgs vortex in dilatonic gravity for arbitrary coupling of the vortex fields to the dilaton and include both massive and massless dilatons.

In this talk I would like to discuss the impact of superstring gravity on defect theories of structure formation, specifically cosmic strings[1]. Many of the constraints on the cosmic string scenario of galaxy formation assume Einstein gravity, however, it seems likely that gravity is modified at sufficiently high energy scales, and the most promising alternative seems to be that offered by string theory, where the gravity becomes scalar-tensor in nature. It is therefore interesting to question the impact of introducing a dilaton.

Calculations involving cosmic string networks generally make use of a 'worldsheet-approximation' in which the string is treated as an infinitesimally thin source which moves according to, and has an energy momentum tensor appropriate for a two-dimensional worldsheet governed by the Nambu action. Although this has only been proved in the absence of gravity[2], the fact that the self-gravitating infinite local vortex has a relatively small effect on spacetime lends credence to the worldsheet approximation for the string. In the presence of a dilaton, the worldsheet approximation may no longer be appropriate. If the dilaton is massless, there is no reason to expect that the string will not have a long range effect on the dilaton, and even if the dilaton is massive, it introduces an additional length scale which may still have significant impact. Indeed, it has recently been argued[3] that a TeV mass dilaton is incompatible with a GUT scale cosmic string network.

In this talk, I summarize work[4] aimed towards resolving this issue by examining the gravi-dilaton field of a self-gravitating cosmic string in dilaton gravity. A reasonably general form for the interaction with the dilaton was considered, assuming that the abelian-Higgs lagrangian coupled to the dilaton via an arbitrary coupling, $e^{2a\phi}\mathcal{L}$, in the string frame; both massive and massless dilatons were considered. The overall action in the string frame is therefore:

$$\hat{S} = \int d^4x \sqrt{-\hat{g}} \left[e^{-2\phi} \left(-\hat{R} - 4(\hat{\nabla}\phi)^2 - \hat{V}(\phi) \right) + e^{2a\phi}\mathcal{L} \right] \qquad (1)$$

For convenience, and comparison with the Einstein gravity analysis[5], the analysis was in fact performed in the Einstein frame, which is related to the string frame via a conformal transformation: $g_{ab} = e^{-2\phi}\hat{g}_{ab}$. We looked for static solutions representing isolated straight cosmic strings i.e. a cylindrically symmetric solution with an additional boost symmetry along the vortex, the metric, dilaton and vortex fields all depend only on the radial variable. We used a perturbative argument to derive the solution, taking the flat space vortex as a zeroth approximation, deriving the gravitational correction from the full equations of motion to order $\epsilon = \eta^2/2$, where η is the symmetry breaking scale, of order 10^{-3} in Planck units for a GUT string. For the full analysis we refer the reader to the paper [4], and here quote the results.

The solutions qualitatively depend on three parameters (four for the massive dilaton): ϵ, already defined, is the gravitational strength of the string. $\beta = (m_{\text{Higgs}}/m_{\text{vec}})^2$ is the Bogomolnyi parameter, and refers to whether the string is type I, type II, or supersymmetric ($\beta = 1$). a, the coupling to the dilaton also indicates qualitatively different behaviour for the solutions, $a = -1$ in this case being a special point.

The results are the following. For the massless dilaton, the asymptotic solution for the vortex in suitably rescaled coordinates in the string frame is:

$$ds^2 = \hat{r}^{\frac{(a+1)\epsilon\hat{\mu}}{2} + \frac{(a+1)^2\epsilon^2\hat{\mu}^2}{4}} \left[d\hat{t}^2 - d\hat{r}^2 - d\hat{z}^2 - (1-\epsilon\hat{\mu})^2\hat{r}^{2-\frac{2(a+1)^2\epsilon^2\hat{\mu}^2}{4}} d\theta^2 \right] \quad (2)$$

$$e^{2\phi} = \hat{r}^{\frac{(a+1)\epsilon\hat{\mu}}{2}} \quad (3)$$

which is almost, but not quite, a conformally rescaled cone. $\hat{\mu} = \mu/4\pi\epsilon$ is used instead of μ, the energy per unit length, so as to make the ϵ-dependence of the solution explicit. This metric is consistent with the results derived for a cosmic string in Brans-Dicke gravity [6]. Note that the radius at which non-conical effects become important is when $r \simeq \sqrt{\lambda}\eta e^{\frac{2}{(a+1)\epsilon\hat{\mu}}}$, therefore, for a typical GUT string, $r = O(10^{50\text{billion}})$! i.e. well beyond any reasonable cosmological scale. For $a = -1$, the metric is conical in both Einstein and string frames, and the dilaton is shifted in the core relative to infinity, the direction of the shift depending on whether the cosmic string is type I or II, no alteration in the dilaton occurring for $\beta = 1$.

For the massive dilaton, we will use the potential $\hat{V}(\phi) = 2m^2\phi^2e^{-2\phi}$, the slightly unusual appearance reflecting the fact that the analysis is performed in the Einstein frame. Of course, we do not expect that this will be the exact form of the dilaton potential, however, a quadratic approximation will be valid provided ϕ remains close to the minimum of the potential. We expect the unknown dilaton mass to lie in the range 1TeV - 10^{15}GeV. It turns out that the dilaton equation can be solved implicitly using its Green's function, and outside the core the solution for the cosmic string to order ϵ is:

$$ds^2 = e^{\gamma_E} \left[dt^2 - dr^2 - dz^2 \right] - \alpha_E^2 e^{-\gamma_E} d\theta^2 \qquad (4)$$

$$e^{2\phi} = e^{-(a+1)\epsilon \hat{\mu} K_0(mr)} \qquad (5)$$

where subscript 'E' indicates the Einstein string solution. As expected, the metric asymptotes a conical metric, in both string and Einstein frames, however, the string does generate a dilaton 'cloud', approximately of width m_H/m_ϕ. For $a = -1$ the dilaton is only perturbed away from its vacuum value in the core of the string, and for $\beta = 1$, it is not affected at all.

Finally, as these results suggest, a Bogomolnyi argument can be derived which shows that a topological bound for the energy can only be saturated and first order equations for the system derived if $a = -1$ and $\beta = 1$.

These results support a Nambu approximation for the string, since they show that the metric is little affected on cosmological length scales. Our results indicate that the calculations of Damour and Vilenkin[3] ought to be renormalized by factors of $(a + 1)$. Provided a is not close to -1, their conclusion that a TeV mass dilaton was incompatible with a GUT string network will hold. However, for $a = -1$, such as might be the case if the fields are derived from the NS-NS sector of type II string theory for example, there will be little dilatonic radiation from the cosmic string network, and hence a much weaker constraint.

Acknowledgments

It is a pleasure to thank my collaborator Caroline Santos. This work was supported by a Royal Society University Research Fellowship.

References

1. M.B.Hindmarsh & T.W.B.Kibble, *Rep. Prog. Phys.* **58** 477 (1995).
   A.Vilenkin & E.P.S.Shellard, *Cosmic strings and other Topological Defects* (Cambridge Univ. Press, Cambridge, 1994).
2. K.I.Maeda & N.Turok, *Phys. Lett.* **202B** 376 (1988).
   R.Gregory, *Phys. Lett.* **206B** 199 (1988). *Phys. Rev.* **D43** 520 (1991).
3. T.Damour & A.Vilenkin, gr-qc/9609067.
4. R.Gregory & C.Santos, gr-qc/9701014.
5. D.Garfinkle, *Phys. Rev.* **D32** 1323 (1985).
   R.Gregory, *Phys. Rev. Lett.* **59** 740 (1987).
6. C.Gundlach & M.Ortiz, *Phys. Rev.* **D42** 2521 (1990).
   L.O.Pimental & A.N.Morales, *Rev. Mex. Fis.* **36** S199 (1990).

TOWARDS NUMERICAL AND ANALYTICAL STUDIES OF FIRST ORDER PHASE TRANSITIONS

Hans-Reinhard MÜLLER

Department of Physics and Astronomy, Dartmouth College,
Hanover, NH 03755-3528, USA

Discrete lattice simulations of a one-dimensional ϕ^4 theory coupled to an external heat bath are being carried out. Great care is taken to remove the effects of lattice discreteness and finite size and to establish the correct correspondence between simulations and the desired, finite-temperature continuum limit.

1 Introduction

To be able to study numerically certain properties of cosmological first order phase transitions such as the nucleation rate of bubbles, we investigate first the effects that finiteness and discreteness of a lattice have on results derived from simulations. For this basic study we limit ourselves to one spatial dimension and to a real scalar field $\phi(x,t)$ subjected to a potential $V_0(\phi)$ and an environmental temperature T. The dynamics obey a Langevin equation as given in Borrill and Gleiser [1]. This 2D-paper indicates that for a given tree-level potential, results obtained numerically on the lattice can't straightforwardly be identified with analytical results. They employ a renormalization procedure to get rid of the lattice spacing dependence and to identify the correct continuum limit of the simulations. Similar problems in 1D are treated in what follows.

2 Method

For classical field theories, the one-loop corrected effective potential is given by a momentum integral [2], and evaluated to

$$V_{1L}(\phi) = V_0(\phi) + \frac{T}{2} \int_0^\infty \frac{dk}{2\pi} \ln\left[1 + \frac{V_0''(\phi)}{k^2}\right] = V_0(\phi) + \frac{T}{4}\sqrt{V_0''(\phi)}. \quad (1)$$

The discretization δx of the lattice and its finite size L introduce short and long momentum cutoffs $k_{\min} = 2\pi/L$ and $\Lambda = \pi/\delta x$. Therefore the simulation only sees $\tilde{V}_{1L} = V_0(\phi) + (T/2)\int_{k_{\min}}^\Lambda \dots dk$. If one neglects the effect of $k_{\min}$ which is possible for sufficiently large L [a], integration and expansion in powers

---

[a]Lattice simulations only know one size parameter, the number of degrees of freedom $N = L/\delta x$. With a given N it is always possible to choose L big enough for the effects of δx to dominate over those of L.

of V_0''/Λ^2 (possible for sufficiently large Λ) yields

$$\tilde{V}_{1L} = V_0 + \frac{T}{4}\sqrt{V_0''} - \frac{T}{4\pi}\frac{V_0''}{\Lambda} + \Lambda \ O\left(\frac{V_0''^2}{\Lambda^4}\right) . \tag{2}$$

As is to be expected for a one dimensional system, the limit $\Lambda \to \infty$ exists and is well-behaved; there is no need for renormalization due to divergences. However, the effective one-loop potential is lattice-spacing dependent through the explicit appearance of Λ, and so are the corresponding numerical simulations, as evidenced in the tree-level potential cases of Fig. 1 (left graphs).

Similar to the renormalization procedure for 2D systems given by Borrill and Gleiser [1] we remove this dependence on δx by adding counterterms to the tree-level potential V_0. In contrast to higher-dimensional systems, these counterterms are *finite*, namely $V_{CT}(\phi) = (T/4\pi)(V_0''(\phi)/\Lambda)$. Hence the lattice simulation works with the corrected potential

$$V_{Latt}(\phi) = V_0(\phi) + \frac{TV_0''(\phi)}{4\pi^2}\delta x \tag{3}$$

and simulates the continuum limit to one loop, $\tilde{V}_{1L}(\phi) = V_0 + (T/2)\int_0^\Lambda \ldots dk = V_{1L}(\phi)$ (where V_{Latt}'' is employed in the integrand), just as it should be.

3   Application

Since the numerical extraction of bubble nucleation rates is a contrived process the ideas of Section 2 are tested initially with the *symmetric* double well potential $V_0(\phi) = (\lambda/4)\left(\phi^2 - \phi_0^2\right)^2$. We compare simulations using V_0 alone with those employing $V_{Latt}(\phi) = V_0(\phi) + 3\lambda T\delta x\phi^2/4\pi^2$ (eq. 3). One set of runs investigates the mean field value $\langle\phi\rangle$ of the metastable equilibrium before the first kink-antikink pair occurs $(\bar{\phi}(t) = (1/L)\int \phi(x,t)dx)$. Another set of runs measures the kink-antikink pair density n_p (proportional to the number of zeroes of the low-pass filtered field). Fig. 1 shows the comparison for different lattice spacings δx. Apart from a discrepancy for very coarse grids ($\delta x \approx 1$) the average field value is clearly lattice-spacing independent in the right panel (V_{Latt}), in contrast to the use of V_0. The effect is even more striking in the case of n_p where the addition of the finite counterterm removes any δx dependence.

In summary it was demonstrated that even in a field theory without divergences finite counterterms play a role. Their inclusion gets rid of the dependence of simulations on size and lattice spacing. This can be observed in the averaged field value and in the density of kink-antikinks. Further studies of this renormalization procedure [3] identify the correct continuum limit of simulations, thus matching theory and numerical results.

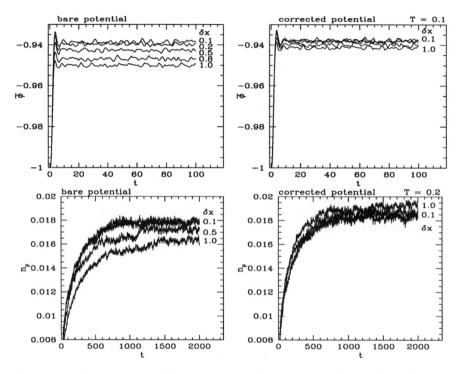

Figure 1: Average field value $\bar{\phi}(t)$, top, and density of kink-antikinks (half of density of zeroes), bottom, using the tree-level potential, left, and the corrected potential, right.

Acknowledgments

The author thanks Marcelo Gleiser for suggesting the topic and valuable discussions. Support from NASA grant NAGW-4270 and NSF grant PHY-9453431 is gratefully acknowledged.

References

1. J. Borrill and M. Gleiser, hep-lat/9607026, in press (Nucl. Phys. B).
2. P. Ramond, *Field Theory: A Modern Primer*, 2nd Ed. (Addison-Wesley, New York, 1990).
3. M. Gleiser and H.-R. Müller, *How to count kinks: a lattice renormalization study*, submitted to Phys. Rev. Lett.

NUMERICAL ANALYSIS OF
A WEAKLY FIRST-ORDER PHASE TRANSITION

JUN'ICHI YOKOYAMA

Yukawa Institute for Theoretical Physics, Kyoto University,
Kyoto 606-01, Japan

MASAHIDE YAMAGUCHI

Department of Physics, School of Sciences,
The University of Tokyo, Tokyo 113, Japan

We report the results of a series of numerical calculation of a phenomenological Langevin equation of an interacting scalar field in order to investigate the dynamics of a weakly first-order phase transition such as the electroweak phase transition in the standard model.

The dynamics of the electroweak phase transition in the early universe is still a subject of controversy. Although the one-loop effective potential of the Higgs field ϕ,

$$V_{\text{eff}}(\phi, T) = D(T^2 - T_2^2)\phi^2 - ET\phi^3 + \frac{1}{4}\lambda_T\phi^4 , \tag{1}$$

where $D = 0.17$, $E = 0.0097$, $T_2 = 92.64$ GeV for the Higgs mass $M_H = 60$ GeV which is the current experimental lower bound, shows it is of first order, the potential barrier between the two minima at the critical temperature, $T_c = 93.39$ GeV (again for $M_H = 60$ GeV), is so shallow that it has been doubted if the conventional picture of nucleation of critical bubbles in the homogeneous false-vacuum background really works. Some people claim that it is valid, others stress the importance of subcritical fluctuations which induce the "phase mixing" of false and true vacua. Here we approach the issue [1] in terms of numerical calculations of the Langevin equation of the expectation value of the scalar field. We discuss the discrepancy originates from the spatial scale on which one estimates the magnitude of field fluctuation.

Whether the universe was in a homogeneous state of $\phi = 0$ at the critical temperature has been studied by Borrill and Gleiser [2] who solved a simple phenomenological equation for a real scalar ϕ,

$$\Box\phi(\boldsymbol{x}, t) + \eta\dot{\phi}(\boldsymbol{x}, t) + V_{\text{eff}}'(\phi, T_c) = \xi(\boldsymbol{x}, t) , \tag{2}$$

where an overdot denotes time derivation and $\xi(\boldsymbol{x}, t)$ is a random Gaussian noise with the correlation function

$$\langle \xi(\boldsymbol{x}_1, t_1)\xi(\boldsymbol{x}_2, t_2) \rangle = \mathcal{D}\delta(\boldsymbol{x}_1 - \boldsymbol{x}_2)\delta(t_1 - t_2) , \tag{3}$$

satisfying the fluctuation-dissipation relation $\mathcal{D} = 2\eta T_c$. They have solved the above equation on a lattice with the spacing $\delta x = 1/\sqrt{2DT_2}$ and concluded that the phase mixing is manifest for all the experimentally allowed values of M_H.

We have started with re-analysis of equation (2) and (3) but with various values of δx and found that the result depends on δx rather strongly. The phase mixing is manifest only if we take $\delta x \lesssim 6/\sqrt{2DT_2}$ for $M_H = 60$ GeV which we assume throughout. Thus in order to draw a sensible conclusion from numerical analysis we should determine δx from a physical point of view.

One often adopts the so-called correlation length or the inverse-mass length at $\phi = 0$, which is about $8/\sqrt{2DT_2}$ here, as the coarse-graining scale. In the case of an interacting field with a nontrivial potential, however, it does not make sense since the spatial correlation function has a different behavior from a massive scalar field with the same mass. Here we propose to consider the correlation length of the noise term as the coarse-graining scale which is the only source of inhomogeneous evolution in the present system.

For this purpose we recall how a Langevin-like equation is derived in thermal field theory. There the noise term is naturally introduced as an auxiliary field which makes the equation of motion derived from the effective action real. The correlation function of the noise is therefore related with the imaginary part of the effective action.

For example, consider a simpler model in which ϕ interacts with a massless scalar field χ and a massless fermion ψ through

$$\mathcal{L}_{\text{int}} = -\frac{1}{4}g^2\chi^2\phi^2 - f\phi\bar{\psi}\psi. \tag{4}$$

The former mimics interactions with gauge bosons while the latter with quarks and leptons. Then there arise two noise terms, $\xi_1(\boldsymbol{x}, t)\phi$ and $\xi_2(\boldsymbol{x}, t)$, from these interactions at the one-loop level.

The spatial correlation function of the bosonic noise, $\xi_1(\boldsymbol{x}, t)$, originating from χ is given by

$$\langle \xi_1(\boldsymbol{x}, t)\xi_1(0, t) \rangle = \frac{g^4}{2}\,\text{Re}\left[G_\chi^F(\boldsymbol{x}, 0)^2\right] = \frac{g^4}{32\pi^2\beta^2 r^2}\coth^2\left(\frac{r}{\beta}\pi\right), \tag{5}$$

with $r \equiv |\boldsymbol{x}|$ and β is the inverse temperature. That is, the correlation function damps with a power-low.

On the other hand, the fermionic noise, $\xi_2(\boldsymbol{x}, t)$, originating from ψ has the spatial correlation

$$\langle \xi_2(\boldsymbol{x}, t)\xi_2(\boldsymbol{0}, t) \rangle = -f^2\,\text{Re}\left[\text{Tr}\left(S_\psi^F(\boldsymbol{x}, 0)S_\psi^F(-\boldsymbol{x}, 0)\right)\right]$$

$$= \frac{f^2}{4\pi^4 r^2} \left[\frac{\pi}{\beta r} \frac{1}{\sinh(r\pi/\beta)} + \frac{\pi^2}{\beta^2} \frac{\cosh(r\pi/\beta)}{\sinh^2(r\pi/\beta)} \right]^2 \simeq \frac{f^2}{4\beta^4} \frac{e^{-2\pi r/\beta}}{r^2}. \quad (6)$$

The last expression is valid for $r \gtrsim \beta/\pi$ and it damps exponentially. The temporal correlation is unimportant in the thermal background and so neglected.

In writing down the master equation, it is much more difficult to determine the other essential ingredient or the dissipation term from field theoretic approach in which non-perturbative effects must be taken into account in some way. Here we instead determine them from the properties of the noises so that the fluctuation dissipation relation is satisfied. For example, in the case only one species of the noise with the correlation $\langle \xi(x) \xi(x') \rangle = B(x - x')\delta(t - t')$ is present, the master equation reads,

$$\Box\phi(x) + V'_{\mathrm{eff}}(\phi) + \frac{1}{2T} \int d^3 x' B(x - x')\dot{\phi}(x', t) = \xi(x). \quad (7)$$

Since the properties of bosonic noise and fermionic noise are very different from each other, we have examined their effects on the dynamics of the phase transition separately.

First we have solved the Langevin equation starting from $\phi = \dot{\phi} = 0$ everywhere at $T = T_c$ taking only the bosonic noise into account. Since ξ_1 is a multiplicative noise we have also incorporated an additive noise arising at the two-loop level whose correlation function damps in proportion to r^{-3}. As a result we have found that the field configuration remains homogeneous and phase mixing does not occur.

Next we have examined the effect of fermionic noise which has no spatial correlation beyond $r \gtrsim \beta/2\pi$ and white noise approximation (3) is valid as long as we take δx larger than this value. In this case phase mixing is manifest.

In the actual electroweak phase transition both types of noises are present with the similar magnitudes, and so the latter result is expected to apply. In short, although gauge interaction plays the essential role to induce the cubic term in the effective potential, the non-equilibrium dynamics is dominated by fermionic interactions. As a result, the conventional picture of first-order phase transition based on the one-loop potential is suspect and we must deal with the effective action directly to clarify the real dynamics of the phase transition.

References

1. M. Yamaguchi and J. Yokoyama, (1996) Preprint YITP-96-61, and papers cited therein.
2. J. Borrill and M. Gleiser, *Phys. Rev.* D **51**, 4111 (1995).

Cosmic Connections

N.J. Cornish

DAMTP, Cambridge University, Silver Street, Cambridge CB3 9EW, England

G.D. Spergel

Princeton University Observatory, Princeton, NJ 08544, USA

G.D. Starkman

Dept of Physics, Case Western Reserve University,
10900 Euclid Ave, Cleveland OH 44106, USA

The universe may be multiply-connected on observable scales. This multiple-connectedness could explain the homogeneity of the universe on the very largest scales. Experiments in the next few years would detect this cosmic topology.

"In fourteen hundred and ninety two, Columbus sailed the ocean blue," goes the children's rhyme. As the seafarers of yore expanded their horizons beyond their local shores, they realized that the surface of the earth was curved. Testing the boundaries of the world they knew, they sought to determine, and exploit, its topology.

Today we may find ourselves at a similar juncture. Several observations [1] seem to suggest that the spatial sections of the universe are curved, though with the opposite sign to the surface of the earth, and that the scale of curvature has recently come within our horizon. If this is true, then we too can begin to explore the topology of our universe.

The universe appears very homogeneous and isotropic on large scales. Its overall spatial geometry is then determined by the ratio, Ω, between the average energy density and a critical energy density. Geometry constrains but does not dictate topology. Positively-curved geometry ($\Omega > 1$) is naturally compact, but flat ($\Omega = 1$) and negatively-curved ($\Omega < 1$) geometries can also be compact. The difference is that the compact topologies of geometries with $\Omega \leq 1$ are not simply connected. In two dimensions, for example, they are topologically equivalent to doughnuts with one or more holes.

If Ω is measurably less than unity, then not only is the curvature of the universe observable, but we may also expect to see its topology. The photons of the east may be arriving from the west.

Why do we expect to see topology today? First, for $0.1 < \Omega < 0.5$, the range which current observations prefer, the ratio of the curvature scale, R_c, to the horizon scale, the distance to which we can see, is between 0.3 and 0.6.

That is, the curvature scale is comparable to, but smaller than, the horizon size.

Second, we have reason to believe that the scale on which the universe is multiply-connected, the topology scale, is roughly the curvature scale. For a homogeneous negatively-curved universe with compact topology, the spatial volume in units of R_c^3 is equal to a topological invariant [2], α^3. The topology scale will be approximately αR_c. Some quantum cosmologists believe that the birth of the universe involves its nucleation out of nothing. Their calculations suggest that the smaller a universe, the more likely it is to be nucleated [3]. Thus, universes with small values of α are preferred. Recently we suggested another reason why universes with small α are preferable [4]: compact topology causes chaotic mixing and hence homogenization of the contents of the universe [5]. This allows inflation [6] to occur if $\alpha \lesssim 10$. If $\alpha < 3$, the chaotic homogenization also helps explain why the universe looks smooth on large scales even if measurably $\Omega < 1$.

For Ω and α in this expected range, the scale of topology is within the diameter of our horizon. Consequently, the topology of the universe should be observable, and can be searched for generically by looking at fluctuations in the cosmic microwave background radiation (CMBR). Below, we will explain how all small topologies lead to correlated circles in the microwave sky, and how we will soon be able to search for these circles [7].

We are not alone in favoring a finite universe [8]. Einstein and Wheeler [9] both favored compact spatial slices for the universe on the basis of Mach's principle. Others have expressed distaste for the infinite replication of events in an infinite volume universe [10]. A universe with a compact three-space and whose Euclidean time is compact would satisfy the Hartle-Hawking no-boundary proposal [11] for the wavefunction of the universe.

What do these philosophical motivations say about the type and scale of topology? In three dimensions, there is believed to be a rather remarkable connection between topology and geometry [2]. All three-dimensional spaces are conjectured to be uniquely divisible into pieces, each of which is topologically equivalent to a homogeneous space. Remarkably, it is found that most [2] spaces are constructed of only one such piece, and that piece has constant negative curvature and non-trivial topology. Thus, in the space of all possible universes, most have three-manifolds with multiply-connected topologies and average negative curvature. All things being equal, we expect a universe to be created with negatively curved (hyperbolic) spatial sections.

We also expect universes with small volume to be favored. Quantum mechanical tunneling is a well-understood phenomenon. Its extension to the nucleation of the universe out of nothing via Euclidean instantons in semiclassical quantum gravity is admittedly speculative. Nevertheless, some such calculations indicate that the action for a universe of homogeneous geometry increases in proportion to the volume[3]. The probability of creating our universe with a particular hyperbolic topology is

$$P(\alpha)d\alpha \propto N(\alpha) \exp\left[-\alpha^3\right] d\alpha, \tag{1}$$

where $N(\alpha)d\alpha$ is the number of universes with $V^{1/3}/R_c$ between α and $\alpha + d\alpha$. So long as the entropy, $N(\alpha)$, does not overwhelm the action, $\exp(-\alpha^3)$, the nucleation of small universes will dominate. This is analogous to the process of bubble nucleation, where the action is minimized for perfectly spherical bubbles, but the entropy is larger for irregular bubbles. Typically action wins over entropy and highly spherical bubbles are formed. Action also wins over entropy in the case of topology, with the proviso that some physical process rules out the formation of non-orientable topologies. If we allow non-orientable topologies, N_V grows very quickly with V and overwhelms the action. However, non-orientable topologies are generally excluded[12] as two-component fermion fields cannot be constructed in such spacetimes.

Another nice feature of small hyperbolic universes is the natural chaotic homogenization they provide. Classical trajectories and quantum fields are known to be strongly mixed in such spacetimes[13]. The mixing smooths out inhomogeneities at a rate given by $\exp(-\kappa d)$, where $\kappa \simeq 1/(\alpha R_c \ln 2)$ is the Kolmogorov-Sinai entropy due to the topology, and d is the comoving distance traveled by the inhomogeneous mode. Between the nucleation of the universe, which we expect to be an energy density of $M_p^4 \simeq (10^{19}\text{GeV})^4$, and the start of inflation when the universe reaches a density of $\lambda M_p^4 \simeq (10^{15}\text{GeV})^4$, the mixing can reduce inhomogeneities in the inflaton potential to at most a few parts in 10^5 so long as $\alpha < 3$.

If the universe is too inhomogeneous then inflation can never get started[14]. Thus, a small hyperbolic universe can solve this inflationary initial conditions problem[4]. Moreover, if Ω is measurably less than unity today, then inhomogeneities on the scale of the curvature, which were present when inflation began, would not have been stretched away by a single period of inflation. If the universe is already compact due to the process of nucleation, it makes sense to make use of the mixing this provides, and avoid the need for multiple periods of inflation[15]. While chaotic mixing does not explain why α is small, it does show that small universes can lead to a simple cosmological model.

The cosmic microwave background radiation is the primary evidence supporting the Big Bang theory of cosmology. Some effort has been made to search for topology by looking for a mode cutoff in the spectrum of CMBR fluctuations[16]. This cutoff does arise in a compact flat universe, albeit at wavelengths as much as six times the topology scale. In hyperbolic compact universes, if there is a cutoff it may be at fifty, one hundred or more times the topology scale [13]. CMBR fluctuations are, however, still the best place to look for topology. In any geometry, the photons of the CMBR that we receive on Earth today emanated from a last-scattering surface which is a sphere centered on us and with radius slightly smaller than radius of the horizon. In a multiply-connected universe, that sphere self-intersects once the diameter of the last scattering surface, $\simeq 2R_h$, is larger than the topology scale $\simeq \alpha R_c$. The self-intersections of the sphere are circles. The mapping from the surface of last scatter to the night sky is conformal. Since conformal maps preserve angles, the identified circles at the surface of last scatter would appear as identified circles on the night sky. Thus, topology produces circles of equal radius on the sky on which the fluctuations in the CMBR are correlated [7]. The existence of the circles is generic, only the size, number and location of the circles depends on the particular topology.

Currently available data [17] from the Cosmic Background Explorer's Differential Microwave Radiometer (COBE/DMR) is being analyzed to search for these circles in the sky. Unfortunately, because of the low signal-to-noise and angular resolution of the DMR, COBE probably does not definitively test the idea of topology. However, these correlations will be measurable with the next generation of CMBR satellites, planned for launch in the late 1990's. By the end of the millennium, we will know if there is horizon-scale topology. If the universe is visibly multiply-connected, we might be able to determine which topology it has. Since a multiply-connected spacetime is generically only *locally* homogeneous and isotropic, we may also learn where we are in the universe and in what direction we are headed!

Acknowledgments

DNS is supported by the NASA MIDEX MAP grant and by the NSF. GDS is supported by an NSF CAREER award, by the DOE, the Ohio Board of Regents and by funds from CWRU. NJC thanks CWRU for their hospitality while this work was in progress.

1. J. A. Peacock & S. J. Dodds, Mon. Not. R. Astr. Soc. **267**, 1020 (1994); M. Kamionkowski & D.N. Spergel, Ap. J., 431, 1 (1994); A.R. Liddle, D.H. Lyth, D. Roberts, & P.P.T. Viana, to appear in Mon. Not. R. Astr. Soc. (1995) (astro-ph/9506091).
2. W. P. Thurston and J. R. Weeks, Scientific American, July '84, 108; W. P. Thurston, *The Geometry and Topology of 3-Manifolds* (Princeton University Press, Princeton, 1978).

414

3. D. Atkatz and H. Pagels, Phys. Rev. **D25**, 2065 (1982); Ya. B. Zel'dovich and A. A. Starobinsky, Sov. Astron. Lett. **10**, 135 (1984); Y.P. Goncharov and A.A. Bytsenko, Astrophys. **27**, 422 (1989); S. Carlip, Phys. Rev. **D46**, 4387 (1992).
4. N. Cornish, D. Spergel and G. Starkman, astro-ph/9601034.
5. C.N. Lockhart, B.Misra and I.Prigogine, Phys. Rev. **D25**, 921 (1982).
6. A. Linde, *Particle Physics and Inflationary Cosmology*, (Hardwood Academic Publishers, Chur, Switzerland, 1990), E. W. Kolb and M. Turner, *The Early Universe*, (Addison-Wesley Publishing, New York, 1990) and references therein.
7. N. Cornish, D. Spergel and G. Starkman, gr-qc/9602039
8. M. Lachieze-Rey, J.P. Luminet, Phys. Reports **254**, 135 (1995).
9. A. Einstein *The Meaning of Relativity* (Princeton University Press, Princeton, 1955), p. 108; J.A. Wheeler, *Einstein's Vision* (Springer, Berlin, 1968).
10. G.F. Ellis, Q.J.R.Astron. Soc. bf 16, 245 (1975).
11. J. Hartle and S. Hawking, Phys. Rev. **D28**, 2960 (1983).
12. R. Penrose and W. Rindler, *Spinors and Spacetime*, Cambridge University Press (1984).
13. N. L. Balazs and A. Voros, Phys. Rep. **143**, 109 (1986).
14. T. Piran and D.S. Goldwirth, Phy. Rep. **214**, 223 (1992) and references therein.
15. M. Sasaki, T.Tanaka, K. Yamamoto, J. Yokoyama, Prog. Theor. Phys., 90, 1019 (1993); K. Yamamoto, M. Sasaki & T. Tanaka, ApJ, 455, 412 (1995); B. Ratra and P.J.E. Peebles, Phys. Rev. **D52**, 1837 (1995), M. Bucher, A.S. Goldhaber and N. Turok, Phys.Rev. **D52**, 3314 (1995); A. Linde, Phys.Lett.B351:99-104,1995; A. Linde and A. Mezhlumian, Phs. Rev. **D52**, 6789 (1995).
16. I.Y. Sokolov, JETP Lett **57**, 617 (1993); A.A. Starobinsky, *ibid.*, p. 622; D. Stevens, D. Scott and J. Silk, Phys. Rev. Lett **71**, 20 (1993).
17. C.L. Bennett, A.J. Banday, K.M. Gorski, G. Hinshaw, P. Jackson, P. Keegstra, A. Kogut, G.F. Smoot, D.T. Wilkinson, and E.L. Wright, astro-ph/9601067, submitted to Ap. J. Lett.

HOW KALUZA-KLEIN SPACE
BECAME THREE DIMENSIONAL

JOHN J. DYKLA

Department of Physics, Loyola University Chicago
6525 N. Sheridan Rd., Chicago, IL 60626, USA

The existence of a unique universal gravitational attraction which is associated with dynamical degrees of freedom of the curvature tensor requires the inflation of three dimensions of space, independent of the total number of spacetime dimensions in any "theory of everything". Exploration of a non-linear sigma model in which the stress-energy-momentum tensor of scalar fields triggers inflation and compactification at equal rates indicates that in these spacetimes inflation of three dimensions is dynamically favored in a 10-dimensional Kaluza-Klein scenario, and the inflation of 10 dimensions is favored in a 26-dimensional picture. Preliminary evidence suggests the suppression of the complementary possible inflations (6 dimensions from 10, or 15 dimensions of 26).

1 Background

1.1 Motivation and Observational Constraints

The reason for considering space with other than three dimensions in the early universe is that a ten-dimensional spacetime allows the supersymmetric unification of bosons and fermions in a quantum field theory of all fundamental interactions. The only other anomaly-free "theory of everything" is a purely bosonic theory in twenty six dimensions. In outer and inner space, observations tightly constrain the current number, m, of dimensions: since the perihelion advance of Mercury's orbit agrees with general relativity, $|m-3| < 10^{-9}$; since the Lamb Shift of Hydrogen agrees with quantum electrodynamics, $|m-3| < 3.6 \times 10^{-11}$. Anthropic arguments can be given that imply $m = 3$ exactly.

1.2 Geometrodynamics and Teleology

Consider a spacetime with $1 + m$ dimensions after inflation. If there is a unique universal gravitation associated with dynamical degrees of freedom of a curvature tensor, then [1] space must have m dimensions, where

$$m^4 + 4m^3 - 7m^2 - 34m - 24 = 0 \qquad (1)$$

The four roots are non-zero integers, but only $m = 3$ is positive. In an early universe "theory of everything", there will be $D = 1 + m + n$ spacetime dimensions, where n remain compactified while m inflate. Post-inflation space has "achieved the goal" of 3 macroscopic dimensions so gravity can exist. Kaluza-Klein dynamics yields insight into the process by which this came to be.

2   Dynamics of Inflation and Compactification

2.1   Results Assuming Isotropir Subspaces

To study Kaluza-Klein cosmologies, consider the line element

$$ds^2 = c^2 dt^2 - R_m^2 g_{ij} dx^i dx^j - R_n^2 h_{kl} dy^k dy^l \tag{2}$$

where $R_m(t)$ and $R_n(t)$ are the scale factors of the inflating m-dimensional and compactifying n-dimensional subspaces. In an obvious generalization of the Robertson-Walker metric, the tensors g_{ij} and h_{kl} are assumed independent of cosmic time, t. Gell-mann and Zweibach proposed [2] introduction of n scalar fields (non-linear sigma model) in $(4 + n)$-dimensional spacetime to spontaneously compactify n dimensions. This study assumes n scalar fields in $(1 + m + n)$-dimensional spacetime, and selects a division into subspaces by requiring that inflation and compactification rates be equal, so that scalar field interactions not distinguish one direction of time from another. The scale factors for the subspaces of dimensions m and n evolve as

$$R_m(t) = R(t - t_0)^\alpha \quad , \quad R_n(t) = R(t - t_0)^\beta \tag{3}$$

where R is the size of all space dimensions at $t = t_0$ and

$$\alpha = \frac{n + \sqrt{mn(m + n - 1)}}{n(m + n)} > 0 \quad , \quad \beta = \frac{m - \sqrt{mn(m + n - 1)}}{m(m + n)} < 0 \tag{4}$$

give inflation and compactification rates. These exponents have equal absolute values for spacetimes and subspaces shown in Table 1. Implications for theories of everything are clear. Anomalies in unified quantum field theories of fundamental interactions may be eliminated if and only if $D = 26$ (purely bosonic spacetime) or $D = 10$ (basis of the supersymmetric model of bosons and fermions). Table 1 shows both numbers related through gravitational dynamics, which "predicts" macroscopic space dimensionality to be $m = 3$.

Table 1: Dimensions of Spacetimes and Inflating Subspaces.

Spacetime, D	2	5	10	17	26	37	50	65	82	101	...
Inflating, m	0	1	3	6	10	15	21	28	36	45	...

2.2 A Limitation of the Isotropic Argument

It can be objected that the principle that directions of time not be distinguished requires consideration of the possibility of exchanging the roles of inflating and compactifying subspaces. Indeed the field equations are satisfied for isotropic subspaces if we exchange m and n in Equations 4. Since $n = 6$ for $D = 10$, and $n = 15$ for $D = 26$, neither of the values of D of interest in "theories of everything" lead to gravity or supersymmetry for this choice. It is unsatisfying to reject dynamical options on the anthropic ground that we could not exist in a universe where they were realized. The solution must be a mechanism which breaks the symmetry between past and future.

2.3 Coupled Anisotropic Fluctuations

Breaking time symmetry requires coupling inflating and compactifying subspaces. Coupled dimensions must evolve anisotropically, else either all will inflate or all will compactify. With no symmetry imposed, a complete solution would satisfy 55 second-order coupled non-linear equations for $D = 10$, and 351 for $D = 26$. With no hope of analytic solution, preliminary numerical study of a "toy model" (for insight if not correct numbers) is fruitful. Investigations began with my former student Ethan P. Honda, now at UT Austin. Table 1 reveals that $D = 5$ implies $m = 1$ and $n = 3$, so that the goal is to find conditions under which 1 dimension inflates and 3 compactify (conjectured robust to perturbations from isotropy), versus 3 inflating and 1 compactifying (conjectured unstable to such perturbations). Suppose the metric is

$$g_{00} = 1 \; , \; g_{0i} = 0 \; , \; g_{ii} = -R_i^2(1 + e_i)^2 \; , \; g_{ij} = -R_i R_j e_i e_j \tag{5}$$

where spatial indices i and j are distinct, no sum is implied in the third form, and $R_i(t)$ and $e_i(t)$ are time-dependent scale and anisotropy/coupling factors satisfying 8 independent field equations, containing nonlinearities through 7th degree. After some examination, 4 intermediate coupling factor products were removed on the grounds that they would not affect the outcome of inflation and compactification but merely uninteresting intermediate details of the process. Results supported the conjectures, and directions for future work are evident.

References

1. J.J. Dykla in *12th International Conference on General Relativity and Gravitation: Abstracts of Contributed Papers*, 371 (1989).
2. M. Gell-mann and B. Zweibach, *Nucl. Phys.* B **260**, 569 (1985).

ADJUSTMENT OF VACUUM ENERGY BY TENSOR FIELD

A. D. DOLGOV

Teoretisk Astrofysik Center, Juliane Maries Vej 30,
DK-2100, Copenhagen, Denmark
and
ITEP, 117259, Moscow, Russia

A mechanism of cancellation of vacuum energy by the back reaction of a massless second rank tensor field is considered. The field is supposed to be minimally coupled to gravity with only derivative terms in the Lagrangian. The isotropic components of this field are unstable in De Sitter background and the induced energy-momentum of the unstable components cancels out the original vacuum energy or, in other words, the cosmological constant down to (almost) zero.

Any theoretical estimate of the cosmological constant or, what is the same, of the vacuum energy gives the result which is by 50-100 orders of magnitude larger than the astronomical upper bound. The latter roughly speaking is [1]:

$$\rho_{vac} < 10^{-47}\,\mathrm{GeV}^4 \qquad (1)$$

while theoretically found contributions into vacuum energy vary from $m_{Pl}^4 = 10^{76}\,\mathrm{GeV}^4$ to approximately $10^{-2}\,\mathrm{GeV}^4$. The higher contribution comes from the Planck scale physics (supergravity, superstrings), while the lower one comes from very well theoretically (and, in a sense, experimentally) established contributions to vacuum energy from the condensates of quarks and gluons. While the the high energy contributions may be rather speculative, the contribution from Quantum Chromodynamics surely exists.

Thus one has to conclude that there is an unknown term in the vacuum energy which cancels out all known contributions with the fantastic precision better than at least one part in 10^{45}. There are several proposals to find a natural mechanism of such a cancellation like modification of gravity, adjustment by a new field, anthropic principle, etc (for a review see e.g. refs. [2,3]). It is not easy (if possible) to modify gravity in such a way that general covariance is preserved, energy-momentum tensor is covariantly conserved and simultaneously the vacuum part of this tensor, which is proportional to the metric tensor $g_{\mu\nu}$, does not gravitate. The second interesting possibility, anthropic principle, was recently discussed in ref. [4]. It may be the only viable solution if nothing better is found. The situation with the problem of the cosmological constant now reminds the one that existed in the Friedman cosmology before the inflationary scenario has been found.

To my mind the best possibility for the solution of the cosmological constant problem is the dynamical adjustment mechanism. The idea is quite simple. One assumes that there exists a new light or massless field which interacts with gravity in such a way that a classical condensate of this field is developed. The energy-momentum tensor of this condensate cancels out the original vacuum energy-momentum tensor generically with the precision m_{Pl}^2/t^2, so that the non-compensated amount of the vacuum energy is parametrically the same as the critical energy density at any given time moment. Unfortunately no known model of this kind gives a realistic cosmology. Moreover the no-go theorem has been proven [2,4], that a scalar field cannot realize such adjustment. However higher rank tensor fields are more promising and with relatively simple Lagrangians can do the necessary cancellation. This talk is based on ref. [5] where one can find more detail and the list of references.

Let us assume that there exists a massless symmetric tensor field $S_{\alpha\beta}$ with the simple Lagrangian

$$\mathcal{L} = -S_{\alpha\beta;\gamma}S^{\alpha\beta;\gamma} \tag{2}$$

In the spatially flat Robertson-Walker metric $ds^2 = dt^2 - a^2(t)d\vec{r}^{\,2}$ the homogeneous components of this field, $S_{\alpha\beta}(t)$ satisfy the equations:

$$(\partial_t^2 + 3H\partial_t - 6H^2)S_{tt} - 2H^2 s_{jj} = 0 \tag{3}$$

$$(\partial_t^2 + 3H\partial_t - 6H^2)s_{tj} = 0 \tag{4}$$

$$(\partial_t^2 + 3H\partial_t - 2H^2)s_{ij} - 2H^2\delta_{ij}S_{tt} = 0 \tag{5}$$

where $s_{tj} = S_{tj}/a(t)$, $s_{ij} = S_{ij}/a^2(t)$ and $H = \dot{a}/a$, it is determined by the expression: $3H^2 m_{Pl}^2/8\pi = \rho^{(vac)} + \rho_S$ where ρ_S is the energy density of the field $S_{\alpha\beta}$. One can check that there exists a rising with time solution which at large t behaves as $S_{tt} = Ct$, $s_{ij} = \delta_{ij}Ct/3$, and $s_{tj} = 0$, where $C \sim \sqrt{\rho^{vac}} = const$. The condition of vanishing of s_{tj} is not stable but with a more complicated Lagrangian [5] the stability or vanishing of non-isotropic components S_{tj} can be naturally realized. The energy-momentum tensor of this solution has the vacuum form, i.e. it is proportional to $g_{\mu\nu}$ and it cancels the original vacuum energy-momentum tensor $T_{\mu\nu}^{(vac)} = \rho^{(vac)}g_{\mu\nu}$ down to asymptotically vanishing terms $\delta\rho \sim m_{Pl}/t^2$. The original exponential expansion, which was induced by the dominating vacuum energy, when $S_{\alpha\beta}$ was small, ultimately turns into the power law one with $H = 3/8t$. This is not a realistic cosmology but at least it is not too far from it. The energy density of usual matter in this model decreases rather slowly, $\rho_{rel} \sim t^{-3/2}$ and $\rho_{nr} \sim t^{-9/8}$. Corresponding values of the parameter $\Omega = \rho_{matter}/\rho_c$ would be much larger than 1. Though the energy density of the usual matter may be the dominant one, the Hubble parameter does not depend on it. So the results obtained above remain valid also in the presence of normal matter. This is connected with the fact that the expression for the energy density of the field $S_{\alpha\beta}$ contains terms proportional to H^2. In particular for the solution presented above we have:

$$\rho = -\left[\frac{1}{2}(\dot{S}_{tt}^2 + \dot{s}_{ij}^2) + H^2(3S_{tt}^2 + s_{ij}^2 + 2S_{tt}s_{jj})\right] \tag{6}$$

One can get different asymptotics of expansion regime adding to the original Lagrangian (2) a few more derivative terms:

$$\Delta\mathcal{L} = \eta_1 S_{\alpha\beta;\gamma}S^{\alpha\gamma;\beta} + \eta_2 S_{\beta;\alpha}^{\alpha}S_{;\gamma}^{\gamma\beta} + \eta_3 S_{\alpha;\beta}^{\alpha}S_{\gamma}^{\gamma;\beta} \tag{7}$$

With a particular choice of constants η_j, which possibly corresponds to the selection of a certain spin states in $S_{\alpha\beta}$, one get either the relativistic expansion regime $H = 1/2t$ or the nonrelativistic one $H = 2/3t$, but it is not evident if one could make a cosmological model with a change from the relativistic regime at an early stage to the nonrelativistic one at a later stage.

The model, as it is, does not seem to give a realistic cosmology but at least it is quite simple and gives a natural cancellation of the vacuum energy. It remains to be seen if a realistic cosmology can be found along these lines. The model may encounter serious theoretical problems if one tries to quantize the non-gauge tensor field $S_{\alpha\beta}$. Even if quantization can be successfully done, quantum corrections may change the form of the original Lagrangian introducing dangerous potential terms like $U(S^2)$ or non-minimal coupling to gravity. Hopefully the symmetry with respect to the transformation $S_{\alpha\beta} \to S_{\alpha\beta} + const\, g_{\alpha\beta}$ does not permit generation of such terms.

Acknowledgments

This paper was supported in part by the Danish National Science Research Council through grant 11-9640-1 and in part by Danmarks Grundforskningsfond through its support of the Theoretical Astrophysical Center.

References

1. See e.g. talk by W. Freedman at this Conference.
2. S. Weinberg, *Rev. Mod. Phys.* **61**, 1 (1989).
3. A.D. Dolgov in *The Very Early Universe*, ed. G.Gibbons, S.W.Hawking, and S.T.Tiklos (Cambridge University Press, 1982).
4. S. Weinberg, astro-ph/9610044; UTTG-10-96.
5. A.D. Dolgov, TAC-1996-021; astro-ph/9608175; submitted to Phys.Rev.

DECOHERENCE OF VACUUM FLUCTUATIONS IN COSMOLOGY

MILAN MIJIĆ

Department of Physics and Astronomy,
California State University Los Angeles, CA 90032,
and, Institute for Physics, P.O. Box 522, 11001 Belgrade, Yugoslavia

A calculation of the one-mode occupation numbers for vacuum fluctuations of massive fields in De Sitter space shows that their use for the generation of classical density perturbations in inflationary cosmology very much depends on their masses and conformal couplings. A new mechanism for the decoherence of relatively massive fields, $2 < m^2/H_0^2 < 9/4$, has been identified. Similar analysis of the power law inflation shows that production of adiabatic density perturbations may not take place in models with power $p < 3$.

Decoherence of vacuum fluctuations in inflationary cosmology [1] is at the foundation of the current paradigm for structure formation. General arguments for classical behavior of fluctuations larger than the Hubble radius [2] usually appeal to the existence of Gibbons-Hawking temperature of De Sitter space, or to the rapid stretching of these modes which collects them in a kind of infrared condensate. Attractive as they are, such arguments were never very explicit, and leave some puzzles. For instance, what about the decoherence of fluctuations of very massive fields? Both mentioned arguments seem independent on mass, yet stochastic approach to inflationary phase [3] apparently shows that the random walk of coarse-grained fields takes place only for fields with mass low compared to the Hubble scale.[4] It seems therefore appropriate to look for a more explicit insight into this process.

Results reported here are based on the investigation of particle production of free fields in expanding universe.[5] Similar lines of reasoning were pursued in Ref.'s 6-7. The key physical mechanism in all of these cases is rapid expansion of the spacetime background, so that classical behavior emerges even in the case of free fields. This is in contrast with the more frequent studies of the decoherence based on the construction of a suitable reduced density matrix for the subsystem of interest. For application of this later idea to fluctuations in inflationary universe see Ref. 8.

Our method is to evaluate the one-mode occupation number when the wavelength of the mode exceeds the Hubble radius. If the occupation number is large we will have classical behavior. For fluctuations of massive fields in spatially flat De Sitter space the results are as follows.

(a) $m^2/H_0^2 < 2$, and $2 < m^2/H_0^2 \leq 9/4$. The one-mode occupation number at times after the mode crosses the Hubble radius grows as,

$$n(z) \sim \left(\frac{\lambda_{phys}(z)}{H_0^{-1}}\right)^{2\nu} . \tag{1}$$

Here, $z \equiv -k\eta$ is a new time variable (k is fixed). $\lambda_{phys} \equiv S(\eta)k^{-1}$ is the physical wavelength, and parameter ν measures the mass in Hubble units: $\nu^2 \equiv 9/4 - m^2/H_0^2$. Since the occupation number diverges each mode forms a classical condensate. For $m^2/H_0^2 < 2$ this happens because modes outside the Hubble radius roll along the upside-down potential.[6] For $2 < m^2/H_0^2 \leq 9/4$ the potential is upside-right at all times, including the times later than the Hubble crossing time, but the occupation number still diverges due to the dominance of just one mode (as opposed to simultaneous presence of both) as the amplitude of this upside-right oscillator settles to the minimum. In both cases the same criteria for classical behavior are satisfied.

(b) $m^2/H_0^2 = 2$. Minimally coupled field with this mass is equivalent to conformally coupled massless field, so there is no particle production whatsoever.

(c) $m^2/H_0^2 > 9/4$. In this case the late time behavior of $n(z)$ is purely oscillatory,

$$n(z) = A_0 + A_c \cos\left[2|\nu|\log(z/2) - 2\Phi_\Gamma\right] + A_s \sin\left[2|\nu|\log(z/2) - 2\Phi_\Gamma\right] . \tag{2}$$

Φ_Γ is phase fixed by the mass. The potential is upside right in this case, but both oscillatory modes must be kept. This leads merely to oscillations in some finite particle number, not to its divergence. One can show that the amplitude is bounded as $n(z) \leq f \equiv A_0 + (A_c + A_s)^{1/2} \ll 1$, for $m^2/H_0^2 - 9/4 \geq 1$.

(d) *Non-minimal coupling.* The effect of adding an $\xi R\phi^2/2$ term is particularly simple in case of a flat De Sitter background: in all the expressions above one should replace m^2 with $m^2 + 12\xi H_0^2$. The behavior of the occupation number as parameterized by different values of this quantity is the same as before. The basic physics of the phenomena is unchanged.

For quantum fluctuations of inflaton that drives the power law inflation $a \sim t^p, p > 1$, one finds that the convenient characteristic parameter is the power of expansion p. The corresponding characteristic values are as follows: for $m^2/H_0^2 = 9/4$ in De Sitter case we have $p = 3$ in case of power law inflation, while $m^2/H_0^2 = 2$ corresponds to $p = p_+ \equiv (7 + \sqrt{33})/4 \approx 3.186$. The equation of motion, boundary conditions and solutions are the same. From the analysis of the De Sitter case one deduces the following decoherence properties of vacuum fluctuations in power law inflation:

(a) $p > p_+$, and $3 \leq p < p_+$. The occupation number diverges as the wavelength exceeds the Hubble radius. For $p > p_+$ the oscillators are upside-down, and the argument of Guth and Pi [6] applies. For $3 \leq p < p_+$ the oscillators are always upside-right but the occupation number nevertheless diverges due to the dominance of just one mode.

(b) $p = p_+$. Fluctuations in this case are the same as that for a massless minimally coupled field, and there is no particle production.

(c) $p < 3$. The oscillators are upside-right, but classical solutions are oscillatory and both modes must be kept. The occupation number is finite, but oscillatory, and always smaller then unity.

Therefore, only the power law inflation of type (a) has classical adiabatic perturbations. These results may also have some consequence for the suspected increase of power towards large scales in models of extended inflation.

To conclude, we find the traditional particle production to be a simple and explicit mechanism, sufficient to describe evolution of vacuum fluctuations into classical perturbations. The same method may be applied to study of the decoherence in any Robertson-Walker spacetime.

References

1. A. Guth, *Phys. Rev.* D **23**, 347 (1981).
2. E. Kolb and M.S. Turner, *The Early Universe*, (Addison Wesley, 1990); A. Linde, *Particle Physics and Inflationary Cosmology*, (Harwood Academic Press, 1990).
3. A.A. Starobinsky, in *Current Topics in Field Theory, Quantum Gravity and Strings,* H.J. de Vega and N. Sanchez (eds.), Springer (1986).
4. M. Mijic, *Phys. Rev.* D **49**, 6434 (1994).
5. M. Mijić, Cal State LA report, (to appear in *Phys Rev* **D**).
6. A. Guth and S-Y. Pi, *Phys. Rev.* D **32**, 1899 (1985).
7. D. Polarski and A.A. Starobinsky, Class. Quan. Grav. **13**, 377 (1996); J. Lesgourges, D. Polarski and A.A. Starobinsky, report gr-qc/9611019.
8. M. Sakagami, Prog. Theor. Phys. **79**, 443 (1989); R. Brandenberger, R. Laflamme, and M. Mijić, Mod. Phys. Lett. **A28**, 2311 (1990); Y. Nambu, *Phys. Lett.* B **276**, 11 (1992); B.L. Hu, J.P. Paz, and Y. Zhang, in *The Origin of Structure in the Universe*, 227, E. Gunzig and P. Nardone (eds.), (Kluwer, 1993); H. Kubotani, T. Uesugi, M. Morikawa, and A. Sugimoto, Ochanomizu University report, (1996).

MILLIMETER AND SUBMILLIMETER WAVELENGTH ARRAYS

J. M. MORAN

Harvard-Smithsonian Center for Astrophysics, 60 Garden St., Cambridge, MA 02138, USA

The characteristics of the existing four millimeter wavelength arrays and four of the arrays that have been proposed or that are under construction are described. The properties of various sites and the scientific objectives of the new projects are also discussed.

1 Introduction

A major focus in instrumentation for radio astronomy is in the area of high resolution ground–based arrays at millimeter and submillimeter wavelengths. There are several reasons for this emphasis. First, with the development of quantum mixer technology, receiver temperatures have improved dramatically recently (e.g. from about 1000 K to less than 50 K between 1985 and 1995 at 345 GHz). Secondly, the scientific importance of the band has become increasingly apparent, especially for the study of galactic thermal sources and molecular gas at high redshift. The sensitivity of a radio array is

$$\Delta S \propto \frac{T_s}{A_e} \frac{1}{\sqrt{B\tau}} \qquad (1)$$

where ΔS is the rms fluctuation in flux density, T_s is the system temperature referred outside the atmosphere, A_e is the effective collecting area, B is the bandwidth, and τ is the integration time. For thermal dust emission the emissivity increases about as ν^2 so that the flux density in the Rayleigh–Jeans regime is proportional to ν^4. Hence, whereas dust emission is difficult to detect longward of 1 cm wavelength, and it is often opaque at optical wavelengths, it can be partially transparent and easily detectable at millimeter and submillimeter wavelengths. For spectral line observations it is more useful to express the sensitivity in terms of fluctuations in brightness temperature. In the Rayleigh–Jeans regime,

$$\Delta T_b \propto T_s A_e^{-1} (\Delta v \tau)^{-0.5} \nu^{-2.5} \theta^{-2} \qquad (2)$$

where Δv is the velocity resolution, θ is the angular resolution, and ν is the frequency. The power emitted in an optically thin spectral line is $P = h\nu A n_u V$ where h is Planck's constant, A is the Einstein coefficient for spontaneous emission, n_u is the upper level population, and V is the volume. Since $A \propto \nu^3$, and $n_u \propto \nu$ for temperatures below the peak of the Boltzmann distribution, the power scales as ν^5, the flux density scales as ν^4 and the brightness temperature scales as ν^2.

2 Instruments

There are currently four millimeter arrays in operation [The Berkeley-Illinois-Maryland Array (BIMA), the Owens Valley Radio Observatory (OVRO) array, the Institute for Millimeter Radio Astronomy (IRAM) array and the Nobeyama Radio Observatory (NRO) array]. Their characteristics are listed in Table 1. These sites could support some operation at wavelengths as short as 0.8 mm. The Submillimeter Array (SMA) is a joint project of the Smithsonian Astrophysical Observatory and the Institute of Astronomy and Astrophysics of the Academia Sinica (Taiwan). The Millimeter Array (MMA) is a project of the National Radio Astronomy Observatory. Funds for its initial design are currently in the FY 1998 budget. No site has been chosen but extensive testing at a high site in the Atacama desert of Chile has been undertaken. The Large Millimeter and Submillimeter Array (LMSA) is a project of the National Astronomical Observatory of Japan (NAOJ) These two instruments may be located close to each other for enhanced resolution and sensitivity. The European initiative, the Large Souther Array (LSA), is currently oriented towards an instrument of very large collecting area, operating at a larger and more accessible 3 km elevation site, that would concentrate on longer wavelengths.

Table 1: Properties of Millimeter and Submillimeter Arrays

Inst	Elements	Area m	Elev m	min λ mm	Funds	Complete
BIMA	9 × 6 m	250	1043	1.4	complete	now
OVRO	6 × 10 m	470	1236	1.4	complete	now
IRAM	5 × 15 m	880	2552	1.4	complete	now
NRO	6 × 10 m	470	1350	2	complete	now
SMA	8 × 6 m	225	4084	0.3	secure	1999
MMA	40 × 8 m	2000	5050	0.3	dev/FY98?	>2006
LMSA	50 × 10 m	3900	5050	0.3	pending	>2008
LSA	50 × 16 m	10050	3300	0.8	pending	?

426

3 Atmosphere

Atmospheric water vapor is a major problem at millimeter and submillimeter wavelengths. It absorbs signals, increases background level, and introduces phase fluctuations that distort the images. For 1 mm of precipitable water vapor, the atmospheric transmission is about 0.95 at 230 GHz, 0.8 at 345 GHz, and 0.4 at 460 GHz, 690 and 805 GHz. Beyond 1000 GHz the opacity increases even more but there is some possibility of ground based operations up to 1050 GHz from sites above 6000 m. The quality of a site is primarily determined by its altitude since the water vapor is approximately exponentially distributed with a scale height of about 2 km. There is a lesser dependence on latitude and local weather conditions (see Table 2). Phase fluctuations can be a severe problem. The atmosphere below 1000 GHz is largely non-dispersive and hence phase fluctuations are proportional to frequency. For strong sources the techniques of self–calibration can be used. There is expectation that the atmospheric phase shifts can be removed by rapid switching or precise measurement of the atmospheric emission.

4 Science

An overview of scientific applications can be found in the proceedings of a conference entitled *Science with Large Millimeter Arrays*[1]. Among the goals of the MMA are: (1) image anisotropies in the CMB (2 μK in one hr; (2) image S–Z effect in distant clusters, (3) image dust emission in galaxies to z = 20; (4) detect CO in starburst galaxies to z = 3; (5) achieve resolution of 0.25 pc on Magellanic Clouds; (6) trace star formation conditions from cores (0.1 pc) to circumstellar disk (5 pc); (7) detect gaps in circumstellar disks due to planets; (8) detect stars in every part of H-R diagram; (9) study dust formation around evolved stars; and (10) obtain orbits for near Earth objects.

Table 2: Percentage of Time that Precipitable Water Vapor is less than 1 mm

Site	Possible Instrument	Elevation m	Time %	Ref
Chajnantor (Chile)	MMA/LMSA	5050	55	[2]
Rio Frio (Chile)	LMSA	4100	32	[2]
Mauna Kea (Hawaii)	SMA	4100	25	[3]
El Chino (Chile)	LSA	3300	18?	[1]

5   The Submillimeter Array

The only array currently under construction is the SMA. It will be located on Mauna Kea approximately 200 m west of the JCMT. The array will consist of eight 6–meter diameter antennas arranged on the sides of Reuleaux triangles. Four configurations will be available with diameters of approximately 24, 64, 171 and 470 m. Each antenna will be equipped with a cryostat at its Nasmyth focus that will accept eight receivers covering all usable bands from 230 to 850 GHz. The maximum angular resolution will vary from 0.4 to 0.1 arcseconds over the frequency range. Signal processing will be performed on a special purpose XF correlator, which is based on a chip developed at the Haystack Observatory and the NASA Space Engineering Research Center for VLSI Design. The correlator will accept two channels (for either dual polarization or dual frequency operation) from each antenna of 2 GHz bandwidth each. The subchannel bandwidth is 104 MHz. Spectral resolutions as fine as 0.6 km/s will be available with full processing capacity.

The antennas have reflector backup structures constructed of carbon fiber tubes and steel nodes. Each primary reflector consists of 72 machined aluminum panels. These panels have an rms accuracy of about 5 microns and the overall reflector surface is expected to have an rms accuracy of 12 microns. In the laboratory, receiver temperatures of 25, 30 and 65 K(DSB) have been achieved at 230, 345 and 460 GHz (1.4, 0.9 and 0.7 mm), respectively, for SIS mixer receivers with junctions fabricated at JPL. The first antenna has been assembled and preliminary receiver tests have been performed at 230 GHz. The initial interferometric tests on celestial sources are planned for late 1997 at the assembly site on the grounds of the Haystack Observatory in Westford, MA. Preparations are also being made to include the 15-m JCMT and 10-m CSO telescope in the array on a part time basis.

428

References

1. P. A. Shaver, Science with Large Millimeter Arrays, ESO Astrophysics Symposia, Springer, 1996.
2. M. A. Holdaway, M. Ishiguro, S. M. Foster, R. Kawabe, K. Kohno, F.N. Owen, S.J.E. Radford, M. Saito, NRAO Millimeter Array Memo Series No. 152, 1996.
3. C. R. Masson, in Astronomy with Millimeter and Submillimeter Wavelength Interferometry, ASP Conference Series, vol. 59, 87, 1994.

VERITAS: the Very Energetic Radiation Imaging Telescope Array System

T.C.WEEKES

Whipple Observatory, Harvard-Smithsonian CfA, P.O.Box 97, Amado, AZ 85645-0097, USA

A next generation atmospheric Cherenkov telescope (ACT) is described based on the Whipple Observatory γ-ray telescope. A total of nine such imaging telescopes will be deployed in an array that will permit the maximum versatility and minimum flux sensitivity in the 100 GeV - 50 TeV band.

1 Introduction

Recent successes in Very High Energy (VHE) γ-ray astronomy using the atmospheric Cherenkov technique have triggered a spate of projects aimed at extending and improving the detection technique. Although these projects (at various stages of conceptual design, detailed planning or actual construction) have radically different experimental approaches, they have important features in common: (i) all are based on the belief that the scientific benefits gained by an increase in sensitivity justify a major effort; (ii) all are agreed that major improvements in sensitivity are physically possible and technically straightforward; (iii) by the standards normally used in this field all the projects are expensive (in the $3M to $15M range).

State-of-the-art imaging ACTs are best represented by the Whipple Collaboration telescope in southern Arizona; the French CAT telescope in the French Pyrenees, the Armenian-German-Spanish HEGRA telescope array in the Canary Islands, the Durham telescope in Narrabri, Australia, the Australian-Japanese telescopes at Woomera, Australia and the Japanese Telescope Array in Utah, USA.

The Whipple Observatory γ-ray telescope consists of a 10m diameter optical reflector focussed onto an array of 109 PMTs. The energy threshold is ~300 GeV and the telescope obtains a 5σ excess from the Crab Nebula in one-half hour of on-source observations. In 1998 it is planned to replace the 109 pixel camera with a 541 pixel camera.

The features of VHE ACTs that can be improved include: (a) energy threshold; (b) flux sensitivity; (c) energy resolution; (d) angular resolution; (e) field of view. Of these energy threshold is the easiest to achieve since energy threshold scales as $(\text{mirror area})^{-1}$. The proposed detector, VERITAS, would make improvements in all these parameters.

430

2 VERITAS

The philosophy underlying VERITAS comes from 30 years of development of ACTs at the Whipple Observatory; the objective is to build a VHE γ-ray observatory which will have a useful lifetime well into the next century. The initial aim is to have the maximum sensitivity in the 100GeV-10TeV range but to have significant sensitivity down to 50 GeV (and lower as new technology photo-detectors become available) and as high as 50TeV (using the low elevation technique). The detection technique will be the so- called "imaging" atmospheric Cherenkov technique which was originally demonstrated at the Whipple Observatory but is now under considerable development at a number of centers. The basic telescopes will be modelled on the Whipple 10m telescopes with wide field cameras of 331 to 541 pixels. The array will consist of nine such telescopes, all capable of independent or coincident operation. The telescope layout will have three telescopes at the vertices of an equilateral triangle of side 20 m surrounded by six telescopes on a hexagonal of side 50m.

The proposed location of VERITAS is a flat area at the Whipple Observatory Basecamp (elevation 1.3km) where there is ample space for development as well as easy access to roads, power, etc. Southern Arizona has been shown to be an excellent site for these kinds of astronomical investigation with an impressive record of clear nights. The dark site is not environmentally sensitive nor is there the potential for conflict with other astronomical activities.

The parameters of the array are chosen to give the optimum flux sensitivity in the 100GeV-10TeV range which has proven to be rich in scientific returns. The predicted flux sensitivity is shown in Figure 1; it is seen to be a factor of ten better than any other detector in this range. In these two decades of energy the major background comes from hadron-initiated air showers for which successful identification methods have been developed. At the lower end single muons become the major background but these can be removed by the coincident requirement in the separated telescopes. At lower energies the cosmic electron background constitutes an irreducible isotropic background. Over these two decades of energy the angular and energy resolutions will be pushed to their limits (0.05° and 8% respectively).

3 Conclusions

There are a number of alternative projects designed to increase the sensitivity of telescopes in the 10GeV-10TeV range. These include the solar farm projects (STACEE in the USA and CELESTE in France), the single dish approach (MAGIC in Germany), the large water Cherenkov air shower detector (MILAGRO in New Mexico, USA), the next generation space telescope (GLAST

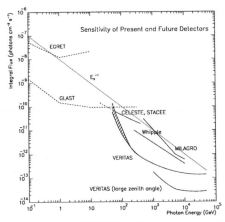

Figure 1: Predicted sensitivity of VERITAS. Also shown are the known sensitivities of EGRET and Whipple and the predicted sensitivities of the MILAGRO, STACEE, CELESTE and GLAST experiments, all of which are at various stages of planning and construction. The exposure for VERITAS, STACEE and CELESTE is 50 hours. The exposure for EGRET, GLAST and MILAGRO is one year of sky survey operation.

which will not be launched before 2004). All of these have merit in particular areas; the relative merits of these projects are compared with VERITAS for different experimental parameters in Table 1. In this table the ranking system is *** = excellent; ** = very good: * = good and blank = not good. Since both the choice of parameters and the ranking are assigned by the author it is not surprising that VERITAS compares favorably with all the other projects.

Table 1: *Comparison of Proposed Projects*

Concept	Solar Farm	Single Dish	Array	Particle	Space
Project	STACEE	MAGIC	VERITAS	MILAGRO	GLAST
En. Thres.	***	**	*	*	***
Dynamic Range	*	**	**	**	***
Flux Sens.	*	**	***	*	***
Energy Res.	*	*	***	*	**
Angular Res.	*	**	***	*	**
FOV		**	***	***	***
Cost	***	**	**	**	*
Timeliness	***	**	**	***	*

DEVELOPMENT OF THE SOLAR TOWER ATMOSPHERIC CHERENKOV EFFECT EXPERIMENT (STACEE)

René A. Ong

The Enrico Fermi Institute, The University of Chicago,
5640 S. Ellis Ave, Chicago, IL 60637, USA

(For the STACEE Collaboration[1])

STACEE is a proposed telescope for ground-based gamma-ray astrophysics between 25 and 500 GeV. The telescope will make use of large mirrors available at a solar power plant to achieve an energy threshold lower than existing ground-based instruments. This paper describes recent development work on STACEE.

1 Introduction

Discoveries from the Compton Gamma Ray Observatory (CGRO)[2] and from ground-based experiments[3] indicate that the high energy sky is rich with interesting astrophysics. Yet, there is a gap in experimental coverage between 20 and 250 GeV. Satellite instruments, such as GLAST,[4] may eventually extend their reach above 20 GeV, but the experiments with the most promise to explore the gap in the near future are ground-based detectors using the atmospheric Cherenkov technique.

The energy threshold of atmospheric Cherenkov detectors is governed by a number of parameters, of which the easiest to control is the mirror collection area.[5] Large collection area translates into lower energy threshold, and large solar mirrors (heliostats) are readily available at existing power facilities. Since early 1994, we have been developing an experiment (STACEE) to use heliostat mirrors for Cherenkov astronomy. A similar experiment (CELESTE) is also under development in France.[6]

2 STACEE Development

Our development work has concentrated on the key issues associated with building an innovative atmospheric Cherenkov detector. We have carried out tests using prototype detector equipment at two solar heliostat fields, the Solar Two Power Plant (Barstow, CA) and the National Solar Thermal Test Facility (NSTTF) at Sandia National Laboratories (Albuquerque, NM). The results from work at Solar Two have been published[7] and in 1996, the successful tests at the NSTTF encouraged us to develop a complete instrument design using

48 heliostats at Sandia. **Documents describing the Test Results and the STACEE design can be found on the Web.**[8] Here we *very briefly* summarize these documents.

3   Results from the Sandia Tests

We carried out two tests at Sandia (Aug. and Oct. 1996). To summarize:

- we verified that the site is suitable for Cherenkov astronomy by measuring the clarity of the sky and the ambient flux of night sky photons, and

- we determined that the heliostats have excellent pointing accuracy ($\sim$ 0.04°) and stability ($\sim$ 0.05°), and typical spot sizes of 1.5 m and reflectivities of $\sim$ 80%.

We built a complete detector prototype consisting of a 2 m secondary mirror, support structure, photomultiplier tube (PMT) camera, electronics and data acquisition system.[8] The detector prototype performed extremely well, and it proved easy to detect atmospheric Cherenkov radiation from cosmic ray showers with little accidental background. Using these showers:

- we measured the trigger rate dependence on zenith angle, the effect of tilting the heliostats to the interaction point, and the cosmic cosmic ray spectral index, and

- from the trigger rate (5 Hz) and simulations, we determined a cosmic ray energy threshold of $\sim$ 290 GeV for vertical showers.

The cosmic ray threshold translates into an effective gamma-ray threshold of $\sim$ 75 GeV, indicating that the prototype instrument operated at a lower energy threshold than any atmospheric Cherenkov detector to date.

4   Overall Detector Design

The full experiment will use 48 heliostats at Sandia, corresponding to a total mirror area of $\sim$ 1770 m^2. The heliostats will be divided into three sectors and Cherenkov light from each sector will be reflected onto a separate 2 m diameter secondary mirror. Each secondary will image the light onto a 16-element camera, consisting of PMTs equipped with Winston cones.

The PMT signals will be amplified and discriminated. The discriminated signals will be delayed and combined to form an overall multiplicity trigger. The amplified PMT signals will be continuously sampled by a digitizer which

434

will store a waveform for each PMT upon receipt of a trigger. The PMT arrival times and pulse-heights will be determined from the digitized waveforms.

Simulations show that STACEE will have a substantial collection area $(10,500\,\mathrm{m}^2)$ for 50 GeV gamma-ray primaries, and that the experiment will be fully efficient by 75 GeV. In addition, the experiment should possess substantial capability to reject hadronic cosmic rays (rejection factor of $\sim$210 at 50 GeV and $\sim$ 95 at 100 GeV), as a result of the rapid decease in the Cherenkov yield for cosmic rays below 200 GeV, the narrow field-of-view of each heliostat, the multiplicity trigger condition, and the measured lateral distribution of the Cherenkov light. ¿From simulations we expect that STACEE will have excellent point source sensitivity ($\sim 8\sigma$ significance on the Crab in one hour).

5 Summary

We have completed the design of an innovative atmospheric Cherenkov detector sensitive to gamma-rays in an unexplored energy region. The complete experiment can be built on a two year timescale.

Acknowledgments

We thank Richard Fernholz, G.H. Marion, Antonino Miceli, Heather Ueunten, Patrick Fleury, Eric Paré, David Smith, and staff of the National Solar Thermal Test Facility. This work was supported by the National Science Foundation, the Natural Sciences and Engineering Research Council, the California Space Institute, and the University of Chicago.

References

1. M.C. Chantell, C.E. Covault, M. Dragovan, R.A. Ong, S. Oser (Univ. of Chicago), D.S. Hanna, K. Ragan (McGill Univ.), O.T. Tumer, D. Bhattacharya (Univ. of California, Riverside), D.A. Williams (Univ. of California, Santa Cruz), R. Mukerjee (USRA), P. Coppi (Yale Univ.), and D.T. Gregorich (California State Univ., Los Angeles).
2. D.J. Thompson *et al.*, Ap. J. Suppl. **101**, 259 (1995).
3. See, for example: J. Gaidos *et al.*, Nature **383**, 319 (1996).
4. GLAST Home Page: http://www-glast.stanford.edu/
5. T.C. Weekes, Phys. Rep. **160**, 467 (1988).
6. Celeste Home Page: http://wwwcenbg.in2p3.fr/Astroparticule/celeste/
7. R.A. Ong *et al.*, Astroparticle Phys. **5**, 353 (1996).
8. STACEE Home Page: http://hep.uchicago.edu/~stacee/

STATUS REPORT OF THE KASCADE EXPERIMENT TO MEASURE COSMIC RAY COMPOSITION

KASCADE Collaboration: G. Schatz[a], W.D. Apel[a], K. Bekk[a], E. Bollmann[a],
H. Bozdog[c], I.M. Brancus[c], M. Brendle[d] A. Chilingarian[e], K. Daumiller[b], P. Doll[a],
J. Engler[a], M. Föller[a], P. Gabriel[a], H.J. Gils[a], R. Glasstetter[a], A. Haungs[a],
D. Heck[a], J. Hörandel[a], K.-H. Kampert[a,b], H. Keim[a], J. Kempa[f], H.O. Klages[a],
J. Knapp[b], H.J. Mathes[a], H.J. Mayer[a], H.H. Mielke[a], D. Mühlenberg[a],
J. Oehlschläger[a], M. Petcu[c], U. Raidt[d], H. Rebel[a], M. Roth[a], H. Schieler[a],
G. Schmalz[a], H.J. Simonis[a], T. Thouw[a], J. Unger[a], G. Völker[a], B. Vulpescu[c],
G.J. Wagner[d], J. Wdowczyk[f], J. Weber[a], J. Wentz[a], Y. Wetzel[a], T. Wibig[f],
T. Wiegert[a], D. Wochele[a], J. Wochele[a], J. Zabierowski[f], S. Zagromski[a],
B. Zeitnitz[a,b]

[a] *Institut für Kernphysik, Forschungszentrum Karlsruhe, D 76021 Karlsruhe*
[b] *Institut für Experimentelle Kernphysik, Universität Karlsruhe, D 76021 Karlsruhe,
Germany*
[c] *Institute of Physics and Nuclear Engineering, RO 7690 Bucharest, Romania*
[d] *Physikalisches Institut, Universität Tübingen, D 72076 Tübingen, Germany*
[e] *Cosmic Ray Division, Yerevan Physics Institute, Yerevan 36, Armenia*
[f] *Inst. for Nuclear Studies and Dept. of Experimental Physics, University of Lodz,
PL 90950 Lodz, Poland*

A new extensive air shower (EAS) experiment has been installed at the site of
the Forschungszentrum Karlsruhe. The main aim of the KASCADE project is the
determination of the chemical composition in the energy range around the knee
of the primary cosmic ray spectrum by the simultaneous measurement of a large
number of observables for each individual event. Data taking with a large part of
the experiment has started. First preliminary results are presented.

1 Introduction

The chemical composition of ultrahigh energy cosmic ray particles is an important clue for the modelling and understanding of cosmic ray origin, acceleration and transport. Especially in the energy range around the knee, a distinct
change of the spectral index in the primary spectrum at about 5×10^{15} eV,
the determination of the relative abundances of light and heavy nuclei and
the change of this ratio with energy is of prime interest. Direct measurements
using detector systems on satellites, space craft or high altitude balloons are
limited in detector area and exposure time. Therefore, data above 10^{14} eV are
sparse and lack statistical accuracy [2,4]. The KASCADE experiment [1] attempts
to determine primary composition in an energy range around the knee in the
spectrum by observing extensive air showers (EAS) on ground level.

436

It is well known that a number of characteristics of EAS depends on the energy per nucleon of the primary nucleus, notably the ratio of electron to muon numbers, the energies of hadrons in the shower, and the shapes of the lateral distributions of the various components of the shower.

2 Layout and Status of the Experiment

KASCADE (Karlsruhe Shower Core and Array Detector) is located on the site of the Forschungszentrum Karlsruhe, Germany, at 8° E, 49° N, 110 m above sea level. It consists of three main components: detector array, central detector system, and muon tunnel. An extensive description of the experiment has recently been published by Klages et al. [1] to which the reader is referred for experimental details. Data taking has started in late 1995 with large parts of the experiment in separate mode. Correlated data have been registered first in April 1996 with 100 m^2 of the calorimeter operational.

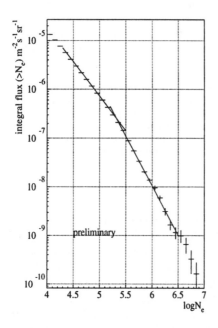

Figure 1: Event rate vs. reconstructed shower size for EAS in the zenith angle range 15° to 25° (one week of data).

Figure 2: Distribution of hadronic energy in a shower core as measured by the calorimeter. The total energy amounts to c. 43 TeV.

This has been increased to 200 m² in the mean time. The remaining part of the calorimeter and the muon tunnel are expected to start operation in 1997. Up to January 1997 c. 20 million correlated events have been registered.

3 Preliminary Results

Owing to the ongoing installation and calibration work data analysis is in a very preliminary stage. The lateral dependence of the signal amplitudes in the scintillators of the array can be well described by the NKG function from which the shower size can be determined. When we plot (fig. 1) the event rate against the reconstructed electron size of the showers in the zenith angle range 15° to 25° we find a distinct change in slope near log $N_e = 5.5$ in agreement with previous experiments. Fig. 2 displays the hadron energy distribution in a shower core as measured by the calorimeter. The shading indicates the energy of individual hadrons on a logarithmic scale. The total hadronic energy of the event displayed was around 43 TeV. The first measurements with the central detector system include the determination of the flux spectrum of single cosmic hadrons. The results are in good agreement with previous experiments and especially those with the prototype of the calorimeter[3].

4 Conclusions

The installation of the KASCADE experiment is nearing completion. Data taking with a large part of the experiment, including 2/3 of the calorimeter area, has started. The rest of the hadron calorimeter and the streamer tube muon detectors will be fully installed in 1997. The detectors perform as expected. Multiparameter analysis procedures are still under further development. Preliminary data look very promising.

References

1. H. O. Klages et al., Proc. 9. Int. Symp. Very High Energy Cosmic Ray Int., Karlsruhe 1996; Nucl. Phys. B Suppl., in press
2. D. Müller et al., Astrophys.J. 374 (1991) 356
3. H.H. Mielke et al., J. Phys. G: Nucl. Part. Phys. 20 (1994) 637 ; H. Kornmayer et al., J. Phys. G: Nucl. Part. Phys. 21 (1995) 439
4. M. Giller, paper presented at the XVth Cracow Summer School of Cosmology, Lodz, July 15 - 19, 1996, to be published

New results on cosmic ray H and He composition from the JACEE collaboration

B.S. NILSEN[1], K. ASAKIMORI[2], T.H. BURNETT[3], M.L. CHERRY[1], K. CHEVLI[4], M.J. CHRISTL[5], S. DAKE[6], J.H. DERRICKSON[5], W.F. FOUNTAIN[5], M. FUKI[7], J.C. GREGORY[4], T. HAYASHI[8], A. IYONO[9], J. IWAI[3], J. JOHNSON[4], M. KOBAYASHI[10], J. LORD[3], O. MIYAMURA[11], K.H. MOON[5a], H. ODA[6], T. OGATA[12], E.D. OLSON[3b], T.A. PARNELL[5], F.E. ROBERTS[5], K. SENGUPTA[1c], T. SHIINA[4], S.C. STRAUSZ[3], T. SUGITATE[11], Y. TAKAHASHI[4], T. TOMINAGA[11], J.W. WATTS[5], J.P. WEFEL[1], B. WILCZYNSKA[13], H. WILCZYNSKI[13], R.J. WILKES[3], W. WOLTER[13], H. YOKOMI[14], E. ZAGER[3].

1. Dept. of Physics and Astronomy, Louisiana State Univ., Baton Rouge, LA 70803; 2. Kobe Women's Junior College, Kobe, Japan; 3. Dept. of Physics, Univ. of Washington, Seattle, WA 98195; 4. Dept. of Physics, Univ. of Alabama, Huntsville, AL 35899; 5. NASA Marshall Space Flight Center, Huntsville, AL 35812; 6. Kobe Univ., Kobe Japan; 7. Kochi Univ., Kochi, Japan; 8. Waseda Univ., Tokyo, Japan; 9. Okayama Univ. of Science, Okayama, Japan; 10. KEK, Tsukuba, Japan; 11. Hiroshima Univ., Hiroshima, Japan; 12. Inst. for Cosmic Ray Research, Tokyo, Japan; 13. Inst. for Nuclear Physics, Krakow, Poland; 14. Tezukayama Univ., Nara, Japan.

Results for the cosmic ray hydrogen and helium spectra up to 800 TeV, near the "knee" region, are presented. There is no sign of a break in either the hydrogen or helium spectra. The differential power law slopes are 2.72±0.10 for hydrogen and 2.56±0.06 for helium. With these new H and He measurements, together with earlier reported results for the heavier elements, the sum of the spectra give an all-particle spectrum that is in good agreement with the all-particle spectrum measured using extensive air showers.

By making cosmic ray composition measurements through the all-particle spectrum's "knee" region (10^{14}–10^{16} eV), we hope to learn why the all-particle spectrum changes slope from ~2.6 to ~3.0[1]. Below the knee, instruments on balloons and satellites make direct measurements of the charge of individual cosmic rays; but near the knee and above, the all-particle measurements rely on indirect extensive air shower techniques. The JACEE (Japanese-American Cooperative Emulsion Experiment) collaboration has now measured the hydrogen and helium cosmic ray spectra up to 800 TeV for hydrogen and 400 TeV/n for helium.

JACEE uses electron-sensitive nuclear emulsion and x-ray film as the sensitive elements in a large-area balloon-borne thin electromagnetic sampling calorimeter. (See reference [2] and the references therein for a complete discussion of the techniques used.) JACEE has now completed fifteen flights, including two long duration flights (>70 m^2 hrs) from Australia to South America and four flights (>300 m^2 hrs) in Antarctica. We have now completed the analysis of the hydrogen and helium cosmic ray spectra through JACEE flight 12, thereby nearly doubling the total exposure reported previously. When we finish the analysis of JACEE flights 13 and 14, the exposure will double again.

The hydrogen and helium differential spectra are shown in Fig. 1. The JACEE results agree well with earlier measurements. There are now 25 hydrogen events above 90 TeV and 37 helium events above 25 TeV/n. The differential spectra are, however, sensitive to the binning used. Therefore, we fit to the integral spectra shown in Fig. 2.

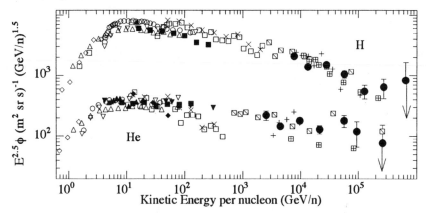

Fig. 1 Differential spectra for hydrogen and helium (●). Also shown for comparison are other measurements, see references in (2).

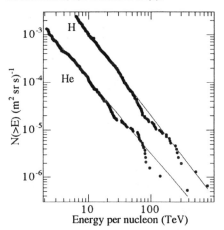

Fig. 2 Integral spectra for hydrogen and helium. Each point corresponds to one more event than the point to the right. At high energies the correlated nature of an integral spectrum and the low statistics are responsible for the characteristic "waviness".

There is no sign of a break in either distribution shown in Fig. 2. This is contrary to earlier results[3] which suggested a break in the hydrogen spectrum at about 40 TeV but no comparable break in the helium spectrum. Fitting the spectra to a single power law gives $N(>E)=$ 0.051 ± 0.014 $(E/1TeV)^{-1.72\pm0.10}$ $(m^2$ sr s$)^{-1}$ for hydrogen and $(3.90+0.24/ -0.58)\times10^{-3}$ $(E/1TeV/n)^{-1.56\pm0.06}$ $(m^2$ sr s$)^{-1}$ for helium with a $\chi^2_{d.o.f.}$ of 0.66 and 1.12 respectively.

In 1992 the Cosmic Ray Program Working Group issued a report advocating composition measurements through the knee region[4]. The JACEE results now make possible a direct comparison of the balloon-borne and air shower results at these energies. Using these new hydrogen and helium spectra, along with fits to preliminary JACEE CNO, Ne-S, and Fe group data[5], an estimate of the all-particle spectrum can be made (Fig. 3). The cutoff in the spectra are chosen here to be at Zx1000 TeV/nucleus, where Z is the particle charge, as suggested by supernova shock wave acceleration models. We assume no flattening

440

in the Ne-S and Fe group spectra (as might be expected when the rigidity-dependent escape mean free path begins to dominate the interaction mean free path[6]). The all-particle spectrum obtained from summing these spectra (assuming power laws with no flattening and a cutoff at Zx1000 TeV/nucleus) is remarkably close to the observed all-particle spectrum.

With these new results, JACEE is beginning to determine the composition approaching the all-particle knee region. With the analysis of JACEE flights 13 and 14 we will either find a knee in the hydrogen and/or helium spectra or push up the position of the knee to still higher energies.

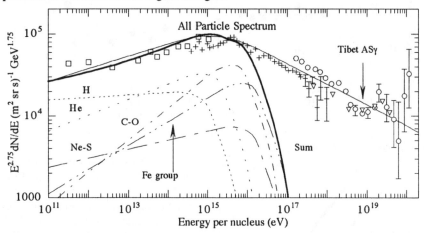

Fig. 3 The all-particle spectrum plus the elemental spectra measured by JACEE. The thick solid line is the sum of the elemental spectra.

References

a Deceased

b Present Address: WRQ Inc., 1500 Dexter Ave. N., Seattle, WA 9809

c Present Address: Horizon Computer Corp., 5 Lincoln Hwy., Edison, N.J. 08820

1 Amenomori, M. *et al.*, Ap. J. **461**, 408 (1996)

2 Asakimori, K. *et al.*, submitted to Ap. J. (1997)

3 Asakimori, K. *et al.*, 1993, Proc. 23rd Intl. Cosmic Ray Conf. (Calgary) **2**, 21 and **2**, 25

4 Waddington, C. J. *et al.*, "Galactic Origin and the Acceleration Limit", NASA report, September 1992

5 Takahashi, Y., Proc. IInd Rencontres du Vietnam, "The Sun and Beyond", Oct. 21-28, 1995, Ho Chi-Minh City, ed. Tran Thanh Van (World Scientific 1996)

6 Esposito, J. A. *et al.*, Ap. J. **351**, 459 (1990)

TRAJECTORIES OF ULTRA HIGH ENERGY COSMIC RAYS

G.A.Medina Tanco, E.M.de Gouveia Dal Pino and J.E.Horvath

Instituto Astronômico e Geofísico, Universidade de São Paulo Av. M.Stéfano 4200, 04301-904 São Paulo SP, BRAZIL

1 Where do Ultra High Energy Cosmic Rays come from ?

Several events corresponding to primary energies above $10^{20}\,eV$, the so-called Ultra-High Energy regime (or UHE) have been recently shown to exist beyond any reasonable doubt[1]. Those particles pose a big challenge for the acceleration models and may point to the sources under certain conditions on the deviating effects. In order to help the UHECR source identification we have performed 3-D simulations of proton trajectories propagated through the IGM and halo coponent. We have calculated the arrival direction and the line-of-sight to the source and the time delay $\Delta\tau$ of the UHECR with respect to photons. We wish to address how strongly do these particles point to their sources.

Given the energy range we are interested in, we have included only photomeson production losses due to interactions with the CMBR photons in the code. Our assumed model for the magnetic field configuration is one in which B is uniform on scales smaller than its correlation length L_c, that is, the halo and IGM consist of bubbles without voids from the source to the Earth.

Several numerical experiments have been performed by injecting $\sim 8 \times 10^5$ particles having a $N(E) \propto E^{-2}$ spectrum in the energy interval $2 \times 10^{19}\,eV - 1 \times 10^{23}\,eV$. Although the propagated particles are tracked down to energies below $10^{20}eV$, our analysis is strictly valid for $E > 10^{20}eV$. The simulations were performed for sources of UHECR at extended halo ($100\,kpc$) and nearby extragalactic ($50\,Mpc$) distance scales (different propagation models and physical conditions are in progress). The simulations allow an accurate explicit determination of α and $\Delta\tau$, an information which is lost in diffusion-type schemes.

2 Results and Discussion

We have checked first that (in agreement with the GZK cutoff expectation) an extragalactic origin is possible if $d \leq 50\,Mpc$, otherwise the injection energy must be unreasonably high[2]. The full results of our work can be appreciated in the following figures:

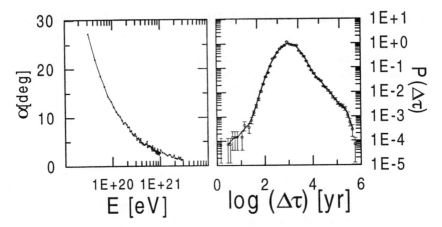

Figure 1: a) Arrival angle α of a proton of energy E. Extended halo case for a power-law injected spectrum (see text),source distance $d = 100\,kpc$, magnetic field strength $B_H = 10^{-6}\,G$ and $L_c = 1.5\,kpc$. b) Probability density distribution $P(\Delta\tau)$ as a function of the proton delay with respect to photons $\Delta\tau$, assuming simultaneous injection.

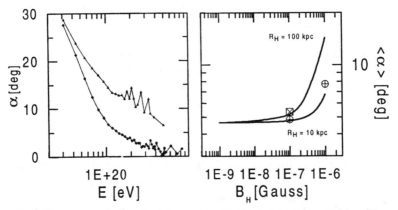

Figure 2: a) The same as in Fig. 1a for an extragalactic source at $d = 50\,Mpc$. The curves are given for $B_{IGM} = 10^{-9}\,G$, $L_c = 1\,Mpc$ without considering the effects of B_H (lower curve) and, for the same values of the IGM, with the inclusion of a maximally magnetized halo having $B_H = 10^{-6}\,G$, $L_c = 1.5\,kpc$ and size $R_H = 100\,kpc$ (upper curve). b) Influence of the halo on α of extragalactic protons injected at $d = 50\,Mpc$ with a power-law spectrum and traveling through an IGM ordered on $1\,Mpc$ scale. Two halo models are considered : a standard R_H of $10\,kpc$ (lower curve) and an extended one with $R_H = 100\,kpc$ (upper curve). The curves span all cases ranging from asymptotically negligible halos (for $B_H < 10^{-7}\,G$) to maximally deviating halos having $B_H = 10^{-6}\,G$ (end of the curves), an extreme and unlikely situation. The symbols show three simulations in which L_c has been doubled for $R_H = 10\,kpc$ (crossed circles) and $R_H = 100\,kpc$ (square).

From the results we conclude that, within this class of models not including a large-scale symmetry of the magnetic field, the association of the UHECR with sources in the supergalactic plane[3] is not supported unless the halo is very extended/strongly magnetized. The Yakustsk-Fly's Eye events are consistent with a single source. From Fig.1b we also claim that a $\sim$ few month delay of UHECR with respect to GRB has a very low probability $<\ 10^{-4}$ for extended halo sources[4]. More work on this problem is clearly needed to identify the origin of these primaries[5].

References

1. See for example D.Bird et al. ApJ 441, 144 (1995) and references therein.
2. F.A.Aarhonian and J.W.Cronin, Phys.Rev.D 50, 1892 (1992).
3. T.Stanev et al., Phys.Rev.Lett. 75, 3056 (1995).
4. M.Milgrom and V.V.Usov, Astropart. Phys. 4, 365 (1996).
5. A full version of this work can be retrived at astro-ph/9610172.

ON COSMIC RAY ACCELERATION BY PULSAR WIND

MARIA GILLER, WOJCIECH MICHALAK

Division of Experimental Physics, University of Lodz,
Pomorska 149/153, 20-236 Łódź, Poland

The pulsar driven SNRs contain large-scale magnetic and electric fields. If cosmic ray particles, accelerated by the shock in the interstellar medium, could get into the pulsar wind region, they would be further acccelerated. We show that the situation when $\vec{\mu} \cdot \vec{\Omega} > 0$ ($\vec{\mu}$ – pulsar magnetic moment, $\vec{\Omega}$ – its angular velocity of rotation) is much more favourable than the opposite case ($\vec{\mu} \cdot \vec{\Omega} < 0$) considered earlier.

1 Introduction

The bulk of cosmic ray (CR) particles is believed to be accelerated by the SN induced shocks in the interstellar medium. The maximum attainable energy is however of the order of $10^{14} - 10^{15}$ eV (for protons). The question arises, what mechanisms are responsible for CR acceleration to higher energies. Here we consider a possibility that in the pulsar wind driven SNRs high energy particles (accelerated by the outer shock) can penetrate inside the SN cavity and be further accelerated by the large scale electric field of the pulsar wind. This idea was put forward by Bell[1] and Berezhko[2,3] and developed further by us[4]. Bell considered the situation when the pulsar magnetic moment has a negative component along the direction of its rotational velocity $\vec{\Omega}$. In consequence, the large scale electric field $\vec{E}$ (directed along meridians) points from poles towards equator. High energy particles can drift inwards to the nebula due to the zenith angle gradient of the (azimuthal) magnetic field $\vec{B}$. For a particular choice of $B(\theta)$ only particles which enter at the poles with zenith angles $\theta < \theta_{max} = (cp_\perp /3 \cdot \Delta E_{max})^{1/3}$ (where ΔE_{max} is the maximum energy gain, $p_\perp$ is particle momentum perpendicular to $\vec{B}$) will be drawn inside and accelerated.

In our earlier work [4] we have calculated the effect that the pulsar wind acceleration would have on the particle energy spectrum for the case $\vec{\mu} \cdot \vec{\Omega} < 0$. The numerical values for nebula size, magnetic field and expansion velocity were adopted as for the Crab Nebula. The incident energy spectrum corresponded to the most probable spectrum obtained after diffusive shock acceleration acting for $\sim 10^3$ years [5] (age of Crab Nebula). We showed that the acceleration was significant only for a small fraction of particles (less than 10^{-2}). Here we consider the situation when $\vec{\mu} \cdot \vec{\Omega} > 0$ i.e. particles enter at equator (with zenith angles θ closer to the equator than $\theta_{max} = (cp_\perp / 6 \cdot \Delta E_{max})^{1/2}$)

and exit at zenith angles farther from the equator. We shall show that for this case fraction of accelerated particles becomes considerably larger.

2 Model of the nebula

The details of the pulsar wind model, adopted here, are described by Bell [1] and us [4], being based on earlier works by Rees and Gunn [6] and Kennel and Coroniti [7]. We only remind here that according to it, the pulsar wind fills the whole SN remnant (nebula) of radius $R_n = 5 \cdot 10^{18}$ cm, forming an inner shock at radius $R_s = 0.1 \cdot R_n$. The azimuthal magnetic field, frozen in the wind, increases towards the outer regions, reaching values $< 10^{-3}$ G at R_n. The maximum energy gain $\Delta E_{max} = 4 \cdot 10^{15}$ eV for the above parameters.

3 Method of calculation

We have calculated the effect of the existence of the wind electric field on the energy spectrum of particles accelerated first by the diffusive mechanism at the outer shock at R_n for the case $\vec{\mu} \cdot \vec{\Omega} > 0$. We followed numerically the particle trajectories. In conditions where it could be applied (usually on the particle way outward) we used the guiding centre approximation. Particles, incident on the surface $R = R_n$, were isotropically distributed, with the energy spectrum $\sim E^{-2} \cdot \exp(-E^2/E_{cut}^2)$, corresponding roughly to the spectrum obtained from the diffusive shock acceleration going on for a finite time. We expect $E_{cut}/\Delta E_{max} \sim 5 \cdot 10^{-3}$ for the Crab Nebula. Only particles entering the nebula near the equator can be drawn inside (the inward component of the $\vec{B} \times grad B$ drift overcomes the outward $\vec{E} \times \vec{B}/B^2$ drift) and leave the nebula at larger latitudes (on both hemispheres). The most effective acceleration occurs however, when a particle reaches the inner shock at R_s and enters inside it.

4 Results and discussion

Fig.1 shows the initial differential spectrum ($\times E^2$) together with the final one obtained after particles gained energy after one cross of the nebula – curve A. We see that the effect is not large, neither in the energy change, nor in number of the accelerated particles. For this case however, we have not allowed for field irregularities. Particle trajectories (and thus their energy gains) are sensitive to the field near the equator (where it tends to zero) and a scattering on field irregularities may increase the particle fraction reaching the inner shock at R_s. To estimate this effect we considered two other cases: case B - all particles with initial distances from equator less than their local Larmor radius enter the inner shock, case C - all particles which enter the nebula at R_n, enter the inner shock at R_s (this case can be considered as an upper limit to the modified spectrum). It can be seen from fig.1 that for these two cases the final spectrum differs dramatically from the case A. The cut-off of the modified spectrum even for the case B extends up to ΔE_{max} for large fraction of particles. This effect is much bigger here than for the case $\vec{\mu} \cdot \vec{\Omega} < 0$ considered by Giller and Michalak [4]. Fig.2 shows a comparison of the modified energy spectra for the two situations (entry at pole and equator) for the cases B and C. These results show that pulsar wind may play an important role (in the considered sort of objects) in the particle acceleration. Of course, a more detailed study, taking into account the time evolution of a pulsar driven SNR, is needed.

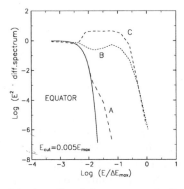

Fig.1. Particle spectra before (solid) and after (dashed) wind acceleration: case A – no scattering, case B and C – estimations of the scattering effect (see text). Entry at equator.

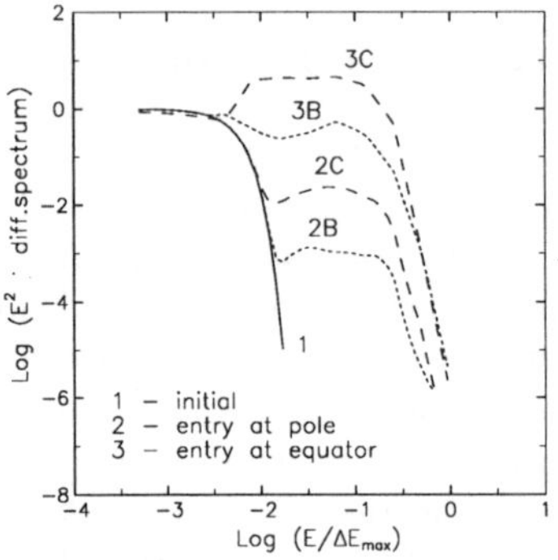

Fig.2. Comparison of the possible scattering effect. Particle spectra before (line 1) and after wind acceleration: 2 – entry at pole, 3 – entry at equator.

References

1. Bell A.R., *MNRaS* **257**, 493 (1992)
2. Berezhko E.G., *Proc. 22nd ICRC, Dublin* **2**, 436 (1991)
3. Berezhko E.G., *Proc. 23rd ICRC, Calgary* **2**, 348 (1993)
4. Giller M. and Michalak W., *Proc. 24th ICRC, Rome* **3**, 249 (1995)
5. Drury L.O'C, *Proc. 21st ICRC, Adelaide* **12**, 85 (1990)
6. Rees M.J. and Gunn J.E., *MNRaS* **167**, 1 (1974)
7. Kennel C.F. and Coroniti F.V., *Ap.J.* **283**, 694 (1984)

Propagation of UHE cosmic rays in a structured universe

Jörg P. Rachen

Pennsylvania State University, Astronomy Department
525 Davey Lab, University Park, USA-PA 16802. jorg@astro.psu.edu

In a gravitationally unstable universe, the structure of dark matter and galaxies, intergalactic gas and magnetic field can have severe impact on the propagation of ultra high energy cosmic rays (UHECR)[2] The possible effects include spatial confinement and directional focusing along the supergalactic matter sheets, as well as universal re-acceleration at large scale shock fronts, and spectral modification due to energy dependent leakage into cosmic voids. As a result, the GZK-cutoff may be less pronounced and occur at a higher energy, where the stochastic nature of both acceleration and energy loss processes has to be taken into account.

1 Supergalactic magnetic field structure and cosmic ray confinement

Very little is known about the strength and orientation of the magnetic field outside our galaxy. For cosmic ray transport calculations, one mostly uses the assumption of a nanogauss field, which is homogeneous over cells with some reversal scale of order 1 Mpc. In such fields, the highest energy cosmic ray protons have a gyro-radius $r_g \sim 300$ Mpc, thus propagate almost in straight lines; this opens the possibility of an "UHECR astronomy", as anticipated by the Pierre Auger Project[1]

Models of structure formation in cosmology, however, draw a quite different picture: The magnetic field is aligned with the matter sheets, where it can reach a field strength up to $\sim \mu$G, while in the large cosmic voids the field drops to its primordial value of $\lesssim$ pG;[2] this scenario is fully consistent with existing observations[3] In the sheets, which have a typical thickness of $\sim$10 Mpc, the highest energy cosmic rays have $r_g \sim 1$ Mpc, and are thus confined. Outside the sheets, the accretion flow of intergalactic gas drives the cosmic rays back, but the rapidly decreasing magnetic field may allow diffusive losses in upstream direction, which can imply spectral modifications due to a "leaky box" mechanism. Fringe field effects may additionally focus and align the cosmic rays with the field direction in the sheets; this might explain the apparent correlation of UHECR arrival directions with the local sheet, the "supergalactic plane"[4,5] Since the universe needs no longer to be homogeneously filled with cosmic rays, the total energy budget for UHECR sources is strongly diminished.

2 Large scale shocks and universal acceleration

Another prediction in a structured universe is the existence of large scale shock fronts,[2] providing the possibility of cosmic ray acceleration by the very effective shock-drift acceleration mechanism[6] In a global picture, the matter sheets form the collective downstream region, and the voids the collective upstream region in a foam-like shock topology. The cosmic rays, sliding sideways along the shock, never effectively leave the acceleration region. The spatial extension of the acceleration region in the direction of the flow can be estimated by the diffusion length, l_D, which depends on particle energy for quasi-perpendicular shocks and a Kolmogoroff turbulence spectrum as $l_D \propto E^{5/3}$. At the highest energies, l_D can become comparable to the sheet thickness, and the particles scatter freely between the boundary shock fronts. In this case, a stationary particle spectrum will no longer be obtained by the balance of diffusion over the shock front and downstream advection, but rather by the balance of energy gains and losses due to MBR interactions; here, the stochastic nature of the loss process turns out to be important.

3   The stochastic nature of MBR pion production losses

The transport of a proton in the MBR which is subject to pion production losses has to be described by a Markov point process, where the energy loss occurs randomly in distinct steps of random-distributed width. We may simplify the process to a pure counting process of unit steps, which is in case of a constant interaction rate known as a Poisson process. In photopion interactions, particles lose energy fractionally, i.e $\Delta E/E \approx \Delta \ln E = \text{const} \approx 0.2$. For a Poisson process, one can show that an initial spectrum power law spectrum, $f \propto E^{-a}$, of a source at distance D, suffers an energy independent reduction by a factor $M = \exp[-(D/\lambda)(1 - e^{-\alpha})]$, if λ is the mean interaction length and $\alpha = a(\Delta \ln E)$. For a linearly increasing interaction rate, $\rho = c/\lambda = \rho' \ln(E/E_0)$, the modified spectrum can be approximately described as a power law steepened by $\Delta a = (D\rho'/c)(1 - e^{-\alpha}).$[7]

The interaction rate in the microwave background can be best modeled relative to the maximum rate ρ_1, which is reached for $E > E_1 \approx 1\,\text{ZeV}$ and corresponds to $\lambda_1 \approx 4\,\text{Mpc}$. For $E_0 \approx 30\,\text{EeV} < E < E_1$ it is linearly increasing, $(\rho/\rho_1) \approx 0.3 \ln(E/E_0)$, and $\rho = 0$ for $E < E_0$. A continuing initial power law spectrum maps then to a piecewise power law with index a for $E < E_0$ and $E > E_1$, and $a + \Delta a$ in between. A spectral cutoff in the source maps to an exponential decline of the observed spectrum somewhat below the source cutoff energy.[7] We may give two numerical examples: A radio galaxy at $D = 30\,\text{Mpc}$, producing a spectrum $f \propto E^{-2}$ with a sharp cutoff at $1\,\text{ZeV}$, is observed with a power law index $a' \approx 2.75$ between 30 and 300 EeV, followed by an exponential cutoff. A topological defect at $D = 100\,\text{Mpc}$, producing a $f \propto E^{-1.3}$ spectrum, is observed with a power law index $a' \approx 3$ between 30 EeV and 1 ZeV, flattening back to $a=2$ for higher energies.

4   Consequences for the GZK cutoff and cosmic ray observatories

The time scale of large scale shock acceleration, t_a, is generally larger than the time scale for MBR photopion losses, t_π; depending on magnetic field strength and shock velocity, we may find ratios $t_a/t_\pi \sim 1-100$ at $\gtrsim 100$ EeV.[8] Thus the acceleration is not really effective in the ordinary sense; however, considering the stochastic nature of energy losses and the breakdown of advection at the highest energies, the resulting stationary spectrum can still be relatively flat for $E \gtrsim 100$ EeV: In the simple case of a constant acceleration time scale, interaction loss balanced shock acceleration leads to power laws $f \propto E^{-b}$, and the relation $t_a/t_\pi = b[1 - \exp(-0.2b)]^{-1}$ holds. Spectral indices as observed in the UHECR spectrum are obtained for $t_a/t_\pi \approx 4$, but steepen very fast for larger values; under realistic conditions, the equilibrium spectrum is probably concave and too steep to explain the highest energy event rates.

Therefore, the existence of large scale shocks in the universe does not make cosmic ray point sources unnecessary; radio galaxies, AGN, gamma ray bursters or topological defects may still contribute as UHECR sources. Clusters of galaxies, which are the sites of the strongest large scale shocks and well located in the universe, can play an intermediate role between point sources and large scale acceleration.[8] The importance of large scale shocks is rather that they provide a *re-acceleration mechanism* which is *as universal as the GZK process*, and thus may lead to revised estimates of the maximum distance of the possible sources of highest energy cosmic rays. Consequently, the pros and cons for the various source models have to be reconsidered in a structured universe. For the Pierre Auger UHECR observatory, the large values of the magnetic field arising from large scale structure simulations give little hope to see point sources of charged cosmic ray particles; however, UHECR events are likely to occur in clusters and map the local large scale structure of the magnetic field.

Acknowledgements

Work of JPR is funded by NASA grant 5-2857. This work is based on a PhD thesis supervised by P.L. Biermann at the MPIfR Bonn, and a collaboration with H. Kang. T. Stanev is acknowledged for discussions.

References

1. J.E. Horvath, this proceedings; and *Pierre Auger Project Design Report*, Chapter 3.
2. D. Ryu & H. Kang, this proceedings, and references therein.
3. P.P. Kronberg. Rep. Prog. Phys. **57**, 325 (1994); and private communication.
4. T. Stanev, et al. Phys. Rev. Lett. **75**, 3056 (1995).
5. N. Hayashida, et al. Phys. Rev. Lett. **77**, 1000 (1996).
6. J.R. Jokipii. ApJ **313**, 842 (1987).
7. J.P. Rachen, PhD-th., MPIR Bonn (1996). http://www.astro.psu.edu/users/jorg/PhD.html
8. H. Kang, J.P. Rachen & P.L. Biermann. MNRAS in press (1997).

A UNIFIED MODEL FOR ULTRA HIGH ENERGY
COSMIC RAYS

ERELLA OPHER[a], REUVEN OPHER[b]

Instituto Astronômico e Geofísico - IAG/USP, Av. Miguel Stéfano, 4200
CEP 04301-904 São Paulo, S.P., Brazil

Ultra high energy cosmic rays (UHECR), $> 10^{20}$ eV, and gamma ray bursts (GRB) are suggested to have a common origin: a current circuit (CC), where the current generator is a rotating magnetized neutron star. The CC produces a double layer (DL) which accelerates particles and produces a jet. The jet is identified with the GRB and has a Lorentz factor $\Gamma_1 \sim 10^5 - 10^8$. Another DL entrained in the jet produces particles with $\Gamma_2' \sim \Gamma_1$. In the laboratory frame, $\Gamma \sim \Gamma_1 \Gamma_2' \sim 10^{10} - 10^{16}$, corresponding to UHECR with energies $\sim 10^{21}$ eV.

We suggest that ultra high energy cosmic rays (UHECR), $> 10^{20}$ eV, are produced by rotating magnetic neutron stars with accretion discs at cosmological distances. In an intermediary step, gamma ray bursts (GRB's) are also produced by the rotating magnetic neutron stars. We assume that a current circuit (CC) with current $I \sim 10^{20}$ A is produced by the magnetic field which is twisted due to the anchoring of the field lines of the rotating neutron star in the accretion disc. The magnetic field has the same shape as that of CC's of extragalactic jets. The current flows out along the axis of rotation ($r = 0$ in cylindrical coordinates) up to a maximum distance, z_{max}, and flows back at a radius $r \sim r_c$, the cocoon radius. It is assumed that $r_c \sim z_{max} \sim R_\star$, the neutron star radius. The self-inductance of the CC of length $\sim l$ is $L \sim \mu_0 l$. The CC has a magnetic energy $W = 1/2 L I^2 \sim 1/2 \mu_0 l I^2$. For $I \sim 10^{20}$ A and $l \sim R_\star \sim 10^6$ cm, we obtain $W \simeq 1/2 \times 10^{45}$ ergs. This energy is comparable to the magnetic field energy of the neutron star.

It is assumed that the CC creates double layers (DL's). The classic double layer [1] is an electrostatic structure which can appear within a current carrying plasma and can sustain a significant potential drop, ϕ_{DL}. For a strong DL, $e\phi_{DL} >> k_B T$, the DL thickness, according to Borovsky [2], is $d \sim 0.9 \lambda_D (e\phi_{DL}/k_B T)^\alpha$, where T is the temperature of the plasma, α is a constant and λ_D is the Debye length. $\lambda_D \propto (T/n)^{1/2}$. For example, for $k_B T \sim 1$ MeV and the numerical density of the plasma $n \sim 5 \times 10^{14}$ cm^{-3}, $\lambda_D \sim 0.2$ cm. For $e\phi_{DL} \sim 10^{14}$ eV, $(e\phi/k_B T)^\alpha \sim 10^6$ with $d \sim 10^5$ cm and $\alpha = 0.75$ as suggested by Borovsky [2]. $(e\phi/k_B T)^\alpha \sim 10^8$ for $d \sim 10^7$ cm and

---

[a]email: erella@orion.iagusp.usp.br
[b]email: opher@orion.iagusp.usp.br

$\alpha = 1.0$. Since d is not accurately known, we assume that $\alpha \sim 0.75 - 1$. With the above parameters, $d \sim 10^6$ cm $\sim R\star$.

For d comparable to the radius of the current carrying region, the value of ϕ_{DL} is [3]

$$\phi_{DL} = \left(\frac{\phi_i}{4\pi\epsilon_0}I\right)^{1/2}$$

With $e\phi_i = m_i c^2$ and m_i, the ion mass, $e\phi_i \sim 0.939 \times 10^4$ eV for protons. The power dissipated by the DL is $P_{DL} \simeq \phi_{DL}$. The characteristic time for the release of the energy is $\tau = W/P_{DL}$. Thus $\tau = \mu_0 lI/2\phi_{DL}$.

In our model, we envision that the neutron star with an accretion disc first creates the CC with a current I. The CC is then interrupted by the DL over a time τ, creating the induced voltage ϕ_{DL} across the DL. For $I \sim 3 \times 10^{20}$A and $l \sim 10^6$ cm, $e\phi_{DL} \sim 10^{14}$ eV and $\tau \sim 10^4$ s. This time is comparable to the observed time for GRB's. We suggest that GRB's are formed in the scenario outlined above: A CC of $I \sim 3 \times 10^{20}$A is produced by a rotating neutron star with an accretion disc which then forms a DL which is active over a time τ. The potential drop across the DL is $\phi_{DL} \sim 10^{14}$ V.

A very high energy density ("fireball") is formed near the DL. At this energy, the Lorentz factor for a proton is $\Gamma_p \sim 10^5$ and for an electron, $\Gamma_e \sim 2 \times 10^8$. Many electron-positron pairs are created in the fireball which acts to cool the plasma. The fireball expands with an effective Lorentz factor $\Gamma_1 \sim 10^5 - 10^8$. The expansion of the fireball is primarily along the rotation axis due to the confining action of the electric current, $I \sim 3 \times 10^{20}$A, which is sufficient to "pinch" a jet with a plasma energy $\leq$ magnetic energy of an axial magnetic field $\sim 10^{12}$G. Thus, a jet with a Lorentz factor Γ_1 is formed. We assume that a DL exists in the jet plasma. For simplicity, we assume that the DL is perpendicular to the jet motion and that the effective Γ of the particles in the reference frame of the jet is $\Gamma_2' \sim \Gamma_1$. In the laboratory frame, the Γ is $\Gamma \sim \Gamma_2'\Gamma$. Since $\Gamma_2' \sim \Gamma_1 \sim 10^5 - 10^8$, $\Gamma \sim 10^{10} - 10^{16}$. For protons, this Γ corresponds to energies $\sim 10^{19} - 10^{25}$ eV, comparable to the highest energy UHECR observed, $> 10^{20}$ eV.

Acknowledgments

R.O. would like to thank the Brazilian agency CNPq for partial support.

1. I. Langmuir, *Phys. Rev.*, **33**, 954 (1929).
2. J.E. Borovsky, *Phys. Fluids*, **26**, 3273 (1983).
3. A.L. Peratt, *Physics of the Plasma Universe*, (Springer-Verlag, New York, 1992).

THE NUCLEOSYNTHESIS OF THE GALACTIC COSMIC RAY Fe AND Ni ISOTOPES

J.A. SIMPSON[1,2] AND J.J. CONNELL[1]

[1]The Enrico Fermi Institute
[2]The Department of Physics
The University of Chicago, Chicago, IL 60637

We report the relative abundance measurements of ^{54}Fe, ^{55}Fe, ^{56}Fe, ^{58}Fe, ^{58}Ni, ^{58}Ni, ^{60}Ni, ^{62}Ni in the overall energy range between 300 and ~ 420 MeV-nucleon^{-1}. The cosmic ray source abundances are derived from the measurements using models for propagation from distributed sources in the galaxy. Overall, except for ^{54}Fe/^{56}Fe and ^{57}Fe/^{56}Fe, we show that the principal Fe and Ni isotopic source ratios have values close to the solar system ratios derived from meteorites. In particular, we note that ^{58}Fe and ^{62}Ni display no evidence of neutron enrichment.

1 Introduction

The nuclei accelerated in the galaxy to become the cosmic radiation carry information on their origin in nucleosynthesis processes. Thus, the determination of cosmic ray source composition provides a critical test for nucleosynthesis theories. In order to avoid elemental biases we have measured the cosmic ray isotopic composition, which is determined by the nucleosynthetic history of the cosmic ray source.[1,2] Among the tests for the origin of the source matter are the source abundances of Fe and Ni isotopes. The following brief report is based on our recent publication[3] of isotopic measurements of Fe and Ni.

2 Measurements

The measurements in the overall energy range ~30 to ~500 MeV per nucleon are obtained from the University of Chicago's High Energy Telescope (HET)[4], carried on the *Ulysses* spacecraft. Figure 1 shows mass histograms for Fe and Ni. Table 1, Column 2 lists the principal measured isotopic ratios for Fe and Ni.

3 Source Abundances

We have calculated cosmic ray propagation from their sources in the galaxy to the observer via the interstellar medium and the heliosphere. We note that Fe is a factor of $> 10^3$, and Ni > 10 more abundant than the elements beyond Ni. This ensures that, aside from ^{56}Fe spallating into ^{54}Fe and ^{55}Fe, secondary nuclear species produced during interstellar propagation are not major contributors to the observed Fe and Ni isotopes.

The cosmic ray source abundance ratios derived from these calculations are listed in Table 1, Column 2. The current accepted values for the corresponding solar system nuclidic abundance ratios are shown in Column 3. Based on the reviews by Anders and

Ebihara[5] and Cameron[6] we conclude that, for the Fe and Ni solar system nuclides, the ratios probably have errors of the order of ± 1 to 2 percent, smaller than the statistical uncertainties in our galactic source determinations.

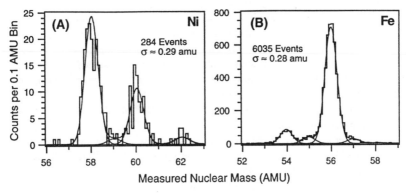

Figure 1: Mass Histogram of Ulysses HET Fe and Ni Isotopes.

Table 1: Ulysses HET Isotopic Abundance Measurements.

Isotopic Ratio	(1) Measured (%)	(2) Source (%)	(3) SS (%)	(4) GCRS/SS Ratio
$^{54}Fe/^{56}Fe$	11.4 ± 0.6	9.3 ± 0.6	6.3	1.5 ± 0.1
$^{55}Fe/^{56}Fe$	5.4 ± 0.4	1.6 ± 0.5	0	—
$^{57}Fe/^{56}Fe$	3.9 (+0.35, -0.38)	3.7 (+0.33, -0.36)	2.3	1.6 (+0.14, -0.16)
$^{58}Fe/^{56}Fe$	0.34 (+0.10, -0.14)	0.18(+0.10, -0.14)	0.32	0.6 (+0.3, -0.4)
$^{59}Ni/^{58}Ni$	4.6 (+2.6, -2.1)	2.6 (+2.1, -1.7)	0	—
$^{60}Ni/^{58}Ni$	45.0 (+6.8, -6.5)	43.2 (+6.7, -6.4)	38.2	1.1 ± 0.2
$^{61}Ni/^{58}Ni$	< 1.6	< 1.2	1.7	< 0.7
$^{62}Ni/^{58}Ni$	5.8 (+2.3, -1.9)	5.4 (+2.2, -1.8)	5.3	1.0 (+0.4,-0.3)

4   Discussion and Conclusions

The principal isotopic ratio $^{60}Ni/^{58}Ni$ clearly shows no significant neutron-rich enhancement that would be predicted by nucleosynthesis models requiring neutron enrichment in Ni[7], but is close to its solar system value. The cosmic ray source ratio $^{54}Fe/^{56}Fe$ is approximately 50% higher than the solar system ratio (Column 4). Part of the excess ^{54}Fe may be secondary nuclides produced during the propagation of ^{56}Fe, as indicated by the presence in Figure 1 and excess in Table 1 of ^{55}Fe, which does not exist in the solar system. ^{55}Fe is an electron capture nuclide ($\tau_{1/2}$ = 2.68 years) that is not present in the cosmic ray source.

^{54}Mn is also a secondary product of ^{56}Fe spallation. We find that a small part of the ^{54}Fe/^{56}Fe excess appears to arise from both spallation of ^{56}Fe and from ^{54}Mn decay by non-electron capture branches. We know of no propagation models, including re-acceleration models, wherein source abundances could assumed to be enhanced significantly with heavier isotopes of Fe and Ni, but, after propagation, could be reduced to simulate the observed ratios of ^{58}Fe/^{56}Fe, ^{60}Ni/^{58}Ni or ^{62}Ni/^{58}Ni. From Table 1 we conclude--with the exception of ^{54}Fe/^{56}Fe and ^{57}Fe/^{58}Fe--that their source matter is close to solar systems values.

For supernovae Type I (helium burning white dwarfs) Thielemann[8] find that the Fe-group nuclides do not show a composition close to solar; ^{57}Fe, ^{58}Ni and ^{62}Ni are over-produced by more than a factor 2.

Since the interstellar medium isotopic composition appears from present investigations to be more solar system-like than the composition from explosive nucleosynthesis[7], present evidence suggests that cosmic ray particle acceleration occurs mainly in the interstellar medium.

For additional conclusions derived from our analysis, see Connell and Simpson[3].

Acknowledgements

We appreciated discussions with J. Truran and F. Timmes. This research was supported in part by NASA/JPL Contract 955432, NASA Grant NGT-51300 and Argonne National Laboratory-University of Chicago Grant 95-021.

References

1. Simpson, J.A. in *Ann. Rev. of Nucl. & Part. Sci.*, (Palo Alto: Annual Rev. Inc.) **33**, 323, Chap. 9 (1983).
2. Mewaldt, R.A. in A.I.P. Conf. Prof. No.183, Cosmic Abundances of Matter, ed. C.J. Waddington (New York: AIP), 124 (1989).
3. Connell, J.J. and Simpson, J.A. *Ap. J. Lett.* **475**, L61(1996).
4. Simpson, J.A. et al. Astron. and Astrophys. Suppl. 93, 365 (1992).
5. Anders, E. and Grevesse, N. Geochim. Cosmochim. Acta 53, 197 (1989).
6. Cameron, A.G.W. in *Essays in Nuclear Astrophysics*, ed. C.A.Barnes, D.D.Clayton & D.N.Schramm (Cambridge: Cambridge Univ. Press; 23, 1982).
7. Arnett, D. *Supernovae and Nucleosynthesis*, (Princeton,NJ: Princeton Univ. Press; 1996).
8. Thielemann, F.-K., Nomoto, K., Iwamoto, K. & Brachwitz, F., 1995, Submitted for publication.

Magnetic Monopole Cosmic Rays at $E \geq 10^{20}$ eV

Thomas W. Kephart and Thomas J. Weiler

Dept. of Physics & Astronomy, Vanderbilt University, Nashville, TN 37235

We suggest that the highest energy $\geq 10^{20}$ eV cosmic ray primaries may be relativistic magnetic monopoles. Motivations for this hypothesis are twofold: (i) conventional primaries are problematic, while monopoles are naturally accelerated to $E \sim 10^{20}$ eV and above by galactic magnetic fields; (ii) the observed highest energy cosmic ray flux is just below the Parker limit for monopoles. By matching the cosmic monopole production mechanism to the observed highest energy cosmic ray flux we estimate the monopole mass to be $\leq 10^{10}$ GeV.

****

The discoveries by the AGASA, Fly's Eye, Haverah Park, and Yakutsk collaborations(see Ref. [1] for detailed references) of cosmic rays with energies above the GZK cut–off at $E_c \sim 5 \times 10^{19}$ eV present an intriguing challenge to particle astrophysics. For every mean free path ~ 6 Mpc of travel, a proton loses 20% of its energy on average. So if protons are the primaries for the highest energy cosmic rays they must either come from a rather nearby source or have an initial energy far above 10^{20} eV. Neither possibility seems likely, although the suggestion has been made that radio galaxies in the supergalactic plane may be origins [2].

A primary nucleus with energy above $\sim 10^{19}$ eV is photo–dissociated by the 3K background [3]. Gamma–rays and neutrinos are other possible primaries but the gamma–ray hypothesis appears inconsistent [4] with the time–development of the Fly's Eye event, while for neutrinos since the Fly's Eye event occured high in the atmosphere, the expected event rate for early development of a neutrino–induced air shower is down from that of an electromagnetic or hadronic interaction by six orders of magnitude [4]. Moreover, the acceleration problem for γ and ν primaries is just as daunting as for hadrons.

Given these problems we have suggested that the primaries of the ultra high energy cosmic rays may be relativistic magnetic monopoles [1,5], hence requiring the monopole mass M to be $\leq 10^{10}$ GeV. Energies above the GZK cut–off are naturally attained by monopoles when accelerated by known cosmic magnetic fields and the observed cosmic ray flux above the cut–off is of the same order of magnitude as the theoretically allowed "Parker limit" monopole flux. The Kibble mechanism for monopole generation in an early–universe then leads to an upper bound on the monopole mass, which turns out to be similar to the above limit. Thus, we arrive at a flux of monopoles with mass

$M \leq 10^{10}$ GeV as a viable explanation the highest energy cosmic ray data. This hypothesis has testable signatures.

The minimum monopole charge is $q_M = e/2\alpha$, and our local galactic B field is about $5\mu G$ with a coherence length[7] $L \sim 1kpc$. Thus, since the Lorentz force on a magnetic monopole due to a magnetic field is $\mathbf{F} = q_M\mathbf{B}$, a galactic monopole will typically have kinetic energy $E_K \sim q_M BL\sqrt{N}$ where $N \sim 20$ is the number of magnetic domains encountered by a typical monopole as it traverses the galaxy. Note that these energies are generally above the GZK cut-off. Thus, the "acceleration problem" for $E \geq 10^{20}$ eV primaries is naturally solved with the monopole hypothesis.

Now let us briefly comment on detection. Magnetic monopoles in typical grand unified models have hadronic interactions, since the classical monopole solution to the field equations have a nonvanishing color components. The monopole mass, the form of the monopole's QCD form factor and hence the hadronic crossection[6] for monopole-baryonic matter (e.g., air) scattering are all model dependent, but it is a strong constraint that the monopole be light. Some well known models in this class are based on SO(10)[8,9] or SU(15)[10,11].

The electromagnetic energy loss of a relativistic monopole traveling through matter is very similar to that of a heavy nucleus with similar γ–factor and charge $Z = q_M/e = 137/2$. One result is a $\sim 6\,\mathrm{GeV}/(\mathrm{g\,cm}^{-2})$ "minimum–ionizing monopole" electromagnetic energy loss. Integrated through the atmosphere, this totals $\sim (6.2/\cos\theta_z)$ TeV, for zenith angle $\theta_z \leq 60°$. For a horizontal shower the integrated energy loss is ~ 240 TeV. Cerenkov radiation is at the usual angle but enhanced by $(137/2)^2 \sim 4700$ compared to a proton primary, and may help in the identification of the monopole primary.

If the monopole has $M \leq \sqrt{2mE_M}$, i.e. $\leq 10^6$ GeV for $E_M \sim$ few $\times 10^{20}$ eV, it may transfer an O(1) fraction of its energy in a forward–scattering event, possibly mimicking a standard air shower; but a relativistic monopole primary with $M > 10^6$ GeV will retain most of its energy per each scattering, and so will continuously "initiate" the shower as it propagates through the atmosphere. The smaller energy transfer per collision for a $M > 10^6$ GeV monopole as compared to that of the usual primary candidates may constitute a signature for heavy monopole primaries. Moreover, the back–scattered atmospheric particles in the center–of–mass system are forward–scattered in the lab frame into a cone of half–angle $1/\gamma_M$; at the given energy of $E \sim 10^{20}$ eV, this angle will be large for a heavy monopole primary compared to the angle for a usual primary particle, possibly offering another monopole signature. Finally, monopole primaries should be asymmetrically distributed on the sky, showing a preference for the direction of the dominant (perhaps local galactic, perhaps extragalactic) magnetic field.

We look forward to more cosmic ray data at these highest energies with the ongoing experiments and the "Auger Project", an international effort[12] to instrument two 5,000 km^2 detectors and collect thousands of events per year above 10^{19} eV.

Acknowledgments

This work was supported in part by the U.S. Department of Energy grant no. DE-FG05-85ER40226.

References

1. T. W. Kephart and T. J. Weiler, *Astropart. Phys.* **4**, 271 (1996).
2. T.Stanev, P. L. Biermann, J. Lloyd-Evans, J. Rachen and A. Watson, *Phys. Rev. Lett.* **75**, 3056 (1995).
3. F. W. Stecker, *Phys. Rev.* **180**, 1264 (1969).
4. F. Halzen, R. A. Vazquez, T. Stanev, and V. P Vankov, *Astropart., Phys.*, **3**, 151 (1995).
5. T. J. Weiler and T. W. Kephart, *Nucl.Phys. B (Proc. Suppl.)* **52B**, 218 (1996).
6. T. W. Kephart, T. J. Weiler, and S. D. Wick, work in progress.
7. P. P. Kronberg, *Rep. Prog. Phys.* **57** , 325 (1994).
8. G. 't Hooft, *Nucl.Phys.* **B105**, 538 (1976).
9. N.G. Deshpande, B. Dutta, and E. Keith, *Phys. Lett.* **B385**, 116 (1996).
10. P. H. Frampton and B.-H. Lee, *Phys. Rev. Lett.* **64**, 619 (1990).
11. P. H. Frampton and T. W. Kephart, *Phys. Rev.* **D42**, 3892 (1990).
12. J. W. Cronin and A. A. Watson, announcement from the Giant Air Shower Design Group (recently renamed the "Auger Project"), October 1994.

Evolution of the morphological luminosity distributions within rich clusters $(0.0 < z < 0.55)$

S.P.Driver, W.J.Couch

School of Physics, University of New South Wales,
Sydney, NSW 2052, AUSTRALIA

S.C.Odewahn, R.A.Windhorst

Department of Physics and Astronomy,
Arizona State University, Tempe, AZ 85287-1504, USA

We demonstrate the ability to recover morphological luminosity distributions (LDs) within medium redshift clusters ($z \sim 0.55$) based on *Hubble Space Telescope* WFPC2 observations. We postulate that a detailed survey of the morphological LDs in local, low and medium redshift clusters may provide strong constraints on the modes of galaxy evolution in rich clusters. Preliminary results suggest that in clusters, as also seen in the field, very strong evolution (*i.e.* $\Delta m \approx 2.5$ mags since $z = 0.55$) is occurring in the late-type spiral and irregular populations.

1 Introduction

The Canada-France-Redshift-Survey (CFRS) of field galaxies has demonstrated that by recovering the optical luminosity distributions (LDs) of galaxies over a wide range of epochs it is possible to place strong constraints on the amount of stellar evolution for various galaxy types[1]. Through these surveys, the results of morphological number count studies[2][3][4] and the surface photometry of MgII absorption systems[5], we now have a consistent picture of the passive evolution of elliptical and early-type galaxies and the rapid evolution of late-type systems over the redshift range $0.0 < z < 1.0$.

Here we introduce an analogous program to trace the evolution of the galaxy LDs within rich cluster environments. Recovering the LDs within rich clusters is much simpler than for field galaxies as clusters represent the ideal prepackaged volume-limited sample. Essentially to recover a cluster's LF all that is required is the redshift of the cluster and the subtraction of the mean field number-counts from the number-counts along the cluster sight-line. This technique has now been applied to a number of clusters and more details on the subtleties of the basic technique are described elsewhere[6].

2   Morphological LDs with HST WFPC2

With the advent of the high-resolution imaging capability of the Wide Field
Planetary Camera 2 (WFPC2) on the Hubble Space Telescope (HST) the
technique can now been taken a step further. Instead of simply recovering
the overall luminosity function, HST's high-resolution imaging can be used to
determine number-counts and LDs according to morphology. The critical step
is the accuracy with which the morphological field number-counts are known.
In previous papers [2][4][7] we presented detailed morphological number-counts
from deep random field images observed by WFPC2 (including results from
the Hubble Deep Field). Using these fields as our mean background sample we
have recovered the morphological luminosity distributions for three clusters
also observed with WFPC2. These clusters are: CL0016 (z=0.54), CL0054
(z=0.56) and CL0412 (z=0.51), the images for which were kindly provided
by the MORPHS collaboration [8]. The data were re-analysed and morphologically classified using the same Artificial Neural Network-based method and
software as used for the field sample. This approach to morphological galaxy
classification is described in more detail by Odewahn et al. [7].

Figure 1 shows the resulting *mean* luminosity distributions for the three
clusters divided into three morphological categories — E/S0, Sabc and Sd/Irr.
The LDs have been converted to rest frame magnitudes by adopting a mean
K-correction per type and assuming a standard flat cosmology with $H_o = 50$
$kms^{-1}Mpc^{-1}$.

3   Discussion

Overlaid on Figure 1 are the local schematic morphological LDs [9]. The indication is for purely passive/no-evolution for early-types and the strong evolution
($\Delta m \approx 2.5$ mags since $z = 0.55$) of late-type systems. This result is analogous
to that seen in the field [4] and begs the question as to whether the Butcher-
Oemler effect [10] seen in clusters and the Faint Blue Galaxy problem are one
and the same ?

Acknowledgments

SPD acknowledges DITAC for providing the funding to attend this conference
and thanks the MORPHS group for the early release of their HST WFPC2
cluster data.

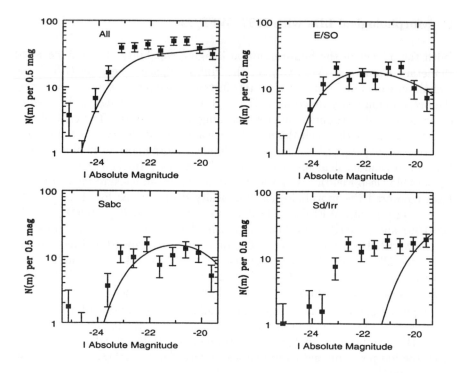

Figure 1: The mean morphological luminosity distributions observed in three medium red-shift clusters. The panels represent (a) all galaxy types, (b) E/S0s (c) Sabc's and (d) Sd/Irr's. The solid lines show the equivalent z=0 distributions based on population studies of the Virgo cluster.

References

1. S.J. Lilly *et al.* ApJ **455**, 108 (1995)
2. S.P. Driver, R.A. Windhorst and R.E. Griffiths ApJ **453**, 48 (1995)
3. K.G. Glazebrook *et al.* MNRAS **275**, 19pp (1995)
4. S.P. Driver *et al.* ApJ **449**, L23 (1995)
5. C.C. Steidel, M. Dickinson and S.E. Persson, ApJ **437**, L75 (1994)
6. S.P. Driver *et al.* MNRAS **000**, 000 (in press)
7. S.C. Odewahn *et al.* ApJ **472**, L13 (1996)
8. I. Smail *et al.* ApJ **000**, 000 (in press)
9. B. Binggeli, A. Sandage and G.A. Tammann, ARA&A **26**, 509 (1988)
10. H. Butcher, and A. Oemler, ApJ **219**, 18 (1978)

THE EFFECT OF CLUSTER MERGERS ON GALAXY EVOLUTION

M.J. HENRIKSEN

University of North Dakota, Department of Physics, Box 7129, Grand Forks, ND 85202-7129, USA

Recent X-ray and optical observations have established that clusters of galaxies evolve through violent mergers. We have found that the cluster gravitional potential will substantially increase star formation within the disk of a spiral galaxy during a merger through enhancing the number of high velocity interstellar cloud collisions. Applying the Jeans criterion to the post-collision cloud gas density derived from isothermal shock conditions suggests that $> 25\%$ of the clouds form stars in a disk galaxy as it follows a radial orbit toward the cluster center. This increase in star formation from cloud collsions may be responsible for the high level of past and present star formation found in spiral galaxies within clusters at the Z $= 0.2$ epoch.

1 Cluster Evolution

The importance of mergers in the formation of clusters of galaxies is apparent in the large fraction of clusters which show substructure or asymmetry in their galaxy distribution and/or X-ray images. Hydrodynamical simulations predict that the signature of an ongoing merger is high temperature gas which is shock heated from collision and a bimodal or asymmetric surface brightness distribution.

The Advanced Satellite for Astrophysics and Cosmology (ASCA) observed the A754 cluster of galaxies providing an unprecedented temperature distribution which shows clear evidence of a merger [1]. A temperature map of the clusters has a high > 12 keV temperature component near the NW galaxy clump. A low temperature component is also found near the SW galaxy component. The pattern of heating predicted by simulations is perpendicular to the merger axis. In this case the merger axis is along the line of sight and the heating is in the plane of the sky. The location of the galaxy clumps indicates that the subclusters have slightly missed having a head on collision. The location of the cool component is coincident with the ROSAT surface brightness peak which appears to be trailing behind the SW subcluster. It is possible that this cool material is the remnant of a cooling flow which has been removed by the impact of the collision.

2 Dynamical Modeling of Cluster Mergers

The A3395 cluster of galaxies appears to be in the early stage of a merger, suggested morphologically by the asymmetry of the ROSAT X-ray surface brightness distribution and a dynamical modeling of the galaxy positions and velocities[2]. This cluster has several hundred galaxy redshifts available which were used to measure the projected velocity dispersion and radial separation of the subclusters. Radial infall or outflow is assumed with two angles describing the true 3-dimensional location of the subclusters. The differential equation is solved with the projected quantities constraining the solutions. We find that bound solutions exist in which the subclusters are infalling with high velocity $\simeq 1500$ km/sec. They have a separation of about 1 Mpc indicating a collision of centers in less than 1 Gyr. This is consistent with the X-ray morphology in which the atmospheres are just beginning to merge.

3 The Effect of a Merger of Subclusters on Galaxies

There are three mechanisms by which galaxies are believed to evolve: interactions and mergers of individual galaxies, ram-pressure of the intergalactic medium, and the tidal effect of the cluster potential. The tidal field of the cluster should dominate ram-pressure[3] near the cluster center even for the very high merger velocities reported in section 2. Galaxy interactions lead to inflow of interstellar gas which drives the starburst phenomenon. However, recent Hubble Space Telescope Observations show that $z\simeq0.2$ clusters are dominated by late type spirals rather than starbursts. Furthermore, a simple simulation of two galaxies in a subcluster potential indicates that the galaxies may have their merger delayed or interrupted within the cluster potential. The delay can be long enough so that the galaxies, assuming radial infall from a zero initial velocity, will experience the tidal effect of the cluster potential before an interaction or merger. Our simulations show that an increase in star formation from cloud collisions for two interacting galaxies begins when the galaxies are separated by about 60 kpc, for a disk radius of 20 kpc. This results in the galaxies only experiencing a significant interaction or merger outside of approximately 2 - 3 Mpc from the cluster center. For initial pair separation of 100 or 200 kpc, the pair will experience the cluster tidal field before the interaction unless it is greater than approximately 3 Mpc from the center, initially. The volume of the cluster is very large and the mean spacing of clusters is much greater than 100-200 kpc at this radius for a King Model. Indeed, optical observations of clusters at a $z=0.2$ typically show very few pairs.

The implications of this study is a contradiction to the claim that finding

E+A galaxies outside of clusters indicates that pair interactions are sufficient to explain the phenomenon within clusters. Since the cluster environment restricts interactions of galaxies anywhere but far from the center (i.e., and isolated cluster) it is more probable that the tidal effect of the cluster potential triggers star formation in gas poor systems and in all systems within approximately 100 - 160 Kpc.

4 Simulations

We have simulated a disk galaxy on a radial orbit into a cluster of galaxies. The cluster potential is given by a static, King model with a core radius of 50 Kpc, derived from gravitational lensing in galaxies at the epoch of approximately 0.2. The mass is 1000 times the disk galaxy mass. There are approximately 12,000 gas clouds in the disk. When a collision occurs we apply the shock conditions. A typical upstream density is assumed for clouds comparable to neutral Hydrogen clouds as well as a typical mass, 300 Solar Masses, and temperature. After the collision, the Jeans criterion is applied to determine whether the cloud becomes unstable to collapse, eventually forming stars. For two equal mass galaxies, the star formation rate (no. stars formed/no. collisions) is 3 - 5%. For the cluster, the star formation rate is 25%. The star formation begins at 160 kpc for the cluster and 60 kpc for a perturbing galaxy. For an isolated galaxy, there are hundreds of cloud collisions, r < 10 pc, but no stars are formed with this star formation scenario. If the mass of the perturbing cluster is higher, 10,000 times the spiral galaxy mass, which might be the potential a late infalling galaxy would experience, the fraction of star forming collisions increases, however, the total number of stars formed is less. This is because the galaxy spends less time being perturbed, since it infalls at a higher velocity and transits the central region over a shorter period of time.

Acknowledgments

I would like to acknowledge support from the National Science Foundation through a CAREER grant, NASA ADP, and ND EPSCOR.

References

1. M. Henriksen and M. Markevitch, *Astrophys. J* **466**, L79 (1996).
2. M. Henriksen and C. Jones, *Astrophys. J* **465**, 666 (1996).
3. M. Henriksen and G. Byrd, *Astrophys. J* **459**, 82 (1996).

CLUSTERS OF GALAXIES IN THE RIXOS SURVEY

F.J. CASTANDER

University of Chicago, Department of Astronomy and Astrophysics, 5640 S Ellis Ave., Chicago, IL 60637, USA

In the simplest case of cluster evolution, changes in cluster properties are linked directly to the gravitational potential and self-similar scaling laws apply. Alternatively, the detailed thermodynamic history of the intracluster plasma may play an important role. Here we present results from a faint flux-limited sample of X-ray selected clusters compiled as part of the ROSAT International X-ray and Optical Survey (RIXOS). Very few distant clusters have been identified and we show that the form of the redshift distribution obtained is inconsistent with standard models in which the X-ray properties of clusters evolve in self-similar fashion.

1 Introduction

The most accessible way to observationally study the evolution of the population of cluster of galaxies out to large redshifts is to investigate the evolution of the X-ray luminosity function (XLF). The first results on cluster X-ray evolution showed that the number of X-ray luminous clusters was significantly lower in the past (*negative evolution*) [1,2]. In particular, the EMSS survey [2] found significantly fewer luminous clusters at $z > 0.3$ than locally. However, these results have been recently questioned [3,4]. In order to explain the EMSS data a steeper slope of the XLF is required at high redshift. If this steep slope is maintained to lower luminosities, surveys that reach fainter fluxes would reveal a large increase in sub-luminous clusters at intermediate redshifts.

2 The RIXOS survey

The sample presented here was derived from the ROSAT International X-ray Optical Survey (RIXOS) [5], a project aimed at producing a catalogue of optical identifications for $\simeq$400 X-ray sources found in 81 northern ROSAT fields observed with the Position Sensitive Proportional Counter (PSPC). Fields were selected with exposure times > 8 ksec and high galactic latitude ($b > +28°$). In total, 385 X-ray sources were catalogued to a limiting flux of $f_X \geq 3.0 \times 10^{-14}$ erg cm^{-2} s^{-1} in the 0.5–2.0 keV energy band. Optical imaging and spectroscopic follow-up was carried at La Palma using time allocated under the international programme supported by the Comité Científico Internacional. The results presented here were obtained by concentrating on a subset of 59 completely identified fields selected randomly [6].

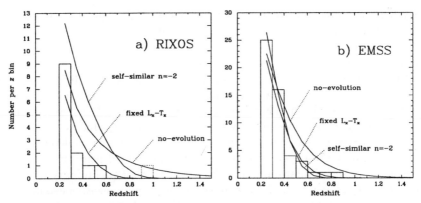

Figure 1: Observed redshift distribution in the (a) RIXOS and (b) EMSS surveys and comparison to three models (see text).

3 Discussion

Fig. 1 shows the observed redshift distribution together with predictions for three physically-interesting cases computed as described elsewhere [6,7].

Our first prediction is that expected for a non-evolving population. Although it matches the data to $z \simeq 0.3$, the faint limiting flux of our survey implies clusters would be visible to very high redshifts and these are not observed. The RIXOS survey therefore agrees with the EMSS results showing a marked decline in the volume density of luminous clusters with redshift.

The next model we consider assumes self-similarity with a shallow slope of initial fluctuations ($n = -2$). Although such a model can account for the evolutionary trends seen in the EMSS survey, for RIXOS far too many intermediate redshift clusters are predicted.

Our final model examines the evolution that can be expected if self-similarity is broken. This particular model reproduces qualitatively the 'preheated' intracluster medium model [8]. A good fit to both the RIXOS and EMSS data can be obtained. Although by no means a unique description of the datasets obtained so far, the improvement over the self-similar model consistent only with the EMSS data is striking.

Recently, new groups are examining the public PSPC ROSAT archives to study cluster X-ray evolution based on searching algorithms maximized to detect extended sources: the SHARC, RDCS and WARPS surveys (Romer, Rosati and Scharf, these proceedings). Their data indicate that the comoving number volume density of clusters has undergone only mild negative evolution or no evolution.

The RIXOS survey used a detection algorithm optimized for the detection of point sources (as AGN were the main goal of the survey). Our theoretical predictions correct for the flux lost by our detection algorithm. However it is difficult to quantify the fraction of extended objects that were missed. In view of the results being presented by other groups, we[9] have started a re-analysis of the RIXOS X-ray fields with a detection algorithm optimized for extended sources. Preliminary indications show that almost all the diffuse objects we find in this way do not reach the flux threshold required for detection by the point-source algorithm. There are, however, a small number of exceptions. We are currently working to quantify the extent of this possible incompleteness.

Summarizing, our data (even including possible imcompleteness effects) does not show the expected increase in the number density of faint clusters predicted in self-similar models. Taking into account preliminary results from other surveys[10,11,12], the XLF appears to show mild negative evolution. One appealing possibility to explain such evolution is that the intracluster gas was heated at an early epoch (perhaps as the result of the galaxy formation process) and has only recently become sufficiently cool to be trapped by the gravitational potential of clusters.

Acknowledgments

The RIXOS project has been made possible by the the award of International Time on the La Palma telescopes by the Comité Científico Internacional. We thank all the numerous individuals who have contributed to the RIXOS project and whose work has made this paper possible. We thank the SHARC survey.

References

1. A.C. Edge *et al*, MNRAS **252**, 414 (1990).
2. J.P. Henry *et al*, ApJ **386**, 408 (1992).
3. H. Ebeling *et al* in *Wide Field Spectroscopy and the Distant Universe*, ed. Maddox & Aragón-Salamanca (World Scientific, 1995).
4. R.C. Nichol *et al*, ApJ , in press (1997).
5. K.O. Mason *et al*, in preparation.
6. F.J. Castander *et al*, Nature **377**, 39 (1995).
7. F.J. Castander, PhD Thesis, University of Cambridge (1996).
8. N. Kaiser, ApJ **383**, 104 (1991).
9. F.J. Castander, R.G. Bower, A.K. Romer, R.C. Nichol *et al*, in prep.
10. C.A. Collins *et al*, ApJLett , in press (1997).
11. P. Rosati *et al*, ApJ **445**, L11 (1995).
12. C.A. Scharf *et al*, ApJ , in press (1997).

NO EVOLUTION IN THE X-RAY CLUSTER LUMINOSITY FUNCTION OUT TO $z \simeq 0.7$

ROMER, A.K., NICHOL R.C.,

Department of Physics, Carnegie Mellon University, 5000 Forbes Avenue, Pittsburgh, PA 15213, USA

COLLINS, C.A., BURKE, D.J.,

Astrophysics Group, School of Electrical Engineering, Electronics and Physics, Liverpool John Moores University, Byrom Street, Liverpool, L3 3AF, UK

HOLDEN, B.P.,

Department of Astronomy and Astrophysics, University of Chicago, 5640 S. Ellis Avenue, Chicago, IL 60637, USA

METEVIER, A., ULMER, M.P., PILDIS, R.

Department of Physics and Astronomy, Northwestern University, 2145 Sheridan Road, Evanston, IL 60208, USA

In contrast to claims of strong negative evolution by other groups, we find no evidence for evolution of the X-ray cluster luminosity function out to $z \simeq 0.7$.

The Serendipitous High-redshift Archival ROSAT Cluster (SHARC) survey (http://www.astro.nwu.edu/sharc) is a project to identify $\gtrsim$ 100 X-ray clusters at redshifts greater than $z = 0.3$. Although similar in aims and approach, it is much larger than the RDCS[12], WARPS[13] and RIXOS[2] surveys: when complete it will cover $\gtrsim$ 200 square degrees. The SHARC survey X-ray data pipeline is fully automated and is based on the EXAS[14] reduction package and a wavelet transform source detection algorithm. The pipeline has been thoroughly tested[10] and has already been applied[11] to 530 high galactic latitude PSPC pointings. A further $\simeq$ 500 PSPC and $\simeq$ 200 HRI pointings will be added to the survey in 1997. Optical follow-up is underway at the ARC 3.5m, ESO 3.6m and AAT 3.9m telescopes.

The primary research goal of the SHARC survey is the robust quantification of X-ray cluster evolution. Reports of rapid negative evolution in the X-ray cluster luminosity function (XCLF) seen in the EMSS[7][8] and RIXOS[2] cluster samples have attracted much attention because they are not consistent with standard hierarchical models of structure formation[9]. Unfortunately, both the EMSS and RIXOS samples suffer from small number statistics. With the availability of thousands of deep pointings in the ROSAT archive, the subject is ripe for re-examination. Nichol *et al.* (1997) discusses a re-analysis of the

468

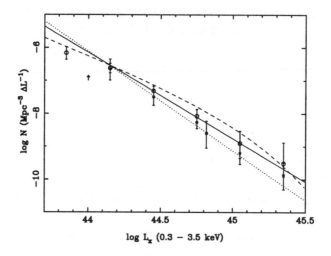

Figure 1: High redshift XCLF (solid line and symbols) and low redshift XCLF (dotted lines, open symbols) for the EMSS cluster sample. Also shown is the low redshift XCLF[6] for RASS clusters (dashed line). (See Nichol *et al.* 1997 for more details.)

properties of the EMSS cluster sample. A large fraction of the EMSS clusters were observed during ROSAT pointed observations. By applying the SHARC data analysis techniques to these observations, we were able to determine the cluster fluxes more accurately than was possible with the original Einstein data. Using these revised fluxes, and incorporating new optical data where available, we produced the XCLFs shown in Figure 1. As can be seen from that figure, the EMSS low and high redshift XCLFs are very similar, indeed they differ by only 1σ. A comparison between the EMSS high redshift XCLF and a local XCLF[6] based on ROSAT All-Sky Survey (RASS) data yields no significant evidence for evolution out to $z \simeq 0.5$.

Collins *et al.* (1997) describes the results of a redshift survey[1] of extended sources found in 66 southern PSPC pointings. The survey yielded a sample of 36 clusters over 17 square degrees to a flux limit of $\simeq 3.9 \times 10^{-14}$ergs sec^{-1} cm^{-2}. Sixteen of the 36 clusters lie in the redshift range $0.30 < z < 0.67$. This is in stark contrast to the RIXOS survey[2], which found only four $z \geq 0.3$ clusters over a similar area (see Figure 2). Using predictions based on extrapolation of two local[4 5] XCLFs, we conclude that our sample shows no evidence for significant evolution out to $z \simeq 0.7$.

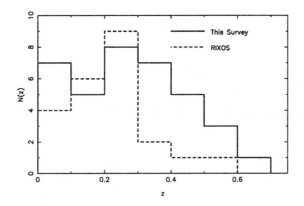

Figure 2: Comparison of the redshift histograms for southern SHARC[1] (solid line) and RIXOS[2] (dotted line) cluster samples. (See Collins *et al.* 1997 for details.)

The studies described above challenge previous claims of a rapidly evolving XCLF. They provide momentum to our continuing ambition to provide the public with the largest distant X-ray cluster sample possible. This sample will have a well understood selection function and will be vital not only to evolution studies but also to studies of superclustering, the Butcher-Oemler effect, the cluster temperature function and the Sunyaev-Zeldovich effect.

This work was partially supported by NASA ADP grant NAG5-2432.

1. Burke, D.J., *et al.*, 1996, MPE Report 263 (Munich), 569.
2. Castander, F.J., *et al.*, 1995, Nature, 377, 39.
3. Collins, C.A., *et al.*, 1997, ApJ Lett, in press. (astro-ph/9701143)
4. De Grandi, S., 1996, MPE Report 263 (Munich), 577.
5. Ebeling, H., *et al.*, 1996, MPE Report 263 (Munich), 579.
6. Ebeling, H., *et al.*, 1997, ApJ, in press.
7. Gioia, I.M., *et al*, 1990, ApJ, 356, L35.
8. Henry, J.P., *et al.*, 1992, 1990, ApJ, 386, 408.
9. Kaiser, N., 1986, MNRAS, 22, 323.
10. Nichol, R.C., *et al.*,1997, ApJ in press. (astro-ph/9611182)
11. Romer, A.K., *et al.*, 1997, in preparation.
12. Rosati, P., *et al.*, 1995, ApJ 445, L11.
13. Scharf, C.A., *et al.*, 1997, ApJ, in press.
14. Snowden, S.L., *et al.*, 1994, ApJ, 424,714.

CLUSTER EVOLUTION IN THE WIDE ANGLE ROSAT POINTED SURVEY (WARPS)

C. A. SCHARF (GSFC/UMD), L. R. JONES (U. Birm.), E. PERLMAN (STScI),
H. EBELING (IfA), G. WEGNER (Dart.), M. MALKAN (UCLA)
& D. HORNER (GSFC/UMD)
*Contact address: Lab. for High Energy Astrophysics, Code 662,
NASA/Goddard Space Flight Center, Greenbelt, MD 20771, USA*

A new catalogue of low luminosity ($L_x \leq 10^{44}$erg s^{-1}) X-ray galaxy clusters covering a redshift range of $z \sim 0.1$ to $z \sim 0.7$ has been produced from the WARPS project. We present the number counts of this low luminosity population at high redshifts ($z > 0.3$). The results are consistent with an unevolving population which does not exhibit the evolution seen in the higher luminosity cluster population. These observations can be qualitatively described by self-similarly evolving dark matter and preheated IGM models of X-ray cluster gas, with a power law index for the spectrum of matter density fluctuations $n \geq -1$.

1 Galaxy cluster evolution

The dynamical timescales of clusters of galaxies are of the order t_0, the Hubble time. Clusters are therefore still young systems. Measurements of the evolution of the X-ray luminosity function (XLF) of the most luminous ($L_x \geq 10^{44}$) clusters in the Einstein Medium Sensitivity Survey [4] (EMSS) show evidence for some negative evolution at redshifts $z \gtrsim 0.3$. Although these results allow some constraints to be put on cosmological and structure formation models they sample only the high end of the cluster XLF. The WARPS [5] was designed to extend this measurement to the faint end of the XLF, at $z > 0.3$ (c.f. other similar projects [6,7,8,9]), and to further test cosmological models.

2 The WARPS cluster sample

Serendipitous X-ray sources were detected in ROSAT PSPC archived fields in the 0.5-2 keV band using the Voronoi Tessellation and Percolation (VTP) method [1,5]. From the 16.6deg^2 currently surveyed (~ 90 fields) a sub-sample of sources with detected flux $> 3.5 \times 10^{-14}$ erg s^{-1} cm^{-2} (total flux $> 5.5 \times 10^{-14}$ erg s^{-1} cm^{-2}) was extracted, extents and corrected fluxes were determined [5] and complete optical followup was performed. Redshifts were obtained for most of the 34 cluster candidates which confirmed them as groups and clusters of galaxies with $10^{42} < L_x \leq 10^{44}$ erg s^{-1} (0.5-2 keV) and $0.1 \lesssim z \lesssim 0.7$.

Since the detection efficencies, exposure maps and flux corrections are well understood the statistical weight of each cluster can be accurately calculated [5].

3   Testing for evolution

In Figure 1 the differential number counts of WARPS clusters (corrected to a uniform sky coverage) with $z > 0.3$ are presented. For comparison, at $z > 0.3$ the RIXOS survey [6] found a surface density of 0.33 ± 0.15 clusters deg^{-2} (to a similar, slightly lower flux limit). The WARPS finds 0.84 ± 0.22 clusters deg^{-2}, a factor of 2.5 times higher. This discrepancy is almost certainly due to different detection efficiencies and data modeling.

We have compared our results with models constructed from the low redshift cluster XLF of the ROSAT BCS [3]. The model predictions plotted in Figure 1 are the expected number counts obtained by integrating the $z = 0$ BCS XLF [3] to high redshift and low luminosity with $q_0 = 0.5$ under varying assumptions about the cluster evolution.

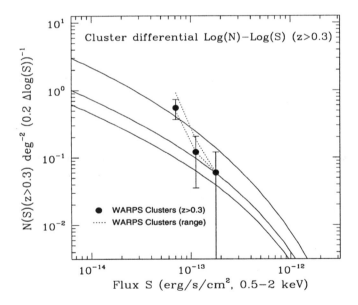

Figure 1: Cluster differential number counts at high redshift. Points indicate WARPS cluster counts, dotted lines indicate possible spread in these counts due to remaining gaps in spectroscopic followup. The uppermost curve is the model prediction with no evolution ($q_0 = 0.5$), the middle curve is the prediction of negative evolution modeled as number density evolution, $\propto (1 + z)^{-2}$, and the lowest curve is the prediction from the minimum amount of evolution seen in the EMSS (approximated as number density evolution, $\propto (1 + z)^{-3}$). From Jones et al, in preparation.

Our conclusion is that the WARPS results show no significant evolution in the low luminosity cluster population at $z > 0.3$ (mean sample redshift $\simeq 0.4$). The amount of negative evolution allowed by the EMSS result is not seen. Interestingly, this is in qualitative agreement with the model described by Kaiser [10], where self-similar dark matter evolution combined with an initially hot IGM can reproduce the basic cluster observations. The low luminosity data here would suggest that Kaiser's model could succeed if the power law index of the spectrum of matter density fluctuations is $n \geq -1$. Recent X-ray observations of element abundances in cluster gas [11] also suggest a preheated IGM.

Acknowledgments

CAS and LRJ acknowledge NRC fellowships during the major period of this work. This work was made possible by the HEASARC at NASA/Goddard Space Flight Center.

References

1. H. Ebeling & G. Wiedenmann, Phys. Rev. E, 47, 704 (1993).
2. H. Ebeling et al, *MNRAS* **submitted**, (1997).
3. H. Ebeling et al, *ApJ* **in press**, (1997).
4. J. P. Henry et al, *ApJ* **386**, 408 (1992).
5. C. A. Scharf et al, *ApJ* **March 1**, (1997).
6. F. J. Castander et al, *Nature* **377**, 39 (1995).
7. P. Rosati et al, *ApJ* **445**, L11 (1995).
8. D. J. Burke et al, Proc. Roentgenstrahlung from the Universe, Wurzburg 1996.
9. A. Vikhlinin et al, BAAS, 188, 06.10, (1996).
10. N. Kaiser, *ApJ* **383**, 104 (1991)
11. M. Loewenstein, R. F. Mushotzky, *ApJ* **466**, 695 (1996).

A catalogue of galaxy cluster models

Eelco van Kampen

Theoretical Astrophyics Center, Juliane Maries Vej 30, DK-2100 Copenhagen, Denmark

We build a model cluster catalogue for the $\Omega_0 = 1$ CDM scenario that is designed to mimic the *ENACS* sample of rich Abell clusters. We use the distribution of richness, corrected for incompleteness, to fix the present epoch. We find $\sigma_8 = 0.4 - 0.5$, which is consistent with other determinations. The catalogue is 70 per cent complete for a richness larger than 50, but we do have a *complete* subsample of cluster models for richnesses larger than 75.

1 Introduction

The traditional approach for obtaining a sample of model clusters is to extract clusters from large-scale N-body simulations [4,9,1]. The advantage of this approach is that completeness of the resulting catalogue is automatically guaranteed. However, a disadvantage is that the resolution of the extracted cluster models is relatively poor since voids and filaments, which are simulated as well, take up most of the volume. This means that the fraction of particles in clusters is only about 10 per cent. We avoid this problem by running an ensemble of individual, high-resolution cluster simulations. This technique has been used by various authors in order to study the influence of the choice of initial conditions on the final cluster properties [2,3,10]. However, because these authors did not aim to construct a fair sample, a statistical comparison with observed catalogues was not feasible. This is exactly the goal we aim for here: to construct a statistically fair catalogue of individual cluster models, each of which is simulated at relatively high resolution. In this context the term 'fair sample' is meant to indicate a sample which is representative of a volume-limited sample.

2 Assumptions and techniques

In order to construct a sample of individually simulated cluster models, we need to make a few assumptions. We assume that clusters of galaxies, as we observe them at the present epoch, formed from peaks in the initial density field smoothed at the length-scale appropriate for clusters. These initial peaks have several characteristics which can be used to predict whether they will evolve into rich Abell clusters. We also assume that the initial density fluctuations are Gaussian distributed. This allows the use of the theory of Gaussian random fields, which provides the probability distributions for the peak parameters in the early (linear) stages of the evolution of the density field.

The construction of a sample of cluster models involves the generation of a set of initial conditions that produce a set of simulated clusters with the same *statistical* properties as for an observed sample. Such realisations of initial matter distributions that produce a cluster with specific properties can be obtained by means of the Hoffman-Ribak method of constructing constrained random fields[5]. We used the implementation of this method by van de Weygaert & Bertschinger[8]. An important limitation of this method is that it can only constrain *linear* functionals of a field. This means that the defining quantity of the catalogue has to be a linear functional as well. Given this limitation on the constrained random field method of generating initial conditions, the final cluster mass will be the best choice for the defining quantity.

Under the assumption that rich clusters of galaxies originate from peaks in the initial density field Gaussian smoothed at $4h^{-1}$Mpc, we select a catalogue on the basis of characteristic parameters of these peaks. Given the fact that we need to use the constrained random field method to generate initial conditions for the individual numerical models, we need to select the catalogue on the basis of linear functionals of the initial density field. With this in mind, we argued that cluster mass is the best catalogue defining quantity, because it can be predicted reasonably well from a linear function of the amplitude *and* the curvature of the initial density peak which is the progenitor of the cluster. Besides this practical argument, total cluster mass *is* a basic property of a galaxy cluster.

For the individual models a numerical code was used that contains a prescription to form galaxies, and also allows galaxies already formed to grow (i.e. accrete particles) and merge with other galaxies[11,12]. Not only does this allow us to directly compare to optical catalogues of galaxy clusters, but the properties of the galaxy population also directly set the present epoch for the cosmological scenario in which the model catalogue is embedded, i.e. they set the amplitude of the initial cosmological fluctuation spectrum.

3 Results

We used a traditional, single large-scale simulation in order to test several issues related to the construction of the catalogue and the reliability of the individual models. We traced peaks in these simulations, and found that they do not merge during the entire run. So we indeed have a one-to-one mapping of initial peaks to final rich Abell clusters. Note that this may not be so in other cosmological scenarios. We also found that both the distribution and the evolution of the peak parameters in the single large-scale low-resolution simulation are similar to those of the set of individual high-resolution simulations. Finally we found that the expected mass relation, a linear function of peak amplitude and peak curvature, is almost equal to that obtained from the clusters extracted from the single simulation.

Having built a catalogue selected on expected final cluster mass, we found that is was 70 per cent complete for richness $C_{3D} \geq 50$. This incompleteness mostly concerns the low-mass end and is mainly due to the noise still present in the relation between final cluster mass and initial peak parameters. But note that the richness measure itself has a rather large intrinsic noise originating from its very definition. To allow a comparison of statistical distribution functions of model clusters properties to those observed we devised a method to corrected for this incompleteness. However, we do have a complete subsample for $C_{3D} \geq 75$, which permits a fully unbiased comparison to observations.

We use the richness distribution of both samples to set the present epoch in the models, i.e. fix σ_8. We find that $\sigma_8 = 0.46$ matches best to the observed $ENACS$ cluster sample [7,6]. This is in the range of allowed values derived from a match of the galaxy-galaxy autocorrelation function for a set of field models similar to the clusters models discussed here to the observed autocorrelation function [12]. The elimination of the free parameter σ_8 of the $\Omega_0 = 1$ CDM scenario allows us to test it with other measures.

References

1. Eke V.R., Cole S., Frenk C.S., 1996, MNRAS, 282, 263
2. Evrard A.E., 1989, ApJ, 341, L71
3. Evrard A.E., Mohr J.J., Fabricant D.G., Geller M.J., 1993, ApJ, 419, L9
4. Frenk C.S., White S.D.M., Efstathiou G., Davis, M., 1990, ApJ, 351, 10
5. Hoffman Y., Ribak E., 1991, ApJ, 380, L5
6. Katgert P., Mazure A., Jones B., den Hartog R., Biviano A., Dubath P., Escalera E., Focardi P., Gerbal D., Giuricin G., Le Fèvre O., Moles O., Perea J., Rhee G., 1996, A&A, 310, 8
7. Mazure A., Katgert P., den Hartog R., Biviano A., Dubath P., Escalera E., Focardi P., Gerbal D., Giuricin G., Jones B., Le Fèvre O., Moles O., Perea J., Rhee G., 1996, A&A, 310, 31
8. van de Weygaert R., Bertschinger E., 1996, MNRAS, 281, 84
9. van Haarlem M.P., Frenk C.S., White S.D.M., 1996, MNRAS, in press
10. van Haarlem M.P., van de Weygaert R., 1993, ApJ, 418, 544
11. van Kampen E., 1995, MNRAS, 273, 295
12. van Kampen E., 1997, submitted to MNRAS

GALAXY COLLISIONS AND THE FORMATION EPOCH OF HUBBLE TYPES

C. BALLAND [a]

Center for Particle Astrophysics and Astronomy Department,
University of California, Berkeley CA 94720, USA

Some aspect of a semi-empirical model of galaxy formation currently under construction is presented. In this model, galaxy formation proceeds through a series of rapid non-merging collisions with surrounding objects. From a semi-analytical treatment of galaxy collisions in a $\Omega_0 = 1$ universe, rules for the formation of morphological types are defined. These rules are coupled to the Press & Schechter mass function for a Cold Dark Matter spectrum normalized to the present distribution of X-ray clusters. The model reproduces the observed morphology-density relation and predicts the formation redshift of field ellipticals to be $z \geq 2$, while spirals form at $z \leq 1.5$. Predictions are made for the redshift evolution of morphological populations in the field as well as in clusters

1 Introduction

There is growing evidence that gravitational interactions between galaxies have a dramatic impact on galaxy properties such as star formation histories and morphologies. In this paper, we assume that the life of a galaxy is dominated by a series of rapid non-merging collisions with surroundings objects rather than by one or two major merger events. Based on the results of N-body simulations [1], a collision cross-section is assigned to any given galaxy. This cross-section depends mainly on the relative velocity between colliding galaxies, the impact parameter and the density of the environment. From this semi-analytic approach, we build the collision history of galaxies in various environments for a given cosmology. We then propose a phenomenological definition of Hubble types based on the galaxy collision history. The collision model is incorporated into a more general scenario of galaxy formation through the use of the Press & Schechter (PS) mass function (Sec. 2). Predictions for the redshift evolution of morphological fractions in various environments are made for the case of a Cold Dark Matter (CDM) universe (Sec. 3).

2 The collision model

We first consider the case of a test galaxy interacting with a perturbing object. In a cluster, the relative velocity between galaxies is of the order of the cluster

---

[a]In collaboration with Joseph Silk (UC Berkeley) and Richard Schaeffer (Saclay, France)

velocity dispersion, that is much higher than the internal velocity dispersion of the test galaxy. The perturbation δV in the stellar velocity field of the test galaxy can be modeled in this case using the impulse and straight line approximations[2]. In the field, these are no longer valid: the gravitational potential of the test galaxy affects the pertuber's trajectory and a focussing factor is introduced[3]. The numerical results of Aguilar & white[1] are then used to define a cross-section $\dot{\Delta}$ for energy exchanges during the collision, which scales as $(\delta V)^2$. We extend this cross-section to the case of many-bodies interactions by integrating $\dot{\Delta}$ over a background population of perturbing objects of various masses. Integrating $\dot{\Delta}$ over time in a $\Omega_0 = 1$ universe, we trace the collision history of the test galaxy since its formation epoch. The integration is performed either in the field or in clusters and uses the relevant scalings with redshift of the physical quantities involved in $\dot{\Delta}$. The resulting collision factor Δ contains quantitative information on the total amount of energy exchanged by the test galaxy during collisions over its lifetime.

We propose a phenomenological definition of Hubble types based on the following line: as soon as a protogalactic cloud collapses, collisions occur and tend to inhibit the gentle infall of the gas needed to allow for the formation of a disk. The condition for a spiral galaxy to form in our model is thus to experience few, if any, collisions between its formation redshift and the epoch z considered. This translates into a condition on the collision factor such as $\Delta < \Delta_{spi}$ where Δ_{spi} is a threshold to be determined. Similarly, galaxies such as $\Delta > \Delta_{ell}$ will end up as ellipticals while galaxies for which $\Delta_{spi} < \Delta < \Delta_{ell}$ will define S0s. These conditions are equivalent to conditions on the formation redshift of galaxies, ellipticals forming at $z > z_{ell}$ in order to experience a sufficient amount of collisions while spirals form at $z < z_{spi}$ with $z_{spi} < z_{ell}$.

3 Results

We use the PS mass function for a CDM spectrum to evaluate the relative abundance of morphological types at any epoch in the field or in clusters. The CDM spectrum is normalized to the present day distribution of X-ray clusters, i.e. $\sigma_8 \approx 0.6$[4]. We impose the observed luminosity function to fit the theoretical mass function on the scale of present day L_* galaxies. This fixes M/L and scales the whole model. The redshift cuts z_{ell} and z_{spi} are determined by requiring the model to reproduce the abundance of morphological types in the field today[5]. This leads to ellipticals forming at $z \geq 2$ and spirals at $z \leq 1.5$. These values are sensitive to the normalization of the fluctuations spectrum considered. The model reproduces the morphology-density relation[5] (see Balland et al. 1997[3] for details). We then make predictions for the redshift

478

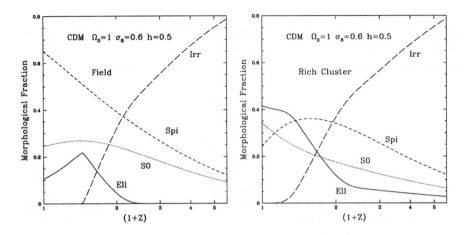

Figure 1: Redshift evolution of morphological types in the field (left) and in clusters (right)

evolution of the morphological fractions in the field and in cluster environments (Fig. 1). The left panel is for field galaxies while the right panel is the case for a rich cluster. Irregulars are defined as galaxies having experienced substantial collisions but which did not have time to relax[3]. In the field, the spiral fraction decreases as one goes towards high redshift due to the increasing density of the environment. Conversely, the irregular fraction increases at higher redshift. Note the difference in the two panels at low redshift. In clusters, the high density of the environment causes the spiral fraction to decrease at a roughly constant rate from $z \sim 0.6$ to $z = 0$, a Butcher-Oemler like effect. The elliptical fraction is dominant in clusters due to the large number of collisions at high density. We refer the interested reader to Balland $et\ al.$ [3] for an extensive discussion of these results. The model predictions should be tested against observations in the near future.

References

1. L. A. Aguilar, S. D. M. White, ApJ **295**, 374 (1985)
2. L. Spitzer, ApJ **127**, 17 (1958)
3. C. Balland, J. Silk, R. Schaeffer, ApJ, *submitted*, (1997)
4. S. D. M. White, G. Efstathiou, C. S. Frenk, $MNRAS$ **262**, 1023 (1993)
5. M. Postman, M. J. Geller, ApJ **281**, 95 (1984)

KILOPARSEC SCALE CORRELATIONS IN THE HUBBLE DEEP FIELD

J.E. RHOADS

Kitt Peak National Observatory, P.O. Box 26732, Tucson, AZ 85726-6732

W.N. COLLEY, J.P. OSTRIKER, O.Y. GNEDIN, D.N. SPERGEL

Princeton University Observatory, Peyton Hall, Princeton, NJ 08544-1001

Sources in the Hubble Deep Field are positionally associated, with a two-point angular correlation function $w(\theta)$ that peaks at $w(0.4'') \approx 2$. These angular correlations imply physical correlations on scales of a few kiloparsecs or less for any source redshift. Such correlations might be expected from the detection of substructure in galaxies similar to those in the local universe. Alternative hypotheses involving merging or accretion scenarios have dynamical implications. Ongoing exploration of these implications indicates that merging or accretion cannot easily explain the observed correlation.

1 Introduction

This work was motivated by three observations: First, the Hubble Deep Field[8] shows a large number density of faint sources. Second, a large fraction of faint galaxies in this and earlier deep HST images show peculiar morphologies. Third, normal galaxies in the local universe look considerably less regular in ultraviolet light.

We therefore asked if individual galaxies at high redshift are being counted as multiple sources in the Hubble Deep Field. To test this possibility, we studied the 2 point angular correlation function of objects in the Hubble Deep Field (HDF).

The appearance of a normal galaxy at cosmological distances is determined by several factors. First, observed optical light was emitted at UV wavelengths for $z \gtrsim 1$. This increases the prominence of young stellar populations relative to old ones: The rest wavelength $0.27\mu m - 0.55\mu m$ color of a stellar population is -1.5 mag for a single burst 10^7 yr old, -0.1 for a population with exponentially declining star formation over the past 3×10^9 yr, and 1.0 for a single burst 4×10^9 yr old.[1] O'Connell and Marcum[5] illustrate this effect using Ultraviolet Imaging Telescope data. Second, angular size is only a weak function of redshift for $z \gtrsim 1$, so the Hubble Space Telescope can resolve kiloparsec scale structures at any redshift. And third, the $(1+z)^4$ dimming of bolometric surface brightness may hide the outer regions of galaxies from view.

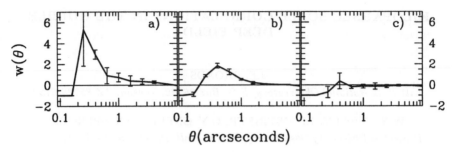

θ(arcseconds)

Figure 1: Angular correlation functions for HDF catalogs: (a) Objects having $U - B >$ $B - R + 1.2$. (b) Other sources. (c) Cross-correlation of the two color bins.

2 Methods and Results

The 2-point angular correlation function $w(\theta)$ is defined such that the probability of a given object having a neighbor in an infinitesimal solid angle $d\,\Omega$ located angle θ away is $(1 + w(\theta))\,\Sigma d\,\Omega$, where Σ is the mean number density of objects on the sky.

We have measured $w(\theta)$ for object catalogs derived from the Hubble Deep Field, using both the entire data set and color-selected subsamples. (E.g., $U - B > B - R + 1.2$, which selects for $2.4 \lesssim z \lesssim 3.4$ [Steidel et al [7]].)

The correlation functions for a representative object catalog are shown in figure 1. (Additional correlation functions are shown in Colley et al 1996.[2]) The objects are substantially correlated ($w \geq 1$) on subarcsecond scales. The correlation strength is somewhat greater for the subset of objects with the colors expected at high redshifts. Cross-correlations between objects of different colors vanish. The results were essentially unchanged when a we used a different object catalog generated by Warrick Couch [4] using a different object finding algorithm. They were also unchanged when we restricted our attention to the objects for which Steidel et al [7] measured spectroscopic redshifts $z \geq 2.59$.

3 Discussion

Because the correlation function is constructed to account for random alignments of sources, the observed correlations must reflect physical associations among identified sources. Also, the scale of these correlations is *at most* a few kiloparsecs, comparable to the size of present galaxies. The vanishing cross-correlations between color bins suggest that associated sources have similar stellar populations. These results are consistent with the substructure hypothesis.

Other hypotheses for the observed correlations include merging scenarios and satellite accretion scenarios. These are subject to dynamical tests.[3]

Merging model: We construct an average radial density profile in neighbor objects from the HDF catalogs. Assuming a mass-to-light ratio converts this into a radial mass density profile, $\rho \sim r^{-3.1}$ for $\theta \lesssim 1''$ ($r \lesssim 6\,\text{kpc}$). If this profile represents a steady state, we can estimate the accreted mass by integrating $M_{acc} = 4\pi r_{in}^2 \rho(r_{in}) v_r$ over the Hubble time at $z \approx 1$. The result is of order $10^{12} M_{\odot}$, which is large compared to the stellar mass of present galaxies or to the central mass concentration inferred from the radial light profile.

Accretion model: Here we consider accretion of satellite galaxies by dynamical friction. If the radial density profile of satellites is in a steady state, it should be a power law $n(r) \propto 1/r$ (Ostriker & Turner [6]). However, the profile for the HDF catalogs is more like $n \propto r^{-2.5}$.

Conclusions: Sources in the Hubble Deep Field show strong angular correlations on subarcsecond scales ($w(0.4'') \approx 2$). These angular correlations imply physical correlations on galaxy-sized scales (a few kiloparsecs). Substructure within galaxies offers a plausible explanation of the observed correlations. Merging and accretion scenarios face dynamical difficulties.

Acknowledgments

This work has been supported by the Fannie and John Hertz Foundation; NSF grants AST-9529120, AST 91-17388, and AST-9424416; NASA ADP grant 5-2567; and NSF traineeship DGE-9354937.

References

1. Charlot, S., & Bruzual A., G. GISSEL computer program; see *ApJ* **405**, 538 (1993)
2. Colley, W., Rhoads, J.E., Ostriker, J.P., & Spergel, D.N. *ApJ* **473**, L63 (1996)
3. Colley, W., Gnedin, O.Y., Ostriker, J.P., & Rhoads, J.E. in preparation
4. Couch, W., http://ecf.hq.eso.org:80/hdf/catalogs/ (1996)
5. O'Connell, R. W. & Marcum, P., 1996,in *HST and the High Redshift Universe*, eds. N.R. Tanvir et al (1996)
6. Ostriker, J.P., & Turner, E.L. *ApJ* **234**, 785 (1979)
7. Steidel, C., Giavalisco, M., Dickinson, M., & Adelberger, K. *AJ* **112**, 352 (1996)
8. Williams et al in *Science with the Hubble Space Telescope— II*, ed. P. Benvenuti, F. D. Macchetto, & E. J. Schreier (STScI, Baltimore, 1996)

BEYOND HDF - SEARCHING FOR EARLY STAR FORMATION IN THE INFRARED

D. THOMPSON

Max-Planck-Institut für Astronomie
Königstuhl 17
D-69117 Heidelberg, Germany

The success of the Hubble Deep Field (HDF) data in identifying galaxies at red-shifts up to ~3 has been quite spectacular. It is possible to extend this to even higher redshifts using infrared techniques, several of which are briefly described in this paper.

1 Introduction

Other papers in this volume describe some of the results which have come from the HDF data. I describe here several types of observations in the infrared which can take us beyond the redshifts probed by the HDF. Although I have split this paper into two main sections, one covering emission-line techniques and the other covering continuum-based techniques, it should be noted that all of these survey methods make use of some strong spectral feature in order to select galaxies out of a background of field galaxies at lower redshift.

2 Emission-Line Surveys

Stretching the intended meaning of *infrared* to include CCD-based narrowband surveys at wavelengths beyond 7000 Å includes searches for Lyα-bright galaxies up to a redshift approaching 7. Field surveys for such objects[1] have, to date, been unsuccessful at identifying any significant population of galaxies, though observations targeting existing structures,[1] such as quasar absorption-line systems of known redshift, have identified a number of interesting objects. Because the Lyα line is resonant, and can suffer multiple scattering off atomic hydrogen as it passes through the ISM, the chances of absorption by dust grains is proportionally higher. This dust *quenching* of the Lyα line is generally thought to explain the lack of success in these field surveys.

Recent, detailed models by Thommes & Meisenheimer,[2] including both dust formation and a scatter in the time when massive star formation begins, give considerably more pessimistic predictions on the volume density of forming galaxies than the canonical results of Baron & White[3] Even so, it should still be possible to identify high-redshift galaxies by targetting their Lyα emission redshifted into the CCD infrared. The Calar Alto Deep Imaging Survey[4] (CADIS) is attempting to cover sufficient volume and depth to reach these new limits, and several $z > 5$ candidate objects have been identified to date.

If sufficient dust is generated early on in a starburst to completely destroy the Lyα photons, then it would still be possible to detect the starbursts through other emission lines, most notably the restframe optical lines of [O II] 3727 Å, Hβ 4861 Å, [O III] 5007Å, and Hα 6563 Å. At high redshift, these lines are shifted into the near infrared JHK bands, where they can be imaged through narrowband filters.

This technique has only been practical since the development of reasonably large infrared arrays.[5,6] There are several groups currently engaged in surveys using this technique,[1,7,8] with a number of objects having already been identified. These surveys primarily target the Hα line at $z \simeq 2.4$, where it is redshifted into the K band, but the method is sensitive to any emission lines, and can be used to image [O II] 3727 Å at redshifts as high as five.

3 Continuum Surveys

Complementary to the emission-line surveys are those based on continuum features. Steidel et al.[9] have used the Lyman limit feature at a restframe wavelength of 912 Å with remarkable success to identify a population of galaxies at $z \simeq 3.25$, many without strong emission lines despite relatively high star formation rates.

If we try to push this technique into the near infrared, we run into problems with the continuum depression across the Lyα line, which increases dramatically at redshifts greater than four.[10] Extrapolating to $z \simeq 7$, which places the Lyα line between the I and J bands, one would expect that virtually all of the flux blueward of the Lyα line would be absorbed. These I-band drop-out objects would appear around $J \simeq 23.4$ (Johnson), assuming little extinction from dust, for star formation rates of $\simeq 50\,M_\odot\,\mathrm{yr}^{-1}$. Such limits are technologically feasible with the current generation of infrared arrays on 4m-class telescopes.

Rather than searching directly for forming galaxies at high redshift, another method of probing the earliest epoch of galaxy formation is to look for old, evolved objects at lower redshift. Evolved, or passively evolving, stellar populations, such as found locally in elliptical galaxies, can develop a strong break in their spectra around 4000Å a few hundred million years after their last burst of star formation, giving another spectral feature on which to base a survey. Sufficiently evolved objects at $z > 1$ can imply formation redshifts of $z > 3$.

As this 4000 Å-break feature is redshifted through the optical bands, the optical-to-infrared colors of these objects becomes increasingly red. At $z \simeq 1.5$, this break lies between the I and J bands. Deep, multicolor imaging in the near infrared can thus distinguish these "extremely red objects" from foreground galaxies. This is potentially a new population of objects; while relatively easy to detect at near infrared wavelengths, no significant numbers of these objects would have been included in even the deepest, optically-selected redshift surveys. Several such galaxies have recently been identified[11,12] from serendipitous observations, though the true extent of any field population is largely unknown. A large-scale field survey would thus be valuable, to determine the space density of such objects and produce a sample for further study.

4 Conclusions

The development of large infrared arrays has opened up several possibilities for pushing beyond the redshifts probed by the HDF data. Current ground-based efforts prehaps presage what we might expect from the NICMOS camera on the Hubble Space Telescope.

References

1. Space limitations prevent listing of references.
2. E. Thommes and K. Meisenheimer in *Galaxies in the Young Universe*, ed. H. Hippelein et al. (Springer, Heidelberg, 1995).
3. E. Baron and S. D. M. White, *Astrophys. J.* **322**, 585 (1987).
4. K. Meisenheimer et al. in *Galaxies in the Young Universe*, ed. H. Hippelein et al. (Springer, Heidelberg, 1995).
5. D. Thompson, S. Djorgovski, and S.V.W. Beckwith, *Astron. J.* **107**, 1 (1994).
6. F. Mannucci and S.V.W. Beckwith, *Astrophys. J.* **442**, 569 (1995).
7. D. Thompson, F. Mannucci, and S.V.W. Beckwith, *Astron. J.* **112**, 1794 (1996).
8. M.A. Malkan, H. Teplitz, and I.S. McLean, *Astrophys. J. Lett.* **468**, L9 (1996).
9. C.C. Steidel et al., *Astrophys. J. Lett.* **462**, L17 (1996).
10. J.D. Kennefick, S.G. Djorgovski, and R.R. deCarvalho, *Astron. J.* **110**, 2553 (1995).
11. J. Dunlop et al., *Nature* **381**, 581 (1996).
12. J.R. Graham and A. Dey, *Astrophys. J.* **471**, 720 (1996).

Formation and Signatures of the First Stars [†]

Z. Haiman & A. Loeb

Harvard University, 60 Garden Street, Cambridge,
MA 02138, USA

We use a spherical hydrodynamics code to show that in cold dark matter cosmologies, the first stars form at $z \sim 50$ through the direct collapse of gas in low–mass systems ($\sim 10^4 M_\odot$). Photons from the first stars easily photodissociate H_2 throughout the universe and so molecular cooling does not affect the subsequent fragmentation of gas clouds into stars. We examine observable signatures of the pre–galactic population of stars. These include the detected metallicity and photo-ionization of the intergalactic medium, and the soon to be detected damping of microwave anisotropies on small angular scales ($\lesssim 10°$). The Next Generation Space Telescope will be able to directly image the pre–galactic star clusters, while the DIMES experiment could detect their Bremsstrahlung emission.

In popular Cold Dark Matter (CDM) cosmologies, the first baryonic objects form at redshifts as high as $z \sim 50$. Although these redshifts are well beyond the current horizon of direct observations, $z \sim 5$, the existence of pre–galactic stars has a number of observable consequences. The stars could reionize the intergalactic medium (in accordance with the lack of the Gunn–Peterson effect out to $z \sim 5$), and the resulting optical depth to electron scattering would damp the microwave background anisotropies on angular scales $\lesssim 10°$. The latter signature will be searched for by future satellite experiments such as MAP or COBRAS/SAMBA. Pre–galactic stars could also produce the amount of carbon necessary to explain the roughly universal $\sim 1\%$ solar metallicity that was detected recently [1,2] in Lyα absorption systems at $z \sim 3$. Finally, an old population of low–mass ($M \lesssim 3 M_\odot$) stars would behave as collisionless matter during galaxy formation and populate the diffuse halos of galaxies. Such stars might account for some of the events observed by ongoing microlensing searches in the halo of the Milky Way [3] and could also be detected in the future through their lensing of distant quasars [4].

We have quantified these observational signatures in CDM cosmologies [5] using a simple semi–analytic approach in which clouds virialize according to the Press–Schechter theory and fragment into stars with an efficiency that yields the observed C/H ratio. This approach is complimentary to more detailed, but computationally expensive and less versatile 3–D numerical simulations [6]. The main ingredients of our model are:

(i) *The Collapsed Fraction of Baryons.* We use the Press–Schechter theory to find the abundance and mass distribution of virialized dark matter halos.

[†] To appear in the Proceedings of the 18th Texas Symposium

However, the collapse of the baryons is delayed relative to the dark matter in low mass objects where gas pressure force resists gravity. We obtained the exact collapse redshifts of spherically symmetric perturbations by following the motion of both the baryonic and the dark matter shells with a one dimensional hydrodynamics code [7]. We find that due to shell–crossing with the cold dark matter, baryonic objects with masses $10^{2-3}M_\odot$, well below the linear–regime Jeans mass, are able to collapse by $z \sim 10$. In calculating the abundance of virialized objects, we take into account the increase in the effective Jeans mass due to photoionization.

(ii) *Star Formation.* We calibrate the fraction of the gas which is converted into stars in each virialized cloud based on the inferred C/H ratio in the Lyα absorption forest. We use tabulated ^{12}C yields of stars with various masses [8], and consider three different initial mass functions (IMFs). In addition, we include a negative feedback on star–formation due to the photodissociation of molecular hydrogen by photons with energies in the range 11–13.6 eV. We find that soon after the appearance of the first few stars, molecular cooling is suppressed even inside dense objects [9]. Due to the lack of any other cooling agent in the metal–poor primordial gas, the bulk of the pre–galactic stars form due to atomic line cooling and fragmentation inside massive objects ($\gtrsim 10^8 M_\odot$) with virial temperatures $\gtrsim 10^4$K.

(iii) *Propagation of Ionization Fronts.* The composite spectrum of ionizing radiation which emerges from the star clusters is determined by the stellar IMF and the recombination rate inside the cluster. We follow the time-dependent spectrum of a star of a given mass based on standard spectral atlases [10] and the evolution of the star on the H–R diagram as prescribed by theoretical evolutionary tracks [11]. To calculate the fraction of the ionizing photons lost to recombinations inside their parent clouds, we adopt the equilibrium density profile of gas inside each cloud according to our spherically–symmetric simulations. We then use the time–dependent composite luminosity of each star–forming region to calculate the propagation of a spherical ionization fronts into the surrounding IGM.

Table 1 summarizes our results for a range of parameters. We varied the cosmological power spectrum ($\sigma_{8h^{-1}}$, n), the baryon density (Ω_b), the star formation efficiency (f_{star}), the escape fraction of ionizing photons (f_{esc}), the IMF, and whether or not the negative feedback due to H_2 is present. For almost the entire range of parameters, the universe is reionized by a redshift $\gtrsim 10$. The optical depth to electron scattering is in the range ~ 0.1–0.2. The resulting damping of ~ 10–20% for the amplitude of microwave anisotropies will be detectable with the MAP or the COBRAS/SAMBA satellites. The only exception occurs when the IMF is strongly tilted towards low–mass stars (by a power law index of 1.7), and reionization is suppressed due to the absence of massive stars which ordinarily dominate the ionizing flux. In this case, the

Table 1: Reionization redshift and electron scattering optical depth for a range of parameters.

Parameter	Standard	Range Considered	z_{reion}	$\tau_{e.s.}$
$\sigma_{8h^{-1}}$	0.67	0.67–1.0	25–32	0.14–0.20
n	1.0	0.8–1.0	18–25	0.08–0.14
Ω_b	0.05	0.01–0.1	24–30	0.03–0.26
f_{star}	4%	4%–90%	20–35	0.10–0.20
f_{esc}	$f_{esc}(z)$	3%–100%	14–31	0.08–0.18
IMF tilt (β)	0	0–1.69	<25	0.01–0.14
H_2 feedback	yes	yes/no	25–29	0.14–0.19

large density of low–mass stars could account for the MACHO events detected towards the LMC. [3]

Another detectable signature of reionization is free–free emission. We find [5] that the contribution of free–free emission from star–forming clouds at $z > 10$ to the brightness temperature of the microwave sky is well above the proposed sensitivity of the DIMES experiment [12] in the frequency range $\nu = 1$–100 GHz.

Finally, the Next Generation Space Telescope [13] will be able to directly image and resolve the high redshift star clusters with its projected sensitivity of ~ 1 nJy in the range 1–3.5μm, and angular resolution of 0.06". Based on our calculated luminosity function of star clusters, we find [5] that NGST will probe $\sim 10^4$ objects per field of view at $z \gtrsim 5$.

We thank P. Höflich, M. Rees, and D. Sasselov for useful discussions.

References

1. L.L. Cowie, A. Songaila, T.-S. Kim and E.M. Hu *A. J.* **109**, 1522 (1995).
2. D. Tytler, et al. in *QSO Absorption Lines*, ed. G. Meylan Springer: Heidelberg, (1995)
3. C. Alcock, et al. 1996, preprint astro-ph/9606165
4. A. Gould, *Ap. J.* **455**, 37 (1995).
5. Z. Haiman and A. Loeb, 1997, ApJ, in press, astro-ph/9611028
6. Gnedin, N. Y., Ostriker, J. P., 1996, preprint astro-ph/9612127
7. Z. Haiman, A. Thoul and A. Loeb, *Ap. J.* **464**, 523 (1996).
8. A. Renzini and M. Voli, *A&A.* **94**, 175 (1981).
9. Z. Haiman, M.J. Rees and A. Loeb, *Ap. J.* **476**, 458 (1997).
10. R. Kurucz, CD-ROM No. 13, ATLAS9 Stellar Atmospheres (1993)
11. G. Schaller, et al., *A&ASS.* **96**, 269 (1992).
12. A. Kogut, preprint astro-ph/9607100 (1996)
13. P. Stockman, this volume (1997).

SPECTRAL EVOLUTION OF GAMMA-RAY BURSTS

EDISON P. LIANG

Rice University, Houston, Texas, 77005-1892 USA

We review the spectral evolution properties of gamma-ray bursts and discuss
a specific physical model which appears most promising in explaining some
unique signatures of GRB spectral evolution.

1 Introduction

As we approach the 30th anniversary of the discovery of gamma-ray bursts (GRBs),
two areas of GRB research emerge as the most promising: multiwavelength
counterpart searches and spectral evolution studies. A combination of these two
approaches may lead to some major advances in the near future. As an example the
recent detection of a bright burst by the Wide Field Camera of the Beppo/SAX
satellite (GB970111) led to its prompt localization to a relatively small error box,
within which a weak ROSAT/SAX x-ray source is discovered[1,2]. Subsequently a
new radio source is found in the x-ray error circle[3]. At the same time the combined
BATSE-SAX measurements provide unprecedented broad-band spectral coverage of
this burst. Together they will shed new light on the nature of this burst. Here we
review some of the recent developments on the spectral evolution of GRBs.

2 GRB Spectral Evolutions

Spectral evolution can in principle reveal valuable informations on the burst emission
mechanisms. However the difficulty lies in identifying quantitative trends since the
apparent patterns are often complex and chaotic. It has long been known that many
GRBs evolve from "hard to soft" in which the hardness peaks before the flux[4].
However, there are also pulses in which the spectral hardness "tracks" the flux[5].
Until recently few systematic patterns have be firmly established.

Recently Liang and Kargatis[6] discovered a remarkable new property of
GRB spectral evolution. They found that at least for sufficiently bright long smooth
GRB pulses, the spectral break of the continuum (defined as the peak of the νF_ν
distribution where ν is the photon energy and F_ν is the specific energy flux) often
decreases exponentially as a function of the photon fluence (running time integral of
photon flux). Even more tantalizing is their discovery that the exponential decay
constant Φ_0 (measured in units of photons/cm^2) often remains the same between
different pulses of the same burst, even when the pulses have very different
hardnesses, intensities and temporal profiles. This hints that the underlying physical
mechanism for such multiple-pulse bursts is likely a regenerative process.

Liang and Kargatis[6] then proposed that the exponential decay behavior is
consistent with simple radiative cooling of a plasma in which the characteristic
energy of the emerging photons is simply proportional to the instantaneous average
energy of the emitting particles. While many radiative mechanisms may be

consistent with this condition, Liang and Kargatis[6] pointed out that multiple or saturated Compton upscattering of soft photons is one such mechanism since the upscattered photons reach "Wien equilibrium" with the emitting plasma before escaping.

3 Recent Developments

The idea that saturated Comptonization of soft photons may be the underlying GRB emission mechanism got another boost recently when Crider et al[7] discovered that the low energy spectral slope below the spectral break (α as defined in the Band et al[8] model) of many hard-to-soft pulses does not stay constant in time as was previously assumed, but generally decreases with time. Moreover α is often positive at the beginning of a pulse, becoming negative only at late times. Again this behavior is consistent with the saturated Compton upscattering scenario in which the Thomson depth of the plasma is decreasing with time, possibly due to expansion.

Another telltale signature of the saturated Comptonization scenario is the late-time soft x-ray excess[9], provided x-ray absorption attenuation by the ISM and circumburster matter is not too severe. Preliminary results from Preece[10] and Piro et al[11] seem to be consistent with such predictions.

4 The S^3C Model

After having exhaustively considered a large number of options, Liang et al[9] concluded that the most likely source of soft photons is internal synchrotron photon emitted by the same leptons doing the upscattering, but in a moderate magnetic field (0.1 Gauss to tens of Gauss for a typical bright burst). The anticorrelated evolution of hardness and flux in the beginning of hard-to-soft pulses is then naturally understood in terms of lepton cooling combined with optical thinning of the highly self-absorbed synchrotorn source. This saturated-synchrotron-self-Compton (S^3C) model makes a number of predictions that are currently being systematically tested with BATSE and other data.

If the emitting leptons are nonthermal with a power law index $p \sim 2\text{-}5$ ($\sim$ slope of the typical BATSE-COMPTEL-EGRET spectra), then the S^3C model predicts that the spectral break should occur at $\geq$ a few hundred keV. While the spectral breaks are indeed usually $\geq$ a few hundred keV at the beginning, at late times they often decrease to 10's of keV. Such low values can only be accommodated in the S^3C model if the leptons have a substantial thermal (Maxwellian) component. This suggests that as the impulsively accelerated nonthermal leptons cool (via Compton upscattering) below a certain energy, they become thermalized. Monte Carlo simulations show that this thermalization energy is $\sim$ couple of hundred keV.

5 Astrophysical Implications of the S^3C Model

If the source is stationary relative to the observer, the S^3C model predicts [9]:

(a) The average magnetic field of the emission region is B~ 0.1-10 G.

(b) The ratio of GRB distance d to emission region size R is d/R $\sim 10^{13}(\tau_T 4\pi/\Phi_o\Omega)^{1/2}$ where τ_T is Thomson depth of the emission region (typically ~ few) and Ω is the emission solid angle.

Result (a) says that the source is not in a very strong or very weak field region. If GRBs are somehow associated with strongly magnetized compact objects, then the burst is occuring far out in the magnetosphere of the compact object.

Result (b) is consistent with the extended Galactic halo distance if R is ~ light seconds. Such an emission region size is motivated by two separate considerations: causality and thermalization. Since τ_T changes slowly during the burst (varying by factors of <10 over seconds) the initial size of the source cannot be much smaller than light seconds as $\tau_T \sim R^{-2}$ in spherical expansion. On the other hand causality requires its size to be not > light seconds. Independently, the deduced thermalization energy (cf. Sec.4) requires the lepton density to be ~ 10^{14}-10^{15}/cm^3 if coulomb collision is the only thermalization process. When this is combined with τ_T ~ few, we again obtain R ~ light seconds.

If however the burst source is moving towards the observer with bulk Lorentz factor $\Gamma \gg 1$ (e.g. in a cosmological blast wave scenario[12]), the S^{3}C model in principle would still be applicable in the emitter rest frame. Hence provided Γ~constant for the duration of the burst all of the above results can be taken over and Lorentz boosted to the observer frame. In particular, the formulae for B and d/R have to be modified with appropiate factors of Γ. For example $\Omega \sim \Gamma^{-2}$ in the observer frame. We find d ~ Gpc if Γ~ 100's. In this case the observed spectrum must be down shifted by Γ to obtain the intrinsic emitter spectrum and spectral break.

Acknowledgements

This work was partially supported by NASA Grant NAG5-1515.

References

1. R.C. Butler, et al., IAUC 6539 (1997).
2. W. Voges, et al., IAUC No. 6539 (1997).
3. D. Frail, et al., IAUC No. 6545 (1997).
4. J. Norris, et al., Ap. J. 301, 213 (1986).
5. S.V. Golenetskii, et al., Nature 306, 451 (1983).
6. E. Liang, and V. Kargatis, Nature 381, 49 (1996).
7. A. Crider, et al., Ap. J. in press (1997).
8. D. Band, et al. , Ap. J. 413, 281 (1993) .
9. E. Liang, et al., Ap. J. in press (1997).
10. R. Preece, contribution to this volume (1997).
11. L. Piro, et al., contribution to this volume (1997).
12. P. Meszaros, and M.J. Rees, Ap. J. 405, 278 (1993).

OBSERVATIONS AND IMPLICATIONS OF X-RAY EXCESSES IN GRBS

R. D. PREECE

University of Alabama in Huntsville,
Dept. of Physics, Huntsville, AL 35899, USA

There is a rich history to the observations of GRBs in the X-ray band. Despite the general observation of X-ray paucity in GRBs, a consistent picture is emerging that about 15 percent of all bursts exhibit some sort of X-ray excess, either in the time history or in the spectra. This has implications for the currently accepted (and disputed!) models.

1   Previous Missions

Gamma-ray bursts (GRBs) typically emit the peak of their power in the 50 – 2000 keV energy band. Burst spectra typically fall off towards the lower energies, so that only $\sim 0.02\%$ of their flux is emitted below 10 keV. However, a persistent thread in the observations of GRBs has been that, whenever the detector has sufficient low-energy response (or where there was a complementary low-energy detector observing simultaneously), a fraction of the observed events have detectable X-ray emission. Among the many missions which have published observations of this phenomenon are: *Apollo 16 & Vela 6A*[1], *OSO7*[2], *P78-1*[3], *HEAO-1*[4], *Ginga*[5], and WATCH-*Granat*[6].

This X-ray excess has historically appeared in the form of soft precursors or tails in the burst time history. That is, the burst time profile is considerably different in the X-ray band than it is in conventional gamma-ray energies. Depending on the energy resolution of the detector observing the X-ray flux, spectra have been published that are at least consistent with thermal emission of some sort. However, due to the size of the error bars, none of these spectral fits is compelling evidence for thermal or blackbody emission from GRBs. At least one report has appeared showing marginal X-ray excesses in the low-end of gamma-ray spectra[3].

2   BATSE Observations

The BATSE Spectroscopy Detectors (SDs) were designed with the inclusion of a beryllium window that covers the face of each detector in order to observe energies down to ~ 7 keV. The on-board telemetry design allows for a single low-energy data channel to be collected every 2.048 s, as well as simultaneously

492

with higher time-resolution triggered data. The energy coverage of this channel varies with the gain of the detector, but for our purposes falls between 7 and 20 keV. It is now very well calibrated, using observations of solar flares, background spectra and Earth occultation measurements of the Crab nebula.

In the BATSE study[7], there was a 5 sigma or greater excess emission in the low-energy channel for 14 out of 86 GRBs. For one case, 3B 920517, the excess was on the order of 20 sigma, or $6\times$, greater than the fitted one-component model rate in the low-energy channel. For 3B 920517, two detectors viewed the burst with favorable geometries and slightly different energy coverage, so an additional spectral form with two free parameters could be fit to the excess; however the exact solution which results from fitting a two-parameter model to essentially two data points must be viewed with caution. This can be seen in figure 1, where an OTTB model was assumed for the excess component. The fitted values were $kT = 0.5 \pm 0.1$ keV and A (at 5 keV) $= 1056 \pm 862$ photons s^{-1} cm^{-2} keV^{-1}, with $\chi^2 = 189$ for 207 degrees of freedom. At best, one can say the fit is consistent with the data.

3 Implications

There are several explanations for the existence of X-ray excesses in the observations, given the current state of theoretical modeling. Although a thermal excess naturally leads one to consider fireball models, the Compton attenuation model (CAM)[8] is an example of a cosmological-origin model that predicts a non-thermal excess. In addition, if the time-history of the excess were considered, the CAM predicts both co-evolution of the spectral hardness with the amplitude of the excess as well as a correlation of the excess with the burst gamma-ray flux. Under the Galactic halo umbrella, we have crustquakes and accretion events on neutron stars, each of which most naturally predict independent X-ray and gamma-ray fluxes. Blackbody emission of roughly 1 keV from a neutron star-sized body is almost certainly ruled out in any case, since the sources would have to be closer than ~ 100 pc, a distance which is not supported by isotropy constraints. Blackbody spectral fits to the X-ray precursors or tails reported by *Ginga* are far from being required by the data. Most cosmological scenarios start out with an initial, impulsive event, such as a binary neutron star merger, that is certainly thermal in nature. The gamma-ray flux is an afterthought, arising in secondary shocks of material accelerated by the fireball, which becomes optically thin after it cools to ~ 20 keV. The fireball flux, red-shifted to the observer's frame, might be the origin of the excess component and should be independent of the gamma-ray flux history. Under unsteady-wind models[9], the central engine of the GRB is variable, thus

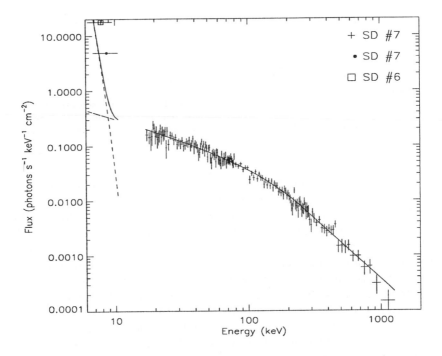

Figure 1: Joint fit of SD7 high spectral resolution data with two low-energy points from SD6 & 7 for 3B 920517. The two components of the model are 'GRB' plus OTTB.

the time-variablility of the excess might be a clue to its behavior. In addition, a blast wave of relativistically-moving material is expected to boost ambient photons at other lower wavelengths into the X-ray band by Comptonization.

References

1. J. I. Trombka et al, ApJ 194, L27 (1974).
2. W. A. Wheaton et al, ApJ 185, L57 (1973).
3. J. G. Laros et al, ApJ 286, 681 (1984).
4. A. Conners & G. J. Hueter, in prep. (1997).
5. A. Yoshida et al, PASJ 41, 509 (1989).
6. A. Castro-Tirado, Ph. D. thesis, Univ. Copenhagen (1994).
7. R. D. Preece et al, ApJ 473, 310 (1996).
8. J. J. Brainerd, ApJ 428, 21 (1994).
9. H. Papthanassiou & P. Mészáros, these proceedings (1997).

Implications of Temporal Structure in GRBs

Tsvi Piran and Re'em Sari

The highly variable temporal structure observed in most GRBs provides us with unexpected clues. We show that variable GRBs cannot be produced by *external shocks* models and consequently cannot be produced by an "explosive" *inner engine*. The observed temporal structure must reflect the activity of the *inner engine* that must be producing unsteady and irregular "wind" which is converted to radiation via *internal shocks*.

Cosmological and even distant galactic halo[1] GRBs models must overcome the compactness problem. Relativistic effects provide the only known solution to this problem. According to the generic picture[2] GRBs arise from a three stages processes: (i) A compact *inner engine* produces a relativistic energy flow (relativistic particles or electromagnetic Poynting flux - but not the observed photons) which (ii) transport the energy outwards to an optically thin region where (iii) it is converted to the observed radiation. Step (iii) could occur due to *external shocks*[3]- resulting from the deceleration of the energy flow into some external medium (e.g. the ISM). Alternatively, *internal shocks*[4,5] that would arise in an irregular flow with non uniform velocities could convert the kinetic energy to radiation. We show here that *internal shocks* rather than *external shocks* convert the energy in GRBs.

Extensive efforts have been devoted to the question of how the observed spectrum is produced, probably because a single spectrum seems to be a universal characteristic of GRBs. The temporal structure, which varies drastically from one burst to another, was practically ignored until recently[6,7,8,9]. Most bursts have a highly variable temporal profile with a rapid variability, on a time scale $\delta T \ll T$, T being the burst's duration. We suggest here that this temporal structure may provide the clue to this mystery[9].

The *inner engine* produces the energy flow which cannot be observed directly. The observed γ-rays emerge only from the outer energy conversion regions. This poses an additional difficulty in deciphering the origin of GRBs. Different engines could produce GRBs provided that they can produce the required relativistic energy flux. Variability on a time scale δT in the bursts dictates an upper limit to the size of this *inner engine* $\sim c\delta T$. *Inner engines* can be "explosive"[10] producing a single outgoing shell whose width Δ is comparable to the size of the *inner engine*. An *inner engine* can also produce a "wind"[11]: an outgoing flows on scales longer than the size of the source. We show here that the observed temporal structure cannot be produced within the energy conversion regions and it must reflect the activity of the source.

Consequently "winds" rather than "explosions" power GRBs.

Consider a relativistic shell that converts its energy to radiation. Let Δ be the width of the shell and let the conversion take place between R_E and $2R_E$. The emitting material moves with a Lorenz factor [a], γ. There are three generic time scales. First is the difference in arrival time between two photons emitted at R_E and $2R_E$, $T_R \approx R_E/\gamma^2 c$. Angular spreading, that is blending of emission from regions from an angle θ from the line of sight leads to a second time scale $T_\theta \approx R_E\theta^2/c$. Finally, Δ/c the light crossing time of the relativistic shell, corresponds to the time difference between the photons emitted from the shell's front and from its back.

Examine now a very thin shell. A typical source that produces a thin shell is an "explosive" fireball for which the width of the shell is comparable to the size of the *inner engine*. However, sources that produces a short wind are also of this kind. Now more specifically require that the shell satisfies: $\Delta \leq R_E/\gamma^2$. It is remarkable that even arbitrarily thick shells will satisfy this conditions if the emission is due to *external shocks* [b]. Since $\Delta < R_E/\gamma^2$ the duration of the burst is determined by the energy conversion region and not by the duration that the *inner engine* operates (which determines Δ).

Because of relativistic beaming an observer detects radiation from an angular scale γ^{-1} around the line of sight. Thus the angular size of the observed regions always satisfies $\theta \leq \gamma^{-1}$. If the system is "spherical" (that is spherical on a scale larger than γ^{-1}) $\theta \approx \gamma^{-1}$ and then $T_\theta \approx R_E/\gamma^2 c \approx T_R \approx T$. Thus angular spreading will erase all temporal structure on scales shorter than T_θ resulting in $\delta T \approx T$. In order to produce variable bursts with $\delta T \ll T$, within the external shock scenario, one must break the spherical symmetry on scales smaller than γ^{-1}.

It is useful to define a variability parameter $\mathcal{V} \equiv T/\delta T \sim 100$. Detailed analysis [9] shows that the emitting regions must have an angular size smaller than $(\gamma\mathcal{V})^{-1} \leq 10^{-4}$ to produce such a variability. A sufficiently narrow jet can bypass this restriction. But it is not clear how one can accelerate and collimate it. Hydrodynamic acceleration, for example, cannot produce an angular width smaller than γ^{-1}. A second possibility is an emitting region made of numerous small size bubbles. The number of bubbles (emitting regions) should be smaller than $\mathcal{V}$, otherwise the contribution from different bubbles will average out to a smooth signal. The maximal solid angle of each bubble is $(\gamma\mathcal{V})^{-2}$. Therefor the total solid angle of all bubbles is smaller than $(\gamma^2\mathcal{V})^{-1}$, which is only $\mathcal{V}^{-1}$ of the observed solid angle. This leads to an intrinsic inefficiency in conversion of energy to radiation of magnitude $\mathcal{V}^{-1}$, ruling out models based on this idea.

---

[a]Note that γ could be smaller than the initial Lorentz factor of the shell.
[b]Strictly speaking this was shown only for hydrodymanic shocks [9].

Let's turn now to a wide relativistic shell in the form of a wind. If the wind is irregular the energy conversion would be due to *internal shocks* and the condition $T_R < \Delta/c$ would be satisfied. This will produce a burst whose overall duration is Δ/c and the observed variability scale is ^{c} $\delta T \approx T_\theta \approx T_R$. The variability scale could be much shorter than the duration. The duration is determined now by the activity of the *inner engine* and not by the emitting regions. The observed temporal structure reflects the activity of the inner engine, which must be producing a relatively long and highly irregular wind.

We find that only this second possibility, of a "wind" like *inner engine* and energy conversion by *internal shocks*, can produce the observed temporal structure. This conclusions have several direct implications. First it tells us that the emitting regions operate with the *internal shock* mechanism ^{d}. This would have direct implications for any attempts to calculated the observed spectrum from these events. The implications for the *inner engine* are even more dramatic: It must operate for a long duration, up to hundred of seconds in some cases, and it must produce highly variable winds as required to form *internal shocks* and the observed variable activity. This directly rules out all explosive models. We will discuss elsewhere the implication of this conclusion for some specific models.

We thank Ramesh Narayan and Jonathan Katz for helpful discussions. The research was supported by a US-ISRAEL BSF grant and by NASA grant NAG5-3516.

1. Piran, T., & Shemi, A. 1993, ApJ, **403**, L67.
2. For a review see Piran, T. in *Some Unsolved Problems in Astrophysics* Eds. Bahcall J. N. and Ostriker, J. P., Princeton University Press, 1997.
3. Mészáros, P., & Rees, M. J. 1992, MNRAS, **258**, 41p.
4. Narayan, R., Paczyński, B., & Piran, T. 1992, ApJ, **395**, L83.
5. Rees, M. J., & Mészáros, P. 1994, ApJ, **430**, L93.
6. Sari, R., & Piran, T. 1995, ApJ, **455**, L143.
7. Sari, R., Narayan, R., & Piran, T. 1996, ApJ, **473**, 204.
8. Fenimore, E. E., Madras, C., & Nayakshin, S. 1996, ApJ, **473**, 998.
9. Sari, R., & Piran, T. astro-ph: 9701002, 1997.
10. Goodman, J. 1986, ApJ, **308**, L47.
11. Paczyński, B. 1986, ApJ, **308**, L51.
12. Sari, R., & Piran, T. 1997, MNRAS, in press.

---

^{c}This is provided, of course, that the cooling time is shorter than T_θ [12].

^{d}In fact we have shown here that *external shocks* cannot produce the observed temporal structure. We have not shown yet that *internal shocks* can produce it. This work is in progress now.

MERGING NEUTRON STARS
AS CENTRAL ENGINES FOR GAMMA-RAY BURSTS

M. RUFFERT, H.-Th. JANKA,

MPA, Postfach 1523, D-85740 Garching, Germany

We continue our numerical investigations of the dynamics and evolution of coalescing neutron stars by integration of the three-dimensional Newtonian equations of hydrodynamics with the "Piecewise Parabolic Method". Our extensions to previously published models include lower minimum densities, nested grids to cover a larger volume, different masses of the two neutron stars and of both binary components, and studies of the effects of a central black hole formed by the collapse of the merged system.

1 Numerics, Initial Conditions, and Neutrino Emission

The scenario of coalescing neutron stars is a three-dimensional problem with the orbital plane being a plane of symmetry. We nest grids 4 levels deep, each grid being Cartesian and equidistant with a resolution of 32^3 or 64^3. The three-dimensional Newtonian equations of hydrodynamics are integrated by the 'Piecewise Parabolic Method' (PPM; Colella and Woodward, 1984). We include the effects of the emission of gravitational waves and the corresponding backreaction on the hydrodynamics in a way originally proposed by Blanchet, Damour and Schäfer (1990).

The properties of neutron star matter are described by the equation of state of Lattimer and Swesty (1991). In addition to the fundamental hydrodynamic quantities, density, momentum, and energy, we follow the time evolution of the electron density in the stellar gas. Energy loss and changes of the electron abundance due to the emission of neutrinos are taken into account by an elaborate "neutrino leakage scheme", which includes a careful calculation of the lepton number and energy source terms of all neutrino types (Ruffert et al, 1996). Neutrinos are produced via thermal processes and lepton captures onto baryons. Matter is rendered optically thick to neutrinos due to neutrino-nucleon scattering and absorption of neutrinos onto baryons.

We simulate the coalescence of two cool neutron stars with different mass combinations: several models consider equal mass neutron stars with baryonic masses of $\approx 1.2\,M_\odot$ and $\approx 1.6\,M_\odot$, respectively. Another model investigates the case of unequal neutron stars with masses $\approx 1.2\,M_\odot$ and $\approx 1.8\,M_\odot$. The radii of the neutron stars vary only slightly about approximately ≈ 15 km. Initially the stars are spherical and orbit around the common center of mass. The orbital velocities of the coalescing neutron stars are prescribed according

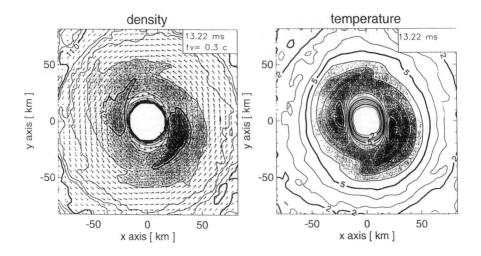

Figure 1: Cut in the orbital plane showing the density distribution together with the velocity field in the left panel and the temperature distribution in the right panel. The density contours are spaced logarithmically with intervalls of 0.2 dex and the bold contours are labeled with their respective values, in units of g/cm³. The temperature contours are spaced linearly with intervalls of 1 Mev and the bold contours are labeled with their respective values. The darker shades of gray indicate higher densities and temperatures. The central circle indicates the sphere with a semidiameter of two Schwarzschild-radii (see text). The legend at the top right corner gives the scale of the velocity vectors and the time elapsed since the beginning of the simulation.

to the motions of point masses as computed from the quadrupole formula. For different models, different spins are added to the neutron stars to take into account rotations around their axes vertical to the orbital plane.

The orbit decays due to gravitational wave emission, and after one revolution the stars are so close that dynamical instability sets in. They merge into a rapidly spinning ($P \approx 1$ ms), high-density body ($\rho \approx 10^{14}$ g/cm³) with a surrounding thick disk of material with densities $\rho \approx 10^{10} - 10^{12}$ g/cm³ and rotational velocities of 0.3–0.5 c. Because of its high mass of ≈ 3 M$_\odot$ we expect the central object to collapse into a black hole essentially instantaneously.

2   Extensions to Previously Published Models

We present preliminary results obtained with our nested grid code. This enables us to model a volume 4^3 times larger than in our previous simulations (Ruffert et al, 1996, 1997), while at the same time keeping a fine resolution in the central regions. The new model equivalent to model B64 from Ruffert

et al (1996) was additionally evolved for a further 5 ms in order to study the
effects of a black hole which forms at the center of the merged binary after a
few ms. In Fig. 1 we show the density distribution together with the velocity
field and the temperature distribution for this case.

A point mass was placed at the center of mass position and given a mass
of roughly $\approx 3\,M_\odot$. This point mass acts gravitationally on the surrounding
matter via the Paczyński-Wiita pseudo-potential (Paczyński & Wiita, 1980).
Because of the divergence of this potential at the Schwarzschild radius, we reset
to a very small finite value the density (and other state variables of matter)
within *two* Schwarzschild radii of the position of the point mass, for numerical
reasons. In this way we hope to mimic the gravitational and absorptive features
of a black hole at the central position of the merged neutron stars. We aim
at determining the amount of matter that remains in the disk surrounding
the black hole and the state this matter is in, which temperatures it has, etc.
We find that the luminosity of neutrinos emitted from the disk to be within
factor of two equal to what has already been reported for the central regions
previously (Ruffert et al, 1996, 1997). This result confirms the difficulty that
prompt (within 10 ms) emission of neutrinos during neutron star mergers could
power gamma-ray bursts.

Acknowledgments

The calculations were performed at the Rechenzentrum Garching on a Cray-
YMP 4/64 and a Cray-EL98 4/256. It is a pleasure to thank Wolfgang Keil
for transforming Lattimer and Swesty's FORTRAN equation of state into a
usable table and for providing the initial neutron star models. MR would like
to thank Sabine Schindler for her patience in our office. One of us (M.R.) was
acknowledges partial support by the Deutsche Forschungsgemeinschaft (Mittel
des Auswärtigen Amts).

References

1. L. Blanchet, T. Damour, and G. Schäfer, *MNRAS* **242**, 289 (1990).
2. P. Colella, and P.R. Woodward, *J.Comput. Phys.* **54**, 174 (1984).
3. J.M. Lattimer and F.D. Swesty, *Nucl. Phys.* **A535**, 331 (1991).
4. B. Paczyński and P.J. Wiita, *A&A* **88**, 23 (1980).
5. M. Ruffert, H.-T. Janka, and G. Schäfer, *A&A* **311**, 532 (1996).
6. M. Ruffert, H.-T. Janka, K. Takahashi, and G. Schäfer, *A&A* , submitted
(1997).

PHASE TRANSITIONS OF ACCRETING NEUTRON STARS AND GAMMA-RAY BURSTS

K.S. Cheng[1] and Z.G. Dai[2]

[1] *Department of Physics, University of Hong Kong, Hong Kong*
[2] *Department of Astronomy, Nanjing University, Nanjing 210093, China*

When neutron stars in low-mass X-ray binaries accrete sufficient matter, they may undergo phase transitions to strange stars. These transitions may result in cosmological γ-ray bursts, and strong gravitational wave bursts if the periods of the resulting strange stars of the order of milliseconds.

1. Introduction

The recent observational results from the BATSE detector on the *CGRO*[1] strongly suggest that the sources of weak γ-ray bursts are at cosmological distances. Many mechanisms including the mergers of binaries consisting of either two neutron stars or a neutron star and black hole have been proposed (for a detailed review see [2]). We here argue that during the evolution of neutron stars in low-mass X-ray binaries, these stars may undergo phase transitions to strange stars. We discuss the astrophysical implications of the phase transitions. We suggest that the transitions not only result in cosmological γ-ray bursts[3] which are consistent with the observations but also produce gravitational wave bursts[4] which may be detected by the advanced LIGO.

2. Model

According to the standard scenario of evolution of a low-mass X-ray binary[5], mass is transferred from the companion to the neutron star, which is spun up to a millisecond period. The processes such as orbital gravitational radiation or magnetic braking keep the system in a steady mass-transfer state throughout the evolutionary timescale of the companion while the accretion rate of the neutron star is near the Eddington value. Thus, the neutron star can accrete mass $\geq 0.5 M_\odot$ in $\sim 10^8$ years and become a millisecond pulsar[6]. If we make an assumption that the mass of the neutron star before accretion is $1.4 M_\odot$, which is supported by the current theories of Type II supernova explosion and observations of masses of pulsars (e.g., the Hulse-Taylor binary system), then the stars in an evolutionary timescale $\geq 10^8$ yrs must become rather massive ones ($\geq 1.8 M_\odot$). Now we ask a question: what could possibly occur in the interiors of these massive neutron stars?

To discuss this question, we first analyze possible equations of state (EOSs) for neutron stars. So far there have been many approaches to determine an EOS for dense matter through the many-body theory of interacting hadrons. Unfortunately, these approaches have given EOSs with different stiffnesses and in turn very different structures of neutron stars. However, the EOSs should be constrained by the observations as follows. First, Link, Epstein & van Riper[7] used a model-independent approach to analyze the postglitch recovery in four isolated pulsars (Crab, Vela, PSR 0355+54 and PSR 0525+21) which are likely to be isolated $1.4 M_\odot$ neutron stars, and concluded that soft EOSs at high densities are ruled out. Second, if the EOSs in cores of neutron stars with mass $\sim 1.4 M_\odot$ were soft, the massive compact objects after the accretion phase of low-mass X-ray binaries could be black holes[8]. In fact, these objects have been identified as millisecond pulsars. This means that soft EOSs may not occur in neutron stars with mass $\sim 1.4 M_\odot$.

Furthermore, the existence of isolated strange stars with $\sim 1.4 M_\odot$ is doubtful for two reasons. First, the postglitch behavior of pulsars is well described by the neutron-superfluid vortex creep theory[9], but current strange-star model cannot explain the observed pulsar glitches. Second, the conversion of a neutron star to a strange star requires the formation of a strange-matter seed, which is produced through the deconfinement of neutron matter at a density ~ 7–$9\rho_0$ (where ρ is the nuclear matter density), much larger than the central density of a $1.4_\odot$ neutron star with a moderately stiff to stiff EOS. Therefore, isolated strange stars with $1.4_\odot$ may not exist. But, as argued by Cheng and Dai recently[3], strange stars can be formed in low-mass X-ray binaries. This is because when the neutron star in a low-mass X-ray binary accretes sufficient mass (perhaps $\geq 0.4 M_\odot$) its central density can reach the deconfinement density and subsequently the whole star will undergo a phase transition to become a strange star in a timescale less than 1 s.

The resulting strange star has a thin crust with mass $\sim 2 \times 10^{-5} M_\odot$, thickness ~ 150m, and internal temperature $\sim 10^{11}$ K. Thus, the nuclei in this crust may decompose into nucleons. The star will cool by the emission of neutrinos and antineutrinos, and because of the huge neutrino number density, the neutrino pair annihilation process $\nu\bar{\nu} \to e^+ e^-$ operates in the region close to the strange star surface. On the other hand, the processes for $n + \nu_e \to p + e^-$ and $p + \bar{\nu}_e \to n + e^+$ play an important role in the energy deposition. The process, $\gamma\gamma \leftrightarrow e^+ e^-$, inevitably leads to creation of a fireball. However the fireball must be contaminated by the baryons in the thin crust of the strange star. If we define $\eta = E_0 / M_0 c^2$, where E_0 is the initial radiation energy produced ($e^+ e^-$, γ) and M_0 is the conserved rest mass of baryons with which the fireball is loaded, then, since the amount of the baryons contaminating the fireballi cannot exceed the mass of the thin crust, we have $\eta \geq 5 \times 10^3$ (see [3]) and the fireball will expand outward. The expanding shell (having a relativistic factor $\Gamma \sim \eta$) interacts with the surrounding interstellar medium and its kinetic energy is finally radiated through non-thermal processes in shocks[10,11]. As estimated from the observed numbers of both low-mass X-ray binaries and millisecond pulsars in [3], a burst rate is about 3–10 events per day.

3. Discussion

We now discuss the other implications of the phase transitions from neutron stars to strange stars:
(1) The phase transition may excite stellar radial oscillations, which are damped not only due to dissipation of the vibration energy of stellar matter into heat by bulk viscosity but also due to conversion of this energy into gravitational radiation. We

have shown that if the periods of strange stars are of the order of milliseocnds, the radial osicillations are mainly damped by gravitational radiation instead of internal viscosity[4]. The resulting gravitational wave bursts can be detected by the advanced LIGO[12] at a rate of about three per year.

(2) The phase transitions might be supported by evidence that there is a gap in the magnetic field distribution of pulsars[13]. This is because neutron stars and strange stars have different crusts which support different minimum magnetic fields.

(3) Hard X-ray bursts from the transient source GRO J1744-28 were detected recently. This source has been further identified as a X-ray pulsar with a low-mass companion. In [14], we show that if this pulsar is a strange star from conversion of a neutron star, the observed bursts may be due to deconfinement of accreted matter when the thin crust of the strange star breaks.

Acknowledgments

This work was supported by a RGC grant of Hong Kong and the National Natural Science Foundation of China.

References

1. G.J. Fishman, and C.A. Meegan, *Ann. Rev. Astr. Astrophys.* **33**, 425 (1995).

2. D. Hartmann, in *The Lives of the Neutron Stars* eds. M.A. Alpar, Ü. Kiziloglu and J. van Paradijs, p. 495 (Kluwer : Dordrecht, 1995).

3. K.S. Cheng and Z.G. Dai, *Phys. Rev. Lett.* **77**, 1210 (1996).

4. K.S. Cheng and Z.G. Dai, *Astrophys. J.*, submitted (1996).

5. D. Bhattacharya and E.P.J. van den Heuvel, *Phys. Rep.* **203**, 1 (1991).

6. E.P.J. van den Heuvel and O. Bitzaraki, *Astron. Astrophys.* **297**, L41 (1995).

7. B. Link, R.I. Epstein and K.A. Van Riper, *Nature* **359**, 616 (1992).

8. G.E. Brown, *Nature* **336**, 519 (1988).

9. D. Pines, and M.A. Alpar, *Nature* **316**, 27 (1985).

10. P. Mészáros and M. Rees, *Astrophys. J.* **405**, 278 (1993).

11. K.S. Cheng and D.M. Wei, *Mon. Not. R. Astron. Soc.* **283**, L133 (1996).

12. A. Abramovici *et al.*, *Science* **256**, 325 (1992).

13. K.S. Cheng and Z.G. Dai, *Astrophys. J. Lett.* **476**, 39 (1997).

14. K.S. Cheng, Z.G. Dai, D.M. Wei and T. Lu, preprint (1996).

SPECTRAL PROPERTIES OF INTERNAL AND EXTERNAL SHOCKS FOR GAMMA RAY BURSTS

H. PAPATHANASSIOU

525 Davey Lab, Pennsylvania State Un., University Park, PA 16801, USA

Two types of shocks that develop during the evolution of relativistic flows appropriate for the description of Gamma Ray Bursts (GRBs)have been discussed in the literature. Those are the "external" shocks that form during the deceleration of a relativistic fireball by the surrounding medium and the "internal" shocks that develop during the dissipation of an unsteady relativistic wind. We model the spectra in terms of parameters that describe the physical quantities in the shocks and explore the parameter space that allows spectra like the observed GRB ones. We discuss the general spectral properties of the two kinds of shocks.

1 Internal and External Shocks

A cataclysmic event that a creates relativistic outflow can develop two kinds of shocks. If the flow is an unsteady wind, internal shocks will dissipate a substantial amount of the wind's bulk kinetic energy [1]. Irrespective of the unsteady character of the flow, its deceleration by the surrounding medium will create a blast wave and a reverse shock (external shocks) that will radiate efficiently the remaining of the energy [2]. Both kinds of shocks produce bursts with non thermal spectra. The main difference between the two is that the internal shocks are nonrelativistic in the unshocked fluid frame, while the blast wave is extremely relativistic and the reverse nonrelativistic.

2 Parameterization and Parameter Search Criteria

The ultrarelativistic wind is the result of the activity of a "central engine". The central engine is assumed to deposit a total amount of energy $E = 10^{51} E_{51}$ with entropy $\eta = E/M_o c^2$ content, on a timescale t_w, and variability on t_{var} -which is related to the initial wind volume ($\sim (ct_{var})^3$). The wind expands in a jet of opening angle $\theta = 0.1$. The unsteady character of the wind is parameterized by $q = \Delta\eta/\eta$. In table 1 we summarize the quantities that describe the dynamics of such flows.

The spectral characteristics of the bursts are largely determined by the conditions in the shocks (magnetic field, particle acceleration). We parameterize the physical processes and quantities involved, thus bypassing our lack of knowledge about their specifics, as follows: the magnetic field by the fraction of its equipartition value λ, the fraction of the electrons that are injected

504

Table 1: Dynamic quantities for "internal" and "external" shocks for $E_{51} = 1, \theta = 0.1$

	Unsteady wind	Impulsive fireball
type of shocks	"Internal"	"External"
burst duration [s]	t_w	$170 n_e x^{-1/3} \eta_2^{8/3}$
burst variability	t_{var}	undetermined
shocks develop at [cm]	$r_{dis} \approx 3\ 10^{14} q^2 \eta_2^2 t_{var}$	$r_{dec} \approx 10^{16} n_{ex}^{-1/3} \eta_2^{-2/3}$
Lorenz factor of shocks	$\Gamma_r = \Gamma_f \approx 1$	$\Gamma_r \approx 1, \Gamma_f \approx \eta$

in the shocks ζ, and the low end of the electron power law ($dN/d\gamma \sim \gamma^{-3}$, $\gamma \geq \gamma_{min}$) by κ ($\gamma_{min} = \Gamma_{shock}\kappa$, $1 \leq \kappa \leq m_p/m_e\zeta$). All these parameters have a wide dynamic range. We perform a parameter search with the objective of determining those values of the parameters that pertain to GRBs.

The dissipation of a large part of the flow's energy should happen in an optically thin environment[15]. In the unsteady wind case, this constrains η.

Taking one of the allowed values for η, we search the parameters that are related to the radiation production (through synchrotron emission and its inverse Compton scattering) in the shocks, λ, ζ, κ. The criteria for this parameter search are set by the characteristics of the bursts that are observed by BATSE, i.e. peak fluence higher than $10^{-9} erg/cm^2$, spectral break in the $80 - 800 keV$ range[4], spectral slope in νF_ν below the break of ~ 1[3], X-ray paucity. There have been reports of excess emission around the $10 keV$ region[6]. We allow for this possibility in our search.

3 Comparative presentation of parameter space search and spectra

For the internal shocks the radiation parameter space search reveals two regions in the ζ, λ, κ space for any given set of t_w, t_{var}. One is the narrow stripe of low κ values (L branch) and the other is a wider high κ region (H branch). Extreme values of κ are excluded for all reasonable ζ. Values in the L branch give spectra that have a lower frequency component (the dashed symbols even correspond to spectra with X-ray excess), they are relatively wide and fall off exponentially at $\sim \eta$ MeV due to e^-e^+ pair production. Combinations of values in the H branch give spectra that have absolutely no low frequency emission, but may have considerable extremely hard emission. The highest allowed values of κ, for a given ζ value, give narrow spectra that show a high frequency roll over and may pick up again at very high frequencies.

The parameter search for external shocks goes along the same lines. Apart from the greater freedom in the choice of η, the radiation parameters show similar (but richer) patterns as in fig 1. The spectra though show a greater

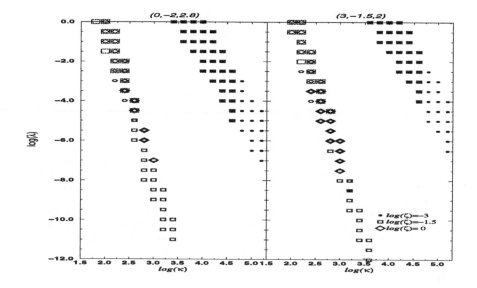

Figure 1: Radiation parameter search for internal shocks. Each of the two graphs is for a different set of values of $(logt_w, logt_{var}, log\eta)$.

variety of shapes and broad band behavior. In general, they do not show a pair production cut-off, are wide, and may have several components.

If one adheres to such a parameterization and believes it has some merit in the physical description of the radiation processes in the shocks, one could use simultaneous multiwavelegnth observations (such as those possible with the now lost *HETE* mission) to further constrain the parameter space [5] and use this insight in the particle acceleration modeling and magnetic field turbulent generation mechanism in the shocks.

References

1. M.J. Rees and P. Mészáros *ApJ* **430**, L93 (1994).
2. P. Mészáros and M.J. Rees *ApJ***405**, 278 (1993).
3. D. Band *et al ApJ* **413**, 281 (1993).
4. G.J. Fishman and C.A. Meegan, *ARA&A***33**, 415 (1995).
5. H. Papathanassiou and P. Mészáros *ApJ***471**, L91 (1996).
6. R. Preece *et al ApJ***473**, 310 (1996).

Limits on Expanding Relativistic Shells from Gamma-Ray Burst Temporal Structure

E. E. FENIMORE

Los Alamos National Laboratory, MS D436,
Los Alamos, NM 87544, USA

We calculate the expected envelope of emission for relativistic shells under the assumption of local spherical symmetry. Gamma-Ray Burst envelopes rarely conform to the expected shape, which has a fast rise and a smooth, slower decay. Furthermore, the duration of the decay phase is related to the time the shell expands before converting its energy to gamma rays. From this, one can estimate the energy required for the shell to sweep up the ISM. The energy greatly exceeds 10^{53} erg unless the bulk Lorentz factor is less than 75. This puts extreme limits on the "external" shock models. However, the alternative, "internal" shocks from a central engine, has one extremely large problem: the entire long complex time history lasting hundreds of seconds must be postulated at the central site.

The temporal structure of long complex Gamma-Ray Bursts (GRBs) presents a myriad of problems for models that involve a single central release of energy, as in many cosmological scenarios. Bursts with 50 peaks within 100 seconds are not uncommon, and there is the recent report [1] of a burst which might have lasted from 10^3 to 10^5 seconds. In Fenimore, Madras, & Nayakshin [2] (hereafter FMN), we used kinematic limits and the observed temporal structure of GRBs to estimate the characteristics of the gamma-ray producing regions. The bulk Lorentz factor of the shell, Γ, must be 10^2 to 10^3 in order to avoid photon-photon attenuation [3,4]. Since the emitting surface is in relativistic motion, the simple rule that the size is limited to $\sim c\Delta T$ does not apply. The high Γ factor implies that visible shells are moving directly towards the observer: if the material of the shell is a narrow cone, it is unlikely that the observer would be within the radiation beam yet outside the cone of material (see FMN).

Surprisingly, the curvature of the shell within Γ^{-1} is just as important in determining the envelope of emission as the overall expansion. This is understood by distinguishing the arrival time of the photons at the detector from the detector's rest frame time. We denote the former as T, and the latter as t. Assume the shell expands at velocity v and emits for time t. Because the emitting surface keeps up with the emitted photons, the photons will arrive at the detector within time $T = (c - v)t/c \approx t/(2\Gamma^2)$. In contrast, the curvature of the shell causes photons emitted from the material at angle $\theta = \Gamma^{-1}$ to arrive after the photons emitted on axis by $T = vt(1 - \cos\theta) \approx t/(2\Gamma^2)$. Thus, both the overall expansion (which might last 10^7 sec) and the delays caused by the curvature spread the observed signal over arrival times by about $t/(2\Gamma^2)$. Envelopes should, therefore, be estimated under the assumption of "local spherical symmetry": local because only $\theta \sim \Gamma^{-1}$ can contribute, symmetric because the material is seen head on, and spherical because curvature effects are important.

One can calculate the expected envelope of emission from an expanding shell. Let $P(\theta, \phi, R)$ give the rate of gamma-ray production for the shell as a function of spherical coordinates. Motivated by the "external shock" models[5], we assume a single shell, $R = vt$, which expands for a time (t_0) in a photon quiet phase and then emits from t_0 to t_{max} (i.e., $P(\theta, \phi, R) = P_0$ from $R = vt_0$ to $R = vt_{max}$, and zero otherwise). In terms of arrival time, the *on-axis* emission will arrive between $T_0 = t_0/(2\Gamma^2)$ and $T_{max} = t_{max}/(2\Gamma^2)$. However, because the curvature is important, off-axis photons will be delayed, and most emission will arrive much later. The expected envelope, $V(T)$, is (see Eq. 11 in FMN):

$$V(T) = 0 \qquad\qquad\qquad \text{if } T < T_0 \qquad (1a)$$

$$= KP_0 \frac{T^{\alpha+4} - T_0^{\alpha+4}}{T^{\alpha+2}} \qquad \text{if } T_0 < T < T_{max} \qquad (1b)$$

$$= KP_0 \frac{T_{max}^{\alpha+4} - T_0^{\alpha+4}}{T^{\alpha+2}} \qquad \text{if } T > T_{max} \qquad (1c)$$

where α is a typical number spectral index (~ 1.5) and K is a constant.

This envelope is similar to a "FRED" (fast rise, exponential decay) where the fast rise depends mostly on the duration of the photon active phase ($T_{max} - T_0$) and the slow, power law decay depends mostly on the duration of the photon quiet phase. The decay phase is due to photons delayed by the curvature.

Often, GRBs do not have a FRED-like shape, implying that something must break the local spherical symmetry. Perhaps $P(\theta, \phi, R)$ is patchy on angular scales smaller than Γ^{-1}, with each patch contributing an observed peak. If so, we define the "filling factor", f, to be the ratio of the observed emission to what we would expect under local spherical symmetry (see Eq. 32 in FMN):

$$f = \frac{\int P(\theta, \phi, t)(1 - \beta \cos\theta)^{-3} dA}{\int (1 - \beta \cos\theta)^{-3} dA} \qquad (2)$$

Thus, we propose the "shell symmetry" problem for cosmological GRBs: models incorporating a single release of energy that forms a relativistic shell must somehow explain either how the material is confined to pencil beams narrower than Γ^{-1} or how a shell can have a low filling factor with a correspondingly higher energy requirement.

From Eq. 1, we find that the half-width of a GRB, $\sim T_{dur}/2$, is $\sim T_0/5$. Thus the shell expands to about $R \sim 5\Gamma^2 T_{dur}$ before becoming active. In previous work [5], the photon quiet phase was estimated from $E_0 = (\Omega/4\pi)R_{dec}^3 \rho_{ISM}(m_p c^2)\Gamma^2$ where E_0 is the energy required to sweep up the ISM with density ρ_{ISM}, m_p is the mass of a proton, Ω is the total angular size of the shell, and R_{dec} is the radius of the photon quiet phase where the shell decelerates and begins to convert its energy to gamma-rays. (Note that one cannot solve E_0 for R_{dec} with an assumed Γ because R is related to Γ through the curvature effects.) Using $R = 5\Gamma^2 T_{dur}$, we find that E_0 is an extremely strong function of Γ: $E_0 \sim 10^{32}\Gamma^8 T_{dur}^3 \Omega \rho_{ISM}$ erg. Unless E_0 is much larger than 10^{53} erg, Γ is quite small (~ 75) for bursts with $T_{dur} \sim 100$ s.

Piran [6] has suggested that the filling factor is $\sim 1/N$, where N is the number of peaks in a burst, and that this filling factor is so small that it rules out single relativistic shells in favor of central engines. However, it is possible to create many peaks and have a large filling factor (as in Eq. 2) by allowing for variations in $P(\theta, \phi, R)$ (work in progress). Thus, we believe it is too premature to "rule out" single relativistic shells. Also, there are other ways to overcome inefficiencies. For example, Ω might be small.

Shaviv [7] has suggested that a single shell sweeps over a cluster of stars with each star contributing a peak to the time history. However, in such a scenario, T_0 is effectively zero so the envelope should have a rise that scales as T^2 (cf. Eq. 1), which is not often seen. In addition, the Shaviv model requires $\Gamma \sim 10^3$, so the energy to sweep up the ISM is extremely large: $10^{62}\rho_{ISM}\Omega$. Globular clusters will have small ρ_{ISM}, but not small enough. Other issues related to the time history and emission process have been raised by Dermer [8].

We conclude that GRBs do not show the signature of a single relativistic shell, and models must, therefore, explain how local spherical symmetry is broken enough to produce the chaotic time histories.

References

1. V. Connaughton, In Preparation.
2. E. E. Fenimore, C. D. Madras, and S. Nayakshin, *Ap. J.* **473**, 998 (1996).
3. M. G. Baring, *Ap. J.* **418**, 391 (1993).
4. S. Nayakshin and E. E. Fenimore, to be submitted, *Ap. J.*.

5. P. Meszaros and M. Rees, *Ap. J.* **405**, 278 (1993).

6. T. Piran, These Proceedings and R. Sari and T. Piran, Preprint.

7. N. Shaviv, These Proceedings and N. Shaviv and A. Dar, *Mon. Not. Royal. Astron. Soc.* **269**, 1112 (1995).

8. C. Dermer, submitted to *Ap. J.*.

WHAT DOES IT TAKE TO SOLVE THE GRB MYSTERY ?

D. HARTMANN
Department of Physics and Astronomy,
Clemson University, Clemson, SC 29634-1911, USA

K. HURLEY
University of California Berkeley,
Space Sciences Laboratory, Berkeley, CA 94720-7450, USA

The phenomenon of cosmic gamma-ray bursts (GRBs) is arguably the most challenging problem of modern astrophysics. The isotropic angular distribution of GRBs on the sky and their non-uniform brightness distribution has ruled out the old paradigm of local sources in the Galactic disk. The current debate over the distance scale considers bursts from either cosmological distances or an extended Galactic halo. The ultimate goal of understanding the nature of GRBs implies identification of their sources and underlying energy generation- and transport mechanisms. Establishing the distance scale can only be a first step, and we must advance our observational capabilities to the point at which the physical nature of GRBs can be revealed. Finding counterparts or host objects is the holy grail of GRB studies and future efforts should be directed to accomplish this crucial step. Just more data is not good enough. New phase space must be explored, including extensions to fainter flux limits, observations of low-, and high-energy spectra beyond the current range of ~ 10 keV $- 10$ GeV, and much better positions.

1   Where do we stand ?

The current debate about the nature of GRBs and their distance scale is well summarized in several papers collected in PASP, Vol. **107**. While establishing the burster distance scale has only second priority, much of our efforts have gone towards the analysis of global statistical properties of GRBs in order to decide whether they are of cosmological or Galactic origin. The most direct arguments are derived from angular- and brightness distributions, and also from their lightcurves (e.g., cosmological time dilation). Many arguments are indirect, such as constraints on burst recurrence. If bursts were found to repeat, most cosmological models would be ruled out. Another issue is the presence of cyclotron lines, perhaps originating near strongly magnetized neutron stars, which are hard to generate if sources are more distant than a few kpc. There are many outstanding questions and debates about GRBs that bear on the quest for the nature and distances of GRBs: How isotropic are bursts ? Are they clustered ? Do they repeat ? Are they correlated with other populations ? Do they show evidence for cosmological evolution ? Is there evidence for time dilation ? Do they exhibit cyclotron lines or pair annihilation lines ?

What causes spectral breaks ? What is the highest photon energy GRBs can generate ? Do spectra extend into the optical/UV range ? Are neutrino bursts or gravity waves associated with GRBs ? Why don't the small error boxes of bright GRBs contain bright host galaxies ? Shouldn't there be an excess of bursts from M31 and nearby galaxies, if they are in halos ? What is their duration distribution ? How many bursts have precursors and delayed emission ? And on and on, filling at least one proceedings volume per year.

2 Location, Location, Location

While future missions could be designed to improve our understanding of most of these questions, tight funding for studies of this challenging astrophysical problem forces us to select a few crucial experiments with high prospects for a breakthrough. While some missions currently under development (such as GLAST) will undoubtedly contribute much to our understanding of GRBs, only dedicated burst missions hold the promise for a solution of this mystery within the coming decade. We argue that counterpart identification is the widely reckognized avenue of choice (Vrba 1996 provides a comprehensive review on counterpart searches). To accomplish this task, it is clear that much smaller error boxes must be obtained. How small ? The answer to this important question is model dependent, but while it may not be neccessary to reach the 1 arcsec level, positions much better than 1 arcminute are clearly required if one wishes to avoid source confusion down to optical magnitudes accessible to HST or ground-based telescopes of the 10 m class. There are basically two approaches to obtaining this kind of accuracy. The first method relies on gamma-rays directly and uses either the burst lightcurves to triangulate the position, or coded mask techniques to image the burst. The latter method is subject of ongoing concept studies. Alternatively, arcsec locations may be obtained from "optical" images of GRBs. This approach requires the existence of detectable emission from bursts, which is only a hypothesis at this point in time. The optical transient (OT) could be found if bursts repeat and if serrendipituous coverage of the burst location exists. This is the method of archival OTs (Schaefer 1981), but the reality of archival OT images is under debate. The more promising approach involves instantaneous burst localization, followed by a rapid response with wide FOV instruments on the ground. Below we discuss this method in greater detail. The HETE experiment (http://space.mit.edu/HETE/), a multi-wavelength burst detector, would have played a major role in this approach, but this mission was lost shortly after launch. HETE-rebuild options (within two years) are currently under consideration. While the OT method is based on the unproven concept of "optical" emission from GRBs, it is the cheapest way to make progress.

3 Inter Planetary Network (IPN)

The currently most accurate burst locations are derived from triangulation using widely separated spacecraft in interplanetary space. A minimum of three long baselines is required to determine a unique position, but additional detectors would not only improve the accuracy of the method but also provide redundancy. The IPN3 network (e.g., Hurley *et al.* 1996) utilizing BATSE, Ulysses, PVO, produced typical error box dimensions of 10" − 1000" (with an average of $\sim$ 6', see Cline *et al.* 1992 for some examples). Most of these positions became available within $\sim$ days of the burst, and optical surveys subsequently produced inventories of these regions to magnitudes of $m_v \sim 24$ (e.g., Vrba, Hartmann, and Jennings 1995). Although "unusual" objects were found in some error boxes, no unambiguous counterpart emerged. The loss of PVO terminated IPN3, and because both Mars94 (Mitrofanov *et al.* 1995) and HETE were lost, there is presently no full network. While future planetary missions could provide a fourth IPN with arcsecond resolution (Hurley & Cline 1994), no committment has been made to place the required detectors aboard any of the planned missions. Concept studies for a multi-satellite network in solar orbit (Energetic Transient Array, ETA; Ricker 1990) show that arcsecond triangulations can be achieved.

4 Optical Transient Searches

Detection of a flaring optical counterpart would yield the desired position accuracy. To obtain simultaneous "optical" data, one must either monitor a large solid angle at all times, or "instantly" determine the burst position and communicate it to the ground for follow-up observations. The capability of rapid burst localisation was first provided by the BACODINE system (Barthelmy *et al.* 1994). However, its accuracy (several degrees) is not good enough for large aperture telescopes, which require arcmin FOVs. In addition, large telescopes usually can not point within seconds to a random position. To address these problems, special optical systems were developed that cover large error boxes, reach meaningful flux limits, and have pointing delays of $\sim$ seconds. The first of these systems was the Gamma-Ray Optical Counterpart Search Experiment (GROCSE) at LLNL, which recently reported limits of $m_v \sim 7$ for several bursts (Park *et al.* 1997). Another system is the BATSE/COMPTEL/NMSU rapid response network, involving a large number of widely distributed telescopes (e.g., McNamara *et al.* 1995), but the response of this network is usually long compared to the burst duration. The Explosive Transient Camera (ETC) at Kitt Peak (e.g., Krimm, Vanderspek, and Ricker 1994) has obtained several simultaneous exposures, without detecting optical flux. This experiment places magnitude limits similar to those reported for GROCSE. Other efforts include the TAROT experiment (http:// www.astrsp-mrs.fr/www/tarot01.html), which covers a 2 degree FOV and slews in less than 3 seconds. Magnitude limits of $m_v \sim$ 13-14 have now been reached with the Livermore Optical Transient Imaging System (LOTIS), which replaced the GROCSE system and is located at a dark site near LLNL. Given the potential breakthrough OT detections might provide, it is not surprising that worldwide efforts in this direction are now underway. If GRBs indeed produce optical emission, these efforts may soon be rewarded with discovery. Several theoretical models suggest that GRBs should produce flux detectable with current technology, but sofar nature has not lifted the veil. Observers must push the limits until she eventually yields the secret.

References

1. S. Barthelmy, *et al.* 1994, in *Gamma-Ray Bursts*, eds. G. J. Fishman, J. J. Brainerd, and K. Hurley, AIP **307**, 643.
2. T. Cline, *et al.* 1992, in *Gamma-Ray Bursts*, ed. W. Paciesas and G. J. Fishman, AIP **265**, 72.
3. K. Hurley, & T. Cline 1994, in *Gamma-Ray Bursts*, eds. G. J. Fishman, J. Brainerd, and K. Hurley, AIP **307**, 653.
4. K. Hurley, *et al.* 1996, in *Gamma-Ray Bursts*, eds. C. Kouveliotou, M. Briggs, and G. Fishman, AIP **384**, 427.
5. H. A. Krimm, R. Vanderspek, & G. Ricker 1996, in *Gamma-Ray Bursts*, eds. C. Kouveliotou, M. Briggs, and G. Fishman, AIP **384**, 661.
6. B. J. McNamara, T. E. Harrison, & C. L. Williams 1995, ApJ, **452**, L25.
7. I. Mitrofanov, *et al.* 1995, Acta Astronautica Suppl. 35, 119. (World Scientific, Singapore, 1988).
8. H.S. Park, *et al.* 1997, ApJ, submitted.
9. G. Ricker 1990, in *High Energy Astrophysics in the 21st Century*, AIP, ed. P. Joss, p. 375.
10. B. E. Schaefer 1981, Nature, **294**, 722.
11. F. J. Vrba, D. Hartmann, & M. Jennings 1996, ApJ, **446**, 115.
12. F. J. Vrba 1996, in *Gamma-Ray Bursts*, eds. C. Kouveliotou, M. Briggs, and G. Fishman, AIP **384**, 565.

AN UNUSUAL CLUSTER OF BATSE GAMMA-RAY BURSTS

V. Connaughton[a], C. Meegan, G. Fishman
NASA Marshall Space Flight Center, Huntsville AL 35812

R. M. Kippen, G. N. Pendleton
University of Alabama in Huntsville

C. Kouveliotou
Universities Space Research Association

K. Hurley
University of California at Berkeley

T. Cline, D. Palmer, S. Barthelmy, P. Butterworth,
B. Teegarden, H. Seifert, J. in 't Zand
NASA Goddard Space Flight Center

E. Mazets, S. Golenetskii
Ioffe Physical-Technical Institute, St Petersburg

We report the detection by BATSE over a 2 day period of 4 triggered gamma-ray bursts which may have a common origin. The locations, time profiles and spectra of the events are discussed along with the probability that these four events are unrelated and clustered spatially and temporally by chance.

1   Introduction

In nearly thirty years of observations of classical gamma-ray bursts (GRBs), no conclusive candidates for a repeating source of cosmic bursts have been seen. The Burst And Transient Source Experiment (BATSE)[2] on the Compton Gamma-Ray Observatory (CGRO) provides unprecedented sensitivity to GRBs. BATSE detects cosmic GRBs at a rate of about 0.8 per day, triggering in the energy range 50-300 keV with a limiting fluence sensitivity $\sim 10^{-8}$ erg cm^{-2}.

The location for each burst is derived from the detected counts in the eight BATSE large area detectors using an algorithm which accounts for the detector response and both atmospheric and spacecraft scattering[6]. An InterPlanetary Network (IPN) location[4] can be obtained using triangulation methods if a burst is seen by more than one spacecraft. Currently the only distant spacecraft is Ulysses[3]. The instruments near Earth include Konus[5] and TGRS[7] on the Wind spacecraft.

---

[a]National Research Council Research Associate

514

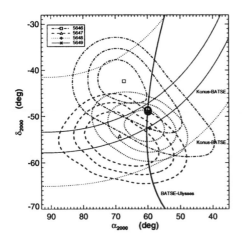

Figure 1: Location of four BATSE bursts detected between October 27 and 29 1996. Only statistical errors are shown. The IPN arcs are obtained from measurements with BATSE and at least one other instrument and the line style of each arc is indicative of the burst to which it applies.

2 Observations

Between 27 and 29 October 1996 BATSE detected 4 bursts from a small ($\sim 9°$ radius cone) region of the sky - the first two on 27 October, separated by 1100 seconds, then two more on 29 October, 650 seconds apart.

Figure 1 is a sky map in equatorial coordinates of the four burst error boxes seen by BATSE and the associated IPN arcs. Statistical errors for each burst are shown as 1,2 and 3 σ χ^2 contours. A systematic error (due to uncertainties in detector response and scattering effects) of 1.6° is associated with three of the events. Trigger 5647, however, is more poorly located and its position can be shifted by 8.5° from that obtained in the standard BATSE analysis (in the direction of burst 5648) with an equally good fit, suggesting a larger systematic error for this burst.

Time histories of each of the four BATSE bursts are shown in Figure 2. The temporal structures of the events are quite different, apparently ruling out a gravitational lens interpretation. By itself, trigger 5649 is one of the longest BATSE events, and among the top 10% in peak flux. Its total fluence is difficult to determine because of telemetry gaps in the BATSE data, but extrapolating from the Konus, TGRS and Ulysses data, one obtains an approximate fluence between 20 and 1800 keV of 3×10^{-4} erg cm^{-2} (accurate to within a factor of 2) which is twice the fluence of the most luminous BATSE burst. The third and fourth events are very probably separate triggers from a single burst, making the combined event the longest burst (1420 sec) ever seen in this energy range. The energy spectra of 5647 and 5649 can be fit using the Band function [1] while

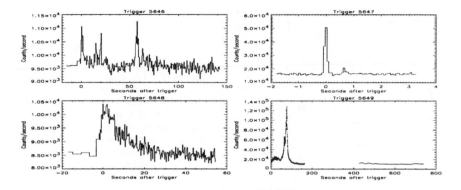

Figure 2: Lightcurves for four BATSE events seen between October 27 and 29 1996

data paucity prevents spectral fitting for 5646 and 5648. A soft gamma repeater (SGR) can be excluded as a common source by the spectral hardness of the events, as well as by their durations.

The *a posteriori* probability of seeing 4 bursts from separate sources within the solid angle subtended by a $10°$ cone in a 2 day window is 10^{-4} per six years. If one considers the third and fourth events to be one long burst from the same source, the chances of seeing 3 unrelated events in this cone in two days are 5%. Although it is likely that one might see two unrelated events in two days from the same direction, combining the events as two bursts of 1200 and 1420 seconds makes them unusual - the probability of seeing two such long bursts from this small cone in two days is 10^{-6} - lower than the four-fold coincidence.

References

1. D. Band *et al.*, *Ap.J.* **413**, 281 (1993).
2. G.J. Fishman *et al.*, Proc. GRO Science Workshop, ed. W.N. Johnson (Greenbelt, NASA 1989), 39.
3. K. Hurley *et al.*, *Astron.Astrophys.Supp* **92**, 401 (1992).
4. K. Hurley *et al.*, AIP Proc. 307, ed. G.J. Fishman, J.J. Brainerd and K. Hurley, (1994), 27.
5. E.P. Mazets *et al.*, AIP Proc. 384, ed. C. Kouveliotou, M.S. Briggs and G.J. Fishman (1996), 492.
6. G.N. Pendleton, M.S. Briggs and C.A. Meegan, AIP Proc. 384, ed. C. Kouveliotou, M.S. Briggs and G.J. Fishman (1996), 877.
7. H. Seifert *et al.*, *Astrophys.Space Science* **231**, 475 (1995).

Astrophysical Effects of Extreme Gravitational Lensing Events

Yun Wang and Edwin L. Turner
Princeton University Observatory, Peyton Hall, Princeton,
NJ 08544, USA

Every astrophysical object (dark or not) is a gravitational lens, as well as a re-ceiver/observer of the light from sources lensed by other objects in its neighbor-hood. For a given pair of source and lens, there is a thin on-axis tubelike volume behind the lens in which the radiation flux from the source is greatly increased due to gravitational lensing. Any objects which pass through such a thin tube or beam will experience strong bursts of radiation, i.e., Extreme Gravitational Lensing Events (EGLEs). We have studied the physics and statistics of EGLEs. EGLEs may have interesting astrophysical effects, such as the destruction of dust grains, ignition of masers, etc. Here we illustrate the possible astrophysical effects of EGLEs with one specific example, the destruction of dust grains in globular clusters.

We propose a new way of looking at gravitational lensing by noting that, for any given pair of source and lens, there is a thin on-axis tubelike volumn behind the lens in which the radiation flux from the source is greatly increased due to gravitational lensing. Any objects which pass through such a thin tube or beam will experience strong bursts of radiation, i.e., Extreme Gravitational Lensing Events (EGLEs).

Inside an EGLE beam, the flux from the source is *greater* than a given value f; the larger f, the thinner the EGLE beam. The characteristic cross section of an EGLE beam is given by

$$\pi r_{EGLE}^2(f) = \frac{8\pi R_S}{27 D_{ds}^3} \left(\frac{L_S}{4\pi f} \right)^2 ,$$

(1)

where $R_S = 2GM/c^2$ is the Schwarzschild radius of the lens of mass M, D_{ds} is the distance between the lens and the source, and L_S is the luminosity of the source.

For a point source, the EGLE beam is infinitely long; it tapers off at infinity. For a finite source, the EGLE beam ends at a distance behind the lens, where the maximum magnification of the source is just enough to bring the unlensed flux up to f, the minimum flux inside the EGLE beam. Let us define a dimensionless parameter

$$\alpha(f) \equiv \frac{8 R_S D_c}{\rho^2} \left(\frac{L_S}{4\pi D_c^2 f} \right)^2 ,$$

(2)

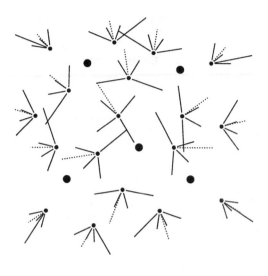

Figure 1: A cartoon of the network of EGLE beams.

where D_c is the maximum separation between the lens and the source, and ρ is the physical radius of the source. $\alpha(f)$ measures the maximum magnification of the source relative to the flux f. $\alpha \to \infty$ is the point source limit; the length of the EGLE beam increases with α.

EGLEs may have interesting astrophysical effects. For an astrophysical system with a population of sources and a population of lenses, space is criss-crossed by a complex network of very bright but narrow beams of light and other forms of radiation. Fig.1 shows a cartoon of the network of EGLE beams for five sources (larger dots) and 18 lenses (smaller dots); the EGLE beams produced by the center source are indicated by dotted lines to illustrate the orientation of the EGLE beams. Without considering EGLE, the hot regions in this system are just the several small spheres around each source. Taking into consideration EGLE, each source produces a thin hot beam which comes out of each lens and points away from the source. Additional hot regions which are thin beams coming out of each lens are thus produced.

In a globular cluster, there are several very bright and hot x-ray sources (believed to be accreting neutron stars) and a large number of stars which act as lensing objects. In a typical globular cluster there would be millions of EGLE beams, one produced by each ordinary star acting on x-rays from each of the cluster's sources. These EGLE regions form a complex network of very hot and narrow beams. Dust grains drift about in this system. For dust grains to be destroyed by EGLE, the following two conditions are sufficient: (1) One EGLE heats up a dust grain to sufficiently high temperature for a sufficient amount of time to destroy the dust grain; (2) Time between two EGLEs is less than the typical lifetime of a dust grain in the absence of EGLE. Both these conditions are satisfied in a typical globular cluster. Whenever a dust grain crosses into an EGLE beam, it will be evaporated and hence destroyed. EGLE may in fact explain why no dust has been observed in globular clusters; they could all have been destroyed by EGLE.

In conclusion, we note that extreme gravitational lensing can be a source of perturbation in astrophysical systems of all scales; it may eventually provide the simplest explanation for some unexplained astrophysical phenomena. Extreme Gravitational Lensing Events can be very dramatic when the source is extremely bright (for example, a gamma-ray burst), and the lens is very massive (for example, a giant black hole). EGLE could provide the trigger for threshold phenomena; it means that objects in interstellar space are subjected to much larger variations in their radiation environment than had been realized previously.

Acknowledgments

We gratefully acknowledge support from NSF grant AST94-19400.

References

1. Y. Wang and E.L. Turner, Astrophys. J., **464**, 114 (1996).
2. Y. Wang and E.L. Turner, Astrophys. J., in press (1997).

Evidence for dust in gravitational lenses

S. Malhotra

IPAC, MS 100-22, Caltech, Pasadena, CA 91125; san@ipac.caltech.edu

J.E. Rhoads

Kitt Peak National Observatory, P.O. Box 26732, Tucson, AZ 85726-6732

E.L. Turner

Princeton University Observatory, Peyton Hall, Princeton, NJ 08544-1001

In an near-infrared survey of known lensed systems we find that the lensed systems identified in radio and infrared searches have redder optical-IR colors than optically selected ones. This could be due to a bias against selecting extincted and reddened quasars in the optical surveys, or due to the differences in the intrinsic colors of optical and radio quasars. Comparison of the radio-selected lensed and unlensed quasars shows that the lensed ones have redder colors. We therefore conclude that at least part of the color difference between the two lens samples is due to dust in the lensing galaxy. Extinction by dust in lenses could hide the large number of lensed systems predicted for universe with a large value of the cosmological constant Λ. These results substantially weaken the strongest constraint on models with a large cosmological constant. They also raise the prospect of using gravitational lenses to study the interstellar medium in high redshift galaxies. (The paper submitted to MNRAS can be found on the astro-ph preprint archives: astro-ph/9610233)

1   Introduction

The statistics of gravitational lensing provide some of the strongest upper limits on the cosmological constant Λ (Turner 1990, Fukugita, Futamase, & Kasai 1990), one of the three parameters (Ω, Λ, H_0) that determine the geometry and the age of the universe (Carroll, Press & Turner 1992). Models with large values of Λ predict many more lensed objects than are presently observed in optical surveys (Maoz & Rix 1993, Kochanek 1993).

The upper limits on Λ from lensing may be too strict if there is enough dust in lensing galaxies to hide the large number of lensed systems predicted for a flat universe with a large value of Λ (Fukugita & Peebles 1993).

2   Method

We report an empirical test to determine whether there is dust in lenses by comparing the optical-IR colors of radio-selected lens systems with those of optically selected systems. Dust transmits redder wavelengths preferentially, so objects seen through large amounts of dust are reddened and dimmed in optical wavelengths and may easily be missed by optical surveys. At radio wavelengths, dust is transparent, so radio surveys should contain lens systems with all degrees of reddening. The radio selected sample should therefore be redder if there is a significant amount of dust in the lensing galaxies. The difference in colors between the optical and radio lensed samples could also be due to the difference in the intrinsic colors of the background objects. We address this issue by comparing the optical-IR colors of lensed and unlensed radio and optical quasars.

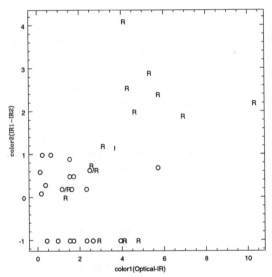

The optical-infrared colors of known multiply imaged gravitationally lensed objects. The optical band is mostly R and sometimes r or i. The near infrared bands are J for IR1 and H or K for IR2. The radio selected objects are represented by R and optically selected objects are denoted by O. The color distribution of optical and radio samples are almost disjoint, and the radio sample is much redder than the optical. Where we have a measurement of only one infrared band we denote the IR1-IR2 color as -1.

3 Results

Figure 1 shows the distribution of the optical-IR2 colors and the IR1-IR2 colors of the two samples. The optical band is mostly R and the two infrared bands are J(IR1) and H or K (IR2). The inhomogeniety in the observed wavelength is negligible compared to the range of emission wavelengths due to the redshift distribution of the sources.

It is clear from Figure 1 that the radio selected lensed images are redder than the optically selected ones. The optically selected and radio-selected quasars have almost disjoint distributions in optical-IR colors. The Wilcoxon test for the means of the two distributions shows that the mean optical-IR color of the radio selected sample is different from the color of the optical sample at 99.99% confidence level. The IR1-IR2 colors are likewise very different in the different samples.

Figure 2 shows the comparison of lensed images with unlensed radio and optical quasar samples. Since the different lensed objects lie at different redshifts and are measured in a slightly inhomogeneous set of bands, the best way to compare their colors is by means of the spectral index α, where $f_\nu \propto \nu^\alpha$.

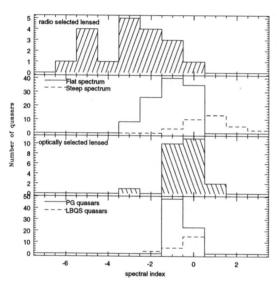

A histogram of the distribution of spectral indices in various samples of quasars. Going from the bottom panel to the top we plot optical unlensed quasars (PG sample [solid line] and LBQS sample [dashed line]), optical lensed quasars, radio unlensed quasars (with both flat radio spectra [solid line] and steep radio spectra [dashed line]), and radio selected lensed systems. While there is not much difference in the spectral indices of lensed and unlensed optical quasars, the radio samples of lensed and unlensed quasars differ significantly.

Comparing the spectral indices derived from these two samples separately by Wilcoxon test, we conclude that the radio selected lensed quasars have steeper spectral indices than the radio selected quasars at the 99.99% confidence level. The optically selected lensed quasars show no significant difference in their spectral indices as compared to the unlensed sample.

References

1. Carroll, S.M., Press, W.H. & Turner, E.L. *ARA&A* **30**, 499 (1992)
2. Fukugita & Peebles, 1993, preprint, astro-ph/9305002.
3. Fukugita, Futamase, & Kasai *MNRAS* **246**, 25p (1990)
4. Kochanek, C.S., *ApJ* **419**, 12 (1993)
5. Maoz, D., & Rix, H.-W., *ApJ* **416**, 425 (1993)
6. Turner, E. L.,*ApJ* **365**, L43 (1990)

Weighing a galaxy bar in the lens Q2237+0305

Robert W. Schmidt

Astrophysikalisches Institut Potsdam, An der Sternwarte 16, 14482 Potsdam, Germany

Rachel L. Webster

School of Physics, University of Melbourne, Parkville, Victoria 3052, Australia

Geraint F. Lewis

SUNY at Stony Brook, Stony Brook, NY11794-2100, USA

We present a new lens model for the quadruple lens Q2237+0305 that includes the lens effect of the bar in this system.

1 A barred lens

Quadruply imaged quasars offer a unique way to weigh galaxies at cosmological distances and to learn about their mass distribution. In the case of Q2237+0305, four images of the quasar are observed superimposed on the core of a barred spiral galaxy [1,2]. The situation is shown in Figure 1. Also shown in this Figure are the inclination axis of the galaxy disk and the direction of the bar. In the following, bulge and disk are treated as a composite component simply called 'bulge'.

If one uses realistic elliptical lensing models to model this system, the major axis has to be aligned with the axis through images C and D [3,6,7,8]. This axis is marked in Figure 1 with the corresponding position angle of 67°. The direction of the major axis of the elliptical isophotes of the bulge, however, is identical with the inclination axis of the galaxy and differs by $\approx 10°$ from the direction of the major axis of a lensing model; there is a discrepancy between the lensing models and the observations.

The motivation for this work was the idea to use the observed bar component to explain this discrepancy. The bar then twists the axis through images C and D by ten degrees in its direction in order to comply with the observed quasar image geometry. By constructing a two component model with bulge and bar, we can measure properties of both components.

2 Models

We used a power-law elliptical model for the bulge and an elliptical Ferrers profile [9] for the bar. A detailed description of these components and their

Figure 1: Image geometry and labelling of Q2237+0305 [3,4]. The relative sizes of the circles correspond to the radio flux ratios [5]. The arrows indicate the galaxy inclination axis [2], the direction of the bar [2] and the direction corresponding to a position angle of 67°. Also shown is the position of the galaxy core.

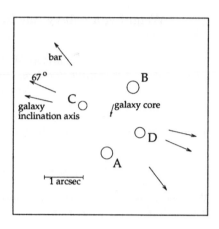

lensing properties as well as our modelling procedure can be found in a forthcoming paper [10]. The model is only fitted to the quasar positions, so that the predicted magnifications can be compared with observed flux values.

The models are degenerate with respect to the ellipticity of the bulge and the length of the major as well as minor axis of the bar. If one uses the observed [11] ellipticity of the bulge (0.31 ± 0.02), the bulge can be restricted to a narrow range of models around a singular isothermal model. The model is insensitive to the length of the major bar axis (≈ 20 arcsec), but the minor bar axis directly determines the steepness of the mass profile of the bar. The length of the minor bar axis ($2 - 4$ arcsec) has not yet been determined precise enough to break this degeneracy.

3   Results: mass, magnification and microlensing

In agreement with previous results [7,12] for this system, we obtain a total mass of $1.49 \pm 0.01 \times 10^{10} h_{75}^{-1} \mathcal{M}_\odot$ for bulge and bar within 0.9 arcsec ($0.6 \, h_{75}^{-1}$ kpc) of the galaxy centre. For all possible bar parameters, the bar mass in this region amounts to 5% of this mass. The total magnification of the quasar ranges from 13 to 21 in the allowed range of bulge ellipticities. The predicted magnification ratios relative to image A for the best bar model are given in Table 1. In this Table, the predictions are compared with radio flux measurements. Although we did not fit for the magnification ratios, the model predicts two out of the three flux ratios correctly within the observational errors.

An explanation for the observed low magnification of image D could be microlensing. In the optical, flux variations due to microlensing have been ob-

Table 1: Magnification ratios relative to image A. The best bar model is compared with measurements from 3.6cm VLA observations by Falco et al.[5]

	predicted	observed
B/A	1.04±0.03	1.08±0.27
C/A	0.60±0.02	0.55±0.21
D/A	1.24±0.02	0.77±0.23

served in all images of Q2237+0305[13]. Falco et al.[5] argue that microlensing is unlikely in the radio, but they can not rule it out completely. Since microlensing in the radio would place strong constraints on the radio emitting region, further observations are needed to explore the radio properties of the quasar. Our model is discussed in detail in a forthcoming paper[10].

Acknowledgments

RWS thanks the Studienstiftung des deutschen Volkes and the organisers of the Texas Symposium for travel support to Chicago. During this work, RWS was supported by a Melbourne University Postgraduate Scholarship and an Australian Overseas Postgraduate Research Scholarship.

References

1. J. Huchra, M. Gorenstein, E. Horine, S. Kent, R. Perley, I. I. Shapiro, G. Smith, 1985, AJ, 90, 691
2. H. K. C. Yee, 1988, AJ, 95, 1331
3. S. M. Kent, E. E. Falco, 1988, AJ, 96, 1570
4. P. Crane, et al., 1991, ApJ, 369, L59
5. E. E. Falco, J. Lehár, R. A. Perley, J. Wambsganss, M. V. Gorenstein, 1996, AJ, 112, 897
6. C. S. Kochanek, 1991, ApJ, 373, 354
7. J. Wambsganss, B. Paczyński, 1994, AJ, 108, 1156
8. H. J. Witt, S. Mao, P. L. Schechter, 1995, ApJ, 443, 18
9. N. M. Ferrers, 1877, Quart. J. Pure Appl. Math., 14, 1
10. R. W. Schmidt, R. L. Webster, G. F. Lewis, 1997, in preparation
11. R. Racine, 1991, AJ, 102, 454
12. H. W. Rix, D. P. Schneider, J. N. Bahcall, 1992, AJ, 104, 959
13. R. Østensen, et al., 1996, A&A, 309, 59

Gravitational Lensing of Quasars by Spiral Galaxies

Abraham Loeb

Harvard Astronomy Department, 60 Garden Street, Cambridge, MA 02138

Gravitational lensing by a spiral galaxy occurs when the line-of-sight to a background quasar passes within a few kpc from the center of the galactic disk. Since galactic disks are rich in neutral hydrogen (cf. Fig. 1), the quasar spectrum will likely be marked by a damped Lyα absorption trough at the lens redshift. Therefore, the efficiency of blind searches for gravitational lensing with sub-arcsecond splitting can be enhanced by 1–2 orders of magnitude [1] by selecting a subset of all bright quasars which show a low-redshift ($z \lesssim 1$) damped Lyα absorption with a high HI column density, $N \gtrsim 10^{21}$ cm^{-2}.

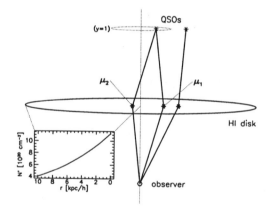

Figure 1. Lensing geometry of a spiral galaxy. Quasars are multiply imaged if their projection lies within the dotted circle. The magnification factors of the two images of the central quasar are μ_1 and μ_2. The insert displays the characteristic HI column density as a function of radius on the face of the HI disk of the lens.

The magnification bias due to lensing changes the statistics of damped Lyα absorbers (DLAs) in quasar spectra by bringing into view quasars that are otherwise below the detection threshold. For optical observations this effect is often counteracted by dust extinction in the lensing galaxy. The combination of lensing and dust extinction results in a net distortion of the intersection probability of HI column densities for spirals. [2] The distortion shows a peak which rises above the unlensed probability distribution. The peak disappears at high column densities because lensing bends the light rays and prevents the brightest quasar image from crossing the HI disk at an arbitrarily small impact parameter. The width of this peak is ultimately determined by the variance of HI profiles and potential wells in the absorber population. [3]

Spiral lenses are difficult to find because their characteristic image separation is a fraction of an arcsecond. The multiple image signature of lensing could, however, be identified spectroscopically and without a need for high-resolution imaging. In the case of the double image system illustrated in Figure 1, the HI absorption spectrum of the quasar would show a generic double–step profile due to the superposition of the two absorption troughs of the different images. This profile is shown schematically in Figure 2. The different images cross the absorbing disk at different impact parameters and therefore have different damped Lyα widths, $W_i \propto N_i^{1/2}$, $(i = 1, 2)$. The metal absorption lines would probe different kinematic components in the two images. In the absence of extinction by dust, the depth of a given step is fixed by the magnification of the corresponding image, μ_i, which reflects the fraction of all detected photons that probe the column density associated with that image. This method for identifying lenses might be contaminated by intrinsic fluctuations in realistic absorption troughs. As a first step towards establishing the feasibility of this technique, one might measure the absorption spectrum of quasars which *are known* to be lensed by a spiral galaxy. The spiral lens B0218+357 constitutes a generic example for such a case. The lensed quasar was discovered in the radio and found to be an Einstein ring of radius $\sim 0.3''$ with two compact components.[4,5] Observations of 21 cm absorption in this lens[6] indicate a high HI column density, $N = 4 \times 10^{21}$ cm^{-2} $(T_s/100\,\text{K})/(f/0.1)$, where T_s is the spin temperature of the gas and f is the HI covering factor. The structure of the Lyα absorption trough in this lens will be measured spectroscopically in the near future with the Hubble Space Telescope[7]. Systems similar to B 0218+357 should be common among quasars with strong damped Lyα absorption.

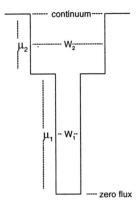

Figure 2. Structure of the Lyα absorption spectrum of a quasar lensed by a spiral galaxy. The double trough profile allows *spectroscopic* identification of a gravitational lens without the need for sub–arcsecond imaging.

Recent microlensing searches [8] indicate that a non–negligible fraction of the halo mass of the Milky–Way galaxy might be in the form of Massive Compact Halo Objects (MACHOs). As a supplement to local searches, it would be interesting to examine whether MACHOs populate the halos of high–redshift spirals. While the microlensing probability is only $\sim 10^{-6}$ in the Milky-Way halo (~ 10 kpc), its value increases up to unity in the cores of halos at cosmological distances (~ 5 Gpc). This follows from the linear dependence of the lensing cross–section on the observer–lens distance for a source at infinity. In particular, one could predict the expected microlensing probability (due to disk stars or MACHOs) in distant spiral galaxies which show up as DLAs in quasar spectra [9]. The obvious advantage of DLAs is that they are selected based on their proximity on the sky to a quasar.

The characteristic Einstein radius of a solar mass lens at a cosmological distance is $\sim 5 \times 10^{16}$ cm, comfortably in between the scales of the continuum–emitting accretion disk ($\lesssim 10^{15}$ cm) and the broad line region ($\sim 3 \times 10^{17}$ cm) of a bright quasar. This implies that a microlensing event would magnify the continuum but not the broad lines emitted by the quasar. As a result, the equivalent width distribution of broad lines [10] will be systematically shifted towards low values in a sample of microlensed quasars. [11, 12] The microlensing probability of a spiral lens is strongly enhanced if the halo of the galaxy is made of MACHOs. [9] In this case, the distortion imprinted by microlensing on the equivalent width distribution of quasar emission lines could be detected in a relatively small sample of only ~ 10 DLAs with HI column densities $N \gtrsim 10^{21}$ cm^{-2} and absorption redshifts $z_{abs} \lesssim 1$. [9] In addition, about a tenth of all quasars with DLAs ($N \gtrsim 10^{20}$ cm^{-2}) might show excess variability on timescales of order 1–10 years. [9] A search for these signals would complement microlensing searches in local galaxies and calibrate the MACHO mass fraction in galactic halos at high redshifts.

Finally, we note that a highly inclined disk could contribute substantially to the lensing cross–section of spiral galaxies. In particular, a quasar behind a razor–thin edge–on disk with a projected mass-per-unit-length μ, would always acquire multiple images with separation $\sim G\mu/c^2$, irrespective of its radial distance from the lens center. Even after averaging over all disk inclinations, the cross-section for lensing by spirals could still be substantially larger than the standard singular-isothermal-sphere value. [13] The lens system B 1600+434 was recently reported [14] to have an edge-on disk in between the lensed quasar images. It therefore provides an excellent candidate for a disk–lensing configuration. Since extinction by dust is common in spiral lenses such as B 0218+357 or B 1600+434, the incidence of spiral lenses could be more common in radio surveys such as CLASS [15].

Acknowledgments

I thank Matthias Bartelmann and Rosalba Perna for many discussions on the subjects mentioned in this contribution.

References

1. M. Bartelmann, & A. Loeb, *Ap. J.* **457**, 529 (1996)
2. R. Perna, A. Loeb, & M. Bartelmann, submitted to *Ap. J.* (1996)
3. A. H. Broeils, & H. van Woerden, A&A S **107**, 129 (1994)
4. C. P. O'Dea, et al.,AJ **104**, 3120 (1992)
5. A. R., Patnaik, et al. MNRAS **261**, 435 (1993)
6. C. L. Carrili, M. P. Rupen, & B. Yanni, *Ap. J.* **412**, L59 (1993)
7. E. Falco, A. Loeb, & M. Bartelmann, "A Search for Multiple Images of QSOs Seen Through Damped Ly-alpha Absorbers", approved proposal for cycle-6 observations with HST (1997)
8. C. Alcock et al., submitted to *Ap. J.*, astro-ph/9608036 (1996)
9. R. Perna, & A. Loeb, to be submitted to *Ap. J.* (1997)
10. P. J. Francis,*Ap. J.* **405**, 119 (1992)
11. C. R., Canizares,*Ap. J.* **263**, 508 (1982)
12. J. J. Dalcanton, C. R. Canizares, A. Granados, & J. T. Stocke, *Ap. J.* **568**, 424 (1994)
13. A. Loeb, & M. Bartelmann, in preparation (1997)
14. A. O. Jaunsen, & J. Hjorth, astro-ph/9611159 (1996)
15. S. Myers, these proceedings (1997)

GRAVITATIONAL LENS MAGNIFICATION AND THE MASS OF ABELL 1689

A.N. TAYLOR

Institute of Astronomy, University of Edinburgh,
Royal Observatory, Blackford Hill, Edinburgh, UK

The absolute mass of a cluster can be measured by the gravitational lens magnification of a background galaxy population. Taking into account the uncertainty introduced by intrinsic clustering of background galaxies, shot noise, and the magnification effect in the strong lensing regime, I measure the mass profile of the cluster Abell 1689. I find that the mass interior to $1h^{-1}$Mpc is $M(< 1h^{-1}\text{Mpc}) = (1.6 \pm 0.3) \times 10^{15}h^{-1}M_\odot$, implying for an isothermal profile a velocity dispersion of $\sigma = (1.7 \pm 0.3) \times 10^3$kms^{-1}. However the profile appears somewhat shallower than isothermal at small radii and steeper beyond a core radii.

1 The Magnification Effect

The observed number of galaxies seen in projection on the sky is[1, 2]

$$n' = n_0 A^{\beta-1}(1 + \Theta), \tag{1}$$

where n_0 is the expected mean number of galaxies in a given area at a given magnitude. Variations about the mean arise from angular perturbations in galaxy density, Θ, due to galaxy clustering, and gravitational lens magnification. The lensing amplification due to a cluster is $A = |(1 - \kappa)^2 - \gamma^2|^{-1}$, where κ is the desired surface mass density and γ is the shear. The galaxy luminosity function can be locally approximated by $\Phi(L) \sim L^{-\beta}$. The index of amplification, $\beta - 1$, then accounts for a decrease in number counts due to the expansion of the background image and for an increase as any faint sources come into view.

2 Galaxy Clustering Noise

The main sources of noise in measuring the amplification, and hence the surface mass density κ, are due to intrinsic clustering of the background source population introducing correlated fluctuations in the angular counts, and shot-noise. We account for this by modeling the angular counts by a Lognormal–Poisson model[1, 2] – a random point–process sampling of a lognormal density field. We find that this reproduces well the distribution of counts in deep fields. The only parameter is the variance of the lognormal field, the amplitude and evolution of which we can estimate from, e.g., the CFRS[3].

3   The Strong Lensing Regime

To transform from the amplification to the surface mass density we shall assume that the lens is smooth over some scale. In this case, for a sufficiently smooth lens, $\gamma \leq \kappa$. The equality holds in the case of a highly isotropic lens, for instance the isothermal lens, while the inequality holds for any anisotropic lens. Van Kampen (in prep.) has recently suggested that for realistic clusters this puts useful constraints on κ from the amplification –

$$A = \frac{1}{|1 - 2\kappa^-|}, \qquad\qquad A = \frac{1}{(1 - \kappa^+)^2} \qquad (2)$$

where κ^- is the isotropic limit and κ^+ is the anisotropic limit. These two estimates bound the true value $\kappa^- \leq \kappa \leq \kappa^+$, and before caustic crossing it can be shown that $\kappa^- \leq \kappa^+ \leq \kappa_{\text{weak}}$, where $A = 1 + 2\kappa_{\text{weak}}$ is the weak lensing limit [2]. Hence the weak lensing approximation will overestimate the cluster mass in the strong regime.

4   The Mass Profile of Abell 1689

Fig. 1 shows the radial mass profile of the cluster Abell 1689. The number counts, observed to $I < 24$, were binned in 13 annuli about the peak of the cluster light distribution. Cluster members were removed by colour selection and masked over. Red galaxies have relatively few faint counts, so that the expansion term in Eq. 1 dominates and there is a net underdensity of red galaxies behind the cluster. Conversely, faint blue galaxies are numerous and cancel the expansion effect. The observed uniformity of blue galaxies across the cluster indicated that the effect is not due to obscuration by dust.

The points in Fig. 1 are calculated using a Lognormal–Poisson likelihood estimator with each of the two strong lensing relations. The dark grey indicates the region between the two extreme estimators. Away from the cluster center the estimators agree and are equal to the weak lensing estimator. The light grey region indicates the uncertainty due to both shot-noise and clustering effects. In the weak lensing regime these effects dominate the uncertainty. Closer to the cluster center the uncertainty due to the unknown shear becomes important and is dominant for $\theta < 2'$. However the cluster mass profile is significantly detected between $1' < \theta < 3.5'$. Converting this to a mass we find $M(< 1h^{-1}\text{Mpc}) = (1.6 \pm 0.3) \times 10^{15} h^{-1} M_\odot$, consistent with the mass inferred from the shear effect [4]. We also appear to see a deviation from an isothermal profile.

532

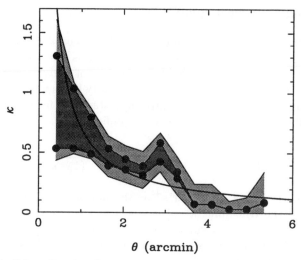

Figure 1: Radial profile of surface mass density of cluster Abell 1689. The uncertainty is due to different strong lensing estimators (dark shaded region) and clustering and shot-noise effects in the background (lighter shaded region). The solid line is an isothermal profile.

5   Conclusions

I have shown that the surface mass profile of a cluster can be measured in the strong lensing regime via the magnification effect on background sources. This takes into account both the uncertainty due to intrinsic galaxy clustering and shot-noise in the background distribution and the uncertainty due to the effect of shear on the amplification.

Acknowledgments

I thank the PPARC for a research associateship and my collaborators Tom Broadhurst, Eelco van Kampen, John Peacock, Nick Kaiser and Narciso Benítez.

References

1. A.N. Taylor and S. Dye, submitted MNRAS (1997).
2. T.J. Broadhurst, A.N. Taylor and J.A. Peacock, ApJ, **438**, 49 (1995).
3. Hudon J.D. and Lilly S.J., preprint (1996).
4. N. Kaiser, in *"Gravitational Dynamics"*, eds O. Lahav, E. Terlevich, R.J. Terlevich, Cambridge University Press (1996).

CLUSTER POTENTIAL RECONSTRUCTION

M. BARTELMANN

Max-Planck-Institut für Astrophysik, P.O. Box 1523, D-85740 Garching, Germany

Gravitational lensing provides a unique way to determine the mass distribution of galaxy clusters. It does not rely on any assumptions about the physical state and the nature of the matter in clusters. I first discuss what kind of information can be obtained from weak and strong lensing by galaxy clusters. I then proceed to compare mass estimates from gravitational lensing and X-ray emission and discuss biases in mass determinations. Finally, I describe how the three-dimensional structure of clusters can be constrained by combining X–ray with lensing information.

1 Weak and Strong Cluster Lensing

The principle underlying all weak-lensing cluster reconstruction techniques is that the *convergence* κ of a cluster and its *shear* γ are related through an effective scalar potential ψ. κ is the surface mass density Σ in units of a critical value Σ_{cr}, and γ measures the tidal field of the mass distribution. ψ is proportional to the projected Newtonian potential Φ of the cluster, $\psi(x_1, x_2) \propto \int \mathrm{d}x_3\, \Phi(\vec{x})$. κ satisfies the Poisson equation $\kappa = \Delta\psi/2$, and the two components of γ are $\gamma_1 = (\psi_{11} - \psi_{22})/2$ and $\gamma_2 = \psi_{12}$, where a subscript i on ψ denotes the derivative of ψ with respect to x_i.

The shear field coherently distorts the images of galaxies in the background of clusters. The distortion is measurable at sufficient resolution because the number density of faint, distant galaxies is as high as $\sim 40\,\mathrm{arcmin}^{-2}$. The intrinsic ellipticities of these galaxies cause noise that can be reduced by averaging, assuming that the intrinsic orientations of the galaxies are random. The finite number density and the need to average limit the resolution of weak-lensing mass maps to $\sim 30''$. In the linear limit, $|\gamma| \ll 1$ and $\kappa \ll 1$, the measured distortion is an estimator of the shear. The intrinsic relation between κ and γ therefore allows to reconstruct κ from the distortion pattern created by a cluster. The first non-parametric reconstruction technique was suggested by Kaiser & Squires (1993).

I cannot go into any more detail of the reconstruction techniques here. There are numerous difficulties in detail, which were extensively discussed in the literature, and led to modifications of the original reconstruction scheme. Recent results demonstrate that both observational and theoretical problems were solved to a degree that allows to infer reliable mass maps of galaxy clusters (N. Kaiser, these proceedings). A shortcoming of all reconstruction techniques based on shear measurements alone is that they yield κ only up to a one-parameter family of transformations. This so-called *mass-sheet degeneracy* can be broken when magnification information is used in addition to shape information (cf. A. Taylor, these proceedings). Converting κ to the surface mass density requires knowledge of the redshift distribution of the distorted sources and the redshift of the cluster.

When $\kappa + |\gamma| \rightarrow 1$, lensing causes strong distortions. The lens mapping can then become singular on closed curves. Arcs can be formed close to such *critical curves*. They are images of background galaxies that are stretched either radially or tangentially with respect to the cluster center. Arcs therefore trace either *radial* or *tangential* critical curves.

For axially symmetric matter distributions, the average κ enclosed by a tangential critical curve is unity, hence the average surface mass density is Σ_{cr}, which is determined by the redshifts of arc source and cluster. The location of tangentially distorted arcs therefore provides a straightforward mass estimate. In the presence of asymmetries, this estimate can be shown to be biased high. More accurate mass estimates from large arcs thus require detailed modeling.

The appearance (e.g. the curvature or thinness) of arcs depends on the structure of the matter distribution in the cluster core. Arc statistics is highly sensitive to cluster asymmetries. Asymmetric matter distributions can increase the probability for arc formation by more than an order of magnitude. Weak lensing thus yields mass maps of clusters as a whole, and strong lensing can be used to infer the structure of cluster cores. For more details on lensing, see, e.g. Schneider et al. (1992) or Narayan & Bartelmann (1997).

2   X–ray Emission vs. Lensing

With X–ray luminosities within $10^{43-45} \, \mathrm{erg \, s^{-1}}$, clusters are the most luminous X–ray sources known. X–ray spectra show that the radiation process is thermal bremsstrahlung. Being a two-body process, it follows that the X–ray emissivity is proportional to the square of the intracluster gas density. X–ray emission by clusters therefore emphasizes cluster cores, where the gas density is highest.

The X–ray emission by clusters is usually interpreted based on the assumptions that the cluster potential is spherically symmetric and that the gas is in hydrostatic equilibrium. Then, the total cluster mass can be inferred from the X–ray flux profile and the spectral temperature. Even if the symmetry and equilibrium assumptions hold, such mass estimates are likely to be biased in the presence of substructure and background noise. Noise limits the radial range within which the X–ray flux profiles can accurately be measured. In the likely case that they do not reach their asymptotic fall-off, X–ray mass estimates are biased low. In addition, if a cluster is substructured, part of the intracluster gas is incompletely thermalized. The spectral temperatures are then also biased low, adding to the bias in the X–ray mass estimates (Bartelmann & Steinmetz 1996 and references therein).

The high sensitivity of strong lensing to cluster asymmetries implies that clusters selected for their ability to produce arcs are likely to be dynamically active and thus out of equilibrium. X–ray mass estimates are then biased low, and the X–ray luminosities are lower than they would be if the clusters were in equilibrium. Consequently, a large fraction of the arc optical depth is contributed by clusters with moderate rather than high X–ray luminosity. Thus, selecting clusters *for* high X–ray luminosity is likely to bias *against* the ability to form arcs.

Systematic discrepancies between mass estimates based on X–ray emission and lensing provide information about the structure of clusters along the line-of-sight. For a spherically symmetric, relaxed cluster, the two mass estimates are expected to agree. If a cluster consists of multiple lumps along the line-of-sight, lensing measures the total, projected mass, while the X–ray emission is determined by the cores of the individual lumps. In such cases, the X–ray emission yields systematically lower mass estimates than lensing. In turn, such discrepancies can be used to constrain the three-dimensional structure of the clusters (Bartelmann & Steinmetz 1996).

Both lensing and X–ray emission measure weighted projections of the same Newtonian potential. The marked difference between the weightings should allow to determine the shape of the potential along the line-of-sight. Ultimately, cluster potentials should be determined such that they simultaneously fit all kinds of information available (Bartelmann & Kolatt 1997, in preparation).

Acknowledgements

This work was supported in part by the Deutsche Forschungsgemeinschaft.

References

1. M. Bartelmann and M. Steinmetz, MNRAS **283**, 431 (1996)
2. N. Kaiser and G. Squires, ApJ **404**, 441 (1993)
3. R. Narayan and M. Bartelmann, *Lectures on Gravitational Lensing* in *Proc. 1995 Jerusalem Winter School*, eds. A. Dekel and J.P. Ostriker (University Press, Cambridge, 1997)
4. P. Schneider, J. Ehlers and E.E. Falco, *Gravitational Lenses* (Springer, Heidelberg, 1992)

MASSIVE BLACK HOLES AND THE LASER INTERFEROMETER SPACE ANTENNA (LISA)

PETER L. BENDER, DIETER HILS, AND ROBIN T. STEBBINS
JILA, University of Colorado, Boulder, CO 80309, USA

The goals of the LISA mission include both astrophysical investigations and fundamental physics tests. The main astrophysical questions concern the space density, growth, mass function, and surroundings of massive black holes. Thus the crucial issue for the LISA mission is the likelihood of observing signals from such sources. Four possible sources of this kind are discussed briefly in this paper. It appears plausible, or even likely, that one or more of these types of sources can be detected and studied by LISA.

The Laser Interferometer Space Antenna (LISA) mission was proposed in 1993 by scientists from both Europe and the United States as a candidate for the third medium-sized mission (M3) of ESA. The proposal was for a joint ESA-NASA mission, but the brief ESA mission study that was carried out in 1993-1994 was for an ESA mission. LISA was not chosen for M3, but was considered and chosen in 1994 as the third new Cornerstone mission under the ESA Horizons 2000 Programme. Since the time scale for LISA as an ESA-led Cornerstone mission appears to be very long, there also has been interest recently in the US in the possibility of flying a NASA-led version of LISA on a considerably shorter time scale. A joint ESA-NASA mission would of course be even more attractive, if a way can be found in the future to accomplish this in a timely fashion.

The ESA Cornerstone version of LISA has been described by Danzmann ,[1] Thorne,[2] and Danzmann et al.[3] In that version, two spacecraft are located at each corner of a triangle 5 million km on a side, and located 50 million km behind the Earth in orbit around the Sun. In the NASA-led version, and probably in a planned revised version of the ESA Cornerstone mission, there will be only one spacecraft at each corner, with two separate optical assemblies pointing along the two arms of the triangle. The on-orbit mass of each spacecraft is about 200 kg. The sensitivity for one year of integration and a S/N ratio of 5 is 1×10^{-23}.

Most sources observable by LISA would be galactic binaries consisting of two compact stars. At least hundreds and probably thousands of individual sources of this kind are certain to be observable. At frequencies below about 3 millihertz, there will be so many such sources that most of them cannot be resolved. Their signals therefore will give a confusion noise level that prevents other weaker signals from being observed.

Supermassive black holes with masses larger than 10^7 M_o are now believed to exist in the centers of most large galaxies. However, much less is known about the abundance of massive black holes (MBHs) with masses between roughly 100 M_o and 10^7 M_o, except for those believed to exist in our galaxy and in M32. If the fraction of smaller or medium sized elliptical and spiral galaxies containing such MBHs is substantial, there are several types of signals associated with them that LISA may well be able to observe (see e.g., Ref. 4 for some additional discussion). The essential assumption we have to make if we expect LISA to see such signals is that roughly 10^9 to 10^{10} galaxies contain MBHs.

One important question is how the MBHs formed. If they grew from seed MBHs, and the seed MBHs were formed by stellar collisions in dense galactic nuclei, then coalescences of multiple seed MBHs in each such galaxy are likely to be observable by LISA. Quinlan and Shapiro[5] have simulated the collisional formation of seed black holes with masses of up to roughly 100 M_o, and find that typically a number of such seeds are formed. If tens of seeds survive and grow to roughly 500 M_o or larger before coalescing, then the resulting event rate could be substantial. LISA could observe the gravitational wave signals over the last year before coalescence out to a redshift of z=5.

On the other hand, Rees[6-9] and Haehnelt and Rees[10] have suggested that very large black holes are more likely to have formed by collapse of supermassive stars produced by rapid contraction of gas and dust clouds in galactic centers. If this is true for most 10^5 or 10^6 M_o black holes as well as for supermassive black holes, then the issue becomes whether signals from the supermassive star collapse are likely to be observable. Although the collapse would be quite slow if the supermassive star is rotating slowly (see e.g., Rees[9]), fast rotation could lead to a bar instability and more efficient gravitational wave radiation at frequencies LISA is sensitive to.

Another promising source of signals for LISA is highly unequal mass binaries. The capture rate for compact stars or stellar mass black holes orbiting MBHs in galactic nuclei has been discussed by Hils and Bender,[11] Sigurdson and Rees,[12] and Sigurdson.[13] For neutron stars or white dwarfs the expected capture rate is high, but often scattering by other stars near the MBH will make the duration of gravitational wave emission much less than a year, and such events difficult to observe. However, for 5 or 10 M_o black holes in the cusp surrounding the MBH, the fraction of captures giving observable signals will be a lot higher. An event rate of higher than once per year appears to be quite plausible, if our assumption on the total number of MBHs is correct.

A fourth possible source of gravitational waves is coalescence of galactic center MBHs following mergers of galaxies or of pregalactic structures. Mergers of galaxies are observed to be occurring today, and the merger rate is believed to have been higher in the past. However, how frequently such events occurred is not known (see e.g., Haehnelt[14]; Rees[8]; and Vecchio[15]). If MBHs were produced only when galactic size structures were formed, it seems difficult to see how enough coalescences could have taken place to give an event rate higher than roughly once per decade. However, if the formation of MBHs took place during the condensation of substantially smaller pregalactic structures that later merged to form the present galaxies, a higher coalescence rate may be possible.[14]

Acknowledgments

Support for this work under NASA Grants NAGW-4772 and NAGW-4865 is appreciated. We also thank many of the authors of the references given below and our colleagues on the LISA Science Team for useful information that they have provided.

References

1. K. Danzmann, Proc. 17th Texas Symp. on Relativistic Astrophys., Annals N. Y. Acad. Sci., Vol. 759, p. 481 (1995).
2. K. Thorne, *ibid.*, p. 127 (1995).
3. K. Danzmann *et al.*, LISA Pre-Phase A Report, Max-Planck-Institut fur Quantenoptik, Report MPQ 208, Garching, Germany (1996).
4. P. L. Bender and D. Hils, in *Advances in Space Research*, submitted (1997).
5. G. D. Quinlan and S. L. Shapiro, *Astrophys. J.* **356**, 483 (1990).
6. M. J. Rees, *Proc. Natl. Acad. Sci. USA* **90**, 4840 (1993).
7. M. J. Rees, in Gravitational Dynamics: Proc. 36th Herstmonceux Conf. (Cambridge Univ. Press, 1996).
8. M. J. Rees, *Class. & Quantum Gravity*, in press (1997).
9. M. J. Rees, in *Reviews of Modern Astronomy*, Vol. 10, in press (1997).
10. M. G. Haehnelt and M. J. Rees, *Mon. Not. R. Astron. Soc.* **263**, 168 (1993).
11. D. Hils and P. L. Bender, *Astrophys. J.* **445**, L7 (1995).
12. S. Sigurdsson and M. J. Rees, *Mon. Not. R. Astron. Soc.*, in press (1997).
13. S. Sigurdsson, *Class. & Quantum Gravity*, in press (1997).
14. M. G. Haehnelt, *Mon. Not. R. Astron. Soc.* **269**, 199 (1994).
15. A. Vecchio, *Class. & Quantum Gravity*, in press (1997).

PERSPECTIVES FOR GRAVITATIONAL WAVE DETECTION WITH RESONANT DETECTORS

M. CERDONIO, L. TAFFARELLO, J.P. ZENDRI

Department of Physics, University of Padova, and I.N.F.N., Sezione di Padova, via Marzolo 8, 35131 Padova, Italy

R. MEZZENA, G.A. PRODI, S. VITALE

Department of Physics, University of Trento, and I.N.F.N., Gruppo Coll. di Trento, 38050 Povo, Trento, Italy

M. BONALDI, P. FALFERI

CeFSA, ITC-CNR Trento and I.N.F.N. Gr. Coll. di Trento, I-38050 Povo, Trento, Italy

L. CONTI, V. CRIVELLI VISCONTI, A. ORTOLAN, G. VEDOVATO

I.N.F.N. Legnaro National Laboratories, via Romea 4, I-35020 Legnaro, Padova, Italy

P. FORTINI

Department of Physics, University of Ferrara and I.N.F.N. Sez. di Ferrara, I-44100 Ferrara, Italy

Ultracryogenic resonant detectors will play a crucial role in the detection of gravitational wave bursts when they will approach their quantum limited sensitivity. We briefly present what has still to be done on the AURIGA type bar detectors to meet this requirement and what advantages it could give as for a confident detection of gravitational waves.

1 Current Situation

The best operating gravitational wave detectors are presently cryogenic resonant bars: ALLEGRO[1], EXPLORER[2] and NIOBE[3]. They demonstrated the capability of long term operation over years with stable noise performance and very high duty cycle. Two ultracryogenic detectors, NAUTILUS[4] and AURIGA[5] are close to operation and work is in progress to improve significantly their sensitivity towards the quantum limit for bursts detection. At present, all these resonant bar detectors are oriented parallel, so to maximize the probability of detection in coincidence. The typical strain noise spectral density at bar input shows minimum values $\sim 10^{-21}/\sqrt{Hz}$ and effective bandwidths $\Delta\nu_{eff} \sim 1\ Hz$ per each resonant mode. The achieved energy sensitivity for millisecond pulses is well above the quantum limit of the present configurations, in the range of $10^3 \div 10^4$ *quanta*, and corresponds to a minimum detectable strain amplitude of a millisecond pulse h_{min} in the range of $\sim 10^{-19}$ at Signal to Noise Ratio $SNR = 1$. A Galactic supernova collapse with large efficiency of conversion to gravitational waves should therefore be detectable by the present bar detectors operating in coincidence. However, the perspectives for an actual detection are limited by the following facts: i) each detector shows events in excess of the the high energy tail of the modeled thermal noise, for best detectors about ~ 1 *events/day*, and thus the rate of accidental coincidences may be greater than the rate of Galactic supernovae, masking the real gravitational wave events unless a signature for the burst signals is provided; ii) even if the detectors duty cycle are high, the fraction of time during which the detectors are directed away from potential Galactic sources is about one third.

2 Perspectives of Improvements

We believe that significant improvements are necessary both on single detectors performances and on the data analysis capabilities of the network. In particular, it is crucial i) to approach the quantum limited sensitivity of each detector, ii) to improve internal vetoes of spurious events in the detectors, and iii) to make a correlation analysis among detectors such that distinctive properties of a gravitational wave signal can be determined.

A resonant bar detector like AURIGA or NAUTILUS operating close to its quantum limit for burst detection would gain a factor of about 10^3 over the best present energy sensitivity. In this case, the minimum detectable amplitude of a burst would be $h_{min} \sim 10^{-21} \div 10^{-20}$ and the strain noise spectral density would show a minimum of $\sim 10^{-23} \ Hz^{-1/2}$ with a bandwidth of $\sim 50 \ Hz$ around 920 Hz. The same technical achievements developed to approach the quantum limit of a bar could be used as well on a massive spherical detector, with the further gain of a factor of $20 \div 50$ in energy sensitivity in respect to the optimally oriented cylinder and, of course, with the advantage of isotropic figure pattern [6].

Our model of the AURIGA detector predicts that to approach its quantum limit many stringent requirements must be met together: i) the SQUID amplifier energy resolution should be lowered to a few $\hbar$, from the actual $N_{SQUID} \sim 5 \times 10^3$; ii) the brownian noise of the bar and transducer should be lowered to the same level, therefore the ratio between the mechanical quality factors to thermodynamic temperature should be $Q/T \geq 10^{10}/N_{SQUID}$, to be compared with the achieved $Q/T \sim 5 \times 10^7$; iii) the capacitive transducer and the impedance matching transformer should be optimized to reach the widest effective bandwidth, $\sim 50 Hz$. The mechanical suspension system of the bar must be improved as well, because additional $\geq 60 \ dB$ to the present attenuation of $\sim 245 \ dB$ in the bandwidth are necessary to make negligible the seismic noise contribution.

To improve the confidence of gravitational wave detection, it is crucial to reject false alarms and false dismissals and to get a signature of the signal. In this respect we implementeded a data acquisition and analysis procedure for the AURIGA detector which enables new important features: i) the high resolution measurement of arrival time of an impulsive excitation [7] and ii) the test on consistency of the detected event with a g.w. excitation [8]. The latter consists in a goodness of fit test on the output of the optimal data filter and can discriminate between events caused by direct mechanical excitation of the bar and those caused by electromagnetic pick-up or other kinds of mechanical excitations. We are beginning experiments to check what part of the non modeled detector noise can be rejected with this method. The former allows to measure time delay between pulse detection at different sites with an accuracy $\sim 170 \ \mu s/SNR$, provided that the effective bandwidth be wide enough, $> 10 \ Hz$. With at least four detectors in coincidence, the direction of the incoming wavefront can be determined together with its propagation speed. Even with fewer detectors, the consistency of the delays with speed of light can provide strong evidence for a g.w. signal.

Resonant detectors with these performances will be complementary to the planned long baseline interferometers for all g.w. signals which do not show distinctive shapes inside the detector's bandwidth. In this case, a confident detection of g.w. signals must rely on a correlation of the outputs of the detectors aimed at the solution of the inverse problem and at the discrimination of distinctive g.w. properties, such as the propagation speed. Other distinctive properties can be provided only by resonant detectors, as the tracelessness of the Riemann tensor. Moreover, resonant bars can be easily oriented in order to ensure a uniform sky coverage of the future network of detectors.

1. E. Mauceli *et al.*, Phys. Rev. **D54** (1995) 1264.
2. P. Astone *et al.*, Phys. Rev. **D47** (1993) 2.
3. D.G. Blair *et al.*, *First E. Amaldi Conf. on G.W. Experiments*, (World Scientific, Singapore, 1995) p.194.
4. P. Astone *et al.*, Europhys. Lett. **16** (1991) 231.
5. M. Cerdonio *et al.*, *First E. Amaldi Conf. on G.W. Experiments*, (World Scientific, Singapore, 1995) p.176.
6. E. Coccia *et al.*, Phys. Rev. **D52** (1995) 3735.
7. S. Vitale *et al.*, *First E. Amaldi Conf. on G.W. Experiments*, (World Scientific, Singapore, 1995) p. 220.
8. S. Vitale *et al.*, Nuclear Phys. **B48** (1996) 104.

THE STATUS OF THE VIRGO EXPERIMENT

The VIRGO Collaboration: C. Boccara[1], J.B. Deban[1], O. Germain[1], P. Gleyzes[1], M. Leliboux[1], V. Loriette[1], R. Nahoum[1], J.P. Rogier[1], Y. Acker[2], L. Dognin[3], P. Ganuau[3], B. Lagrange[3], J.M. Mackowski[3], C. Michel[3], M. Morgue[3], M. Napolitano[3], L. Pinard[3], C. Arnault[4], G. Barrand[4], J.L. Beney[4], R. Bihaut[4], F. Bondu[4], V. Brisson[4], F. Cavalier[4], R. Chiche[4], M. Dialinas[4], A. Ducorps[4], P. Hello[4], P. Heusse[4], A. Hrisoho[4], P. Marin[4], M. Mencik[4], A. Reboux[4], P. Roudier[4], R. Barrillet[5], M. Barsuglia[5], J.P. Berthet[5], A. Brillet[5], J. Cachenaut[5], F. Cleva[5], H. Heitmann[5], L. Latrach[5], C.N. Man[5], M. Pham-Tu[5], V. Reita[5], M. Taubman[5], J.Y. Vinet[5,5a], F. Bellachia[6], B. Caron[6], T. Carron[6], D. Castellazzi[6], F. Chollet[6], A. Dominjon[6], C. Drezen[6], R. Flaminio[6], C. Girard[6], X. Grave[6], J.C. Lacotte[6], B. Lieunard[6], F. Marion[6], L. Massonet[6], C. Mehemel[6], R. Morand[6], B. Mours[6], P. Mugnier[6], V. Sannibale[6], M. Yvert[6], E. Bougleux[7], R. Cecchini[7], F. Ciuffi[7], M. Mazzoni[7], P.G. Pelfer[7], R. Stanga[7], D. Babusci[8], S. Bellucci[8], S. Candusso[8], H. Fang[8], G. Giordano[8], G. Matone[8], F. Barone[9], E. Calloni[9], L. Di Fiore[9], F. Garufi[9], A. Grado[9], M. Longo[9], M. Lops[9], L. Milano[9], A. Aragona[10], F. Cagnoli[10], L. Gammaitoni[10], J. Kovalik[10], F. Marchesoni[10], M. Punturo[10], M. Beccaria[11], M. Bernardini[11], S. Braccini[11], C. Bradaschia[11], G. Cella[11], A. Ciampa[11], E. Cuoco[11], G. Curci[11], V. Dattilo[11], R. De Salvo[11], R. Del Fabbro[11], A. Di Virgilio[11], D. Enard[11], I. Ferrante[11], F. Fidecaro[11], A. Gaddi[11], A. Gennai[11], A. Giassi[11], A. Giazotto[11], P. La Penna[11], L. Holloway[11,11a], G. Losurdo[11], M. Maggiore[11], S. Mancini[11], F. Palla[11], H.B. Pan[11], A. Pasqualetti[11], D. Passuello[11], R. Poggiani[11], P.Popolizio[11], F. Raffaelli[11], Z. Zhang[11], Bronzini[12], V. Ferrari[12], E. Frasca[12], E. Majorana[12], P. Puppo[12], P. Rapagnani[12], F. Ricci[12].

[1] ESPCI, Paris, France, [2] INSU, Paris, France, [3] IPNL, Lyon, France, [4] LAL, Paris–Sud, France, [5] CNRS–Orsay Laser Optics, Paris, France, [5a] Ecole Polytechnique, Paliseau–Cedex, France, [6] LAPP and ESIA, Annecy, France, [7] University of Firenze and INFN, Firenze, Italy, [8] LNF, Frascati, Italy, [9] University of Napoli and INFN, Napoli, Italy, [10] INFN, Perugia, Italy, [11] INFN, Pisa, Italy, [11a] University of Urbana, Urbana, USA, [12] INFN, Roma, Italy.

The status report of the VIRGO experiment is presented. The experiment has been approved in September 1993 and is now in the construction stage. Its aim is the detection of gravitational waves over a broad frequency range (from 10 Hz to 10 kHz) using a Michelson interferometer equipped with 3 km-long Fabry–Perot cavities. The experiment is being installed in Cascina near Pisa and is planned to be operative during year 2000. The planned sensitivity is $\tilde{h} \approx 10^{-21}~Hz^{-1/2}$ at 10 Hz and $\tilde{h} = 3 \times 10^{-23}~Hz^{-1/2}$ at 500 Hz.

presented by A. Giazotto

1   Introduction

Gravitational waves (GW) are predicted by the theory of General Relativity [1] and actually there is only their indirect evidence [2]. The direct observation of GW, beside being a relevant test of General Relativity, will start a new picture of the Universe. In fact, GW carry complementary information with respect to electromagnetic waves since GW are not absorbed by matter and are emitted from massive sources in strong gravity conditions. The aim of the VIRGO experiment is the direct detection of GW and, in joint operation with other similar detectors, to perform gravitational waves astronomical observations. VIRGO is designed for a broad band detection, from 10 Hz to 10 kHz. The capability of low frequency operation should give the best possibilities for detecting gravitational radiation from monochromatic sources and coalescing binaries. VIRGO consists of a recycled Michelson interferometer having the arms constituted by 3 Km long Fabry–Perot cavities (Fig.1). The long arms are in high vacuum conditions. VIRGO is a large international collaboration with physicists and engineers from nine different laboratories in France and in Italy.

2   Gravitational waves sources

In the weak field approximation the GW perturb the metric of the space–time as follows [3]:

$$g_{\mu\nu} = \eta_{\mu\nu} + h_{\mu\nu} \tag{1}$$

where $h_{\mu\nu}$ is the perturbation to the flat space–time metric $\eta_{\mu\nu}$ and $|h_{\mu\nu}| \ll 1$. The GW have two states of polarization h_+, $h_\times$ with quadrupolar patterns. The perturbation h causes a change ΔL of the distance L between two free masses, such that:

$$\Delta L = \frac{hL}{2} \tag{2}$$

in the hypothesis that the wavelength λ of the GW is much larger than the mass separation L. Mainly five source types are considered promising. The collapse of type II supernovae into neutron stars or black holes is predicted to produce short radiation bursts. For a supernova exploding in the VIRGO cluster (10 Mpc) a value of $h \sim 10^{-21}$ is expected, with a rate of a few per year [3]. The coalescence of a binary system, in which one of the components is a neutron star or black hole, is a GW source. A few events per year are expected within a 100 Mpc radius [4]. Asymmetries in the mass distribution of pulsars are believed to produce GW emission at twice the rotation frequency with h ranging [3] from 10^{-28} to 10^{-24}. The relic radiation which carry the inprinting of the Big Bang event, due to GW low interaction cross section with matter. The detection of relic radiation can only be made in coincidence with other detectors. The predicted number of rotating neutron stars is very large $(10^8 \div 10^{10})$; it has been proposed to detect the collective GWs emitted by the whole ensemble using a non–linear analysis [5].

3   VIRGO: the detector

The main features of the laser interferometer VIRGO will be described here, referring the reader to [6,7] for a detailed discussion. A GW propagating orthogonally to the plane of the Michelson interferometer will alternatively stretch one arm and shrink the other one: thus there will be an oscillating change of the relative phase of the light in the two arms. With the use of Fabry–Perot cavities in the arms, the optical path length is increased from 3 km to 120 km. A feedback system controls the positions of the mirrors in such a way that the interferometer is kept in destructive interference condition, that is operating on the dark fringe. The GW signal is extracted from the current circulating in the feedback loop. Various sources of noise affect the sensitivity of VIRGO. The sources and the proposed approaches to reduce the noise will be discussed now.

3.1   Shot noise

The shot noise is the limiting factor to the sensitivity in the frequency region above a few hundreds Hz. It is due to the statistical fluctuations in the number of detected photons. It is inversely proportional both to the square root of the efficiency of the photodetector η and of the beam power P:

$$\tilde{h}_{sn} = \frac{\lambda}{8FL}(1 + (\frac{2\omega FL}{\pi c})^2)^{\frac{1}{2}}\sqrt{\frac{h\nu}{\eta P}} \tag{3}$$

where λ, ν are the wavelength and frequency of the laser, L the interferometer arm length, h the Planck constant, F the cavity finesse. In VIRGO the use of a 20 W Nd:Yag laser ring at 1.06 μm is planned. The recycling technique[8] is used to increase the power in the interferometer, by adding a further mirror between the laser and the beam splitter.

3.2   Laser frequency and power fluctuations

Since small asymmetries between the two arms of the interferometer are unavoidable, the fluctuations in the laser frequency must be kept as small as possible. For an asymmetry of 1% beetwen the arms the laser frequency stabilization must be kept around[7] $10^{-6} Hz/\sqrt{Hz}$. The laser is locked by the injection–locking technique to an ultra–stable laser with 1 W power. The power fluctuations are also reduced by an active control system to $\frac{\Delta P}{P} \leq 10^{-8} Hz^{-1/2}$.

3.3   Vacuum requirements

The VIRGO interferometer will be operated under vacuum. The statistical fluctuation of the residual gas density can induce a fluctuation of the index of refraction thus of the interferometer phase difference. A partial pressure of $10^{-9} \div 10^{-10}$ mbar is planned for typical gases, while for hydrocarbon this limit is lowered to 10^{-14} mbar, to prevent the pollution of optical elements.[7] These requirements will contribute a noise level of $10^{-24}\ Hz^{-1/2}$.

3.4 Seismic noise

The ground vibrations are strongly reduced by suspending each optical component to a cascade of mechanical oscillators, the *superattenuator* (SA). A new SA equipped with a pre isolator plus five pendulum stages has replaced the old one with seven cascade pendulum. The pre isolator stage consists[7] in an inverted pendulum with a very low resonant frequency ($\approx 30mHz$). The use of a pre isolator stage has three advandages: a) seismic isolation; b) active control loop simplification; c) the required dynamics for the suspension point is obtained using noiseless electromagnetic forces applied on soft flexure joints. To improve the seismic noise isolation VIRGO will use digital servo loops.

3.5 Thermal noise

The thermal noise is related to the dissipation in the system through the Fluctuation–Dissipation theorem, which states that stochastic forces arise with spectral density: $\tilde{F}^2(\omega) = 4k_B T R(\omega)$ where $R(\omega)$ is the real part of the mechanical impedance $Z(\omega) = F(\omega)/v(\omega)$. The force causes Brownian motion of the system. Due to operation in vacuum, the dissipation in VIRGO is coming from the internal friction in the suspension materials. The dominant contribution in such a multistage system is given by the dissipation in the last stage named *marionetta*. The chosen design of the marionetta and of the clamps and wires system supporting the mirrors allow to reduce the vertical thermal noise[9]. A R&D program to improve geometries and materials for the mirror suspension is in progress[10]. The Fabry–Perot cavities mirrors are kept as massive as possible (20 to 40 Kg) to reduce the thermal noise.

3.6 The Test interferometer

The construction of the VIRGO interferometer in Cascina will require about 5 years. The buildings construction was started and the central area of VIRGO is planned to be ready at the end of September 1997. During 1998 all the critical subsystems of VIRGO will be installed. While the construction of the 3 km arms is going on, we will have, in 1999, the opportunity to work with a recycled Michelson interferometer in the final position in the central building, and to test all the VIRGO subsystems and their final functions. In this way, we expect to start data acquisition with full sensitivity by the beginning of 2001, a few months only after the construction is completed.

3.7 Conclusions

The overall VIRGO sensitivity curve is shown in Fig. 2. It is evident that the use of the SA should decrease the seismic noise down to negligible levels. Various mechanical resonances contribute to the thermal noise in the frequency region of interest: the high frequency tail of the pendulum mode, the low frequency tail of the mirror internal modes and the narrow resonances of the violin modes. The contribution of the vertical modes is well below the pendulum one at low frequencies. We point out that the lowest detectable frequency in VIRGO is 4 Hz. At this frequency the contribution of the seismic noise equals the contribution from Newtonian noise. It is evident that there is no advantage in further improvements of the seismic attenuation below this frequency. The sensitivity in the high frequency region is of the order of $3 \times 10^{-23} \ Hz^{-1/2}$.

The most critical problems for VIRGO construction have been solved. The beginning of the interferometer operation is planned in year 2001.

1. C. W. Misner, K, S. Thorne and J. A. Wheeler, *Gravitation*, W. H. Freeman and Company, San Francisco (1973)
2. J. H. Taylor and J. M. Weisberg, *Astrop. J.* **345** (1989) 434
3. *Three hundreds years of gravitation*, Ch. 9: Gravitational radiation by K. S. Thorne, ed. by S. W. Hawking and W. Israel, Cambridge University Press (1987)
4. B. F. Schutz, *Nature* **323** (1986) 310
5. S. Bonazzola, A. Giazotto and E. Gorgoulhon, *Phys. Rev. D* **55**, 2014 (1997)
6. *VIRGO: proposal for the construction of a large interferometric detector of gravitational waves* (1989); *VIRGO: Final Conceptual Design* (1992); *VIRGO: Final Design* (1995); A. Giazotto, *Phys. Rep.* **182** (1989) n.6
7. *VIRGO: Final Design* (1997)
8. C. N. Man et al., presented at the "7th Marcel Grossman Meeting", San Francisco, July 1994
9. S. Braccini et al., *Phys. Lett.* **A199** (1995) 307
10. F. Marchesoni et al., *Phys. Lett.* **A187** (1994) 359
11. B. Caron et al., presented at "Frontier Detectors for Frontier Physics", La Biodola, May 1994, *Nucl. Instr. and Meth.* **A360** (1995) 375

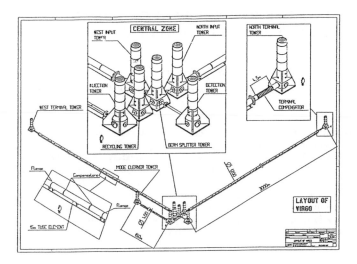

Figure 1: The layout of VIRGO with the towers containing the suspended optical elements

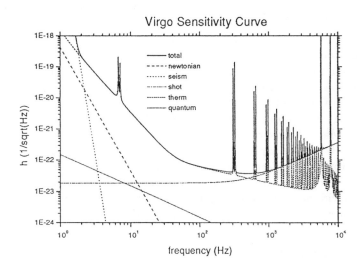

Figure 2: The VIRGO expected sensitivity

MODELLING THE PERFORMANCE OF AN INITIAL-LIGO DETECTOR WITH REALISTICALLY IMPERFECT OPTICS

B. BOCHNER

*LIGO Project (MIT), Room 20F-109, 18 Vassar Street,
Cambridge, MA 02139, USA*

The satisfactory performance of interferometric detectors in the Laser Interferometer Gravitational-Wave Observatory (LIGO) project will depend upon exceedingly high-quality optical components. In order to accurately predict the response of real detectors, we have written a grid-based simulation program which models the steady-state laser fields in a complete LIGO interferometer with multiply-coupled-cavities, for a wide variety of possible optical imperfections. Using measurements of exceptionally smooth mirrors obtained from industry, we show how feasibly obtainable levels of mirror deformations may degrade the sensitivity of the LIGO detector to astrophysically-generated gravitational waves.

1 Introduction and Background

The LIGO project [1] will use long-baseline interferometers (IFO's) to detect astrophysically-generated gravitational waves (GW's) via their perturbing forces on the interferometer mirrors. In order to detect these extraordinarily small GW-induced mirror motions ($\Delta L/L \sim 10^{-21}$), several limiting noise sources must be controlled, especially seismic, thermal, and photon shot noise.

Imperfections in the IFO optics will inhibit the detection of GW's by causing a reduction in the amount of resonating power available for the sensing of mirror positions, and by increasing the amount of unmodulated (i.e. non-signal-bearing) light which emerges from the IFO signal port and contributes to the shot noise. The net result is a degradation of the *shot-noise-limited* part ($\nu_{GW} \geq 100$ Hz) of the expected LIGO sensitivity envelope, $h(f)_{SN}$.

To quantify these effects, we have developed a computer code to perform detailed numerical simulations of an initial-LIGO IFO, with the capability of simulating a wide variety of IFO imperfections. Here we introduce the code and present a selection of results; a detailed description may be found elsewhere [2].

2 The LIGO Simulation Program

The program is a Fortran code, adapted for execution on the massively-parallel Paragon supercomputers at Caltech. As a grid-based program, the fundamental objects it manipulates are complex, 2-D maps representing mirror profiles and transverse slices of the laser beam electric field, sampled at various points in the IFO. Using the *paraxial approximation*, beam propagations become simplified procedures primarily involving FFT's[3]. Reflections and transmissions at mirrors are performed using a small-distance approximation[4], reducing them to pixel-by-pixel multiplications of an electric field map with a mirror map.

The program can incorporate many different optical imperfections, including, for example: (i) Deformations in surface figure and substrate homogeneity profiles, (ii) Finite mirror apertures and realistic beam clipping, (iii) Mirror displacements, tilts, curvature errors and beam mismatch, (iv) Pure losses into which we lump our estimates of high-angle scattering and power absorption.

We simulate a static IFO, neglecting the dynamics of control systems and power buildup. An iterative procedure relaxes the electric fields to their steady-state distributions. The code simultaneously implements a number of parameter optimizations to ensure that all of the proper resonance conditions are achieved, and that the interferometer is optimally configured for GW detection. Finally, sideband frequency beams for the LIGO heterodyne detection scheme are modelled, so that we can explicitly compute $h(f)_{SN}$ for the IFO.

We have obtained 2 maps of real mirror deformations from industry: a fused-silica substrate homogeneity map from Corning, and a surface figure map (of a polished but uncoated substrate) from Hughes-Danbury. To create enough substrate and surface maps for all of the IFO mirrors, Fourier transform techniques were used to convert each of the source maps into a family of maps with identical power spectra but different, uncorrelated structure. Finally, the initial family of surface maps (w/RMS deformations of $\sim .6$ nm $\sim \lambda_{YAG}/1800$ in the central portion of the source map) were scaled up by constant factors to create surface profile families of $\lambda/1200$, $\lambda/800$, and $\lambda/400$, for conservative estimation given poorly known mirror coating homogeneity limitations.

3   Results and Discussion

Five baseline runs are presented here to characterize the effects of realistically deformed optics upon LIGO's GW sensitivity: one (control) run with perfectly smooth surfaces and substrates, and 4 runs with (respectively) $\lambda/1800$, $\lambda/1200$, $\lambda/800$, or $\lambda/400$ surface maps on all mirrors, plus the deformed substrates.

Fig. 1 plots strain noise spectral density, $h(f)$, vs. GW frequency f. The five shot-noise curves are computed[2] for each of the simulation runs, and are shown against the overall *GW-strain-equivalent noise requirement envelope*[5] for the initial-LIGO interferometers. All of the runs except for the very worst case ($\lambda/400$ surfaces) meet the LIGO requirement.

Fig. 2 shows the effects of deformed mirrors upon LIGO's sensitivity to the periodic GW's emitted by a non-axisymmetric pulsar. The noise curves

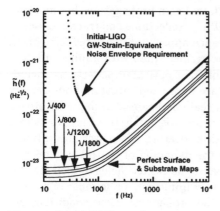

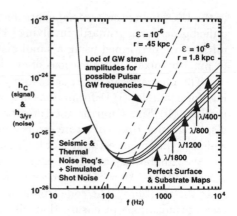

Figure 1: Initial-LIGO requirement vs. simulation-derived shot noise curves.

Figure 2: Effects of deformed mirrors on LIGO's sensitivity to GW's from pulsars.

are the quadratic sums of LIGO's seismic and thermal noise *requirements* plus the simulation-derived shot noise curves of Fig. 1. They have been converted to $h_{3/yr}(f)$ [6], representing high-confidence coincident detection in all 3 LIGO IFO's. The dashed lines represent the characteristic signal strengths, $h_c(f)$ [6], for pulsars with ellipticity $\epsilon = 10^{-6}$ at different distances (r) from the earth. Estimating roughly by setting $h_c = h_{3/yr}$ at the frequency of peak sensitivity, going from worst case to best gains a factor of ~ 4 in 'lookout distance' r, yielding a potential event rate increase ($\propto r^2$ in galactic disk) of ~ 16.

To summarize: (1) Our simulation program can be used to drive specifications for LIGO optics [7], (2) The sensitivity goals of the initial IFO's can be met with feasibly obtainable mirrors (pending acceptable coatings), (3) Significant benefits to LIGO science are gained with extremely high quality optics.

This work was supported by NSF Cooperative Agreement PHY-9210038.

References

1. A. Abramovici, *et al.*, *Science* **256**, 325 (1992)
2. B. Bochner and Y. Hefetz, *in preparation*.
3. A.E. Siegman, *Lasers* (University Science Books, California, 1986).
4. J.-Y. Vinet, *et al.*, *J. Phys.* I *(Paris)* **2**, No. 7, 1287 (1992).
5. A. Lazzarini and R. Weiss, Internal doc. **LIGO-E950018-02-E** (1996).
6. K.S. Thorne, in *300 YEARS OF GRAVITATION*, eds. S.W. Hawking and W. Israel (Cambridge University Press, Cambridge, England, 1987).
7. S. Whitcomb, *et al.*, for the *TAMA Workshop on GW Detection* (1996).

INFALL OF A PARTICLE INTO A BLACK HOLE AS A MODEL FOR GRAVITATIONAL RADIATION FROM THE GALACTIC CENTER

Carlos O. Lousto

Department of Physics, University of Utah, Salt Lake City, UT 84112, USA

I present here the results of the study of the gravitational radiation generated by the infall (from rest at radius r_0) of a point particle of mass m_0 into a Schwarzschild black hole of mass M. We use Laplace's transform methods and find that the spectra of radiation for $\sim 5M < r_0 < \infty$ presents a series of evenly spaced bumps. The total radiated energy is not monotonically decreasing with r_0, but presents a *joroba* (hunch-back) at around $r_0 \approx 4.5M$. I finally discuss the detectability of the gravitational radiation coming from the black hole in the center of our galaxy.

1 Perturbative approach

Here I will report on work made in collaboration with R. Price (see Ref. 1 for further details.) The problem of a particle falling into a non-rotating black hole can be treated in the regime where the particle contributes perturbatively to the Schwarzschild metric. Thus, it was not surprising that soon after Zerilli (1970) wrote down his equation describing the propagation of gravitational waves on the Schwarzschild background, the case of a particle falling from infinity, both at rest and with a finite velocity, was solved using Fourier transform techniques. The key to the resolution to the problem of the infall from a finite distance (that had to wait 25 years) is the use of the Laplace's transform method instead. In this case, Zerilli's equation reads

$$\frac{\partial^2 \Psi_\ell}{\partial r*^2} + \left[\omega^2 - V_\ell(r)\right]\Psi_\ell = -\dot{\psi}_0(r) + i\omega\psi_0(r) + S(r,\omega) , \qquad (1)$$

where the ℓ-multipole of the waveform is related to the metric components (in the Regge–Wheeler gauge) by

$$\psi_\ell(r,t) = \frac{r}{\lambda+1}\left[K + \frac{r-2M}{\lambda r + 3M}\left\{H_2 - r\partial K/\partial r\right\}\right] , \qquad (2)$$

where $\lambda \doteq (\ell+2)(\ell-1)/2$. An important ingredient in Eq. (1) is the source term $S(r,\omega)$ that we had to compute for the particle infalling along a geodesic. We also had to consider the initial waveform $\psi_0(r)$ corresponding to a solution of the hamiltonian constraint. Here we have taken the particle limit of the Brill-Lindquist initial solution (the results do not differ notably had we chosen the Misner initial solution), and a time-symmetric situation ($\dot{\psi}_0(r) = 0$).

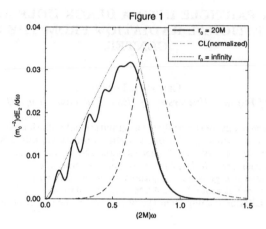

Figure 1

2 Results

To solve Eq. (1) we use the Green's function method, integrating over the total effective source, i.e. the right hand side of Eq. (1), to obtain the complex amplitude $A(\omega)$ of the outgoing radiation. The main results we have obtained can be summarized in the two figures: In Fig. 1 the new feature is the appearance of bumps in the spectrum whose spacing decreases as r_0 increases. This can be understood as a consequence of the interference between the initial burst of radiation, soon after $t = 0$, and the final one due to the infall of the particle into the black hole, In Fig. 2 the novelty is the *joroba*, with a local maximum at $r_0 = 4.5M$. This purely general relativistic effect can be attributed to the higher efficiency in the generation of radiation by the initial burst close to the maximum of the potential (located at $r \approx 3.1M$.)

3 Observational estimates

The confirmation[2] of the presence of a black hole with $M = 2.4 \times 10^6 M_\odot$ in the center of our galaxy provides an astrophysical scenario for testing our computations. Let us first consider the amplitude of the metric perturbations as the gravitational radiation reaches the earth. From the waveforms given in Ref. 1 and Eq. (2), we find that $\Delta K \sim h \sim 10^{-17}(m_0/M_\odot)(8kpc/r)$, independent of M, the mass of the black hole[a]. The duration of the final burst of radiation

---

[a] Rotation of the hole and angular momentum of the infalling particle may increase this value in two orders of magnitude

Figure 2

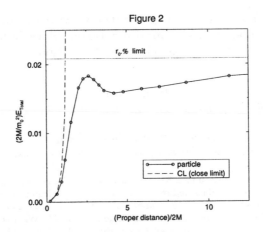

can be estimated as $\Delta t \approx 3 \times 10^{-4}(M/M_\odot)sec \approx 12min$. The next important issue is the frequency of such events. As an order of magnitude estimate we can study the quantity $f = \rho\sigma\nu$, where $\rho \approx 6.5 \times 10^9 M_\odot/pc^3$, is the density of stars near the hole, $\sigma_{cap} = 16\pi M/v^2$ is the capture cross section, and $\nu \approx 400km/sec \approx v$ is the mean velocity of the stars relative to the hole. All this together gives $f = 7.5 \times 10^{-24}(M_\odot/m_0)(M/M_\odot)^2(c/v)sec^{-1} \approx 1\,\text{event}/yr$. Finally, from the spectrum of gravitational radiation we see that its maximum takes place at a frequency $\omega_{max} \approx 7 \times 10^4 (M_\odot/M)\text{Hz} \approx 0.03\text{Hz}$. With all this numbers[b] in hand we conclude that the galactic center may well be among the first positively detected sources of gravitational radiation (by LISA).

Acknowledgments

I would like to thank A. Giazotto, J. Horvath, and R. Price for discussions on this problem, and NSF Grant No. PHY0507719 for financial support.

References

1. C.O. Lousto and R.H. Price, *Phys. Rev.* D**55**, No. 4 (1997).
2. A. Eckart and R. Genzel, *Nature* **383**, 415 (1996).

---

[b]The tidal forces of the black hole will disrupt the upper atmosphere of the infalling star (generating a burst of electromagnetic radiation accompanying the gravitational one), but leave its core (where most of the star's mass resides) practically untouched. This justifies our particle approximation.

Post-Newtonian expansion of gravitational waves from a particle in circular orbits around a rotating black hole – Effect of black hole absorption –

Hideyuki Tagoshi

National Astronomical Observatory, Mitaka, Tokyo 181, Japan

Shuhei Mano and Eiichi Takasugi

Department of Physics, Osaka University, Toyonaka, Osaka 560, Japan

When a particle moves around a Kerr black hole, it radiates gravitational waves. Some of those waves are absorbed by the black hole. We calculate such absorption of gravitational waves produced by a motion of a particle of mass μ in circular orbit on a equatorial plane around a Kerr black hole of Mass M. We assume that the velocity of the particle is much smaller than the speed of light and calculate the gravitational waves using a post-Newtonian expansion technique of the Teukolsky equation

1 Introduction

Among the possible sources of gravitational waves, coalescing compact binaries are the most promising candidates for detection by near-future, ground based laser interferometric detectors such as LIGO, VIRGO, GEO600, and TAMA. It is very important to investigate detailed wave forms from coalescing compact binaries. It was pointed out[1] that we have to perform the post-Newtonian expansion up to extremely high order. Then the post-Newtonian wave generation formalisms have been developed to very high order.

On the other hand, a study of gravitational waves which are absorbed by a black hole is developed only a little in the context of the post-Newtonian approximation. These effect was calculated by Poisson and Sasaki using the post-Newtonian expansion in the case when a test particle is in circular orbit around a Schwarzschild black hole[2]. In that case, the effect of the black hole absorption is small since that effect appears at $O(v^8)$ beyond the Newtonian quadrupole formula. However, as we shall see in this paper, the black hole absorption appear at $O(v^5)$ when a black hole is rotating. Then it is important to investigate the effect of black hole absorption to the orbital evolution of coalescing compact binaries when at least one of the stars is a rotating black hole. However, it is difficult to extend their method to higher order. It is also difficult to apply their method to a rotating black hole case. On the other hand, another analytic techniques for the Teukolsky equation was found by Mano, Suzuki, and Takasugi[3]. Since this method is very powerful for the calculation of the post-Newtonian expansion of Teukolsky equation, we adopt this method in this paper.

2 Methods

In the Teukolsky formalism, gravitational perturbations of the Kerr black hole are described by a function ψ_4. We decompose ψ_4 into Fourier-harmonic components. Then, the radial function $R_{\ell m \omega}$ satisfies the Teukolsky equation

$$\Delta^2 \frac{d}{dr}\left(\frac{1}{\Delta}\frac{dR_{\ell m\omega}}{dr}\right) - V(r)R_{\ell m\omega} = T_{\ell m\omega}, \tag{1}$$

We solve the Teukolsky equation using the Green function method. We calculate two kind of homogeneous solutions of the Teukolsky equation $R_{\ell m\omega}^{\text{in}}$ and $R_{\ell m\omega}^{\text{up}}$. A non-trivial step in calculating gravitational waves is to obtain these homogeneous solutions. New analytic representations of the homogeneous Teukolsky equation were found by Mano, Suzuki and Takasugi[3]. In this method, two kind of expansion are used. One is a series of hypergeometric functions which has good convergence property around the horizon and and the other is a series of Coulomb wave functions which has good convergence property at infinity. We can match those two series *analytically* where both of two series converge. Then we can obtain various solutions of homogeneous Teukolsky equation in a region from the horizon to infinity by this method. It can also be shown that the the order of the the post-Newtonian approximation increase when the order of above expansions increase. Then this method is very useful in performing the post-Newtonian expansion.

3 Results

The total luminosity becomes up to $O(v^9)$ as

$$\begin{aligned}\left\langle \frac{dE}{dt}\right\rangle =\ & \left(\frac{dE}{dt}\right)_N \left[\left(\frac{-q}{4}-\frac{3\,q^3}{4}\right)v^5 + \left(-q-\frac{33\,q^3}{16}\right)v^7\right.\\ & + \left(\frac{1}{2}+2\,\alpha(q)\,q-\frac{115\,q^2}{12}+6\,\alpha(q)\,q^3-\frac{93\,q^4}{2}+\right.\\ & \left.\frac{\sqrt{1-q^2}}{2}+\frac{13\,q^2\,\sqrt{1-q^2}}{2}+3\,q^4\,\sqrt{1-q^2}\right)v^8\\ & + \left(\frac{-187\,q}{28}-\frac{2369\,q^3}{126}-\frac{1901\,q^5}{336}\right)v^9\Bigg],\end{aligned} \tag{2}$$

556

q	$(1.4M_\odot,10M_\odot)$	$(1.4M_\odot,40M_\odot)$	$(1.4M_\odot,70M_\odot)$
0	$< 10^{-2}$	0.01	0.1
0.9	1.1	3	4.5
-0.9	0.2	0.33	0.28

Table 1: The error of the total cycle N caused by neglecting the black hole absorption effect in the formulas for dE/dt for typical neutron star(NS)-black hole(BH) binaries with mass (M_{NS}, M_{BH}). The initial frequency is 10Hz and the final orbital radius is $6(M_{NS} + M_{BH})$.

where $(dE/dt)_N = 32/5(\mu/M)^2 v^{10}$, $q = a/M$, a is the black hole spin, $v = (M/r_0)^{1/2}$, r_0 is the orbital radius, $\alpha(q) = \text{Im}\left[\psi^{(0)}\left(3 + 2iq/\sqrt{1-q^2}\right)\right]$, and $\psi^{(n)}(z)$ is polygamma functions.

If we set $q = 0$ above formula reduced to $dE/dt = (dE/dt)_N v^8$ which was derived by Poisson and Sasaki[2]. We see that the absorption effect is more important in the case $q \neq 0$.

Using the above results, we estimate the effect of the black hole absorption to the orbital evolution of the coalescing compact binaries (Table 1). We find that the black hole absorption is more important when the mass of the black hole is large. This is because that convergence property of the post-Newtonian expansion is slow when black hole mass is large. We also find that the black hole absorption is more important in cases $q > 0$ (say, co-rotating cases).

All the details of this paper will be presented elsewhere[5].

Acknowledgments

We thank M.Sasaki for discussion. Tagoshi was supported by Research Fellowships of the Japan Society for the Promotion of the Science for Young Scientists. This work is also supported in part by the Japanese Grant-in-Aid for Scientific Research of Ministry of Education, Science, Sports and Culture, No. 06640396 and 08640374.

References

1. C. Cutler et al., Phys. Rev. Lett. **70**, 2984 (1993). references therein.
2. E.Poisson and M.Sasaki, Phys. Rev.**D51**, 5753(1995).
3. S. Mano, H. Suzuki and E. Takasugi, Prog. Theor. Phys. **95**, 1079 (1996).
4. S.A. Teukolsky, Astrophys. J. **185**, 635 (1973).
5. H. Tagoshi S. Mano, and E. Takasugi, in preparation.

A SEARCH FOR LARGE-SCALE STRUCTURE AT HIGH REDSHIFT

A.J. CONNOLLY, A.S. SZALAY

Department of Physics and Astronomy, The Johns Hopkins University, Baltimore,
MD 21218

A. K. ROMER, R.C. NICHOL

Department of Physics, Carnegie Mellon University, 5000 Forbes Avenue,
Pittsburgh, PA 15213, USA

B. HOLDEN

Department of Astronomy and Astrophysics, University of Chicago, 5460 S. Ellis
Ave, Chicago, IL 60637

D. KOO

University of California Observatories, Lick Observatory, University of California
Santa Cruz, CA 95064

T. MIYAJI

Max-Planck-Institut für Extraterrestrische Physik Postf. 1603, D-85740, Garching,
Germany

We present new and exciting results on our search for large-scale structure at high redshift. Specifically, we have just completed a detailed analysis of the area surrounding the cluster CL0016+16 ($z = 0.546$) and have the most compelling evidence yet that this cluster resides in the middle of a supercluster. From the distribution of galaxies and clusters we find that the supercluster appears to be a sheet of galaxies, viewed almost edge-on, with a radial extent of 31 h^{-1}Mpc, transverse dimension of 12 h^{-1}Mpc, and a thickness of ~ 4 h^{-1}Mpc. The surface density and velocity dispersion of this coherent structure are consistent with the properties of the "Great Wall" in the CfA redshift survey. Full details and followup observations can be found at http://tarkus.pha.jhu.edu/~ajc/papers/supercluster/sc.html

Understanding the clustering of galaxies as a function of redshift provides important constraints on the initial perturbation spectrum of the Universe. While in the local Universe extensive redshift surveys (e.g. CfA) have detected coherent structures of 100 h^{-1}Mpc in extent, wide angle surveys at $z > 0.2$ have been precluded due to the extensive observational resources required. Consequently, while overdensities in the redshift distribution of galaxies have been detected their angular distribution and, therefore, their spatial extent have yet to be well determined. We consider here the distribution of galaxies in the direction of Selected Area 68[3]. We identify this region as a potential site

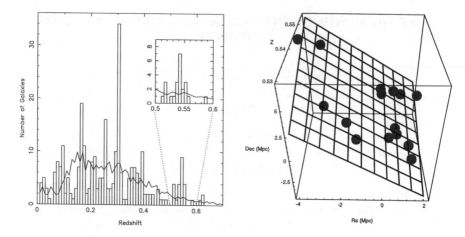

Figure 1: (a) The redshift distribution of galaxies in the Selected Area SA68. The solid line represents the expected distribution of redshifts for this sample assuming no clustering. The expectation value for the number of galaxies between $0.530 < z < 0.555$ is 3.54 ± 3.7. The observed number of galaxies within this redshift range is 14, a factor of four larger. (b) Combining the optical and X-ray data the distribution of galaxies and clusters appears planar. The galaxies appear to be part of a 2-dimensional sheet-like structure which we are viewing edge on (as opposed to a 1-dimensional filament). The extent of this structure, as defined by the current data set, is 31 h^{-1}Mpc radially, 13 h^{-1}Mpc in the transverse direction and with a "thickness" of 433 km s^{-1} orthogonal to the plane.

of an intermediate redshift supercluster because of a coherent distribution of very red galaxies, estimated to be at $z \sim 0.5$, and the presence of two nearby X-ray clusters at the same redshift[2].

Figure 1a shows the redshift distribution for those galaxies in SA68 with $B_J < 23.0$. The solid line represents the expected redshift distribution of galaxies assuming an unclustered universe. In the redshift range $0.530 < z < 0.555$ we find that the observed number of galaxies exceeds the expected value by a factor of 4 (the expectation value is 3.54 ± 3.7 and the number detected 14). The projected angular distribution of these galaxies forms a linear structure passing from the South-West of the SA68 field through to the North-East.

Along this direction, a little over half a degree from the center of SA68, lie the X-ray clusters CL0016+16 and J0018.3+1618[2] with redshift $z = 0.54$. Further, as part of the SHARC survey (see Romer et al. in these proceedings) we have identified an additional X-ray cluster in the vicinity of CL0016+16[1]

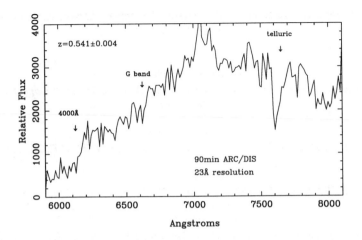

Figure 2: The spectrum for the central elliptical galaxy of J0018.8+1602 observed with the ARC 3.5m telescope at Apache Point. The spectrum is consistent with an elliptical at a redshift of $z = 0.541$.

(RX J0018.8+1602). A recent observation with the ARC 3.5m telescope has confirmed the redshift of this cluster as $z = 0.541$ (Figure 2).

Combining the optical and X-ray data we, therefore, have a coherent structure, at a redshift of $z = 0.54$, extending about one degree across the sky from the survey field SA68 through the cluster CL0016+16 (see Figure 1b). The positions of the galaxies and clusters within this volume are not randomly distributed but appear to lie in a planar distribution (i.e. their redshifts and angular distribution are strongly correlated). Fitting a two dimensional surface to the spectroscopic redshifts we determine an orientation $40° \pm 10°$ East of North and an angle $12° \pm 2°$ from the line of sight. The surface density and velocity dispersion of this supercluster are consistent with the measurements of the "Great Wall" from the CfA survey.

1. Connolly A.J., Szalay, A.S., Koo, D.C., Romer, A.K., Holden, B., Nichol, R.C. & Miyaji, T., 1996 ApJ 473, L67
2. Hughes J.P., Birkinshaw, M., Huchra, J.P., 1995, ApJ, 448, L93
3. Kron, R.G., 1980, ApJS, 43, 305

Superlarge-Scale Structure in the Universe: Observations and Simulations

Andrei G. Doroshkevich

Theoretical Astrophysics Center, Juliane Maries Vej 30, DK-2100 Copenhagen Ø, Denmark

Richard Fong

Dept. of Physics, University of Durham, Durham, DH1 3LE, England

Douglas Tucker

FERMILAB, MS 127, P.O. Box 500, Batavia, IL 60510, USA

Huan Lin

Department of Astronomy, University of Toronto, 60 St. George St., Toronto, ONT M5S 3H8, Canada

The spatial galaxy distributions in the Las Campanas Redshift Survey and in one simulation are examined with core-sampling, clustering, inertia tensor, and Minimal Spanning Tree analyses. A bimodal galaxy distribution is established in both samples: Superlarge-Scale Structure (SLSS), formed by a system of wall-like elements, and Large-Scale Structure (LSS), formed by galaxy filaments. The SLSS incorporates about 50% of all galaxies in overdensities which are $\sim 10\times$ denser than the mean, and therefore constitutes the major component of the observed structure in the Universe. The main physical parameters of LSS and SLSS are found to be similar for the observed and simulated catalogues. The possible interpretation of the SLSS parameters requires either the matter density to be $\Omega_m \approx 0.2$ or the bias on the SLSS scale to be $b_{SLSS} \approx 2$, or both, the likeliest possibility.

1 Introduction

Large maps of the galaxy distribution like the Durham/UKST Galaxy Redshift Survey (Ratcliffe et al. 1996) and the Las Campanas Redshift Survey (LCRS) (Shectman et al. 1996) have recently become available, and it now is possible to investigate properties of the galaxy distribution in the Universe on truly large scales - out to $200\text{--}300\,h^{-1}$Mpc (where $h \equiv H_0/100$ km/s/Mpc). The analysis of both these catalogues shows that $\approx 50\%$ of galaxies are concentrated within the Superlarge-Scale Structure (SLSS) which surrounds huge underdense regions (UDRs) on scales of $50\text{--}100\,h^{-1}$Mpc. In many respects the main properties of SLSS structures are similar to those of the Great Wall. In particular, SLSS elements are most easy to discern, since they have a galaxy density $\geq 10\times$ that of the mean observed density.

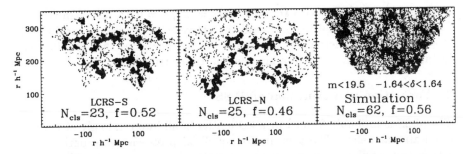

Figure 1: The SLSS, indicated by the heavily emphasized points, in two slices of the LCRS (south, S, and north, N) and in the simulation. N_{cls} and f denote the number and fraction of galaxies within SLSS elements.

Large-Scale Structure elements (galaxy filaments, LSS) bridge and fill the huge 'voids' between the SLSS elements; such 'voids' are more properly designated as underdense regions (UDRs). Our analysis finds this bimodal distribution into two morphological classes of structure to be most striking, possessing as they do very different properties and physical characteristics. It would seem that the formation of structure in the Universe must have been quite intricate, likely driven by a number of different physical processes.

Theoretically, one expects the deep wells of the initial gravitational potential to have collapsed to form the SLSS, whilst the relatively underdense regions expanded to give rise to the 'voids' we see today. We report here on an N-body simulation produced to mimic the forthcoming 2dF Galaxy Redshift Survey, and in which we see, for the first time in a DM simulation, SLSS similar to that in the observations.

2   Analyses of the LCRS and an N-body simulation

We have used *three* different analyses in our studies of LSS and SLSS in order to address their physical properties, such as their 'thickness', t, 'width', w, 'length', L, and separation, D_{wall}, directly: to wit, the core sampling (Doroshkevich et al. 1996), 3D clustering, and Minimal Spanning Tree techniques, supplemented with an inertia tensor analysis of the resulting clusters. The sizes of the SLSS elements are found to fall in the ranges $L \approx 30$–$70 h^{-1}$Mpc, $w \approx 20 h^{-1}$Mpc, and $t \approx 6$–$10 h^{-1}$Mpc. The distribution of wall separations appears Poissonian with a sample dependent mean of $< D_{wall} > \approx 40$–$60 h^{-1}$Mpc or more and a dispersion of $\sigma_{wall} \approx 20$–$30 h^{-1}$Mpc.

A similar bimodal LSS and SLSS structure in a DM distribution is found in a simulation performed by Cole et al. (1997) using a CDM power spectrum and

562

a box size $345\,h^{-1}$Mpc with $\Omega_m = 0.3$, $\Omega_\Lambda = 0.7$, $h = 0.83$, although the DM SLSS overdensity is only $\sim 5.5\times$ and the 1D velocity dispersion within SLSS is $\sim 600\,$km/s. Nonetheless, the simulation demonstrates that gravitational instability can by itself yield a bimodal matter distribution.

3 Formation and evolution of SLSS

The SLSS is probably formed by a 2D matter infall into randomly distributed gravitational wells. The typical sizes of SLSS proto-elements both for observed and simulated catalogues are estimated to be ~ 15–$20\,h^{-1}$Mpc, which corresponds to an infall velocity $v_{inf} \geq 1500\,$km/s and a velocity dispersion $\sigma_v \approx v_{inf}/\sqrt{3} \geq 800\,$km/s for a universe with $\Omega_m = 1$, clearly exceeding the observed value of 300–400 km/s. These values could be lowered by using cosmological models with $\Omega_m \leq 0.2$–0.3. However, even for the observed σ_v, DM SLSS elements would still be short-lived, since soon after formation much of the DM in a SLSS wall would pass through and move quickly away from the centre of the gravitational well and the wall would break up into a more disparate system of galaxy clumps. Thus, if $\Omega_m \geq 0.3$ and/or the SLSS is observed in the galaxy distribution at $z \geq 2$, as some authors have claimed, some other physical mechanism for the formation of SLSS would be required.

A biasing of the galaxy distribution relative to the DM distribution can permit smaller SLSS proto-elements, since the bias itself contributes to the observed overdensity of galaxy SLSS elements. A natural physical mechanism for such a bias is the radiational reheating of gas at high redshift, inhibiting the formation of galaxies in the UDRs where galaxies would otherwise have formed at a later time than in the overdense regions. In this way, the galaxy SLSS could manifest itself before SLSS forms in the underlying DM distribution. If such reheating is responsible for the observed SLSS, then it would imply an epoch of galaxy formation, $5 \lesssim z_f \lesssim 10$.

Acknowledgments

This paper was supported in part by Denmark's Grundforskningsfond through its support for an establishment of Theoretical Astrophysics Center.

1. Cole S., Weinberg D.H., Frenk C.S., Bharat R, MNRAS, submitted
2. Doroshkevich A.G. et al, MNRAS, **283**, 1281, (1996)
3. Ratcliffe A. et al, MNRAS, **281**, L47, (1996)
4. Shectman S.A. et al, ApJ., **470**, 172, (1996)

Higher Order Galaxy Correlations

István Szapudi
Fermilab, Theoretical Astrophysics Group,
Batavia, IL 60510

Moments and joint moments of cell counts are used to estimate the amplitudes of higher order correlation functions in galaxy catalogs. The results are compared with theoretical predictions by perturbation theory and N-body simulations. It is found that the picture of gravitational instability with Gaussian initial conditions is qualitatively correct in both catalogs (EDSGC, APM) analyzed, without the need of significant biasing. These methods in conjunction with future data and N-body simulations will be able to pin down the amplitudes of the higher order correlations with unprecedented accuracy.

$N-$point correlation functions [3] provide provide a wealth of information about linear and non-linear gravitational growth, the Gaussianity of the initial conditions, and the degree and nature of biasing, while being fairly insensitive to Ω. However, they are burdened with a combinatorial explosion of terms, which severely complicates their direct measurement and subsequent interpretation. Thus indirect methods became increasingly popular for high precision extraction of higher order correlations. The simplest of these methods consists of calculating the (factorial) moments of the distribution of counts in cells, and from that, the cumulants, S_N's, of the underlying distribution [5]. These quantities measure the amplitude of the N-point correlation function averaged in a particular window. Since the averaging causes a significant loss of information, alternative methods based on moment correlators use a pair of cells [4]. A new set of reduced functions can be introduced, Q_{NM}: the cumulant correlators [6]. Both methods are illustrated next with two state of the art measurements: counts in cells measured in the The Edinburgh/Durham Southern Galaxy Catalogue [1] (EDSGC), and the cumulant correlators extracted from the joint moments in the Automatic Plate Measuring [2] (APM) survey.

In the following measurements both catalogs had approximately the same $b_j^{APM} = 17 - 20$ magnitude cut. The available surface area was about 1000 square degrees for the EDSGC, and roughly four times that for the APM. The consequential reduction in cosmic errors allowed to perform the slightly more demanding joint moment analysis on the latter catalog. Details on the properties of the catalogs and methods used can be found in Szapudi *et al.* [4,5,6], as well as key references on the phenomenology of higher order functions, perturbation theory, extended perturbation theory, the hierarchical assumption, and N-body simulations, which were omitted here due to space limitations.

(i) EDSGC Counts in cells were estimated by calculating the results corresponding to an infinite [5] number of square cells with sizes in the range $0.015125° - 2°$. From the probability distribution of counts in cells, P_N, the

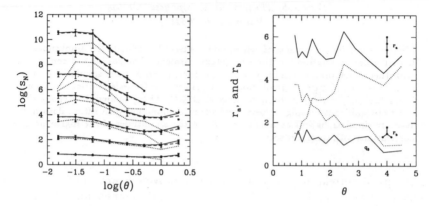

Figure 1: Figure 1. a. The solid line is the measurement of the s_N's over the entire survey area with infinite sampling, the dotted line is the same with low sampling. Under-sampling results in a systematic underestimate of the coefficients. The squares show the mean of the measurements in four equal parts of the survey, and the errors are calculated from the dispersion. The triangles display the s_N's corresponding to the best fitting formal $n_{\rm eff}$. b. The hierarchical amplitudes as calculated from the fully non-linear cumulant correlators are displayed. The solid lines to the estimator of r_a, and r_b, the amplitudes of the fourth order snake, and star graphs, respectively. The dotted lines show the linear approximation, which breaks down at smaller scales at this level of precision.

factorial moments, $F_k = \sum P_N(N)_k$ were calculated, where $(N)_k = N(N - 1)..(N-k+1)$ is the k-th falling factorial of N. Factorial moments take Poisson noise into account automatically, since their ensemble average is equal to the continuum moments [4]. This facilitates the calculation of the cumulants, which are proportional to the N-th connected moment of the underlying distribution,

$$S_N \propto \langle \delta_1^N \rangle_c. \tag{1}$$

The results shown on Figure 1.a. exhibit two plateaus, one at small scales ($< 0.03°$) and a second at large ($> 0.5°$). The large scale plateau is approaching the width of the survey, and so may merely reflect edge effects. The plateau at small separations, however, may indicate that the strongly nonlinear clustering limit has been reached, in which case the hierarchical form for the angular correlations should apply, for which the coefficients appear to converge.

(ii) APM The factorial moment correlators are simple two-point generalization of the factorial moments. Similarly, the cumulant correlators are

natural two-point extension of the cumulants obtained from the logarithm of the moment generating function [4,6]. Their physical meaning is simplified by the continuum properties of the factorial moments:

$$Q_{NM} \propto \langle \delta_1^N \delta_2^M \rangle_c. \tag{2}$$

When the different tree topologies and all the non-linearities are taken into account, it is possible to show that the hierarchical assumption is an excellent approximation to the statistics of cumulant correlators at the twenty percent level. This is illustrated on Figure 4.b. for the four-point correlations. The dotted lines show the linear results for r_a, and r_b, the weights for the snake and star topologies. The solid lines display the fully non-linear calculation: the corrections remove most of the scale dependence, as expected if the hierarchy is satisfied. So far, these results represent the highest quality confirmation of the hierarchical ansatz motivated by the BBKGY equations in the highly non-linear regime. The actual values can provide a clue for a solution.

The above results are strikingly similar to corresponding results from N-body simulations, despite possible projection effects. De-projection and further detailed comparison performed in Szapudi $et\ al.$ [5,6] reveals excellent agreement with the predictions of perturbation theory, extended perturbation theory, N-body simulations, and the theory of the gravitational BBKGY equations: all based on Gaussian initial conditions, gravitational instability and no biasing. Although present measurement errors prevent establishing a strict constraint on biasing, this will be possible in future second generation catalogs, such as the Sloan Digital Sky Survey. The author was supported by DOE and NASA through grant NAG-5-2788 at Fermilab.

References

1. Heydon-Dumbleton, N. H., Collins, C. A., & MacGillivray, H. T. *MN-RAS*, **238**, 379 (1989)
2. Maddox, S. J., Efstathiou, G., Sutherland, W. J., & Loveday, L. *MNRAS*, **242**, 43P (1990)
3. Peebles, P.J.E. *The Large Scale Structure of the Universe* (Princeton: Princeton University Press, 1980)
4. Szapudi, I., Dalton, G., Efstathiou, G.P., & Szalay, A. *ApJ*, **444**, 520 (1995)
5. Szapudi, I., Meiksin, A., & Nichol, R.C. *ApJ*, **473**, 15 (1996)
6. Szapudi, I. & Szalay, A. submitted to *ApJ* (1997)

Understanding Gravitational Clustering with Non-Linear Perturbation Theory

Román Scoccimarro

CITA, McLennan Physical Labs, 60 St George Street, Toronto ON M5S 3H8

I discuss new results concerning the evolution of the bispectrum due to gravitational instability from gaussian initial conditions using one-loop perturbation theory (PT). Particular attention is paid to the transition from weakly non-linear scales to the non-linear regime at small scales. Comparison with numerical simulations is made to assess the regime of validity of the perturbative approach.

1 Introduction

This work is based on results to be reported in [1]. Here, I briefly present results on the one-loop corrections to the bispectrum and compare them to numerical simulations for CDM initial spectra.[a] In particular, we work in terms of the hierarchical amplitude Q defined from the bispectrum, $B(\mathbf{k_1}, \mathbf{k_2})$, and power spectrum, $P(k)$, as follows (superscripts denote, tree-level, one-loop PT, and so on):

$$Q \equiv \frac{B(\mathbf{k_1}, \mathbf{k_2})}{\Sigma(\mathbf{k_1}, \mathbf{k_2})} = \frac{B^{(0)}(\mathbf{k_1}, \mathbf{k_2}) + B^{(1)}(\mathbf{k_1}, \mathbf{k_2}) + \cdots}{\Sigma^{(0)}(\mathbf{k_1}, \mathbf{k_2}) + \Sigma^{(1)}(\mathbf{k_1}, \mathbf{k_2}) + \cdots} = Q^{(0)} + Q^{(1)} + \cdots, \quad (1)$$

with:

$$\langle \delta(\mathbf{k}) \delta(\mathbf{k'}) \rangle = \delta_\mathbf{D}(\mathbf{k} + \mathbf{k'}) \, P(k), \quad (2)$$

$$\langle \delta(\mathbf{k_1}) \delta(\mathbf{k_2}) \delta(\mathbf{k_3}) \rangle = \delta_\mathbf{D}(\mathbf{k_1} + \mathbf{k_2} + \mathbf{k_3}) \, B(\mathbf{k_1}, \mathbf{k_2}). \quad (3)$$

$$\Sigma(\mathbf{k_1}, \mathbf{k_2}) \equiv P(k_1) \, P(k_2) + P(k_1) \, P(k_3) + P(k_2) \, P(k_3) \quad (4)$$

where we have used that the bispectrum is defined for closed triangle configurations, $\sum_{i=1}^{3} \mathbf{k_i} = 0$. The perturbative quantities in Eq. (1) can be calculated from the standard machinery of PT.[1,2] In the following, we consider Q for configurations where $k_1/k_2 = 2$, as a function of θ, the angle between $\hat{\mathbf{k}}_1$ and $\hat{\mathbf{k}}_2$.

---

[a]with $\Omega = 1$, $\Gamma = 0.25$. The simulation data is publically available through the Hydra Consortium Web page ().

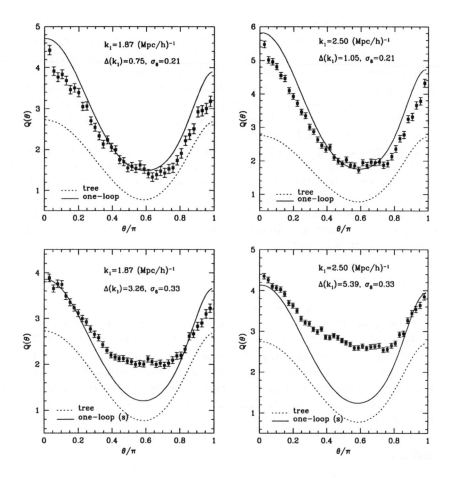

Figure 1: The hierarchical amplitude Q for triangle configurations with $k_1/k_2 = 2$ as a function of the angle θ between $\hat{k}_1$ and $\hat{k}_2$ in the Hydra CDM numerical simulations (symbols), tree-level PT (dotted lines) and one-loop PT (solid lines).

568

2 Results

Figure 1 shows Q in numerical simulations compared to tree-level PT and one-loop PT. Error bars in these plots are estimated from the number of independent Fourier modes contributing to each configuration assuming gaussianity[1]. The degree of non-linearity in each case can be inferred from the dimensionless power spectrum, $\Delta(k) \equiv 4\pi k^3 P(k)$. In the top panels ($\sigma_8 = 0.21$), we note a clear deviation of the N-body results from the tree-level PT prediction for Q, and a good agreement with the one-loop correction. At this stage of non-linear evolution, the dynamics is dominated by large-scale power and therefore an enhancement of Q at collinear configurations ($\theta = 0, \pi$) develops[2]. In the bottom panels ($\sigma_8 = 0.33$), where already $\Delta(k_1) > 1$, we use the ratio of one-loop quantities in Eq. (1) (denoted as "one-loop (s)" in Fig. 1) for the one-loop prediction[2]. We see very good agreement for configurations close to collinear, and a progressively flattening of $Q(\theta)$ as we look at smaller scales. The flattening is due to configurations close to equilateral becoming more probable due to random motions at small scales[1]. At even more non-linear scales, Q becomes configuration independent, in rough agreement with the hierarchical ansatz for the three-point function[1].

Acknowledgments

This work is based on a project in collaboration with S. Colombi, J.N. Fry, J.A. Frieman, E. Hivon, & A. Mellot. I would like in addition to thank F. Bernardeau, E. Gaztañaga, B. Jain, R. Juszkiewicz, C. Murali for conversations, and especially Hugh Couchman for numerous helpful discussions. The CDM simulations analyzed in this work were obtained from the data bank of cosmological N-body simulations provided by the Hydra consortium (http://coho.astro.uwo.ca/pub/data.html) and produced using the Hydra N-body code[3].

References

1. R. Scoccimarro, S. Colombi, J.N. Fry, J.A. Frieman, E. Hivon, and A. Melott, in preparation (1997).
2. R. Scoccimarro, submitted to *Astrophys. J.*, astro-ph/9612207, (1996).
3. H. Couchmann, P.A. Thomas, and F. Pearce, *Astrophys. J.* **452**, 797 (1995).

WEAK LENSING BY LARGE-SCALE STRUCTURE

BHUVNESH JAIN

Max-Planck-Institut für Astrophysik, 85740 Garching, Germany

UROŠ SELJAK

Center For Astrophysics, Harvard University, Cambridge, MA 02138 USA

Weak lensing by large scale structure induces correlated ellipticities in the images of distant galaxies. The two-point correlation is determined by the matter power spectrum along the line of sight. We use the fully nonlinear evolution of the power spectrum to compute the predicted ellipticity correlation. Nonlinear effects significantly enhance the ellipticity for $\theta < 10'$ — on $1'$ the rms ellipticity is $\simeq 0.05$, which is nearly twice the linear prediction. We discuss the dependence of the predicted ellipticity correlation on the matter power spectrum and the cosmlogical parameters Ω and Λ.

1 Introduction

Weak lensing magnifies and shears the images of distant galaxies. The shear induces an ellipticity in the image of an intrinsically circular galaxy. By averaging over the observed ellipticities of a large number of faint galaxies, the induced ellipticity can be measured, and related to the mass fluctuations along the line of sight and to the spatial geometry of the universe. Ellipticities of distant background galaxies averaged over several arcminute windows are sensitive to the mass power spectrum on scales of $1 - 10h^{-1}$ Mpc. For a given spectrum of mass fluctuations, it is sensitive to the cosmological parameters Ω_m and Ω_Λ. Thus while strong lensing which leads to multiple images probes non-typical regions of the universe which contain massive halos, weak lensing provides a different and a more direct measure of the mass fluctuations on large scales.

The first calculations of the shear signal due to weak lensing that used modern models for the large scale structure power spectrum were performed in the early 1990s [2,8,6]. Our work generalizes these results to include the effects of nonlinear evolution of the matter fluctuations for flat and for open and $\Lambda-$dominated cosmologies [3,4,9]. Detailed results from our calculations can be found in Jain & Seljak (1996) [5]. Related aspects have been considered recently by other authors as well [13,1,7,12].

We follow the formalism of Seljak (1995,1996) [10,11] to relate the shear due to gravitational lensing to the perturbations in the gravitational potential along the line of sight. These potential perturbations are related to density perturbations via the cosmological Poisson equation. We are thus led to expressions relating two-point statistics that can quantify the induced ellipticites in galaxy images to a line of sight integral over the matter power spectrum $P_\delta(k,\chi)$ [5]. The simplest such statistic is the rms ellipticity $\bar{p}(\theta) \equiv \langle \bar{p}(\theta)\bar{p}(\theta)^* \rangle^{1/2}$, where the overbar indicates the average within a circular aperture of radius θ. In the weak lensing regime $p = 1 - b/a$, where b/a is the ratio of the minor to major axis of an elliptical image. The ellipticity variance is given by

$$\bar{p}^2(\theta) = 36\pi^2 \, \Omega_m^2 \int_0^\infty kdk \int_0^{\chi_S} a^{-2}(\chi) \, P_\delta(k,\chi) \, g^2(\chi) W_2^2[kr(\chi)\theta] \, d\chi \,, \quad (1)$$

where χ is the radial coordinate, $W_2(x) = 2J_1(x)/x$ with $J_1(x)$ being the Bessel function of first order, and a is the expansion factor normalized to unity today. The angular diameter distance is denoted $r(\chi)$ and defines the line of sight window function $g(\chi) = r(\chi) \, r(\chi_S - \chi)/r(\chi_S)$ if all the galaxies lie at the same redshift z_s corresponding to χ_S.

2 Results for CDM-like models

It is evident from examining the above equation that $\bar{p}(\theta)$ depends on Ω, Λ (via the distance and growth factors in the line of sight integral), σ_8 and on the shape of $P_\delta(k)$. A measurement of shear due to large scale structure can thus constrain these parameters. Figure 1 shows the expected signal-to-noise

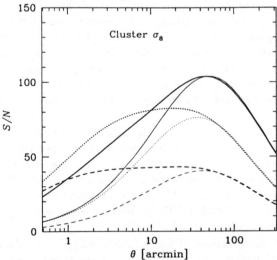

Figure 1: The quantity $(\theta/5')\,\bar{p}^2(\theta)\,10^5$ is shown for three cosmological models (solid – Einstein-de Sitter, dotted – Λ, dashed – open). This provides an estimate of the signal to noise in the measurement of $\bar{p}$. Nonlinear evolution causes the curves to be much less sensitive to θ in the range $2' < \theta < 2°$ compared to the linear prediction, which shows a significant peak around $\theta = 1°$. The linear curves for the three models are shown by the thin curves.

for a one square degree field with 2×10^5 galaxies, as obtained from a typical observation with a limiting magnitude $I = 26$. The galaxies are assumed to be at a redshift of one. The theoretical power spectrum used is the $\Gamma = 0.25$ CDM-like spectrum, normalized to the abundance of galaxy clusters [14]. A detailed comparison of model predictions for $\bar{p}(\theta)$ and the ellipticity power spectrum can be found in Jain & Seljak (1996)[5].

References

1. Bernardeau, F., van Waerbeke, L., & Mellier, Y. 1996, astro-ph/9609122
2. Blandford, R. D., Saust, A. B., Brainerd, T. G., & Villumsen, J. V. 1991, *MNRAS*, 251, 600
3. Hamilton, A. J. S., Kumar, P., Lu, E., & Matthews, A. 1991, *ApJ*, 374, L1
4. Jain, B., Mo, H. J., & White, S. D. M. 1995, *MNRAS*, 276, L25
5. Jain, B., & Seljak, U. 1996, preprint, astro-ph/9611077
6. Kaiser, N. 1992, *ApJ*, 388, 272
7. Kaiser, N. 1996, astro-ph/9610120
8. Miralda-Escude, J. 1991, *ApJ*, 380, 1
9. Peacock, J. A., & Dodds, S. J. 1996, *MNRAS*, astro-ph/9603031
10. Seljak, U. 1995, Ph.D. thesis, MIT, unpublished
11. Seljak, U. 1996, *ApJ*, 463, 1
12. Stebbins. A. 1996, preprint astro-ph/9609149
13. Villumsen, J. V. 1996, *MNRAS*, 281, 369
14. White, S. D. M., Efstathiou, G., & Frenk C. S. 1993, *MNRAS*, 262, 1023

COSMIC SHOCK WAVES ON LARGE SCALES OF THE UNIVERSE

DONGSU RYU

Dept. of Astronomy, Univ. of Washington, Seattle, WA 98195-1580
Dept. of Astronomy & Space Science, Chungnam Nat. Univ., Korea

HYESUNG KANG

Dept. of Astronomy, Univ. of Washington, Seattle, WA 98195-1580

In the standard theory of the large scale structure formation, matter accretes onto high density perturbations via gravitational instability. Collisionless dark matter forms caustics around such structures, while collisional baryonic matter forms accretion shocks which then halt and heat the infalling gas. Here, we discuss the characteristics, roles, and observational consequences of these accretion shocks.

The simulations of large scale structure in the universe, which include the evolution of baryonic matter as well as that of dark matter, have shown the formation of accretion shocks around the nonlinear structures such as super-galactic sheets, filaments, and clusters of galaxies (see, for example, Kang *et al.* 1994). The upper panel of Figure shows the density contours of baryonic matter in one of those simulations (Kulsrud *et al.* 1997; Ryu, Kang, & Biermann 1997). Accretion shocks exist around the high density structures of clusters, filaments, and sheets in the density contours.

The properties of the shocks and the accreting matter outside the shocks depend upon the power spectrum of the initial perturbations on a given scale as well as the background expansion in a given cosmological model. To study them, we calculated the accretion of dark matter particles around clusters in one-dimensional spherical geometry under various cosmological models (Ryu & Kang 1997). The velocity of the accreting matter around clusters of a given temperature is smaller in a universe with smaller Ω_o, but only by up to $\sim 24\%$ in the models with $0.1 \leq \Omega_o \leq 1$. It is given as $v_{acc} \approx 0.9 - 1.1 \times 10^3$km s$^{-1}[(M_{cl}/R_{cl})/(4 \times 10^{14}M_\odot/\text{Mpc})]^{1/2}$. However, the accretion velocity around clusters of a given mass or a given radius depends more sensitively on the cosmological models.

Considering that these accretion shocks are very big with a typical size $\gtrsim$ a few Mpc and very strong with a typical velocity jump $\gtrsim$ a few 1000 km s^{-1}, they could serve as possible sites for the acceleration of high energy cosmic rays by the first-order Fermi process (Kang, Ryu, & Jones 1996; Kang, Rachen, & Biermann 1997). With Jokipii diffusion, the observed cosmic ray spectrum near 10^{19}eV could be explained with reasonable parameters if about 10^{-4} of the infalling kinetic energy can be injected into the intergalactic space as the high energy particles.

The shocks could serve also as sites for the generation of weak seeds of cosmic magnetic field by the Biermann battery mechanism. Then, these seeds could be amplified to strong (up to a few μG) and coherent (up to the galaxy scale) magnetic field by the Kolmogoroff turbulence endemic to gravitational structure formation (Kulsrud et al. 1997). The lower panel of Figure shows the vectors of seed magnetic field generated by the Biermann mechanism in the simulation. In the highest density regions of clusters, the magnetic field is chaotic since the flow motion is turbulent. However, in the regions which are identified as filaments or sheets, the magnetic field is aligned with the structures due to the streaming flow motion along the structures.

If there is aligned magnetic field in filaments and sheets, this would induce the Faraday rotation in polarized radio waves from extra-galactic sources. Then, an upper limit in its strength can be placed by comparing the expected rotational measure with the observed limit of rotational measure RM = 5 rad m^{-2} at $z = 2.5$ (Kronberg 1994) due to the intergalactic magnetic field. We performed this calculation using the data of the simulation in Figure. The result indicates that, with the present value of the observed limit in rotational measure, the existence of magnetic field of $\lesssim 1\mu$G in filaments and sheets can not be ruled out (Ryu, Kang, & Biermann 1997). It is interesting to notice that the equipartition magnetic field strength in filaments and sheets, $B = 0.77h\sqrt{T_7}\sqrt{\rho_b/\rho_c}$ μG, is close to this limit. Here, T_7 is the temperature in unit of 10^7K and ρ_b/ρ_c is the baryonic density in unit of critical density.

One interesting implication of such strong magnetic field in filaments and sheets is its effects on the propagation of high energy cosmic rays through the universe. The discoveries of several reliable events of high energy cosmic rays at an energy above 10^{20}eV raise questions about their origin and path in the universe, since their interaction with the cosmic microwave background radiation limits the distances to their sources to less than 100 Mpc, perhaps within our Local Supercluster. In Biermann, Kang & Ryu (1996), we noted that if the magnetic field of $\sim 1\mu$G or less exists inside our Local Supercluster and there exist accretion flows infalling toward the supergalactic plane, it is possible that the high energy cosmic rays above the so-called GZK cutoff ($E > 5 \times 10^{19}$eV) can be focused in the direction perpendicular to the supergalactic plane, analogously but in the opposite direction to solar wind modulation. This would explain naturally the correlation between the arrival direction of the high energy cosmic rays and the supergalactic plane. Also, focusing means that for all the particles captured into the sheets, the dilution with distance d is $1/d$ instead of $1/d^2$, increasing the cosmic ray flux from any source appreciably

574

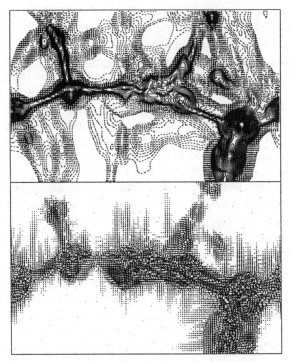

Figure 1: Two-dimensional cut of the simulated universe. The plot shows a region of $32h^{-1} \times 20h^{-1}\mathrm{Mpc}^2$ with a thickness of $0.25h^{-1}\mathrm{Mpc}$, although the simulation was done in a box of $(32h^{-1}\mathrm{Mpc})^3$ volume. The upper panel shows baryonic density contours, and the lower panel shows magnetic field vectors.

with respect to the three-dimensional dilution.

References

1. Biermann, P. L., Kang, H. & Ryu, D., 1996, in Proc. ICRR Symposium on Extremely High Energy Cosmic Rays, ed. M. Nagano & K. Tanashi, in press.
2. Kang, H., Cen R., Ostriker, J. P., & Ryu, D., 1994, ApJ, 428, 1.
3. Kang, H., Rachen, J. P., & Biermann, P. L., 1997, MNRAS, in press.
4. Kang, H., Ryu, D., & Jones, T. W., 1996, ApJ, 456, 422.
5. Kronberg, P. P., 1994, Rep. Prog. Phys. 57, 325.
6. Kulsrud, R. M., Cen, R., Ostriker, J. P., & Ryu, D., 1997, ApJ, in press.
7. Ryu, D., & Kang, H., 1997, MNRAS, in press.
8. Ryu, D., Kang, H., & Biermann, P. L., 1997, MNRAS, submitted.

THE STRUCTURE OF COSMIC STRING WAKES

A. SORNBORGER

DAMTP, University of Cambridge, 3 Silver Street,
Cambridge CB2 1ND, UK

R. BRANDENBERGER

Department of Physics, Brown University,
Providence, RI 02912, USA

B. FRYXELL and K. OLSON

Institute for Computational Science and Informatics,
George Mason University, Fairfax, VA 22030, USA

We present results of a cosmological hydrodynamical study of gravitational accretion in cosmic string wakes and filaments. Cosmic string wakes are formed by fast moving ($v \sim c$) strings. A conical deficit angle in the string spacetime induces a velocity perturbation in the background matter and a two-dimensional wake accretes in the path of the string. Filaments are formed by slow moving strings with a large amount of small scale structure. The major gravitational perturbation from slow moving strings is due to the Newtonian field induced by the effective mass of the wiggles. In cosmic string wakes, cool streams of baryons collide and are trapped at the center of the wake causing an enhancement of baryons versus dark matter by a factor of 2.4. In filaments, a high pressure is induced at the filament core and baryonic matter is expelled leading to a baryon deficit in the center of the filament.

1 Introduction

Strings are an inevitable result of symmetry breaking in many theories in particle physics as well as in numerous condensed matter systems. Once the temperature decreases below the symmetry breaking scale, the field undergoes a phase transition and a string network forms. In an expanding universe, the string distribution quickly approaches a scaling solution in which there are (statistically) the same number of strings per horizon volume at all times. Numerical simulations show that the number of strings per horizon volume is in the range $10 - 30$. Potential systematic errors due to coarse graining in simulations may tend to favor a lower number of strings per horizon volume.

If symmetry breaking occurs at the GUT scale, the strings are massive enough to be responsible for large-scale structure in the universe. In this case, the scaling solution leads to a Harrison-Zel'dovich spectrum consistent with that observed by COBE. Calculations of the power spectrum give good agreement with the observed power spectrum with a bias factor around 2. Normalization of the spectrum with COBE gives a value of $G\mu \sim 10^{-6}$, a dimensionless quantity proportional to the mass per length of the string μ. This value is within observational bounds, the most restrictive being that from timing analyses of the millisecond pulsar.

2 Large-Scale Structure Formation

Large-scale structure in the cosmic string model is due to the growth of density perturbations resulting from the gravitational field of the string network. The gravitational field of a string can conceptually be broken into two parts. The first part is due to purely general relativistic effects: the string leads to a conical spacetime. Geodesics in such a spacetime converge after passing the string. This means that a moving string will cause a velocity perturbation as it passes through matter. For strings with little small-scale structure such as waves and kinks moving along the string, the velocity perturbation from the conical spacetime is the predominant perturbing mechanism. Large-scale structure formed by fast moving strings will thus be in large sheet-like objects called wakes.

Strings with a large amount of small-scale structure move slowly. The predominant perturbation mechanism for these strings is the Newtonian field resulting from the effective mass of the waves and kinks on the string. These strings move slowly accreting long cylindrically shaped overdensities called filaments.

3 Non-linear Matter Evolution (Our Study)

What has been missing in the study of large-scale structure in the context of the cosmic string model is an understanding of non-linear evolution of matter accretion in wakes and filaments. In particular, what is the detailed distribution of the baryonic and dark matter?

To answer this question, we have developed a high-resolution PPM/PIC cosmological hydrodynamical numerical code for investigating the formation and evolution of baryonic and dark matter as they accrete in cosmic string wakes and filaments.

4 Results

We have investigated the matter distribution in the case of wakes and filaments. The matter distribution in the wake is markedly dissimilar to that of the filament.

First, we investigated the matter distribution in a wake caused by a long straight string. As the string passes through matter, its conical spacetime gives a velocity kick to the matter on either side of its path and a wake forms. Two cool coherent streams of matter flow toward the center of the wake. The baryonic matter in the streams collides at the center forming an overdensity bounded by a shock. The dark matter streams flow through each other giving rise to an overdensity bounded by caustics. The baryonic matter is more concentrated at the center of the wake than the dark matter leading to an enhancement of baryonic matter to dark matter of about 2.4. Temperatures at the wake center are of the order of 100 degrees Kelvin.

Next, we examined the evolution of a filament caused by a slowly moving string. Here, the matter distribution was quite different than that of the wake. In the filament, the kinetic energy of the inflow is higher than that of the wake leading to high post-thermalization temperatures at the filament core. The associated high pressures force the baryons out of the filament core resulting in a net baryon deficit inside the filament. Temperatures at the filament core are of the order a few times 10^6 degrees Kelvin.

5 Conclusions

From our results we conclude that biasing in the cosmic string model is perturbation dependent. In wakes, biasing of a factor of 2 to 3 can be expected, while in filaments, antibiasing is expected. The amount of biasing in wakes is consistent with calculations of the amount of biasing required by power spectrum calculations.

We find wakes to be relatively cool, with temperatures of the order 100 degrees Kelvin. Filaments are hotter, with temperatures of the order 10^6 degrees Kelvin.

The baryon enhancement that we find in wakes is a possible mechanism for explaining baryon overabundances in objects such as the Coma cluster. It has been suggested that clusters will form at triple wake intersections. If this is the case, the enhancement will be tripled to a factor of 7, giving a measured baryon fraction of 0.35.

References

1. All references may be found in: A. Sornborger, R. Brandenberger, B. Fryxell & K. Olson, 'The Structure of Cosmic String Wakes', accepted for publication in the Astrophysical Journal, June 10, 1997, also in the archives at Los Alamos: http://xxx.lanl.gov/ps/astro-ph/9608020

ABUNDANCE RATIOS IN LOW METALLICITY STARS; FAMILIAR TRENDS AND NEW SURPRISES

CHRISTOPHER SNEDEN

Department of Astronomy and McDonald Observatory,
University of Texas, Austin, TX 78712

Recent high resolution spectroscopic analyses of old, very low metallicity stars of the galactic halo show that some elemental abundance ratios have extremely large (and generally unexpected) departures from the ratios in higher metallicity stars. We comment on the trends with metallicity of the iron-peak, α-capture, and neutron-capture elements.

1 Introduction

Abundance ratios of stars as functions of stellar metallicity provide the basic data set for probing stellar nucleosynthesis throughout the galaxy. Reviews by *e.g.* Wheeler *et al.*[1] have summarized the first few decades of observational work in this field. Recently however, extensive analyses of larger samples of relatively bright stars in the range $-2.5 \leq$ [Fe/H] $\leq +0.3^a$ and similar studies of newly discovered halo stars with [Fe/H] < -2.5 (hereafter "ultra-metal-poor" or UMP stars) have found surprising changes in the abundance patterns of the lowest metallicity stars. Such changes are now revealing the story of early galactic nucleosynthesis. In this brief review we highlight some of the major results from these recent studies.

2 Fe-Peak Elements: Large Shifts from Solar Abundance Ratios

The Fe-peak elements are those with $21 \leq Z \leq 30$, except the partial α-capture element Ti. Nearly all Fe-peak abundance ratios were thought to be approximately solar (*e.g.* [Ni/Fe] $\simeq 0.0$) at all metallicities. However, there are large observed departures from this assumption in UMP stars. In Fig. 1 we summarize the current observational knowledge for these elements, combining free-hand sketches of the mean trends for Sc, Mn, Co and Ni (Ryan *et al.*[2]), for V (Gratton and Sneden[3]), and for Cu and Zn (Sneden *et al.*[4]). Each curve is truncated at the approximate metallicity limit of trend reliability; see the cited papers for more complete discussions. Several of the abundance ratios for stars with metallicities [Fe/H] ≥ -2.5 are well matched by recent chemical evolution models (*e.g.* Timmes *et al.*[5]). The apparent upward trend in Ni

---

^{a}[A/B] $\equiv \log_{10}(N_A/N_B)_{star} - \log_{10}(N_A/N_B)_\odot$.

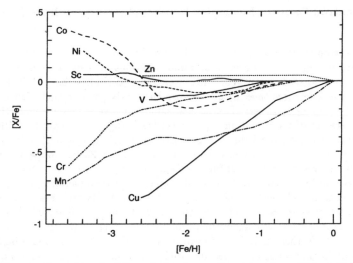

Figure 1: General trends of Fe-peak abundance ratios as functions of Fe metallicity.

is not yet observed in enough UMP stars to be certain of the effect. Radical declines (substantially in excess of observational uncertainties) are seen in the abundances of the odd-Z elements Mn and Cu; but such underabundances were suspected in some early analyses of very metal-poor stars. The remaining surprises are the large and theoretically unexpected underabundance of Cr and overabundance of Co in UMP stars (see also the discussion of McWilliam *et al.* [6]). Nucleosynthesis models of low metallicity supernovae must begin to address these metallicity trends; when these data can be matched it will be of great interest to see what other anomalous abundance ratios are predicted by the models.

3 Alpha-Capture Elements: Evidence for Inhomogeneities

α-capture elements (O, Mg, Si, S, Ca, and Ti) seem always to be overabundant in metal-poor stars: $[\alpha/\text{Fe}] \sim 0.3$ to 0.5 when $[\text{Fe/H}] \leq -1$. The earlier-cited review discusses this point; it is often suggested that the small star-to-star scatters of these overabundances are dominated by observational uncertainties. Theoretically the effect is understood by the dominance of Type II supernova nucleosynthesis in the early galactic halo. However, a few exceptions to α-rich metal-poor stars are appearing. First, Brown *et al.* [7] have demonstrated that two moderately metal-poor outer halo globular clusters have $[\alpha/\text{Fe}] \sim 0.0$ for all observed α-capture elements. Second, Carney *et al.* [8] have found a field halo star with $[\text{Fe/H}] \sim -2.0$ but $[\alpha/\text{Fe}] \sim -0.2$. What fraction of metal-poor stars are deficient in these elements relative to the vast majority in the galactic halo. How did such stars get their anomalous compositions? How chemically homogeneous is the galactic halo?

580

4 Neutron-Capture Elements: Individual Nucleosynthesis Events Seen?

The neutron-capture elements are all those beyond the Fe-peak (Z > 30) that are manufactured via slow or rapid additions of neutrons (the s- and r-processes) to seed nuclei. For UMP stars, several studies (*e.g.* McWilliam *et al.*[6] and references therein) have shown that early galactic nucleosynthesis favored r-process synthesis of these elements. Moreover, the very large star-to-star scatter of the overall [n-capture/Fe] levels suggest strongly that UMP stars were formed from halo ISM gas enriched with the ejecta of individual supernovae. The most prime example is the star CS 22892-052, which has overabundances of neutron-capture elements as large as +1.5 dex. Sneden *et al.*[9] have subjected this star to a detailed analysis, deriving abundances for 20 neutron-capture elements. These abundances match very well a scaled solar system set of r-process abundances, and allow the use of a detected thorium feature in a new galactic age estimate of 15.2±3.7 Gyr. This value has been subjected to further analysis by Cowan *et al.*[10]. CS 22892-052 should be the target of further analyses, especially with new near-UV spectra to uncover other n-capture elements; at the present time there is no other metal-poor star with such a contrast between neutron-capture elements and Fe-peak elements. It is one of the best neutron-capture laboratories known, and deserves further scrutiny.

Acknowledgments

We thank John Cowan and Jim Truran for helpful discussions. This work has been supported by NSF grants AST-9315068 and AST-9618364.

References

1. J.C. Wheeler *et al.*, *Ann. Rev. Ast. Ap.* **27**, 279 (1989).
2. S.G.Ryan *et al.*, *Ap. J.* **471**, 254 (1996).
3. R.G. Gratton and C. Sneden, *Ast. Ap.* **241**, 501 (1991).
4. C. Sneden *et al.*, *Ast. Ap.* **246**, 354 (1991).
5. F.X. Timmes *et al.*, *Ap. J. Supp.* **98**, 617 (1995).
6. A. McWilliam *et al.*, *Ast. J.* **109**, 2757 (1996).
7. J.A. Brown *et al.*, *Ast. J.*, submitted (1997).
8. B.W. Carney *et al.*, *Ast. J.*, submitted (1997).
9. C. Sneden *et al.*, *Ap. J.* **467**, 819 (1996).
10. J.J. Cowan *et al.*, *Ap. J.* **467**, 819 (1996).

NUCLEOSYNTHESIS IN TYPE IA SUPERNOVAE AND CONSTRAINTS ON PROGENITORS

KEN'ICHI NOMOTO , NOBUHIRO KISHIMOTO

Department of Astronomy and Research Center for the Early Universe
School of Science, University of Tokyo, Tokyo 113

Among the major uncertainties involved in the Chandrasekhar mass models for Type Ia supernovae are the companion star of the accreting white dwarf and the flame speed after ignition. We present nucleosynthesis results from relatively slow deflagration to constrain the rate of accretion from the companion star. Because of electron capture, a significant amount of neutron-rich species are synthesized. To avoid the too large ratios of $^{54}Cr/^{56}Fe$ and $^{50}Ti/^{56}Fe$, the central density of the white dwarf at thermonuclear runaway must be as low as $\lesssim 2 \times 10^9$ g cm^{-3}. Such a low central density can be realized by the accretion as fast as $\dot{M} \gtrsim 1 \times 10^{-7} M_\odot$ yr^{-1}. These rapidly accreting white dwarfs might correspond to the super-soft X-ray sources.

1 Introduction

Supernovae of different types have different progenitors, thus producing different heavy elements on different time scales during the chemical evolution of galaxies. Type II supernovae (SNe II) and Type Ib/Ic supernovae cause the heavy-element enrichment in the early phase of the galactic evolution. In contrast, Type Ia supernovae (SNe Ia) produce heavy elements on a much longer time scale in the later phase of the galactic evolution. There are strong observational and theoretical indications that SNe Ia are thermonuclear explosions of accreting white dwarfs. However, the exact binary evolution that leads to SNe Ia has not been identified (e.g., Branch *et al.* 1995 for recent reviews). In this paper, we provide some important constraints on the progenitor system from the viewpoint of nucleosynthesis, namely, the carbon ignition density which is translated into the accretion rate for the Chandrasekhar mass models.

2 Nucleosynthesis in Slow Deflagration

For the Chandrasekhar mass white dwarf model, carbon burning in the central region leads to a thermonuclear runway. A flame front then propagates at a subsonic speed as a *deflagration wave* (e.g., Nomoto *et al.* 1996 for a review). Here the major uncertainty is the flame speed which depends on the development of instabilities of various scales at the flame front. Multi-dimensional hydrodynamical simulations of the flame propagation have been attempted by several groups (see Niemeyer in these Proceedings). These simulations have suggested that a carbon deflagration wave propagates at a speed v_{def} as slow as 1.5 - 3 % of the sound speed v_{s} in the central region of the white dwarf. Though the calculated flame speed is still very preliminary, it is useful to examine nucleosynthesis consequences of such a slow flame speed.

In the deflagration wave, electron capture enhances neutron excess, which depends on both v_{def} and the central density of the white dwarf $\rho_9 = \rho_c/10^9$ g cm^{-3}. We calculate explosive nucleosynthesis for two cases with $\rho_9 = 1.37$ (C) and 2.12 (W) at the thermonuclear runaway. For the slow (S) deflagration, we adopt $v_{\text{def}}/v_{\text{s}} = 0.015$ (WS15, CS15) and 0.03 (WS30). The central region behind the slower deflagration undergoes electron capture for a longer time than in the carbon deflagration model W7 (Nomoto et al. 1984), thereby having significantly reduced Y_e. In general it can be recognized that small burning front velocities lead to steep Y_e-gradient which flatten with increasing velocities. Lower central ignition densities shift the curves up (CS15).

In the abundance distribution, $Y_e \sim 0.46 \approx 26/56$ leads to a large abundance of stable ^{56}Fe (not from ^{56}Ni decay). $Y_e = 0.44$ - 0.46 result also in ^{48}Ca, ^{50}Ti, ^{54}Cr, and ^{58}Fe. Of these nuclei ^{48}Ca with Z/A ≈ 0.42 is only produced if Y_e approaches values close to and smaller than 0.44 (Woosley 1996; Meyer et al. 1996).

3 Nucleosynthesis in Delayed Detonation

The deflagration might induce a detonation at low density layers as in the *delayed detonation* model (Khokhlov 1991; Woosley & Weaver 1994). We study explosive nucleosynthesis with a large reaction network (Thielemann et al. 1996b) for various delayed detonation (DD) models, which has not been discussed in detail before (Nomoto et al. 1996). We artificially transform the slow deflagration WS15 into detonation when the density ahead of the flame decreases to 3.0, 2.2, and 1.7 $\times$ 10^7 g cm^{-3} (WDD3, WDD2, and WDD1), respectively. Then the carbon detonation propagates through the layers with $\rho < 10^8$ g cm^{-3}. The explosion energy of three WDD models is 1.5 $\times$ 10^{51} ergs s^{-1} and the mass of ^{56}Ni is 0.73 $M_{\odot}$ (WDD3), 0.58 $M_{\odot}$ (WDD2), and 0.45 $M_{\odot}$ (WDD1).

4 Yields of Type Ia Supernovae

Total isotopic compositions of WDD2 are compared with the solar abundances. We note:

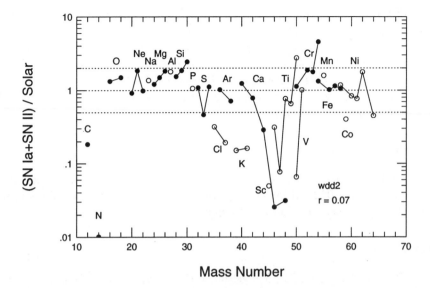

Figure 1: Solar abundance pattern based on synthesized heavy elements from a composite of SNe Ia and SNe II explosions with the most probable ratio. Relative abundances of synthesized heavy elements and their isotopes normalized to the corresponding solar abundances are shown by circles. Here WDD2 is adopted for the SNe Ia model.

1) Among the WDD models, WDD2 produces almost the same amount of ^{56}Ni as W7 ($\sim 0.6\ M_\odot$) but more Si-Ca than W7 by a factor of ~ 2, since more oxygen is burned in the outer layers.

2) Some neutron-rich species such as ^{54}Cr and ^{50}Ti are overproduced with respect to ^{56}Fe. To see the degree of overproduction, we combine nucleosynthesis products of SNe Ia and SNe II with various ratios and compare with solar abundances of heavy elements and their isotopes.

Nucleosynthesis products of SNe II as a function of stellar masses are taken from the calculations by Nomoto & Hashimoto (1988), Hashimoto et al. (1989, 1996) and Thielemann et al. (1996a) as summarized in Tsujimoto et al. (1995). SNe II yields are integrated over $m_l = 10\ M_\odot$ to $m_u = 50\ M_\odot$ with the Salpeter IMF. Nucleosynthesis products of SNe Ia are those from WDD2.

Aided with a reasonable chemical evolution model (Yoshii et al. 1996), the number of SNe Ia ever occurred relative to SNe II is determined to be $N_\mathrm{Ia}/N_\mathrm{II} = 0.12$ in order to reproduce the solar abundances as seen in Figure 1. This is consistent with the fact that the observed estimate of SNe Ia frequency is as low as 10 % of the total supernova occurrence (van den Bergh & Tammann 1991; Tsujimoto et al. 1995).

With this relative frequency, ^{56}Fe from SNe Ia is about 50 % of total ^{56}Fe. For WDDs, ^{54}Cr and ^{50}Ti are overproduced as seen in Figure 1.

The above results imply that the central density of the Chandrasekhar mass white dwarf at thermonuclear runaway must be as low as $\lesssim 2 \times 10^9$ g cm^{-3}. Such a low central density can be realized by the accretion as fast as $\dot{M} > 1 \times 10^{-7} M_\odot$ yr^{-1}. Such rapidly accreting white dwarfs might correspond to the super-soft X-ray sources.

Acknowledgments

This work has been supported in part by the grant-in-Aid for Scientific Research (05242102, 06233101) and COE research (07CE2002) of the Ministry of Education, Science, and Culture in Japan.

References

1. Branch, D., Livio, M., Yungelson, L.R., Boffi, F.R., & Baron, E. 1995, PASP 107, 717
2. Hashimoto, M., Nomoto, K., & Thielemann, F.-K. 1989, A&A 210, L5
3. Hashimoto, M., Nomoto, K., Tsujimoto, T., & Thielemann, F.-K. 1996, in IAU Colloquium 145, Supernovae and Supernova Remnants, ed. R. McCray & Z. Wang (Cambridge University Press), p. 157
4. Khokhlov, A.M. 1991, A&A 245, 114
5. Meyer, B.S., Krishnan, T.D., & Clayton, D.D. 1996, ApJ 462, 825
6. Nomoto, K., & Hashimoto, M. 1988, Phys. Rep. 163, 13
7. Nomoto, K., Thielemann, F.-K., & Yokoi, K. 1984, ApJ 286, 644
8. Nomoto, K., Iwamoto, K., Nakasato, N., Thielemann, F.-K., Brachwitz, F., Young, T., Shigeyama, T., Tsujimoto, T., & Yoshii, Y. 1996b, in Thermonuclear Supernovae, ed. R. Canal et al. (NATO ASI, Kluwer), p. 349
9. Thielemann, F.-K., Nomoto, K., & Hashimoto, M. 1996a, ApJ 460, 408
10. Thielemann, F.-K., Nomoto, K., Iwamoto, K., & Brachwitz, F. 1996b, in Thermonuclear Supernovae, ed. R. Canal et al. (NATO ASI, Kluwer).
11. Tsujimoto, T., Nomoto, K., Yoshii, Y., Hashimoto, M., Yanagida, S., & Thielemann, F.-K., 1995, MN 277, 945
12. van den Bergh, S., & Tammann, G. 1991, ARA&A 29, 363
13. Woosley, S.E. 1996, preprint
14. Woosley, S.E., & Weaver, T.A. 1994, in Supernovae (Les Houches, Session LIV), ed. J. Audouze et al. (Elsevier Sci. Publ.) p. 63
15. Yoshii, Y., Tsujimoto, T., & Nomoto, K. 1996, ApJ 462, 266

NUCLEAR PHYSICS AND ASTROPHYSICS OF THE R-PROCESS

WOLFGANG HILLEBRANDT

Max-Planck-Institut für Astrophysik, Karl-Schwarzschild-Straße 1,
D-85740 Garching, Germany

A summary of recent progress towards a better understanding of r-process nucleosynthesis is given. In particular, new results concerning solar-system r-
abundances as well as abundances observed in very metal-poor stars are discussed
in the context of searches for constraints on the still rather uncertain nuclear
physics input data and the astrophysical models.

1 Introduction

It is a common belief that the neutron-rich isotopes of heavy elements are
produced by a very intense and short neutron burst occuring, presumably,
in a core-collapse supernova. This belief rests on the fact that in the course
of hydrostatic stellar evolution most isotopes shielded from β-stable ones by
a short-lived nucleus cannot be reached by a sequence of neutron captures.
Morover, the observed abundance maxima at $A = 80, 130$ and 195 indicate
(interpreted as being due to closed neutron shells) that these nuclei were formed
very far from stability. In order to prove or disprove this general concept we
have to know the answers to several questions which are addressed in the
following sections.

2 What are the (solar system) r-process abundances?

The general procedure to determine r-process abundances is to subtract from
observed ones s- (and p-) process contributions which are thought to be better understood. The residuals are attributed to the r-process and attempts
are made to fit them by certain models. The most widely used data set is
due to Käppeler et al.[1] and is based on an large number of n-capture cross-
sections measured by the Karlsruhe group, and a careful analysis of s-process
conditions obtained from certain branchings where n-captures and β-decays
compete. Assuming a 2-component s-process modelled by a set of two exponential exposures, for s-only isotopes they could fit the solar (σN_s)-relation
very well. The same set was then used to get the r-process residuals.

Although this approach seems to be conclusive, some aspects introduce
uncertainties. Besides the question what exactly the "solar system" abun-

dances are and how universal (or special) they can be assumed to be another difficulty is that with the exception of nuclei at the peaks s-abundances are not much different from the total, and after subtraction the uncertainties of the residuals can be large. In fact, Wisshak *et al.*[2] have presented a new compilation of r-process residuals based on n-capture cross-sections obtained with the new Karlsruhe 4π BaF_2 detector and more recent solar abundances in the range from $A = 110$ to 170 and found rather big changes. In general, the new r-abundance curve is more smooth then before and the r-contributions to the Sn and Nd isotopes have come down by up to a factor of 8. Interestingly, for elements beyond Ba the elemental r-abundances are not much changed. Therefore the conclusion remains valid that in the extremely metal-poor star CS22892-052 the abundances resemble a pure r-process[3,4].

A second uncertainty relates to the assumption of a 2-component s-process, because it is by no means clear that the solar s-process was indeed the result of a superposition of two exponential exposures. Goriely[5] has recently investigated this question. He made fits to the solar (σN_s)-curve allowing for more general distributions of neutron exposures. He could show that, depending on the nuclei he incorporated in his fit, several nuclei do not even need r-process contributions anymore. Again, this result should be taken as a warning not to overemphasize details of the commonly accepted r-process abundances.

3 Do we know the nuclear physics data sufficiently well?

In the canonical model of the r-process it is assumed that for most of its duration (n, γ) and (γ, n) reactions are in equilibrium, and that the time-scale is set entirely by the (slow) β^--decays. For an appropriate combination of neutron densities and temperatures one obtaines the correct abundace peaks.

The nuclear input needed for the canonical model includes nuclear masses (or reaction Q-values), single particle levels to compute (n, γ) -rates during freeze-out, β -decay rates and, possibly, inelastic neutrino-scattering cross-sections if the presently favored scenario of a neutrino-heated hot bubble on the surface of a newly-born neutron star is the site, and all these data have to be known for some 2000 unstable nuclei in total.

The most important nuclear masses are those near closed neutron shells. In fact, it has been argued that the "observed" solar r-process abundances give evidence for shell-quenching for very neutron-rich nuclei[6]. But given the still existing uncertainties both in the abundances and in the astrophysical models, this conclusion may not be secure. Certainly, the question of shell-quenching is an important one. It is related to the isospin dependence of the spin-orbit interaction and it should be answered by nuclear physics[7].

Similar concerns are in order for single-particle levels and β-decay rates. It is probably fair to state that none of the existing nuclear models can predict single-particle levels for nuclei beyond the sd-shell and in particular for those with low neutron separation energies, and that the uncertainties in the β-decay half-lives are still of the order of a factor of 10 or more[8]. One may hope for a new generation of experiments to cure the deficiencies, such as the upgraded ISOLDE facility at CERN or post-accelerators in conjunction with high neutron flux nuclear reactors, but these experiments will give informations for a few selected isotopes only.

4   Do we have good astrophysical models?

In general terms, three kinds of scenarios have been suggested over the years none of which, however, is fully convincing. One can characterize them by their temperature (or entropy) and their lepton fraction (or neutron excess).

Firstly, one can form r-nuclei as secondaries from pre-existing seeds, presumably s-processed matter, at moderate temperatures and rather "normal" y_e. An example is explosive He-burning in a supernova shock passing the He shell af a massive star. However, this model gives mainly nuclei in the A=130 peak but cannot reproduce the others[9]. However, there is room for improvements. For example, massive stars are rapid rotators and rotationally induced mixing of H into the He-shell would encrease the ^{13}C content and thereby the neutron flux. Modifications of this sort should be analysed carefully.

Secondly, and this scenario is now favored, one can think of a high entropy, moderately low y_e environment where the neutron density is not hugh but due to the high entropy also the abundance of (in-situ produced) seed nuclei is low. Here the "hot bubble" created by an intense neutrino flux near a newly born neutron star is the standard example. However, as was shown in various numerical studies, supernova models do not give entropies high enough to make it work[10,11]. Short-comings include enormous overproductions of Sr and Zr, and the failure to reach the A=195 peak. Again, there are necessary modifications. Until recently, inelastic neutrino interactions have been ignored during the operation of the r-process in the hot bubble and after freeze-out. This, however, does not seem to be justified[12]. Moreover, the mean neutrino energies obtained in core-collapse supernova models appear to be too low for powerful explosions[13]. Therefore the difference between $\bar{\nu}_e$ and ν_e energies (which determine y_e) may also not be predicted correctly.

Finally, one can think of expanding low entropy, low y_e matter, found in the crusts of neutron stars[14,15,16]. The "cold decompression" idea has never been pursued intensively because of the lack of realistic models. This situation

588

has changed, however. Neutron star mergers are discussed because they might serve as sources of gravity waves and also as candidates for γ -ray bursters, and it was found that during the merging indeed some neutron-rich matter will be ejected with fairly low entropy[17]. Such a scenario predicts that mainly the very heavy r-nuclei form, in contrast to what was found for most of the other models.

5 Conclusions

At present there is no unique answer to the questions where and how the n-rich isotopes of the heavy elements formed. Nevertheless there are many sites known where a r-like nucleosynthesis does happen. Maybe, the solution to the problem is that there is no unique site, but that we observe in the solar system the result of mixing and superposition of different sites. A test of this hypothesis is to find in very metal-poor stars r-nuclei below Ba and to show that their abundance pattern is rather non-solar. Also, it would be useful to incorporate those sites which are reasonably well understood into a chemical evolution code and to see what fraction of the r-process abundances is already reproduced by them. Only the residuals might call for an exotic explanation.

1. F. Käppeler, H. Beer and K. Wisshak, *Rep.Prog.Phys.* **52**, 945 (1989).
2. K. Wisshak, F. Voss, F. Käppeler, in *8th Workshop on Nuclear Astrophysics*, eds. W. Hillebrandt, E., Müller (MPA/P9, 1996), p 16.
3. C. Sneden *et al.*, *ApJ* **467**, 819 (1996).
4. J.J. Cowan *et al.*, *ApJ* **460**, L115 (1996).
5. S. Goriely, in *NIC-96*, *Nucl. Phys. A*, in press.
6. K.-L. Kratz *et al.*, *ApJ* **403**, 216 (1993).
7. G.A. Lalazissis *et al.*, in *Proc. ENAM-95*, eds. M. de Saint Simon, O. Sorlin, (Edition Frontieres, 1995), p.71.
8. I.N. Borzov, S. Goriely and J.M. Pearson, in *NIC-96*, *Nucl. Phys. A*, in press.
9. J.J. Cowan, A.G.W. Cameron and J.W. Truran, *ApJ* **265**, 429 (1983).
10. K. Takahashi, J. Witti and H.-Th. Janka, *A & A* **286**, 857 (1994).
11. S.E. Woosley *et al.*, *ApJ* **433**, 229 (1994).
12. Y.-Z. Qian *et al.*, preprint nucl-th/9611010, *Phys. Rev. D*, in press.
13. A. Mezzacappa *et al.*, preprint astro-ph/9612107, *ApJ*, submitted.
14. J.M. Lattimer and D.N. Schramm, *ApJ* **192**, L145 (1974).
15. E.R. Hilf *et al.*, *Phys.Scripta* **10A**, 132 (1974).
16. M. Colpi, S.L. Shapiro and S.A. Teukolsky, *ApJ* **414**, 717 (1993).
17. M. Ruffert *et al.*, preprint MPA957, *Astron. Astrophys.*, in press.

THE RADIUS OF THE NEUTRON STAR RXJ185635-3754 AND IMPLICATIONS FOR THE EQUATION OF STATE

JAMES M. LATTIMER

Department of Earth & Space Sciences
State University of New York, Stony Brook, NY 11794-2100, USA

MADAPPA PRAKASH

Department of Physics
State University of New York, Stony Brook, NY 11794-3800, USA

Combined Hubble optical and Rosat X-ray data of the bright soft X-ray source RXJ185635-3754 are consistent with thermal emission from an isolated (non-pulsing, single) neutron star. The stellar radius is apparently limited to $R < 9$ km, which imposes severe constraints upon the equation of state.

1 Observations

Walter, Wolk and Neuhäuser[1] identified the bright soft X-ray source RXJ185635-3754 as a nearby isolated neutron star. The X-ray source is nearly blackbody with $T_\infty \approx 57$ eV and little extinction ($A_V \approx 0.07$ mag). The observed X-ray flux and temperature, if originating from a uniform blackbody, correspond to an effective stellar radius of $R_\infty = R/\sqrt{1 - 2GM/Rc^2} \approx 7.3 (D/120 \text{ pc})$ km. Here, R is the Schwarzschild r-coordinate stellar radius, normally called the neutron star radius. The low extinction and the location of the star in the direction of the R CrA molecular cloud limits the distance to $D < 120$ pc. The emission is therefore consistent with nearly blackbody radiation from at least part of the surface of a neutron star. The fact that the source is not observable in radio and its lack of variability implies that it is not a pulsar. It may give us the clearest view so far of the surface of a neutron star without complications from non-thermal emission in a magnetosphere. This, coupled with a known upper limit to its distance, severely limits its radius and the equation of state of matter above nuclear density.

In August, 1996, we used the WFPC2 on the Hubble Space Telescope to detect the optical counterpart[2], which has a V magnitude of 25.8 on the Vega system. The object is very blue with F300W-F606W ≤ -1.22 mag. The optical fluxes are about a factor of two above the extrapolated X-ray black body, and the optical flux ratios yield an extinction $A_V \approx 0.5$ mag, incompatible with the X-ray extinction. These observations may suggest the existence of spectral features in the optical similar to those inferred for Geminga[3], but the total optical fluxes can be partially reconciled with the X-ray flux if the neutron star has a temperature differential across its surface. The best-fit model with temperature distribution $T(\theta) = T_p\sqrt{\sin\theta}$ yields $T_p \approx 62$ eV and $R_\infty \approx 11(D/120 \text{ pc})$ km. The best-fit model with two regions of uniform temperature yields nearly equal temperatures and $R_\infty \approx 7.5(D/120 \text{ pc})$ km. This does not fit the optical observations as well as the sinusoidal model. Thus, the reconciliation of the optical and X-ray data seems to imply an increase in R, although the increase is probably smaller than 50% relative to the best-fit X-ray value $R_\infty = 7.3$ km for a distance of $D = 120$ pc.

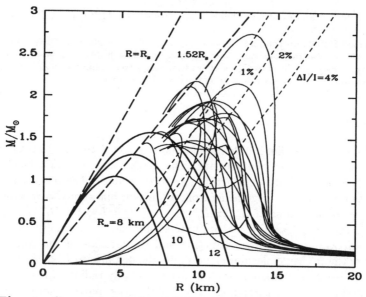

Fig. 1. Curve labeled $R = 1.52R_s$ shows limit imposed by GR + causality. Curves with $R_\infty = 8, 10$ and 12 km are shown, as are $M - R$ curves for several recent equations of state (see text). Also displayed are $M - R$ relations for $\Delta I/I = 1, 2$ and 4 %.

2 Restrictions in the Mass-Radius Diagram

General relativity (GR) restricts the radii of objects with gravitational mass M to be larger than the Schwarzschild radius $R_s = 2GM/c^2 = 2.954(M/M_\odot)$ km. In addition, GR hydrostatic equilibrium coupled with causality non-violation, i.e., $dP/d\rho \leq c^2$, restricts[4] the radius further: $R \geq 1.5R_s$. Coincidentally, the largest permitted mass M for a given value of $R_\infty = R/\sqrt{1 - (R_s/R)}$ occurs when $R = 1.5R_s$ and is $R_{s,max} = (2/3^{3/2})R_\infty$. Since supernova models suggest that neutron stars are formed with $M > 1$ $M_\odot$, this implies $R_\infty > 7.6$ km. This is consistent with the observations of RXJ185635-3754, if a temperature differential across the neutron star's surface is posited. These relations are illustrated in Fig. 1.

The small values apparently allowed for R severely restrict the high-density equation of state. In Fig. 1 are shown typical $M - R$ curves for several equations of state [5], including a potential model, a field-theoretical model, a kaon condensate model and a model containing self-bound strange quark matter. It is apparent that if $R_\infty < 11$ km, this star must have some form of enhanced softening in its high-density matter.

Such a small radius for a neutron star would be inconsistent with many glitch models. The inferred change in the star's moment of inertia I during a glitch is assumed to be at least partially due to the crustal moment of inertia ΔI. $\Delta I/I$ is a function of M and R with only slight sensitivity to the equation of state [6]. In fact, the position of a $\Delta I/I$ contour in $M - R$ space depends chiefly upon the density n_{cc} and enthalpy $\mathcal{H}$ at the core-crust boundary [7]. Results for $\Delta I/I$ are shown in Fig. 1 assuming $n_{cc} = 0.074\,\mathrm{fm}^{-3}$ and $\mathcal{H} = 24.1$ MeV. The largest glitches observed imply $\Delta I/I > 0.01$, which is incompatible with $R_\infty < 10$ km and $M > 1 M_\odot$.

Acknowledgments

This research was supported by NASA grant NAG52863 and by the USDOE grants DOE-FG02-87ER-40317 and DOE-FG02-88ER-40388. We thank Manju Prakash and Penghui An for discussions.

References

1. F.M. Walter, S.J. Wolk and R. Neuhäuser, *Nature* **379**, 233 (1996).
2. F.M. Walter, L.D. Matthews, P. An, J.M. Lattimer and R. Neuhäuser, in *Proceedings of the 18th Texas Symposium*, ed. A. Olinto, J. Frieman, and D. Schramm (World Scientific, Singapore, 1997).
3. G.F. Bignami, P.A. Carravo, R. Mignani, J. Edelstein and S. Bowyer, *Astrophys. J. Lett.* **456**, L111 (1996).
4. J.M. Lattimer, M. Prakash, D. Masak and A. Yahil, *Astrophys. J.* **355**, 241 (1990)
5. M. Prakash, I. Bombaci, P.J. Ellis, Manju Prakash, J.M. Lattimer and R. Knorren, *Phys. Rep.* **280**, 1 (1996).
6. D.G. Ravenhall and C.J. Pethick, *Astrophys. J.* **424**, 846 (1994).
7. J.M. Lattimer and M. Prakash, In preparation (1997).

BINARY NEUTRON STARS IN QUASI-EQUILIBRIUM CIRCULAR ORBIT: A FULLY RELATIVISTIC TREATMENT

T. W. BAUMGARTE, S. L. SHAPIRO

Department of Physics. University of Illinois at Urbana-Champaign, Urbana, Il 61801

G. B. COOK, M. A. SCHEEL and S. A. TEUKOLSKY

Center for Radiophysics and Space Research, Cornell University, Ithaca, NY 14853

We present a numerical scheme that solves the initial value problem in full general relativity for a binary neutron star in quasi-equilibrium. While Newtonian gravity allows for a strict equilibrium, a relativistic binary system emits gravitational radiation, causing the system to lose energy and slowly spiral inwards. However, since inspiral occurs on a time scale much longer than the orbital period, we can adopt a quasi-equilibrium approximation. In this approximation, we integrate a subset of the Einstein equations coupled to the equations of relativistic hydrodynamics to solve the initial value problem for binaries of arbitrary separation, down to the innermost stable orbit.

Neutron star binaries are of interest for several reasons. They exist, even within our own galaxy, and are among the most promising sources for gravitational wave detectors like LIGO, VIRGO and GEO. More fundamentally the two-body problem is one of the outstanding unsolved problems in classical general relativity.

So far most researchers have treated binary neutron stars in Newtonian theory[1]. In post-Newtonian treatments[2] the stars are usually treated as point-sources, so that hydrodynamical effects are absent. More recently, Nakamura[3] and Wilson and Mathews[4] have initiated studies of binary neutron stars in general relativity.

In our work we assume that the two stars have equal mass, are co-rotating and obey a polytropic equation of state, $P = K\rho_0^\Gamma$. In Newtonian gravity, a strict equilibrium solution for two stars in circular orbit can be found. Since this solution is stationary, the hydrodynamical equations reduce to the Bernoulli equation,

$$\frac{\Gamma}{\Gamma - 1}\frac{P}{\rho_0} + \Phi - \frac{1}{2}\Omega^2(x^2 + y^2) = C, \qquad (1)$$

where C is a constant, Φ the gravitational potential, Ω the angular velocity, and where the rotation is about the z-axis. The gravitational potential satisfies Poisson's equation, $\nabla^2\Phi = 4\pi\rho_0$. These equations comprise a coupled system, containing a linear elliptic PDE in 3D, which must be solved numerically.

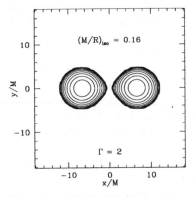

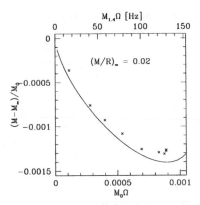

Figure 1: Rest density contours of a close, highly relativistic neutron star binary in the equatorial plane. The contours span densities logarithmically between the central density and 1 percent of that value.

Figure 2: Binding energy of a $\Gamma = 2$ polytrope. The solid line is a result of a post-Newtonian calculation for compressible ellipsoids [5]; crosses are the result of this work.

Because of the emission of gravitational waves, a binary in general relativity cannot be in strict equilibrium. However, up to the innermost stable circular orbit (ISCO) the timescale for orbital decay by radiation will be much longer than the orbital period so that the binary can be considered to be in "quasi-equilibrium". This allows us to neglect both gravitational waves and wave-induced deviations from a circular orbit to good approximation.

To minimize the gravitational wave content we choose the 3-metric to be conformally flat [6]. The field equations then reduce to a set of coupled, quasi-linear elliptic equations for the lapse, shift, and the conformal factor. Neglecting small deviations from circular orbit, the fluid flow is again stationary, and the hydrodynamical equations reduce to the relativistic Bernoulli equation. Solving these equations yields a valid solution to the initial value (constraint) equations and an approximate solution to the full Einstein equations at any given moment, prior to plunge.

As in the Newtonian case, a system of coupled elliptic equations must be solved, and since the two problems have a very similar structure, they can both be solved with very similar numerical methods. We have developed parallel FAS multigrid solvers for both applications. Because of the symmetries of the problem, is is sufficient to work in one octant. The codes are written in cartesian coordinates and use the DAGH infrastructure that has been developed as part of the Binary Black Hole Grand Challenge project. For code development we typically run on 8 processors on the IBM SP2 parallel cluster at Cornell. We use up to 5 levels of refinement, which gives a $(64)^3$ grid on the finest level. The matter is covered by about 20 gridpoints in each direction.

We are interested in quasi-equilibrium models in their own right, but we also plan to use the models as initial data for fully relativistic evolution codes. We show a density profile of a close neutron star binary in Figure 1. In isolation, each star would have a compaction of $(M/R)_\infty = 0.16$, showing that this configuration is highly relativistic. The maximum mass configuration for this Γ satisfies $(M/R)_\infty = 0.22$.

We construct quasi-equilibrium sequences for binaries of fixed rest mass. Up to the ISCO, these sequences approximate evolutionary sequences. We plot in Figure 2 the binding energy $(M - M_\infty)/M$ versus the separation, parameterized by the angular velocity, for a mildly relativistic sequence $((M/R)_\infty = 0.02)$. The turning point of this curve indicates the onset of orbital instability at the ISCO and the angular velocity there. Note that we are restricted to co-rotating sequences. Sequences of conserved circulation are probably more realistic, since maintaining co-rotation would require excessive viscosity [7].

In the near future we plan to implement adaptive mesh refinement (AMR) and increase the accuracy of our calculation. We will then explore the physics of fully relativistic binary neutron stars for different polytropic indices, separations, and values of $(M/R)_\infty$.

1. See, for example: I. Hachisu and Y. Eriguchi, *Publ. Astron. Soc. Japan* **36**, 239 (1984); M. Shibata, T. Nakamura and K. Oohara, *Prog. Theor. Phys.* **88**, 1079 (1992); F. A. Rasio and S. L. Shapiro, *Ap. J.* **401**, 226 (1992); X. Zughe, J. M. Centrella, S. L. W. McMillan, *Phys. Rev.* D **50**, 6247 (1994); M. Ruffert, H.-T. Janka and G. Schäfer, *Astrophys. Sp. Sci.* **231**, 423 (1995);

2. L. Blanchet, T. Damour, B. R. Iyer, C. M. Will and A. G. Wiseman, *Phys. Rev. Lett.* **74**, 3515 (1995) and references therein

3. T. Nakamura, *in Proceedings of the Fourth Workshop on General Relativity and Gravitation*, edited by K. Nkao et. al., pp. 302, 1994

4. J. R. Wilson and G. J. Mathews, *Phys. Rev. Lett.* **75**, 4161 (1995)

5. J. C. Lombardi, F. A. Rasio and S. L. Shapiro, 1997, in preparation

6. J. R. Wilson and G. J. Mathews, in *Frontiers in Numerical Relativity*, edited by C. R. Evans, L. S. Finn, and D. W. Hobill (Cambridge University Press, Cambridge, England, 1989), pp. 306; G. B. Cook, S. L. Shapiro and S. A. Teukolsky, *Phys. Rev.* D **53**, 5533 (1996)

7. C. S. Kochanek, *Ap. J.* **398**, 234 (1992); L. Bildsten and C. Cutler, *Ap. J.* **400**, 175 (1992)

LESSONS FROM 3D HYDRODYNAMIC CALCULATIONS OF BINARY NEUTRON STAR COALESCENCE

FREDERIC A. RASIO

Dept of Physics, M.I.T. 6-201, Cambridge, MA 02139

The main lessons learned from several years of Newtonian (and post-Newtonian) numerical hydrodynamic calculations of neutron-star binary coalescence are briefly summarized and some important test calculations are suggested for fully relativistic 3D hydrodynamic codes now being developed.

1 Summary of Recent Work

Coalescing binary neutron stars (NS) are important sources of gravitational waves that should become detectable with the laser interferometers now being built as part of the LIGO, VIRGO and GEO projects. For recent reviews and references on the detection and sources of gravitational radiation, see Thorne[16].

Post-Newtonian (PN) approximation methods have been used to calculate waveform templates in the low-frequency, slow-inspiral phase of the binary evolution [3]. In this phase the two stars are still well separated and can be treated as point masses. Near the end of the coalescence, however, hydrodynamic effects and the interior structure of the stars play an increasingly important role. Hydrodynamics becomes dominant when the two stars finally merge together into a single object. The shape of the corresponding burst of gravitational waves provides a direct probe into the interior structure of a NS and the nuclear equation of state (EOS). In the Newtonian limit, the inspiral phase is quasi-hydrostatic, and the final merging is driven entirely by a global hydrodynamic instability caused by tidal effects [11]. After a brief episode of mass shedding, most of the fluid settles down into a new, nearly axisymmetric, hydrostatic equilibrium configuration [10].

Numerical hydrodynamic calculations of binary NS coalescence have now been performed by a number of different groups, using a variety of numerical methods. Unfortunately, up until very recently, no direct comparison between the different published results was possible because different groups focused on different aspects of the problem. Nakamura and collaborators[8] were the first to perform 3D calculations of binary NS coalescence, using a traditional Eulerian finite-difference code. Rasio and Shapiro[10][11] have been using the Lagrangian method SPH and have focused on determining the stability properties of initial binary models in strict hydrostatic equilibrium and calculating the emission of gravitational waves from the coalescence of unstable binaries. Many of their results have now been independently confirmed in the work of New [9], who used completely different numerical methods but also focused on stability questions. Zhuge et al. [17] have also used SPH and studied the dependence of the gravitational waveforms on the initial NS spins. Davies et al. [4] and Ruffert et al. [12] have incorporated a treatment of the nuclear physics in their calculations (done using SPH and PPM codes, respectively) and focus on NS mergers as sources of gamma-ray bursts.

For NS binaries, and particularly if the NS EOS is fairly soft, relativistic effects combine nonlinearly with Newtonian tidal effects so that close binary configurations can become dynamically unstable earlier during the spiral-in phase (i.e., at larger binary separation and lower orbital frequency) than predicted by Newtonian hydrodynamics alone. The combined effects of relativity and tidal interactions on the stability of close compact binaries have only very recently begun to be studied. Preliminary results have been obtained using both analytic approximations [15] [6] [7] (basically PN generalizations of the work done by Lai et al. [5]) as well as numerical 3D calculations incorporating simplified treatments of relativistic effects [18] [13] [1]. A NASA Grand Challenge project is under way [14] that will ultimately attempt a fully relativistic calculation of the final coalescence, combining the techniques of numerical relativity and numerical hydrodynamics in 3D.

2 Main Lessons for Future Fully Relativistic Calculations

In the context of future 3D relativistic calculations, some important lessons to be remembered from existing Newtonian and PN calculations are as follows.

- **Relevance of the Newtonian limit**. For stiff NS EOS (NS radius $R \sim 12 - 15$ km, $M/R \sim 0.1 - 0.2$), relativistic effects may in fact remain relatively unimportant during the entire hydrodynamic merger. Thus it should be kept in mind that fully relativistic codes may not be entirely necessary to calculate NS mergers numerically. Much simpler PN codes may be sufficient, even to obtain quantitatively accurate results.

- **Importance of the Newtonian limit as a test case**. Even if they were not so relevant for real NS binaries (e.g., because the NS EOS turned out to be rather soft), Newtonian or PN results would remain important to test relativistic codes. These codes, when run for very large R/M (cf. Ref. 1), should be able to reproduce all the features of the Newtonian results. In particular, they should be able to maintain binaries in stable quasi-equilibrium circular orbits at large binary separations $r \gg R$, and they should be able to identify the onset of dynamical instability at $r/R \sim 3$.

- **Sensitivity to the mass ratio**. Newtonian calculations[10] have revealed a high sensitivity of the gravitational waveforms to the binary mass ratio, even for very small departures from the equal-mass case. Thus, although the equal-mass case may seem to be the most natural test case, it is *not* a realistic case.

- **Synchronized vs Nonsynchronized Binaries**. Hydrodynamic mergers of initially *nonsynchronized* binaries are far more challenging to calculate numerically, compared to mergers of initially synchronized (uniformly rotating) NS. This is because an unstable vortex sheet can form at the interface between the two stars during the merger[10]. This leads to small-scale Kelvin-Helmholtz instabilities and turbulence, which are basically intractable (particularly in 3D). Therefore, although they may not be the most realistic[2], synchronized initial conditions should be preferred, at least for test calculations.

1. T. Baumgarte, et al. in this volume.
2. L. Bildsten and C. Cutler, *ApJ* **400**, 175 (1992).
3. L. Blanchet et al., *Class. Quant. Grav.* **13**, 575 (1996).
4. M.B. Davies et al. *ApJ* **431**, 742 (1994).
5. D. Lai, F.A. Rasio, and S.L. Shapiro, *ApJ* **437**, 742 (1994). [LRS]
6. D. Lai and A.G. Wiseman, *Phys. Rev.* D **54**, 3958 (1997).
7. J.C. Lombardi, F.A. Rasio, and S.L. Shapiro, submitted to ApJ (1997).
8. T. Nakamura, in *Relativistic Cosmology*, ed. M. Sasaki (Universal Academy Press, 1994), and references therein.
9. K. New, PhD Thesis, Louisiana State University (1996).
10. F.A. Rasio and S.L. Shapiro, *ApJ* **432**, 242 (1994).
11. F.A. Rasio and S.L. Shapiro, in *Compact Stars in Binaries*, Proc. of IAU Symp. 165, eds. J. van Paradijs et al. (Kluwer, Dordrecht, 1996).
12. M. Ruffert, H.-T. Janka, and G. Schäfer, *A&A* **311**, 532 (1996).
13. M. Shibata, *Prog. Theor. Phys.* **96**, 317 (1996).
14. W.-M. Suen, in this volume.
15. K. Taniguchi and T. Nakamura, *Prog. Theor. Phys.* **96**, 693 (1996).
16. K.S. Thorne, in *Compact Stars in Binaries*, Proc. of IAU Symp. 165, eds. J. van Paradijs et al. (Kluwer, Dordrecht, 1996).
17. X. Zughe et al., *Phys. Rev.* D **50**, 6247 (1994); **54**, 7261 (1996).
18. J.R. Wilson et al., *Phys. Rev.* D **54**, 1317 (1996).

Numerical General-Relativistic Hydrodynamics: Riemann Solvers

José Mª. Ibáñez[1], José A. Font[2], José Mª. Martí[1] and Juan A. Miralles[1]

[1] *Departament d'Astronomia i Astrofísica*
Universitat de València, E-46100 Burjassot (València), Spain
[2] *Max-Planck-Institut für Gravitationsphysik*
Schlaatzweg 1, D-14473 Potsdam, Germany

The network of gravitational-wave laboratories under construction in Europe (GEO600, VIRGO), Japan (TAMA) and USA (LIGO) encourages theorists to continue developing powerful analytical and numerical tools allowing the correct interpretation of the gravitational waveforms coming from astrophysical (and cosmological) sources of gravitational radiation. The accurate knowledge of the features of the gravitational radiation released in astrophysical events, such as the stellar core collapse or the coalescing compact binaries at the final merging stage, requires the state-of-the-art in numerical relativity and numerical hydrodynamics. Strong shocks forming and interacting in relativistic flows, which are evolving in a dynamical spacetime, are common to the above astrophysical scenarios. In this contribution we summarize the research carried out in our group in order to extend modern high-resolution shock-capturing schemes (HRSC) –well experimented in Newtonian hydrodynamics– to the multidimensional relativistic (special and general) hydrodynamics as well as the present status of these techniques and further developements.

In the pioneering works by May & White (1967) and Wilson (1971,1972), which laid the foundations of numerical relativistic hydrodynamics, the authors used classical numerical techniques of computational fluid dynamics (CFD). Since then, the field of CFD has evolved extraordinarily and, in particular, a special branch is, nowadays, devoted to exploit the hyperbolic character of the equations (Godunov-type or HRSC methods). To take advantage of the conservation properties of the system, modern HRSC algorithms are written in *conservation form*, in the sense that –in the absence of sources– one gives the variation of the mean values of the conserved quantities, within the numerical cells, by the fluxes accross the cell boundaries. Furthermore, these fluxes can be obtained from solutions of discontinuous initial problems (*Riemann problems*) between neighboring numerical cells. In this way, physical discontinuities appearing in the flow are treated consistently (*shock-capturing property*). The use of the solution of Riemann problems in numerical codes comes from the original idea of Godunov (1959). It was not until the late seventies when, thanks to the development of *cell-reconstruction* procedures (see references in, e.g., Martí 1997), HRSC techniques were recognized as the most effective way to describe complex flows accurately. Since then, efficient *Riemann solvers* based on exact or approximate solutions of the initial value problem have been developed.

In the frame of relativistic fluid dynamics, the use of HRSC techniques is more recent and still does not cover the full set of possible astrophysical applications. Nowadays, half a dozen special relativistic Riemann solvers (SRRS) have been proposed. A provisional classification is (Martí 1997): 1) *The Exact Riemann Solver* (Martí and Müller, 1994). This is the analogous one to the original Godunov's Riemann Solver for the Newtonian case. It is interesting not only from the purely theoretical point of view but also from the practical one. It allows one to build the exact solution of any relativistic Riemann problem and, consequently, to check the numerical solution obtained with a hydrocode. 2) *Linearized Riemann Solvers* which are based on the original work by Roe. 3) *Other SRRS* are the natural extension of the classical ones: i) Harten-Lax-van Leer Method, ii) Colella's two-shock approximation, iii) Glimm's method.

In Martí, Ibáñez and Miralles (1991) we proposed how to extend HRSC to the relativistic case (in one-dimensional calculations) and made use of a linearized SRRS based on the *spectral decomposition* of the Jacobian matrices of the system. In Font et al. (1994) we extended them to multidimensional special relativistic flows. The extension of HRSC methods to multidimensional general relativistic hydrodynamics has been accomplished in Banyuls et al.(1997); it has been carried out in the $\{3 + 1\}$ *formalism* well suited for the solution of the Einstein field equations and by defining the *conserved variables* according to measurements by Eulerian observers.

In the last years we have carried out an exhaustive study of the performances of our numerical procedure for describing relativistic flows (for a summary see Martí 1997). In Banyuls et al. (1997), numerical experiments of standard 1D shock-tube problems in Minkowski space-time have been extended to multidimensional problems, involving more complex metrics, by means of changes of coordinates. A battery of numerical experiments (relativistic shock-tube tests, relativistic spherical shock reflection, Oppenheimer-Snyder collapse, spherical accretion onto a compact object, detection of the zero value for the fundamental mode against radial oscillations of a compact object, etc) allows us to assure the quality of our procedure. Present astrophysical applications have mainly been made in the simulation of extragalactic relativistic jets, spherically symmetric general-relativistic stellar core collapse and non-spherical accretion onto a Schwarzschild black hole (see references in Banyuls et al., 1997).

To end let us conclude that a general relativistic hydrocode –such as the

600

one having the main features described here– can be used to solve numerically the evolution of matter in a *dynamical spacetime*, when *appropiately coupled with Einstein equations*. A very promising approach might be the one of solving at each timestep – of a given spacetime foliation– the matter part keeping the geometry as a background at that timestep. An iterative procedure would warranty the accuracy of this approach. Such a marriage between numerical relativity and numerical relativistic hydrodynamics could be useful to deepen into the knowledge of the dynamics (and the physics) of, for example, coalescing compact binaries. Our study on the spherically symmetric general relativistic stellar core collapse allow one to be confident in that such a coupling be *attainable* in the future.

Acknowledgments

This contribution relies on extensive analytical and numerical work done in collaboration with F.Banyuls, A.Marquina and J.V.Romero. This work has been supported by the Spanish DGICYT (grant PB94-0973). José M$^{\underline{a}}$. Martí has benefited from a European contract (nr. ERBCHBICT930496).

References

1. F. Banyuls, J.A. Font, J. M$^{\underline{a}}$. Ibáñez, J. M$^{\underline{a}}$. Martí and J.A. Miralles, *Astrophys. J.* , in press (1997).
2. J.A. Font, J.M$^{\underline{a}}$. Ibáñez, A. Marquina and J. M$^{\underline{a}}$. Martí, *Astronomy and Astrophysics* **282**, 304 (1994).
3. S.K. Godunov, Mat. Sb. 1959, 47, 271 *Mat. Sb.* **47**, 271 (1959).
4. J. M$^{\underline{a}}$. Martí in *Relativistic Gravitation and Gravitational Radiation*, ed. J-A. Marck and J-P. Lasota (Cambridge University Press, Cambridge, UK, 1997).
5. J. M$^{\underline{a}}$. Martí, J.M$^{\underline{a}}$. Ibáñez and J.A. Miralles, *Phys. Rev.* D **43**, 3794 (1991).
6. J. M$^{\underline{a}}$. Martí and E. Müller, *J. Fluid Mech.* **258**, 317 (1994).
7. M.M. May and R.H. White, *Math.Comp.Phys.* **7**, 219 (1967).
8. J.V. Romero, J. M$^{\underline{a}}$. Ibáñez, J. M$^{\underline{a}}$. Martí and J.A. Miralles, *Astrophys. J.* **462**, 839 (1996).
9. J.R. Wilson, *Astrophys. J.* **163**, 209 (1971).
10. J.R. Wilson, *Astrophys. J.* **173**, 431 (1972).

HYPERBOLIC FORMULATION OF GENERAL RELATIVITY

ANDREW ABRAHAMS[a], ARLEN ANDERSON, YVONNE
CHOQUET-BRUHAT[b], and JAMES W. YORK, JR.

Dept. Physics and Astronomy
Univ. North Carolina
Chapel Hill, NC 27599-3255 USA

Two geometrical well-posed hyperbolic formulations of general relativity are described. One admits any time-slicing which preserves a generalized harmonic condition. The other admits arbitrary time-slicings. Both systems have only the physical characteristic speeds of zero and the speed of light.

Einstein's theory, viewed mathematically as a system of second-order partial differential equations for the metric, is not a hyperbolic system without modification and is not manifestly well-posed, though physical information propagates at the speed of light. A well-posed hyperbolic system admits unique solutions which depend continuously on the initial data and seems to be required for robust, stable numerical integration. The well-known traditional approach achieves hyperbolicity through special coordinate choices. The formulations[1] described here permit much greater coordinate gauge freedom. Because these exact nonlinear theories incorporate the constraints, they are natural starting points for developing gauge-invariant perturbation theory.

Consider a globally hyperbolic manifold of topology $\Sigma \times R$ with the metric

$$ds^2 = -N^2 dt^2 + g_{ij}(dx^i + \beta^i dt)(dx^j + \beta^j dt), \qquad (1)$$

where N is the lapse, β^i is the shift, and g_{ij} is the spatial metric. The derivative, $N^{-1}\hat{\partial}_0 = N^{-1}(\partial/\partial t - \mathcal{L}_\beta)$, where $\mathcal{L}_\beta$ is the Lie derivative in a time slice Σ along the shift vector, is the derivative with respect to proper time along the normal to Σ. This implies that $\hat{\partial}_0$ is the natural time derivative for evolution. The extrinsic curvature K_{ij} of Σ is defined as

$$\hat{\partial}_0 g_{ij} = -2N K_{ij}. \qquad (2)$$

Together with (2), the dynamical part of Einstein's theory (in vacuum) can be expressed in 3+1 language as

$$R_{ij} = -N^{-1}\hat{\partial}_0 K_{ij} + J_{ij} + \bar{R}_{ij} = 0, \qquad (3)$$

[a]Permanent address: NCSA, Beckman Institute, University of Illinois, Champaign IL 61820 USA

[b]Permanent address: Gravitation et Cosmologie Relativiste, t.22-12, Un. Paris VI, Paris 75252 France.

where $\bar{R}_{ij}$ is the spatial Ricci curvature and j_{ij} consists of terms at most zeroth order in derivatives of $\mathbf{K}$, first order in derivatives of $\mathbf{g}$, and second order in derivatives of N. The spatial Ricci curvature is second order in derivatives of $\mathbf{g}$ in such a way as to spoil the hyperbolicity of the combined equations (2) and (3). To achieve hyperbolicity for the 3+1 equations, we proceed as follows.

By taking a time derivative of (3) and subtracting appropriate spatial covariant derivatives of the momentum constraints, we obtain an equation with a wave operator acting on the extrinsic curvature. In vacuum, one finds

$$\hat{\partial}_0 R_{ij} - \bar{\nabla}_i R_{0j} - \bar{\nabla}_j R_{0i} = N\hat{\Box}K_{ij} + J_{ij} + S_{ij} = 0, \tag{4}$$

where $\hat{\Box} = -N^{-1}\hat{\partial}_0 N^{-1}\hat{\partial}_0 + \bar{\nabla}^k\bar{\nabla}_k$, J_{ij} consists of terms at most first order in derivatives of $\mathbf{K}$, second order in derivatives of $\mathbf{g}$, and second order in derivatives of N, and

$$S_{ij} = -N^{-1}\bar{\nabla}_i\bar{\nabla}_j(\hat{\partial}_0 N + N^2 H) \tag{5}$$

($H = K^k{}_k$). The term S_{ij} is second order in derivatives of $\mathbf{K}$ and would spoil hyperbolicity of the wave operator $\hat{\Box}$ acting on $\mathbf{K}$. Hyperbolicity is achieved by imposing "generalized harmonic slicing"

$$\hat{\partial}_0 N + N^2 H = g^{1/2}\hat{\partial}_0(g^{-1/2}N) = f(x,t; g^{-1/2}\alpha N, g^{1/2}\alpha^{-1}), \tag{6}$$

where f is an arbitrary scalar function of its arguments and α is an arbitrary scalar density of weight one.

After using (2) to replace $\mathbf{K}$ in (4) and designating g_{ij} and $g^{-1/2}\alpha N$ as variables, the resulting equation and (6) form a quasi-diagonal hyperbolic system with principal operators $\hat{\partial}_0\hat{\Box}$ and $\hat{\partial}_0$. This system can also be put in first order symmetric hyperbolic form by introducing sufficient auxiliary variables and by using the equation for R_{00} (thus incorporating the Hamiltonian constraint). The Cauchy data for the system (in vacuum) are: i) $(\mathbf{g}, \mathbf{K})$ such that the constraints $R_{0i} = 0$, $G^0{}_0 = 0$ hold on the initial slice; ii) $\hat{\partial}_0\mathbf{K}$ such that $R_{ij} = 0$ on the initial slice; and iii) $N > 0$ arbitrary on the initial slice. Note that the shift $\beta^k(x,t)$ is arbitrary. Using the Bianchi identities, one can prove that this system is fully equivalent to the Einstein equations. The constraints of this system propagate in a well-posed causal hyperbolic way. The restriction on the initial value of $\hat{\partial}_0\mathbf{K}$ is what prevents the higher derivative from introducing spurious unphysical solutions.

All variables propagate either with characteristic speed zero or the speed of light. The only variables which propagate at the speed of light have the dimensions of curvature, and one sees that this is a theory of propagating curvature.

In the above formulation, the shift is arbitrary, but the lapse is determined by (6). The theory is therefore covariant under coordinate transformations which preserve the slicing. Can the lapse be made arbitrary so that one can simply drop the generalized harmonic condition (6)? The answer is yes.

By taking another time derivative and adding an appropriate derivative of R_{00}, one finds (in vacuum)

$$\hat{\partial}_0 \hat{\partial}_0 R_{ij} - \hat{\partial}_0 \bar{\nabla}_i R_{0j} - \hat{\partial}_0 \bar{\nabla}_j R_{0i} + \bar{\nabla}_i \bar{\nabla}_j R_{00} = \hat{\partial}_0 (N \square K_{ij}) + \mathcal{J}_{ij} = 0, \quad (7)$$

where $\mathcal{J}_{ij}$ consists of terms at most third order in derivatives of $\mathbf{g}$ and second order in derivatives of $\mathbf{K}$. Together with (2), these form a system for $(\mathbf{g}, \mathbf{K})$ which is hyperbolic non-strict in the sense of Leray-Ohya.[2] Both the lapse and the shift are arbitrary $(N > 0)$ and are not dynamical variables. The Cauchy data of the previous form (in vacuum) must be supplemented for (2) and (7) by $\hat{\partial}_0 \hat{\partial}_0 \mathbf{K}$ such that $\hat{\partial}_0 R_{ij} = 0$ on the initial slice. This guarantees that the system is fully equivalent to Einstein's theory. This system does not have a first order symmetric hyperbolic formulation.

Acknowledgments

A.A., A.A., and J.W.Y. were supported by National Science Foundation grants PHY-9413207 and PHY 93-18152/ASC 93-18152 (ARPA supplemented).

References

1. Y. Choquet-Bruhat and J.W. York, *C. R. Acad. Sci. Paris* t. **321**, Série I, 1089 (1995); A. Abrahams, A. Anderson, Y. Choquet-Bruhat, and J.W. York, *Phys. Rev. Lett.* **75**, 3377 (1995); Y. Choquet-Bruhat and J.W. York in *Gravitation, Electromagnetism and Geometric Structures*, ed. G. Ferrarese (Pitagora Editrice, Bologna, Italy, 1996) 55 (gr-qc/9601030); A. Abrahams and J.W. York, to appear in *Astrophysical Sources of Gravitational Radiation*, ed. J-A. Marck and J-P. Lasota (North Holland, Amsterdam, 1997), gr-qc/9601031; A. Abrahams, A. Anderson, Y. Choquet-Bruhat, and J.W. York, to appear in *Cl. Q. Grav.*, gr-qc/9605014 (1996); Y. Choquet-Bruhat and J.W. York, gr-qc/9606001 (1996); A. Abrahams, A. Anderson, Y. Choquet-Bruhat, and J.W. York, *C. R. Acad. Sci. Paris*, t. **323**, Série IIb, 835 (1996).
2. J. Leray and Y. Ohya, *Math. Ann.* **162**, 228 (1966).

A SLIGHTLY LESS GRAND CHALLENGE: COLLIDING BLACK HOLES USING PERTURBATION TECHNIQUES

H.-P. NOLLERT

Center for Gravitational Physics and Geometry, and
Department of Astronomy and Astrophysics, 104 Davey Lab,
The Pennsylvania State University, University Park, PA 16802, USA

Perturbation techniques can be used as an alternative to supercomputer calculations in computing gravitational radiation emitted by colliding black holes, provided the process starts with the black holes close to each other. We give a summary of the method and of the results obtained for various initial configurations, both axisymmetric and without symmetry: Initially static, boosted towards each other, counter-rotating, or boosted at an angle (pseudo-inspiral). Where applicable, we compare the perturbation results with supercomputer calculations.

1 The perturbation approach

The grand challenge for supercomputers in numerical relativity is the simulation of the inspiral and merger of two black holes, and the computation of the gravitational radiation emitted in the process. An alternative to this approach is the use of perturbation theory: If the colliding black holes start out so close to each other that they have already formed a common horizon, they can be treated as a linearized perturbation of a single, spherically symmetric black hole (Fig. 1). This method is less complicated and requires considerably less computer resources (CPU time, memory), than the full, non-linear approach; it can also yield additional analytic insight.

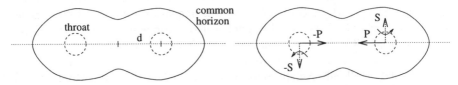

Figure 1: Close collision of two black holes, interpreted as a perturbation of the final black hole. Left: Initially static. Right: Boosted towards each other, and carrying non-axisymmetric, antiparallel spins.

We assume that the collision leads to a final Schwarzschild black hole, and therefore use the Schwarzschild metric as the background metric. Linearizing the field equations around the background metric and using the spherical symmetry of the background to separate angular variables finally reduces the perturbation equations to a single, one-dimensional wave equation for the scalar Zerilli function.

In order to obtain initial data for the collision, we follow Bowen and York[1], solving the constraint equations for a conformally flat initial spatial slice. The conformal factor for several initially static wormholes was found by Misner[2]; the extrinsic curvature vanishes in this case. Bowen and York[1] gave solutions for the extrinsic curvature of one black hole with momentum P or spin S; it is linear in P or S. The extrinsic curvature for several black holes can be obtained by combining contributions from each black hole.

Initial data for the Zerilli function itself is derived from the initial spatial metric[3], the initial derivative is obtained from the extrinsic curvature[4].

The total energy of the gravitational radiation emitted in the close collision is completely determined by the time evolution of all relevant Zerilli functions[4].

2 Results

Price and Pullin[3] first used the perturbation approach for two initially static black holes. The results (Fig. 2) show excellent agreement with supercomputer calculations even for rather large separations (a common apparent horizon begins to form at $\mu_0 = 1.36$).

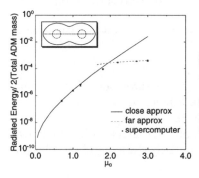

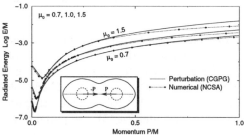

Figure 2: Total radiated energy for a collision of two initially static black holes. μ_0 parametrizes the initial separation.

Figure 3: Total radiated energy for two black hole initially boosted towards each other.

In the case of two black holes initially boosted towards each other (Fig. 3), it turns out that the agreement with fully non-linear, numerical calculations extends to rather large values of the initial momentum as well[5].

Two counter-rotating black holes, with antiparallel spins aligned along the direction of the collision ("Cosmic Screw"), still form an axisymmetric

606

system and result in a final Schwarzschild black hole. The radiated energy increases strongly as a function of spin (Fig. 4). No comparison with numerical calculations is available so far.

Axisymmetry does not play a major role in the perturbation approach. Two black holes with antiparallel spins, aligned perpendicularly to the line of collision, form a simple *non-axisymmetric* system. Contrary to common expectations, the violation of axisymmetry does not increase the amount of radiation for this particular situation; rather, the radiation contributed by the spins is *lower* by a factor of 3/4, compared to the axisymmetric case.

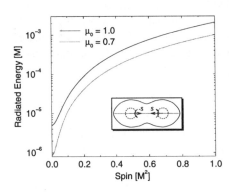 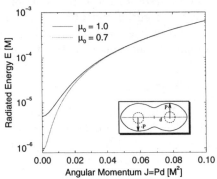

Figure 4: Cosmic Screw: Antiparallel spins along the direction of collision.

Figure 5: Pseudo-Inspiral: Initial boost perpendicular to the connecting line.

Two black holes with momenta perpendicular to the line connecting them can be regarded as the start of a collision from a slowly decaying circular orbit. Since this case results in a rotating black hole, we can only treat it as a perturbation of a Schwarzschild black hole if the final angular momentum is small. For $J = 0.1M^2$, the emitted radiation already increases by about two orders of magnitude (Fig. 5), compared to the initially static, head-on collision.

Acknowledgments

This work was supported in part by the NSF and the DFG.

1. J.M. Bowen and J.W. York, Jr., *Phys. Rev.* D **21**, 2047 (1980)
2. C.W. Misner, Ann. Phys. **24**, 102 (1963)
3. R.H. Price and J. Pullin, *Phys. Rev. Lett.* **72**, 3297 (1994)
4. A.M. Abrahams and R.H. Price, *Phys. Rev.* D **53**, 1963 (1996).
5. J. Baker *et al*, *Phys. Rev.* D **55**, 829 (1997).

Dynamics of spin-2 fields in Kerr background geometries

W. KRIVAN, P. LAGUNA, P. PAPADOPOULOS

Dept. of Astronomy & Astrophysics and
Center for Gravitational Physics & Geometry
Penn State University, University Park, PA 16802

We have developed a numerical method for evolving perturbations of rotating black holes. Solutions are obtained by integrating the Teukolsky equation written as a first-order in time, coupled system of equations, in a form that explicitly exhibits the radial characteristic directions. We follow the propagation of generic initial data through the burst, quasi-normal ringing and power-law tail phases. Future results may help to clarify the role of black hole angular momentum on signals produced during the final stages of black hole coalescence.

At first instance, a direct derivation of the equations governing the perturbations of Kerr spacetimes is to consider perturbations of the metric. This path, however, leads to gauge-dependent formulations. A theoretically attractive alternative is to examine *curvature* perturbations. Using the Newman-Penrose formalism, Teukolsky [1] derived a master equation governing not only gravitational perturbations (spin weight $s = \pm 2$) but scalar, two-component neutrino and electromagnetic fields as well. For the case $s = 0$, it yields the equation for a scalar wave propagating in a Kerr background, a system which we had studied previously [2].

To our knowledge, most of the work on the dynamics of perturbations of Kerr spacetimes has been performed under the assumption of a harmonic time dependence. Here we are interested in the time integration of physical initial data, possibly from the inspiral collision of binary black holes. Fourier transformation of the data and subsequent evolution of such data in the frequency domain approach is, in principle, possible but numerically cumbersome. The main complication one faces by keeping the equation in the time domain is that one cannot longer benefit from the reduction of dimensionality implied by the separation of variables. We have chosen the option to evolve one single 2+1 PDE instead of the equivalent approach of solving the set of ODEs corresponding to the Fourier spectrum. The computational burden of both approaches is likely similar. The resulting evolution equation is a hyperbolic, linear equation which is quite amenable to numerical treatment, provided suitable coordinates, variables and boundary conditions are chosen.

The two key factors in successfully solving the Teukolsky equations were: first, to carefully select the evolution field and its asymptotic behavior, and second to rewrite the Teukolsky equation in a form that explicitly exhibits the radial characteristic directions. On the analytical level, one obtains bounded solutions for any direction of propagation by choosing $s = -2$ and rescaling by an appropriate function of r. A convenient choice is simply r^3, a factor that is regular at the horizon. Regarding the choice of spatial coordinates, we use the Kerr tortoise coordinate r^* and the Kerr $\tilde{\phi}$ coordinate instead of the Boyer-Lindquist coordinate ϕ. Then the ansatz for the solution to the Teukolsky equation is: $\Psi(t, r^*, \theta, \tilde{\phi}) \equiv e^{im\tilde{\phi}} r^3 \Phi(t, r^*, \theta)$. After a series of unsuccessful numerical experiments with this second-order in time and space equation for Φ, we found that numerical instabilities due to the first order in time derivatives in the Teukolsky equation were suppressed by introducing an auxiliary field $\Pi \equiv \partial_t \Phi + b \, \partial_{r^*} \Phi$, that converts the Teukolsky equation to a coupled set of first-order equations in space and time[3], where $b \equiv (r^2 + a^2)/\Sigma$ and $\Sigma^2 \equiv (r^2 + a^2)^2 - a^2 \Delta \sin^2 \theta$. The resulting first order system is hyperbolic in the radial direction. Stable evolutions were achieved using a modified Lax-Wendroff method, discretizing the equation on a uniform two dimensional polar grid. Typically we used a computational domain of size $-100M \le r_i^* \le 500M$ and $0 \le \theta_j \le \pi$ with $0 \le i \le 8000$ and $0 \le j \le 32$. Details of the numerical scheme are described in [3]. The stability of the code was verified with long-time evolutions, of the order of $1000M$. Its accuracy in turn was tested using standard convergence tests.

The evolution of generic initial data consists of three stages, as seen from an observer located away from the hole. During the first stage, the observed signal depends on the structure of the initial pulse and its reflection from the angular momentum barrier (burst phase). This phase is followed by an exponentially decaying quasinormal ringing of the black hole (quasinormal phase). In the last stage, the wave slowly dies off as a power-law tail (tail phase). The numerical algorithm described in the previous section is used to obtain the time evolution of generic perturbations impinging on the rotating black hole. The broad features of the evolution are demonstrated in Fig. 1, where quasinormal ringing and tail behavior are clearly manifested. We then performed a series of simulations with $m = 0$ for different values of a. Focusing on the later part of the ringing sequence, which clearly depicts the least damped mode of oscillation, we read off the oscillation frequencies. A comparison of our values with those given by Kokkotas[6] and Leaver[7] shows an agreement of better than 1%.

We have shown[2], that the late time evolution of scalar fields in the background of rotating black holes is qualitatively similar to the non-rotating case. We extend here this result to the physically more interesting spin-2 field evolution. Our calculations show that the exponents governing the behavior for $a \ne 0$ do not exhibit a significant change when compared to the Schwarzschild

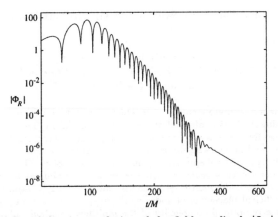

Figure 1: Log-log plot of the time evolution of the field amplitude $|\Phi_R|$ for an observer located at $r^* = 20M$, $\theta = \pi/2$ ($a = 0.9M$, $m = 0$). The exponentially damped oscillation of the scattered signal, and the late time power-law tail are clearly exhibited.

case, if the initial data pulse is given by the lowest allowed mode for a particular value of m. That is, the power-law tail behavior is basically determined by the dominant asymptotic form of the potential. When the initial data pulse is not given by the lowest allowed mode for a particular value of m, mixing of modes occurs [2].

We thank H.-P. Nollert, R. Price and J. Pullin for helpful discussions. This work was supported by the Binary Black Hole Grand Challenge Alliance, NSF PHY/ASC 9318152 (ARPA supplemented) and by NSF grants PHY 96-01413, 93-57219 (NYI) to PL. WK was supported by the Deutscher Akademischer Austauschdienst (DAAD).

References

1. S. A. Teukolsky, *Phys. Rev. Lett.* **29**, 1114 (1972).
2. W. Krivan, P. Laguna, and P. Papadopoulos, *Phys. Rev.* D **54**, 4728 (1996).
3. W. Krivan, P. Laguna, and P. Papadopoulos, *Dynamics of perturbations of rotating black holes*, to be submitted to *Phys. Rev.* D (1997).
4. R.H. Price, *Phys. Rev.* D **5**, 2419 (1972).
5. C. Gundlach, R.H. Price, and J. Pullin, *Phys. Rev.* D **49**, 883 (1994).
6. K.D. Kokkotas, *Class. Quantum Grav.* **8**, 2217 (1991).
7. E.W. Leaver, *Proc. R. Soc. Lond.* A **402**, 285 (1985)

A 3D Finite Element Solver for the Initial Value Problem

D. Bernstein and M. Holst

*Applied Mathematics 217-50, California Institute of Technology,
Pasadena, CA 91125, USA*

We describe a new finite element code which solves the fully general initial-value problem using the York conformal decomposition formalism. The finite element approach allows for domains with complex topology, natural representation of boundary surfaces, and provides a solid theoretical framework for analyzing the accuracy of the resulting numerical approximations. The code employs adaptive mesh refinement based on robust *a posteriori* error estimation and simplex bisection, coupled with CG-like methods, multilevel methods, and Gummel/Newton methods. The extremely efficient placement of nodes by the adaptive refinement procedure along with the nearly optimal complexity of the multilevel methods allows these computations to be performed on workstations rather than supercomputers. As an example we compute the initial data corresponding to two "stars" in an approximately circular orbit in the presence of a gravitational wave and on a slice on which $\mathrm{tr}K$ is non-zero.

1 The General Initial-Value Problem

We work in the ADM formalism and use the standard York conformal decomposition of the 3–metric and extrinsic curvature [1]. The initial value equations in this form appear as four coupled quasi-linear elliptic equations for the conformal factor ϕ and a vector potential W^a. In the standard notation and with the usual conformal weighting these are

$$\hat{\gamma}^{ab}\hat{D}_a\hat{D}_b\phi = \frac{1}{8}\hat{R}\phi - \frac{1}{8}\phi^{-7}(\hat{A}_{ab}^* + (\hat{l}W)_{ab})^2 + \frac{1}{12}(\mathrm{tr}K)^2\phi^5 - 2\pi\hat{\rho}\phi^{-3}, \quad (1)$$

$$\hat{D}_b(\hat{l}W)^{ab} = \frac{2}{3}\phi^6\hat{D}^a\mathrm{tr}K + 8\pi\hat{j}^a, \quad (2)$$

$$(\hat{l}W)^{ab} = \hat{D}^aW^b + \hat{D}^bW^a - \frac{2}{3}\hat{\gamma}^{ab}\hat{D}_cW^c. \quad (3)$$

The equations have been solved numerically many times in a variety of physical settings (e.g., Mathews and Wilson [2]). However, because of the complexity of the operators and the nonlinear coupling of the equations it has been common to make various simplifying assumptions, the most frequent of which are to force $\hat{\gamma}_{ab}$ to be flat and to let the initial slice be maximal ($\mathrm{tr}K = 0$). In addition, the majority of previous work has been done with finite differences which create well-known difficulties when the boundaries of the domain do not lie along constant coordinate surfaces or when the coordinate system has singularities.

The code which we describe here allows us to solve the equations in three dimensions without recourse to any of the above assumptions. The nature of the finite element method is such that the placement of the nodes in space is independent of the coordinate system in which the equations are written. This permits the use of a non-singular coordinate system (if one exists) which covers the entire computational domain and also makes handling boundaries with complicated geometry or topology natural and automatic. Thus, supplied with the necessary freely specifiable data and appropriate boundary conditions on ϕ and W^a, the code is, in principle, able to solve *any* initial value problem in three dimensions. (In practice, on desktop machines, we are limited to problems which require 50,000 nodes or less.)

2 Finite Element Discretization and Mesh Refinement

We employ a standard Galerkin finite element method in which the basis and test functions are taken to be piecewise polynomials with local support over disjoint polyhedral subregions of the underlying spatial domain. A weak formulation of the original strong form equations (1)–(2) is required, obtained by multiplication of the strong form equations by test functions u and v^a and integration by parts. The functions ϕ and W^a are expanded in terms of the basis, yielding linear and nonlinear algebraic equations for the expansion coefficients (matrices which occur are very sparse due to the local support property of the basis).

The underlying spatial domain is broken into disjoint polyhedra; we employ simplices (tetrahedra in this setting), and the natural piecewise-linear basis. A simplex may be divided into two (bisection) or eight (octasection) child simplices in a recursive fashion such that all progeny remain non-degenerate [3,4]. Adaptive mesh refinement consists of marking certain simplices which are deemed too large, and performing bisection or octasection to produce smaller simplices covering the same region in space. The marking procedure is driven by *a posteriori* error estimation [5]. A conforming finite element method requires a conforming mesh; a requirement for such a mesh is that it contains no hanging vertices (vertices which lie along the middle of edges of one or more simplices). Adaptive octasection always produces such hanging nodes, and must be supplemented with bisection to produce conforming meshes.

The discrete coupled nonlinear algebraic system which arises from discretizing (1)–(2) with the finite element method is solved using a damped, inexact-Newton procedure for linearization, coupled with a multilevel preconditioned conjugate gradient iteration on the normal equations for solving the resulting linear systems.

612

3 Numerical Results

We compute initial data for the following (admittedly artificial) problem: two spherical masses with constant density and velocity fields which put them in an approximately circular orbit. The conformal metric is an axisymmetric Brill wave with moderate amplitude and $\mathrm{tr}\,K$ is a small, slightly oscillatory function which is nonzero in the neighborhood of the masses and vanishes rapidly as one moves towards the outer boundary. The masses have coordinate radius 1, central separation 8, and the outer boundary has coordinate radius 100. The following table gives results for a 200MHz Pentium Pro running Linux.

Error Indicator	Unknowns	CPU Seconds	Newton Steps	RAM
1.0e-1	516	55	18	1Mb
1.0e-2	3,312	399	24	4Mb
1.0e-3	25,022	4,385	31	19Mb
5.0e-4	48,126	13,725	43	34Mb

Acknowledgments

This work was supported in part by the NSF under Cooperative Agreement No. CCR-9120008.

References

1. J. W. York. Kinematics and dynamics of general relativity. In L. L. Smarr, editor, *Sources of Gravitational Radiation*, pages 83–126. Cambridge University Press, Cambridge, MA, 1979.
2. J. R. Wilson, G. J. Mathews, and P. Marronetti. Relativistic numerical model for close neutron-star binaries. *Phys. Rev. D*, 54:1317–1331, 1996.
3. A. Mukherjee. *An Adaptive Finite Element Code for Elliptic Boundary Value Problems in Three Dimensions with Applications in Numerical Relativity*. PhD thesis, Dept. of Mathematics, The Pennsylvania State University, 1996.
4. S. Zhang. *Multi-level Iterative Techniques*. PhD thesis, Dept. of Mathematics, Pennsylvania State University, 1988.
5. R. Verfürth. A posteriori error estimates for nonlinear problems. finite element discretizations of elliptic equations. *Math. Comp.*, 62(206):445–475, 1994.

Bifunctional adaptive mesh (BAM) for 3d numerical relativity

Bernd Brügmann

Max-Planck-Institute for Gravitational Physics
Schlaatzweg 1, 14473 Potsdam, Germany
bruegman@aei-potsdam.mpg.de

We report on an adaptive mesh package for 3d numerical relativiy, BAM, that performs both adaptive mesh refinement for hyperbolic (evolution) problems and adaptive multigrid for elliptic problems.

Numerical methods for the solution of partial differential equations often involve discrete grids, and the error introduced is directly related to the resolution of the grids. Adaptive mesh refinement refers to the natural idea that one should adapt the resolution locally to the local discretization errors. On the one hand, one ensures in this way that the local errors stay below a given threshold. On the other hand, using non-uniform resolution one can avoid wasting computer resources on regions where the global maximum of resolution is not required.

Natural as this idea may seem, adaptive mesh refinement has only recently found its way into numerical relativity [1]. The main reason certainly is that numerical relativity is still far from being a mature field. To mention just one great outstanding problem, not even the two-body problem of vacuum general relativity in three dimensions has been solved. In its "cleanest" form this refers to the evolution of two black holes, which in the general case amounts to a difficult numerical simulation problem.

Adaptive mesh techniques are relevant in this case because one could hope that the boost in efficiency can lift the numerical problems into the reach of current supercomputers, which would allow numerical experiments to examine the at least equally severe conceptual problems related to the choice of coordinates and such.

To this end we are working on an adaptive mesh package that combines (3+1)-dimensional adaptive mesh refinement for the hyperbolic (evolution) part of the Einstein equations with 3-dimensional adaptive multigrid for elliptic problems such as solving the constraints of the Einstein equations. We refer to it as BAM for bifunctional adaptive mesh.

The hyperbolic part has been described in Ref. 2. The (3+1)-dimensional ADM equations are implemented in a standard explicit finite difference form. The core of the evolution code, however, is produced by a Mathematica script so that other formulations should be simple to implement. Encouraging results

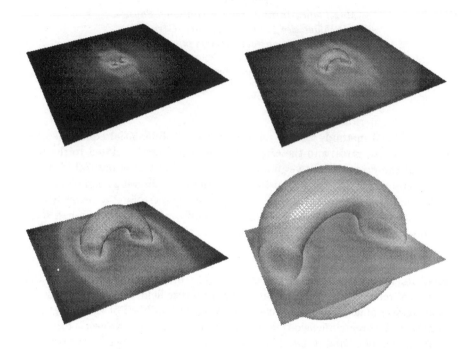

Figure 1: Conformal factor $\psi(x, y, z)$ for a 3d time-symmetric Brill wave obtained in BAM. Four nested grids of size 33^3 with edge lengths 16, 8, 4, and 2 (in appropriate units), respectively, centered at the origin are shown. The grids are not allowed to grow beyond 33^3 points, which in this case we ensured by a refinement schedule. The Euclidean norm of the residual has dropped to below 10^{-10} on all grids, which is far below the discretization error of about 10^{-3}. The conformal factor tends to 1.0 at the outer boundary. In each plot the same isosurface $\psi = 1.25$ is displayed. "Brill wave" refers to a particular way to specify the free data for a gravitational wave in vacuum without singularities. Here a toroidal distribution of a certain metric coefficient around the z-axis is chosen. The conformal factor ψ is then obtained numerically as a solution to the Hamiltonian constraint equation. Using standard notation for Brill-waves, we choose for the free metric function q, $q = A\rho^2 \sin^2(\theta) \exp(-(\rho - \rho_0)^2/\sigma^2)$, where θ is the usual polar angle, and ρ is the radial coordinate in the x-y plane. In the figure, $A = -1.5$, $\rho_0 = 0.5$, and $\sigma^2 = 0.2$.

were obtained for the evolution of a single static black hole in a dynamic slicing (geodesic slicing). A key feature is the use of the apparent horizon boundary condition which requires some effort in conjunction with the adaptive mesh technique.

The elliptic part is based on a full approximation storage adaptive multigrid code built around a non-linear Gauss-Seidel relaxation [3]. This type of relaxation method turns out to have the required smoothing properties for typical elliptic operators of the initial data problem of general relativity. In figure 1 we show Brill wave initial data. In Ref. 4, general black hole initial data sets are discussed.

We have performed first tests combining the elliptic and hyperbolic parts of BAM. Both operate on identical data structures, so it was straightforward to, for example, evolve the initial data sets just mentioned for a brief period of time. However, many issues related to the boundary conditions and the choice of coordinates (incidentally often also implemented by solving elliptic equations) still have to be sorted out. We hope that these quite non-trivial problems can be addressed in the technical framework created by BAM.

It is a pleasure to thank Steven Brandt, Bernd Schmidt, Bernard Schutz, Ed Seidel, and Paul Walker for helpful discussions.

References

1. M. Choptuik, *Phys. Rev. Lett.* **70**, 9 (1993).
2. Bernd Brügmann, *Phys. Rev.* D **54**, 7361 (1996).
3. A. Brandt, in *Multigrid Methods*, ed. W. Hackbusch and U. Trottenberg (Springer Lecture Notes in Mathematics No. 960, 1982); W. Press, S. Teukolsky, W. Vetterling, and B. Flannery, *Numerical Recipes in C* (Cambridge University Press, 1992)
4. Steven Brandt and Bernd Brügmann, AEI-preprint (1/1997).

Finding apparent horizon for a special family of 3D space in numerical relativity

Masaru Shibata

Department of Earth and Space Science, Graduate School of Science,
Osaka University, Toyonaka, Osaka 560, Japan

We have developed a method to find the apparent horizon(AH) on a special family of 3D spacelike hypersurfaces which has the π-rotation symmetry around z axis as well as the reflection one with respect to the equatorial plane. We demonstrate that the method works well by determining the AH for the triaxial compact ellipsoid of dust.

One of the most important issues in numerical relativity is to clarify formation process of black hole(BH). Standard method for investigation of the formation process of BH's in numerical relativity is to find the apparent horizon(AH) on each 3D spacelike slice[1]. When we denote two surface of the AH as $h(\theta, \varphi)$, where θ and φ are the spherical coordinates, h obeys the 2D elliptic-type equation as

$$h_{,\theta\theta} + \cot\theta\, h_{,\theta} + \frac{h_{,\varphi\varphi}}{\sin^2\theta} = S(h, g_{\mu\nu}), \qquad (1)$$

where "," and $g_{\mu\nu}$ denote the ordinary derivative and the 4D metric. S is the source term which depends on the non-linear terms of h and $g_{\mu\nu}$. Since Eq. (1) is the elliptic-type equation, usually, it is solved as the boundary value problem imposing the boundary conditions at $\theta = 0$, π and $\varphi = 0$, 2π. In the general 3D case, however, we can not specify the boundary condition at $\theta = 0, \pi$, and up to now, Eq. (1) is solved by the harmonic expansion method[1][2] although it is not so efficient.

We propose an efficient method which can be used for the case when the 3D space has π-rotation symmetry around the z axis as well as the reflection one with respect to the equatorial plane. Assumption that the 3D space has such symmetries is appropriate for many problems in 3D numerical relativity; for example, merging of binary neutron stars of equal mass to be a BH, collapse of rotating ellipsoid to be a BH, and so on. In this case, we can impose the boundary conditions at $\theta = 0$ and $\pi/2$ as $h_{,\theta} = 0$, and as a result, Eq.(1) can be solved as the boundary value problem. In ref.3, we describe the detailed numerical method and show that it works very well for a wide variety of test problems. Here, we determine the AH for the time symmetric initial conditions of the triaxial compact ellipsoidal dust of its axes a_1, a_2 and a_3 $(a_1 > a_2 > a_3)$ as an example.

The time symmetric initial conditions in general relativity are obtained by solving the Hamiltonian constraint. If we assume the conformal flat 3D metric, the equation is

$$\Delta \Psi = -2\pi \rho \Psi^5 \qquad (2)$$

where Δ is the Laplacian for the flat space, Ψ is the conformal factor and ρ is the mass density of the dust. We set $\rho \Psi^5 = \rho_h$ =constant, and in this case, the solution for Ψ is well known[4]. In the following, we fix the mass of system as $M = 4\pi \rho_h a_1 a_2 a_3 / 3 = 2$. Thus, for the case $a_1 = a_2 = a_3 < 1$, the AH appears at $r = 1$.

In numerical calculation, we first fix a_3, and then changing a_1 and a_2, we try to find the AH. In figure, we show results for $a_3 = 0.05$; if the AH is found, we plot the circle, and if not, we plot the cross. As for the case when $a_1 = a_2$ or $a_2 = a_3$, the results agree well with a previous study[5]. We calculate the area of the AH A when it is determined. (Hereafter, we use the normalized one $\hat{A} \equiv A/16\pi M^2$.) For the case $a_1 = a_2 = a_3$, $\hat{A} = 1$, and we check and confirm that. The area for the spindle like configuration is slightly smaller than that for the oblate like configuration, but $\hat{A}$ does not change so much irrespective of the axis ratio($0.99 < \hat{A} \leq 1$ in the present case). Thus, if we perform numerical simulation starting from these initial conditions, gravitational radiation is hardly emitted.

Interesting question is whether the naked singularity is formed or not in numerical simulation starting from the initial condition where $a_1 \gg a_2 \sim a_3$. For initial conditions of $a_2 \neq a_3$, the singularity does not appear even for $a_3 = 0$(i.e., geodesics are complete). To produce the singularity in simulation, a_2 must be equal to a_3. Thus, it is one of the interesting problems in 3D numerical relativity to investigate whether such a configuration is realized or not.

The author thanks K. Nakao and T. Nakamura for useful discussions. This work was in part supported by Japanese Grant-in-Aid for Scientific Research of Ministry of Education, Culture, Science and Sports, No. 08NP0801.

References

1. T. Nakamura, K. Oohara and Y. Kojima, Prog. Theor. Phys. suppl. **90**, 21(1987).
2. J. Libson, J. Masso, E. Seidel, and W.-M. Suen, gr-qc/9412058.
3. M. Shibata, Phys. Rev. **D55**(1997).
4. S. Chandrasekhar, *Ellipsoidal Figure of Equilibrium*, (Dover, 1987).
5. T. Nakamura, S. L. Shapiro, and S. A. Teukolsky, Phys. Rev. **D38**, 2972(1988).

618

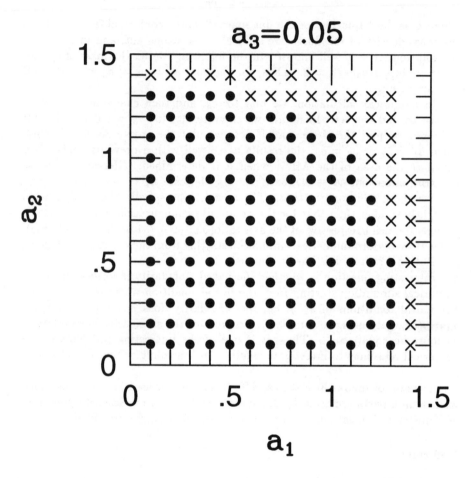

CAN WE DETECT BRANS-DICKE SCALAR GRAVITATIONAL WAVES IN GRAVITATIONAL COLLAPSE?

HISA-AKI SHINKAI

Department of Physics, Washington University, St. Louis, MO 63130-4899, USA

MOTOYUKI SAIJO and KEI-ICHI MAEDA

Department of Physics, Waseda University, Shinjuku, Tokyo 169, Japan

Using a model of a non-spherical dust shell consists of test particles falling into a Kerr black hole, we study gravitational waves in Brans-Dicke theory of gravity. We assume that each test particles plunge with a constant azimuthal angle, and calculate the emitted energy of both scalar and tensor modes of gravitational radiation. We figure out when the scalar modes dominate, changing Kerr parameter a and shell configuration. When a black hole is rotating, the tensor modes do not vanish even for a "spherically symmetric" shell, instead a slightly oblate shell minimizes their energy but with non-zero finite value, which depends on a. As a result, we find that the scalar modes dominate only for highly spherical collapse, but they never exceed the tensor modes unless the Brans-Dicke parameter $\omega_{BD} \lesssim 750$ for $a/M = 0.99$ or unless $\omega_{BD} \lesssim 20,000$ for $a/M = 0.5$, where M is mass of black hole. We conclude that the scalar gravitational waves with $\omega_{BD} \lesssim$ several thousands do not dominate except for very limited situations. Therefore presice observation of polarization is also required when we determine the theory of gravity by the observing of gravitational waves.

Currently, a number of worldwide projects for detecting gravitational radiation directly using a km-scale laser interferometer such as LIGO[2], VIRGO, GEO, and TAMA are progressing. These direct detections of gravitational waves will enable us to see strong gravitational phenomena such as coalescence of compact stars or black holes. By observing gravitational waves, we expect to extract many parameters of sources. Here, we examine an alternative information in waveform of gravitational waves; whether we can find some evidence for alternative theories of gravity instead of general relativity. Especially, we consider the Brans-Dicke (BD) theory[3] of gravity, which is the simplest and proto-type among all scalar-tensor theories of gravity. [We know the recent strictest bound[4] for BD theory is to the BD parameter $\omega_{BD} \gtrsim 500$.]

In BD theory, gravitational waves have three modes, i.e., a scalar mode, which we call a scalar gravitational wave (SGW), as well as two tensor modes (+ and × modes) of conventional gravitational waves, which we shall call tensor gravitational waves (TGWs) here. To find the expected templates of SGWs, Shibata, Nakao and Nakamura[5] calculated the waveform of SGWs from spherically symmetrical dust fluid collapse and concluded that the advanced LIGO may detect SGWs even if the BD coupling constant ω_{BD} is $10^4 \sim 10^6$.

Since there is no TGWs emission in a spherically symmetric collapse, their case is idealized situation for SGWs. We extend their studies in more realistic cases and figure out when the SGWs can dominate.

We analyze a 'non-spherical' dust shell collapsing into a rotating black hole with mass M and Kerr parameter a. We assume a dust shell consists of test particles, which plunges with a constant azimuthal angle θ into a rotating black hole. The gravitational radiation from a dust shell is derived from a linear combination of test particles falling into a black hole. Along with the early studies of such a distorted shell[6], we define the shape of a dust shell at $r_0 = 10M$ by a function $r_1(\theta) = r_0(1+\delta\cos^2\theta)$(prolate) or $r_2(\theta) = r_0(1+\delta\sin^2\theta)$ (oblate), where $\delta \geq 0$, and sum up emitted radiations from each test particles.

We show our results in Figure 1 for $\omega_{BD} = 500$. Since the system is not completely spherically symmetric, the TGW is generated for any values of δ. There exists some value of δ_0, which is not zero (e.g. $\delta_0 \sim 0.02$ for $a = 0.5$, $\delta_0 \sim 0.08$ for $a = 0.9$, and $\delta_0 \sim 0.10$ for $a = 0.99$), at which the emitted energy of TGWs takes a minimum value. As a increases, its minimum energy and the corresponding deformation parameter δ_0 increase.

The existence of a non-vanishing minimum value of TGW is important for observation. The emitted energy of SGW does not depend much on a deformation parameter δ nor a, because the $l = 0$ mode dominates in the SGW. We also find both energies of TGW and SGW decrease for large δ because of a phase cancellation effect. Therefore we conclude that the SGW dominates the TGWs for $\delta_1(= -0.05,\ 0.03$ and 0.05 for $a = 0.5,\ 0.9$ and $0.99) < \delta < \delta_2(= 0.09,\ 0.11$ and 0.12 for the same as).

On the other hand, the emitted SGW is proportional to $1/\omega_{BD}$ if $\omega_{BD} \gg 1$. Therefore if ω_{BD} is larger than some critical value ω_{BD}^{cr}, which depends on a, the SGWs never dominate TGWs. Therefore we may speculate that the SGW cannot be dominant unless $\omega_{BD} \lesssim 750$ for $a/M = 0.99$ or unless $\omega_{BD} \lesssim 20,000$ for $a/M = 0.5$.

Finally, we remark on the detectability of SGWs. From analysis of maximum amplitude of emitted gravitational radiation fixing the BD coupling constant $\omega_{BD} = 500$, we find that the limiting distance of SGWs in First LIGO for NS-NS or BH-BH collision is $r=100$kpc (for advanced LIGO, 20Mpc). Conversely, if the SGW is observed in a gravitational wave detector with polarization information, we can discuss the upper bound of constraint on ω_{BD}. For example, we can easily say that if NS-BH collision occurred in our Galaxy and the SGW were observed, we could determine definitely the BD coupling constant up to 10^5.

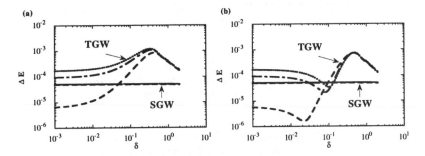

Figure 1: Total emitted energy by non-spherical dust shell collapsing (with deforming parameter δ) into a Kerr black hole ((a) prolate shell and (b) oblate shell) for Brans-Dicke parameter $\omega_{BD} = 500$. Dashed, dash-dotted and dotted lines represent Kerr parameter $a = 0.5, 0.9$ and 0.99 respectively for both tensor modes (TGW) and scalar modes (SGW) of gravitational radiation.

The above results require observations of polarizations and extractions of information of the scalar mode from observed data, for which we need several observational sites in the world. It will, however, open a new window in gravitational astronomy to test theories of gravity by direct observation of gravitational radiation.

This work was supported partially by the Grant-in-Aid for Scientific Research Fund of the Ministry of Education, Science and Culture (Specially Promoted Research No. 08102010), NSF 96-00507 and by a Waseda University Grant for Special Research Projects.

References

1. The detail of this article is in, M. Saijo, H. Shinkai and K. Maeda, WU-AP/64/96 and gr-qc/9701001. Submitted to *Phys. Rev. D*.
2. A. Abramovici, *et al.*, *Science* **256**, 325 (1992).
3. C. Brans and R. H. Dicke, *Phys. Rev.* **124**, 925 (1961).
4. R. D. Reasenberg, *et al.*, *Astrophys. J.* **234**, L219 (1979).
5. M. Shibata, K. Nakao and T. Nakamura, *Phys. Rev.* D **50**, 7304 (1994).
6. T. Nakamura and M. Sasaki, *Phys. Lett.* B **106**, 69 (1981).
   S. L. Shapiro and I. Wasserman, *Astrophys. J.* **260**, 838 (1982).

THE NATURE OF THE GENERIC COSMOLOGICAL SINGULARITY

B.K. BERGER

Department of Physics, Oakland University
Rochester, MI 48309, USA

V. Moncrief

Department of Physics, Yale University
New Haven, CT 06511, USA

A symplectic PDE solver is used to evolve spatially inhomogeneous cosmologies toward the singularity in order to test the Belinskii *et al* conjecture that the generic singularity is locally Mixmaster-like.

1 Introduction

Singularities in spatially homogeneous cosmologies can be velocity dominated or Mixmaster-like. Belinski, Khalatnikov, and Lifshitz (BKL)[1] conjectured that the generic cosmological singularity is locally of the Mixmaster type. So far, a proof of this conjecture is beyond analytic techniques. Analytic and numerical studies have shown that generic Gowdy plane-symmetric cosmologies have velocity dominated singularities with the approach to the velocity dominated regime governed by nonlinear terms in the equations for the gravitational wave amplitudes.[2] The more general $U(1)$ symmetric cosmologies are sufficiently generic that they might have Mixmaster-like singularities although polarized $U(1)$ models appear to have velocity dominated singularities.[3]

2 Numerical Methods

Einstein's equations for all these models can be derived from a Hamiltonian of the form $H = H_1 + H_2$ where H_1's variation yields a velocity dominated solution (only time derivatives are involved) while H_2 contains all the spatial derivatives. To evolve from t to $t + \Delta t$, the evolution operator may be approximated through order $(\Delta t)^2$ as $U(t + \Delta t) = U_1(t + \Delta t/2)U_2(t + \Delta t)U_1(t + \Delta t/2)$ where U_1, U_2 are associated respectively with H_1, H_2.[4,5] This symplectic approach is useful when the subhamiltonians are exactly solvable as is the case here.[2] However, spatial differencing problems in the $2 + 1$ $U(1)$ models limit the duration of the simulation. A differencing scheme and spatial averaging algorithm provided by A. Norton (private communication) are used to extend the evolution. The constraint and initial value problems are trivial for Gowdy models. In $U(1)$ models, a partial solution to the initial value problem is found and the constraints are subsequently monitored. Spatial averaging can create artifacts which are detected by increasing spatial resolution. The Gowdy evolution provides a test case for algorithms and for signatures of velocity dominance.[2]

3 Cosmological Models

Homogeneous cosmologies can be described as dynamical systems in minisuperspace where the spatial scalar curvature produces a potential. In a velocity dominated singularity there is a final bounce of the system off the potential while in a Mixmaster one there is no last bounce.

In the vacuum Gowdy cosmology on $T^3 \times R$, the dynamical degrees of freedom are amplitudes of the two polarizations of gravitational waves P and Q which depend on spatial variable θ and time τ. These satisfy nonlinearly coupled equations which decouple from the constraints. Initial values for P and Q and their time derivatives are essentially arbitrary while the constraints may be satisfied by construction of a background variable using the known evolution of P and Q. The wave equations are obtained by variation of

$$ H = \oint d\theta \tfrac{1}{2} \left(\pi_P^2 + e^{-2P} \pi_Q^2 \right) + e^{-2\tau} \oint d\theta \tfrac{1}{2} \left(P_{,\theta}^2 + e^{2P} Q_{,\theta}^2 \right). $$

The kinetic, K, and potential, V, parts of H naturally become H_1 and H_2 in the symplectic PDE solver described above. The velocity dominated regime is characterized as $\tau \to \infty$ by $P \to v\tau$, $Q \to Q_0$ and $V \to 0$ if $v < 1$.

The more general $U(1)$ symmetric cosmologies[6] are described by five degrees of freedom $\{x, z, \Lambda, \phi, \omega\}$ which depend on spatial variables u, v, and time τ. Einstein's equations are obtained by variation of

$$ H = K(x, z) + K(\phi, \omega) + K_{free}(\Lambda) + V(x, z, \Lambda, \phi, \omega) $$

where K has the same functional form as in the Gowdy case, K_{free} is a free particle kinetic energy, and V is a complicated potential containing all the spatial derivatives. Constraints are non-trivial but may be solved algebraically at the initial time by setting some variables to zero. Four free functions remain. Polarized $U(1)$ models are obtained by setting the degree of freedom associated with ω (analog of Q) to zero. This is preserved by the evolution. These polarized models seem to be velocity dominated. Adding back even a small amount of the missing degree of freedom completely changes the character of the solution. The differences are easily seen in the behavior of V (or terms in

624

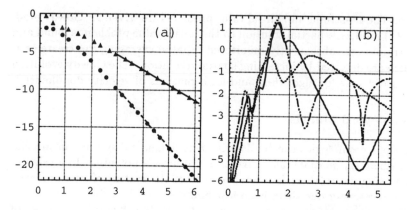

Figure 1: (a) Decay of V in the polarized $U(1)$ model. $\log_{10} V(u,v,\tau)$ vs τ is plotted for two locations in the u-v plane. The points are from the simulation. The lines represent the asymptotic best fit. (b) Decay of the $(\nabla\omega)^2$ term in V in a generic $U(1)$ model. $\log_{10} V_{\nabla\omega}(u,v,\tau)$ vs τ is plotted for three locations in the u-v plane.

V) at particular points in the u-v plane. Although persistant nonzero terms could arise from Mixmaster-like oscillations, they could also be due to numerical artifacts, nonlinear wave interactions which would eventually die off, or a foolish choice of τ (part of the space hits the singularity early on). Work in progress involves the exploration of ways to run the simulations longer to distinguish among these possiblities.

Acknowledgments

This work was supported in part by NSF PHY-9507313 to Oakland University and PHY-9503133 to Yale University.

References

1. V.A. Belinskii *et al*, Sov. Phys. Usp. **13**, 745 (1971).
2. B.K. Berger and V. Moncrief, *Phys. Rev.* D **48**, 4676 (1993).
3. B.K. Berger in *Relativity and Scientific Computing*, edited by F.W. Hehl, R.A. Puntigam, H. Ruder (Springer-Verlag, Berlin, 1996)
4. J. A. Fleck *et al*, Appl. Phys. **10**, 129 (1976).
5. M. Suzuki, Phys. Lett. **A146**, 319 (1990).
6. V. Moncrief, Ann. Phys. (N.Y.) **167**, 118 (1986).

EMISSION PHYSICS IN RADIO PULSAR NANOSTRUCTURE: THEORY AND OBSERVATIONS

T. H. HANKINS and J. C. WEATHERALL

Physics Department, New Mexico Institute of Mining and Technology, Socorro, NM 87801

We are conducting theoretical and observational studies of the pulsar emission mechanism to compare theoretical predictions of emission process time signatures and polar cap plasma dynamics with observed intensity nanostructure. We show that simulations of beam-driven plasma turbulence produce extremely short pulse structure on the order of a few nanoseconds. We show some 10-ns resolution data recently obtained on the Crab "giant" pulses that demonstrates that there is significant unresolved structure at that resolution.

1 Objectives

Our strategy is to identify the primary inputs to the pulsar polar cap environment, then apply several approaches to the plasma microphysics in the polar cap which can produce radio emission. We then note that the radio signals are affected by magnetospheric and interstellar medium propagation before being observed on the earth. We identify the observables of spectral bandwidth and intensity nanostructure as important for discerning which of the viable emission mechanisms is correct. This we view as the "Pulsar Problem".

2 Plasma Turbulence Modelling

Following the premise that temporal and spectral behavior of the radio emission are determined by the microphysics of the emission process, we develop a model[1] for plasma turbulence emission[2] based on wave-mode equations coupled through a cubic nonlinearity. Numerical solution of the nonlinear wave behavior exhibits modulational instabilities of the electrostatic plasma beam wave generated by the polar cap current flow, and a strongly turbulent state characterized by the multi-dimensional collapse of wave packets. Inasmuch as the beam wave has no Poynting flux, the turbulence plays the essential role of generating electromagnetic modes which escape the plasma.

The quantitative features of the radio emission in this model are almost entirely controlled by the microphysics of the source plasma. The saturated turbulent energy density is $E^2/8\pi \sim nk_BT = 1\,\mathrm{erg\,cm^{-3}}$. The energy in the 10^4-cm source region can be released into escaping radiation by turbulent processes over the time of light transit. Taking into account the relativistic boosting at $\gamma \sim 300$, the received radiative flux calculated from this model is ~ 20 Jy, consistent with Crab giant pulse fluxes.

The numerical simulations show characteristic temporal structure which may be unique to this model and can be tested against observation. The intrinsic pulse length is characteristic of the escape time, $R/(\gamma c) \sim 2$ ns. The pulse also exhibits structure due to mode coupling, which causes energy to transfer back and forth in a resonant interaction between the plasma beam mode and electromagnetic modes. This characteristic time scale is $(2\pi/\gamma\omega_p)(8\pi n k_B T/E^2) \sim 5$ ns.

3   Observations

To test our emission model we have made ultra-high time resolution measurements of the total intensity of the Crab "giant" pulses. By applying the coherent dedispersion technique developed by one of us[3], we have obtained a time resolution of 10 ns. We find intensity structures unresolved at this resolution as shown in Figure 1. If we make the conventional assumption that the duration of a pulse corresponds to the light travel time across the emitting region, we are detecting radiation entities that are no larger than 3 meters in extent.

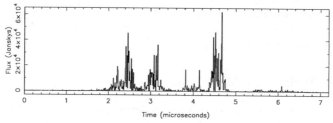

Figure 1: A single "giant" pulse from the Crab pulsar recorded at 4.8 GHz is shown containing unresolved structure even with the 10 ns resolution afforded by the VLA.

Rickett[4] developed the Amplitude Modulated Noise (AMN) model for pulsar signals where the emitted signal is modelled as a white Gaussian noise process modulated by a more slowly varying function that produces the envelope of the nano-, micro- and sub-pulses observed. We tested our data for consistency with the AMN model by computing the normalized ACF of the pulse intensities, and we found no departures from the model predictions. One interpretation is that we have not yet fully resolved the intensity structure time scale.

We have detected the *same* "giant" pulses at several narrow and widely

spaced frequencies. First, the dynamic spectrum shows that the pulse amplitudes are not strongly modulated across the 50-MHz bandpass of the VLA
at 1.4 and 4.9 GHz. We see "giant" pulses simultaneously at the following
frequency pairs (allowing for the dispersion delay): 1.3 and 1.6 GHz, 0.6 and
1.4 GHz plus 1.4 and 4.9 GHz. These measurements indicate that the intrinsic
bandwidth of the emission spans at least two octaves, and a single pulse may
cover the whole of the observed radio spectrum.

4 Conclusions

Our simulations predict that we should see intensity structure about 2 to 4
times shorter than our current observational resolution limits. The predicted
bandwidths of nanostructure are narrower than those we measure for the Crab
"giant" pulse. It could be that the pulses we see are an ensemble of narrower-
bandwidth short pulses, each of which has a ≈10% instantaneous bandwidth,
making up a wider-band pulse.

Both the theoretical and observational limits must be pushed further to
try to close the loop on the pulsar emission problem. We have developed tools
that are permitting us to look into regimes never before examined. We expect
to develop tests of the emission signatures that can be compared both with
observations and simulations. The simulations are the direct connection with
the theoretical models. We expect to be able to validate or refute several of the
theories that have been proposed and close the loop with observational tests.

Acknowledgments

This work has been supported by NSF Grant AST9315285 to New Mexico
Institute of Mining and Technology. We also acknowledge help from the Very
Large Array, which is a facility of the National Radio Astronomy Observatory,
operated by Associated Universities, Inc., under a cooperative agreement with
the National Science Foundation.

References

1. J. Weatherall, *Astrophys. J*, in press, (1997).
2. E. Asseo, G. Pelletier, and H. Sol, *MNRAS* **247**, 529 (1990).
3. T. H. Hankins, *Astrophys. J.* **169**, 487 (1971).
4. B. J. Rickett, *Astrophys. J.* **197**, 185 (1975).

GENERATION OF PULSAR GLITCHES: A SUPERFLUID CORE MODEL

A. SEDRAKIAN & J. CORDES

Center for Radiophysics and Space Research, Cornell University,
Ithaca, NY 14853

We show that the neutron star's crust-core interface acts as a potential barrier on the peripheral neutron vortices approaching this interface and thus prevents their continuous decay on it in the course of equilibrium state deceleration. The barrier arises due to the interaction of vortex magnetic flux with the Meissner currents set up by the crustal magnetic field at the interface. While the friction force decreases due to the spin-down of the star, the Magnus force remains fixed. When the non-balanced part of the Magnus force reaches the value at which the vortices are able to annihilate at the interface, the rapid transfer of angular momentum from the superfluid spins-up the observable crust on short dynamical coupling times.

Models of generation of pulsar glitches are supposed to explain the following observational facts: (i) the short spin-up time scales, which are less than 120 s in the Vela pulsar 0833–45 and less than an hour in the Crab pulsar 0531+21; (ii) the magnitudes of the jumps in the rotation and spin down rates, $\Delta\nu/\nu \sim 10^{-8} - 10^{-6}$ and $\Delta\dot{\nu}/\dot{\nu} \sim 10^{-3} - 10^{-2}$, respectively; and (iii) the origin of the instability driving a glitch along with characteristic intervals between glitches (typically of the order of several months to years). A number of existing generic models invoke as trigger crust- and core-quakes [1] (discontinuous adjustments of the solid crust to the gradually changing oblateness of the star as it spins down); spontaneous quantum transitions of rotating superfluid between quasistationary eigenstates corresponding to different eigenvalues of total angular momentum [2]; collective unpinning of a large number ($\sim 10^{13}$) of vortices in the neutron star crusts [3,4] and thermal instabilities [5]. The increasing bulk of observational evidence provides good chances of discrimination between the theoretical models in the future.

Here we shall give a brief account of a new trigger mechanism for generation of pulsar glitches in neutron star's superfluid core. The complete discussion will be given in ref. 6. In the interjump epoch a neutron star is decelerating; consequently the vortex lattice in the superfluid core is expanding and the peripheral vortices attempt to cross the crust-core boundary. The crust-superfluid core interface acts as a potential barrier on the vortices in the superfluid core that approach this boundary. The barrier arises due to the magnetic interaction between the crustal magnetic field H_0 (which penetrates the superconducting core exponentially within a scale δ_p - the penetration depth) and quantum vortices whose magnetic field is governed by the generalized London equation

$$\delta_p^{-2} \vec{\nabla} \times (\vec{\nabla} \times \vec{B}_v) + \vec{B}_v = \vec{\nu}_p \, \Phi_0 \delta^{(2)}(\vec{r} - \vec{r}_p) + \vec{\nu}_n \, \Phi_1 \delta^{(2)}(\vec{r} - \vec{r}_n). \qquad (1)$$

Here $\vec{\nu}$'s are circulation unit vectors; subscripts p and n refer to protons and neutrons; Φ_0 is the flux quantum carried by proton vortices; and Φ_1 is the non-quantized flux of neutron vortices due to the entrainment effect (i.e. the effect of superconducting proton mass transport by the neutron superfluid circulation, see ref. 7). The total field $\vec{B}$ is the superposition of $\vec{B}_v$ of eq. (1) and crustal filed $\vec{B}_{cr} = \vec{H}_0 \exp(-\vec{r} \cdot \vec{n}/\delta_p)$, ($\vec{n}$ being the normal of the interface).

Suppose that the interface is the (yz)-plane of a Cartesian system of coordinates and the vectors of vortex circulation are in the positive z-direction. The half-plane $x < 0$ corresponds to the crust while $x > 0$ corresponds to the superfluid core. The knowledge of magnetic field distribution with the boundary condition at the interface $B_z = H_0$, allows one to calculate the relevant part of the Gibbs free energy of the system $G = F - (4\pi)^{-1} \int \vec{B} \cdot \vec{H}_0 dV$, where the free-energy is

$$F = \frac{\delta_p^2}{8\pi} \int \left[\vec{B} \times (\vec{\nabla} \times \vec{B}) \right] \cdot d\vec{S} + \frac{\delta_p^2}{8\pi} \int \vec{B} \cdot [\, \delta_p^{-2} \vec{B} + \vec{\nabla} \times (\vec{\nabla} \times \vec{B}) \,] \, dV. \qquad (2)$$

The force per single vortex of effective flux $\Phi_* (\equiv \Phi_0; \Phi_1)$ derived from G is

$$f(x) = \frac{\Phi_*}{4\pi} \left[\frac{H_0}{\delta_p} e^{-x/\delta_p} - \frac{\Phi_*}{2\pi\delta_p^3} K_1 \left(\frac{2x}{\delta_p} \right) \right]. \qquad (3)$$

Here K_1 is the modified Bessel function. The first term in equation (3) is the repulsive force acting between the vortex magnetic flux and the crustal magnetic field. It can also be interpreted as a Lorentz force resulting from superposition of velocity fields of the vortex and the surface Meissner currents. The second term is the attractive force acting between the vortex and the interface. For large distances the repulsive term dominates; (the attractive one goes to zero faster since $K_1(2x) \propto \sqrt{\pi/4x} \, e^{-2x}$ for $x \to \infty$). For small distances the second (attractive) term in eq. (3) dominates. Thus the repulsive part of the vortex - crust-core interface interaction, which dominates at large distances, acts as a potential barrier on a vortex approaching the boundary. The Magnus force remains locally fixed, while the friction force decreases due to the spin down of the star. When the disbalance between this forces reaches the value at which the vortices annihilate at the interface, the resultant rapid transfer of angular momentum from the superfluid spins-up the crust.

Let us next estimate the interaction. For relevant densities, $\rho \simeq 2 \times 10^{14}$ g cm^{-3}, we have $\delta_p \simeq 100$ fm (e. g. ref. 7) and assuming a conventional value for the crustal magnetic field $H_0 = 10^{12}$ G, we find a maximal repulsive force $f^{\max} = 3.13 \times 10^{12}$ dyn cm^{-1}. In general, the magnitude of the crustal magnetic field at the crust-core interface for different objects can vary in a reasonable range $10^9 < H_0 < 10^{13}$ G. We find that, the maximal force increases (decreases) by two orders of magnitude when the crustal magnetic field is increased (decreased) by an order of magnitude.

Further progress needs to specify the ground state structure of vortices in the superfluid core. For the vortex cluster model of ref. 7, 8 (no residual field when the superconducting state sets on or a complete Meissner expulsion of residual field) the number of proton vortices per neutron vortex interacting with the interface is $N \leq 283$ and the maximal force is $f_C^{\max} = N f^{\max} = 8.9 \times 10^{14}$ dyn cm^{-1}. It should sustain the Magnus force excess (i.e. the part which is non-balanced by the friction along the normal $\vec{n}$) acting on a neutron vortex $\delta f^M = 3.23 \times 10^{17}$ $(\delta \omega_s / \mathrm{s}^{-1})$ dyn cm^{-1}, where $\delta \omega_s$ is the angular velocity difference between the superfluid and the normal components.[a] ¿From the balance condition $f_C^{\max} = \delta f^M$, the maximal departure that can be sustained by the boundary force on the cluster is $\delta \omega_s^{\max} \simeq 0.003$ s^{-1}. Then the angular momentum conservation in a Vela-type glitch implies that the ratio of the moment of inertia of the superfluid region to the normal component should be $I_s / I_n = 0.023$. This value is close to a previous estimate of the ratio $I_s / I_n = 0.02$ for a superfluid shell at the crust-core boundary with short dynamical coupling times $(\leq 120$ s$)$ [8].

Acknowledgments

One of us (A.S.) greatfully acknowledges a research grant from the Max-Kade-Foundation, NY.

References

1. M. Ruderman, *Nature* **223**, 597 (1969); *ApJ* **382**, 576 (1991).
2. R. E. Packard, *Phys. Rev. Lett.* **28**, 1080 (1972).
3. P. W. Anderson and N. Itoh, *Nature* **256**, 25 (1975).
4. A. Alpar, P. Anderson, D. Pines, and J. Shaham, *ApJ* **249**, L33 (1981).
5. B. Link and R. I. Epstein, *ApJ* **457**, 844 (1996).
6. A. D. Sedrakian and J. M. Cordes, in preparation.
7. A. D. Sedrakian and D. M. Sedrakian, *ApJ* **447**, 305 (1995).
8. A. D. Sedrakian, D. M. Sedrakian, J. M. Cordes, and Y. Terzian, *ApJ* **447**, 324 (1995).

---

[a]The neutron star model used in our estimates is discussed i detail in ref. 8.

Entrainment in a mixture of two superfluids and its effect on the dynamics of neutron stars

Majid Borumand, Robert Joynt & Włodzimierz Kluźniak[1]
*1150 University Ave. Physics Department, Madison,
WI 53706, USA*

[1] *and Copernicus Astronomical Center
ul. Bartycka 18, 00-716 Warszawa, Poland*

The correct expressions for superfluid densities responsible for coupling the motion of two superfluids are derived. Implications of this coupling in relaxation between normal and superfluid components of a neutron star are discussed. The relaxation time scale is of importance in the context of post-glitch behaviour and gravitational radiation instability of neutron stars.

1 Introduction

Ninety percent of the neutron star interior is made of superfluid neutrons. A superfluid rotates by forming vortices. The amount of circulation around a closed curve depends on how many vortices are enclosed by the curve (Alpar *et al.* 1984),

$$\oint_C \vec{v} \cdot \vec{dl} = N\kappa,$$

where N is the number of vortices inside the curve and κ is the quantum of vorticity. For a superfluid rotating with angular velocity of Ω we have:

$$R\Omega \cdot 2\pi R = N\kappa \text{ or } \Omega = \frac{N\kappa}{2\pi R^2}.$$

So by changing N one can change Ω. There are two relevant questions regarding the dynamics of the core of the neutron star: a) What is the mechanism of interaction between the charged component of the star and these vortices? b) What is the time scale for relaxation between the superfluid and the normal (crust) component of the star?

2 Entrainment

In 1984 Alpar *et al.* introduced the "drag" effect as the most efficient mechanism for coupling between normal and charged components of the star. Because of entrainment, velocity of one component in a mixture of two superfluids will create mass currents of both spices. For the mixture of superfluid neutrons and superconducting protons in the core of a neutron star:

$$\vec{g}_p = \rho_{pp}\vec{v}_p + \rho_{pn}\vec{v}_n,$$

$$\vec{g}_n = \rho_{np}\vec{v}_p + \rho_{nn}\vec{v}_n.$$

Entrainment causes the neutron vortices to become magnetized. Scattering of degenerate, relativistic electrons off these vortices couples the superfluid and crust of the star. The superfluid densities must satistfy the following constraints:

$$\rho_{pp} + \rho_{pn} = \rho_p, \rho_{nn} + \rho_{np} = \rho_n$$

$$\rho_{pn} = \rho_{np}$$

Alpar *et al.* (1984), Lindblom & Mendell (1994) and Sedrakian & Sedrakian (1995) have used different expressions for superfluid densities but none of them satisfies the above constraints. Here we use the Fermi liquid theory to derive the exact expressions for these densities.

3   Fermi liquid theory

In a Fermi gas particles do not interact and energy of a particle in state $\vec{k}$ is independent of the other particles, $\epsilon(\vec{k}) = k^2/2m$. In a Fermi liquid particles are strongly interacting with each other. It is assumed that to every energy state of a Fermi gas there correspondes one state of the Fermi liquid. Energy of quasiparticle in a Fermi liquid is a functional of the distribution function,

$$\varepsilon = \varepsilon[n(\vec{k})],$$

$$\varepsilon(\vec{k}) = \varepsilon_0(\vec{k}) + \sum_{\vec{k}'} f(\vec{k}, \vec{k}')\delta n(\vec{k}'),$$

in which $\varepsilon_0(\vec{k})$ is the energy of the quasi particle corresponding to the ground state distribution.

4   Derivation of densities

Let us assume protons and neutrons are in the ground state,

$$n_0^n = \theta(k - k_{fn}),$$

$$n_0^p = \theta(k - k_{fp}).$$

Now let's move neutrons with velocity $\vec{v}_n$, $(v_n \ll v_{fn})$,

$$\delta n^n = m_n \vec{v}_n \cdot \nabla_{\vec{k}} \theta(k - k_{fn}),$$

$$\delta \epsilon^p(\vec{k}) = \Sigma_{\vec{k}'} f^{np}(\vec{k}, \vec{k}') \delta n^n(\vec{k}').$$

Putting this in a linearized Landau equation :

$$\frac{\partial \delta n^p(\vec{k}, \vec{r}, t)}{\partial t} + \nabla_{\vec{r}} \delta n^p(\vec{k}, \vec{r}, t) \cdot \nabla_{\vec{k}}(\epsilon_0^p) - \nabla_{\vec{k}} n_0^p \cdot \nabla_{\vec{r}}(\delta \epsilon^p) = 0$$

and summing over $\vec{k}$ we arrive at the continuity equation. We can extract $\vec{g}_p$ from the continuity equation and as a result we get:

$$\rho_{np} = \rho_{pn} = \frac{1}{9\pi^4} m_n m_p k_{Fn}^2 k_{Fp}^2 f_1^{np}.$$

One can derive the other two densities in a similar manner (Borumand et al. 1996), they are:

$$\rho_{pp} = \frac{m_p^2}{m_p^*} n_p (1 + F_1^{pp}/3),$$

$$\rho_{nn} = \frac{m_n^2}{m_n^*} n_n (1 + F_1^{nn}/3).$$

Using these correct expressions for the densities and the relation for the magnetic field of a vortex given by Alpar et al. 1984, we find the coupling time (Borumand et al. 1996) between core and crust in the case of Vela pulsar to be 56 seconds. Considering a factor of 2 uncertainty in the value of the Landau parameters (F_1) and the effective masses one would say entrainment is responsible for the crust-core coupling of Vela.

This work supported in part by KBN grant No. 2-1244-91-01.

References

1. A. Alpar, S. Langer & J. Sauls, 1984, ApJ. **282**, 533.
2. M. Borumand, R. Joynt & W. Kluźniak, 1996, Physical Review C **54**, 2745.
3. L. Lindblom & G. Mendell, 1994, ApJ. **421**, 689.
4. A. Sedrakian, & D. Sedrakian, 1995, ApJ. **447**, 305.

DYNAMICS OF PSR J0045-7319/B-STAR BINARY AND NEUTRON STAR FORMATION

DONG LAI

Theoretical Astrophysics, 130-33, California Institute of Technology
Pasadena, CA 91125

Recent timing observations have revealed the presence of orbital precession due to spin-orbit coupling and rapid orbital decay due to dynamical tidal interaction in the PSR J0045-7319/B-star binary system. They can be used to put concrete constraints on the age, initial spin and velocity of the neutron star.

One of the fundamental questions in the studies of pulsars concerns the physical conditions of neutron star at birth. Of particular interest is the initial spin periods and velocities of pulsars, as they are related to such issues as supernova explosion mechanism and gravitational wave emission from core collapse. The PSR J0045-7319 binary (containing a 0.93 s radio pulsar and a massive B-star companion in an eccentric, 51 days orbit [1]) is unique and important in that it is one of the two binary pulsars discovered so far that have massive main-sequence star companions (The other one is PSR B1259-63). These systems evolve from MS-MS binaries when one of the stars explode in a supernova to form a neutron star. Thus the characteristics of such pulsar binaries can potentially be used to infer the physical conditions of neutron star formation. The PSR J0045-7319 system, in particular, owing to its relatively small orbit and "clean" environment (the mass loss from the B-star is negligible), exhibits interesting dynamical orbital behaviors, which allow for concrete constraints on the initial spin and kick of the pulsar.

Dynamics of PSR J0045-7319/B-star Binary

Spin-Orbit Coupling: Based on earlier timing data, it was suggested that classical spin-orbit coupling is observable in the PSR J0045-7319/B-star binary [2]. This spin-orbit coupling results from the flattening of the B-star due to its rapid rotation: the dimensionless distortion is related to the spin rate Ω_s by $\varepsilon \sim \Omega_s^2/(GM_c/R_c^3)$, and the resulting quadrupole moment is $Q \sim k M_c R_c^2 \varepsilon$, where M_c, R_c are the mass and radius of the B-star, k is a constant measuring the mass concentration inside the star. There are two effects associated with this spin-induced quadrupole moment: (i) The *advance of periastron* due to perturbation in the interaction potential $\Delta V \sim GM_p Q/r^3$ (where M_p is the pulsar mass, r is the separation); (ii) When the B-star's spin $\mathbf{S}$ is misaligned with the orbital angular momentum $\mathbf{L}$, there is an interaction torque $N \sim (GM_pQ/r^3)\sin\theta$ (where θ is the angle between $\mathbf{S}$ and $\mathbf{L}$) between the spin and the orbital motion, giving rise to *precessions of $\mathbf{S}$ and $\mathbf{L}$* around a fixed $\mathbf{J} = \mathbf{L} + \mathbf{S}$. Both of these effects have been confirmed by recent observation[3].

Rapid Orbital Decay: Recent timing data also reveal that the orbit is decaying[3], on a timescale of $P_{\rm orb}/\dot{P}_{\rm orb} = -0.5$ Myr (shorter than the lifetime of the $8.8M_\odot$ B-star and the characteristic age of the pulsar). Since mass loss from the B-star is negligible (as inferred from dispersion measure variation), the orbital decay must have a dynamical origin. It was suggested that dynamical tidal interaction can do job[4]: Each time the pulsar passes close to the B-star, it excites internal oscillations (mainly g-modes) in the star, transferring orbital energy to the stellar oscillations. An interesting prediction of this theory is that in order for the energy transfer to be sufficiently large to explain the observed orbital decay rate, $\mathbf{S}$ and $\mathbf{L}$ must be not only misaligned but also more or less anti-aligned. The reason that *retrograde rotation* can significantly increase the tidal strength is the following: During a periastron passage, the most strongly excited modes are those (i) propagating in the same direction as the orbital motion, (ii) having frequencies in the inertial frame comparable to the "driving frequency" (equal to twice of orbital frequency at periastron), and (iii) coupling strongly to the tidal potential. Since the higher-order (lower frequency) g-modes have smaller couping coefficients than the low-order ones, the trade-off between (ii) and (iii) implies that the dominant modes in energy transfer are those with frequencies higher than the resonant mode. If the B-star were nonrotating, the dominant modes would be g_5-g_9, and the inferred $\dot{P}_{\rm orb}$ would be two orders of magnitude too small to explain the observed value. However, a retrograde rotation "drags" the wave modes backwards and reduces the mode frequencies in the inertial frame. As a result, energy transfer is dominated by lower-order modes, which couple much more strongly to the tidal potential. At $\Omega_s = -0.4(GM/R^3)^{1/2}$ (projected along the $\mathbf{L}$ axis), for example, the dominant modes are g_3-g_5, and the energy transfer increases by two orders of magnitude as compared to the nonrotating value. A recent analysis[5] of the radiative damping of g-modes indicates that an additional ingredient, i.e., differential rotation, is needed to compensate for the longer damping times of lower-order g-modes.

Constraints on the Initial Conditions of PSR J0045-7319

Evidence for Supernova Kick: As mentioned before, the PSR/B-star binary evolves from a MS-MS binary. At this earlier stage, the B-star's spin is most likely to be aligned with the orbital angular momentum. The only way to

636

transform this aligned configuration into the current misaligned configuration is that the supernova was asymmetric and gave the pulsar a kick [2,3], and the kick velocity must have nonzero components (i) in the direction out of the original orbital plane, and (ii) in the direction opposite to and with magnitude larger than the original orbital velocity. Let the total mass, semimajor axis, eccentricity of the system before and after the supernova be $(M_i, a_i, 0)$ and (M_f, a_f, e_f). The kick velocity is given by

$$|V| = (GM_f/a_f)^{1/2} \left[2\xi - 1 + \xi\eta^{-1} - 2(1 - e_f^2)^{1/2}\xi^{3/2}\eta^{-1/2}\cos\theta\right], \quad (1)$$

where $\eta = M_f/M_i < 1$ and $\xi = a_f/a_i$. With $(1 + e_f)^{-1} \leq \xi \leq (1 - e_f)^{-1}$ and $125° \leq \theta \leq 155°$ (as constrained by the measurement of precessions [3] and for retrogarde rotation; tighter constraint can be obtained using the observed surface velocity of the B-star, but it depends on the precession phase), we get $|V| \gtrsim (GM_f/a_f)^{1/2} \simeq 125\ \mathrm{km\,s^{-1}}$, where we have used the current observed values for a_f and e_f. Orbital evolution since the supernova tends to make a_f and e_f larger, hence decreases this lower limit.

Age of the Binary and Initial Spin of the Pulsar: The theory of dynamical-tide induced orbital decay gives a scaling relation [4]

$$\dot{P}_{\rm orb} \propto P_{\rm orb}^{-7/3-4\nu}(1 - e)^{-6(1+\nu)}, \quad (2)$$

and a similar relation for $\dot{e}$, where ν lies in the range $0.2 - 1.0$, reflecting the uncertainty in the rotation rate. The proportional constant depends on the (uncertain) mode damping time. Using the observed value of $\dot{P}_{\rm orb}$ for the current system, this constant can be fixed, and the equations can be integrated backward in time. It was found that regardless of the uncertainties, the age of the binary since the supernova is less than 1.4 Myr. This is significantly smaller than the characteristic age (3 Myr) of the pulsar, implying that the latter is not a good age indicator. The most likely explanation for this discrepancy is that the initial spin period of the pulsar is close to its current value. Thus the pulsar was either formed rotating very slowly, or has suffered spin-down due to accretion in the first $\sim 10^4$ years (the Kelvin-Holmholtz time of the B-star) after the supernova (E. van den Heuvel, private communication).

This research is supported by the Richard C. Tolman Fellowship at Caltech and NASA Grant NAG 5-2756.

1. V.M. Kaspi, et al. ApJ, 423, L43 (1994).
2. D. Lai, L. Bildsten, and V.M. Kaspi, ApJ, 452, 819 (1995).
3. V.M. Kaspi, et al. Nature, 381, 584 (1996).
4. D. Lai, ApJ, 466, L35 (1996).
5. P. Kumar and E.J. Quataert, ApJ, submitted (1997).

OPTICAL OBSERVATIONS OF ISOLATED NEUTRON STARS

PATRIZIA A. CARAVEO

Istituto di Fisica Cosmica del CNR, Via Bassini, 15 - 20133 Milano, ITALY

Owing also to recent HST observations, nine isolated neutron stars have been associated with an optical counterpart, the emission of which can be recognized as being either non-thermal or thermal.

1 The Identification Methods

In order to pinpoint the counterpart of an Isolated Neutron Star (INS) in the optical domain one must use a signature capable of unambigously identify it. Traditionally, pulsation is the distinctive character most widely used to identify the counterpart of a pulsar at wavelengths other than radio. However, in recent years the availability of new ground based as well as HST instruments has added several potential candidates too faint to be studied through fast photometry. Therefore, most of the recently proposed identifications are based on positional coincidence. While more accurate positioning, possibly leading to a better coincidence, can corroborate an identification, the use of some additional distinctive source parameter, such as its proper motion, can definitely secure it. This method is unambigous as timing and can be easily applied also to faint targets such as the newly proposed INS counterparts.

2 The Data

Table 1 lists the optical identifications proposed so far for classical (i.e. non-msec) pulsars. For each object we specify the method used for identification (timing, proper motion or positional coincidence), the value of the proper motion, either measured in the radio (r) or in the optical (o), and the distance, either measured through radio or optical parallactic displacement or estimated from the radio dispersion measure or from absorption of the soft X-ray flux. However, as mentioned above, not all the identifications have the same degree of certainty. Apart from Crab, Vela and PSR0540-69, identified through pulsation, only the identification of Geminga is certain (see Bignami and Caraveo, 1996). The remaining five identifications are based just on positional coincidence: while the last three FOC identifications are new, the study of PSR1509-58 and PSR0656+14 (now observed also with HST) has been pursued at length with the ESO NTT (New Technology Telescope). Table 1 shows the optical-vs-radio displacements as a measure of the goodness of the positional

Table 1: INSs with an optical identification

PULSAR	ID	PM(mas/yr)	Distance (pc)
CRAB	Timing	13 ± 4 (o)	2,000
PSR0540-69	Timing	-	55,000
PSR1509-58	≤ 0.35" (NTT)	-	4,400
VELA	Timing	52 ± 5 (o)	500
GEMINGA	Proper Motion	169 ± 6 (o)	160 (HST parallax)
PSR0656+14	≤ 0.5" (WFPC2)	70 ± 10 (r)	200-600 (X-ray abs.)
PSR1055-52	≤ 0.1" (FOC)	-	500-1000 (X-ray abs.)
PSR0950+08	≤ 1.83" (FOC)	34 ± 9 (r)	130 (Radio parallax)
PSR1929+10	≤ 0.39" (FOC)	106 ± 7 (r)	Conflicting Radio values

coincidence. This parameter ranges from 0.1" for PSR1055-52 (Mignani et al, 1997), whose identification of which is quite solid, to 1.8" for PSR0950+08 (Pavlov et al. 1996), which is a much more doubtful case.

Table 2 summarizes the data collected so far. Here, for each object we give the measured V magnitude and the different kind of data available (timing, polarization, spectrum, colors). Unfortunately, INSs are faint optical emitters and our knowledge on their optical phenomenology is hampered by their low brigthness. ¿From the relatively bright Crab, for which medium resolution spectroscopy as well as phase resolved photometry and polarization are possible, we have to step down to V $\sim$ 22-23 to find PSR 0540-69, similar to Crab but in the LMC, Vela and the proposed counterpart of PSR 1509-58. For these objects, too faint for spectroscopy, timing analysis together with polarization and multicolor photometry are the tools to be used. For the remaining 5 objects, all fainter than V $\sim$ 25, only multicolor photometry is possible.

Thermal vs. non-thermal radiation

The examination of this limited sample shows that in optical, like in X-rays, there are two classes of mechanisms at work: thermal and non-thermal. Non-thermal processes are powered by the star rotation through its electromagnetic energy loss. In the optical range, non-thermal magnetospheric emission has been firmly detected for Crab, Vela and PSR0540-69, and possibly for PSR1509-58. For older objects, thermal emission takes over the non-thermal one. Owing to temperatures in the region of $10^5 - 10^6$ K, thermal emission from neutron stars are Planckians, from either the star's solid crust or its surrounding atmosphere. This emission peaks in the EUV-soft X-ray band, as

Table 2: Summary of optical observations available for all the INSs identified so far

PULSAR	V mag	Timing	Pol	Photometry
CRAB	16.6(*)	Ground and HSP	Y	J to U, spectrum
PSR0540-69	22.4	Ground and HSP	Y	I,R,V,B,U
PSR1509-58	22.0	negative	-	R,V,B
VELA	23.6	Ground	-	R,V,B,U
GEMINGA	25.5	-	-	I,R,V,B 555,432,342,190
PSR0656+14	25.1	Ground? Shearer et al.	-	I,R,V,B, 555,130L
PSR1055-52	24.9 (U)		-	342
PSR0950+08	27.1 (130L)		-	130L
PSR1929+10	25.7 (U)		-	130L,342

(*)a hint of secular decrease at a rate of 0.008 +/- 0.004 mag/y has been reported by Nasuti et al. (1996).
Three digit numbers refer to HST filters

recently shown by ROSAT PSPC observations (Becker, 1996). Indeed, the first evidence for a neutron star atmosphere comes from the detection of a broad line, interpreted as ion-cyclotron emission (Bignami et al. 1996), superimposed to the black-body optical emission from Geminga. Also the behaviour of PSR0656+14 is interesting, with optical/UV points well above the extrapolation of the X-ray Planckian. On the other hand, the recent detection of PSR1055-52 places its flux on the extrapolation of the soft X-ray Planckian. Thus, three INSs of similar age and overall energetics seem to show a markely different behaviour in the optical/UV band: predominantly thermal for Geminga, most probably non-thermal for PSR0656+14 and little known (may be thermal) for PSR1055-52. The thermal scenario could also fit the data of PSR0950+08 and PSR 1929+10, the older INSs in our sample.

References

1. W. Becker, MPE report 263,103 1996
2. G.F. Bignami and P.A. Caraveo, Ann. Rev. A&A 34,331, 1996.
3. G.F. Bignami et al. Ap.J. Lett. 456, L111 1996.
4. R. Mignani, P. Caraveo and G.F. Bignami Ap.J. Lett. 474,L51, 1997.
5. F. Nasuti et al. A&A 314.849, 1996.
6. G.Pavlov et al.Ap.J.,467,370, 1996.
7. A.Shrearer et al.IAU Circ.6502, 1996

THE SPECTRAL ENERGY DISTRIBUTION OF THE ISOLATED NEUTRON STAR RXJ185635-3754

F.M. WALTER, L.D. MATTHEWS, P. AN, J. LATTIMER
Department of Earth and Space Sciences, University at Stony Brook
NY 11794-2100, USA

R. NEUHÄUSER
Max-Planck-Institut für extraterrestrische Physik
85740 Garching, Germany

The evidence suggests that the bright soft X-ray source RXJ185635-3754 is an isolated neutron star. We have now identified the optical counterpart. The optical and X-ray data are inconsistent with a pure blackbody, but the spectral energy distribution appears consistent with a model having sinusoidal temperature variations with latitude.

1 Introduction

Walter, Wolk, and Neuhäuser[1] identified the bright soft X-ray source RXJ185635-3754 as a nearby isolated neutron star. The X-ray source has a lightly absorbed (A_V=0.07 mag) blackbody spectrum (kT_∞=57 eV). There is no evidence for variability. The lack of an optical counterpart to V~23 mag, and the consequent $\frac{f_X}{f_V}$ >7000, rules out most other types of astrophysical objects as plausible counterparts. The low extinction and the location in the direction of the R CrA molecular cloud limits the distance to less than 120 pc. At 100 pc, the X-ray source has a radius R_∞ of 6.1 km. The temperature and the limit on the distance and luminosity are consistent with either accretion onto the surface of a neutron star, or emission from the hot surface of a young neutron star. Since there is no evidence that the target is a pulsar (it is not a radio source), this object may give us a clean view of the surface of a neutron star without complications from non-thermal emission in a magnetosphere.

2 The Optical Counterpart

Extrapolating the blackbody fit from the X-rays gives a predicted V magnitude of 26.8. We obtained time on the Hubble Space Telescope to use the WFPC2 to detect the optical counterpart and obtain V and U magnitudes. The observations consist of a 4400 sec exposure in the wide-V (F606W) filter and a 2400 sec exposure in the wide-U (F300W) filter. We identified a blue object at right ascension $18^h 56^m 35.48^s$, declination -37° 54' 36.9" (J2000). This

Ferlet *et al.* (1996) have reviewed D/H measurements obtained with the Copernicus, IUE, and GHRS instruments. Since the analysis of new LOS observed with the GHRS is proceeding very rapidly, the conclusions that one can now draw from the data have changed. The previous studies of Lyman line absorption toward both hot and cool stars with the Copernicus and IUE satellites left a confused picture in which the uncertainties in D/H for individual LOS were large and the possiblilty of spatial variations in D/H by a factor of 2 or larger was consistent with the data. The flood of beautiful new GHRS spectra has changed this picture dramatically. The first clear indication of this paradigm shift was the measurement of D/H = $(1.60^{+0.14}_{-0.19}) \times 10^5$ for the Capella LOS (Linsky *et al.* 1995). Since this GHRS result lies outside of the published error bars for all previous results for this LOS to a bright star, the older results are likely unreliable because of systematic errors.

GHRS echelle spectra have far higher S/N and spectral resolution than Copernicus and IUE. Since the core of Lyman-α is highly saturated, high S/N and spectral resolution are critical for inferring the H column density which is more uncertain than the D column density. This alone may explain much of the previous scatter in the D/H values. Another critical issue is the presence of many velocity components in the LOS. Two or more velocity components are often observed even for short LOS. Ultra-high resolution spectra of the NaI and CaII lines for many LOS show many closely spaced narrow velocity components (e.g., Welty *et al.* 1996). Components with column densities orders of magnitude smaller than the main absorber would not be detected in metal lines but could be optically thick in the H Lyman-α line and change the inferred H column densities if not taken into account. For example, the analysis of GHRS spectra of the nearest (1.3 pc) stars, α Cen A and B (Linsky & Wood 1996), required a second absorption component in the LOS with a column density 1/1000 that of the main component, a red shift of about 4 km s^{-1}, and a high temperature (30,000 K). The most likely explanation is a "hydrogen wall" around the Sun created by charge exchange reactions between the inflowing interstellar gas and the solar wind leading to a pileup of neutral hydrogen atoms (cf. Baranov and Malama 1995). Given the highly saturated nature of the main component, the inclusion of this extra component raises the inferred D/H ratio from $\approx 6 \times 10^{-6}$ to $(1.2 \pm 0.7) \times 10^{-5}$. We have now identified hydrogen walls around other stars. Another potential source of systematic error is the unknown stellar Lyman-α emission line, but we have learned to minimize this problem by observing high radial velocity stars and spectroscopic binary systems at opposite quadratures.

Figure 1 shows the derived D/H ratios for 12 stars with interstellar radial velocities indicating that their LOS pass through the Local Interstellar Cloud

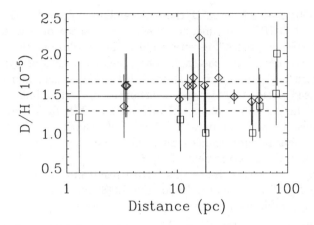

Figure 1: D/H ratios for interstellar gas toward all nearby stars observed with the GHRS. Diamond symbols are for gas in the LIC and square symbols are for other warm clouds.

(LIC) and 7 stars with LOS that pass through other warm clouds. The mean value for the LIC is $(D/H) = (1.50 \pm 0.10) \times 10^{-5}$ and the 1σ error bars for all 12 data points are consistent with the errors in the mean value. The data for other clouds are more scattered with $(D/H) = (1.28 \pm 0.36) \times 10^{-5}$ and the scatter may indicate real D/H variations. The figure shows the mean relation for all data points, $(D/H) = (1.47 \pm 0.18) \times 10^{-5}$. These results will be described in detail elsewhere. We conclude that the value of D/H in the tiny region of the Galaxy occupied by the LIC is now known, but we are just beginning to sample more distant lines of sight. STIS and FUSE data will allow us to study D/H further out in the Galactic disk and in the halo.

References

1. D. Tytler, S. Burles, & D. Kirkman, *Ap.J.* in press.
2. R. Lallement *et al.*, *A&A* **304**, 461 (1995).
3. R. Ferlet *et al.* in *Science with the HST-II*, ed. B. Benvenuti *et al.* (Space Telescope Science Institute, Baltimore, 1996).
4. J.L. Linsky *et al.*, *Ap.J.* **451**, 335 (1995).
5. D. Welty *et al.*, *Ap.J. Suppl.* **106**, 533 (1996).
6. J.L. Linsky & B.E. Wood, *Ap.J.* **463**, 254 (1996).
7. V.B. Baranov & Y.G. Malama, *JGR* **100**, 14755 (1995).

CONSTRAINTS ON THE MASS AND RADIUS OF PULSARS FROM X-RAY OBSERVATIONS OF THEIR POLAR CAPS

G.G. PAVLOV

Pennsylvania State University, 525 Davey Lab,
University Park, PA 16802, USA

V.E. ZAVLIN, J. TRÜMPER

Max–Planck–Institut für Extraterrestrische Physik,
D-85740 Garching, Germany

The properties of X-ray radiation from the polar caps predicted by the radio pulsar models depend on the surface chemical composition, magnetic field and star's mass and radius as well as on the cap temperature, size and position. Fitting the radiation spectra and light curves with the neutron star atmosphere models enables one to infer these parameters. We present here results obtained from the analysis of the soft X-ray radiation of PSR J0437−4715. In particular, with the aid of radio polarization data, we put constraints on the pulsar mass-to-radius ratio.

1 Introduction

Current models of radio pulsars [1,2,3] predict a typical polar cap (PC) size $R_{\rm pc} \sim (2\pi R^3/Pc)^{1/2}$ and PC temperatures $T_{\rm pc} \sim 3 \times 10^5 - 6 \times 10^6$ K, depending on the model adopted. The spectra of the thermal PC radiation are mainly determined by the temperature, gravitational acceleration, magnetic field and chemical composition of emitting layers (atmospheres). The light curves of the pulsed PC radiation depend not only on the orientation of the magnetic and rotation axes, but also on the magnetic field and chemical composition which affect the angular distribution of radiation, and the neutron star (NS) mass-to-radius ratio, M/R, which determines the gravitational bending of the photon trajectories. The best candidates for the investigation of the PC radiation are nearby, old pulsars for which both the nonthermal radiation from relativistic particles and thermal radiation from the entire NS surface are expected to be negligibly faint. Here we report results obtained for PSR J0437−4715 ($P = 5.75$ ms, $\tau = P/2\dot{P} = 5 \times 10^9$ yr, $B \sim 3 \times 10^8$ G, $d \approx 180$ pc). *ROSAT* observations of this object have revealed [4] smooth pulsations with the pulsed fraction $\sim 25 - 50\%$ growing with the photon energy. Such behavior has been predicted [5] for the radiation emergent from the NS atmospheres with low magnetic fields $B < 10^{10}$ G. Making use of the low-field NS atmosphere models [5], we model PC radiation with allowance for the gravitational effects [6] and compare the results with observational data.

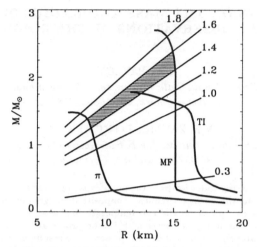

Figure 1: NS mass-radius diagram with the lines of constant values of M/R (the numbers in units of $M_\odot/10$ km) and the $M(R)$ curves for soft (π) and hard (TI and MF) equations of state of superdense matter[10]. The shaded region indicates the mass-radius domain compatible with $\zeta = 40°$ and $\alpha = 35°$.

2   Results

Our analysis has shown[7] that both the spectra and the light curves obtained with *ROSAT* and *EUVE* can be interpreted as thermal radiation from two PCs with radii $R_{\mathrm{pc}} = 0.8 - 0.9$ km (comparable to 1.9 km expected from the simple estimate) covered with hydrogen or helium at a temperature $T_{\mathrm{pc}} = (8 - 9) \times 10^5$ K. Neither the blackbody nor iron atmosphere models give acceptable spectral fits. Those results were obtained for fixed $M = 1.4 M_\odot$ and $R = 10$ km, and the angles between the pulsar rotation axis and the line of sight, $\zeta = 40°$, and between the rotation and magnetic axes, $\alpha = 35°$, evaluated from the phase dependence of the radio polarization position angle[8]. We checked that these parameters do not affect significantly the results of the spectral fits (PC temperature and size). However, the values of α, ζ and M/R drastically affect the shape of the light curve and the pulsed fraction.

To constrain these parameters, we (i) fitted the *ROSAT* PSPC count rate spectrum with the hydrogen atmosphere models on a grid of ζ, α and M/R values and obtained the corresponding set of R_{pc} and T_{pc}; (ii) used this set to compute the model *ROSAT* PSPC light curves and evaluated their deviations (χ^2 values) from the observed light curve; (iii) calculated the confidence regions in the ζ-α plane at trial values of M/R. If there were no observational information about the α, ζ values, the only constraint on the M/R ratio would be $M < 1.6M_\odot(R/10$ km$)$, or $R > 8.8(M/1.4M_\odot)$ km, at the 99% confidence level. If, however, we adopt $\zeta = 40°$ and $\alpha = 35°$, the M/R ratio lies in the range $1.4 < (M/M_\odot)/(R/10$ km$) < 1.6$ (Fig. 1). This means that if the pulsar mass is $M = 1.4M_\odot$, its radius is in the range $R = 8.8 - 10.0$ km. An alternative set of angles[9], $\zeta = 24°$ and $\alpha = 20°$, yields $(M/M_\odot)/(R/10$ km$) < 0.3$, leading to very low masses, $M < 0.5M_\odot$, at any R allowed by the equations of state.

Thus, the analysis of the PC X-ray radiation in terms of the NS atmosphere models provides a new tool to constrain the NS mass and radius and the equation state of the superdense matter. A similar analysis of X-ray radiation from pulsars with strong magnetic fields, $B \sim 10^{12}$ G (e. g., PSR B1929+10) would enable one to additionally constrain the magnetic field strength.

Acknowledgments

The work was partially supported through the NASA grant NAG5-2807, IN-TAS grant 94-3834 and DFG-RBRF grant 96-02-00177G. VEZ acknowledges the Max-Planck fellowship.

References

1. A.F. Cheng and M.A. Ruderman, *Astrophys. J.* **235**, 576 (1980).
2. J. Arons, *Astrophys. J.* **248**, 1099 (1981).
3. V.S. Beskin, A.V. Gurevich and Ya.N. Istomin, *Physics of the Pulsar Magnetosphere* (Cambridge Univ. Press, Cambridge, 1993).
4. W. Becker and J. Trümper, *Nature* **365**, 528 (1993).
5. V.E. Zavlin, G.G. Pavlov and Yu.A. Shibanov, *Astron. Astrophys.* **315**, 141 (1996).
6. V.E. Zavlin, Yu.A. Shibanov and G.G. Pavlov, *Astron. Let.* **21**, 141 (1995).
7. V.E. Zavlin, G.G. Pavlov, W. Becker and J. Trümper, *Astron. Astrophys.*, submitted (1997).
8. R.M. Manchester and S. Johnston, *Astrophys. J.* **441**, L65 (1995).
9. J. Gil and A. Krawczyk, in *Pulsars: Problems and Progress*, IAU Coll. 160, eds. S. Johnston, M.A. Walker and M. Bailes (ASP, San Francisco, 1996).
10. S. Shapiro and S. Teukolsky, *Black Holes, White Drawfs and Neutron Stars* (Wiley, New York, 1983).

ABUNDANCE PATTERNS IN LOW-REDSHIFT DAMPED LYMAN-ALPHA QSO ABSORBERS

DAVID M. MEYER

*Department of Physics and Astronomy, Northwestern University,
Evanston, IL 60208, USA*

Recent observations of species such as Zn II, Cr II, Fe II, Mn II, and Ti II in low-redshift ($z < 1.5$) damped Lyman-alpha absorbers are discussed. The Zn II, Cr II, and Fe II column densities yield Cr/Zn and Fe/Zn gas-phase abundance ratios that are typically much greater than even those of low-density Galactic ISM sightlines and imply that the absorbers are quite dust-poor. There is also evidence in some cases of a Mn underabundance and a Ti overabundance that is consistent with the nucleosynthetic signature of a gas cloud with a base metallicity similar to that of metal-poor Galactic stars.

1 Introduction

Due to their low-ionization metal absorption and large H I column densities, the damped Lyα systems observed at various redshifts in the spectra of QSOs are believed to sample the progenitors of modern galaxies at various epochs.[1] Since the absorption lines are sensitive to the metallicity and dust content of these systems, they can be used to probe galactic chemical evolution back to epochs that are coeval with Population II star formation in the Milky Way.

The Zn II $\lambda\lambda2026$, 2063 doublet is particularly useful as a metallicity indicator for damped QSO absorbers since Zn II is the dominant ion of Zn in H I clouds, Zn is not depleted much into dust in the Galactic ISM, and the Zn II lines are not heavily saturated.[2,3] In a Zn II survey of about 25 damped Lyα systems at $z > 1.5$, Pettini *et al.*[4] find an average Zn metallicity of about 10% solar at $z \approx 2$ with the most metal-poor systems at $z > 2.5$. In addition, based on observations of the nearby Cr II $\lambda\lambda2056$, 2062, 2066 triplet, Cr is much less depleted into dust in the $z > 1.5$ damped absorbers than in the Galactic ISM with implied dust-to-gas ratios that are about 3% of the Milky Way value.[5] If the damped Lyα systems are indeed sampling a population of evolving galaxies, one would expect these metal and dust abundances to increase at lower redshifts. In the first test of this idea at a redshift that roughly corresponds to the epoch of the Sun's formation, Meyer and York[6] found that the metallicity and dust content of the $z = 0.692$ damped system toward the QSO 3C286 is similar to that of the $z > 1.5$ absorbers. With imaging of the 3C286 field revealing a low surface brightness galaxy as the likely absorber[7], the question is whether 3C286 is a special case or the beginning of a trend.

Table 1: Damped Lyα QSO absorber gas abundances at $z < 1.5$.

QSO	z_{abs}	[Fe/H]	[Mn/Fe]	[Ti/Fe]	Reference
0935+417	1.373	−1.22	−0.39	+0.26	Meyer et al.[8]
0449−134	1.267		−0.28		Lu et al.[9]
1247+267	1.223	−1.29	−0.32		Lanzetta et al.[10]
0450−132	1.174		−0.22		Lu et al.[9]
0014+813	1.112		−0.10	+0.42	Roth & Songaila[11]
0454+039	0.860	−1.00	−0.36		Lu et al.[9]
2206−199	0.752		−0.10	+0.27	Prochaska & Wolfe[12]
3C336	0.656	−1.21	−0.32		Steidel et al.[13]

2 The Data

The study of damped Lyα systems at $z < 1.5$ has been accelerating recently
with a number of new identifications obtained through *IUE* and *HST* archival
searches as well as the insightful use of Keck 10 m HIRES data. For example,
Meyer, Lanzetta, and Wolfe[8] have acquired KPNO 4 m echelle observations
of the Zn II, Cr II, Fe II, Mn II, and Ti II absorption from the damped Lyα
system at $z = 1.373$ toward the QSO 0935+417. The Zn metallicity of this
system is about 10% solar and the Cr/Zn and Fe/Zn abundance ratios imply
little, if any, dust. This lack of dust also helps to account for the detection of
the weak Ti II λ1911 doublet which has yet to be seen in the Galactic ISM
due to the heavy depletion of Ti from the gas phase. Most importantly, the
paucity of dust opens a pathway to using ratios such as Mn/Fe and Ti/Fe as
nucleosynthetic markers. Table 1 lists the [Mn/Fe] and [Ti/Fe] ratios measured
to date (where $[X/Y] = \log(X/Y)_{obs} - \log(X/Y)_{solar}$) for damped Ly$\alpha$ systems
at $z < 1.5$. The [Fe/H] ratios are also listed for those systems with measured
H I column densities. The only other system besides the 0935+417 case with
a definitive Zn measurement in the Table 1 sample is that of 0454+039[14] and
it is also consistent with a 10% solar metallicity.

3 Discussion

The most striking feature of Table 1 is the fact that the low-redshift damped
Lyα absorbers consistently exhibit [Mn/Fe] < 0 whereas [Mn/Fe] ≥ 0 in even
the most dust-poor Galactic ISM sites. Along with the fragmentary [Zn/H] and
[Fe/H] measurements, this result implies that there is still no chemical evidence

yet of a metal-rich, dusty damped system at $z < 1.5$. It is possible that this finding is a manifestation of the Fall and Pei [15] selection effect where damped Lyα absorber samples are biased against dusty, metal-rich cases because the background QSOs are preferentially missed in flux-limited spectroscopic surveys. However, the H I column densities of the low-redshift sample in Table 1 are typically low enough that even a Galactic dust-to-gas ratio would not have excluded the QSOs in question from any survey.

Although there does not appear to be much dust in the $z < 1.5$ damped Lyα systems, the abundance data alone cannot rule out the possibilities of either some dust or no dust. However, given the Galactic dust depletion pattern where species such as Fe, Mn, and Ti exhibit similar depletions in dust-poor sightlines [16], this same dust insensitivity makes ratios such as [Mn/Fe] useful as nucleosynthetic probes. Thus, it is interesting that the absorbers in Table 1 exhibit Mn underabundances and Ti overabundances (relative to Fe) that are characteristic of old Galactic stars with $\approx 10\%$ solar metallicities. [17,18] Efforts are currently underway to image the galaxies responsible for these absorbers.

References

1. A.M. Wolfe in *QSO Absorption Lines*, ed. G. Meylan (Springer-Verlag, Berlin, 1995).
2. M. Pettini *et al.*, *Ap. J.* **426**, 79 (1994).
3. D.M. Meyer and K.C. Roth, *Ap. J.* **363**, 57 (1990).
4. M. Pettini *et al.* in *QSO Absorption Lines*, ed. G. Meylan (Springer-Verlag, Berlin, 1995).
5. M. Pettini *et al.*, preprint (1996).
6. D.M. Meyer and D.G. York, *Ap. J.* **399**, L121 (1992).
7. C.C. Steidel *et al.*, *A. J.* **108**, 2046 (1994).
8. D.M. Meyer, K.M. Lanzetta, and A.M. Wolfe, *Ap. J.* **451**, L13 (1995).
9. L. Lu *et al.*, *Ap. J. Suppl.* **107**, 475 (1996).
10. K.M. Lanzetta, D.M. Meyer, and A.M. Wolfe, in preparation (1997).
11. K.C. Roth and A. Songaila, in preparation (1997).
12. J.X. Prochaska and A.M. Wolfe, *Ap. J.* **474**, 140 (1997).
13. C.C. Steidel *et al.*, *Ap. J.*, in press, (1997).
14. C.C. Steidel *et al.*, *Ap. J.* **440**, L45 (1995).
15. S.M. Fall and Y.C. Pei in *QSO Absorption Lines*, ed. G. Meylan (Springer-Verlag, Berlin, 1995).
16. B.D. Savage and K.R. Sembach, *Ann. Rev. Ast. Ap.* **34**, 279 (1996).
17. R.G. Gratton, *Astr. Ap.* **208**, 171 (1989).
18. R.G. Gratton and C. Sneden, *Astr. Ap.* **241**, 501 (1991).

INITIAL CHEMICAL ENRICHMENT IN GALAXIES

Limin Lu[a], Wallace, L. W. Sargent, Thomas A. Barlow

Caltech, 105-24, Pasadena, CA 91125

We present evidence that damped Lyα galaxies detected in spectra of quasars may not have started forming stars until the redshift $z \sim 3$. If damped Lyα absorbers are the progenitors of disk galaxies, then the above result may indicate that star formation in galactic disks first began at $z \sim 3$.

1 Introduction

One definition of the epoch of galaxy formation is when galaxies first began to form stars. Different types of galaxies (e.g., ellipticals and spirals) or different parts of the same type of galaxies (e.g., bulge and disk of spirals) may have formed at different epochs. A important goal of observational cosmology is to identify these different formation epochs.

Conventional wisdom suggests that ellipticals and the spheroidal component of spirals formed very early, followed by disks. The population of galaxies at $z > 3$ identified using the Lyman limit drop-out technique may very well be the progenitors of the spheroidal component of massive galaxies [1,2]. Here we discuss evidence for a sharp rise in the metallicity distribution of damped Lyα absorbers at $z \leq 3$, which may signify the onset of star formation in galactic disks.

2 Results

Damped Lyα (DLA) absorption systems seen in spectra of background quasars are widely accepted to be the progenitors of present-day galaxies [3], although their exact nature (dwarfs, spheroids, or disks?) is still unclear. A program is carried out using the Keck telescopes to study the chemical compositions of the absorbing gas in DLA systems. One of the goals is to (hopefully) identify the epoch of the first episode of star formation in these galaxies, hence constraining theories of galaxy formation.

Figure 1 shows the distribution of [Fe/H] in DLA systems as a function of redshift. Detailed descriptions of the data and analyses are given in refs [4,5]. The low metallicities of DLAs testify the youth of these galaxies: they have yet to make the bulk of their stars. Remarkably, all 6 of the highest redshift absorbers have [Fe/H]≤ -2; while many absorbers have reached ten times

---

[a]Hubble Fellow

649

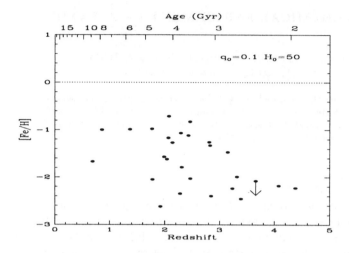

Figure 1: Metallicity distribution of damped Lyα absorbers.

higher metallicity at just slightly lower redshifts. This indicates an epoch of rapid star formation at $z \sim 3$. The effect is likely to be real: if DLA systems at $z > 3$ have [Fe/H] that is uniformly distributed between -1 and -2.5 (i.e.,, similar to the distribution at $2 < z < 3$), then the *posterior* probability for all six of the highest redshift systems to have [Fe/H]≤ -2 by chance is 1.4×10^{-3}.

Coincidentally, the metallicities of DLA systems at $z > 3$, [Fe/H]$= -2$ to -2.5, is identical (within the uncertainties) to those found for the IGM clouds at similar redshifts, as inferred from the C IV absorption associated with Lyα forest clouds [6,7,8]. This coincidence suggests that the metals in DLA galaxies at $z > 3$ may simply reflect those in the IGM, however they were made (e.g., Pop III stars, ejected from protogalaxies); DLA galaxies did not start making their own stars (hence metals) until $z \sim 3$.

3 Discussion

The implications of the above result for the general question of galaxy formation and evolution depend on the nature of the DLA galaxies.

It was suggested [9] that DLA systems may represent the progenitors of disk galaxies. This is supported by the very recent finding [10] that the kinematics of DLA absorbers as inferred from the metal absorption line profiles appears to be dominated by rotation with large circular velocities (> 200 km s^{-1}). However, the mean metallicities of DLAs at $z > 1.6$ are significantly below that of the Milky Way disk at the corresponding epoch [4,11]. The problem with the metallicity distribution may be lessened if DLAs represent a thick disk phase of galaxies [3] or if low surface brightness disk galaxies (which have substantially sub-solar metallicities) make up a significant fraction of DLA absorbers [12]. *If* the disk hypothesis for DLA absorbers is correct, we may have identified the epoch of initial star formation in disk galaxies.

Alternatively, DLAs may represent dwarf galaxies or the spheroidal component of massive galaxies; this conjecture stems from the similarity between the metallicity distribution of DLAs and those in halo globular clusters and local gas-rich dwarf galaxies [4]. In this case, however, one has to explain the kinematics of DLAs [10] by other means.

Acknowledgments

LL appreciates support from a Hubble Fellowship (HF1062-01-94A). WWS was supported by NSF grant AST95-29073.

References

1. Steidel, C.C., Giavalisco, M., Pettini, M., Dickinson, M., & Adelberger, K.L. 1996, ApJ, L17
2. Giavalisco, M., Steidel, C.C., & Macchetto, F. 1996, ApJ, 470, 189
3. Wolfe, A.M. 1995, in *QSO Absorption Lines*, ed. G.Meylan (Springer-Verlag), p13
4. Lu, L., Sargent, W.L.W., Barlow, T.A, Churchill, C.W., & Vogt, S. 1996, ApJS, 107, 475
5. Lu, L., Sargent, W.L.W., & Barlow, T.A. 1997, in preparation
6. Cowie, L.L., Songaila, A., Kim, T.S., & Hu, E.M. 1995, AJ, 109, 1522
7. Tytler, D., Fan, X.-M., Burles, S., Cottrell, L., Davis, C., Kirkman, D., & Zuo, L. 1995, in *QSO Absorption Lines*, ed. G.Meylan (Springer-Verlag), p289
8. Sargent, W.L.W., Womble, D.S., Barlow, T.A., & Lyons, R.S. 1997, in preparation
9. Wolfe, A.M., Turnshek, D.A., Smith, H.E., & Cohen, R.D. 1986, ApJS, 61, 249
10. Prochaska, J.X., & Wolfe, A.M. 1997, ApJ, submitted
11. Pettini, M., Smith, L.J., Hunstead, R.W., & King, D.L. 1994, ApJ, 426, 79
12. Lu, L., Sargent, W.L.W., & Barlow, T.A. 1997, ApJ, submitted

THE SIZE AND SHAPE OF LYMAN ALPHA FOREST CLOUDS

ARLIN P.S. CROTTS

*Dept. of Astronomy, Columbia University, 550 W. 120th St., New York,
NY 10027, USA*

Recent observations have shown that Ly α clouds are much larger than previously thought, while even more recent hydrodynamical/gravitational models indicate that clouds of this size tend to be flattened or elongated. These size estimates came from observations of QSO sightlines adjacent to each other in the sky. Adjacent triplets of sightlines might betray the clouds' shape. Furthermore, QSOs at different redshifts can show how these characteristics might change over time. We discuss new observations of close pairs and triplets of QSOs which give qualitatively new information on the size evolution and shape of absorbers among the Ly α forest. These results indicate that there is little evidence for evolution of Ly α absorber size, but that they are *not* simply unclustered, uniformed-sized spheres. Evidence supports the clouds being flattened disks expanding in the Hubble flow, but argues against them being filaments.

1 Introduction and Observations

This paper summarizes results found in two recently submitted papers: Crotts and Fang [1], and Crotts, Burles and Tytler [2]. They involve spectra from pairs and triplets of QSOs observed in order to study the Ly α forest's behavior in the transverse dimension between sightlines. Furthermore, evidence is claimed for evolution of the cloud size with redshift [3] (from the 1017-0232/0107-0235, with $\langle z \rangle = 0.7$ in the forest), which we test here.

The first paper [1] involves observations taken on the Kitt Peak National Observatory's 4-meter telescope with the RC Spectrograph, covering 3170-4720Å at 1.7Å resolution, with typical signal-to-noise ratio $S/N = 60$ for the QSO triplet (Q1623+2653, $z = 2.526$; Q1623+2651A, $z = 2.467$; Q1623+2651B, $z = 2.605$; separations of 127″, 147″, and 177″) and typical $S/N = 30$ for the 102″-separation QSO pair Q1517+2356 ($z = 1.903$) and Q1517+2357 ($z = 1.834$). Down to 1700Å, the Q1517+2356/Q1517+2357 pair was observed with *HST*'s FOS. The KPNO 4m/RC Spec was also used to observe $S/N \approx 25$, 1.4Å-resolution spectra for the triplet for 4450-5750Å.

The second paper [2] involves 7.9 km s^{-1}-resolution, $S/N \approx 20$ spectra of the triplet, obtained with Keck/HIRES, covering 3872-6299Å (with small gaps beyond 5165Å).

2   Results and Interpretation

The sightline separations, for $q_o = 1/2$, correspond to proper distances of $432\ h^{-1}$ kpc for the Q1517+2356/7 pair and 508, 588 and 702 h^{-1} kpc for the triplet. In comparison, the most recent estimates of the size of Ly α clouds[4] converge to proper diameters $\approx 300\ h^{-1}$ kpc (for sightlines approaching zero separation). A feature is seen[4] when cross-correlation between the Ly α forest distributions for sightlines closer than $\approx 0.8\ h^{-1}$ Mpc is made, showing a surplus of pairs within 200 km s^{-1} of each other. This excess of close pairs is seen only, however, for lines stronger than rest equivalent width $W_o = 0.4$Å, except for the one QSO pair at smallest separation, $40\ h^{-1}$ kpc (QSO pair Q1343+2640A/B[5]), where it is also seen for weaker lines.

We have developed a Bayesian statistical estimator for the cloud radius assuming that the clouds are 1) spherical, 2) of uniform size, and 3) unclustered. We calculate this radius for different sightline pairs (for $W_o > 0.4$Å), including all of the above mentioned pairs as well as 0307-1931/0307-1932 [6], both at $z \approx 2.0$. The results of this calculation of most probable model cloud radius R as a function of sightline separation S and redshift z shows a significant trend with S, $\partial R / \partial S = 0.37 \pm 0.18$, but an insignificant decrease of radius with increasing redshift, $\partial R / \partial z = -13$ kpc$\pm$81kpc. These results allow us to conclude, first, that the model assumption is false, given the lack of a consistent R value. This implies either *the forest clouds are non-spherical, not of uniform size, or clustered* (or some combination thereof). Also, *size evolution is not established*, but indicates that we need further close, low-z pairs.

We use the triplet data to investigate the possibility of non-spherical clouds. Within the 200 km s^{-1} cross-correlation interval defined above, a "hit" refers to a detection in two sightlines. We compute a probability for such occurrences for a simple model of a uniformly circular cross-sectioned cylinder oriented at random with respect to the sightline, for different cross-sectional radii R and length-to-width aspect ratios a. A hit on sightlines A and B affects the conditional probability of a line in A given another line within 200 km s^{-1} in sightline B, $[b|a] = [a|b] \equiv P_{ab}$. A three-way hit is described by $P_{abc} \equiv ([a|bc] + [b|ac] + [c|ab])/3$.

The observed hit statistics constrain P_{ab}, P_{ac}, P_{bc}, and P_{abc} which can then be compared to their model values from the simple rod model. The four P constraints agree at the 95% confidence level only for $a < 4$, with a best value of $a = 1.4$, $R = 300\ h^{-1}$ kpc. (68% confidence intervals correspond to $1 < a < 2.0$ abd 230 h^{-1} kpc$< R < 330\ h^{-1}$ kpc.) This conclusion applies only to lines stronger than $W_o = 0.4$Å; for weak lines there is no clear hit signal on 500 h^{-1} kpc scales. This implies, at least for strong systems, that Ly

α *forest clouds are roughly circular in cross-section,* not elongated as seen for many objects of this size in recent numerical models.

For weaker lines, the lack of a detectable excess contributing to P_{abc} precludes a shape determination, but in the closest pair ($S = 40\ h^{-1}$ kpc) there are three hits and four "misses" for lines weaker than $W_o = 0.4$Å, implying a median probability $R = 63\ h^{-1}$ kpc. While smaller than the implied size for more strongly absorbing clouds, this is still large compared to luminous portions of galaxies and probably C IV absorber sizes.

Since clouds flattened in one dimension, disks, can have a nearly circular cross-section even when accounting for typical inclination effects, three-way hit statistics e.g. P_{abc} have little leverage on cloud shape. If such disks are still expanding with the Hubble flow along their unflattened dimensions, however, adjacent sightlines might detect this as a velocity shear between sightlines. For the four $W_o = 0.4$Å absorbers spanning all three sightlines of the triplet, we find a distribution in shear velocity consistent with the Hubble flow expectation for any of several reasonable cosmologies ($[\Omega_o, \Lambda/3H_o^2] = [1,0], [0.1,0], [0.1,0.9]$). Furthermore, this is supported by the observation that the strongest systems, $W_o > 0.8$Å, cluster within 600 km s^{-1} along adjacent sightlines, as might be expected in a correlation between W_o and Δv, the velocity difference between sightlines, caused by inclination of sheets of absorbing gas expanding in the Hubble flow. This evidence tends to support Ly α clouds being sheet-like in shape, as opposed to filaments as discounted above.

Additionally, we find that two of the four $W_o > 0.4$Å absorbers spanning all three sightlines are associated with C IV absorption, compared to only 1/9 of the overall $W_o > 0.4$Å sample. A third $W_o > 0.4$Å absorber lands at the same redshift as a foreground QSO. There is some evidence that these absorbers spanning the sightlines preferentially include collapsed, star-forming objects.

References

1. A.P.S. Crotts and Y. Fang, *Ap. J.* , submitted (1996).
2. A.P.S. Crotts, S. Burles and D. Tytler, *Ap. J. Let.* , submitted (1997).
3. N. Dinshaw, C.D. Impey, R.J. Weymann and S.L. Morris Nature **373**, 223 (1995).
4. Y. Fang, R.C. Duncan, A.P.S. Crotts and J. Bechtold *Ap. J.* **462**, 77 (1996).
5. J. Bechtold, A.P.S. Crotts, R.C. Duncan and Y. Fang *Ap. J. Let.* **437**, 79 (1994).
6. P.A. Shaver and J.G. Robertson *Ap. J.* **268**, 57 (1983).

LARGE–SCALE STRUCTURE AS SEEN FROM QSO ABSORPTION–LINE SYSTEMS

J.M. QUASHNOCK, D.E. VANDEN BERK, D.G. YORK

Department of Astronomy and Astrophysics, University of Chicago,
5640 S. Ellis Avenue, Chicago, IL 60637, USA

We study clustering on very large scales — from several tens to hundreds of co-moving Mpc — using an extensive catalog of heavy–element QSO absorption–line systems. We find significant evidence that C IV absorbers are clustered on comoving scales of $100 \, h^{-1}$ Mpc ($q_0 = 0.5$) and less. The superclustering is present even at high redshift ($z \sim 3$); furthermore, it does not appear that the superclustering scale (comoving) has changed significantly since then. Our estimate of that scale increases to $240 \, h^{-1}$ Mpc if $q_0 = 0.1$, which is larger than the largest scales of clustering seen at the present epoch. This may be indicative of a larger value of q_0, and hence Ω_0. We identify 7 high–redshift supercluster candidates, with 2 at redshift $z \sim 2.8$. The evolution of the correlation function on $50 \, h^{-1}$ Mpc scales is consistent with that expected in cosmologies with $\Omega_0 =$ ranging from 0.1 to 1. Finally, we find no evidence for clustering on scales greater than $100 \, h^{-1}$ Mpc ($q_0 = 0.5$) or $240 \, h^{-1}$ Mpc ($q_0 = 0.1$).

It has been recognized for some time now that QSO absorption line systems are particularly effective probes of large–scale structure in the universe.[1] This is because the absorbers trace matter lying on the QSO line of sight, which can extend over a sizable redshift interval out to high redshifts. Thus, the absorbers trace both the large–scale structure and its evolution in time, since the clustering pattern can be examined as a function of redshift out to $z \sim$ 4. The evolution of large–scale structure is of great interest, since, in the gravitational instability picture, it depends sensitively on Ω_0.[2]

Here we study clustering by computing line–of–sight correlations of C IV absorption line systems, using a new and extensive catalog of absorbers.[3] (A more complete version of this work has appeared elsewhere.[4]) This catalog contains data on all QSO heavy–element absorption lines in the literature. It is an updated version of the York et al. (1991) catalog,[5] but is more than twice the size, with over 2200 absorbers listed over 500 QSOs, and is the largest sample of heavy–element absorbers compiled to date.

Figure 1 shows the C IV line–of–sight correlation function, ζ_{aa}, as a function of absorber comoving separation, Δr, for the entire sample of absorbers. The results are shown for both a $q_0 = 0.5$ (*left panel*, $25 \, h^{-1}$ Mpc bins) and a $q_0 = 0.1$ (*right panel*, $60 \, h^{-1}$ Mpc bins) cosmology.[a] The vertical error bars

---

[a]Larger bins are required for $q_0 = 0.1$ because, at high redshift, a larger comoving separation Δr arises from a fixed redshift interval Δz.

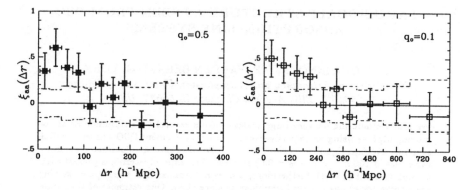

Figure 1: Line–of–sight correlation function of C IV absorbers as a function of absorber comoving separation (from Ref. 4, ©1996 by The American Astronomical Society).

through the data points are $1\,\sigma$ errors in the estimator for ξ_{aa}, which differ from the $1\,\sigma$ region of scatter (*dashed line*, calculated by Monte Carlo simulations) around the no–clustering null hypothesis.

Remarkably, there appears to be significant clustering in the first four bins of Figure 1: The positive correlation seen in the first four bins of Figure 1 has a significance of $5.0\,\sigma$. Therefore, there is significant evidence of clustering of matter traced by C IV absorbers on scales up to $100\ h^{-1}$ Mpc ($q_0 = 0.5$) or $240\ h^{-1}$ Mpc ($q_0 = 0.1$). There is no evidence from Figure 1 for clustering on comoving scales greater than these.

We have investigated the evolution of the superclustering by dividing the absorber sample into three approximately equal redshift sub–samples; namely, low ($1.2 < z < 2.0$), medium ($2.0 < z < 2.8$), and high ($2.8 < z < 4.5$) redshift. We find that the significant superclustering seen in Figure 1 is present in all three redshift sub–samples, so that the superclustering is present even at redshift $z \gtrsim 3$. Furthermore, it does not appear that the superclustering scale, in comoving coordinates, has changed significantly since then.

We have examined the clustering signal more closely and find that a large portion comes from 7 QSO lines of sight that have groups of 4 or more C IV absorbers within a $100\ h^{-1}$ Mpc interval ($q_0 = 0.5$). (From Monte Carlo simulations, we expect only 2.7 ± 1.5 QSOs with such groups.) We have found two potential superclusters, at redshift $z \sim 2.8$, among these groups.

The superclustering is indicative of generic large–scale clustering in the universe, out to high redshift $z \gtrsim 3$, on a scale frozen in comoving coordinates that is — if $q_0 = 0.5$ — similar to the size of the voids and walls in galaxy redshift surveys of the local universe.[6-9] It also appears consistent with the general finding[10,11] that galaxies are clustered in a regular pattern on very large scales, although we have not confirmed that there is quasi–periodic clustering with power peaked at $\sim 128\ h^{-1}$ Mpc.

Our estimate of the superclustering scale increases to 240 h^{-1} Mpc if $q_0 = 0.1$ (see Figure 1), which is larger than the largest scales of clustering known at present. If the structures traced by C IV absorbers are of the same nature as those seen locally in galaxy redshift surveys, the superclustering scale should have a value closer to 100 h^{-1} Mpc . This may be indicative of a larger value of q_0, and hence Ω_0.

We find that the evolution of the correlation function on 50 h^{-1} Mpc scales is consistent with that expected in cosmologies with density parameter ranging from $\Omega_0 = 0.1$ to 1.

Acknowledgments

JMQ is supported by the Compton Fellowship – NASA grant GDP93-08. DEVB was supported in part by the Adler Fellowship at the University of Chicago, and by NASA Space Telescope grant GO-06007.01-94A.

References

1. A.P.S. Crotts, A.L. Melott, and D.G. York, 1985, *Phys. Lett.* B **155B**, 251 (1985).
2. P.J.E. Peebles, *Principles of Physical Cosmology*, (Princeton Univ. Press, Princeton, 1993).
3. D.E. Vanden Berk et al., *Astrophys. J. Suppl.*, submitted, (1997).
4. J.M. Quashnock, D.E. Vanden Berk, and D.G. York, *Astrophys. J. Lett.* **472**, L69 (1996).
5. D.G. York et al., *Mon. Not. Roy. Astr. Soc.* **250**, 24 (1991).
6. R.P. Kirshner, A. Oemler, P.L. Schecter, and S.A. Shectman, *Astrophys. J. Lett.* **248**, L57 (1981).
7. M.J. Geller and J.P. Huchra, *Science*, **246**, 897, (1989).
8. L.N. da Costa et al., *Astrophys. J. Lett.* **424**, L1 (1994).
9. S.D. Landy et al., *Astrophys. J. Lett.* **456**, L1 (1996).
10. T.J. Broadhurst, R.S. Ellis, D.C. Koo, and A.S. Szalay, *Nature* **343**, 726 (1990).
11. J. Einasto et al., *Nature* **385**, 139 (1997).

THE GUNN-PETERSON EFFECT AND THE DENSITY OF INTERGALACTIC HI

S.A. LEVSHAKOV,

Department of Theoretical Astrophysics, Ioffe Physico-Technical Institute, 194021 St.Petersburg, Russia

W.H. KEGEL

Institut für Theoretische Physik der Universität Frankfurt am Main, Postfach 11 19 32, 60054 Frankfurt/Main 11, Germany

We show that in stochastic large scale velocity fields superposed on the general Hubble flow, the formation of the GP-depression in QSO spectra is intimately related to the formation of the absorption-line structure usually called 'Lyα forest'. Therefore the HI-density in the diffuse IGM might be substaintially underestimated if one determines the GP-effect from the apparent continuum in high resolution spectra of QSOs. Our tentative calculations imply a current baryon density ρ_{IGM} that is about 8 per cent of the critical density $\rho_c = 1.1 \times 10^{-29}(H_0/75)^2$ g cm^{-3}, and thus $\rho_{IGM} > \rho_{gal}$.

The diffuse intergalactic medium may be probed by observing absorption features in the spectra of QSOs. The continuum on the blue side of the Lyα emission line is expected to be depressed as compared to that on the red side, since Lyα absorption in the diffuse intergalactic medium leads to an apparently continuous absorption due to the general cosmological expansion [1]. This depression actually has been observed in many low resolution spectra [2].

Gunn and Peterson [1] [GP] estimated the depression in 3C9 to be of the order of 40 per cent, and derived for the HI density a value of the order of 6×10^{-11} cm^{-3} at $z = 2$ (with $q_0 = 0.5$ and $H_0 = 100$ km s^{-1} Mpc^{-1}).

High resolution QSO spectra, however, show numerous narrow absorption lines on the blue side of the Lyα emission line. These have been interpreted as Lyα absorption from intervening clouds [3,4]. Correspondingly the depression of the *continuum* is considerably less than estimated from low resolution spectra. Actually, it has been concluded that there is *no clear evidence* for any GP-effect [5], a result which seems to imply an extremely low HI-density. This conclusion rests on the assumption that the diffuse intergalactic medium is smoothly distributed and that line broadening is purely thermal.

However, if one allows for a large scale stochastic velocity field with a finite correlation length (*mesoturbulence*) superposed on the general Hubble flow, the Lyα absorption by an otherwise smoothly distributed diffuse medium leads to a '*line-like*' structure which is similar to the population of unsaturated lines observed in the Lyα forest [6].

Fig.1 shows an example of our Monte-Carlo [MC] calculations based on a model with $n(\text{HI}) = 4 \times 10^{-11}$ cm^{-3}, $T_{kin} = 10^4$ K, peculiar r.m.s. bulk velocity $\sigma_t = 300$ km s^{-1}, correlation length $l_z = 6$ Mpc at $z = 3$, $H_0 = 70$ km s^{-1} Mpc^{-1} and $q_0 = 0.5$. The top panel shows one realization of the stochastic velocity field along a given line of sight. – The peculiar velocities have the effect that different volume elements may contribute to the absorption at a given value of λ. This is shown in panel b. For comparison also the absorption by a purely thermal gas (no peculiar velocities) is shown by a solid line. Finally panel c shows the resulting "absorption spectrum" and for comparison the classical GP-depression (GP$_{therm}$). In the present example the latter has a value of $\tau_{GP} = 0.3$. On the other hand it is evident that one obtains a much smaller value if one determines the GP-effect from the apparent continuum in the 'Lyα forest' caused by the correlated structure of the large scale velocity field (GP$_{meso}$).

These results and those discussed in some detail in ref. [6] demonstrate that an *essential part* of the Lyα forest lines may be caused by the absorption in the *diffuse* medium *between* optically thick intervening clouds and that therefore the HI-density in the diffuse intergalactic medium may be substantially higher than that estimated from the continuum depression in high resolution QSO spectra. If ionization of the IGM is in equilibrium and the UV background radiation at epoch $z = 3$ is given by $J_\nu \simeq 3 \times 10^{-21}(\nu/\nu_c)^{-1.5}$ ergs s^{-1} cm^{-2} sr^{-1} Hz^{-1} (ref. [7]), then the order-of-magnitude estimation amounts to a current baryon density $\Omega_{IGM}h^2 \simeq 0.04$, i.e. for $h = 0.7$ it is about 8 per cent of the mass needed to close the Universe.

Acknowledgments

This work was supported in part by the Deutsche Forschungsgemeinschaft and by the RFFR grant No. 96-02-16905-a.

1. J.E. Gunn and B.A. Peterson, *ApJ* **142**, 1633 (1965).
2. J.D. Kennefick *et al*, *AJ* **110**, 2553 (1995).
3. R. Lynds, *ApJ* **164**, L73 (1971).
4. W.L.W. Sargent *et al*, ApJSS **42**, 41 (1980).
5. E. Giallongo *et al.* in *QSO Absorption Lines*, ed. G. Meylan (Springer, Berlin, 1995).
6. S.A. Levshakov and W.H. Kegel, *MNRAS* , submit. (1997).
7. H. Vedel *et al*, *MNRAS* **271**, 743 (1994).

660

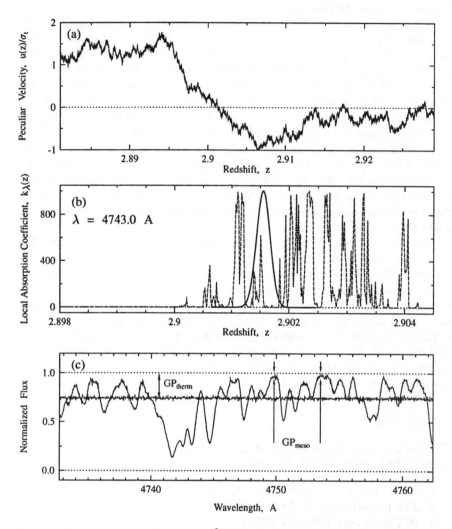

Figure 1: An example of MC simulations[6]. (a) – One realization of the stochastic velocity field $u(z)$ around $z = 2.9$ given in units of its r.m.s. value σ_t. (b) – The corresponding local absorption coefficient for fixed $\lambda = 4743$ Å (dashed curve), and for comparison $k_\lambda(z)$ for a purely thermal gas (solid curve). (c) – A portion of the resulting absorption spectrum and for comparison the GP-depression for purely thermal gas. [The spectrum is calculated with S/N = 100 and convolved with an instrumental function with FWHM = 8 km s^{-1}.]

THE STRUCTURE OF THE LYMAN ALPHA FOREST IN A COLD DARK MATTER COSMOLOGY

A. MEIKSIN

Dept. of Astronomy & Astrophysics, University of Chicago, 5640 S. Ellis Ave.,
Chicago, IL 60637

We report the results of a numerical simulation of the formation of the Lyα forest at high redshift in standard CDM. The H I absorption properties are found to match high spectral resolution observations of the forest extremely well. The absorbers originate from a variety of structures: high column density systems arise in spheroidal minihalos, intermediate systems arise in filaments, and the lowest column density systems arise in unbound structures in underdense minivoids. By contrast, most of the He II absorption is produced in the minivoids. Matching to the measured values of intergalactic He II absorption requires a soft intrinsic spectrum for QSO sources, with $\alpha_Q \approx 1.8 - 2$.

1 Introduction

The most ubiquitous discrete structures at high redshift are systems of intergalactic gas detected as absorption features in the spectra of quasars. These systems are known collectively as the Lyα forest. Because of their large numbers and occurrence over a wide range of redshifts ($0 < z < 5$), they offer a unique opportunity for testing theories of cosmological structure formation over most of the history of the universe. The low metallicity of the systems ensures that the influence of star formation and associated feedback mechanisms, effects that complicate galaxy and cluster formation simulations, are small or absent, simplifying the computational problem to a level that permits comparisons with observations that are relatively assumption free.

The results are presented here for the formation of the Lyα forest in a standard Cold Dark Matter cosmology ($\Omega_0 = 1$, $H_0 = 50 \, \text{km s}^{-1} \, \text{Mpc}^{-1}$, $\sigma_8 = 0.7$). We choose a cosmological baryon density of $\Omega_b = 0.06$, and assume a UV photoionizing background dominated by QSO sources, as determined by Haardt & Madau (1996). The simulation uses a Particle–Mesh code to evolve the dark matter, combined with an Eulerian finite–difference scheme to solve the hydrodynamics. We evolve hydrogen, helium, and their ionization products. The computational box is 9.6 comoving Mpc on a side, with a comoving resolution of 75 kpc in the top grid, and a resolution 4 times greater in a subgrid. More detailed discussions of the simulation results are provided by Zhang, Anninos, Norman, & Meiksin (1997) and Zhang, Meiksin, Anninos, & Norman (1997).

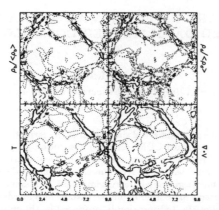

Figure 1: Baryon and dark matter overdensities, baryon temperature, and peculiar velocity divergence at $z = 3$, for a 150 kpc thick slice. The overdensity contours are 0.5 (*dotted*), 1 (*dashed*), 3 (*thin solid*), and 5 (*thick solid*). The temperature contours are at 6, 10, 14, and 20 × 10^3 K. The velocity divergence contours are at 5, 0, -3, and -15, in units of H_0.

2   Results

Figure 1 shows the baryon and dark matter overdensities through a section of the simulation at $z = 3$. The gas and dark matter collapse into a filamentary network, separated by underdense minivoids. The temperature in the minivoids is reduced by adiabatic expansion, as indicated by the positive peculiar velocity divergence in these regions. A sample H I spectrum at $z = 3$ is shown in Figure 2 (left panel). Low column density systems ($N_{HI} < 10^{13}$ cm^{-2}) are associated with fluctuations in underdense minivoids. Both the baryon and dark matter overdensities of the fluctuations are less than unity, showing

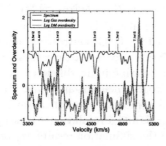

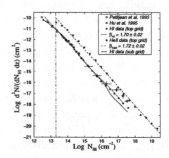

Figure 2: *Left*: Sample H I spectrum. *Right*: N_{HI} and N_{HeII} column density distributions.

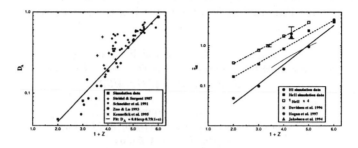

Figure 3: Evolution of D_A (*left*) and $\bar{\tau}_\alpha$ (*right*).

that the systems are gravitationally unbound. Higher column density systems ($N_{HI} > 10^{14}\,cm^{-2}$) are found to originate in filamentary systems, while the highest column density absorbers ($N_{HI} > 10^{16}\,cm^{-2}$) arise in spheroidal halos at the intersections of the filaments. The H I and He II column density distributions are shown in the right panel. Close agreement with the observed H I distribution is found.

The decrement D_A between hydrogen Lyα and Lyβ is shown in Figure 3 (left panel). Although the data show a considerable amount of scatter, the simulation reproduces the measured values reasonably well. The average intergalactic absorption, $\bar{\tau}_\alpha \equiv -\log\langle e^{-\tau}\rangle$, is shown in the right panel. The He II opacity falls below the measured values. A reduction by a factor of ~ 4 of the He II ionizing radiation background is required, indicating a soft intrinsic spectrum for the QSOs, with $\alpha_Q \approx 1.8 - 2$.

Acknowledgments

The author thanks the W. Gaertner fund at the Univ. of Chicago for support.

References

1. F. Haardt and P. Madau, *Astrophys. J.* **461**, 20 (1996).
2. Y. Zhang, P. Anninos, M. Norman, and A. Meiksin, *Astrophys. J.* (1997), in press.
3. Y. Zhang, A, Meiksin, P. Anninos, and M. Norman (1997), in preparation.

Characterization of Lyman Alpha Spectra and Predictions of Structure Formation Models: A Flux Statistics Approach

Rupert A.C. Croft, David H. Weinberg
Department of Astronomy, The Ohio State University,
Columbus, Ohio 43210, USA

Lars Hernquist
Lick Observatory, University of California,
Santa Cruz, CA 95064

Neal Katz
Department of Physics and Astronomy, University of Massachusetts,
Amherst, MA, 98195

In gravitational instability models, Lyα absorption arises from a continuous fluctuating medium, so that spectra provide a non-linear one-dimensional "map" of the underlying density field. We characterise this continuous absorption using statistical measures applied to the distribution of absorbed flux. We describe two simple members of a family of statistics which we apply to simulated spectra in order to show their sensitivity as probes of cosmological parameters (H$_0$, Ω, the initial power spectrum of matter fluctuations) and the physical state of the IGM. We make use of SPH simulation results to test the flux statistics, as well as presenting a preliminary application to Keck HIRES data.

1 Introduction

Hydrodynamical simulations are now providing us with the detailed predictions of structure formation models for what should be seen in quasar absorption spectra[1,2,3]. If we accept the same picture of formation of structure by gravitional instability that is believed to be responsible for the observed galaxy distribution, then in the same sense that galaxy redshift surveys can provide important cosmological constraints, there is a wealth of information to be extracted from the Lyα forest.

We analyse TreeSPH[4] hydrodynamic simulations of three CDM-based cosmological models : SCDM and CCDM have $\Omega_0 = 1$, $h = 0.5$ and $\sigma_8 = 0.7$, $\sigma_8 = 1.2$ respectively, OCDM has $\Omega_0 = 0.4, \Omega_\Lambda = 0, h = 0.65, \sigma_8 = 0.75$. For all, $\Omega_b h^2 = 0.05$ and a photoionizing UV background is included. Details of the simulations are given in [5]. In the models, Lyα absorption arises from a continuous fluctuating medium, the optical depth to absorption, τ, being correlated with the underlying density, which is shown in Figure 1. The empirically measured relationship, $\tau \propto \rho_b^{1.6}$ can be predicted from the interplay

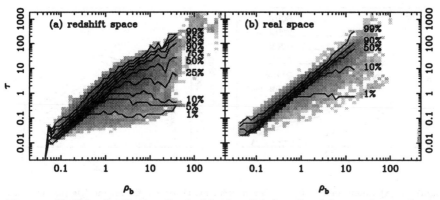

Figure 1: The joint distribution of optical depth and ρ_b (in units of the mean baryon density) for the SCDM model at $z = 3.0$. Panel (a) is in redshift space and in panel (b) peculiar velocities were set to zero. The logarithmic grey scale shows the fraction of pixels in each (ρ_b, τ) bin. Lines show the percentile distribution of τ_{HI} in each bin of ρ_b

between photoionization heating and adiabatic cooling[5,6]. The scatter in 1(a) is mainly due to the effects of peculiar velocities (set to zero in 1(b)).

2 Flux statistics

We apply two sets of statistics to the simulated spectra as well as to a Keck spectrum of Q1422+231 (observations described in [7], $\bar{z}$ of absorption = 3.2):

(1) The fraction of each spectrum (FF, or filling factor) above a given threshold in flux decrement ($D = 1 - e^{-\tau}$). The dotted lines in Figure 2(a) show the results at two different redshifts in the SCDM model. The expansion of the Universe, which results in a lower mean density of neutral hydrogen is the main factor driving the dramatic evolution in the shape of the curve and the mean optical depth. All three models fit the observations reasonably well (see also [8]) , once the strength of the UV background has been adjusted to yield the correct mean amount of absorption, with CCDM giving the best fit.

(2) N, the mean number of times per unit length a spectrum crosses a given threshold in D. We plot N against against FF instead of D (Figure 2(b)) so that the two panels give us entirely independent information. When plotted in this fashion, N is member of a family of statistics such as the distribution of distances between downcrossings and upcrossings ("size of absorbers") which is unaffected by variations in Ω_b and the UV background. N is sensitive to the length scale and therefore H_0, as well as the power spectrum of fluctuations and the temperature of the IGM (thermal broadening results in less crossings).

In Figure 2(b) we see that the three cosmological models have a smaller N

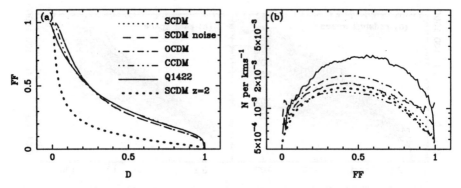

Figure 2: (a) Filling factor, FF vs flux decrement, D for 3 different models (see text) at z=3, as well as SCDM with noise, SCDM at z=2 , and a Keck spectrum of Q1422+231. (b) Number of crossings, N, per kms^{-1} vs FF. A 20 kms^{-1} tophat filter was applied to all the spectra before calculating the statistics for both panels.

than the Keck spectrum of Q1422+231, the difference being worse at high FF (low D). Adding simulated Keck noise to the SCDM model does not resolve the discrepancy. There is genuinely more structure in the QSO spectrum than in low density regions of the simulations, . At the moment it is uncertain whether this discrepancy reflects a failure of these cosmological models (e.g. they have the wrong power spectrum on these scales) or a failure of the simulations to resolve the smallest scale features in the fluctuating intergalactic medium.

Acknowledgments

We thank Toni Songaila and Len Cowie for providing us with the spectrum of Q1422+231 and Jordi Miralda-Escudé for helpul discussions. The simulations were performed at the San Diego Supercomputer Center.

References

1. J. Miralda-Escudé *et al*, ApJ **471**,582(1996)
2. Y. Zhang *et al*, ApJ **453**, L57 (1995)
3. L. Hernquist *et al*, ApJ **457**, L51 (1996)
4. N. Katz *et al* ApJS **105**, 19 (1996)
5. R. Croft *et al*, ApJ *submitted*, astro-ph/9611053 (1996)
6. L. Hui and N. Gnedin, ApJ *in press*, astro-ph/9608157
7. A. Songaila, AJ *to be submitted* (1997)
8. M. Rauch *et al*, ApJ *submitted*, astro-ph/9612245 (1997)

THE BACK REACTION OF GRAVITATIONAL PERTURBATIONS

L.R. ABRAMO, R.H. BRANDENBERGER
Physics Department, Brown University,
Providence, RI 02912, USA

V.M. MUKHANOV
Institut für Theoretische Physik, ETH Zürich,
CH-8093 Zürich, Switzerland

The back reaction of gravitational perturbations in a homogeneous background is determined by an effective energy-momentum tensor quadratic in the perturbations. We show that this nonlinear feedback effect is important in the case of long wavelength scalar perturbations in inflationary universe models. We also show how to solve an old problem concerning the gauge dependence of the effective energy-momentum tensor of perturbations.

1 Back Reaction in Chaotic Inflation

It is a well known fact that gravitational waves carry energy and momentum, and as such are themselves a source of curvature for spacetime. High-frequency gravity waves, in particular, have an equation of state of a radiation fluid ($p = \rho/3$) and this can be used to constrain their amplitude during BBN by taking into account the effect they have on the expansion and cooling rates of the universe.

Back reaction can also be quite important in the inflationary models of the Universe evolution. In the chaotic inflation model with a massive scalar field [1], for example, it is usually supposed that once the inflaton φ drops below the self-reproduction scale $\varphi_{sr} \sim m^{-1/2}$ (in Plack units), the dynamics of the homogeneous FRW background proceeds classically with no influence from metric perturbations created henceforth. Our calculations show, however, that the effect of back reaction may become crucial midway through the period of slow-roll inflation [2,3].

To see that, consider the effective energy-momentum tensor (EEMT) obtained after averaging quadratic terms of the perturbative expansion of Einstein's Equations about the FRW background,

$$\tau_{\mu\nu} = \frac{1}{16\pi} \langle (G_{\mu\nu} - 8\pi T_{\mu\nu})_{,ab} \, \delta q^a \delta q^b \rangle \,, \tag{1}$$

where $\langle \ldots \rangle$ denotes spatial averaging and $\delta q^a \equiv \{\delta g_{\mu\nu}, \delta\varphi\}$ are the perturbations to the metric and matter fields, for which $\langle \delta q^a \rangle = 0$ but $\langle (\delta q)^2 \rangle \neq 0$. This energy-momentum tensor, of second order in the perturbations, serves as an effective source in the RHS of Einstein's equations.

The energy density in long wavelength scalar (density) perturbations (*i.e.*, the 0-0 component of the EEMT) is proportional to $\langle \delta \varphi^2 \rangle$. Making use of the known spectrum of first order perturbations, it can be shown that τ_{00} becomes comparable to the energy density of the background before the end of inflation if the intial value of the scalar field was bigger than $\varphi_0 \sim m^{-1/3}$.

2 Gauge Dependence of the EEMT

Once it is established that back reaction can be relevant and that the EEMT is the proper tool for handling it, we must deal with a problem inherent to that tensor: namely, its gauge dependence.

On a given manifold, a coordinate transformation $x \to \tilde{x} = x + \xi$ (ξ small and $\langle \xi \rangle = 0$) induces a gauge transformation on the tensors of that manifold which is expressed in terms of the Lie derivative: $q \to \tilde{q} = q - \mathcal{L}_\xi q$. In particular, the matter and metric field perturbations are transformed by the same law, which perturbatively reads

$$\delta \tilde{q}^a = \delta q^a - [\mathcal{L}_\xi q_0]^a . \tag{2}$$

Since the EEMT is a function of the explicit perturbations on the fixed background, it is clear by Eq. 1 that it will change accordingly, $\tilde{\tau}_{\mu\nu} \neq \tau_{\mu\nu}$.

The question now becomes, how to calculate back reaction and be sure that it does not include spurious "gauge" effects? In order to answer this we must first clarify the origin of the mystery, *i.e.* why does the EEMT change while the background apparently does not?

3 Finite Gauge Transformations

Back reaction is a second order effect, and therefore any terms of like order should be accounted for. In particular, it must be recognized that a coordinate transformation in fact induces a gauge transformation which involve terms of all orders in perturbation theory. Rather than the simple law Eq. 2, valid only up to first order, tensor fields are transformed by the Lie operator found upon exponentiation of the Lie derivative,

$$\tilde{q} = e^{-\mathcal{L}_\xi} q . \tag{3}$$

Now it becomes clear that, although to first order the background variables do not change under a gauge transformation, to second order they do:

$$\tilde{q}_0^a = q_0^a + \langle \frac{1}{2}\mathcal{L}_\xi \mathcal{L}_\xi q_0^a - \mathcal{L}_\xi \delta q^a \rangle + \mathcal{O}(\epsilon^3) \ , \tag{4}$$

where all first order terms from Eq. 3 vanish by virtue of the spatial average.

We conclude then that, although under a coordinate transformation the EEMT change, so does the background to which it is referred. In this case we would write Einstein's equations in another frame as

$$\langle \tilde{\Pi} \rangle = \langle e^{-\mathcal{L}_\xi} \Pi \rangle = 0 \quad \rightarrow \quad \tilde{\Pi}_0 = -\frac{1}{2}\langle \Pi, ab\delta \tilde{q}^a \tilde{q}^b \rangle \tag{5}$$

where tensor indices have been suppressed for simplicity.

Above, we have shown that the back reaction equation 5 for cosmological perturbations is covariant, that is, takes the same form in any coordinate system. An obvious question is whether a gauge invariant formulation exists. The answer is yes, and the construction is in fact analogous to one that leads to the gauge invariant theory of linear perturbations [2,3].

Acknowledgments

L.R.A. is supported by CNPq (Research Council of Brazil); L.R.A. and R.B. are partially financed by U.S. DOE, contract DE-FG0291ER-40688, Task A; V.M. thanks the SNF and the Tomalla foundation for financial support; the authors also acknowledge support by NSF collaborative research award NSF-INT-9312335.

References

1. A.A. Starobinsky , in "Current Topics in Field Theory, Quantum Gravity and Strings", ed. H.J. de Vega and N. Sánchez, Lecture Notes in Physics Vol. 246 (Springer-Verlag, Berlin 1982.) See also A.S. Goncharov, A.D. Linde and V.F. Mukhanov, *Int. J. Mod. Phys.* **A2**, 561 (1987); A.D. Linde, "Particle Physics and Inflationary Cosmology" (Harwood, Chur 1990); A.D. Linde, *Phys. Scr.* **T36**, 30 (1991).
2. V. Mukhanov, L.R. Abramo and R. Brandenberger, *Phys. Rev. Lett.* (1997), in press (gr-qc/9609026).
3. L.R. Abramo, R. Brandenberger and V. Mukhanov, Brown preprint *BROWN-HET-1046* (1997).

Strong Cosmic Censorship and Causality Violation

Kengo Maeda and Akihiro Ishibashi
Department of Physics,
Tokyo Institute of Technology,
Oh-Okayama Meguro-ku, Tokyo 152, Japan

We investigate the instability of the Cauchy horizon caused by causality violation in the compact vacuum universe with the topology $B \times S^1 \times R$, which Moncrief and Isenberg considered. We show that if the occurrence of curvature singularities are restricted to the boundary of causality violating region, the whole segments of the boundary become curvature singularities. This implies that the strong cosmic censorship holds in the spatially compact vacuum space-time in the case of the causality violation. This also suggests that causality violation cannot occur for a compact universe.

1 Summary

The problem whether singularities are visible from an observer or not is an open question in general relativity. In 1969, Penrose[1] proposed a conjecture which in any realistic gravitational collapse no observer can ever see singularities. This conjecture is called the strong cosmic censorship. The mathematical precise definition of the strong cosmic censorship is as follows. It is said that the strong cosmic censorship holds if a space-time is globally hyperbolic. Namely, there is no Cauchy horizon. The strong cosmic censorship does not hold due to a causality violation because of the appearance of the Cauchy horizon. Over the past few decades a considerable number of studies have been made to elucidate the relations between causality violation and singularities. However, what seems to be lacking is the study of the relation between the strong cosmic censorship and causality violation. In this paper we restrict our attention to causality violation in a spatially compact space-time.

Moncrief and Isenberg[2] showed that causality violating cosmological solutions of the Einstein equations are essentially artifacts of symmetries. They proved that if the Cauchy horizon is compact there exists a Killing symmetry in the direction of the null geodesic generator on the Cauchy horizon by using the Einstein equations. We can easily understand this curious result by inspecting exact solutions which have causality violating regions. For example, the space-time and the Taub-NUT universe, which have compact Cauchy horizons, have Killing symmetries on the Cauchy horizons. However, the physically realistic universe is inhomogeneous and does not admit Killing symmetries. Thus we expect that spatially compact inhomogeneous universe does not have any

Cauchy horizon or, if it does, the Cauchy horizon cannot be compact from the results of Moncrief and Isenberg. In this paper, we will see that such a Cauchy horizon cannot occur by using global method[3].

We shall concentrate on the following Cauchy horizon.

Definition: the chronological Cauchy horizon.

Consider a space-time (M, g) with a partial Cauchy surface S of which Cauchy development $D^+(S)$ has compact spatial sections $\Sigma \approx B \times \mathbf{S}^1$, where B is a compact two-manifold. If the Cauchy horizon satisfies the following condition,

Let $\{q_n\}$ be a sequence of points in $D^+(S)$ which converges to a point p in $H^+(S)$. If there exists an infinite sequence L_n of closed spacelike curves which generate $\mathbf{S}^1$ factors of $D^+(S)$ and each L_n passes through q_n such that, for every point $r_n \in L_n$, the tangent vector K_n of L_n at r_n approaches to null, i.e.

$$\lim_{q_n \to p} g(K_n, K_n)|_{r_n} = 0.$$

then we call such $H^+(S)$ *the chronological Cauchy horizon.*

It is obvious that each null geodesic generator of the chronological Cauchy horizon has no future endpoint in (M, g). This means that the segments of $\dot{D}^+(S, M^+)$ are null and, if exist, singularities are restricted to the boundary of causality violating region. As mentioned above, when we speak of singularity in this paper, it means curvature singularity. The chronological Cauchy horizon is not required in general to be compact and is a generalization of the Cauchy horizon which Moncrief and Isenberg considered. If there is no curvature singularity, the chronological Cauchy horizon is compact and has closed null geodesic generators. The precise definition is given in the reference.[5]

We present a theorem in which no spatially compact space-time can have a chronological Cauchy horizon as follows.

Theorem
Let (M, g) be a spatially compact vacuum space-time which admits a regular partial Cauchy surface S diffeomorphic to $\Sigma \approx B \times \mathbf{S}^1$. If (M, g) satisfies the following conditions,
(i) the generic condition, i.e. every inextendible null geodesic contains a point at which $K_{[a}R_{b]cd[e}K_{f]}K^c K^d \neq 0$, where K^a is the tangent vector to the null geodesic,
(ii) the Cauchy horizon, if any, is the chronological Cauchy horizon $H^+(S)$,

(iii) all occurring curvature singularities are the strong curvature singularities[4],
then, (M, g) is globally hyperbolic.

We give the idea of the proof as follows.
¿From the conditions (i) ,(ii), and (iii) every null geodesic generator of $H^+(S)$ terminates at the strong curvature singularity in both future and past directions. Then, there exist points p and $q(< p)$ for each null geodesic generator such that the expansion $\theta|_p < 0$ and $\theta|_q > 0$ somewhere near the singularity in the future direction from the strong curvature singularity condition. Let us consider an infinite sequence of null geodesics l_n in $D^+(S)$ which converges to the null geodesic generator of $H^+(S)$ and take arbitrary small neighborhoods U_p, U_q of p, q respectively. ¿From the continuity there exists a number N such that all $l_n(n > N)$ have points $p_n \in l_n \cap U_p$, $q_n \in l_n \cap U_q$, $q_n > p_n$ and $\theta|_{p_n} < 0$, $\theta|_{q_n} > 0$ in the future direction. This is a contradiction because the expansion of l_n must decrease monotonously from the Raychaudhuri equation in the future direction. In detail, see the reference[5].

Acknowledgments

We would like to thank Professor A.Hosoya for useful discussions and helpful suggestions. We are grateful to T.Koike, M.Narita, K.Tamai and S.Ding for useful discussions. We are also grateful to T.Okamura and K.Nakamura for kind advice and comments.

References

1. R.Penrose, *Riv.Nuo.Cim.*1,252(1969)
2. V.Moncrief and J.Isenberg, *Com.Math.Phys.*89, 387-413(1983).
3. S.W.Hawking and G.F.R.Ellis, *The large scale structure of space-time,* Cambridge University Press, Cambridge (1973).
4. A.Królak, *J.Math.Phys.28(11)*, 2685-2687 (1987).
5. K.Maeda and A.Ishibashi, gr-qc/9607052, *Strong Cosmic Censorship and Causality Violation.*

QUANTUM INSTABILITIES OF TOPOLOGICAL DEFECTS

Ruth GREGORY

Centre for Particle Theory, University of Durham,
South Road, Durham, DH1 3LE, UK

I review the quantum instability of 'topologically stable' cosmic strings to pair production of black holes, and present selection rules for the decay channel.

Topological defects and other soliton structures have a wide application to many areas of physics. A topological defect is a discontinuity in the vacuum, and in conventional field theory can be classified in terms of the homotopy groups of the vacuum manifold, the idea being that non-trivial vacuum structure can translate into stable finite energy field configurations. Topology protects the defect since to remove it, fields must be lifted away from the vacuum over an infinite region of space. However, this assumes that *spacetime* has no topology. In this talk I want to illustrate, using the cosmic string as an example, that spacetime topology changing processes can destabilize defects and derive a selection rule for when a string can and cannot split.

Consider the usual argument for a cosmic string having no ends. In what follows we will use the U(1) abelian Higgs model for the cosmic string, however, the will be general. For reference, the lagrangian is

$$\mathcal{L}[\Phi, A_a] = D_a \Phi^\dagger D^a \Phi - \frac{1}{4} F_{ab} F^{ab} - \frac{\lambda}{4} (\Phi^\dagger \Phi - \eta^2)^2 \qquad (1)$$

where Φ is a complex scalar field, $D_a = \nabla_a + ie A_a$ is the usual gauge covariant derivative, and F_{ab} the field strength associated with A_a. The presence of a vortex, or cosmic string, is indicated by loops in the vacuum in space which correspond to a non-trivial element of $\Pi_1(M)$, i.e. around which the phase of the Higgs field changes by some integer multiple of 2π. Continuity then demands that $\Phi = 0$ at some point on any surface spanning the loop (this is the locus of the vortex), and then one argues by deforming this surface that the string must be infinite or closed.

This argument is correct in so far as it goes, however, it assumes that spacetime has no topology. If there exist surfaces spanning the loop that cannot be continuously deformed into the initial surface containing the vortex, then it is possible that such surfaces do not have a vortex piercing them, and that the string can therefore 'end'. The condition for the existence of such surfaces is that there is a 'hole' in space, and gravity provides us with such in the guise of black holes. This is all very well in theory, but can a string end *in practise?* For a string to end there are three possible obstacles

(i) *The vortex solution must be able to sit on the event horizon of the black hole. This could potentially be in conflict with 'no hair' theorems.*

The 'no-hair' theorems essentially say that the only long range information that a black hole can carry is its electromagnetic charge, mass and angular momentum. The abelian Higgs no-hair theorems[1] state that the black hole cannot carry any broken U(1)-charge, however, the proofs explicitly exclude the presence of a vortex. This first obstacle was overcome when it was demonstrated[2] that there exists a solution to the Einstein-abelian-Higgs system representing a cosmic string threading a black hole.

(ii) *There must exist a smooth field configuration that consists solely of a vortex at one pole of a sphere.*

For a string to terminate, it must be possible for the string to effectively disappear into the black hole, in other words the string must end on the event horizon which is topologically spherical. This means that there must be a smooth field configuration corresponding to a single vortex sitting on a sphere. This in fact is a well studied problem, and the answer[2,3] is to define the fields on two gauge patches which overlap in an equatorial region, and on this overlap define the gauge transition function $g_{12} = e^{i\phi}$ so that

$$\Phi_1 = g_{12}\Phi_2 \quad ; \quad A_{1\mu} = A_{2\mu} - \frac{g_{12}^{-1}}{e}\partial_\mu g_{12} \qquad (2)$$

on the overlap, and we take $\Phi_2 = A_2 = 0$. Thus we have a vacuum on the southern hemisphere and a vortex on the northern hemisphere connected via a non-singular gauge transformation on the overlap.

(iii) *There must be a solution to the fully coupled euclidean Einstein-abelian-Higgs system of equations which 'splits' the string.*

The instanton[3,4,5] describing this process is based on the C-metric, which is a lorentzian metric representing two black holes undergoing uniform acceleration away from each other connected to infinity by conical singularities. The spatial topology of this solution is $\mathbf{R}^3$ with a handle and euclideanization of this solution gives an interpolation between the initial, infinite vortex, state and the final, split vortex, state, which in the process changes the topology of space. The final step is to show that the conical singularity can be smoothed out by the vortex[6]. The tunneling amplitude for the string decay is approximately $e^{-10^{12}}$ for a GUT string, which indicates this is not a problem for the cosmic string scenario of galaxy formation!

Finally, having reviewed how one actually splits the topologically stable U(1) vortex, we would like to point out that the decay channel can be closed if the vortex is embedded in a larger – still bosonic – theory. Clearly the instanton can still be defined as a classical solution to the euclidean field equations, therefore one would think that the previous conclusions would still hold, however, to calculate a tunneling amplitude one is really using the classical euclidean solution as a saddle point approximation to the functional integral, and therefore one has to consider perturbations around the putative saddle point, and it is here that the argument can break down. If we extend our original model by the addition of extra bosonic fields then these too will have to be defined separately on each gauge patch, and transformed by the gauge transition function according to the particular representation in which they lie. A necessary and sufficient condition for the string to split is that these field configurations can all be consistently defined. This is a relic of the Dirac charge quantization conditions of the unbroken theory. A string ending on a black hole represents confined magnetic flux from a magnetically charged black hole created prior to a phase transition. Such a charged black hole must be made up of particles in the spectrum of the theory, which are known to obey quantization conditions, therefore the string can only terminate on the black hole if its flux corresponds to some magnetic monopole in the unbroken theory. In terms of the underlying symmetries of the theory a necessary condition for a string to break by pair creation of black holes or monopoles is that the group element $\mathbf{g} = \mathbf{P} \exp ie \oint A$ belong to the trivial component in $\pi_0(H)$. In addition, a string can only break by uncharged black holes if $\mathbf{g} \equiv 1$.

Acknowledgments

It is a pleasure to thank my collaborators, Ana Achucarro, Mark Hindmarsh and Konrad Kuijken. This work was supported by a Royal Society University Research Fellowship.

References

1. S.Adler and R.Pearson, *Phys. Rev.* **D18** 2798 (1978).
2. A.Achucarro, R.Gregory and K.Kuijken, *Phys. Rev.* **D52** 5729 (1995).
3. D.Eardley, G.Horowitz, D.Kastor, J.Traschen, *Phys. Rev. Lett.* **75** 3390 (1995).
4. S.W.Hawking and S.F.Ross, *Phys. Rev. Lett.* **65** 3382 (1995).
5. R.Emparan, *Phys. Rev. Lett.* **75** 3386 (1995).
6. R.Gregory and M.Hindmarsh, *Phys. Rev.* **D52** 5598 (1995).
7. A.Achucarro and R.Gregory, *Selection rules for splitting strings*, UG-1/97, DTP/97/3.

COSMOLOGICAL PAIR CREATION OF DILATON BLACK HOLES

RAPHAEL BOUSSO

DAMTP, Cambridge University, Silver Street, Cambridge, CB3 9EW, U.K.

We present a cosmological dilatonic black hole solution with unusual solution space properties. It can be used to describe the pair creation of black holes on the background of dilatonic models of inflation.

1 Einstein-Maxwell Theory

During inflation, the effective cosmological constant changes only slowly. For the purpose of describing black hole pair creation, it can be approximated by a fixed cosmological constant. Therefore, one must consider the nucleation of Reissner-Nordström-de Sitter (RNdS) black holes on the background of de Sitter space. These black holes have spacelike slices of the topology $S^1 \times S^2$, where in general the two-spheres vary in size along the one-sphere. The solution space is parametrised by the cosmological constant, Λ, the black hole mass, M, and their charge, Q.

The RNdS solution space contains three hypersurfaces of solutions that admit Euclidean sections and can therefore be nucleated. They are the extremal solutions, the so-called *lukewarm* solutions, and, finally, the *Charged Nariai* (CN) solutions, the solutions of maximal mass at a given charge. The CN solutions are characterized by the fact that the two-sphere radius is constant along the one-sphere in the spacelike sections.

In the semiclassical approximation, the pair creation rate is given by the exponential of the difference of the Euclidean actions of the black hole and background (de Sitter) solutions. The total number of black holes produced during inflation can be obtained by integrating the pair creation rate, and summing over all three black hole types and all charges. Most of these black holes will evaporate rapidly. Only magnetically charged ones are stable, since there are no magnetically charged particles they could radiate. Taking dilution effects into account, their number in the observable universe today can be estimated. This number turns out to be very small in standard Einstein-Maxwell theory. [1] It depends extremely sensitively on the values of the fine structure constant and the mass of the inflaton field. In models with a dilaton, which derive from higher-dimensional theories such as string theory, the effective values of both of these quantities could have been significantly larger in the early universe than they are today. This means the number of black holes produced should be much higher. Excessive black hole pair creation may even constrain some dilatonic models of inflation. One should therefore examine how the pair creation process can by implemented in such models.

2 Dilatonic Black Holes

We consider a fairly general class of dilatonic theories, in which the dilaton field ϕ couples exponentially to the Maxwell and the cosmological terms, with coupling constants a and b, respectively.[2] Thus the cosmological term forms a Liouville potential for the dilaton. The obvious step would be to find a dilatonic version of the RNdS solutions and go through the pair creation formalism as before. There is a no-go theorem by Poletti and Wiltshire,[3] however, which appears to stand in the way. Essentially, it states that there are no static, spherically symmetric, asymptotically de Sitter dilatonic black hole solutions with a Liouville potential. This is related to the simple fact that there is no de Sitter type solution in such dilatonic models. Black hole pair creation is thus faced with a double problem: neither is there a black hole solution, nor a background on which it could be nucleated.

Fortunately, there are loopholes in this argument. The first loophole will allow us to obtain a black hole solution. It lies in the words "asymptotically de Sitter" in the theorem. Perhaps surprisingly, not all RNdS solutions are asymptotically de Sitter. Most of them are, but the CN family is an exception. Because of the degeneracy of the horizons, the CN solutions look globally like the direct product of $1+1$-dimensional de Sitter space with a round two sphere. They contain no asymptotic $3+1$-dimensional de Sitter region. Thus they can avoid the no-go theorem, and in fact it is easy to find their dilatonic analogues.

First we note that in static, spherically symmetric solutions, the dilaton can only depend on the two-sphere radius. But in the CN solutions this radius is constant. Therefore the dilaton will be constant. Thus the exponential dilaton couplings in the action will just be constant factors and can be absorbed:

$$\tilde{F}^2 = e^{-2a\phi}F^2, \quad \tilde{\Lambda} = e^{2b\phi}\Lambda. \tag{1}$$

With these definitions, the action reduces to the non-dilatonic form. By writing the CN solutions in terms of F^2 and Λ, one can simply obtain their dilatonic cousins by adding the twiddles.

It follows from the dilatonic equation of motion that $\tilde{F}^2/2\tilde{\Lambda} = b/a$. This is an interesting result. In the non-dilatonic case, $F^2/2\Lambda$ is a degree of freedom in the solution space, related to the black hole charge. It determines the shape of the solution, while Λ determines its overall scale. In the dilatonic case, we see that this ratio is completely fixed by the exponential coupling constants of the theory. So one degree of freedom in the metric has been lost.

This degree of freedom resurfaces when one looks at the Maxwell coupling strength, $e^{-2a\phi}$. By adjusting the black hole charge, this quantity can be varied without changing the solution shape. One can thus have solutions with the same geometry, but different electro-magnetic interaction. As a corollary, one finds that the charge is not bounded from above in the dilatonic solutions.

The cosmological dilaton black holes thus obtained admit a Euclidean section. To describe a pair creation process, however, a background is needed on which they can be nucleated. It was mentioned earlier that dilatonic models with a fixed cosmological constant do not admit a de Sitter-like solution. But in inflationary scenarios one does not consider a fixed, but an effective cosmological constant, which will be large only for a finite time. During this time, the universe is similar to de Sitter space. On this background, pair creation is possible and can be described using the dilatonic CN solutions. [4]

In many ways, the pair creation process is similar to the non-dilatonic case: near the beginning of inflation, when the effective cosmological constant is close to the Planck value, pair creation is only weakly suppressed, and the nucleated black holes are very small. Later on, they will be larger, but become exponentially suppressed. There is a difference in the way they can be charged: if there is no dilaton, the maximum charge the black holes can carry increases during inflation, as the black holes grow. With the dilaton, it turns out to decrease, because the black hole growth is more than compensated by the increase in Maxwell coupling as the dilaton decreases during inflation.

While it is satisfying to understand that cosmological dilaton black hole pair creation is possible in principle, the solutions presented above are not sufficient to determine the total number of magnetically charged black holes produced. In the non-dilatonic case most of the black holes are nucleated via the lukewarm solutions. Finding their dilatonic analogue is a more complicated problem than for the CN black holes, but should be attempted in order to make a more reliable estimate of the number of charged primordial black holes.

References

1. R. Bousso and S. W. Hawking, *Phys. Rev.* D **54**, 6312 (1996).
2. R. Bousso: *Charged Nariai black holes with a dilaton*. Accepted for publication in Phys. Rev. D, gr-qc/9608053.
3. S. J. Poletti and D. L. Wiltshire, *Phys. Rev.* D **50**, 7260 (1994).
4. R. Bousso: *Pair creation of dilaton black holes in extended inflation*. Accepted for publication in Phys. Rev. D, gr-qc/9610040.

MODELLING THE DECOHERENCE OF SPACETIME

J.T. WHELAN

Department of Physics, 201 James Fletcher Building
University of Utah, Salt Lake City, UT 84112

The question of whether unobserved short-wavelength modes of the gravitational field can induce decoherence in the long-wavelength modes ("the decoherence of spacetime") is addressed using a scalar field toy model with some features of pertur-bative general relativity. For some long-wavelength coarse grainings, the Feynman-Vernon influence phase is found to be effective at suppressing the off-diagonal elements of the decoherence functional. The requirement that the short-wavelength modes be in a sufficiently high-temperature state places limits on the applicability of this perturbative approach.

1 Environment-induced decoherence in generalized QM

To review briefly, the sum-over-histories implementation of generalized quantum mechanics, as formulated by Hartle[1], assigns to each pair of field histories $\{\varphi, \varphi'\}$ a *fine-grained decoherence functional*

$$D[\varphi, \varphi'] = \rho(\varphi_i, \varphi_i')\delta(\varphi_f' - \varphi_f)e^{i(S[\varphi] - S[\varphi'])} \tag{1}$$

(in units where $\hbar = 1 = c$), where $\varphi_i = \varphi(t_i)$ and $\varphi_f = \varphi(t_f)$ are the endpoints of the path φ, and similarly for φ'. One can *coarse grain* by partitioning the field histories into classes $\{c_\alpha\}$ and summing the fine-grained decoherence functional (1) over all field histories lying in a given class:

$$D(\alpha, \alpha') = \int_\alpha \mathcal{D}\varphi \int_{\alpha'} \mathcal{D}\varphi' \, D[\varphi, \varphi']. \tag{2}$$

If this coarse-grained decoherence functional (which is a Hermitian matrix with indices α and α') is diagonal $[D(\alpha, \alpha') \approx \delta_{\alpha\alpha'}p_\alpha]$, then each diagonal element p_α is the probability that the system follows a history in class c_α. We say then that the coarse graining $\{c_\alpha\}$ *decoheres*. When the alternatives do not decohere, quantum mechanical interference prevents the theory from assigning consistent probabilities (as defined by the classical probability sum rules).

Most decoherence in physical situations arises[2] when the variables φ can be split into "system" varibles Φ which can define coarse grainings of interest and and "environment" ϕ which is only relevant due to its interaction with the system. The action can then be divided: $S[\varphi] = S_\Phi[\Phi] + S_E[\phi, \Phi]$, into a piece S_Φ which would describe the system in the absence of the environment and a

680

piece S_E which describes the enviroment and its coupling to the system. The decoherence functional $D(\alpha, \alpha')$ for a coarse graining based solely on the system variables can be written in terms of $D[\Phi, \Phi'] = \int \mathcal{D}\phi\, \mathcal{D}\phi'\, D[\varphi, \varphi']$. Assuming that the density matrix describing the initial state factors:

$$\rho(\varphi_i, \varphi_i') = \rho_\Phi(\Phi_i, \Phi_i')\rho_\phi(\phi_i, \phi_i'), \tag{3}$$

then $D[\Phi, \Phi']$ can be written

$$D[\Phi, \Phi'] = \rho_\Phi(\Phi_i, \Phi_i')\delta(\Phi_f' - \Phi_f)e^{i(S_\Phi[\Phi] - S_\Phi[\Phi'] + W[\Phi, \Phi'])} \tag{4}$$

where

$$e^{iW[\Phi, \Phi']} = \int \mathcal{D}\phi\, \mathcal{D}\phi'\, \rho_\phi(\phi_i, \phi_i')\delta(\phi_f' - \phi_f)e^{i(S_E[\phi, \Phi] - S_E[\phi', \Phi'])}. \tag{5}$$

$W[\Phi, \Phi']$ is called the Feynman-Vernon influence phase [3]; if the influence functional e^{iW} becomes small for $\Phi \neq \Phi'$, the "off-diagonal" parts of $D[\Phi, \Phi']$ will be suppressed, causing alternatives defined in terms of Φ to decohere.[2]

2 Toy Model

The model which we use to mimic general relativity is that of a scalar field φ on Minkowski space with action

$$S[\varphi] = -\frac{1}{2} \int d^4x \, (\partial_\mu \varphi)(\partial^\mu \varphi)[1 - (2\pi)^{3/2}\ell\varphi]. \tag{6}$$

This action has the same structure as the first few terms of a perturbative expansion of GR: a wave term followed by an interaction term containing two derivatives. In fact it can be obtained[4] from the Nordström-Einstein-Fokker theory[5] theory of a conformally flat metric $g_{\mu\nu} = \Omega^2\eta_{\mu\nu}$ by taking φ proportional to $\Omega^4 - 1$ (motivated by the volume element $\sqrt{|g|} = \Omega^4$) and ℓ proportional to the Planck length, and expanding the action to the first interacting order in ℓ.

3 On the applicability of perturbation theory

As described in section 1, decoherence is enforced if $|e^{iW[\Phi, \Phi']}| \ll 1$ whenever $\Phi - \Phi'$ is appreciable. However, since to zeroth order in perturbation theory, the system and environment are not coupled, a perturbative expansion will yield $|e^{iW}| = 1 + \mathcal{O}(\ell)$. In order to make $|e^{iW}|$ small, the $\mathcal{O}(\ell)$ term must be comparable to unity, making this sort decoherence in some sense an inherently nonperturbative phenomenon.

It is still possible, however, to treat some aspects of the problem perturbatively if there is another small quantity β. In that case, while terms proportional to ℓ can be small and be treated perturbatively, terms proportional to $\frac{\ell}{\beta}$ can be large, producing the seemingly nonperturbative effect. In the present work, the small parameter is the inverse temperature of the thermal state assumed to describe the short-wavelength environment.

4   Results

If the wavenumber k_c dividing the short- and long-wavelength regimes and the inverse temperature β of the environment obey $\beta k_c \ll 1$, one can calculate[4] the upper limit

$$e^{iW[\Phi,\Phi']} \leq \frac{1}{(1 + A^2[\Delta\Phi])^{1/4}}, \qquad (7)$$

which is indeed small for large values of the suppression factor

$$A^2 \gtrsim \frac{\ell^2}{\beta^2} \int\limits_{q<k_c} d^3q \int_{-q}^{q} d\omega \, |\Delta\Phi_{\mathbf{q}\omega}|^2 \frac{(2\pi)^2(q^2 - \omega^2)^2}{32q|\omega|} \ln\left(1 + \frac{|\omega|}{k_c}\right). \qquad (8)$$

Considering, for example, a coarse graining by values of a "field average" $\langle\Phi\rangle = \Delta\omega(\Delta q)^3 \Phi_{\mathbf{q}_0\omega_0}$, the suppression factor is bounded (when $q_0 \gg \omega_0$) by

$$A^2 \gtrsim \frac{\pi^2}{8} \frac{|\langle\ell\Delta\Phi\rangle|^2}{\beta^2 q_0^2} \frac{q_0^4}{\Delta\omega(\Delta q)^3}. \qquad (9)$$

Since the factors $\frac{1}{\beta q_0}$, $\frac{q_0}{\Delta\omega}$, and $\frac{q_0}{\Delta q}$ are all large, A^2 can become large and suppress the off-diagonal components of $D[\Phi,\Phi']$ even when the difference $\ell\Delta\Phi = \ell\Phi - \ell\Phi'$ is still small enough to be within the perturbative regime.

1. See, for example, J.B. Hartle, in *Gravitation and Quantizations,* proceedings of the 1992 Les Houches Summer School, ed. B. Julia and J. Zinn-Justin (North Holland, Amsterdam, 1995).
2. M. Gell-Mann and J.B. Hartle, *Phys. Rev. D* **47**, 3345 (1993).
3. R.P. Feynman and J.R. Vernon, Ann. Phys. (N.Y.) **24**, 118 (1963).
4. J.T. Whelan, preprint gr-qc/9612028
5. G. Nordström, Phys. Z. **13**, 1126 (1912), Ann. d. Phys. **40**, 856 (1913), **42**, 533 (1913), **43**, 1101 (1914); A. Einstein and A.D. Fokker, Ann. d. Phys. **44**, 321 (1914).

BLACK HOLE THERMODYNAMICS AND THE EUCLIDEAN TWO-DIMENSIONAL TEITELBOIM-JACKIW ACTION

JOSÉ P. S. LEMOS

Departamento de Astrofísica, Observatório Nacional-CNPq,
Rua General José Cristino 77, 20921 Rio de Janeiro, Brazil, &
Departamento de Física, Instituto Superior Técnico,
Av. Rovisco Pais 1, 1096 Lisboa, Portugal.

Following indications from several theories that two-dimensional black holes are important and useful to study, we use York's prescriptions for finding the temperature and other thermodynamic quantities of the static black hole in the two-dimensional Teitelboim-Jackiw theory.

Analysis of the behavior of quantum fields in a black hole (BH) background has shown that BHs steadily emit thermal radiation at a given temperature T. Further study by Hawking showed that the thermodynamics comes from a formal calculation of the partition function $Z(\beta)$ as a functional integral over all Euclidean geometries g with period β and Euclidean action $I[g]$. In other words, the probability that a thermodynamic system is in a state of energy E_n is proportional to $e^{-\beta E_n}$. The partition function is then defined as $Z(\beta) = \sum_n e^{-\beta E_n}$, and for a quantum mechanical system this can be written as $Z(\beta) = \sum_n < g_n|e^{-\beta H}|g_n >$ where H is the Hamiltonian and $< g_n|e^{-\beta H}|g_n >$ gives the expectation value of $e^{-\beta H}$ on a state g_n of the field g. Now, from the work of Feynmann one can also write Z as $Z = \int D[g]e^{-I[g]}$, a functional integral over all fields g. In the Lorentzian formulation the integral is a propagator, but by Euclideanizing the time, $t \to \beta = it$, one has a well defined statistical mechanics formalism with $\beta = \frac{1}{T}$. This idea can be extended to include the gravitational field itself. By starting with a BH geometry g, such as the Euclidean-Schwarzschild metric, one obtains through the partition function, an appropriate thermodynamics for the BH.

When applied to Kerr-Newman BHs of small charge and angular momentum, such as the Schwarzschild BH, Hawking's prescription leads to thermodynamic instabilities for the BHs. To circumvent these problems York[1] defined a canonical ensemble for the BH and hot gravity in equilibrium. For the Schwarzschild BH, the ensemble is defined by a spherical heat reservoir with radius R where a temperature $T(R)$ is kept fixed. York's prescription for finding the temperature consists of (i) write the BH metric in static coordinates, (ii) Euclideanize the time $t \to it$ and periodically identify t, (iii) adjust the mass M to remove conical singularities, and (iv) fix the proper period at R, to β. York gets the following reduced action $I(M) = R\beta \left(1 - \sqrt{1 - \frac{2M}{R}}\right) - 4\pi M^2$. The extrema, $\frac{\partial I}{\partial M} = 0$, are given by $\beta = 8\pi M \sqrt{1 - \frac{2M}{R}}$. There are two solutions $M_1 < M_2$, with $\frac{\partial^2 I}{\partial M^2} < 0$ at the extremum M_1, and $\frac{\partial^2 I}{\partial M^2} > 0$ at M_2. Thus, M_2 can be used to approximate the partition function, $Z(\beta) = e^{-I(M_2)}$. Then one obtains, $<E> = R \left(1 - \sqrt{1 - \frac{2M_2}{R}}\right)$ and an entropy $S = \frac{A}{4}$, where A is the area of the black hole. The temperature of the BH is Hawking's temperature $\frac{1}{8\pi M}$ multiplied by the redshift factor $z = \frac{1}{\sqrt{1 - \frac{2M}{R}}}$. One can ask in what sense is the temperature of the BH equal to $\frac{1}{8\pi M}$ as yielded by the original approach using quantum field theory in a BH background geometry. The idea[2] is that Hawking's temperature corresponds to drilling a small hole in the reservoir at R and letting some radiation escape to infinity, where z goes to unity.

Extensions of the 4D analysis to 2D have been of interest, in particular there has been some focus on the Teitelboim-Jackiw theory[3]. Although in this theory the curvature is constant and negative, it has a black hole solution[4]. Here we study the black hole of the Teitelboim-Jackiw theory using York's formalism[1]. For an extended study see[5]. In the Teitelboim-Jackiw 2D theory the action is $I = \frac{1}{2\pi} \int d^2x \sqrt{-g} e^{\Phi}(R - 2\Lambda) + I_B$, where g is the determinant of the metric, R is the curvature scalar, Λ is the cosmological constant (here we put $\alpha^2 \equiv -\Lambda$), and I_B is a boundary term to specify later. This action has got a black hole solution given by $ds^2 = -(\alpha^2 r^2 - b)dt^2 + \frac{dr^2}{\alpha^2 r^2 - b}$, $e^{\Phi} = c\alpha r$ where b and c are positive constants. The curvature scalar of the solution is $R = -\alpha^2$ which is a constant. Therefore, spacetime has constant negative curvature and, in principle, is isomorphic to the anti-de Sitter (ADS) spacetime. One can interpret this solution as a BH solution by truncating spacetime at $r = 0$. This cut off comes from the interpretation of this 2D theory as a theory dimensionally reduced from higher dimensions. The dilaton sets new boundary conditions (b.c.), making two locally indistinguishable solutions topologically and globally different[4]. The mass of the black hole is $M = \frac{\alpha c}{2} b$. $M = 0$ gives the extremal case.

By using the constraints and b.c. one finds the following Euclidean reduced action: $I(h^{-1}) = -(G^{-1})\beta e^{\Phi_B}\alpha \sqrt{1 - e^{2(\Phi_H - \Phi_B)}} - (h^{-1})2\pi e^{\Phi_H} + (G^{-1})\beta e^{\Phi_B}\alpha$, where Φ_H is the value of Φ at the horizon and $I_0 \equiv \beta e^{\Phi_B}\alpha$ was chosen appropriately. We have put back Newton's constant G and Planck's constant h (still puting Boltzmann's constant and the velocity of the light equal to one). Note that in 2D we use the following units for the constants: $[G] = LM^{-1}T^{-1}$ and $[h] = MT^{-1}$. As in 4D[1], a quantum term has appeared in the action, namely the term $2\pi e^{\Phi}$, which is associated with the entropy.

The temperature is implicitly given by $\frac{2M}{\alpha c} = \alpha^2 r_H^2 = \frac{\alpha^2 r_B^2}{1 + \frac{\alpha^2 \beta^2}{4\pi^2}}$, $\beta \equiv \frac{1}{T}$. There is nothing like the instanton solution of the Schwarzschild bath in 4D. The entropy is $S_H = \beta \left(\frac{\partial I}{\partial \beta} \right)_{\Phi_B} - I = 2\pi e^{\Phi_H} = 2\pi \sqrt{\frac{2M}{\alpha c}}$. This functional dependence on the dilaton is the same for all black holes having a simple 2D Brans-Dicke action, in analogy to the 4D case where $S = \frac{A}{4}$. The extreme case has zero entropy. The thermodynamic energy $E \equiv \frac{\partial I}{\partial \beta} \big)_{\Phi_B} = c\alpha^2 r_B \left(1 - \sqrt{1 - \frac{r_H^2}{r_B^2}} \right)$. The zero point is such that for zero mass the thermal energy is zero. The heat capacity is $C_{r_B} \equiv T \left(\frac{\partial S}{\partial T} \right)_{r_B} = 2\pi c\alpha \frac{r_H}{r_B^2}(r_B^2 - r_H^2)$, i.e., it is positive, implying thermal stability always. For the Schwarzschild BH stability exists only within a limited range of the reservoir R.

To know the ground state of the system, one can compare the free energies $F = \frac{I}{\beta}$ of the 2D BH and hot ADS space. One state cannot jump to another classically, but quantum mechanically the topologies can change. The free energy for HADS in 2D can be found from[5] $-I_{HADS} = \frac{\pi}{6\alpha} nT \arctan(\alpha r_B)$, where n is the number of massless spin sates. The ground state is the state of least free energy. We find that HADS dominates whenever $I_{HADS} \leq I_{BH}$, i.e., $T \geq \alpha \frac{12c}{n} \frac{\alpha r_B}{\arctan(\alpha r_B)} \sqrt{1 - \frac{n}{12\pi c} \frac{\arctan(\alpha r_B)}{\alpha r_B}}$. Whenever the number of particle species is relatively large then HADS is favored for sufficiently small r_B. When the boundary $r_B \to \infty$ the black hole is the ground state for finite T.

References

1. J. W. York, *Phys. Rev.* D **33**, 2092 (1986); H. W. Braden, J. D. Brown, B. F. Whiting, J. W. York, *Phys. Rev.* D **42**, 3376 (1990).
2. D. J. Brown, gr-qc/9404006 (1994).
3. C. Teitelboim in *Quantum Theory of Gravity*, ed. S. M. Christensen (Hilger, Bristol, 1984); R. Jackiw in *Quantum Theory of Gravity*, ed. S. M. Christensen (Hilger, Bristol, 1984).
4. D. Christensen, R. B. Mann, *Class. Quantum Grav.* **6**, 9 (1992); A. Achúcarro, M. E. Ortiz, *Phys. Rev.* D **48**, 3600 (1993); M. Cadoni, S. Mignemi, *Nucl. Phys.* B **427**, 669 (1994); M. Cadoni, S. Mignemi, *Phys. Rev.* D **51**, 4139 (1995); J. P. S. Lemos, P. M. Sá, *Mod. Phys. Lett.* A **9**, 771 (1994); J. P. S. Lemos, P. M. Sá, *Phys. Rev.* D **49**, 2897 (1994).
5. J. P. S. Lemos, *Phys. Rev.* D **54**, 6206 (1996).

Chiral ($N = 1$) supergravity

Alfredo Macías[a]

Departmento de Física,
Universidad Autónoma Metropolitana–Iztapalapa,
P.O. Box 55-534, 09340 México D.F., MEXICO.

We consider a purely imaginary *translational Chern–Simons term*, evaluated by using the Cartan relation between torsion and spin. By means of the introduction of this term as boundary term a *chiral reformulation* of ($N = 1$) supergravity is achieved. The translational Chern–Simons term plays the role of a *generating function* for the canonical transformation, which splitts the supergravity field variables into selfdual and antiselfdual parts in a natural way.

The role of *chirality* is fundamental in the gauge interactions of elementary particles and it can be identified with the concept of *selfduality*.

In spite of all the success of the Ashtekar formulation [1], the theory gets afflicted by the complex character of the variables, thus the issue of the reality conditions becomes compulsory. As it is well known, supergravity suffers the same diseases as general relativity. Had we aimed at giving to the supergravity Lagrangian its chiral form, a similar analysis to the one given by Ashtekar should be performed. Jacobson [2] advanced to some extent in this direction. Here we will turn, however, to the Clifford–algebra approach which yields the chiral decomposition from the more fundamental point of view of the *generating function*.

We introduce a purely imaginary *translational* Chern–Simons term as a generating function which induces the *chiral* decomposition into selfdual and anti–selfdual parts in the bosonic sector as well as in the fermionic sector of the ($N = 1$) supergravity theory.

The ($N = 1$) supergravity geometry can be described by means of a very concise formalism employing Clifford algebra–valued exterior differential forms. To this end we use the Dirac matrices γ_α in the Bjorken and Drell convention.

The simplest coupling of a Rarita–Schwinger type spin-$\frac{3}{2}$ field to gravity is *supergravity* theory with one supersymmetry generator, i.e. N=1. The corresponding Hermitian Lagrangian four–form reads

$$L_{\text{Sugra}} = -\frac{1}{2\ell^2}Tr(\Omega \wedge^* \sigma) + -\frac{1}{2}\left(\overline{\Psi} \wedge \gamma_5\gamma \wedge D\Psi - \overline{D\Psi} \wedge \gamma_5\gamma \wedge \Psi\right) \qquad (1)$$

where the Rarita–Schwinger field $\Psi := \Psi_\alpha \vartheta^\alpha$ is a spinor valued one-form.

Because we do *not* want to change the physics of the supergravity action, one is only entitled to add an exact form to the Lagrangian. The translational boundary term arising from

$$dC_{\mathrm{TT}} = \frac{1}{8\ell^2} Tr\left(\Theta \wedge \Theta - 4i\Omega \wedge \sigma\right), \qquad (2)$$

is crucial here in the chiral transition of supergravity Lagrangian.

This term yields parity violating pieces in the gauge field equations, however by adding it with the imaginary unit i as factor to the Lagrangian, one can preserve the more stringent CP-invariance of quantum field theory[1]. This boundary term represents the *generating function* of our canonical transformation to variables involving an *antiselfdual or selfdual* connection.

Bosonic sector of the chiral $N = 1$ supergravity Lagrangian

The the bosonic sector of the complex $(N = 1)$ supergravity Lagrangian reads

$$\overset{(\pm)}{V}_{\mathrm{EC}} := V_{\mathrm{EC}} \pm idC_{\mathrm{TT}} == \pm\frac{1}{\ell^2} Tr\left\{P_{\mp}\,\Omega \wedge \sigma + \frac{i}{8}\,\Theta \wedge \Theta\right\}. \qquad (3)$$

Observe the explicit *chirality projector* obeying $P_{\pm}P_{\pm} = P_{\pm}$. For vanishing torsion, Eq. (3) can be regarded as the *chiral* form of the Hilbert–Einstein Lagrangian. It is possible to rewrite (3) in another form by performing the following identifications: $\Theta = \pm i\ell^2 \overset{(\pm)}{H}_{\mathrm{EC}}$, $\sigma_{\pm} = P_{\mp}\sigma$, and $\Omega_{\pm} = P_{\mp}\Omega$. Thus we find

$$\overset{(\pm)}{V}_{\mathrm{EC}} = \pm Tr\left(\Omega_{\pm} \wedge \sigma_{\pm}\right) \mp \frac{i}{8}\ell^2\, Tr\left(\overset{(\pm)}{H}_{\mathrm{EC}} \wedge \overset{(\pm)}{H}_{\mathrm{EC}}\right), \qquad (4)$$

which shows in an alternative manner the chiral properties of the Lagrangian (3). The first term is the Clifford algebra version of the (anti–) selfdual action introduced simultaneously by Jacobson and Smolin, and Samuel. The last term is related to torsion. By means of the Cartan relation between torsion and spin, it can be shown to be zero in the supergravity case by a Fierz identity.

Fermionic sector of the chiral (N=1) supergravity

Here we focus on the Rarita–Schwinger field contribution to the supergravity Lagrangian L_{Sugra} which will be also transformed into its chiral form. It should be noted that the gravitino field Ψ is a *Majorana spinor* and satisfies the Majorana condition. We will use the real Majorana representation in which all γ^{α} are purely imaginary but the components of the gravitino vector–spinor are consequently all real.

Now according to our results there is in the Lagrangian density $\overset{(\pm)}{V}_{\mathrm{EC}}$ a term quadratic in the torsion: $Tr(\overset{(\pm)}{H}_{EC} \wedge \overset{(\pm)}{H}_{EC}) \sim Tr(\Theta \wedge \Theta)$. This term vanishes for supergravity because of the Fierz identity: $*\left(\Theta \wedge \Psi \wedge \vartheta^{\tau}\right) \cong \left(\overline{\Psi}_{\beta}\gamma^{\alpha}\Psi_{\delta}\right)\left(\gamma_{\alpha}\Psi_{\rho}\right)\eta^{\beta\delta\rho\tau} = 0$.

Had we aimed at giving to the Rarita–Schwinger pice of the supergravity Lagrangian (1) its chiral form, an analysis similar to the previous one for the bosonic sector should be performed. The Rarita–Schwinger vector–spinor is decomposed into left and right–handed pieces $\Psi = \Psi_L + \Psi_R$, where $\Psi_L := P_- \Psi$, and $\Psi_R := P_+ \Psi$, with $D_\pm \Psi_{R,L} := d\Psi_{R,L} + \Gamma_\pm \wedge \Psi_{R,L}$.

In our elegant *Clifford* approach, we note that "on shell", i.e. after using the Cartan relation, the translational Chern–Simons term is given by $C_{TT} \cong \frac{i}{4} \overline{\Psi} \wedge \gamma \wedge \Psi$, which is proportional to the "charge current" of the Rarita–Schwinger field. Similarly, the chiral version of the Rarita–Schwinger Lagrangian is obtained by adding the boundary term dC_{TT}, i.e. $L_{RS\pm} := \frac{1}{2} L_{RS} \pm idC_{TT}$. Acting with the chirality projector on the spinor one–forms as well as on the connection, the chiral Rarita–Schwinger Lagrangian reads explicitly

$$L_{RS\pm} = \frac{i}{2} \left\{ \overline{\Psi_{L,R}} \wedge \gamma \wedge D_\pm \psi_{L,R} + \overline{D_\pm \Psi_{R,L}} \wedge \gamma \wedge \Psi_{R,L} \right\}, \tag{5}$$

where the positive (negative) sign goes with the first (second) of the indices appearing in the vector–spinor $\Psi_{L,R}$. According to this result the chiral and standard actions for supergravity differ by an imaginary boundary term, once the equation of motion for the connection obtained from the chiral one is used. In full, the *chiral* $(N = 1)$ supergravity theory can be related to standard $(N = 1)$ supergravity theory through the identity

$$L_{Sugra}^{Chiral} := \frac{1}{2} \overset{(\pm)}{V}_{EC} + L_{RS\pm} = \frac{1}{2} L_{Sugra} \pm \frac{3}{2} idC_{TT}. \tag{6}$$

More precisely, the imaginary part of the chiral supergravity Lagrangian L_{Sugra}^{Chiral} is the boundary term $L_{Sugra}^{Chiral} - \overline{L}_{Sugra}^{Chiral} = dC_{TT} \cong \frac{i}{4} d \left(\overline{\Psi} \wedge \gamma \wedge \Psi \right)$, whereas the real part is the standard supergravity Lagrangian[2]. Hence, the two theories are dynamically equivalent, but may have different variables in the Hamiltonian formulation.

Acknowledgments

This research was partially supported by CONACyT, Grant No. 3544–E9311.

1. A. Ashtekar, *New Perspectives in Canonical Gravity* (Bibliopolis, Napoli 1988), *Lectures on Non-perturbative Canonical Gravity* (World Scientific, Singapore 1991).
2. T. Jacobson, Class. Quantum Grav. **5** (1988) 923.
3. A. Macías, Class. Quantum Grav. **13** (1996) 3163.

Second–order reconstruction of inflationary dynamics compatible with recent COBE data

Eckehard W. Mielke

Departamento de Física,
Universidad Autónoma Metropolitana–Iztapalapa,
Apartado Postal 55-534, C.P. 09340, México, D.F., MÉXICO

A new method is applied to *reconstruct* the inflationary potential from a more phenomenological point of view. By using the more stringent *nonlinear* second order slow–roll approximation, we found, besides the power law inflation for $n_s < 1$, an *exact* solution for which the spectral index is *necessarily* determined as $n_s \simeq 1.5$. Part of these continous and discrete spectra of the index fits within the 1σ range of recent COBE and CAT data.

1 Introduction

Various *inflationary models* have been proposed in which the *scalar spectral index* n_s takes values between zero and one, whereas those with a *blue* spectrum beyond one are more difficult to obtain. Recently, the general *exact* inflationary solution depending on the Hubble expansion rate H, the "inverse time", has been found [1] and the regime of inflationary potentials allowing a graceful exit is classified. We apply this H–formalism to the more accurate second order perturbation formalism and provide *exact* solutions to the nonlinear equation of Stewart and Lyth [2].

In our second order slow–roll reconstruction of the inflationary dynamics, this nonlinear equation permits exact solutions only for the continous or discrete eigenvalues $-3 < n_s < 1$ or $n_s \simeq \{-6.6, 1.5\}$, respectively, of the spectral index. Although both upper parts are marginally consistent with $n_s = 1.2 \pm 0.3$ from cumulative four years CMB observations of COBE and $n_s = 1.3 \pm 0.4$ of CAT, the blue spectrum is just below the $n_s \leq 1.5$ constraint in order to suppress distortions of the microwave spectrum and the formation of primordial black holes during the phase of reheating.

2 General metric of a spatially flat inflationary universe

Let us consider the Einstein frame for which the Lagrangian density of a rather general class of inflationary models reads

$$\mathcal{L}_{\rm E} = \frac{1}{2\kappa}\sqrt{|g|}\left(R + \kappa\left[g^{\mu\nu}(\partial_\mu\phi)(\partial_\nu\phi) - 2U(\phi)\right]\right). \tag{1}$$

Here R is the curvature scalar, ϕ the scalar field, $U(\phi)$ *the self–interaction potential* and $\kappa = 8\pi G/c^4$ the gravitational coupling constant. We use natural units with $c = \hbar = 1$ and our signature for the metric is $(+1, -1, -1, -1)$ as it is common in particle physics.

For the *flat* $(k = 0)$ Robertson-Walker metric favored by the inflationary paradigm, the evolution of the generic inflationary model (1) is determined by the autonomous first order equations $\dot{H} = \kappa U(\phi) - 3H^2$; $\dot{\phi} = \pm\sqrt{2/\kappa}\sqrt{3H^2 - \kappa U(\phi)}$. By introducing the Hubble expansion rate $H := \dot{a}(t)/a(t)$ as the *new* "inverse time" *coordinate*, it is possible to find the following general metric and scalar field solution [1]:

$$
ds^2 = \frac{dH^2}{\left(\kappa\tilde{U} - 3H^2\right)^2} - a_0{}^2 \exp\left(2\int \frac{H\,dH}{\kappa\tilde{U} - 3H^2}\right) \times
$$
$$
\left[dr^2 + r^2\left(d\theta^2 + \sin^2\theta d\varphi^2\right)\right] , \tag{2}
$$

$$
\phi = \phi(H) = \mp\sqrt{\frac{2}{\kappa}} \int \frac{dH}{\sqrt{3H^2 - \kappa\tilde{U}}} , \tag{3}
$$

where $\tilde{U} = \tilde{U}(H) := U(\phi(t(H))) = (3/\kappa)H^2 + g(H)/\kappa$ is the *reparametrized* inflationary potential and $g(H) < 0$ the graceful exit function excluding the singular de Sitter inflation.

3 Second-order slow–roll approximation

For a reconstruction of the potential $U(\phi)$ under the "umbrella" of *chaotic inflation*, we adopt here the results of Stewart and Lyth [2] for the second order slow–roll approximation, at present the most accurate equations available. In terms of $\epsilon = -g(H)/H^2$, this *nonlinear* second order differential equation [3] involving the *scalar* spectral index n_s reads

$$
2C\epsilon \ddot{\epsilon} - (2C + 3)\epsilon \dot{\epsilon} - \dot{\epsilon} + \epsilon^2 + \epsilon + \frac{n_s - 1}{2} = 0 , \tag{4}
$$

where $\cdot \hat{=} d/d\ln H^2$ and $C := -2 + \ln 2 + \gamma \simeq -0.73$. The corresponding second order equation for the spectral index n_g of *tensor* perturbations turns out to be $n_g = -2\epsilon\left[1 + \epsilon - 2(1 + C)\dot{\epsilon}\right]$.

A rather general solution of (4) is

$$
g(H) = -AH^2 - \epsilon_0 H^{2(A+2)} , \tag{5}
$$

where $A = \frac{1}{2}\left(-1 \mp \sqrt{3 - 2n_s}\right)$ is determined by the empirical n_s, see [4,5].

For $\epsilon_0 = 0$, we reconstructed already in [1] the potential

$$U(\phi) = \frac{3-A}{\kappa} C_3{}^2 \exp(\pm\sqrt{2\kappa A}\,\phi)\,. \tag{6}$$

It belongs to the class of *power–law* models with the expansion factor given by $a(t) \sim t^{1/A}$.

Potentials yielding a blue spectrum are more difficult to obtain. However, for $\epsilon_0 \neq 0$ in our general solution (5), the nonlinearity of (4) requires, in addition, the constraint $A = \left(3 - 2C \pm \sqrt{4C^2 + 4C + 9}\right)/4C$. Consequently, the spectral index n_s is now *completely fixed* by the Euler constant γ via $n_s = 1 - \left(2C + 9 \mp 3\sqrt{4C^2 + 4C + 9}\right)/4C^2 = 1.49575\ (-6.57797)$. The first positive value just fits within the range of recent COBE, Tenerife, and CAT data on the scalar spectral index.

From (3) and (5) we find the *new potential*

$$U(\phi) = \frac{1}{\kappa}\left[-\frac{A}{\epsilon_0}\left(1 + \tan^2(P\phi)\right)\right]^{1/(1+A)} \left(3 + A\tan^2(P\phi)\right) \tag{7}$$

with a fixed scalar spectral index, where $P = \pm\sqrt{-\kappa A/2}\,(1 + A)$. It interpolates between the power–law inflation in the limit $\epsilon_0 \to 0$ and intermediate inflation.

Criteria for a *reheating* of the universe after inflation are analyzed by Kusmartsev et al. [6] using methods of *catastrophe theory*. In a conformal frame, the inflaton can be absorbed in a specific nonlinear curvature scalar Lagrangian parametrized by n_s which is related to effective string models [4].

1. F.E. Schunck and E.W. Mielke, Phys. Rev. **D 50** (1994) 4794; E.W. Mielke and F.E. Schunck, Phys. Rev. **D 52** (1995) 672.
2. E. D. Stewart and D. H. Lyth, Phys. Lett. **B302**, 171 (1993).
3. A. García, A. Macías, and E.W. Mielke: "Stewart–Lyth second–order approach as an Abel equation for reconstructing inflationary dynamics", Phys. Lett **A** (submitted 1997).
4. J. Benítez, A. Macías, E.W. Mielke, O. Obregón, and V.M. Villanueva, Int. J. Mod. Phys. (1996, in print).
5. E.W. Mielke, O. Obregón, and A. Macías, Phys. Lett **B**, in print (1996).
6. F. V. Kusmartsev, E. W. Mielke, Yu. N. Obukhov, and F. E. Schunck, Phys. Rev. D **51**, 924 (1995).

The Design and Status of the Sudbury Neutrino Observatory

D. F. Cowen (for the SNO Collaboration)
University of Pennsylvania, 209 S. 33rd Street,
Philadelphia, PA 19072-6396

A brief overview of the purpose, design and current status of the Sudbury Neutrino Observatory (SNO) is presented. SNO has unique neutrino detection capabilities and has been designed specifically to address the observed deficit of electron neutrinos produced by the sun.

The Sudbury Neutrino Observatory (SNO) is being constructed to study the fundamental properties of neutrinos, in particular the mass and mixing parameters. SNO can accomplish this by measuring the flux of electron type neutrinos, ν_e, which are produced in the sun, and comparing it to the flux of all neutrino flavors detected on earth in an appropriate energy interval. Observation of neutrino flavor transformation through this comparison is compelling evidence that at least one neutrino flavor has non-zero mass. Non-zero neutrino mass is evidence for physics beyond the Standard Model of fundamental particle interactions.[1] Non-zero neutrino mass might also have cosmological implications, and could play a role in the presence of "hot" dark matter implied in some cosmological models by the observations of the cosmic microwave background radiation anisotropy, angular correlations of galaxies, correlations of galaxy clusters, and other cosmological data.[2]

The long distance to the sun makes the search for neutrino mass sensitive to much smaller mass differences than can be studied with terrestrial sources. Furthermore, the matter density in the sun is sufficiently large to enhance the effects of small mixing between ν_e's and ν_μ's or ν_τ's. This "MSW"[3] effect also applies when solar neutrinos traverse the earth, and may cause well defined temporal and spectral modulation of the ν_e signal in the SNO detector.

All solar neutrino experiments performed to date have detected fewer solar neutrinos than expected from models of the energy producing mechanisms in the sun. There are strong hints that this is the result of neutrino flavor transformation between the production point in the sun and the terrestrial detection point. These conclusions are based on a calculated prediction.[4] SNO will be able to confirm existing results, extend them by direct measurement of the ν_e flux, $\Phi(\nu_e)$, and, unlike all other solar neutrino experiments, make the interpretation of the results independent of theoretical calculations by estimating the flavor-blind flux, $\Phi(\nu_l)(l = e, \mu, \tau)$, and the ratio $\Phi(\nu_e)/\Phi(\nu_l)$. Unique mass and mixing parameters can be determined by combining the SNO results

Table 1: Neutrino detection modes and expected data rates for SNO and several other solar neutrino experiments. The addition of NaCl to enhance neutral current detection has been assumed for SNO in the table. The two $\bar{\nu}$ detection modes are also unique to SNO. (In the table, $^\ddagger E_\nu > 5$MeV, $^* \frac{1}{3}$SSM, $^{**} \frac{1}{2}$SSM, and †SSM; "SSM" = Standard Solar Model)

Experiment	ν parent	Detection Reactions	Expected rate‡	Start
SNO	^{8}B	$\nu_e + d \rightarrow p + p + e^-$	10/d*	1997
	^{8}B	$\nu_l + d \rightarrow \nu_l + p + n$	10/d	
	^{8}B	$\nu_l + e^- \rightarrow \nu_l + e^-$	0.5/d*	
		$\bar{\nu}_e + d \rightarrow n + n + e^+$	?	
		$\bar{\nu}_e + p \rightarrow n + e^+$	?	
SuperKamiokande	^{8}B	$\nu_l + e^- \rightarrow \nu_l + e^-$	20/d**	4/1996
Homestake ^{127}I	^{8}B,^{7}Be	$\nu_e + ^{127}$I $\rightarrow ^{127}$Xe $+e^-$	1/d	1997
Borexino	^{7}Be	$\nu_l + e^- \rightarrow \nu_l + e^-$	50/d†	1999

with the existing results of the ^{37}Cl, SuperKamiokande, Kamiokande, and ^{71}Ga solar neutrino experiments. [5]

The SNO experiment is unique in that it utilizes one kiloton of heavy water, D_2O, as a target. This permits detection of the neutrinos through the reactions $\nu_l + e^- \rightarrow \nu_l + e^-$ (1) $\quad \nu_e + d \rightarrow e^- + p + p$ (2) $\quad \nu_l + d \rightarrow \nu_l + n + p$ (3). Each of these interactions is detected when one or more scattered electrons produce Čerenkov light that impinges on a phototube array. The elastic scattering ("ES") interaction of neutrinos with electrons (Eq. 1) is highly directional, and establishes the sun as the source of the detected neutrinos. The charged current ("CC") absorption of ν_e on deuterons (Eq. 2) produces an electron with an energy highly correlated with that of the neutrino. This reaction is sensitive to deviations in the ν_e energy spectrum relative to the expected parent spectrum. The neutral current ("NC") disintegration of the deuteron by neutrinos (Eq. 3) is independent of neutrino flavor and has a threshold of 2.2 MeV. To be detected, the neutron must be absorbed, resulting in the emission of a readily detectable photon. Subsequent Compton scatters of this photon impart enough energy to electrons to create Čerenkov light. Eventually, special–purpose neutral current detectors will be installed which will enable SNO to tag NC events directly. Measurement of the rate of the NC reaction determines the total flux of ^{8}B neutrinos, even if their flavor has been transformed. The ability to measure both the CC and NC reactions will be unique to SNO. Monte Carlo studies have shown that these reactions can be separated on a statistical basis. A summary of all the SNO interaction modes, along with expected data rates in each mode and a comparison with other experiments, is given in Table 1.

The SNO detector is located underground in a large cavity excavated at the 6800 foot level (6000 meter water equivalent) in the INCO, Ltd., Creighton mine near Sudbury, Ontario, Canada. The kiloton D_2O target will be contained in a twelve meter diameter acrylic vessel, the upper hemisphere of which was recently completed. The D_2O is surrounded by a light water shield of minimum diameter twenty two meters. A spherical geodesic shell of diameter seventeen meters surrounds the D_2O and supports almost 10,000 20 cm photomultiplier tubes (PMTs), each with a light collecting reflector. Roughly half of these PMTs have been installed and cabled. Final water fill is scheduled to start in the summer of 1997.

Acknowledgements

This research is financially supported by the Natural Sciences and Engineering Research Council, Industry Canada, and Northern Ontario Heritage Fund Corporation, all in Canada, the Department of Energy in the USA and the Science and Engineering Research Council in the UK. The loan of heavy water from Atomic Energy of Canada Limited is greatly appreciated.

References

1. S. Weinberg, *Phys. Rev. Lett.* **19**, 1264 (1967); A. Salam, in *Elementary Particle Physics*, ed. N. Svartholm (Almqvist and Wiksells, Stockholm, 1968) p. 367; and S. Glashow, J. Iliopolous, and L. Maiani, *Phys. Rev.* D **2**, 1285 (1970).
2. See, for example, C.W. Kim and A. Pevsner, *Neutrinos in Physics and Astrophysics* (Harwood Academic Publishers, 1993).
3. L. Wolfenstein, *Phys. Rev.* D **17**, 2369 (1979); S. P. Mikheyev and A. Yu. Smirnov, Yad. Fiz. **42**, 1441 (1985) (Sov. J. Nucl. Phys. **42**, 913 (1985)); *Nuovo Cimento* **9C**, 17 (1986).
4. John N. Bahcall, *Neutrino Astrophysics*, (Cambridge University Press, 1994), and references therein.
5. R. Davis *et al.*, in *Proceedings of the 21st International Cosmic Ray Conference*, Vol. 12, edited by R. J. Protheroe (University of Adelaide Press, Adelaide, 1990), p. 143; SuperKamiokande Collaboration, Y. Totsuka, in this Conference Proceedings; Kamiokande II Collaboration, K. S. Hirata *et al.*, *Phys. Rev.* D **44**, 2241 (1991); SAGE Collaboration, A. I. Abazov *et al.*, *Phys. Rev. Lett.* **67**, 3332 (1991); GALLEX Collaboration, P. Anselmann *et al.*, *Phys. Lett.* B **285**, 376 (1992).

HELIOSEISMOLOGY AND SOLAR NEUTRINOS

Jørgen Christensen-Dalsgaard

Teoretisk Astrofysik Center, Danmarks Grundforskningsfond, and
Institut for Fysik og Astronomi, Aarhus Universitet, DK-8000 Aarhus C, Denmark

Helioseismology has provided very precise information about the solar internal sound speed and density throughout most of the solar interior. The results are generally quite close to the properties of standard solar models. Since the solar oscillation frequencies do not provide direct information about temperature and composition, the helioseismic results to not completely rule out an astrophysical solution to the discrepancy between the predicted and measured neutrino fluxes from the Sun. However, such a solution does appear rather implausible.

1 Introduction

The persistent discrepancy between the measured fluxes of solar neutrinos and the predictions of solar models (*e.g.* Bahcall;[1] for a recent review, see Bahcall, these proceedings) has led to doubts about the reliability of solar-model calculations. Indeed, a large number of suggestions have been made for changes to the properties of the model which might reduce the predicted neutrino fluxes.

The neutrino flux depends principally on the temperature and composition profiles of the solar core but provide little detailed information about the solar interior. In the last decade far more information on the internal structure of the Sun has been obtained from observations of solar oscillations, providing tight constraints on solar models.[2] Recent "standard" solar models are generally in reasonable agreement with the helioseismic inferences.[3,4,5,6,7] Here I consider the extent to which this agreement argues against an astrophysical solution to the solar neutrino problem. To be definite, I base the analysis on Model S of Christensen-Dalsgaard *et al.*[8] This was computed with OPAL equation of state[9] and opacity,[10] using the Bahcall & Pinsonneault[11] nuclear parameters, and including settling of helium and heavy elements computed with the Michaud & Proffitt[12] diffusion coefficients. The predicted neutrino capture rates in the ^{37}Cl and ^{71}Ga experiments are 8.2 and 132 SNU, respectively, while the flux of high-energy ^{8}B neutrinos is $5.9 \times 10^6 \, \mathrm{cm^{-2} s^{-1}}$, quite similar to, for example, the corresponding model of Bahcall & Pinsonneault.[11]

In common with other current models of the Sun the model is based on significant simplification, neglecting possible hydrodynamical phenomena in the convectively stable part of the Sun, associated *e.g.* with instabilities or motion induced by convective penetration. Indeed, it is well established that the Sun becomes unstable to low-order, low-degree g modes during its evolution;[13] the non-linear development of the instability might conceivable involve mixing of the solar core.[14] Other instabilities, leading to mixing, could be associated with the spin-down from the commonly assumed early rapid rotation.[15]

2 Inversion for the Solar Internal Sound Speed

The observed solar oscillations are adiabatic almost everywhere, to a very good approximation. Thus their frequencies depend on the distribution of mass, pressure p and density ρ, and on the adiabatic compressibility $\Gamma_1 = (\partial \ln p / \partial \ln \rho)_s$, the derivative being at fixed specific entropy s. Assuming that the model is spherically symmetric and in hydrostatic equilibrium, the frequencies are completely determined by specifying $\rho(r)$ and $\Gamma_1(r)$, r being the distance to the solar centre, whereas they do not depend directly on temperature. The observed modes are essentially standing acoustic waves, with frequencies depending predominantly on the adiabatic sound speed c given by $c^2 = \Gamma_1 p / \rho$. Departures from this simple description of the oscillations, occurring very near the solar surface, can be eliminated in the analysis of the frequencies.

The differences $\delta \omega_i = \omega_i^{(\text{obs})} - \omega_i^{(\text{mod})}$ between the observed and computed frequencies of the i-th mode can be linearized around the model, resulting in[16]

$$\frac{\delta \omega_i}{\omega_i} = \int_0^R K^i_{c^2,\rho}(r) \frac{\delta c^2(r)}{c^2(r)} dr + \int_0^R K^i_{\rho,c^2}(r) \frac{\delta \rho(r)}{\rho(r)} dr + \frac{F_{\text{surf}}(\omega_i)}{Q_i} + \epsilon_i . \quad (1)$$

Here the integrals extend to the surface radius R of the Sun, δc^2 and $\delta \rho$ are differences between the Sun and the model in c^2 and ρ, the kernels $K^i_{c^2,\rho}$ and K^i_{ρ,c^2} are known functions, $F_{\text{surf}}(\omega_i)$ results from the near-surface errors in the model, and ϵ_i is the error in the observed frequencies. This equation forms the basis for inferring the corrections to solar models.[17] The principle is to make linear combinations of eqs (1) with coefficients $c_i(r_0)$ chosen such that the sums corresponding to the last three terms on the right-hand side of eqs (1) are suppressed, while in the first term the *averaging kernel* $\mathcal{K}_{c^2,\rho}(r_0,r) = \sum_i c_i(r_0) K^i_{c^2,\rho}(r)$ is localized in the vicinity of $r = r_0$. The corresponding combination of the left-hand sides, $\sum_i c_i(r_0) \delta \omega_i / \omega_i$, then clearly provides a localized average of $\delta c^2 / c^2$ near $r = r_0$.

Several extensive sets of helioseismic data are now available, including initial results from the GONG network[18] and the SOI/MDI project[19] on the SOHO satellite. The results presented here are based on a combination[4] of frequencies from the BiSON network[20] and the LOWL instrument;[21] however,

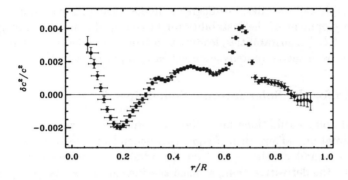

Figure 1: Inferred difference in squared sound speed, in the sense (Sun) – (model).[4] The horizontal error bars mark the first and third quartile points of the averaging kernels, whereas the vertical error bars show 1-σ errors, as progated from the errors in the observed frequencies.

analyses of other, independent sets give results that are generally consistent with those presented here. The inferred difference in squared sound speed between the Sun and the model is shown in Fig. 1. Each point corresponds to an average of $\delta c^2/c^2$, weighted by $\mathcal{K}_{c^2,\rho}(r_0, r)$; however, as indicated these averages are relatively well localized in r. Also, the propagated data errors are small. Thus the procedure has succeeded in providing precise and well resolved measures of the sound-speed errors in the model, even quite close to the centre of the Sun. The differences are evidently highly significant; nonetheless, it is striking that the model reproduces the solar sound speed to within a small fraction of a per cent. This has been achieved without any explicit adjustment of parameters to fit the model to the observations.

From the results shown in Fig. 1 we can evidently reconstruct an estimate of the actual solar sound speed. Tests have shown that this is largely independent of the choice of reference model, even for models differing rather more from the structure of the Sun than the model used here. Thus it is possible from the observed frequencies to determine quite precisely the dependence of sound speed on position in the Sun.

3   Relevance to the Solar Neutrino Problem

The support for the standard solar model provided by the helioseismic results, at least to the level of precision relevant to the current state of neutrino measurements, might argue against solutions to the solar neutrino problem in terms of non-standard solar models.[7] Indeed, such models have generally been ruled out by helioseismology. These include models with partial mixing of the core, or with energy transport in the core from motion of hypothetical weakly interacting massive particles (WIMPs). A measure of the structure of the solar core is provided by the separation $d_{nl} = \nu_{nl} - \nu_{n-1\,l+2}$ in cyclic frequencies ν between modes differing by 1 in radial order n and by 2 in degree l, for low-degree modes. For standard solar models the computed d_{nl} is very close to the observed value. Core mixing increases d_{nl} whereas the inclusion of WIMPs reduces it, such that in both cases models where the neutrino fluxes are reduced to near the measured values are inconsistent with the observed d_{nl}. Thus these models are effectively ruled out.[22,23] However, one might argue that a model combining mixing with WIMPs could perhaps be set up in such a way as to be consistent with both the neutrino and the oscillation data.

This exemplifies the fact that while helioseismology constrains the mechanical properties such as sound speed and density, other aspects of the model cannot be uniquely determined. Indeed, $c^2 \propto T/\mu$, where T is temperature and μ is the mean molecular weight; hence T and μ can be varied individually, as long as their ratio is unchanged. For example, this might be achieved by changing μ and T through mixing and the inclusion of WIMPs, respectively. Only by invoking the physics of stellar interiors, such as information about energy transport and production, is it possible to determine, $e.g.$, the variation of temperature and composition,[24,25,26] and hence to compute the neutrino flux expected from the helioseismically determined model. Since the inferred sound speed is so close to the standard solar model, a substantial reduction in the neutrino flux while keeping the model consistent with helioseismology can be achieved only by modifying several aspects of the model computations.

Which modifications might be contemplated? The most questionable assumption in computations of standard models is probably the absence of motion in the solar interior; thus it is natural to consider the possibility of mixing, modifying the distribution of composition. However, as argued above, any change in μ must be compensated by changes in T. This in turn requires modification of the description of energy generation and transport. It is conceivable that there are significant errors in the opacities; the basic rate of energy generation is probably rather well known, although one cannot entirely exclude problems with the treatment of nuclear screening. Evidently, inclusion of non-standard processes allows substantially more freedom in the modelling; examples are energy transport by WIMPs or waves, or departures from thermal equilibrium in the present Sun, perhaps caused by a recent mixing episode.[14]

Models with such modifications to the physics permit fairly substantial reductions in the computed neutrino flux, while keeping the sound speed consistent with the helioseismic evidence,[27] although typically drastic and perhaps unrealistic reductions in opacity are required. A particularly careful analysis of this nature was carried out by Antia & Chitre,[28] who were able to reduce the computed capture rates to near the observed values. However, in common with other astrophysical attempts at a solution it was not possible to match all neutrino measurements simultaneously.

To account for the details of the measured neutrino fluxes Cumming & Haxton[29] proposed slow mixing of the core at such a rate that ^{3}He is not in nuclear equilibrium, hence shifting the balance between the different components of the pp chains. This would unavoidably lead to homogenization of the hydrogen abundance over a substantial region which, unless other modifications were invoked, would result in a sound-speed profile inconsistent with the helioseismic results.[7] Even allowing opacity reductions by a factor of up to eight it was not possible to obtain an acceptable sound speed. However, further tests taking into account the modified energy generation rate caused by the redistribution of ^{3}He are still required.

4 Conclusions

Given the success of the standard solar model in predicting the solar sound speed, it is tempting to assume that the models are equally successful in determining the temperature and composition. Indeed, it might appear unreasonable if the Sun were to have substantially different temperature and composition profiles, arranged in such a way as to give the same sound speed as in the standard model. Models of this nature can be constructed, although only with very substantial changes to the assumed physics of the solar interior, carefully adjusted to avoid changes in the sound speed. Thus, it is perhaps natural to conclude that the predictions of the neutrino production rates are robust, and that therefore the neutrino discrepancies reflect a need for revision of our understanding of the physics of the neutrino.

However, given the fundamental implications of such a conclusion, it should not be accepted prematurely. Further investigations, taking into account plausible errors in the helioseismic results as well as in the physics of the solar interior, are required to place firmer limits on predicted neutrino fluxes consistent with the helioseismic results.

Acknowledgments

I am grateful for permission to show the results in Fig. 1 before publication.[4] This work was supported in part by the Danish National Research Foundation through its establishment of the Theoretical Astrophysics Center.

1. J.N. Bahcall, *Neutrino astrophysics*, Cambridge University Press (1989)
2. D.O. Gough & J. Toomre, *Ann. Rev. Astron. Astrophys.* **29**, 62 (1991)
3. S. Basu, J. Christensen-Dalsgaard, J. Schou, M.J. Thompson & S. Tomczyk, *Astrophys. J.* **460**, 1064 (1996)
4. S. Basu *et al.*, *Mon. Not. R. astr. Soc.*, submitted (1997)
5. D.O. Gough *et al.*, *Science* **272**, 1296 (1996)
6. O. Richard, S. Vauclair, C. Charbonnel & W.A. Dziembowski, *Astron. Astrophys.* **312**, 1000 (1996)
7. J.N. Bahcall, M.H. Pinsonneault, S. Basu & J. Christensen-Dalsgaard, *Phys. Rev. Lett.* **78**, 17 (1997)
8. J. Christensen-Dalsgaard *et al.*, *Science* **272**, 1286 (1996)
9. F.J. Rogers, F J. Swenson & C.A. Iglesias, *Astrophys. J.* **456**, 902 (1996)
10. C.A. Iglesias, F.J. Rogers & B.G. Wilson, *Astrophys. J.* **397**, 717 (1992)
11. J.N. Bahcall & M.H. Pinsonneault, *Rev. Mod. Phys.* **67**, 781 (1995)
12. G. Michaud & C.R. Proffitt, in *Proc. IAU Colloq. 137: Inside the stars*, eds A. Baglin & W.W. Weiss, *ASP Conf. Ser.* **40**, 246 (1993)
13. J. Christensen-Dalsgaard, F.W.W. Dilke & D.O. Gough, *Mon. Not. R. astr. Soc.* **169**, 429 (1974)
14. F.W.W. Dilke & D.O. Gough, *Nature* **240**, 262 (1972)
15. B. Chaboyer, P. Demarque, D.B. Guenther & M.H. Pinsonneault, *Astrophys. J.* **446**, 435 (1995)
16. D.O. Gough & M.J. Thompson, in *Solar interior and atmosphere*, eds A.N. Cox, W.C. Livingston & M. Matthews, Space Science Series, University of Arizona Press, p. 519 (1991)
17. S. Basu, J. Christensen-Dalsgaard, F. Pérez Hernández & M.J. Thompson, *Mon. Not. R. astr. Soc.* **280**, 651 (1996)
18. J.W. Harvey *et al.*, *Science* **272**, 1284 (1996)
19. P.H. Scherrer *et al.*, *Solar Phys.* **162**, 129 (1996)
20. Y. Elsworth *et al.*, *Astrophys. J.* **434**, 801 (1994)
21. S. Tomczyk *et al.*, *Solar Phys.* **159**, 1 (1995)
22. Y. Elsworth *et al.*, *Nature* **347**, 536 (1990)
23. J. Christensen-Dalsgaard, *Geophys. Astrophys. Fluid Dyn.* **62**, 123 (1991)
24. D.O. Gough & A.G. Kosovichev, in *Proc. IAU Colloquium No 121*, eds G. Berthomieu & M. Cribier, Kluwer, Dordrecht, p. 327 (1990)
25. W.A. Dziembowski, A.A. Pamyatnykh & R. Sienkiewicz, R., *Mon. Not. R. astr. Soc.* **244**, 542 (1990)
26. H. Shibahashi & M. Takata, *Publ. Astron. Soc. Japan* **48**, 377 (1996)
27. I. W. Roxburgh, *Bull. Astron. Soc. India* **24**, 89 (1996)
28. H.M. Antia & S.M. Chitre, *Mon. Not. R. astr. Soc.*, submitted (1997)
29. A. Cumming & W.C. Haxton, *Phys. Rev. Lett.* **77**, 4286 (1996)

TIME VARIATIONS OF THE SOLAR NEUTRINO FLUX & CORRELATIONS WITH SOLAR PHENOMENA

Todor Stanev

Bartol Research Institute, University of Delaware, Newark, DE 19716, USA

Correlations of the solar neutrino flux with various solar phenomena have been often reported since the Homestake detector became operational in the late 1960's. The statistical significance of these correlations reached a peak during the late 1980's. Since then the significance has declined to reach the ~90% confidence level similar to the significance of the 1970's. We may have been fooled by unusually strong statistical fluctuations. An alternative explanation could be that the relevant correlations timescale is the 22 year solar magnetic cycle.

1. Statistical properties of the solar neutrino data

The current world statistics on solar neutrinos, excluding the SuperKamiokande results first presented at this meeting, comes from the 23.6 years of operation of the Homestake detector [1], 5.8 years of Kamiokande II&III [2], 3.8 years of SAGE [3] and 3.1 years of GALLEX [4]. As far as correlations with the ~11 years solar cycle is concerned, the Homestake detector is the only one with long enough exposure to study such effects.

One has to be however very cautious and appreciate the statistical properties of all run by run data samples used to study the time variations. The average rate of ^{37}Ar production in the Homestake detector is 0.48 atoms/day, i.e. a total of 18 atoms per 60 days run and halflife of 35 days. Accounting for all experimental efficiencies this translates into 6 signal events per run over a background of 2 events [1]. The situation with other radiochemical detectors is quite similar. GALLEX measures a ^{71}Ge production of ~14 atoms per 21 day run to end up with similar statistics of 4–5 signal atoms after accounting for all experimental efficiencies. The probability for measuring of a production rate different from the average by a factor of 2 in any particular run is very high, not less than 1/3.

This argument does not apply to the measurement of the average production rate and of the average solar neutrino flux. The events from individual runs are combined, as a function of the time since the end of the exposure, to give a precise value only after ~20 individual runs.

Thus the study of the possible time variations of the solar neutrino flux are dominated by statistical problems which are not present when the average solar neutrino flux is derived. Water Cherenkov detectors have different set of problems, since the signal events are determined statistically from the track direction respectively to the position of the Sun.

2. Brief overwiew of the correlation studies and the statistical methods used

Time variation analyses have searched for periodicities and for correlations with the sunspot numbers, other solar and heliospheric data sets, and with simple models. Even the best of the many papers published on the subject have their specific deficiencies. Periodicity searches on the time scale of one or two solar cycles are not very useful because of the irregularities of the solar cycle itself. The use of simple models is suspect because they certainly oversimplify the complex solar phenomena and could be misleading. The use of data is best, however all data sets are extremely noisy, even noisier than the solar neutrino data sets. Since all studied solar and heliospheric phenomena are highly correlated with the sunspot number, even a high confidence correlation with any particular data set does not guarantee a real physical connection with the solar neutrino production. A partial list of significant papers with indications of the correlation parameter and of the statistical significance found is given in Table I.

Table 1: Partial list of correlation papers. 'moderate' means 2–4 σ.

Authors	Ref	What	Strength
R. Davis Jr.	5	Sunspots	??
Bahcall, Field&Press '87	6	Sunspots	so–so
Filippone&Vogel '90	7	Sunspots	moderate
Bahcall&Press '91	8	Sunspots	probably not
Dorman&Wolfendale '91	9	Sunspots	strong
Bieber et al. '90	10	VVO[11]	moderate
Oakley et al. '94	12	⊙ B field	very strong
Massetti&Storini '95	13	green line	strong for South
Krauss '90	14	p–modes	moderate
Delache et al. '93	15	⊙ dimensions	moderate
McNutt '95	16	solar wind	very strong

Individual solar neutrino run data are so ragged that many correlation attempts start with the smoothing the data set through 3, 5, and 7 points running averages, i.e. 6 months to 2 years for the Homestake detector. Possible shorter term effects are thus eliminated and the statistics becomes very complicated. Correlation studies using running averages [12,13,16] tend to erroneously overestimate, sometimes drastically, the statistical significance of the observed correlations.

χ^2 fits are challenged by the problem of choosing the proper statistics [7] and estimating correctly the errors of the individual runs. Rank order statistics [8] avoid that problem but end up losing the information contained in the errors. The maximum likelihood technique seems to be the best, provided that it represents correctly the experimental data analysis procedure.

It is quite common now to use the F–test, which measures the improvement of the fit when additional parameters are added, to study a correlation. The statistical significance of a correlation is often estimated by a large number of Monte Carlo shuffles of the data sets [10]. This is a practical, if not elegant, solution of the problems stemming out of the unknown small numbers statistics.

3. Current situation

In preparation for this talk I updated the analysis of Ref. [10] of the correlations between the solar neutrino measurements of the Homestake detector (up to and including run #133) with solar phenomena. In addition to the sunspot # and (sunspot # $\times \cos\theta$), where θ is the solar latitude of Earth footprint on the solar surface (to implement the idea of VVO [11], three other data sets were inspected for correlations, These are:
green line intensity. The green line is Fe XIV coronal line, probably excited by magnetic reconnection below the solar surface, as in Ref. [13];
solar wind particle density as in Ref. [16];
solar wind particle flux [16].
The statistical method is to perform χ^2 correlation study using the upper error bars as a measure of error in the individual runs, F–test as a measure of the improvement of the fit and Monte Carlo shuffles as estimate of the statistical significance. Fig. 1 shows the time evolution of the statistical significance of the correlations with the five solar phenomena starting with the Homestake run #22 (end of 1971).

Table II contains a summary of the results of this analysis. Columns 2–4 show the epoch with the highest statistical significance for each correlation, the corresponding value of the F–test and value of $1.- CL$, i.e. the probability that the correlation is caused by constant solar neutrino flux. Columns 5&6 show F and $1.- CL$ for the whole sample, i.e. after run # 133.

There is obviously a very strong tendency for a decrease of the significance of all correlations, which are currently on the pre 1980 level. As noted by Bahcall&Press [8] the correlations first became significant during four runs in 1980, when the Homestake was very low and which coincided with high sunspot numbers. The anticorrelations with the sunspot #, $s \times \cos\theta$ and the green line

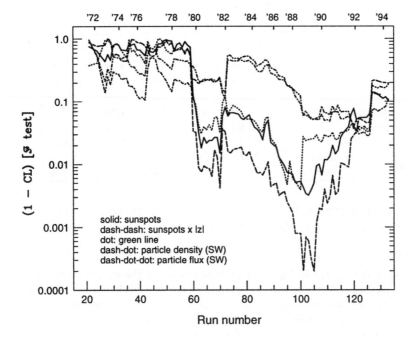

Figure 1: Statistical significance of the correlation of the Homestake neutrino flux with several solar and solar wind parameters.

intensity continued to grow up to the epochs given in Table II, and started decreasing soon afterwards. The analysis summarized above is performed with a Homestake data set provided by R. Davis Jr. The current error estimates [1] (submitted to *Ap. J.*) give smaller upper errors and correspondingly worse anticorrelations.

I have also inspected independently the data of Homestake after 1990 and of SAGE (29 runs in 1990–93) and GALLEX (39 runs in 1991–94). Neither of these data sets shows any correlation with the sunspot numbers. The only data set that supports any anticorrelations is the GALLEX one [17] at a statistically insignificant level. There is also a number of runs by different detectors that are almost coincidental in time. There 14 runs of GALLEX and Homestake with weighted average epochs within 1/2 ^{71}Ge lifetime, 8 such runs for SAGE&Homestake, and 10 for GALLEX&SAGE. There is no correlation between the neutrino fluxes measured by the corresponding detectors. We can conclude that there are no obvious correlations between the solar neutrino flux measured by these three experiments since 1990 and the solar activity.

Table 2: Update of the correlations of the Homestake neutrino flux with some solar and solar wind parameters.

What	Best correlation			After run 133	
	Epoch	F	1.−CL	F	1.−CL
sunspots (s)	89.12	10.9	0.0030	4.58	0.103
s×cosθ	88.60	15.1	0.0002	3.10	0.195
green line	87.48	9.2	0.0050	4.44	0.115
SW particle density	92.38	6.1	0.0300	4.29	0.110
SW particle flux	92.57	5.7	0.0340	2.72	0.189

It is tempting to conclude that unusual fluctuations occurred in 1980's and we are now going back to the statistical average.

4. However

The alternative option is that our understanding of the correlations between the solar neutrino flux and the solar activity is not sufficient and we have not yet even started to comprehend their relation. There are several facts that prevent me from dismissing the correlation hypothesis.

It was noted in Ref.[10] that the four important runs in 1990 actually coincide with the change of the magnetic polarity of the Sun and with very strong magnetic activity in the polar region of the Sun. Furthermore, the epoch when the anticorrelations disappeared coincides with another change of the magnetic polarity. The guess was made that the anticorrelation is only visible when the solar poloidal field is of negative polarity, although we know of no physical process backing up such a guess (with the possible exception of a fine tuned resonant spin–flip transition in twisting magnetic fields[18]).

The correlation studies performed with solar data also reveal some features that are unlikely to be coincidental. Oakley et al.[12] study the correlations with the surface magnetic field measured in bands around the solar equator. Correlations are only strong for bandwidths less than about 20 degrees as it should be if the neutrinos somehow interact with strong subsurface fields. Massetti&Storini[13] also notice similar behaviour, which only occurs when the Earth is south of the solar equator. Strangely enough, Simpson et al[19] derive quite a strong asymmetry of the solar magnetic field toward the Southern hemisphere from the repulsion of cosmic rays measured during the Ulysses flyby. It is after all still possible that in a few years, with the beginning of the new solar magnetic cycle, the anticorrelations of the solar neutrino flux with solar activity will become again observable.

Acknowledgments This is an update of the work performed with John Bieber, David Seckel and Gary Steigman. The author is grateful to Ken Lande and Ray Davis Jr. for sharing the Homestake data and to Marisa Storini and her collaborators for the green line data set. The work of TS is funded in part by the DOE contract DE–FG02–91ER40626.

References

1. B.T. Cleveland, T. Daily, R. Davis Jr., J.R. Distel, K. Lande, C.K. Lee and P.S. Wildenhain, *Ap. J.*, submitted; J.K. Rowley, B.C. Cleveland and R. Davis Jr., *AIP Conf. Proc.* **126**, 1 (1985).
2. K.S. Hirata *et al.*, PLR **65**, 1297 (1990); A. Suzuki, Venice Workshop on Neutrino Telescopes, ed. M. Baldo Ceolin (1996).
3. J.N. Abdurashitov *et al.*, PLB **328**, 234 (1994); J.N. Abdurashitov *et al.*, *Proc. TAUP '95*, eds. A. Morales *et al.*, *Nucl. Phys. B* (Proc. Suppl.) **48**, 299 (1996).
4. P. Anselmann *et al.*, *Phys. Lett.* B **357**, 237 (1995); P. Anselmann *et al.*, *Phys. Lett.* B **314**, 445 (1993)
5. R. Davis Jr., *Proc. TAUP '95*, eds. A. Morales *et al.*, *Nucl. Phys. B* (Proc. Suppl.) **48**, 299 (1996) and references therein.
6. J.N. Bahcall, G.B. Field and W.H. Press, *Ap. J.*, **320**, L69 (1987).
7. B.W. Filippone and P. Vogel, *Phys. Lett.* B **246**, 546 (1990)
8. J.N. Bahcall and W.H. Press, *Ap.J.*, **370**, 730 (1991).
9. L.I. Dorman and A.W. Wolfendale, *J.Phys. G:* Nucl. Part. Phys. **17**, 769 (1991).
10. J.W. Bieber *et al.*, *Nature*, **348**, 407 (1990).
11. L.B. Okun, M.B. Voloshin and M.I. Vysotsky, *Sov. Phys. JETP*, **64**, 446 (1986).
12. D.S. Oakley *et al.*, *Ap. J.*, **437**, L63 (1994).
13. S. Massetti, M. Storini and N. iucci, *Proc. 24th ICRC* (Rome) 4, 1243 (1995); S. Massetti and M. Storini, *Ap. J.*, submitted.
14. L.M. Krauss, *Nature*, **348**, 403 (1990)
15. Ph. Delache *et al. Ap.J.*, **407**, 801 (1993)
16. R.L. McNutt Jr. *Science* **270**, 1635 (1995).
17. T. Kirsten, *Talk at the Topical workshop on Solar Neutrinos*, Gran Sasso National Lab, Italy, May 1996.
18. E.Kh. Akhmedov, P.I. Krastev and A.Yu. Smirnov, *Z. Phys. C* **52**, 701 (1991).
19. J.A. Sipmson, M. Zhang and S. Bame, *Ap. J.*, **465**, L69 (1996).

IS THERE TIME VARIATION OF THE HOMESTAKE NEUTRINO FLUX?

Marcello LISSIA

Istituto Nazionale di Fisica Nucleare, I-09127 Cagliari, Italy

Data are still ambiguous: a constant neutrino flux is consistent with all present data, but the uncertainties do not exclude even strong time variations, *e.g.*, a strong correlation with sunspot number.

For many years possible time variation of the solar neutrino flux has drawn considerable interest because of its physical implications[1]: neutrino magnetic moment[2] (connected to 22-year-long cycle of the solar magnetic activity and to the annual change of the solar latitude of the neutrino path), neutrino oscillations with wave length of the order of the annual variation of the Sun-Earth distance[3], short-time periodic phenomena in the solar production region, etc. The data themselves from the Homestake detector have appeared to suggest time variations over time intervals of the order of 11 years[4,5].

Three questions are in order:

(Q.1) Is a constant solar-neutrino flux consistent with the data of the neutrino experiments?

(Q.2) Does a (physically well-motivated) non constant solar-neutrino flux give a better description of the experimental data?

(Q.3) Are large time variations excluded by the data?

Here we only present a simple analysis of the data just to illustrate what kind of evidence supports our answers to these three questions. Additional statistical analyses[6] and the published literature are in substantial agreement with these results. Answers to Q.2 and Q.3 partially depends on the choice of the model for the time-dependence (linear correlation with sunspot number): for correlation of neutrino signals (mainly Homestake) with other indicators of solar activity, such as different measurements of the surface magnetic field, see talk by T. Stanev in these proceedings and in Ref.[1].

Let assume that the temporal series of the neutrino data and of the sunspot number (as indicator of the solar magnetic activity) be anticorrelated:

$$S(i)^{Cl,Ga,Ka} = A^{Cl,Ga,Ka} \left[1 - B \frac{N_{ss}(i)}{N_{ss}^{max}} \right] \tag{1}$$

where A is a normalization constant that fixes the average signal, N_{ss}^{max} is the maximum number of sunspots and $-1 < B < 1$ measures the (anti) correlation.

The χ^2/d.o.f. is shown in Fig. 1 for each experiment: (i) a constant fit $(B = 0)$ is allowed by all experiments; (ii) very strong (anti)correlation ($|B| \approx 1$) is not excluded by any experiment (only Kamiokande disfavors $B \approx 1$); Kamiokande favors $B \approx 0.35$, but the confidence interval (-0.15, 0.7) includes $B = 0$.

Therefore, our answers to the three questions are:

(A.1) A constant neutrino flux is perfectly compatible with present data. This conclusion holds for the complete set of Homestake data (1970–1994), for several subsets of the same data, for the Kamiokande data and for the gallium data.

(A.2) The question can be answered only within a specific model. If one models the effect of solar activity by assuming linear anticorrelation of the data with the sunspot number, one finds a better, but not significantly better, description of the data. The anticorrelation of the Homestake data and the sunspot number, which appeared very strong for data up to 1990, has significantly decreased both in the same data up to 1990, after their recent recalibration, and even more in the complete set of data up to 1994 (the data in the last few years appear actually to prefer correlation). More precisely, a signal linearly anticorrelated with the sunspot number gives a fit with a better χ^2 relative to the fit with a constant signal, but the magnitude of the χ^2 improvement is not very large and comparable to the one due statistical fluctuations in about 17% of the cases.

(A.3) The statistics of data are such that even strong correlation of neutrino signals and sunspot number cannot be excluded.

The conclusion is that much larger number of events is necessary: compare A.1 and A.3. Larger statistics would also allow to study seasonal variations and further exclude some of the models: there is no evidence of solar latitude dependence of the signal within the present low statistics. In addition it would be desirable that several experiments with different energy windows keep observing the solar neutrino flux in case the time variation is energy dependent.

Acknowledgments

This talk is based on several discussion with V. Berezisky and G. Fiorentini and uses material from the Ph. D. thesis of E. Calabresu[6].

708

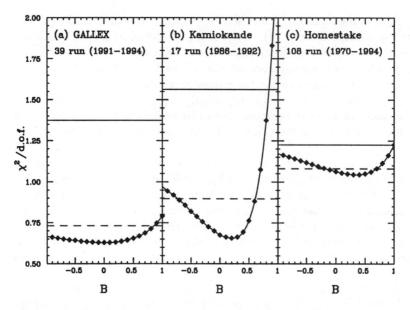

Figure 1: The χ^2 per degree of freedom for a fit to the (a) Gallex, (b) Kamiokande and (c) Homestake temporal series of data with the formula in Eq. (1). For each value of B, A has been chosen to minimize the χ^2. Values of B yielding $\chi^2 \leq \chi^2_{min} + 3.84$ (dashed line) define the 95% confidence interval for B. Value of χ^2 above the solid line, and the corresponding values of B, are excluded at the 95% CL for the appropriate number of degrees of freedom (38, 16 and 107).

References

1. V. Berezinsky and G. Fiorentini (eds.), *New Trends in Solar-Neutrino Physics* (INFN-LNGS, L'Aquila, 1996).
2. E. K. Akhmedov, *Phys. Lett.* B **213**, 64 (1988)
3. S. L. Glashow and L. M. Krauss, *Phys. Lett.* B **190**, 199 (1987)
4. Bieber *et al*, *Nature* **348**, 407 (1990)
5. J. N. Bahcall and Press, *Astr. J.* **370**, 730 (1991)
6. E. Calabresu, Ph. D. Thesis, Cagliari, 1996 (unpublished).

A Critique of Core–Collapse Supernova Theory Circa 1997

Adam Burrows

Department of Astronomy, University of Arizona, Tucson, AZ 85721, USA

There has been a new infusion of ideas in the study of the mechanism and early character of core–collapse supernovae. However, despite recent conceptual and computational progress, fundamental questions remain. In this all–too–brief contribution, I summarize some of the interesting insights achieved over the last few years. In the process, I highlight as–yet unsolved aspects of supernova theory that continue to make it a fascinating and frustrating pursuit.

1 Introduction

It has recently been shown that neutrino–driven Rayleigh–Taylor instabilities between the stalled shock wave and the neutrinospheres ("Bethe" convection) are generic feature of core–collapse supernovae.[1,2,3,4,5,6] Whatever their role in reigniting the stalled explosion, their existence and persistence have altered the way modelers approach their craft. Supernovae must explode *aspherically*, and this broken symmetry is stamped on the ejecta and character of the blast, as well as on its signatures. Consequences of asphericity include significant gravitational radiation,[7,8] natal kicks to nascent neutron stars,[7,9] induced rotation, [4] mixing of iron–peak and r–process nucleosynthetic products, the generation and/or rearrangement of pulsar magnetic fields, and, in extreme cases, jetting of the debris.

However, there is no consensus yet on the centrality of overturn (or "convection") to the mechanism of the explosion itself, with some deeming it either pivotal,[1,2] potentially important,[4,5,10] or diversionary.[6] Nevertheless, *all* agree on the existence of convection in the gain region of the stalled protoneutron star, and this point must be stressed. *A gain region is a prerequisite for the neutrino–driven mechanism.*[11] *For heating to exceed cooling in steady–state accretion, the entropy gradient must be negative, and, hence, unstable. Therefore, a gain region is always convective.* In order to achieve quantitative agreement with the variety of observational constraints (explosion energy, residual neutron star masses, ^{56,57}Ni and "$N = 50$" peak yields, halo star element ratios, neutron star proper motions, etc.), the "final" calculations must be done multi-dimensionally. While if it can be shown that 1–D spherical models do explode after some delay, the true duration of that delay, the amount of fallback, and the energetics of the subsequent explosion must be influenced by the overturning motions that can not be captured in 1–D. Convection changes not only the character of the hydrodynamics, but the entropies in the gain region and the "efficiency" [2] of neutrino energy deposition that is the ultimate driver of the explosion.[4,12,11] Furthermore, implicit in a focus on 1–D calculations is the assumption that multi–D effects could only help, that they do not thwart explosion. Hence, the belief that spherically–symmetric calculations are germane depends upon insights newly obtained from the multi–dimensional simulations. Nevertheless, it will be an important theoretical exercise to ascertain whether 1–D models with the best physics and numerics can explode, if only because such has been a goal for decades. The "viablility" of 1–D models will be influenced in part by the transport algorithm employed (multi–group, flux–limited, full transport, diffusion), the microphysics (opacities and source terms at high and low densities), the effects of general relativity, the equation of state, convection in the inner core that can boost the driving neutrino luminosities,[13,14,10] and the inner density structure

2 Neutrino Transport

Though much of the recent excitement in supernova theory has concerned its multi–dimensional aspects, neutrino heating and transport are still central to the mechanism. The coupling between matter and radiation in the semi–transparent region between the stalled shock and the neutrinospheres determines the viability and characteristics of the explosion. Unfortunately, this is the most problematic regime. Diffusion algorithms and/or flux–limiters do not adequately reproduce the effects of variations in the Eddington factors and the spectrum as the neutrinos decouple. Hence, a multi–group full transport scheme is desirable.

To address the issues surrounding neutrino transport, we have recently created a neutrino transport code using the program **Eddington** developed by Eastman & Pinto.[16] This code solves the full transport equation using the Feautrier approach, is multi–group, is good to order v/c, and does not employ flux limiters. The ν_es, $\bar{\nu}_e$s, and "ν_μs" are handled separately and coupling to matter is facilitated with accelerated lambda iteration (ALI). By default, we employ 40 energy groups from 1 MeV to either 100 MeV ($\bar{\nu}_e$ and "ν_μs") or 230 MeV (ν_e) and from a few to 200 angular groups, depending on the number of tangent rays at the given radial zone, in the Feautrier manner. In this way, the neutrino angular distribution function and all the relevant angular moments (0'th through 3'rd) are calculated to high precision, for every energy group.

The effect of the full Feautrier scheme vis–a–vis previous[4,17,18,6,19,20] calculations will soon be benchmarked and calibrated. However, we have already obtained several interesting results. Since the annihilation of ν–$\bar{\nu}$ pairs into e^+–e^- pairs depends upon the 0'th, 1'st, and 2'nd angular moments of the neutrino angular distribution function, as well as upon the neutrino spectra, we can and have calculated the rate of energy deposition via this process *exactly*, though in the context of previous model runs[4] (still ignoring general relativity). The $\nu_e + \bar{\nu}_e \rightarrow e^+ + e^-$ and $\nu_\mu + \bar{\nu}_\mu \rightarrow e^+ + e^-$ energy deposition rates in the shocked region are no more than 0.01 and 0.001, respectively, those of the dominant charged–current processes, $\nu_e + n \rightarrow e^- + p$ and $\bar{\nu}_e + p \rightarrow e^+ + n$. However, in the unshocked region ahead of the shock, depending upon the poorly–known ν–nucleus absorption rates, the ν–$\bar{\nu}$ annihilation rate can be competitive, though it is still irrelevant to the supernova. These calculations should put to rest the notion that ν–$\bar{\nu}$ annihilation is important in igniting the supernova explosion.

It is thought that neutrino–electron scattering and inverse pair annihilation are the processes most responsible for the energy equilibration of the ν_μ's and their emergent spectra. However, recent calculations imply that the inverse of nucleon–nucleon bremsstrahlung (*e.g.*, $n + n \rightarrow n + n + \nu\bar{\nu}$) is also important in equilibrating the ν_μ's.[21] This process has not heretofore been incorporated in supernova simulations. Our preliminary estimates suggest that inverse bremsstrahlung softens the emergent ν_μ spectrum, since the bremsstrahlung source spectrum is softer than that of pair annihilation. In addition, given the large ν_μ scattering albedo, one must properly distinguish absorption from scattering, in ways not possible with a flux–limiter. Since the relevant inelastic neutral–current processes are stiff functions of neutrino energy, these transport issues bear directly upon the viability of neutrino nucleosynthesis (*e.g.*, of ^{11}B and ^{19}F).[22]

The new code allows us to calculate the difference bewteen the flux spectrum (h_ν) and the energy density spectrum (j_ν). The latter couples to matter and drives the supernova in the neutrino mechanism, while the former, or some variant of it, is frequently substituted for the latter in diffusion codes. Since matter–neutrino cross sections are higher for higher–energy neutrinos, the energy density spectrum is always harder than the flux spectrum. This hardness boosts the neutrino heating rates in the semi–transparent region. To illustrate this effect, in Figure 1 the ratio j_{ν_e}/h_{ν_e} is plotted versus neutrino energy at a time 30 milliseconds after bounce. The shock is then at 124 kilometers. It is clear that the ratio effect can be interesting. However, it is most pronounced in the cooling region below the gain region and tapers off as the shock is approached. Mezzacappa *et al.*,[6] in particular, have highlighted this correction, but self–consistent calculations from collapse to explosion, using the Feautrier or Boltzmann techniques (in principle equivalent), are needed, given the notorious feedbacks in the supernova problem. The same effect may be important in driving the protoneutron star wind[4] thought to be the site of

712

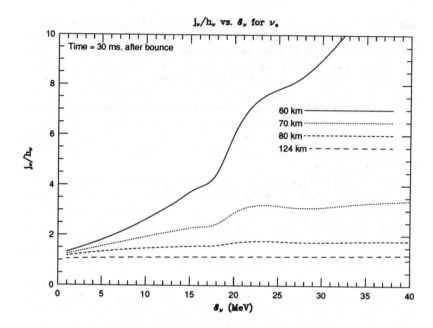

Figure 1: j_ν/h_ν versus ϵ_ν for electron neutrinos 30 milliseconds after the bounce of a 15 $M_\odot$ core, using the code of Burrows & Pinto (1997)

the r–process.[23,24] Indeed, full transport calculations of r–process winds and the supernova, even in 1–D, will be illuminating.

3   Conclusions

In parallel with the ongoing evolution of the numerical tools being brought to bear on the supernova problem is the emerging realization that the systematics of the supernova phenomenon with progenitor is inching closer into view. As we unravel the mechanism, we simultaneously explore the origin of neutron stars and black holes, the birthplace of elements of which we are made, and the source of much of the energy of the ISM. As supernova modelers and the Jayhawks might say, *ad astra per aspera.*

Acknowledgments

Conversations with Willy Benz, Chris Fryer, Tony Mezzacappa, and Phil Pinto that materially altered the content of this squib are gratefully acknowledged, as is the support of the U.S. N.S.F. under grant #AST92-17322.

References

1. H. A. Bethe, *Rev. Mod. Phys.* **62**, 801 (1990).
2. M. Herant, W. Benz, & S. A. Colgate, *Astrophys. J.* **395**, 642 (1992).
3. M. Herant, W. Benz, J. Hix, C. Fryer, & S.A. Colgate, *Astrophys. J.* **435**, 339 (1994).
4. A. Burrows, J. Hayes, & B. A. Fryxell *Astrophys. J.* **450**, 830 (1995).
5. H.-T. Janka & E. Müller, *Astron. & Astrophys.* **306**, 167 (1996).
6. A. Mezzacappa, *et al.*, *Astrophys. J.*, in press (1996).
7. A. Burrows & J. Hayes, *Phys. Rev. Lett.* **76**, 352 (1996).
8. R. Mönchmeyer, G. Schäfer, E. Müller, & R.E. Kates, *Astron. & Astrophys.* **246** (1991) 417.
9. S.E. Woosley, in *The Origin and Evolution of Neutron Stars*, eds. D.J. Helfand & J.-H. Huang (D. Reidel: Dordrecht, 1987), p. 255.
10. R. Mayle & J. R. Wilson, *Astrophys. J.* **334**, 909 (1988).
11. H. A. Bethe, J. R. Wilson, *Astrophys. J.* **295**, 14 (1985).
12. H.-T. Janka, in the *7th Workshop on Nuclear Astrophysics*, eds. W. Hillebrandt and E. Müller (Max-Planck-Institut für Astrophysik, 1993), p. 18.
13. A. Burrows, *Astrophys. J.* **318**, L57 (1987).
14. W. Keil, H.-T. Janka, & E. Müller, preprint (1996).
15. A. Burrows & P. A. Pinto, in preparation (1997).
16. R. Eastman & P. A. Pinto, *Astrophys. J.* **412**, 731 (1993).
17. A. Mezzacappa & S. W. Bruenn, *Astrophys. J.* **405**, 637 (1993).
18. A. Mezzacappa & S. W. Bruenn, *Astrophys. J.* **405**, 669 (1993).
19. J. R. Wilson, in *Numerical Astrophysics* ed. J. Centrella, J. M. LeBlanc, & R. L. Bowers (Jones & Bartlett: Boston, 1985), p. 422.
20. E. Myra & A. Burrows, *Astrophys. J.* **364**, 222 (1990).
21. H. Suzuki, in *Frontiers of Neutrino Astrophysics*, ed. Y. Suzuki & K. Nakamura (Tokyo: Universal Academy Press, 1993), p. 219.
22. S. E. Woosley, D. Hartmann, R. Hoffman, & W. C. Haxton, *Astrophys. J.* **356**, 272 (1990).
23. S. E. Woosley & R. D. Hoffman, *Astrophys. J.* **395**, 202 (1992).
24. Y.-Z. Qian & S. E. Woosley, *Astrophys. J.* **471**, 331 (1996).

SIMULATIONS OF CORE COLLAPSE SUPERNOVAE IN ONE AND TWO DIMENSIONS USING MULTIGROUP NEUTRINO TRANSPORT

A. MEZZACAPPA

ORNL, Physics Division, Building 6010, MS 6354, P.O. Box 2008
Oak Ridge, TN 37831-6354

In one dimension, we present results from comparisons of stationary state multigroup flux-limited diffusion and Boltzmann neutrino transport, focusing on quantities central to the postbounce shock reheating. In two dimensions, we present results from simulations that couple one-dimensional multigroup flux-limited diffusion to two-dimensional (PPM) hydrodynamics.

1 Introduction

Current supernova modeling revolves around the idea that the stalled supernova shock is reenergized by absorption of electron neutrinos and antineutrinos on the shock-liberated nucleons behind it.[1] Key to this process is the neutrino transport in the critical semitransparent region encompassing the neutrinospheres. Potentially aiding this process is convection below the neutrinospheres and the shock,[2,3,4,9] although the verdict has been mixed.[2,4,5,6,8]

In semitransparent regions, neutrino transport approximations such as multigroup flux-limited diffusion (MGFLD) break down. Moreover, neutrino shock reheating/revival depends sensitively on the emergent neutrinosphere luminosities and spectra, and on the neutrino inverse flux factors between the neutrinospheres and the shock. At the very least, computation of these quantities by approximate transport methods have to be checked against exact methods, i.e., against Boltzmann neutrino transport.

We present results from one- and two-dimensional supernova simulations. In one dimension, beginning from postbounce slices obtained from Bruenn's MGFLD simulations, which are subsequently thermally and hydrodynamically frozen, we compute stationary state neutrino distributions with MGFLD and Boltzmann neutrino transport. From these, we compute and compare the neutrino luminosities, RMS energies, and inverse flux factors, and the net neutrino heating rate, between the neutrinospheres and the shock. In two dimensions, beginning with the same postbounce profiles, we couple one-dimensional MGFLD to two-dimensional PPM hydrodynamics to investigate "prompt" convection below the neutrinospheres and "neutrino-driven" convection below the shock. At the expense of dimensionality, we implement neutrino transport that is multigroup and that computes with sufficient realism transport through opaque, semitransparent, and transparent regions.

2 Prompt Convection

Without neutrino transport, prompt convection develops and dissipates in ~15 ms in both our 15 and 25 $M_\odot$ models. When transport is included, there is no significant convective transport of entropy or leptons in either model. Neutrino transport *locally* equilibrates a convecting fluid element with its surroundings, reducing the convection growth rate and asymptotic convection velocities by factors of 4–250 between $\rho = 10^{11-12}$ g/cm^3, respectively.[5] Our transport neglects the finite time it takes for neutrinos to traverse a convecting fluid element; consequently, our equilibration rates are too rapid, and transport's effect on prompt convection is overestimated. Nonetheless, it is difficult to see how prompt convection will lead to significant convective transport when these effects are taken into account.

3 Neutrino-Driven Convection

In our 15 $M_\odot$ model, large-scale semiturbulent convection is evident below the shock and is most vigorous at $t = 225$ ms after bounce (we began our run at 106 ms after bounce when a well-developed gain region was present.) At this time, the maximum angle-averaged entropy is 13.5, and the angle-averaged radial convection velocities exceed 10^9 cm/sec, becoming supersonic just below the shock. Despite this, at $t = 506$ ms after bounce our shock has receded, the convection below it has become more turbulent, and there is no evidence of an explosion or of a developing explosion.[6]

Different outcomes obtained by the various groups are mainly attributable to differences in the neutrino transport approximations, which determine the postbounce initial models and the neutrino luminosities, RMS energies, and inverse flux factors, which in turn define the postshock neutrino heating rates. Most notable are dramatic differences in the RMS energies.[6] For example, for $\eta_{\nu_e} = 2$ and $T_{\nu_e} = T(\tau_{\nu_e} = 2/3)$, which is implemented by Burrows et al. (1995),[2] the electron neutrino RMS energy and matter temperature at the neutrinosphere are related by $< E_{\nu_e}^2 >^{1/2} = 3.6T$, whereas our MGFLD calculations give $< E_{\nu_e}^2 >^{1/2} = 3.0T$. Because the neutrino heating rates depend on the square of the RMS energies, the Burrows et al. rates would effectively be 40–50% higher. In the Herant et al. (1994)[3] calculations, at 100 ms after bounce $< E_{\nu_e} > \sim$ 13 to 14 MeV and $< E_{\bar{\nu}_e} > \sim$ 20 MeV, whereas we obtain the significantly lower values, 10 and 13 MeV, respectively.

716

4 Boltzmann Neutrino Transport

Although differences in the MGFLD and Boltzmann transport luminosities and RMS energies between the neutrinospheres and the shock were seen, the inverse flux factors differed most. The fractional difference relative to MGFLD rose to ~ 20–25% just below the shock. The net heating rate showed an even more pronounced difference, with the Boltzmann rate being 2–3 times higher just above the gain radius.[7] (Small differences in the heating rates lead to large differences in the net heating rate near the gain radii, where heating and cooling balance.) These results have two ramifications: (1) It may be possible to obtain explosions in spherical symmetry in the absence of convection. (2) Improved transport in two and three dimensions will give rise to more neutrino heating, and may give rise to more vigorous neutrino-driven convection. Simulations in both spherical and axi- symmetry with one-dimensional Boltzmann transport are planned. We are also developing realistic multidimensional, multigroup transport.

Acknowledgments

AM is supported at the Oak Ridge National Laboratory, which is managed by Lockheed Martin Energy Research Corporation under DOE contract DE-AC05-96OR22464. This work was also supported at the University of Tennessee under DOE contract DE-FG05-93ER40770.

References

1. H. A. Bethe & J. R. Wilson, *Ap. J.* **295**, 14 (1985).
2. A. Burrows, J. Hayes, & B. Fryxell, *Ap.J.* **450**, 830 (1995).
3. M. E. Herant, W. Benz, W. R. Hix, C. L. Fryer, & S. A Colgate, *Ap. J.* **435**, 339 (1995).
4. H.-Th. Janka & E. Müller, *Astron. Astrophys.* **306**, 167 (1996).
5. A. Mezzacappa, A.C. Calder, S. W. Bruenn, J.M. Blondin, M.W. Guidry, M.R. Strayer, & A.S. Umar *Ap. J.*, accepted (1997a).
6. A. Mezzacappa, A.C. Calder, S. W. Bruenn, J.M. Blondin, M.W. Guidry, M.R. Strayer, & A.S. Umar *Ap. J.*, accepted (1997b).
7. A. Mezzacappa, O. E. B. Messer, S. W. Bruenn, & M.W. Guidry, in preparation.
8. D. S. Miller, J. R. Wilson, & R. W. Mayle, *Ap. J.* **415**, 278 (1993).
9. J. R. Wilson & R. W. Mayle, *Phys. Rep.* **227**, 97 (1993).

CONVECTION IN PROTONEUTRON STARS

W. KEIL, H.-Th. JANKA, E. MÜLLER

Max-Planck-Institut für Astrophysik,
Karl-Schwarzschild-Str. 1, D-85740 Garching, Germany

Convectively enhanced neutrino (ν) luminosities from the protoneutron star (PNS) can provide an essential condition for a Type-II supernova explosion. Very recent two-dimensional, self-consistent, general relativistic simulations of the cooling of a newly-formed neutron star demonstrate and confirm the possibility that quasi-Ledoux convection, driven by negative lepton number and entropy gradients, may encompass the whole PNS within less than 1 s and can lead to an increase of the neutrino fluxes by up to a factor of two.

1 Introduction

Observations of supernova SN 1987A instigated recent work on multi-dimensional numerical modeling of Type-II supernova explosions. Two-dimensional (2D) simulations [1,2,3,4] indicate that overturn instabilities and mixing in the ν-heated region between PNS and supernova shock may indeed help ν-driven explosions to develop for conditions which do not allow explosions in spherical symmetry. Nevertheless, a critical investigation of the sensitivity of the explosion to the value of the ν flux from the neutron star [3,4] cannot confirm claims that a "convective engine" ensures the robustness of the explosion for a wide range of conditions. Instead, the explosion turns out to be extremely sensitive to the ν fluxes from the inner core. Explosions can occur in 2D as well as in 1D, provided the ν luminosity is large enough. Similarly, for too small core ν fluxes, strong convective overturn in the ν-heated region cannot develop and the explosion fizzles. These findings are supported by recent simulations of Mezzacappa and collaborators [5]. In addition, all currently successful numerical models of supernova explosions produce nucleosynthesis yields that are in clear contradiction with observational abundance constraints. These facts indicate severe problems of the numerical modeling of the explosion. The first 2D simulations of the evolution of the nascent neutron star [6] suggest that long-lasting quasi-Ledoux convection *in* the PNS can significantly raise the ν luminosities and may have a positive impact on the supernova explosion and conditions for explosive nucleosynthesis.

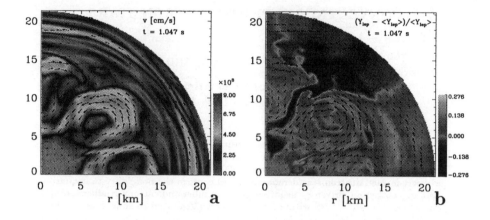

Figure 1: Panel a shows the absolute values of the velocity for the non-rotating PNS model 1.047 s after the start of the simulation in units of 10^8 cm/s. The computation was performed in an angular wedge of 90° between +45° and −45° around the equatorial plane. The PNS has contracted to a radius of about 21 km at the given time. Panel b displays the relative deviations of the lepton fraction Y_{lep} from the angular means $\langle Y_{\text{lep}} \rangle$ at each radius for $t = 1.047$ s. The maximum deviations are of the order of 30%. Lepton-rich matter rises while deleptonized material sinks in.

2 Results of two-dimensional hydrodynamical simulations

Our simulations were performed with an elaborate description of the nuclear equation of state [7] and the ν transport and were started with a PNS model of Bruenn [8] at ~ 25 ms after core bounce. These calculations showed that driven by negative lepton fraction and entropy gradients convective activity can encompass the whole PNS within ~ 1 s and can continue for at least as long as the deleptonization of the PNS takes place. The driving forces for this quasi-Ledoux convection are connected with very specific properties of the nuclear equation of state and the interaction between the diffusive ν transport and the convective transport. In differentially rotating PNSs stabilizing angular momentum distributions seem to suppress convection near the rotation axis.

The convective pattern is extremely non-stationary and has most activity on large scales with radial coherence lengths of several km up to ~ 10 km and convective "cells" of 20°–30° angular diameter, at some times even 45° (Fig. 1a). The maximum convective velocities are $\sim 4 \cdot 10^8$ cm/s, but peak values of $\sim 10^9$ cm/s can be reached. Relative deviations of the lepton fraction

Y_{lep} from the angular mean can be several 10% in rising or sinking buoyant elements (Fig. 1b), and for the entropy (S) they can reach 5% or more. Rising flows always have larger Y_{lep} *and* S than their surroundings.

Since convective activity continues for more than 1 s in a large region of the PNS, convective ν transport shortens the cooling and deleptonization timescales of the PNS compared to results of 1D simulations and enhances the calculated ν fluxes at the surface of the star by up to a factor of 2. The latter can have important influence on the interpretation of the measured ν signal from SN 1987A and might change limits of various quantities in nuclear and elementary particle physics which have been derived from these measurements. Moreover, the enhancement of the ν fluxes can also have important consequences for the explosion mechanism of the supernova, because it aids ν-driven explosions. The faster deleptonization modifies the luminosity ratio of ν_e and $\bar{\nu}_e$ in such a way that the electron fraction Y_e^{ej} in the ν-heated SN ejecta will be raised during the first few 100 ms after core bounce, but it will be lowered for $t \gtrsim 1$ s compared to 1D simulations. This might help to solve the severe problems of the nucleosynthesis in current models of Type-II supernovae which disregard long-lasting convective activities in the PNS. Finally, convection in the PNS causes stochastical asymmetries of the ν flux. The resulting recoil acceleration might explain measured proper motions of pulsars up to several 100 km/s. Both convective mass motions and anisotropic ν emission are a source of gravitational waves.

Acknowledgments

We thank S.W. Bruenn for kindly providing us with the post-collapse model used as initial model in our simulations. Support by the "SFB 375-95 für Astro-Teilchenphysik" of the German Science Foundation is acknowledged.

References

1. M. Herant *et al.*, ApJ **435**, 339 (1994).
2. A. Burrows, J. Hayes, and B.A. Fryxell,ApJ **450**, 830 (1995).
3. H.-Th.Janka and E. Müller, ApJ **448**, L109 (1995).
4. H.-Th.Janka and E. Müller, A&A **306**, 167 (1996).
5. A. Mezzacappa *et al.*, ApJ, submitted (1997).
6. W. Keil, H.-Th. Janka, and E. Müller, ApJ **473**, L111 (1996).
7. J.M. Lattimer and F.D. Swesty, Nucl. Phys. A **535**, 331 (1991).
8. S.W. Bruenn, in *Nuclear Physics in the Universe*, eds. M.W. Guidry and M.R. Strayer, IOP, Bristol, 31 (1993).

NEUTRINO BURST FROM SUPERNOVAE AND NEUTRINO OSCILLATION

K. Sato[1,2] and T. Totani[2]

[1] The Research Center for the Early Universe, School of Science,
the University of Tokyo, Tokyo 113, Japan
[2] Department of Physics, School of Science,
the University of Tokyo, Tokyo 113, Japan

Super Kamiokande has started observations from the April, 1996. We discuss how the neutrino burst can be detected by this detector. We investigate the effects of spin-flavor conversion, in particular. It is pointed out that if the energy spectrum of anti-electron type neutrinos has double peaks, it may imply the existence of this precession.

1   Introduction

In 1987, neutrino burst from Supernova 1987A, which appeared in Large Magellanic cloud, was detected by Kamiokande and IMB groups. This was the first and direct evidence that supernova explosions are induced by gravitational collapse of stellar cores. The event numbers were, however, only 11 and 5, respectively, which were too small to subtract informations on the mechanism of supernova explosions and on neutrino physics. From April, 1996, Super Kamiokande begun observations. If supernova explosion occurs in the center of our Galaxy, it is expected almost 8,000 events will be detected by $\bar{\nu}_e p$ reaction, 230 events by $\nu_e e$ scattering and 130 events by $\nu_e O$ and $\bar{\nu}_e O$ reactions if we take a simulation given by J. Wilson. Then it is greatly expected that not only the informations on the explosion mechanism, but also those on the neutrino physics could be obtained.

In this note, we show our recent investigation on the effects of spin-flavor conversion of Supernova $\bar{\nu}_e$'s.

2   Neutrino Oscillation or Precession of Supernova $\bar{\nu}_e$'s

Some possibilities of non-standard physics of neutrinos may cause the neutrino oscillation of supernova neutrinos. Collapse-driven supernovae emit all the six types of neutrinos in approximately the same luminosity, but the average energy or the effective temperature is different due to the difference of the coupling to dense matter in a proto neutron star: the effective temperatures of ν_e's, $\bar{\nu}_e$'s, and others are $\sim$ 3, 5, and 7–8 MeV, respectively. The most easily detectable flavor is $\bar{\nu}_e$'s, because of the dominantly large cross section of

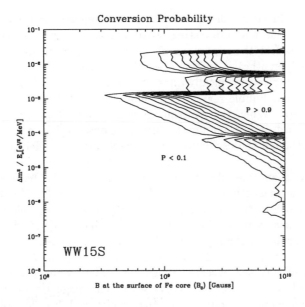

Figure 1:

the $\bar{\nu}_e p$ reaction. Therefore, the oscillation of supernova $\bar{\nu}_e$'s will make clear signals of the non-standard physics of neutrinos in the future gravitational collapse in the Galaxy. The first of such possibilities is the vacuum oscillation of $\bar{\nu}_e \leftrightarrow \bar{\nu}_\mu$, which occurs if the vacuum mixing angle is quite large compared with that of the quark sector. Although such large mixing is suggested from some of the candidates for the solution of the solar neutrino problem or from the atmospheric neutrino anomaly, the conversion probability of $\bar{\nu}_e$'s and $\bar{\nu}_\mu$'s would at most be 1/2, with the maximum vacuum mixing, after the phase averaging.

On the other hand, more efficient conversion process exists, that is, the resonant oscillation induced by the effective matter potential in a star (well known as the MSW effect). However, the ordinary MSW oscillation is relevant only to neutrinos and not to antineutrinos, under the direct mass hierarchy of neutrinos. The most easily detectable flavor, $\bar{\nu}_e$, is not exchanged with $\bar{\nu}_\mu$ by this effect. However, there exist still some possibilities of almost complete conversion of $\bar{\nu}_e$'s and ν_μ-like neutrinos: one is the MSW effect under the inverse mass hierarchy [1,2] and another is the resonant spin-flavor conversion of $\bar{\nu}_e \leftrightarrow \nu_\mu$ induced by the flavor changing magnetic moment of Majorana neutrinos [3,4]. The most interesting possibility is that these exotic oscillation or conversion affect the supernova physics: the explosion mechanism or the r-process nucleosynthesis. In this case, the neutrino mass difference have to be as large as the cosmologically interesting scale, in order that the resonance occurs in the hot bubble region, i.e., below the stalled shock and above the neutrino sphere. When the neutrino mass is smaller than this scale, the $\bar{\nu}_e$ conversion is still important for the future $\bar{\nu}_e$ observation in water Cerenkov detectors, although the oscillation or conversion do not affect the supernova physics.

Figure 1 is an example of the spin-flavor precession between $\bar{\nu}_e \leftrightarrow \nu_\mu$ induced by the magnetic moment of Majorana neutrinos[5]. This figure shows a contour map for the conversion probability of $\bar{\nu}_e \leftrightarrow \nu_\mu$, as a function of the neutrino mass squared difference over neutrino energy and the magnetic field at the surface of the iron core. The magnetic moment is assumed to be $10^{-12}\mu_B$. The magnetic field of about 10^9 Gauss is possible at the iron core surface, considering the observed magnetic field of white dwarfs, which are similar objects to iron cores of massive stars. The magnetic moment sensitive to this phenomenon is about $\sim 10^{-12}\mu_B$, which is 1–2 orders of magnitude lower than the current experimental/astrophysical limits.

Acknowledgments

This work was partially supported by Grant-in-Aids for Scientific Research from the Ministry of Education, Science and Culture of Japan (07CE2002).

1. G. M. Fuller, J. R. Primack, and Y.-Z. Qian, Phys. Rev. D **52**, 1288 (1995).
2. G. G. Raffelt and J. Silk, *Phys. Lett.* **B366**, 429 (1996).
3. C. S. Lim and W. J. Marciano, *Phys. Rev.* **D37**, 1368 (1988)
4. E. Kh. Akhmedov, *Sov. J. Nucl. Phys.* **48**, 382 (1988); Phys. Lett. B **213**, 64 (1988).
5. T. Totani and K. Sato, *Phys. Rev.* D, **54**, 5975 (1996).

TYPE IA SUPERNOVAE: NEWS FROM THE FLAME FRONT

J.C. NIEMEYER AND W. HILLEBRANDT

Max-Planck-Institut für Astrophysik
Karl-Schwarzschild-Str. 1, 85740 Garching, Germany

S. E. WOOSLEY

Board of Studies in Astronomy and Astrophysics
University of California, Santa Cruz, CA 95064, USA

Our current idea of Chandrasekhar mass models for Type Ia supernovae is summarized and two recent developments, a statistcal description for the Rayleigh-Taylor unstable flame front and delayed detonations in the "distributed burning regime", are highlighted.

1   Introduction

Despite increasingly elaborate techniques to use Type Ia supernovae (SN Ia's) as distance indicators [1,2] we are only beginning to grasp the complexity of their explosion mechanism. We will focus here on the historical "standard model" for SN Ia explosions [3], the violent thermonuclear incineration of a carbon and oxygen white dwarf at or near the Chandrasekhar mass $M_{chan} \approx 1.4 \ M_\odot$. In the past few years, this subject has been influenced strongly by chemical combustion research, in particular the theory of turbulent flame propagation [4,5].

Roughly speaking, the dynamics of an exploding white dwarf can be described in terms of the speed u_{eff} of the thermonuclear combustion front that processes carbon and oxygen into (mainly) nickel. It is observationally constrained to be neither too fast nor too slow: a successful explosion requires burning speeds of a fair fraction of the sound speed, while the large amount of intermediate mass elements, inferred from the spectra at peak magnitude, implies some pre-expansion of the star after the runaway has begun. As a consequence of the latter condition it is known that the explosion starts in the so-called deflagration or "flame" mode, i.e. the combustion wave propagates subsonically by means of thermal conduction by degenerate electrons [6]. Owing to the density contrast between the hot ashes produced near the center of the star and the cold unburned material surrounding it, burning bubbles start to rise toward the star's surface (i.e., the situation is Rayleigh-Taylor (RT) unstable) with speeds exceeding that of the laminar flame separating fuel and ashes [7]. This induces strong shear flows along the bubble surfaces and causes the onset of turbulence in the burning regions. As the intensity of turbulent fluctuation increases, the flame surface area and thus u_{eff} grow beyond their laminar values, making the explosion a problem of turbulent combustion similar to combustion engines. Ultimately, u_{eff} is limited by the turnover speed of the fastest (and thus largest) turbulent eddies [8] which, in turn, are limited by the rise velocity of the largest buoyant bubbles.

2 Turbulent Deflagration: the Nonlinear Rayleigh-Taylor Phase

Depending on the time evolution of u_{eff}, one can think of three possible ways to produce a successful SN Ia explosion: pure turbulent deflagrations [9], delayed detonations [10,11,12], or pulsational delayed detonations [11]. The first case occurs if the nonlinear RT instability that drives the turbulent flame is sufficiently effective to burn more than ≈ 0.6 M$_\odot$ of the star to nickel before the flame is quenched by expansion. Three-dimensional simulations by Khokhlov [5] suggest that RT mixing is too inefficient, while 2D simulations of off-center deflagrations [7] and a simple 1D parameterization of the nonlinear RT phase [4] indicate that pure turbulent deflagrations can lead to successful explosions, albeit not extremely strong ones. The latter model uses the so-called Sharp-Wheeler speed,

$$v_{sw} = 0.1\, g_{eff}\, t \, , \qquad (1)$$

with the effective gravitational acceleration $g_{eff} \approx 0.5 g\, \Delta\rho/\rho$ and the time since ignition t, motivated by a statistical model for the merging and rising behavior of an ensemble of RT bubbles [13]. Assuming that all the fuel entrained into the ashes by RT mixing is eventually burned and neglecting large scale freeze-out by expansion, one can use $u_{eff} \approx v_{sw}$ and finds that a healthy explosion is produced [4]. In particular, much more material is burned than needed to unbind the star; u_{eff} would need to be significantly smaller than v_{sw} to allow a pulsational delayed detonation.

3 Delayed Detonation in the Distributed Burning Regime

In the delayed detonation model, the deflagration spontaneously turns into a supersonic detonation during the late stage of the explosion. Only recently specific scenarios have been suggested [4,14] that make physical predictions for the transition density. One of them [4] is based on the idea that after the density has dropped to some 10^7 g cm^3, the flame sheet becomes disrupted by turbulent eddies, and hot ashes is mixed with cold fuel on macroscopic length scales. In the combustion community, this situation is known as the "distributed burning regime". In some regions, burning might be slowed down sufficiently to allow turbulent mixing to smooth out the temperature distribution within a critical volume V_{crit} necessary to trigger a detonation. When this critical volume, heated by slow nuclear burning, later reenters the fast burning mode, it runs away almost simultaneously and may form a sustainable detonation [15]. In this scenario, the transition density ρ_{tr} is fixed by the turbulent flame brush entering the distributed burning regime, which coincides closely with the density inferred from 1D parameterizations of SN Ia events [16]. Owing to the complex physics involved in this problem, most of the current approaches are based on dimensional analysis and phenomenology. Detailed numerical and theoretical work has yet to be done in order to accurately determine ρ_{tr} and V_{crit}, but the underlying ideas appear promising.

Acknowledgments

We would like to acknowledge helpful discussions with A.R. Kerstein and F.X. Timmes.

References

1. A.V. Filippenko, *this volume*
2. S.A. Perlmutter, *this volume*
3. W.D. Arnett, *Astrophys. Space Sci.* **5**, 280 (1969)
4. J.C. Niemeyer and S.E. Woosley, *ApJ*, in press (1997)
5. A.M. Khokhlov, *ApJ* **449**, 695 (1995)
6. F.X. Timmes and S.E. Woosley, *ApJ* **396**, 649 (1992)
7. J.C. Niemeyer, W. Hillebrandt and S.E.Woosley, *ApJ* **471**, 903 (1996)
8. P. Clavin, in *Fluid dynamical aspects of combustion theory*, eds. M. Onofri & A. Tesei (Longman, Harlow, 1990)
9. K. Nomoto, F.-K. Thielemann and K. Yokoi, *ApJ* **286**, 644 (1984)
10. A.M. Khokhlov, *A & A* **246**, 383 (1991)
11. A.M. Khokhlov, *A & A* **245**, L25 (1991)
12. S.E. Woosley and T.A. Weaver, in *Les Houches, Session LIV: Supernovae*, eds. S. Bludman, R. Mochkovitch, and J. Zinn-Justin, (North Holland, Amsterdam, 1994)
13. D.H. Sharp, *Physica D* **12**, 3 (1984)
14. A.M. Khokhlov, *astro-ph/9612226*, unpublished (1996)
15. S.I. Blinnikov and A.M. Khokhlov, *Sov. Astron. Lett.* **12**, 131 (1986)
16. P. Höflich, A.M. Khokhlov, and J.C. Wheeler, *ApJ* **444**, 831 (1995)

A SUPERNOVA BURST OBSERVATORY
TO STUDY μ AND τ NEUTRINOS

R. N. BOYD,[1] R. L. BRODZINSKY,[2] D. B. CLINE,[3] S. A. COLGATE,[4] E. J. FENYVES,[5]
G. M. FULLER,[6] D. KNAPP,[7] D. LEWIN,[8] S. E. LABOV,[7] R. MARSHALL,[9] M. M. NIETO,[4]
D. A. SANDERS,[3] P. F. SMITH[8] K. STEVENS,[9] E. R. SUGARBAKER,[1] W. VERNON,[6]
J. R. WILSON[7]

The Supernova Neutrino Burst Observatory (SNBO) is a proposed dedicated detector for μ and τ neutrinos from a supernova in our Galaxy. Observation of the time profile for 1000 events would enable a non-zero μ or τ neutrino mass greater than 5 eV to be measured through the time-of-flight difference and, in conjunction with the electron antineutrino signal from Super-Kamiokande and other world detectors, would provide unique information on neutrino mixing effects over Galactic distances.

The SNBO is based on the use of natural rock as target, detecting the neutrinos by nuclear excitation, which releases neutrons detected by underground neutron counters. Using loaded scintillator, a new neutron-collection arrangement has been devised which improves the original scheme by a factor of 30. This new scheme is known as SNBO II, and enables 100 tons of target rock to be monitored with only 1 ton of scintillator; and 100 tons of scintillator would be sufficient to observe 1000 events from a supernova at 8 kpc. A suitable underground site would be the DOE-owned salt mine near Carlsbad, New Mexico. Neutron background measurements in this site are consistent with the levels expected from the low U and Th levels in the rock, demonstrating the suitability of this site for a supernova observatory. A coincident detector array could be accommodated in the Boulby Salt mine, UK.

1   The SNBO

The majority of supernovae in our Galaxy are obscured optically, but would be observable as a neutrino burst a few seconds in duration. During the past 1000 years, there have been six visible supernovae within about 5 kpc of the Sun, consistent with a total 30–100 in the Galaxy as a whole. Thus Galactic supernovae occur on average every 20 ± 10 years, the majority emitting most of their energy as a neutrino burst

[1]Ohio State University; [2]Batelle, Pacific Northwest Laboratory; [3]University of California, Los Angeles; [4]Los Alamos National Laboratory; [5]University of Texas, Dallas; [6]University of California, San Diego; [7]Lawrence Livermore National Laboratory; [8]Rutherford Appleton laboratory, UK; [9]University of Manchester, UK.

containing all three neutrino types in approximately equal proportions. For Galactic distances (5–25 kpc), a non-zero neutrino mass would produce time-of-flight differences in the range 0.1–1.0 s, allowing a cosmologically significant mass (> 5 eV) to be extracted from the time profile. Therefore, it would be of considerable interest to measure the relative arrival-time profile of the μ and τ neutrino components.

Existing large water-based detectors (Super-Kamiokande, SNO) and scintillator-based detectors (MACRO, LVD) detect mainly electron antineutrinos through charged-current events; they do not see a sufficient number of neutral-current events from μ and τ neutrinos. Several years ago, a method of observing the latter was devised by members of this collaboration, using any material, including natural rock, as target.[1] Neutral-current excitation of nuclei by the incoming neutrinos releases neutrons that can be detected within ~ 1 ms to give the time profile of the neutrino burst. Moreover, because this process has an (energy)^2 dependence, the signal arises principally from the μ and τ neutrinos, since these are released in the supernova at a higher temperature (~25–27 MeV) than the electron neutrinos (11–15 MeV). The simplicity of this concept suggested that it could be used as the basis of a permanent, dedicated underground observatory for supernova neutrinos.[2]

2 SNBO II

The original scheme (SNBO I) was based on boron trifluoride counters embedded in the rock for which simulations show a count of 0.4 events/m^3 for a supernova at 8 kpc.[1] Thus, despite the zero-cost target, 1000 events would have required the provision of a 2500 m^3 neutron detector. However, new Monte Carlo studies have shown three types of improvements in detection efficiency:[3]

1. Placing detectors in open space in underground caverns, to take advantage of multiple neutron scattering in the cavern (Fig 1);

2. Using a mixture of neutron absorber and thermalizer (*e.g.*, a Li- or B-loaded scintillator) to minimize absorption by the hydrogen thermalizer;

3. Reshaping the detectors for optimum area/thickness.

Each of these changes gains approximately a factor 3. Figure 2 shows the sequence of improvements: Fig. 2A shows an example of improvement No. 1; Fig. 2B shows the additional improvements (Nos. 2 and 3) to produce a typical form of SNBO II.

Further substantial improvements are possible, in principle, through the addition of materials with lower neutron absorption to the cavern (*e.g.*, lead, iron, or their natural mineral compounds). We refer to this extension as SNBO III, but it is not yet known

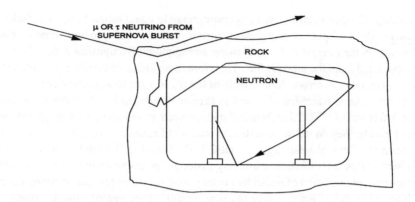

Figure 1: Multiple neutron scattering in the cavern in an open geometry.

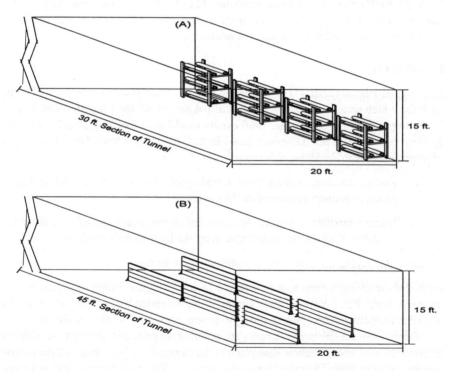

Figure 2: Examples of (A) BF_3 and (B) loaded scintillator (distributed along open tunnels in the rock[3]) detectors in a tunnel.

whether the gain in event collection outweighs the additional cost of adding these materials.

3 Event Numbers and Effective Target Mass

The neutron collection capability is increased, with SNBO II, to ~10–12 events/m ≡ 10–12 events/ton scintillator, compared with the ~0.3 events/ton scintillator obtained by direct detection in MACRO or LVD. Thus SNBO II, in addition to its unique capability of observing μ and τ neutrinos, also provides a very low cost method of supernova detection. Another way of illustrating this is through the effective target mass. The figure of > 10 events detected per ton of scintillator is for a production of 0.1 neutrinos/ton rock (estimated for an 8-kpc supernova). Combining these, it is seen that only 1 ton of scintillator is required to monitor 100 tons of the target rock. In addition, the required minimum of 1000 μ/τ neutrino events can be achieved with only 100 tons of liquid scintillator.

4 Comparison of Detectors

Table 1 compares event numbers in SNBO II with four other world detectors, showing the combined ability to observe all three neutrino types.[2,3]

5 Sites and Background

Neutron background arises from muon interactions and from spallation by alphas from U and Th in the rock. Production by muons is reduced to below the U and Th production at depths > 500 MWe. The major USA site under construction is the DOE-owned salt mine in New Mexico, planned as a Waste Isolation Pilot Plant (WIPP). This site has areas that are some distance from the proposed waste storage region and, thus, which could be used without any background problems. In addition, a supernova array could provide an automatic monitor of the integrity of the nuclear waste over periods of 100s of years – a possibility that exists and has been discussed. Recent measurements (by UCLA–Ohio) of the neutron background flux in the salt tunnels have given results consistent with the known U and Th levels in the rock (*i.e.*, with a continuous neutron production rate of 0.01/tons·s throughout the rock.

This continuous background can be subtracted, leaving only the Poisson fluctuations in the time-binned data. For the first few seconds of a supernova burst, these fluctuations are smaller than the Poisson fluctuations in the signal itself. Similar background levels are obtained at the Boulby salt mine, UK, where an area is currently set aside for dark matter experiments.

Table 1: Comparison of typical SNBO II and III performance with world detectors based on direct detection with water and scintillator targets, showing event numbers for each neutrino type.[2,3]

	Target Material	Fiducial Mass (ton)	Target Elements	ν_e	$\bar{\nu}_e$	$\nu_{\mu,\tau} + \bar{\nu}_{\mu,\tau}$
Combined target–detector						
Super-Kamiokande	H_2O	32000	p,e,O	180	8300	50
LVD	CH_2	1200	p,e,C	14	540	30
MACRO	CH_2	1000	p,e,C	8	350	25
SNO	H_2O	1600	p,e,O	16	520	6
SNO	D_2O	1000[a]	p,e,O	190	180	300[a]
Separated target & detector						
SNBO II:						
rock + neutron detector	Rock	>10 ton/event	All nuclei (nc)	< 50 (nc)	< 50	1000
Direct interaction with scintillator	CH_2	effective[b] + 100 tons CH_2	p,e,C	1	40	3
SNBO III:						
lined caverns + neutron detector	Rock + Fe/Pb	>10 ton/event effective[c] + 10–20 tons CH_2	Fe/Pb	< 50 (nc)	< 50 (nc)	1000

[a]D_2O target not permanently available.
[b]No defined fiducial target mass. Effective mass = events/(production/g).
[c]Optimum rock–metal combination not yet studied.

6 Development Work

Development work is required to optimize the choice of neutron detector and readout, and studies have begun at Ohio, UCLA, and Manchester. Options include panels of loaded scintillator with wavelength-shifting fiber readout, or multi-wire versions of proportional gas detectors with improved thermalization.

7 Conclusion

There is a strong neutrino physics case for constructing SNBO II to observe the μ/τ neutrino profile from the next Galactic supernova, in order to provide a measurement of a cosmologically significant neutrino mass, together with mixing effects, over Galactic distance.

8 References

1. D. B. Cline *et al.*, *Phys. Rev.*, **D 50**, 720 (1994); for earlier discussions, see *Nucl. Phys.*, **B 14A**, 348 (1990) and *Astro. Lett. Comm.*, **27**, 403 (1990).
2. D. B. Cline and G. Fuller, *Neutrino Astrophysics* (Proc. of the 1994 Snowmass Conference), eds. R. Kolb and R. Peccei (*World Scientific*, Singapore, 1996).
3. P. F. Smith, "SNBO II – an Improved Low-Cost Detetor to Measure Mass and Mixing of μ/τ Neutrinos from a Galactic Supernova," in: Proceedings of the Int. Wksp. on the Identification of Dark Matter (The University of Sheffield, UK, Sept. 1996), World Scientific (in press); also see *Astropart. Phys.* (to be published).

THE BLACK-HOLE X-RAY BINARIES

P.A.CHARLES

Department of Physics, Oxford University, Keble Road,
Oxford OX1 3RH, England

The fundamental observed properties of the soft X-ray transient (SXT), black-hole X-ray binaries (BHXRB) are summarised. Particular attention is paid to recent observations of the superluminal transient Nova Sco 1994 (=GRS J1655-40).

1 Key Properties

For a full-scale review of the basic optical properties of the BHXRBs and how they are determined, see Charles [3], Haswell [5] and Casares [2]. For X-ray properties see Tanaka & Shibazaki [14].

- **X-ray** transient outbursts typically reach peak $L_X \sim 10^{37-39}$erg s^{-1}, with outburst/quiescence amplitude $\geq 10^{3-4}$. They exhibit F.R.E.D. light curves with fast rise ($\sim$few days), exponential decay ($\sim$30d) and secondary maximum $\sim$ 60–90d after peak (Chen *et al* [4]), recurring every 10-50yrs. They show complex, variable X-ray spectra with hard power law extending to >100keV, and a possible substantial soft (<10keV) excess.

- **Optical** counterpart brightens visually by ~ 5-7^m. Therefore they are LMXBs (and *not* similar to Cyg X-1 or the harder Be transients). Orbital periods are in range 5h $\longrightarrow$ 6.5d. Quiescent studies reveal companion spectrum (K–M type, except GRS J1655-40) and residual disc emission. Radial velocity curves yield the *mass function* (=*minimum* compact object mass). $\sim$75% are BH candidates, out of $\sim$20 SXTs known (Table 1). Mass ratios $q = M_X/M_C$ (from rotational broadening) are high, $\sim$15 (except GRS J1655-40). Hence expect SU UMa phenomena (seen in 3 SXTs, O'Donoghue & Charles [9]). Lithium line at $\lambda 6707$ seen in 5 SXTs (Martin *et al* [6]). Ellipsoidal modulation of companion (in visual and IR) yields i, and hence (with q and $f(M)$) M_X and M_C (Shahbaz *et al* [11],[12],[13]).

Some exhibit "mini-outbursts" within $\sim$1-2 years of main outburst. e.g. GRS J0422+32 and GRS J1655-40. Two display transient superluminal jets (GRO J1655-40, GRS 1915+105, although latter may not be a LMXB, Mirabel & Rodriguez [8]). The residual accretion disc Hα profile implies disc size $\sim$ 0.4–0.7 primary Roche lobe.

The galactic distribution of SXTs is consistent with population I objects, i.e. massive progenitors (White & van Paradijs [16]). Estimate $\sim$500 in Galaxy.

2   Nova Sco 1994 (=GRS J1655-40)

Has the earliest spectral type (F3-6IV) of any BHXRB and, at V=17, is likely to become a *Rosetta-stone* for the class. Disc effects (even in the optical) are much smaller (or even negligible) in quiescence, and the high γ-velocity (Brandt *et al*[1]) suggested that it might have been formed by the accretion-induced collapse of a neutron star. It also displays superluminal jets, and its high average mass transfer rate suggests that it may be close to being a "steady" LMXB, based on the disc-instability model of van Paradijs [15].

Its low q ($\sim$3) and high i ($\sim$70°, Orosz & Bailyn [10]) makes the ellipsoidal light curve sensitive to both q **and** i, allowing for (future) measurements of the rotational broadening to lead to an independent estimate of q and hence a test of the entire basis for modelling the BHXRB light curves.

References

1. W.N.Brandt, P.Podsiadlowski & S.Sigurdsson, MN **277**, L35 (1995).
2. J.Casares in *Proc. of Spanish Relativity Meeting ERE95* (World Scientific, Singapore, 1997).
3. P.A.Charles in *Compact Stars in Binaries, IAU Symp.* **165**, 341 (1996).
4. W.Chen, C.R.Shrader & M.Livio, *Ap.J.* , in press (1997).
5. C.Haswell in *Compact Stars in Binaries, IAU Symp.* **165**, 351 (1996).
6. E.L.Martin *et al*, *New Astron.* **1**, 197 (1996).
7. J.Miller & T.Shahbaz, *MNRAS* , in preparation (1997).
8. I.F.Mirabel & L.F.Rodriguez *Nature* **371**, 46 (1994).
9. D.O'Donoghue & P.A.Charles, *MNRAS* **282**, 191 (1996).
10. J.A.Orosz & C.D.Bailyn, *Ap.J.* , in press (1997).
11. T.Shahbaz, T.Naylor, P.A.Charles, *MNRAS* **265**, 655 (1993).
12. T.Shahbaz, T.Naylor, P.A.Charles, *MNRAS* **268**, 756 (1994a).
13. T.Shahbaz *et al*, *MNRAS* **271**, L10 (1994b).
14. Y.Tanaka & N.Shibazaki, *Ann.Rev.Astro.Ap.* **34**, 607 (1996).
15. J.van Paradijs, *Ap.J.* **464**, L139 (1996).
16. N.E.White & J.van Paradijs, *Ap.J.* , in press (1997).

Table 1: Dynamical Mass Measurements of Soft X-ray Transients.

Source	V	P (hrs)	f(M) ($M_\odot$)	$v_{rot}\sin i$ (km s^{-1})	q	i	M_X ($M_\odot$)	M_2 ($M_\odot$)
J0422+32	22.4	5.0	1.13±0.09	≤80	>12	20–40	10±5	0.3
G1009-45	20.6	6.9	-	-	-	44±15	-	-
A0620-00	18.3	7.8	2.91±0.08	83	15±1	37±5	10.0±5	0.6
G2000+25	22.5	8.3	4.97±0.10	86	24±10	56±15	10±5	0.7
N Mus 91	20.5	10.4	3.34±0.15	106	7±2	54^{+20}_{-15}	6^{+5}_{-2}	0.8
N Oph 77	21.3	12.5	4.83±0.12	-	-	70±10	7±2	0.4
Cen X-4	18.4	15.1	0.20±0.05	45	5±1	43±11	1.3±0.6	0.4
J1655-40	17	62.6	3.16±0.15	-	3±0.5	70±5	5.5±1	1.2
V404 Cyg	18.4	155.4	6.08±0.06	39	17±1	55±4	12±2	0.6

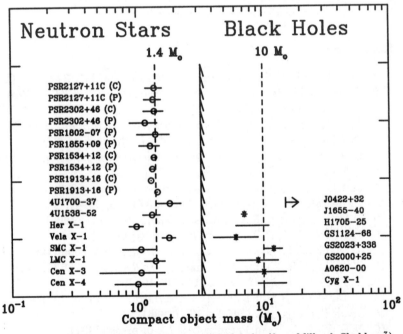

Figure 1: The mass distribution of neutron stars and black holes (from Miller & Shahbaz [7]). Note the remarkably narrow spread of neutron star masses, and the large factor by which the BHXRB masses exceed the maximum mass of $3.2 M_\odot$.

X–RAY VARIABILITY OF CYG X–1 AND 1E1740.7-2942

M.Gilfanov, E.Churazov, R.Sunyaev

MPI für Astrophysik, Karl-Schwarzschild-Str. 1, 85740 Garching, Germany;
Space Research Institute, Profsouznaya 84/32, 117810 Moscow, Russia

The long–term light curves of Cyg X–1 and 1E1740.7-2942 feature extended episodes of low hard X–ray flux accompanied with changes of the hard X–ray spectral properties [1,2,5]. For both sources an approximate correlation between kT^a and L_X exists (Fig.1). For Cyg X–1 the spectral hardness is in general positively correlated with the relative amplitude of short-term variability. The low luminosity end of these approximate correlations (low kT and low rms) corresponds to extended episodes of very low hard X-ray flux which occurred during GRANAT/SIGMA observations in 1993–1994.

The broad band spectra of Cyg X–1 observed in 1993 (low hard X–ray flux) and 1994 (nominal hard state) (ASCA and GRANAT/SIGMA data) are shown in Fig.2. The parameters of the spectral approximation of the ASCA data (0.5–10 keV) are given in Table 1. The spectral changes observed by SIGMA in the high energy domain during low hard X–ray flux episode in 1993–1994 were accompanied with corresponding changes in the standard X–ray band (Fig.2 and columns 2 and 3 of Table 1, see also [4]):

1) The slope of the Comptonized radiation changes from ≈ 1.6 for standard hard state spectrum to ≈ 2.0 (low hard X–ray flux episode). The position of the high energy cut-off of the spectrum shifts towards lower energy indicating that the electron temperature in the Comptonization region is decreasing.

2) Parameters of the soft excess – supposedly emission from optically thick, geometrically thin outer part of the accretion disk [6] – change as well. Qualitatively, mean photon energy, luminosity and relative contribution of the soft excess to the X–ray luminosity increased. In terms of the multicolor disk approximation these changes may be attributed to increase of the disk temperature and, possibly, decrease of the radius of the inner boundary of the optically thick part of the accretion disk.

The fourth column of Table 1 gives the best fit parameters for the May–June 1996 low hard X–ray flux episode. The ASCA observation in 1993 occurred two month before the source reached the lowest value of the hard X–ray flux, whereas in 1996 Cyg X–1 was observed with ASCA nearly at the minimum of the hard X–ray flux. Correspondingly, 1996 observation found steeper power law, higher disk temperature, smaller disk inner radius and higher value

---

[a]The best fit bremsstrahlung temperature was used as a simple one-parameter representation of the spectral hardness in the 40–200 keV energy domain.

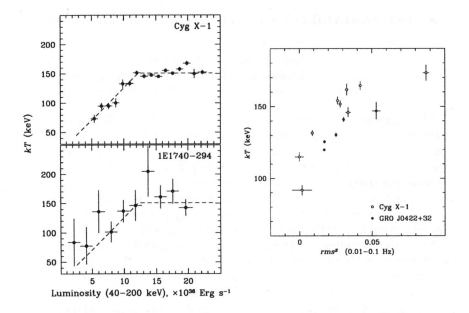

Figure 1: The best-fit bremsstrahlung temperature plotted against the 40-200 keV luminosity (*left*) and the rms^2 of the short-term flux variations (*right*). From ([5]), see also ([1]).

of the mass accretion rate as estimated from the disk parameters (Table 1). It seems very likely, that both 1993 and 1996 events correspond to the same phenomena – transition of the source to the soft spectral state caused by increase of the mass accretion rate by a factor of few.

This research has made use of data obtained through the High Energy Astrophysics Science Archive Research Center Online Service, provided by the NASA/Goddard Space Flight Center.

1. Ballet J. et al., 1996, ASCA Symp. Proc., X-Ray imaging and spectroscopy of cosmic hot Plasma. Tokyo, in press
2. Crary D.J. et al., 1996, ApJ, 462, L71
3. Dotani T. et al., 1996, IAUC 6415
4. Gilfanov M., Churazov E., Sunyaev R., 1997, in the Proceedings of the EARA workshop on Accretion Disks
5. Kuznetsov S. et al., 1997, to be submitted to M.N.R.A.S.
6. Shakura N. & Sunyaev R., 1973 Astron.Astrophys, 24, 337

Table 1: The best fit parameters for the ASCA data approximation with the model consisting of the power law with reflection and multicolor disk emission.

Parameters [1]	25–26/11/94 "nominal" hard state	11-12/11/93 ~beginning of the low hard X–ray flux 1993 episode	30–31/05/96 [4] ~middle of the low hard X–ray flux 1996 episode
photon index, α	1.64 ± 0.01	1.99 ± 0.01	~ 2.2
T_{in}, eV	136 ± 5	156 ± 2	~ 470
R_{in}, km	440 ± 60	420 ± 15	~ 110
F_{disk} [2]	$0.02\ (0.7)$	$0.06\ (1.3)$	$0.2\ (5.7)$
$\dot{M}_{disk}$, 10^{17} g/sec [3]	~ 1.7	~ 2.8	~ 5

[1] $N_H L = 6 \cdot 10^{21}$ cm^{-2}; $D = 2.5$ kpc; $\theta = 70$ degrees

[2] energy flux from the multicolor disk emission component, 10^{-8} erg/sec/cm^2, 0.3–9 keV, the absorption corrected value is given in parenthesis

[3] Estimated using the multicolor disk approximation

[4] The parameters were *roughly* estimated using the spectral parameters from ([3]).

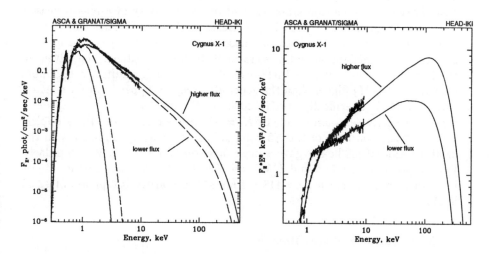

Figure 2: The broad band spectra of Cyg X–1 in the "nominal" hard state (marked as "higher flux") and during 1993 low hard X–ray flux episode (marked "lower flux"). The behavior of the spectral flux density above 30 keV is schematically shown according to ([5]).

KECK-I OBSERVATIONS OF GALACTIC BLACK HOLES

E.T. HARLAFTIS

School of Physics & Astronomy, University of St. Andrews,
North Haugh, St. Andrews, KY16 9SS, U.K.

A.V. FILIPPENKO

Department of Astronomy, University of California,
Berkeley, California 94720-3411, USA

The advent of the large effective aperture of Keck-I has resulted in the determination with unprecedented accuracy of the mass functions of 3 faint ($R \approx 21$ mag) X-ray transients. In addition, the mass ratios of GS 2000+25 and GRO J0422+32 are constrained from the rotational broadening of the companion stars using an optimal subtraction procedure. Finally, important information on the companion star is derived.

1 Mass functions

The search in low-mass X-ray binaries for compact stars which are unambiguously heavier than neutron stars resulted in the discovery of a mass function $f(M)$ of 6.08±0.06 $M_\odot$ in V404 Cyg (Casares *et al.* 1992; Casares and Charles 1994), which is a lower limit to the mass of the compact object and is independent of the inclination and mass ratio of the binary system. The advent of the Keck-I telescope with the LRIS optical spectrograph resulted in two more detections of $f(M)$ higher than the generally accepted upper limit for neutron stars (3.2 $M_\odot$). Specifically, for GS 2000+25 Filippenko *et al.* (1995a) found a radial velocity semi-amplitude of $K = 518 \pm 4$ km s^{-1} from the Doppler-shifted absorption lines of the K5 companion star which, combined with the orbital period of 8.3 hours, gave a mass function of 4.97±0.10 $M_\odot$. Similarly, Filippenko *et al.* (1997) found $f(M) = 4.86 \pm 0.13$ $M_\odot$ for the compact object in Nova Oph 1977 (given $K = 448 \pm 4$ km s^{-1} and an orbital period of 12.55 hours).

2 Mass ratios and star masses

The Keck spectra also allow estimates of the mass ratio $q = M_2/M_1$ by measuring the rotational broadening of the companion, $v \sin i$, through the relation

$$\frac{v \sin i}{K_c} = 0.46 \left[(1 + q)^2 \, q \right]^{1/3} .$$

The rotational broadening is determined using a χ^2 minimization procedure where the solution consists of a set of three variables, namely spectral type, rotational broadening, and veiling factor (see Fig. 1). In particular, spectra of various spectral-type main-sequence stars are processed so as to simulate the observed spectra. Then, optimal subtraction of the main-sequence star from the observed GS 2000+25 spectrum gives a χ^2 minimum for a K5±2 spectral type for the companion star with a rotational broadening of 86±8 km s^{-1} and a contribution of 94±5% to the light (Harlaftis *et al.* 1996). For GRO J0422+32, we similarly derive a spectral type of M2±2, a contribution of 61±4%, and a rotational broadening of 50±30 km s^{-1} at the 68% confidence level (Harlaftis *et al.* 1997). The above results indicate mass ratios of $q = 0.042 \pm 0.012$ for GS 2000+25 and < 0.08 for GRO J0422+32 (using a mass function of 1.21±0.06 $M_\odot$; Filippenko *et al.* 1995b). Using the relation

$$M_1 = f(M_1)\frac{(1+q)^2}{\sin^3 i},$$

the mass of the compact object is $(5.44\pm0.15)\sin^{-3} i\ M_\odot$ for GS 2000+25 and $(1.2\pm0.3)\sin^{-3} i\ M_\odot$ for GRO J0422+32. The mass of the companion star in GS 2000+25 is

$$M_2 = (0.23 \pm 0.02)\sin^{-3} i\ M_\odot$$

which, for an inclination of $44° < i < 75°$ (measured from IR ellipsoidal modulations; Callanan *et al.* 1996; Beekman *et al.* 1996), suggests an undermassive companion star, $0.26 < M_2 < 0.59 M_\odot$ compared to a main-sequence mass of 0.69 $M_\odot$. Work on Nova Oph 1977 is in progress.

References

1. G. Beekman, *et al.*, *MNRAS*, 281, L1 (1996).
2. P. J. Callanan, *et al.*, *ApJ*, 450, L57 (1996).
3. J. Casares, P. A. Charles, and T. Naylor, *Nature*, 355, 614 (1992).
4. J. Casares and P. A. Charles, *MNRAS*, 271, L5 (1994).
5. A. V. Filippenko, T. Matheson, and A. J. Barth, *ApJ*, 455, L139 (1995a).
6. A. V. Filippenko, T. Matheson, and L. C. Ho, *ApJ*, 455, 614 (1995b).
7. A. V. Filippenko, *et al.*, *PASP*, in press (1997).
8. E. T. Harlaftis, K. Horne, and A. V. Filippenko, *PASP*, 108, 762 (1996).
9. E. T. Harlaftis, S. Collier, K. Horne, and A. V. Filippenko, *ApJ*, submitted (1997).

740

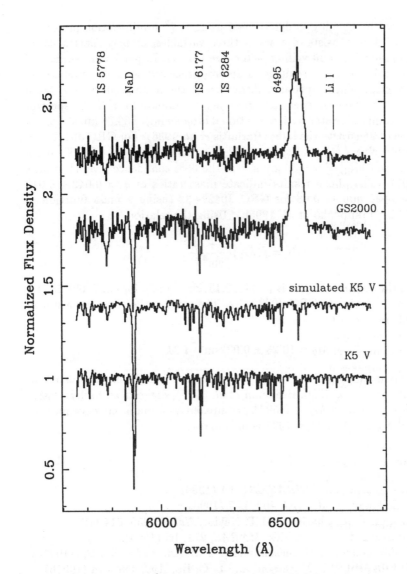

Figure 1: Results of the χ^2 minimization procedure which is used to extract the spectral type, rotational broadening, and light contribution from the companion star of GS 2000+25. A simulated K5 V spectrum (including effects of orbital smearing, rotational broadening, and veiling by the disk light) is subtracted from the Doppler-shifted average spectrum of GS 2000+25. The residual spectrum is dominated by the accretion disk (top), but also shows anomalous line strengths for Hα and Li I 6708 Å from the companion star.

ADVECTION-DOMINATED ACCRETION MODEL OF X-RAY NOVA MUSCAE IN OUTBURST

A.A. Esin, J.E. McClintock, R. Narayan

Harvard-Smithsonian Center for Astrophysics, 60 Garden Street, Cambridge, MA 02138, USA

We present a model for the high-low state transition of the X-Ray Nova GS 1124-68 (Nova Muscae 1991) observed by Ginga[1]. The model consists of an advection-dominated accretion flow (ADAF) near the central black hole surrounded by a thin accretion disk. During the rise phase of the outburst, as the mass accretion rate increases, the transition radius between the thin disk and the ADAF moves closer to the center, until the thin disk extends all the way in to the last stable orbit. The transition radius increases again during decline. We reproduce the basic features of the spectra taken during and after outburst, and the light curves in the soft and hard X-ray bands. We estimate that the accretion rate in Nova Muscae decreased exponentially with a time scale ~ 95 days during decline.

1 Introduction

A bright X-ray nova GS 1124-68 (Nova Muscae 1991) was discovered by the Ginga ASM on January 8, 1991. Subsequent optical observations[2] showed that the system is a short period spectroscopic binary with a mass function of $3.1 M_\odot$, which makes it a strong black-hole candidate.

Like other black hole X-ray novae in outburst, Nova Muscae underwent large changes in its X-ray luminosity and spectral characteristics during outburst[1] (Figs 1(a), 2). These changes are usually described in terms of a succession of spectral states. The highest luminosity corresponds to the "very high" state, characterized by a photon index of ~ 2.5 in the X-ray range 2-20 keV. The "high" state has a lower bolometric luminosity with practically no emission above 5-10 keV. At yet lower luminosity, we have the "low" state where most of the energy comes out in hard X-rays; the X-ray spectrum is usually well characterized by a power-law with a photon index $\sim 1.5 - 1.7$. Finally, at the lowest luminosities, there is the "quiescent state"[3,4].

2 Modeling the State Transitions in Nova Muscae

We model Nova Muscae as a black hole of mass $M = 6 M_\odot$ accreting gas from its companion at a rate $\dot{m}$ (in units of $\dot{M}_{Edd} = L_{Edd}/(0.1 c^2)$). From the outer edge at a radius r_{out} (all radii are in units of $R_{Schw} = 2GM/c^2$) to a transition radius, r_{tr}, the gas accretes via a cool thin disk plus a hot corona formed by

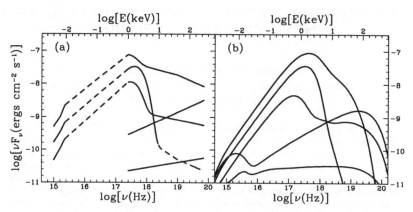

Figure 1: (a) Spectra of GS 1124-68 on days 3 (very high state), 62 (high state), 130, 197 (low state), and 238 of the outburst (in order of decreasing 1 keV flux). (b) Corresponding model spectra with the following parameters: $(\log \dot{m}, \log r_{tr})$ = $(0.2, 0)$; $(-0.4, 0)$; $(-1.0, 0.5)$; $(-1.1, 3.9)$; $(-1.6, 3.9)$. A distance of 5 kpc was assumed.

evaporation of the disk[3,4]. Inside the transition radius, the thin disk is absent and the accretion is via a pure advection-dominated accretion flow[5,6] (ADAF).

This two-zone model has been successfully applied to the quiescent state of two X-ray novae, A0620–00 and V404 Cyg[3,4]. The model reproduces well the observed spectra of these systems and provides strong evidence for the presence of event horizons in the accreting stars. Here we show that the same model also explains observations of X-ray novae in outburst. The various states arise fairly naturally as $\dot{m}$ decreases following the outburst.

Above a critical accretion rate $\dot{m}_{crit}$, it is not possible to have a pure ADAF without a disk[5]. For such $\dot{m}$, we model the flow as a thin disk plus a corona extending all the way from $\log r_{out} = 4.9$ (80% of the Roche radius of the primary) to the last stable orbit at $r = 3$. In our model of Nova Muscae, both the very high state and the high state correspond to this regime of $\dot{m}$. The distinction between the two is that in the former a large fraction of the viscous energy released by the disk is dissipated directly in the corona[7], whereas in the latter the disk and the corona dissipate only their own viscous energy. As a result, in the very high state the system has an active corona which is bright in hard X-rays while in the high state the corona is very quiet and radiates very little hard radiation. The transition between the two states occurs at $\dot{m} \approx 0.5$. The top curve in Fig. 2(b) shows our model spectrum corresponding to the very high state, while the next curve shows the high state.

Approximately 150 days after the peak of the outburst, $\dot{m}$ drops below $\dot{m}_{crit} \approx 0.09$, and at this point the inner regions of the disk evaporate away completely and r_{tr} begins to increase. This corresponds to the high-low state

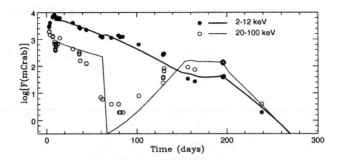

Figure 2: Filled and open circles trace the observed light curve of GS 1124-68 during its 1991 outburst[1]. The lines show the corresponding light curves calculated with our model.

transition which was observed in Nova Muscae between 150 and 200 days after the peak. With increasing time, the transition radius continues to move outward until it reaches its quiescent value of $\log(r_{tr}) = 3.9$ (determined from the width of the H_α line in quiescence[3]). At this point the system is in the low state with $\log \dot{m} \approx -1.1$ and the X-ray spectrum is a pure power law. At later times, as $\dot{m}$ decreases further, the system switches to the quiescent state.

Our model light curve of Nova Muscae is shown in Fig. 2 together with the data. For the model we used the scaling: Time $= 13 \log r_{tr} - 95 \log (\dot{m}/2.51)$ days. The model satisfactorily reproduces the observed variations in both soft and hard X-rays.

In other work, we find that the high-low transition in Cyg X-1 can be explained through variations in r_{tr}, just as in Nova Muscae.

Acknowledgments

This work was supported in part by NASA grant NAG 5-2837 and the Smithsonian Scholarly Studies Program.

References

1. K. Ebisawa *et al*, *P. A. S. J.* **46**, 375 (1994).
2. R.A. Remillard, J.E. McClintock, C.D. Bailyn, *Ap. J.* **399**, L145 (1992).
3. R. Narayan, J.E. McClintock, I. Yi, *Ap. J.* **457**, 821 (1996).
4. R. Narayan, D. Barret, J.E. McClintock, *Ap. J.* **482**, 000 (1997).
5. R. Narayan, I. Yi, *Ap. J.* **428**, L13 (1994), *Ap. J.* **452**, 710 (1995).
6. M. Abramowicz, et al., *Ap. J.* **438**, L37 (1995).
7. F. Haardt, L. Maraschi, *Ap. J.* **380**, L51 (1991).

A NEW VIEW OF ACCRETING BINARY PULSARS

B.A. VAUGHAN, J. CHIU, D.T. KOH, R.W. NELSON, T.A. PRINCE
Space Radiation Laboratory, California Institute of Technology,
Pasadena, CA 91125, USA

M.H. FINGER, B.C. RUBIN, D.M. SCOTT, M. STOLLBERG,
C.A. WILSON, R.B. WILSON
Space Science Laboratory, NASA/Marshall Space Flight Center,
Huntsville, AL 35812, USA

L. BILDSTEN
Dept. of Physics and Dept. of Astronomy,
University of California, Berkeley, CA 94720, USA

D. CHAKRABARTY
Center for Space Research, Massachusetts Institute of Technology,
Cambridge, MA 02139, USA

Continuous observations of accreting binary pulsars with BATSE reveal a qualitatively different picture of their spin behavior than prior sparse observations. Sudden reversals in accretion torque are common among disk accretors, and may explain the long-term behavior of these objects

1   Introduction

Accreting pulsars are rotating, highly magnetized ($B \geq 10^{11}$ G) neutron stars that accrete material from a stellar companion. In the simplest picture of disk accretors, the magnetic field of the neutron star interrupts the quasi-Keplerian flow of plasma through the disk and forces it to corotate at the magnetospheric radius, r_{m}, where magnetic and fluid stresses balance. If the angular momentum of the captured material is transferred to the neutron star, the star will spin up at a rate $\dot{\nu} \simeq \dot{M}/(2\pi I_{\mathrm{ns}})\sqrt{GM_{\mathrm{ns}}r_{\mathrm{m}}}$, where M_{ns} and I_{ns} are the mass and moment of inertia of the neutron star and $\dot{M}$ is the mass accretion rate. The star spins up until it reaches an eqilibrium period, P_{eq}, where the plasma at r_{m} corotates with the star at the corotation radius, r_{co}. It was found that some disk accretors, notably Her X-1 and Cen X-3, spin up ~5–10 times more slowly than the simple theory predicts, and several groups proposed that near equilibrium, negative torques, either from magnetic coupling to the disk beyond r_{co}[1] or from an MHD wind[2] slow *dotν* and can cause spin down. Magnetic coupling has been applied extensively to white dwarfs and T-Tauri stars. Accreting pulsars are the only systems where these ideas can be tested dynamically.

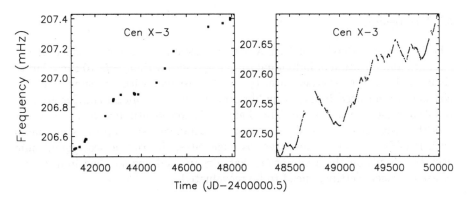

Fig. 1: Long-Term (left) and BATSE (right) frequency history of Cen X-3.

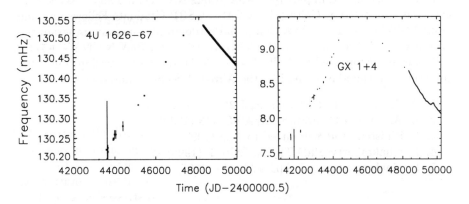

Fig. 2: Frequency histories of 4U 1626−67 (left) and GX 1+4 (right).

2 Observations

The 25-year frequency history of Cen X-3 is shown in Figure 1, along with the 5-year BATSE history. BATSE observations reveal a sequence of 10–100 d intervals of steady spin up and spin down with sudden reversals. The distribution of $\dot{\nu}$ is bimodal[3]. The average spin up rate of $\nu_{su} \sim 7 \times 10^{-12}\,\mathrm{Hz\,s^{-1}}$ is larger than the average spin down rate of $\dot{\nu}_{sd} \sim -3 \times 10^{-12}\,\mathrm{Hz\,s^{-1}}$. The

746

low-mass systems 4U 1626–67 and GX 1+4 have each shown one such reversal, as shown in Figure 2. In both systems $|\dot{\nu}_{su}| > |\dot{\nu}_{sd}|$[4,5].

3   Discussion

The long-term behavior of Cen X-3 is revealed to be a consequence of the strength and duration of individual torque episodes with $|\dot{\nu}|$ ~5–10 times larger than the long-term average rate, $\langle\dot{\nu}\rangle$, and consistent with the simple theory. Neither system need be near equilibrium. Torque switching behavior appears to be common in disk systems The challenge is now identifying the physics that sets the switching time scale. Because instantaneous spin up rates measured with BATSE are ~10 times larger average rates, theories of near-equilibrium accretion, built on the assumption that $\dot{\nu} \simeq \langle\dot{\nu}\rangle$ and now applied in a variety of astrophysical settings, need to be reevaluated.

Acknowledgments

This work was funded in part by NASA grants NAG 5-1458 and NAGW-4517. D.C. was supported at Caltech by a NASA GSRP Graduate Fellowship under grant NGT-51184. The NASA Compton Postdoctoral Fellowship program supported D.C. (NAG 5-3109), L.B. (NAG 5-2666), and R.W.N. (NAG 5-3119). L.B. was also supported by Caltech's Lee A. DuBridge Fellowship, funded by the Weingart Foundation; and by the Alfred P. Sloan Foundation.

1. P. Ghosh and F. K. Lamb, ApJ **234**, 296 (1979)
2. U. Anzer and G. Borner, A&A **83**, 113 (1980)
3. L. Bildsten *et al*, submitted to ApJ (1997).
4. D. Chakrabarty, PhD Thesis, Caltech (1996).
5. D. Chakrabarty *et al*, ApJ **474**, 414 (1997)

Studying Pulsars and X-ray Binaries for Natal Neutron Star Kicks

C. L. Fryer

Steward Observatiory, University of Arizona, Tucson, AZ, 85721

The pulsar velocity distribution provides irrefutable evidence for the existence of natal neutron star (NS) kicks. These kicks are motions placed on neutron stars at birth generally attributed to asymmetries in the explosion. With the most recent proper motion measurements and the newest distance determinations, Lyne & Lorimer (1994) estimated a mean three-dimensional pulsar velocity ~ 450 km s^{-1} which can only be explained with the existence of natal neutron star kicks. They also derived a pulsar velocity distribution which is consistent with the observational data. This three-dimensional velocity distribution is neither equivalent to the natal neutron star kick distribution, nor is it a unique solution to the measured pulsar transverse-velocities. Only by first extracting the effects of binary evolution and motion through the galaxy, and, by using the additional constraints from the formation rates of NS systems, can we uncover the nature of natal NS kicks. The natal neutron star kick distribution which best fits the pulsar transverse-velocity data and the formation rates of NS systems is bimodal, with one peak in the distribution near 0 km s^{-1} and the other near 650 km s^{-1}.

1   Introduction

Natal NS kicks have been invoked by theorists to explain a wide range of objects from a variety of peculiar NS binary systems (Flannery & van den Heuvel 1975; Kaspi et al. 1996; van den Heuvel & Rappaport 1986) to the construction of a working model for galactic gamma-ray bursts (Colgate & Leonard 1994; Lamb 1995). However, to quantitatively determine their application to these systems, the distribution of these kicks must be understood. The three-dimensional velocity distribution of pulsars derived by Lyne & Lorimer (1994) has often been misinterpreted as the only distribution which fits the observed pulsar transverse velocities, and that this distribution is equivalent to the natal neutron star kick distribtuion. These misconceptions are addressed in §2. However, by using additional constraints such as the formation rates of neutron star binaries, the range of allowable kick distributions can be restricted (§3).

2   Extracting a Kick Distribution

To derive a kick distribution from the three-dimensional pulsar distribution, one must first separate the effects of the motion through the galactic potential and binary evolution. Using Monte Carlo techniques to simulate binary/single

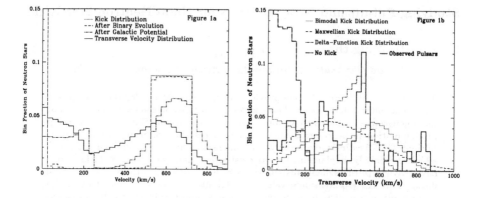

Figure 1: Figure 1a shows the dependence of the pulsar velocity distribution for a double-peaked kick distribution (dotted curve) on binary evolution (dashed curve) and the galactic potential (dot-dashed curve). The solid curve gives the pulsar transverse-velocity distribution which is then compared to the observations. Note that the transverse velocity distribution bears little resemblance to its parent kick distribution, illustrating the importance of including the effects of binary evolution and motion in the galaxy in extracting the actual kick distribution from the observations. Figure 1b emphasizes this by giving the transverse velocity distributions for the best fitting double-peaked, Maxwellian, and delta-function kick distributions along with a smoothed distribution of the observed pulsar transverse velocities. Included for comparison is the transverse velocity distribution for a simulation with no natal neutron star kick. The simulation with no neutron star kicks clearly does not fit the data, but the other distributions can not be ruled out with such small number statistics.

star evolution and then the motion through the galactic potential, Fryer, Burrows, & Benz (1997) study these effects on various possible initial natal NS kick distributions, and calculate the transverse pulsar velocity distribution to compare with the observations. The magnitude of these effects is illustrated for a bimodal kick distribution in Figure 1a. Note that the pulsar velocity distribution is significantly different from the actual kick distribution. Indeed, the mean *pulsar* velocities differ from the mean *kick* velocities by up to 40%.

It is evident from Figure 1a that the observed transverse velocity distribution retains very little information about its progenitor kick distribution. In fact, it is rather difficult to constrain the kick distribution based on the pulsar transverse-velocity data alone. Figure 1b shows 3 different transverse-velocity distributions from 3 drastically different kick distributions (Maxwellian, delta-function, and a bimodal distribution). Superposed on these plots is the actual observed pulsar transverse-velocities. Except for the zero-kick distribution which can be readily ruled out, it is difficult to argue which distribution fits the observed distribution best given the low-number statistics.

3   Constraining the Velocity Distribution

Neutron star kicks do not only affect pulsar velocities, but also the formation rates of X-ray binaries, the formation rates of double neutron star systems, and the retention fractions of neutron stars in globular clusters. By using all of these systems as constraints, we are forced to require a bimodal distribution with one peak in the distribution near 600 km s^{-1} and the rest of the neutron stars receiving very little kick. The details of this result are discussed in Fryer, Burrows, & Benz (1997) but it can be easily understood by realizing that low kicks are required for NS binary systems to remain bound, as well as to retain neutron stars in globular clusters. However, to explain the pulsar velocity distribution, high velocities are necessary.

Hence, despite the fact that a unique natal NS kick distribution can *not* be determined from the observed transverse velocities of pulsars, by including data on NS binaries, we *can* constrain the allowed kick distribution. This bimodal kick distribution, has profound repurcussions on O/B runaway star formation scenarios, galactic gamma-ray burst models, and the kick mechanism itself.

Acknowledgments

It is a pleasure to thank Vicky Kalogera for discussions on the topics addressed here and for a careful reading of the manuscript. This work was funded by NSF grant AST 9206738 and a UA-TAP grant.

References

1. Colgate, S.A., & Leonard, P.J.T., 1993, BATSE Gamma Ray Burst Workshop
2. Flannery, B.P., & van den Heuvel, E.P.J., 1975, A&A, 39, 61
3. Fryer, C.L., Burrows, A., & Benz, W., 1997, submitted to ApJ
4. Kaspi, V.M., Bailes, M., Manchester, R.N., Stappers, B.W., Bell, J.F., 1996, Nature, 381, 584
5. Lamb, D.Q., 1995, PASP, 107, 1152
6. Lyne, A.G., Lorimer, D.R., 1994, Nature, 369, 127L
7. van den Heuvel, E.P.J., & Rappaport, S., 1986, Physics of Be stars, Proc. of the 92nd IAU Coll., 291, Cambridge University Press

Multifrequency Observations of the Galactic Microquasars GRS1915+105 and GROJ1655-40

Ronald A. Remillard, Edward H. Morgan (MIT)
Jeffrey E. McClintock (CFA), Charles D. Bailyn (Yale)
Jerome A. Orosz (Penn State) & Jochen Greiner (AIP, Potsdam)

The two galactic 'microquasars' with superluminal radio jets have been quite active during 1996, generating a variety of studies involving both NASA and ground-based observatories. GRS 1915+105 has displayed dramatic accretion instability in observations with RXTE, revealing X-ray light curves and emission states unlike anything previously seen. Variable QPOs in the range of 0.07–10 Hz have been monitored with the capability to track the individual oscillations. The QPO amplitude is as high as 40% of the mean flux, while both amplitude and phase lag increase with photon energy. The results imply a direct link between the QPO mechanism and the origin of the energetic electrons believed to radiate the X-ray power-law component. GRS1915+105 also displays a transient yet stationary QPO at 67 Hz. The other source, GRO J1655-40, is an optically established black hole binary. Recent optical reports include an excellent model for the binary inclination and masses, while an optical precursor to the April 1996 X-ray outburst has been measured. We report new results from recent RXTE observations. GROJ1655-40 has displayed both the canonical "soft/high" and "very high" X-ray states, with QPOs at 8–22 Hz during the latter state. In addition, there is a high-frequency QPO at 300 Hz. The rapid oscillations in these sources are suspected of providing a measure of the mass and rotation of the accreting black holes, although several competing models may be applied when evaluating the results.

1 Introduction

The two sources of superluminal radio jets[9][15][6] in the Galaxy, GRS1915+105 and GRO J1655-40 have been quite active during 1996. These X-ray sources were originally detected during May 1992 and July 1994, respectively, and they have persisted well beyond the typical time scale for X-ray transients[2]. Optical study of the companion star in GROJ1655-40 has yielded a binary mass function (3.2 $M_\odot$) that indicates an accreting black hole[1][13]. In the case of GRS1915+105, interstellar extinction limits optical/IR studies to weak detections at wavelengths > 1 micron[8]. The compact object in this system is supsected of being a black hole due to the spectral and temporal similarities with GROJ1655-40 and other black hole binaries. Both of these microquasars have now been detected with OSSE[4] out to photon energies of 600 keV.

Investigations of microquasars are motivated by several broad and inter-related purposes: to search for clues regarding the origin of relativistic jets, to probe the properties of the compact objects, and to understand the various spectral components and their evolution as the sources journey through different accretion states. Several research programs are described herein, with emphasis on new results from the Rossi X-ray Timing Explorer (RXTE).

2 RXTE Observations of GRS1915+105

The RXTE All Sky Monitor [7] began operation during 1996 Jan 5-13, and continuous observing with a 40% duty cycle has been achieved since 1996 Feb 20. GRS1915+105 was found to be bright and incredibly active [11], as ASM time series data revealed high amplitude modulations at 10-50 s. These results initiated a series of weekly pointings for the PCA and HEXTE instruments. The yield is approaching ten billion photons in an immensely complex and exciting archive that is fully available as 'public' data.

The ASM light curve of GRS1915+105 (1996 Feb 20 – 1997 Jan 23) is shown in Fig. 1. These results are derived using version 2 (1/97) of the model for the instrumental response to X-ray shadows through the coded masks. The top panel shows the normalized intensity for the full range (2–12 keV) of the ASM cameras, in which the Crab nebula produces 75.5 c/s. The vertical lines in the upper region show the times of the PCA / HEXTE observations in the public archive. Below this light curve, one of the ASM hardness ratios is displayed; $HR2$ is the ratio of normalized flux at 5–12 keV relative to the flux in the 3–5 keV band. The spectrum of GRS1915+105 is harder than the Crab ($HR2 = 1.07$). Since there is an anticorrelation between the count rate and $HR2$ in GRS1915+105, we caution against the presumption that the ASM flux is a direct measure of X-ray luminosity. During 1997, significant progress is expected from efforts to combine the ASM results with those of BATSE and radio monitors, including the newly organized Greenbank Interferometer project. This effort will build on earlier work [5] to investigate the multifrequency evolution of X-ray outbursts and radio flares.

The PCA observations of GRS1915+105 immediately showed dramatic intensity variations [3] with a complex hierarchy of quasi-periodic dips on time scales from 10 s to hours. Complex and yet repeatable 'stalls' in the light curve were preceeded by rapid dips in which the count rate dropped by as much as 90% in a few seconds. These variations were interpreted as an inherent accretion instability, rather than absorption effects, since there was spectral softening during these dips. There were also occasions of flux overshooting after X-ray stalls. These repetitive, sharp variations and their hierarchy of time scales are entirely unrelated to the phenomenology of absorption dips [3]. The dips represent large changes in an absolute sense; the pre-dip or post-dip luminosity in GRS1915+105 is as high as 2×10^{39} ergs cm^{-2} s^{-1} at 2-60 keV, assuming the distance of 12.5 kpc inferred from 21 cm HI absorption profiles [9].

The phenomenology of wild source behavior in GRS1915+105 has expanded since the first series of observations. Three examples are shown in Fig. 2. The Oct 7 display of quasiperiodic stalls preceeded by rapid dips (middle panel) is highly organized and repetitive, while the Jun 16 light curve (top panel) shows complex, interrupted stalls that are not preceeded by rapid dips. In the bottom panel, an entirely new type of oscillatory instability is displayed; hundreds of these ringing features were recorded during Oct 13 and 15 with a recurrence time near 70 s. During Oct 15 the recurrence time increases (see Fig. 2), leading to a long X-ray stall and subsequent flux overshoot. The nature of these astonishing X-ray instabilities is currently a mystery. Note, however, that most of the PCA observations show 'normal' light curves with variations limited to rapid flickering at 10-20 % of the mean rate.

A penetrating analysis of GRS1915+105 was made by investigating the X-ray power spectra and comparing them with the characteristics of the ASM light curve[10]. The shape of the broad-band power continuum and the properites of rapid QPOs (0.01 to 10 Hz) are correlated with the brightness, spectral hardness, and the long-term variations seen with the ASM. Four emission states were found, labelled in Fig. 1 as chaotic (CH), bright (B), flaring (FL), and low-hard (LH). We see QPOs and nonthermal spectral components during all four states, implying that they are new variants of the 'very high state' rarely seen in other X-ray binaries[17][16]. The combination of the intense QPOs and the high throughput of the PCA enabled phase tracking of individual oscillations. Four QPO cases were chosen from three different states [10], with frequencies ranging from 0.07 to 2.0 Hz. The results are remarkably similar: the QPO arrival phase (relative to the mean frequency) exhibits a random walk with no correlation between the amplitude and the time between subsequent events. Furthermore the mean 'QPO-folded' profiles are roughly sinusoidal with increased amplitude at higher energy, and with a distinct phase lag of ≈ 0.03 between 3 and 15 keV. At photon energies above 10 keV, the high amplitudes and sharp profiles of the QPOs are inconsistent with any scenario in which the phase delay is caused by scattering effects. Alternatively, it appears that the origin of the hard X-ray spectrum itself (i.e. the creation of energetic electrons in the inverse Compton model) is functioning in a quasiperiodic manner. These results fundamentally link X-ray QPOs with the most luminous component of the X-ray spectrum in GRS1915+105.

In addition to the frequent X-ray QPOs below 10 Hz, a transient yet 'stationary' QPO at 67 Hz has been discovered [10]. This feature is seen on 6 of the first 31 PCA observations of GRS1915+105. Typically, the amplitude is 1% of the flux and the QPO width is 3.5 Hz. This QPO exhibits a strong energy dependence, rising (e.g. on 1996 May 6) from 1.5 % at 3 keV to 6%

at 15 keV. One may attempt to associate this frequency with the mass and spin rate of an accreting black hole, but the competing models include such concepts as instabilities at the minimum stable orbit of $3Rs$, implying a mass of 33 $M_\odot$ for a nonrotating black hole[10], to relativistic modes of oscillation in the inner accretion disk, implying 10 $M_\odot$ for a nonrotating black hole[12].

3   Recent Observations of GRO J1655-40

During much of 1995 and early 1996, GRO J1655-40 was in a low or quiescent accretion state, permitting a clear optical view of the companion star (near F4 IV). Orosz and Bailyn[13] improved the determinations of the binary period (2.62157 days) and the mass function. They further measured the 'ellipsoidal variations' arising from the rotation of the gravitationally distorted companion star. Their analysis, using B,V,R, and I bandpasses, provide an exceptionally good fit for the binary inclination angle (69.5 deg) and the mass ratio. From these results, they deduce masses of 7.0 ± 0.2 and 2.34 ± 0.12 $M_\odot$ for the black hole and companion star, respectively.

The ASM recorded a renewed outburst from GRO J1655-40 that began on 1996 April 25. The ASM light curve (Feb 1996 to Jan 1997) is shown in the lower half of Fig. 1. With great fortune, our optical campaign had lasted until April 24, and Orosz et al. has shown[14] that optical brightening preceeded the X-ray ascent by 6 days, beginning first in the I band and then accelerating quickly in blue light. These results provide concrete evidence favoring the accretion disk instability as the cause of the X-ray nova. Theorists may now attempt to model the brightness gradients and delay times in the effort to develop a deeper understanding of this outburst.

The ASM $HR2$ measures (Fig. 1) show an initially soft spectrum that becomes brighter and harder for several months during mid outburst. The PCA observations from our GO program confirm this evolution, as the power-law component (photon index ≈ 2.6) dominates the spectrum during the brightest cases. The great majority of PCA measurments of GRO J1655-40 follow single tracks on the intensity:color and color:color diagrams, with a positive correlation between hardness and brightness.

PCA power spectra show transient QPOs in the range of 8–22 Hz that are clearly associated with the strength of the power-law component. Using a PCA-based hardness ratio, PCA_HR2 = flux above 9.6 keV / flux at 5.2–7.0 keV, we detect QPO in the range of 8–22 Hz whenever $PCA\_HR2 > 0.22$. Furthermore, in the 7 'hardest' observations ($PCA\_HR2 > 0.3$), there is evidence of a high-frequency QPO near 300 Hz. In Fig. 3 we show the sum of PCA power spectra in these 3 intervals of PCA_HR2, illustrating the QPO centered at 298 Hz. The Poisson noise has been subtracted, with inclusion of deadtime effects [10]. The integrated feature has a significance of 14σ, a width of 120 Hz, and an amplitude near 0.8%. Applying the 'last stable orbit' model to this feature yields a mass of 7.4 $M_\odot$ for a non-rotating black hole. While this is astonishingly similar to the optically determined mass, we caution that other models can give similar results in the case of significant black hole rotation. We further note that none of the models discussed [10] for the high-frequency QPOs in GROJ1655-40 and GRS1915+105 adequately address the spectral signature of this oscillation, which is more directly associated with the power law component rather than the disk (thermal) component. Nevertheless, the fact of these QPOs, which almost certainly originate very near the accreting compact objects, will remain a vigorous research topic throughout the RXTE Mission.

References

1. C.D. Bailyn, J.A. Orosz, J.E., McClintock, and R.A. Remillard, Nature **378**, 157 (1995).
2. W. Chen, M. Livio, and N. Gehrels, ApJL **408**, L5 (1993).
3. J. Greiner, E.H. Morgan, and R. A. Remillard, ApJL **473**, L107 (1996).
4. E. Grove, this proceedings.
5. B.A. Harmon *et al.* , Nature **374**, 703 (1995).
6. R.M. Hjellming & M.P. Rupen, Nature **375**, 464 (1995).
7. A.M. Levine, H. Bradt, W. Cui, J.G. Jernigan, E.H. Morgan, R.A. Remillard, R.E. Shirey, and D.A. Smith, ApJL **469**, L33 (1996).
8. I.F. Mirabel, *et al.* , A&A **282**, L17 (1994).
9. I.F. Mirabel and L.F. Rodriguez, Nature **371**, 46 (1994).
10. E.H. Morgan, R. A. Remillard, and J. Greiner, ApJL **473**, L107 (1996).
11. E.H. Morgan, R. A. Remillard, and J. Greiner, IAUC 6392, May 2, 1996.
12. M.A. Nowak, R.V. Wagoner, M.C. Begelman, and D.E. Lehr, ApJ, submitted.
13. J.A. Orosz and C.D. Bailyn, ApJ **477**, in press (1997).
14. J.A. Orosz, R. A. Remillard, C.D. Bailyn, & J. E. McClintock, ApJL **478**, in press (1997).
15. S.J.Tingay *et al.* , Nature **374**, 141 (1995).
16. M. van der Klis in *X-ray Binaries*, eds. W. Lewin, J. van Paradijs, and E. van den Heuvel (Cambridge University Press, Cambridge, 1996) p. 252.
17. M. van der Klis, APJS **92**, 511 (1994).

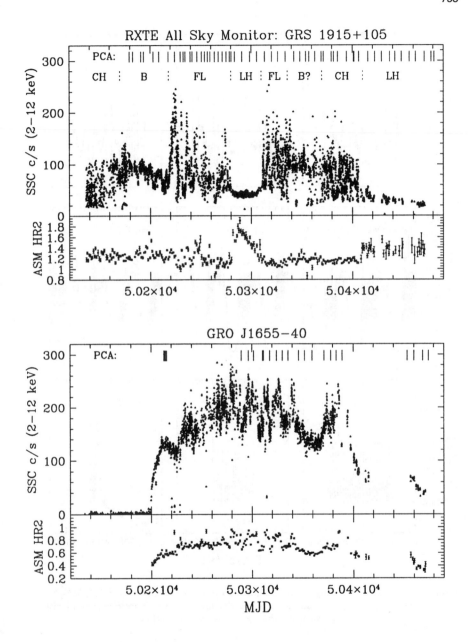

Figure 1: RXTE ASM light curves and hardness ratio of GRS1915+105 and GRO J1655-40.

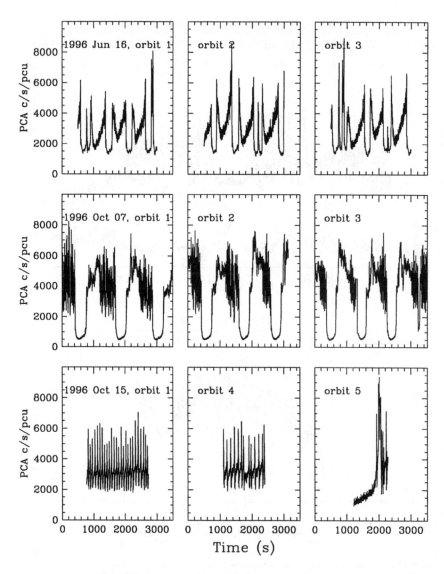

Figure 2: Samples of PCA light curves showing dramatic variability in GRS1915+105 (2–30 keV). The adjacent panels along each row display source measurements from different satellite orbits during the same observation.

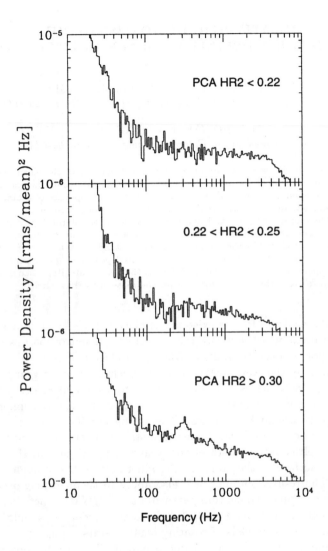

Figure 3: PCA power spectra of GRO J1655-40 in which 20 observations are combined into 3 intervals of PCA_HR2 (see text). A QPO appears at 300 Hz during the 7 observations with the hardest X-ray spectra.

RELATIVISTIC DISKOSEISMOLOGY:
SEARCH FOR A DEFINITIVE BLACK HOLE SIGNATURE

R.V. WAGONER, D.E. LEHR, A.S. SILBERGLEIT
Dept. of Physics, Stanford University, Stanford, CA 94305-4060

M.A. NOWAK, M.C. BEGELMAN
JILA, University of Colorado, Boulder, CO 80309-0440

We will summarize some results of relativistic calculations of the eigenfunctions and eigenfrequencies of normal modes of oscillation trapped within the inner region of thin accretion disks by non-Newtonian gravitational properties of a black hole (or a compact, weakly-magnetized neutron star).The focus will be on the most observable, robust, and best-studied class: the analogue of (internal) gravity modes in stars. The frequency of the peak in the power spectrum of the luminosity modulations produced by the lowest modes is almost independent of the properties of the accretion disk. A feature with the required properties has recently been observed in at least one black hole candidate by the RXTE satellite, which produces a relation between the angular momentum and mass of the black hole.

Presently, evidence for the existence of black holes is based almost entirely upon observations which indicate that a certain amount of mass is contained within a region of a certain radius. However, a black hole is a region of space-time governed by the Kerr metric of general relativity. Our group has been investigating consequences of the realization (by Kato and Fukue [1]) that general relativity can trap normal modes of oscillation near the inner edge of accretion disks around black holes. The strong gravitational fields that are required can also be produced by neutron stars that are sufficiently compact (requiring a soft equation of state) and weakly magnetized that there is a gap between the surface of the star and the innermost stable orbit of the accretion disk. The results obtained will also apply to them to first order in the dimensionless angular momentum parameter $a = cJ/GM^2$, since their exterior metric is identical to that of a black hole to that order. It should be noted that $a < 0.2$ for most models of rotating neutron stars.

These modes of oscillation provide a potentially powerful probe of both strong gravitational fields and the physics of accretion disks, since: (a) they do not exist in Newtonian gravity, (b) their frequencies depend upon the angular momentum as well as the mass of the black hole, and (c) the frequency spread of each mode is proportional to α, the viscosity parameter of the disk.

We have applied the general relativistic perfect fluid perturbation formalism of Ipser and Lindblom [2] to thin accretion disks in the Kerr metric. The radial component of the velocity of the fluid is neglected, which should have little effect on the modes for these thin models. Neglecting the gravitational field of the disk, the adiabatic oscillations of all physical quantities can be expressed in terms of a single scalar potential governed by a second-order partial differential equation. The stationary and axisymmetric unperturbed accretion disk is taken to be described by a relativistic α-disk model.

All fluid perturbations are of the form $f(r, z) \exp[i(\sigma t + m\phi)]$. With angular velocity Ω, the comoving frequency is $\omega(r) = \sigma + m\Omega(r)$. In general, the vertical extent of the modes within the disk is restricted by the requirement that $|\omega|$ be greater than the buoyancy frequency. Numerical results have been obtained for accretion disks which are barotropic (e. g., isentropic) on scales of order their thickness. The effective radial wavelengths are significantly smaller than r, allowing a WKB expansion of the problem and approximate separation of the governing equations. The key ingredient is the relativistic behavior of the radial epicyclic frequency $\kappa(r)$.

The g–modes are trapped where $\omega^2 < \kappa^2$, which for $m = 0$ is between radii on either side of that where $\kappa(r)$ achieves its maximum (at $r_0 = 8GM/c^2$ for $a = 0$). The lowest modes have eigenfrequencies $|\sigma|$ which are close to the maximum possible.

The p–modes are trapped where $\omega^2 > \kappa^2$. The radial ($m = 0$) p–modes that are trapped very near the inner radius of the disk have very little radial extent, and thus will produce negligible direct luminosity modulation. However, for all m these modes will modulate the accretion onto the black hole (or neutron star).

The c–modes are typically nonradial ($m = 1$) incompressible (and therefore less modulating) waves near the inner edge of the disk that slowly precess around the angular momentum of the black hole. The ones that we have studied have a maximum frequency proportional to a^3, only becoming comparable to the g–mode frequencies when $a \to a(max) \cong 0.998$.

The frequencies of all of these modes are proportional to $1/M$, but their dependence on the angular momentum of the black hole is quite different. However, we now turn our attention to the g–modes, which are the most robust and observable of those classes that we have investigated. These (rough) analogs of stellar g–modes have been analyzed most extensively by Perez et al.[3], where references to earlier work can also be found.

The lowest radial g–modes have a frequency $f = -\sigma/2\pi$ given by

$$f = 714(1 - \epsilon_{nj}) \left(\frac{M_\odot}{M}\right) F(a) \text{ Hz} , \qquad \epsilon_{nj} \approx \left(\frac{n + 1/2}{j + \delta}\right) \frac{h}{r_0} .$$

The properties of the accretion disk enter only through the small mode-dependent term ϵ_{nj}, which involves the disk thickness $2h(r_0)$ and the radial (n) and vertical (j) mode numbers, with $\delta \sim 1$. Typically $h(r_0)/r_0 \sim 0.1 L/L_{Edd}$ for a radiation-pressure dominated optically thick disk region. $F(a)$ increases monotonically from $F(0) = 1$ to $F(0.998) = 3.44$. The higher axial modes $(m > 0)$ have higher frequencies and a somewhat different dependence upon the angular momentum of the black hole, which in principle would allow its determination as well as that of the mass.

Nowak and Wagoner [4] have estimated the damping rates of these modes due to turbulent viscosity. The corresponding quality factor Q is given by $1/Q_{jn} \sim [j^2 + (h/r)n^2]\alpha$ for isotropic turbulence. However, the effective width Δf of the g–mode feature will also be determined by what range of mode numbers j and n are sufficiently excited. They later [5] investigated the excitation produced by the turbulence, obtaining maximum displacements of order the thickness of the disk, which should produce maximum luminosity modulations $\delta L_\nu / L_\nu \sim 10^{-2}$, if the photon frequency $\nu > h^{-1}kT(max)$.

Recently, Morgan et al. [6] have detected a feature at $f = 67$ Hz with $Q \approx 20$ in the black hole candidate GRS 1915+105 with RXTE. What distinguishes this peak from others that have recently been detected at high frequencies is the fact that it did not change its frequency as the source luminosity varied by a factor of two, as predicted by the above equation. The amplitude of the peak was greatest ($\sim 6\%$) at the highest x-ray energies. If this feature is produced by a g–mode oscillation, this equation also predicts a black hole mass of $10.6 M_\odot$ if it is nonrotating to $36.3 M_\odot$ if it is maximally rotating. Other aspects of this identification are explored by Nowak et al. [7].

This research was supported by NASA grants NAG 5-3102 (R.V.W.), NAG 5-3225 (M.A.N.), and NAS 8-39225 (GP-B, A.S.S.); by NSF grants AST-95-29175 and AST-91-20599 (M.C.B.); and by a NDSEG Fellowship (D.E.L.).

References

1. S. Kato and J. Fukue, *P.A.S.J.* **32**, 377 (1980).
2. J.R. Ipser and L. Lindblom, *Ap.J.* **389**, 392 (1992).
3. C.A. Perez, A.S. Silbergleit, R.V. Wagoner, and D.E. Lehr, *Ap.J.* **476**, in press (1997).
4. M.A. Nowak and R.V. Wagoner, *Ap.J.* **393**, 697 (1992).
5. M.A. Nowak and R.V. Wagoner, *Ap.J.* **418**, 187 (1993).
6. E.H. Morgan, R.A. Remillard and J. Greiner, *Ap.J.* , in press (1997).
7. M.A. Nowak, R.V. Wagoner, M.C. Begelman and D.E. Lehr, *Ap.J.Lett.* , in press (1997).

SONIC-POINT MODEL FOR HIGH-FREQUENCY QPOs IN NEUTRON STAR LOW-MASS X-RAY BINARIES

M. COLEMAN MILLER

University of Chicago, Department of Astronomy and Astrophysics,
5640 S. Ellis Ave., Chicago, IL 60637, USA

FREDERICK K. LAMB AND DIMITRIOS PSALTIS

University of Illinois, Department of Physics and
Department of Astronomy, University of Illinois at Urbana-Champaign
1110 W. Green St., Urbana, IL 61801-3080, USA

Quasi-periodic brightness oscillations (QPOs) with frequencies in the range 300–1200 Hz have been detected from at least nine neutron star low-mass X-ray binaries. Here we summarize the sonic-point model for these brightness oscillations, which we present in detail elsewhere. If the sonic-point interpretation of kilohertz QPOs is confirmed, measurements of kilohertz QPO frequencies in low-mass X-ray binaries will provide new bounds on the masses and radii of neutron stars in these systems and new constraints on the equation of state of matter at high densities.

1 Introduction

Observations of neutron stars in low-mass X-ray binaries using the Rossi X-ray Timing Explorer have revealed that at least nine produce quasi-periodic X-ray brightness oscillations (QPOs) with frequencies ν_{QPO} ranging from ~ 300 Hz to ~ 1200 Hz (see, e.g., van der Klis et al. 1996; Strohmayer et al. 1996; Berger et al. 1996; Zhang et al. 1996). These high-frequency QPOs are remarkably coherent ($\nu_{QPO}/\Delta\nu_{QPO}$ up to ~ 200) and strong (rms amplitudes up to 20%). In several sources the frequencies are strongly positively correlated with countrate. In four sources, a single highly coherent QPO peak is seen during type I X-ray bursts. High-frequency QPOs have been observed to occur in pairs in six sources.

In the sonic-point model of these high-frequency QPOs, (1) the frequency of the higher frequency QPO in a pair is the Keplerian frequency at the point in the disk flow where the inward radial velocity increases steeply with decreasing radius and typically becomes supersonic, and (2) the frequency of the lower frequency QPO peak is the beat frequency between the sonic-point Keplerian frequency and the stellar spin frequency or its overtones. We emphasize that the sonic-point model is *not* a magnetospheric beat-frequency model (see Miller, Lamb, & Psaltis 1996 [hereafter MLP] for further discussion of the differences between the sonic-point model and magnetospheric models). The sonic-point model is based on earlier work on the effect of radiation forces on disk accretion by neutron stars (Miller & Lamb 1993, 1996).

2   Overview of the Sonic-Point Model

The sonic-point model builds on the standard picture of the Z and atoll sources, in which the neutron star accretes gas from a low-mass companion via a geometrically thin disk and stresses within the disk create only a slow, subsonic inward radial drift. In the sonic-point model, near the neutron star there is a region of the disk flow in which the inward radial velocity increases rapidly with decreasing radius, primarily because (1) radiation from near the star removes angular momentum from the gas in the disk, allowing the gas to accelerate radially, or—if radiation forces are sufficiently weak—(2) the gas drifts inside the marginally stable orbit R_{ms}. Because there is a sharp transition to supersonic radial inflow in a very small radial distance, we refer to this radius as the "sonic point" radius R_{sonic}. We expect that at least some gas in the disk will penetrate in to R_{sonic} provided that the magnetic field of the neutron star is less than $\sim 10^{10}$ G, as in the Z and atoll sources.

We expect that the disk flow will have local density inhomogeneities, or "clumps". Because the radial velocity outside R_{sonic} is small, the inflow time from this region to the stellar surface is large and hence clumps produced outside R_{sonic} are sheared or dissipated before gas from them reaches the surface. In contrast, the inflow time for clumps from near R_{sonic} is small, and hence clumps formed in this region are not completely sheared or dissipated before gas from them impacts the stellar surface. The impact of gas with the stellar surface produces bright, arc-shaped footprints that move around the star. As seen at infinity, the frequency at which the pattern of arcs moves around the star is the Keplerian frequency ν_{Ks} at the sonic point (see MLP). Hence, from the standpoint of a distant observer with a line of sight inclined with respect to the disk axis, the bright arcs are periodically occulted by the star at the frequency ν_{Ks}, and the observer sees brightness oscillations at a frequency ν_{Ks}. If the field is weak but not negligible, some of the accreting gas may be channeled by the magnetic field onto hot spots on the surface that rotate with the star. Because the inflow of gas from the clumps is caused by radiation drag, the enhanced radiation intensity from the hot spots modulates the mass accretion rate at the beat frequency between the sonic-point Keplerian frequency and the stellar spin frequency or its overtones. In this case, a distant observer will also see a modulation of the total luminosity at the beat frequency.

As we explain in detail in MLP, the sonic-point model explains naturally the high frequencies, amplitudes, and coherences of kilohertz QPOs and their changes with countrate, as well as the numerous other observed correlations. The sonic-point model is also consistent with the physical picture constructed previously from observations of the spectra and low-frequency variability of the Z and atoll sources.

3 Predictions and Implications

The sonic-point model makes a number of testable predictions, which are discussed in detail in MLP. In particular, we do not expect kilohertz QPOs of the type discussed here in black hole sources, because in the sonic-point model the presence of a stellar surface is essential. We expect that QPOs of this type will be very weak or undetectable by current instruments from sources which show strong periodic oscillations at their stellar spin frequencies (pulsars), since these sources have strong magnetic fields and hence at most a very a small fraction of the accreting gas will reach the stellar surface without being forced by the magnetic field to corotate with the neutron star.

If the sonic-point model is correct, observation of a high-frequency QPO provides upper limits to the mass and radius of the neutron star that are independent of the equation of state and, for an assumed equation of state, yields estimates of the mass and radius of the source. These constraints can in principle rule out equations of state, particularly if high (>1300 Hz) QPO frequencies are observed, thereby constraining the properties of dense matter.

This work was supported in part by NSF grant AST 93-15133 and NASA grant 5-2925 at the University of Illinois, NASA grant NAG 5-2868 at the University of Chicago, and through the *Compton Gamma-Ray Observatory* Fellowship Program, by NASA grant NAG 5-2687.

References

1. Berger, M., et al., *Astrophys. J.*, 469, L13 (1996).
2. Miller, M. C., & Lamb, F. K., *Astrophys. J.*, 413, L43 (1993).
3. Miller, M. C., & Lamb, F. K., *Astrophys. J.*, 470, 1033 (1996).
4. Miller, M. C., Lamb, F. K., & Psaltis, D., *Astrophys. J.*, submitted (1996).
5. Strohmayer, T. et al., *Astrophys. J.*, 469, L9 (1996).
6. van der Klis, M. et al., *Astrophys. J.*, 469, L1 (1996).
7. Zhang, W., Lapidus, I., White, N. E., & Titarchuk, L., *Astrophys. J.*, 469, L17 (1996).

PARTICIPANTS' E-MAIL ADDRESSES

tabel@ncsa.uiuc.edu	Abel Thomas
abramo@het.brown.edu	Abramo L. Raul
adami@astrsp-mrs.fr	Adami Christophe
jenni@teorfys.uu.se	Adams Jenni
a.albrecht@ic.ac.uk	Albrecht Andreas
alcock@igpp.llnl.gov	Alcock Charles
amendola@oarhp1.rm.astro.it	Amendola Luca
arley@physics.unc.edu	Anderson Arlen
nils@howdy.wustl.edu	Andersson Nils
panninos@ncsa.uiuc.edu	Anninos Peter
annis@fnal.gov	Annis Jim
ansari@lal.in2p3.fr	Ansari Reza
	Anton Theodore
ashtekar@phys.psu.edu	Ashtekar Abhay
aubourg@hep.saclay.cea.fr	Aubourg Eric
audouze@iap.fr	Audouze Jean
dbacker@astro.berkeley.edu	Backer Donald
jnb@sns.ias.edu	Bahcall John
fbaier@aip.de	Baier Frank W.
suchitra@astro.umd.edu	Balachandran Suchitra
audra@as.arizona.edu	Baleisis Audra
balland@physics.berkeley.edu	Balland Christophe
dband@ucsd.edu	Band David
banday@mpa-garching.mpg.de	Banday Anthony
abarger@ast.cam.ac.uk	Barger Amy
barger@phenxi.physics.wisc.edu	Barger Vernon
baron@phyast.nhn.ou.edu	Baron Edward
mbartelmann@mpa-garching.mpg.de	Bartelmann Matthias
r.battye@ic.ac.uk	Battye Richard
thomas@astro.physics.uiuc.edu	Baumgarte Thomas
bayin@newton.physics.metu.edu.tr	Bayin Selçuk
	Baym G.
beacom@nucth.physics.wisc.edu	Beacom John
svwb@mpia-hd.mpg.de	Beckwith Steve
mitch@jila.colorado.edu	Begelman Mitchell
	Belloni T.
sci_quista@vixen.emcmt.edu	Benacquista Matthew
	Bender Peter

alan.bunner@hq.nasa.gov	Bunner Alan N.
burdyuzh@dpc.asc.rssi.ru	Burdyuzha Vladimir
scott@cass154.ucsd.edu	Burles Scott
burrows@as.arizona.edu	Burrows Adam
javier@egret.sao.arizona.edu	Bussons Gordo Javier
byers@physics.ucla.edu	Byers Nina
calder@compsci.cas.vanderbilt.edu	Calder Alan
caldwell@dept.physics.upenn.edu	Caldwell Robert
kcamarda@aei-potsdam.mpg.de	Camarda Karen
	Campana S.
bruce@phys.ualberta.ca	Campbell Bruce
crc@space.mit.edu	Canizares Claude
pat@ifctr.mi.cnr.it	Caraveo Patrizia
ccardall@ucsd.edu	Cardall Christian
carlson@msi.se	Carlson Per
jc@oddjob.uchicago.edu	Carlstrom John
carmelim@bgumail.bgu.ac.il	Carmeli Moshe
fjc@oddjob.uchicago.edu	Castander Francisco
catanese@egret.sao.arizona.edu	Catanese Michael
cavaliere@roma2.infn.it	Cavaliere Alfonso G.
cerdoniam@padova.infn.it	Cerdonio Massimo
jorge@iris14.inin.mx	Cervantes-Cota Jorge L.
chakraba@bose.ernet.in	Chakrabarti Sandip K.
pac@astro.ox.ac.uk	Charles Philip
pcha@xanum.uam.mx	Chauvet Pablo
	Chen K.
hrspksc@hkucc.hku.hk	Cheng Kwong-Sang
masha@hea.iki.rssi.ru	Chernyakova Maria
chiba@yukawa.kyoto-u.ac.jp	Chiba Takeshi
jcd@obs.aau.dk	Christensen-Dalsgaard J.
markc@oddjob.uchicago.edu	Chun Mark
chuvenkov@phys.rnd.runnet.ru	Chuvenkov Vladimir
clarke@fizzle.stanford.edu	Clarke Roland
dcline@physics.ucla.edu	Cline David
cline@lheavx.gsfc.nasa.gov	Cline Thomas
coble@oddjob.uchicago.edu	Coble Kimberly
jdc@cosmos2.phy.tufts.edu	Cohn Joanne
dmc@otto.uchicago.edu	Cole David
comer@newton.slu.edu	Comer Gregory I.
connauv@gibson.msfc.nasa.gov	Connaughton Valerie
ajc@tiamat.pha.jhu.edu	Connolly Andrew

spd@edwin.phys.unsw.edu.au	Driver Simon
duncan@oddjob.uchicago.edu	Duncan Douglas
jdykla@orion.it.luc.edu	Dykla John J.
ehlers@aei-potsdam.mpg.de	Ehlers Jürgen
eichblatt@fnal.gov	Eichblatt Steve
rse@ast.cam.ac.uk	Ellis Richard
repstein@lanl.gov	Epstein Richard
helio@ift.unesp.br	Fagundes Helio V.
am981@torfree.net	Falk Dan
farooqui@het.brown.edu	Farooqui Khurram
feldman@kusmos.phsx.ukans.edu	Feldman Hume
efenimore@lanl.gov	Fenimore Edward
ezbd@utdallas.edu	Fenyves Ervin
ferguson@stsci.edu	Ferguson Henry
ferlet@iap.fr	Ferlet Roger
fsoto@mail.ess.sunysb.edu	Fernandez-Soto Alberto
valeria@roma1.infn.it	Ferrari Valeria
pgf@physics.berkeley.edu	Ferreira Pedro
bfields@cygnus.phys.nd.edu	Fields Brian
alex@astro.berkeley.edu	Filippenko Alexi
dfink@astro.berkeley.edu	Finkbeiner Douglas
fiorentini@vaxfe.fe.infn.it	Fiorentini Giovanni
richard.fong@durham.ac.uk	Fong Richard
eric@astro.columbia.edu	Ford Eric
ford@cosmos2.phy.tufts.edu	Ford Larry
wforman@cfa.harvard.edu	Forman William
fosalba@ieec.fcr.es	Fosalba Pablo
frampton@physics.unc.edu	Frampton Paul
wendy@ociw.edu	Freedman Wendy
ktfreese@umich.edu	Freese Katherine
friedman@mitlns.mit.edu	Friedman Jerome
frieman@fnal.gov	Frieman Joshua
fry@phys.ufl.edu	Fry James
cfryer@as.arizona.edu	Fryer Chris
gaidos@purdd.physics.purdue.edu	Gaidos James
gaisser@bartol.udel.edu	Gaisser Thomas
gaitskell@physics.berkeley.edu	Gaitskell Richard
jgallego@iac.es	Gallegos Julio
aagarcia@fis.cinvestav.mx	Garcia Alberto

gates@oddjob.uchicago.edu	Gates Evalyn
gawiser@physics.berkeley.edu	Gawiser Eric
gaztanaga@ieec.fcr.es	Gaztanaga Enrique
gehrels@gsfc.nasa.gov	Gehrels Neil
gerlach@math.ohio-state.edu	Gerlach Ulrich H.
rgiaccon@eso.org	Giacconi Riccardo
giazotto@pi.infn.it	Giazotto Adalbert
gilfanov@mpa-garching.mpg.de	Gilfanov Marat
mgiller@zpk3.u.lodz.pl	Giller Maria
72734.3072@compuserve.com	Glanz James
physeg@server.uwindsor.ca	Glass Edward
glicens@hep.saclay.cea.fr	Glicenstein Jean-François
gnedin@astron.berkeley.edu	Gnedin Nick
gnedin@pulkovo.spb.su	Gnedin Yuri
	Gold Marvin
alexey@oddjob.uchicago.edu	Goldin Alexey
goldoni@ifctr.mi.cnr.it	Goldoni Paolo
golwala@physics.berkeley.edu	Golwala Sunil
andy@cecs.cl	Gomberoff Andres
p.gondolo1@physics.oxford.ac.uk	Gondolo Paolo
gonthier@physics.hope.edu	Gonthier Peter
	Goodrich Robert
	Gordan Bonnie
jgu@laeff.esa.es	Gorosabel Javier
gorski@tac.dk	Gorski Krzysztof
jgra@loc.gov	Graber James S.
graber@aps.anl.gov	Graber Timothy
	Graves Shandra
carlo@aquila.uchicago.edu	Graziani Carlo
graziani1@llnl.gov	Graziani Frank
lg@seek.uchicago.edu	Grego Laura
dma0rag@gauss.dur.ac.uk	Gregory Ruth
greivel@astrog.physics.wisc.edu	Greiveldinger Chris
gress@mhd1.moorhead.msus.edu	Gress Joseph
hgreyber@capaccess.org	Greyber Howard D.
kgriest@ucsd.edu	Griest Kim
grove@osse.nrl.navy.mil	Grove J. Eric
guendel@bgumail.bgu.ac.il	Guendelman Eduardo
jg@barus.physics.brown.edu	Gunderson Joshua
gursky@ssd0.nrl.navy.mil	Gursky Herbert
gustafso@umich.edu	Gustafson Richard
cgwinn@pelican.physics.ucsb.edu	Gwinn Carl

haiman@cfata5.harvard.edu — Haiman Zoltan
nils@seek.uchicago.edu — Halverson Nils
hanany@physics.berkeley.edu — Hanany Shaul
thankins@nrao.edu — Hankins Timothy
harada@tap.scphys.kyoto-u.ac.jp — Harada Tomohiro
ehh@st-andrews.ac.uk — Harlaftis Emilios
harmon@ssl.msfc.nasa.gov — Harmon B. Alan
bharms@ua1vm.ua.edu — Harms Benjamin
frh@hq.nasa.gov — Harnden F. Rick
hartmann@grb.phys.clemson.edu — Hartmann Dieter
s.w.hawking@dampt.cam.ac.uk — Hawking Stephen
mrsh@roe.ac.uk — Hawkins Mike
heckler@revere.mps.ohio-state.edu — Heckler Andrew
— Hellings R.

helliwell@thuban.ac.hmc.edu — Helliwell Thomas
martin@astro.gla.ac.uk — Hendrie Martin
mahenrik@plains.nodak.edu — Henriksen Mark
heusler@physik.unizh.ch — Heusler Markus
wfh@mpa-garching.mpg.de — Hillebrandt Wolfgang
hirat@phys.h.kyoto-u.ac.jp — Hirata Kouichirou
hirotani@ferio.mtk.nao.ac.jp — Hirotani Kouichi
hiscock@montana.edu — Hiscock William
raph@astro.as.utexas.edu — Hix William Raphael
hobbs@yerkes.uchicago.edu — Hobbs Lewis
hobill@crag.ucalgary.ca — Hobill David
hogan@astro.washington.edu — Hogan Craig
holden@oddjob.uchicago.edu — Holden Bradford
holder@oddjob.uchicago.edu — Holder Gilbert
— Holloway L.
holman@defoe.phys.cmu.edu — Holman Richard
— Holst Michael
deholz@rainbow.uchicago.edu — Holz Daniel
gary@cosmic.physics.ucsb.edu — Horowitz Gary
hoi@astro.psu.edu — Horvath Istvan
hoi@astro.psu.edu — Horvath Jorge
lhui@fnal.gov — Hui Lam
cjh1010@damtp.cam.ac.uk — Hunter Christopher
khurley@sunspot.ssl.berkeley.edu — Hurley Kevin
huterer@oddjob.uchicago.edu — Huterer Dragan
jchan@hanul.issa.re.kr — Hwang Jai-Chan

ibanez@godunov.daa.uv.es — Ibañez Jose-Maria

osamu@th.phys.titech.ac.jp — Iguchi Osamu
— Illarionov A.
misenber@oddjob.uchicago.edu — Isenberg Michael
akihiro@th.phys.titech.ac.jp — Ishibashi Akihiro
ishihara@th.phys.titech.ac.jp — Ishihara Hideki
israel@phys.ualberta.ca — Israel Werner
ani@mpe-garching.mpg.de — Iyudin Anatoli

jaffe@physics.berkeley.edu — Jaffe Andrew
bjain@mpa-garching.mpg.de — Jain Bhuvnesh
cecelia.jarlskog@shogun.matfys.lth.se — Jarlskog Cecilia
jetzer@rzuaix.unith.ch — Jetzer Philippe
joffre@oddjob.uchicago.edu — Joffre Michael
— Joynt R.
visitor1@fnal.gov — Juszkiewicz Roman

kaiser@ifa.hawaii.edu — Kaiser Nick
vicky@astro.uiuc.edu — Kalogera Vassiliki
kamion@phys.columbia.edu — Kamionkowski Marc
— Kang H.
ahya@oddjob.uchicago.edu — Kao Lancelot
s.cheenu.kappadath@unh.edu — Kappadath S. Cheenu
kartje@jets.uchicago.edu — Kartje John
visnja@oddjob.uchicago.edu — Katalinic Visnja
— Kauts Vladimir
skayser@nsf.gov — Kayser Susan
aidan@astro.gla.ac.uk — Keane Aidan
keating@het.brown.edu — Keating Brian
ckeeton@cfa.harvard.edu — Keeton Charles
kephartt@ctrvax.vanderbilt.edu — Kephart Tom
khalat@ccsg.tau.ac.il — Khalatnikov Isack
— Kim Jung Kon
sangkim@knusun1.kunsan.ac.kr — Kim Sang Pyo
— Kim Sung Kyu
kinneyw@fnal.gov — Kinney William
kirshner@cfa.harvard.edu — Kirshner Robert
klose@tls-tautenberg.de — Klose Sylvia
— Kluzniak Wlodzimierz
knox@cita.utoronto.ca — Knox Lloyd
ckochanek@cfa.harvard.edu — Kochanek Christopher
kohnle@aps.anl.gov — Kohnle Antje
— Koike Tatsuhiko

rocky@fnal.gov	Kolb Rocky
m.kolman@elsevier.nl	Kolman Michiel
serguei@amsta.leeds.ac.uk	Komissarov Serguei
dkompan@dpc.asc.rssi.ru	Kompaneets Dmitri
arieh@jets.uchicago.edu	Konigl Arieh
dak@nadn.navy.mil	Konkowski Deborah
sergeikk@cc.nao.ac.jp	Kopeikin Sergei
kosik@tekel.butler.edu	Kosik Dan
rik@mppmu.mpg.de	Kotthaus Rainer
skd@haar.pha.jhu.edu	Kovesi-Domokos Susan
kozlovsky@lheavx.gsfc.nasa.gov	Kozlovsky Ben-Zion
frank@egret.sao.arizona.edu	Krennrich Frank
jkress@maths.newcastle.edu.au	Kress Jonathan
jkrisch@mich1.physics.lsa.umich.edu	Krisch Jean
krivan@astro.psu.edu	Krivan William
kronberg@astro.utoronto.ca	Kronberg Philipp
kuncic@maths.anu.edu.au	Kuncic Zdenka
tomislav@tapir.caltech.edu	Kundic Tomilsav
pablo@astro.psu.edu	Laguna Pablo
dong@tapir.caltech.edu	Lai Dong
f-lamb@uiuc.edu	Lamb Frederick
lamb@oddjob.uchicago.edu	Lamb, Jr. Don Q.
jwlandry@oddjob.uchicago.edu	Landry James
lanou@physics.brown.edu	Lanou Robert
lanzetta@sbastc.ess.sunysb.edu	Lanzetta Ken
uphgfml@montana.edu	Larson Michelle
shane@orion.physics.montana.edu	Larson Shane
lattimer@astro.sunysb.edu	Lattimer James
fil@mir.saclay.cea.fr	Laurent Philippe
	Lee Andrew
chlee@hepth.hanyang.ac.kr	Lee Chul H.
hyongel@nwu.edu	Lee Hyong
wlee@astrog.physics.wisc.edu	Lee William
lemoine@oddjob.uchicago.edu	Lemoine Martin
lemos@obsn.on.br	Lemos José
	Lenz Dawn
leroux@physics.montana.edu	LeRoux Anna
lessard@purdd.physics.purdue.edu	Lessard Rod
lev@astro.ioffe.rssi.ru	Levshakov Sergei
hli@lanl.gov	Li Hui
	Liang Edison

a.liddle@sussex.ac.uk	Liddle Andrew
liebend@quasar.physik.unibas.ch	Liebendörfer Matthias
charley@cdsxb6.u-strasbg.fr	Lineweaver Charles
blink@dante.physics.montana.edu	Link Bennett
jlinsky@jila.colorado.edu	Linsky Jeffrey
	Lissia Marcello
lockitch@alpha1.csd.uwm.edu	Lockitch Keith
aloeb@cfa.harvard.edu	Loeb Abraham
	Lombardy J.
nzamora@tamarugo.cec.uchile.cl	Lopez Carlos
zbylo@camk.edu.pl	Loska Zbigniew
lousto@mail.physics.utah.edu	Lousto Carlos
loveday@fnal.gov	Loveday Jon
ll@astro.caltech.edu	Lu Limin
ernest.ma@ucr.edu	Ma Ernest
celtina@ifcai.pa.cnr.it	Maccarone Maria Concetta
alfredo@janaina.uam.mx	Macias Alfredo
jesm@dfi.aau.dk	Madsen Jes
maeda@th.phys.titech.ac.jp	Maeda Kengo
cmv@hep.saclay.cea.fr	Magneville Christophe
rohan@cfata5.harvard.edu	Mahadevan Rohan
san@ipac.caltech.edu	Malhotra Sangeeta
maller@noether.ucsc.edu	Maller Ari
mannheim@uconnvm.uconn.edu	Mannheim Philip
rmansouri@aip.de	Mansouri Reza
hman@athena.uchicago.edu	Marion Howard
draza@tac.dk	Markovic Draza
craigm@astrog.physics.wisc.edu	Markwardt Craig
marleau@weibel.berkeley.edu	Marleau Francine
hugo@sagredo.as.utexas.edu	Martel Hugo
gmathews@bootes.phys.nd.edu	Mathews Grant
s-matz@nwu.edu	Matz Steven
meiksin@oddjob.uchicago.edu	Meiksin Avery
melott@kusmos.phsx.ukans.edu	Melott Adrian
mendell@phys.ufl.edu	Mendell Gregory
nnp@astro.psu.edu	Meszaros Peter
bmetcalf@physics.berkeley.edu	Metcalf R. Ben
	Metevier Anne
metzler@fnal.gov	Metzler Chris
meyer@elvis.astro.nwu.edu	Meyer David
	Meyer Heinrich

meyer@oddjob.uchicago.edu — Meyer Stephan
mezz@nova.phy.ornl.gov — Mezzacappa Anthony
ekke@janaina.uam.mx — Mielke Eckehard
— Mignani R.

milan@moumee.calstatela.edu — Mijic Milan
miheeva@dpc.asc.rssi.ru — Mikheeva Elena
miller@gamma.uchicago.edu — Miller Coleman
miller@sissa.it — Miller John
mamiller@void.wustl.edu — Miller Mark
mino@vega.ess.sci.osaka-u.ac.jp — Mino Yasushi
mirabel@ariane.saclay.cea.fr — Mirabel Igor Felix
miyahata@vega.ess.sci.osaka-u.ac.jp — Miyahata Keiko
moiseenko@mx.iki.rssi.ru — Moiseenko Sergej
— Moncrief Vincent

prlvm10@amtp.cam.ac.uk — Moniz Paulo V.
moran@cfa.harvard.edu — Moran James
morgan@physics.wm.edu — Morgan David
hiro@phys.ocha.ac.jp — Morikawa Masahiro
masaaki@th.phys.titech.ac.jp — Morita Masaaki
drom@vxcern.ch — Morrison Douglas
leonidas@astro.berkeley.edu — Moustakas Leonidas
karen@kayz.dartmouth.edu — Mueller Karen
ewald@mpa-garching.de — Muller Ewald
hrm@dartmouth.edu — Muller Hans-Reinhard
vmueller@aip.de — Müller Volker
lucia@oddjob.uchicago.edu — Munoz-Franco Lucia
murphy@tekel.butler.edu — Murphy Brian
myers@dept.physics.upenn.edu — Myers Steven T.

nagamine@princeton.edu — Nagamine Kentaro
mnagano@icrr.u-tokyo.ac.jp — Nagano Motohiko
masa@vega.ess.sci.osaka-u.ac.jp — Nagashima Masahiro
akika@cfi.waseda.ac.jp — Nakamichi Akika
takashi@yukawa.kyoto-u.ac.jp — Nakamura Takashi
narita@rikkyo.ac.jp — Nalita Makoto
rnarayan@cfa.harvard.edu — Narayan Ramesh
madeleinen@aol.com — Nash Madeleine
matilda@novelty.tau.ac.il — Ne'eman Yuval
jnelson@ucolick.org — Nelson Jerry
nemiroff@mtu.edu — Nemiroff Robert
netzer@wise.tav.ac.il — Netzer Hagai
— Neuhaeuser Ralph

nichol@oddjob.uchicago.edu	Nichol Robert
jcn@mpa-garching.mpg.de	Niemeyer Jens
nilsen@phepds.dnet.nasa.gov	Nilsen Bjorn
dfnitz@umich.edu	Nitz David
nollert@phys.psu.edu	Nollert Hans-Peter
nollett@oddjob.uchicago.edu	Nollett Kenneth
	Nomoto Haruyo
nomoto@astron.s.u-tokyo.ac.jp	Nomoto Ken'ichi
norman@stsci.edu	Norman Colin
norman@astro.washington.edu	Norman Dara
norman@ncsa.uiuc.edu	Norman Michael
novikov@tac.dk	Novikov Igor
bihkpoh@uxa.ecn.bgu.edu	O'Hara Paul
	Occhionero F.
wco@phys.physics.ucf.edu	Oelfke William
ogelman@astrog.physics.wisc.edu	Ogelman Hakki
olinto@oddjob.uchicago.edu	Olinto Angela
kdo@mit.edu	Olum Ken
rene@hep.uchicago.edu	Ong René
merav@vax.iagusp.usp.br	Opher Merav
opher@vax.iagusp.usp.br	Opher Reuven
oreglia@uchicago.edu	Oreglia Mark
orio@astrog.physics.wisc.edu	Orio Marina
aog@mail.ess.sunysb.edu	Ortiz-Gil Amelia
ortolan@lnl.infn.it	Ortolan Antonello
ovadia@uchicago.edu	Ovadia Jacques
page@pupgg.princeton.edu	Page Lyman
rpain@lpnax1.in2p3.fr	Pain Reynald
nathalie@hep.saclay.cea.fr	Palanque-Delabrouille Nathalie
apanait@astro.psu.edu	Panaitescu Alin
	Papadopoulos Philippos
hara@astro.psu.edu	Papathanassiou Hara
park@llnl.gov	Park Hye-Sook
mep@howdy.wustl.edu	Pati Michael
pavlov@astro.psu.edu	Pavlov George
fpelaez@astro.estec.esa.nl	Pelaez Francois
upen@cfa.harvard.edu	Pen Ue-Li
hilken@maths.ox.ac.uk	Penrose Roger
peop@fnal.gov	Peoples John
perez@goya.ific.uv.es	Perez Armando

saul@LBL.gov

vahe@bigbang.stanford.edu

ravi@cuphyb.phys.columbia.edu
lopr@xanum.uam.mx
piperp@merlo.nwu.edu
tsvi@shemesh.fiz.huji.ac.il
piro@alpha1.ias.fra.cnr.it
gifco94@botes1.tesre.bo.cnr.it
pogosyan@cita.utoronto.ca
poirier@nd.edu
poisson@terra.physics.uoguelph.ca
jpons@schwarz.doa.uv.es
mrp@astro.washington.edu
rob.preece@msfc.nasa.gov
premana@hoffmann.ph.utexas.edu
xavier@cass152.ucsd.edu
zack@mps.ohio-state.edu
pryke@uchicago.edu
demetris@astro.uiuc.edu
pullin@phys.psu.edu
w-purcell@nwu.edu

jmq@gamma.uchicago.edu
quinn@egret.sao.arizona.edu

jorg@astro.psu.edu
rad@rri.ernet.in
ramaty@pair.gsfc.nasa.gov
rasio@mit.edu
rebull@oddjob.uchicago.edu
mir@ast.cam.ac.uk
reese@oddjob.uchicago.edu
refreg@astro.columbia.edu
rr@space.mit.edu
chris@rocinante.colorado.edu
rezzolla@astro.physics.uiuc.edu
srhie@nd.edu
jrhoads@noao.edu
richards@oddjob.uchicago.edu
dor@umich.edu

Perlmutter Saul
Peschke Sybille
Petrosian Vahe
Pierfederici F.
Pilla Ravi
Pimentel Luis O.
Piper Patrick
Piran Tsvi
Piro Luigi
Pizzichini Graziella
Pogosyan Dmitri
Poirier John
Poisson Eric
Pons José
Pratt Mark
Preece Robert
Premadi Premana
Prochaska Jason
Protogeros Zack
Pryke C. L.
Psaltis Dimitrios
Pullin Jorge
Purcell William

Quashnock Jean M.
Quinn John

Rachen Jorg
Radhakrishnan Venkataraman
Ramaty Reuven
Rasio Frederic
Rebull Luisa
Rees Martin
Reese Erik
Refregier Alexandre
Remillard Ron
Reynolds Chris
Rezzolla Luciano
Rhie Sun Hong
Rhoads James
Richards Gordon
Richstone Douglas

grr@space.mit.edu	Ricker George R.
ricker@nu.uchicago.edu	Ricker Paul
riotto@fnal.gov	Riotto Antonio
ronald.roberts@dartmouth.edu	Roberts Ronald
robinson@utdallas.edu	Robinson Ivor
roman@ccsu.ctstateu.edu	Roman Thomas
kromer@nwu.edu	Romer Anita Katherine
rosati@stsci.edu	Rosati Piero
	Rosen S. Peter
	Rosenberg Seth
rosswog@quasar.physik.unibas.ch	Rosswog Stephan
bfr@ferio.mtk.nao.ac.jp	Roukema Boudewijn
rub@physto.se	Rubinstein Hector
mor@mpa-garching.mpg.de	Ruffert Maximilian
ruhl@physics.ucsb.edu	Ruhl John
ryden@mps.ohio-state.edu	Ryden Barbara
ryu@hermes.astro.washington.edu	Ryu Dongsu
pmsa@ualg.pt	Sa Paulo M.
saavedra@to.infn.it	Saavedra Oscar
sadoulet@physics.berkeley.edu	Sadoulet Bernard
samar@astrog.physics.wisc.edu	Safi-Harb Samar
	Saijo Motoyuki
nsakai@mn.waseda.ac.jp	Sakai Nobuyuki
msaniga@auriga.ta3.sk	Saniga Metod
andrea@ifcai.pa.cnr.it	Santangelo Andrea
sato@phys.s.u-tokyo.ac.jp	Sato Katsuhiko
joseph.scanio@uc.edu	Scanio Joseph
caleb@clopper.gsfc.nasa.gov	Scharf Caleb
	Schartman Ethan
schatz@ik3.fzk.de	Schatz Gerd
chschmid@itp.phys.ethz.ch	Schmid Christoph
rschmidt@aip.de	Schmidt Robert
pvs@hep.anl.gov	Schoessow Paul
dns@oddjob.uchicago.edu	Schramm David
bschroed@olivet.edu	Schroeder Brock
fs@astr.maps.susx.ac.uk	Schunck Franz
bschwarz@aip.acp.org	Schwarzschild Bertram
scoccima@cita.utoronto.ca	Scoccimarro Roman
dscott@astro.ubc.ca	Scott Douglas
sedrakia@spacenet.tn.cornell.edu	Sedrakian Armen
	Seely Ron

segreto@ifcai.pa.cnr.it	Segreto Alberto
eseidel@aei-potsdam.mpg.de	Seidel Ed
tom.shanks@durham.dc.uk	Shanks Tom
	Shapiro Maurice
shapiro@astro.as.utexas.edu	Shapiro Paul
shapiro@astro.physics.uiuc.edu	Shapiro Stuart
nir@tapir.caltech.edu	Shaviv Nir
shi@physics.ucsd.edu	Shi Xiangdong
shibata@vega.ess.sci.osaka-u.ac.jp	Shibata Masaru
	Shibley John
shinkai@wurel.wustl.edu	Shinkai Hisaaki
siromizu@utaphp1.phys.s.u-tokyo.ac.jp	Shiromizu Tetsuya
	Shutt Thomas
tsiegfried@dallasnews.com	Siegfried Tom
signore@physique.ens.fr	Signore Monique
steinn@ast.cam.ac.uk	Sigurdsson Steinn
simpson@odysseus.uchicago.edu	Simpson John
singh@tpau.physics.ucsb.edu	Singh Anupam
sivronr@gvsu.edu	Sivron Ran
skillman@astro.spa.umn.edu	Skillman Evan
bob-smither@qmgate	Smither Robert
chris@astro.as.utexas.edu	Sneden Chris
soltan@camk.edu.pl	Soltan Andrzej M.
ats25@damtp.cam.ac.uk	Sornborger Andrew
spaans@pha.jhu.edu	Spaans Marco
spooner1@vz.rl.ac.uk	Spooner Neil
squires@astro.berkeley.edu	Squires Gordon
stanev@bartol.udel.edu	Stanev Todor
gds6@po.cwru.edu	Starkman Glenn
stebbins@fnal.gov	Stebbins Albert
steele@uchicago.edu	Steele Diana
ccs@astro.caltech.edu	Steidel Charles
stockman@stsci.edu	Stockman Hervey
	Stockman P.
norbert@physik.unizh.ch	Straumann Norbert
russell@oddjob.uchicago.edu	Strickland Russell
wms@wugrav.wustl.edu	Suen Wai-Mo
sunyaev@mpa-garching.mpg.de	Sunyaev Rashid
ssurya@suhep.phy.syr.edu	Surya Sumati
dswesty@ncsa.uiuc.edu	Swesty F. Douglas
	Szalay A.
szapudi@fnal.gov	Szapudi Istvan

taam@astro.nwu.edu — Taam Ronald
tagoshi@yso.mtk.nao.ac.jp — Tagoshi Hideyuki
taka@tokyo-rose.uchicago.edu — Takamiya Marianne
letters@astronomy.com — Talcott Rich
tanaka@fys.ruu.nl — Tanaka Yasuo
taniguci@tap.scphys.kyoto-u.ac.jp — Taniguchi Keisuke
ant@roe.ac.uk — Taylor Andy
joe@puppsr.princeton.edu — Taylor Joseph
max@ias.edu — Tegmark Max
saul@spacenet.tn.cornell.edu — Teukolsky Saul
fkt@quasar.physik.unibas.ch — Thielemann Friedrich
th@ifk-mp.uni-kiel.de — Thielheim Klaus
davet@phys.ufl.edu — Thomas David
hcthomas@mpa-garching.mpg.de — Thomas Hans-Christian
djt@mpia-hd.mpg.de — Thompson David
— Thorne Kip S.
serap@amanda.physics.wisc.edu — Tilav Serap
timbie@het.brown.edu — Timbie Peter
— Tipler Frank J.
tkachev@wollaston.mps.ohio-state.edu — Tkachev Igor
olat@fnal.gov — Tornkvist Ola
totsuka@sukip04.icrr.u-tokyo.ac.jp — Totsuka Yoji
jtrefil@gmu.edu — Trefil James
smt@dartmouth.edu — Tribiano Shana M.
trodden@ctpa04.mit.edu — Trodden Mark
— Truemper J.
truran@nova.uchicago.edu — Truran James W.
uphst@gemini.oscs.montana.edu — Tsuruta Sachiko
turkot@fnal.gov — Turkot Frank
mturner@fnal.gov — Turner Michael
n.g.turok@damtp.cam.ac.uk — Turok Neil
turolla@astaxp.pd.infn.it — Turolla Roberto
dtytler@ucsd.edu — Tytler David
uehara@murasaki.scphys.kyoto-u.ac.jp — Uehara Hideya
andrew@astro.princeton.edu — Ulmer Andrew
m-ulmer2@nwu.edu — Ulmer Melville
valinia@rosserv.gsfc.nasa.gov — Valinia Azita
vanbibber@llnl.gov — van Bibber Karl
edvdh@astro.uva.nl — van den Heuvel Edward P. T.
eelco@tac.dk — van Kampen Eelco

mvp@math.mit.edu	van Putten Maurice
devb@oddjob.uchicago.edu	Vanden Berk Daniel
voort@oddjob.uchicago.edu	Vandervoort Peter
brian@srl.caltech.edu	Vaughan Brian
vecchio@aei-potsdam.mpg.de	Vecchio Alberto
aparna@oddjob.uchicago.edu	Venkatesan Aparna
armitage@bologna.infn.it	Venturi Giovanni
visser@kiwi.wustl.edu	Visser Matt
vladilo@oat.ts.astro.it	Vladilo Giovanni
ioav@if.ufrj.br	Waga Ioav
wagoner@leland.stanford.edu	Wagoner Robert
rmwc@midway.uchicago.edu	Wald Robert
fwalter@astro.sunysb.edu	Walter Fred
bdw@ic.ac.uk	Wandelt Benjamin D.
wands@sms.port.ac.uk	Wands David
pyro@ncsa.uiuc.edu	Wang Ed
	Wang John
lifan@astro.as.utexas.edu	Wang Lifan
wqd@nwu.edu	Wang Q. Daniel
ywang@astro.princeton.edu	Wang Yun
morgan@milagro.lanl.gov	Wascko Morgan
a.a.watson@leeds.ac.uk	Watson Alan
mway@newton.umsl.edu	Way Michael
	Weatherall J. C.
	Weber Joseph
rwebster@physics.unimelb.edu.au	Webster Rachel L.
tweekes@egret.sao.arizona.edu	Weekes Trevor
weilertj@ctrvax.vanderbilt.edu	Weiler Tom
ejw@phys.columbia.edu	Weinberg Erick
	Weinberg Steven
weiss@tristan.mit.edu	Weiss Ranier
hanwen@slac.stanford.edu	Wen Han
wentz@ik3.fzk.de	Wentz Juergen
	Whalen John
wheel@astro.as.utexas.edu	Wheeler J. Craig
jwheeler@cc.usu.edu	Wheeler Jim
whelan@physics.utah.edu	Whelan John
white@oddjob.uchicago.edu	White Martin
bernard@phys.ufl.edu	Whiting Bernard
	Wick Stuart
c-wiegert@uchicago.edu	Wiegert Craig

dwigg@amsta.leeds.ac.uk	Wiggins David
ramjw@ast.cam.ac.uk	Wijers Ralph
revak@astro.ufl.edu	Williams Reva K.
wilson@het.brown.edu	Wilson Grant
wilson@ricker.llnl.gov	Wilson James
axw5@email.psu.edu	Wolszczan Alex
	Woosley S. E.
	Wright Edward
yakovlev@sci.fian.msk.su	Yakovlev Vladimir
yess@kusmos.phsx.ukans.edu	Yess Capp
yi@sns.ias.edu	Yi Insu
yokoyama@yukawa.kyoto-u.ac.jp	Yokoyama Jun'ichi
yoneda@cfi.waseda.ac.jp	Yoneda Gen
yoonjh@cosmic.konkuk.ac.kr	Yoon Jong Hyuk
don@oddjob.uchicago.edu	York Donald
young@oddjob.uchicago.edu	Young Yuan-Nan
wenfei@gibson.msfc.nasa.gov	Yu Wenfei
zakharov@vitep3.itep.ru	Zakharov Alexander
nzamora@tamarugo.cec.uchile.cl	Zamorano Nelson
zampieri@donald.physics.uiuc.edu	Zampieri Luca
zane@sissa.it	Zane Silvia
	Zavlin Vyatcheslav
zeh@urz.uni-heidelberg.de	Zeh H. Dieter
	Zepf Steve
zhang@ssl.msfc.nasa.gov	Zhang Shuang-Nan
zhang@xancus10.gsfc.nasa.gov	Zhang William
winfried.zimdahl@uni-konstanz.de	Zimdahl Winfried
zotov@phys.latech.edu	Zotov Natalia